ANIMAL CELL TECHNOLOGY

Animal Cell Technology

From Vaccines to Genetic Medicine

Edited by

Manuel J. T. Carrondo

IBET,
Oeiras, Portugal

Bryan Griffiths

CAMR,
Porton Down, Wiltshire, U.K.

and

José L. P. Moreira

IBET,
Oeiras, Portugal

KLUWER ACADEMIC PUBLISHERS
DORDRECHT / BOSTON / LONDON

A C.I.P. Catalogue record for this book is available from the Library of Congress

ISBN 0-7923-4321-2

Published by Kluwer Academic Publishers,
P.O. Box 17, 3300 AA Dordrecht, The Netherlands.

Kluwer Academic Publishers incorporates
the publishing programmes of
D. Reidel, Martinus Nijhoff, Dr W. Junk and MTP Press.

Sold and distributed in the U.S.A. and Canada
by Kluwer Academic Publishers,
101 Philip Drive, Norwell, MA 02061, U.S.A.

In all other countries, sold and distributed
by Kluwer Academic Publishers Group,
P.O. Box 322, 3300 AH Dordrecht, The Netherlands.

Printed on acid-free paper

Printed in the Netherlands

Table of Contents

SESSION II - Animal Cells as Tools for Discovery and Testing

SESSION IV - Animal Cell Process Engineering

SESSION V - Recombinant Proteins: Biosynthesis and Processing

SESSION VII - Cell and Physiology Engineering

Meeting Sponsors

Companies

Bayer
Baxter
Boehringer Ingelheim
Genentech
Hoffmann-La Roche
Hyclone Europe
Microbiological Associates
Novo Nordisk
Pharmacia Biotech
SmithKline Beecham

Akzo Nobel Pharma
- Organon
- Organon Teknika
- Diosynth

Boehringer Mannheim
Ciba-Geigy
E. Merck
Glaxo-Wellcome
Intervet
Pasteur Mérieux Sérums & Vaccins
Schering

ATCC
Celltech Biologics
IBET

Institutions

European Commission - DGXII - Biology
Calouste Gulbenkian Foundation - F.C.G.
Luso American Development Foundation - FLAD
Junta Nacional de Investigação Científica e Tecnológica - JNICT

Trade Fair Sponsors

Aber Instruments Ltd.
Amicon
Ashby Scientific, Ltd.
B. Braun Biotech International
Bayer Corporation, Diagnostics Division, USA - Kankakee
Bibby Sterilin
BioInvent International AB
Bioprocessing Limited
Boehringer Ingelheim
Boehringer Ingelheim Bioproducts
Cansera International Inc.
Celltech Biologics plc
Corning Hazleton
Costar (Europe), Ltd.
E.C.A.C.C.
Greiner GmbH
Hucoa-Erloss
Hyclone Europe
Inceltech France
Institut Pasteur
Inveresk Research
JRH Biosciences
Life Technologies Ltd.
LSL Biolafitte - Cellex Biosciences
Microbiological Associates, Ltd.
MicroSafe B.V.
Molecular Devices
Nalge/Nunc International
Nature - Macmillan Magazines, Ltd.
PAA Laboratories GmbH
Pall Europe Ltd.
Pharmacia Biotech AB
Q-One Biotech Ltd.
Quality Biotech NV
Sarstedt
Schärfe System GmbH
Sigma BioSciences
Sorvall Centrifuge Products
TC Tech Corporation
Unisyn Technologies Inc

KEYNOTE LECTURES

MOLECULAR BIOLOGY OF PRION DISEASES

C. Weissmann, Institut für Molekularbiologie, Abteilung I, Universität Zürich, 8093 Zürich

The transmissible agent that causes spongiform encephalopathies such as scrapie, the prion, is believed to be devoid of nucleic acid and identical with PrP^{Sc}, a modified form of PrP^{C}. PrP^{C} is a normal host protein encoded by a single copy gene (*Prn-p*) and is found predominantly on the surface of neurons. PrP^{Sc}, in contrast to PrP^{C}, is resistant to protease and accumulates intracellularly. Prusiner proposed that PrP^{Sc}, when introduced into a normal host, causes the conversion of PrP^{C} or its precursor into PrP^{Sc} (*"protein only" hypothesis*). We generated homozygous *Prn-p*$^{o/o}$ ("PrP knockout") mice devoid of PrP^{C}. *Prn-p*$^{o/o}$ mice inoculated with mouse scrapie prions remained free of scrapie symptoms for at least 2 years while wild-type controls all died within 6 months. There was no propagation of prions in the *Prn-p*$^{o/o}$ animals. Introduction of murine PrP transgenes rendered *Prn-p*$^{o/o}$ mice highly susceptible to mouse but not to hamster prions, while the introduction of Syrian hamster PrP transgenes rendered them susceptible to hamster but to a much lesser extent to mouse prions. Our experiments show that PrP^{C}, possibly at close to normal levels, is required for the usual susceptibility to scrapie and that lack of homology between incoming prions and the host's PrP genes decreases susceptibility to disease. They also suggest that cattle or sheep devoid of the PrP gene might be viable and resistant to scrapie.

M.J.T. Carrondo et al. (eds.), Animal Cell Technology, 3–6.

Discussion

Miller: Would you envisage human knockouts?

Weissmann: Not in the immediate future. What your question aims at is: "What would happen if you could suppress the formation of PRP in man by some drug?" Obviously one would like to think of using antisense technology, introduce antisense oligo's into the cerebo-spinal compartment and try to suppress the formation of PRP. It is a possibility but there are many technical difficulties. Not a simple technology, nor is it easy to introduce oligonucleotides to the brain. Another approach to prevent progression of the disease would be to find a drug that prevents the conversion of the normal into the scrapie form. This can be achieved *in vitro* and thus could be followed up.

Onions: There are reports that some mice with the homozygous knockout phenotype kept for a long time are beginning to show signs of ataxia, and other signs associated with a classic spongiform encephalotomy. From the tenor of your comments do I take it that you do not accept those results?

Weissmann: I accept the results. I do not accept that they are due to the PRP knockout. There are two lines of PRP knockout mice that do not show this effect. A knockout group made from a mixed genetic background gives all lines devoid of some genes and just comparing a wild type from a mixed background is elicit. I would believe this if you restore the normal phenotype and make the animals transgenic again for PRP. So far our mice have never shown this effect.

Onions: What do you think PRP is doing? What is its function?

Weissmann: In general terms it could be a signalling molecule or a cell adhesion molecule. In fact there is no good data to show what it does.

Barteling: Prions can penetrate membranes. Can they penetrate into the nucleus?

Weissmann: No evidence that they do. You see them in the endocytotic lysosomal compartment. Sometimes cells release considerable quantities of PRP Scrapie, maybe when they die or in some other process.

Barteling: The point is that in the nucleus they could interfere in DNA transcription.

Weissmann: This hypothesis has been published, but all the experiments speak against it eg messenger RNA has been analysed from infected cells and no difference to normal tissue was found. Chemical analysis of PRP Scrapie, sequence analysis and mass spectrometry has not shown differences in the structure of PRPC and PRP Scrapie. Only difference is conformational.

Barteling: You stated that 1 in 100,000 particles is infectious. Can this be translated that only 1 in 100,000 reaches spots where it infects?

Weissmann: You are right. If there is no difference in chemical structure then 1 particular molecule among 100,000 means you could never see it.

Mannix: Have you considered using the knockout mice with the bovine gene as an assay system?

Weissmann: These experiments are being done. The human gene has been introduced into our knockout mice and obtained mouse lines that are much more susceptible to the human gene so that one can transfer all the human forms of the disease to these transgenic mice carrying a human gene on a knockout background. The same is being tried for the bovine gene but I have not heard any results yet. This could lead to an easier assay for BSE.

Wurm: A provocative question. Would you eat British beef?

Weissmann: Yes, I ate some a few weeks ago. I believe the current measures that have been taken to prevent transmission to humans are quite sufficient. The only place one finds the infectious agent in cells is the brain and spinal cord. In principle, if you do not consume either of these two components, then you are safer than flying to this meeting. The rules are adequate, but have they been followed in an adequate way? It is forbidden to put brain or spinal tissue into food for human consumption, but does it really not happen? The way animals are slaughtered and butchered opens the vertebrate canal leading to a cutting through of the spinal cord which is splashed around. The process is followed by hosing down. This situation exists but is it reason for concern or not? One cannot answer this question. However, there is some evidence that the disease is transmissible from cattle to man as shown by the 10 - 12 youthful patients that have contracted the disease. There is no reason to believe that this cannot happen as it has been clearly shown that feeding bovine brain to experimental animals (mice, sheep) can in fact lead to disease with a relatively low dose. One cannot assume man is different. A dose infective experiment on primates has not been done yet. So we cannot tell what is an infectious dose for a human and how dangerous some splashing around of spinal cord would be. All cattle showing the disease are not processed. But how many animals not showing the disease in a herd which has animals showing the disease are infectious? This is another important value that is needed before the risk can be evaluated. There are a lot of unknowns and if I say I eat English beef you must realise that at my age the risk is not very high either.

Applications of Animal Cell Technology to Drug Discovery and Pharmacology

W.-D. Schleuning
Research Laboratories of Schering AG, D-13342 Berlin, Germany

Introduction

Since the introduction of gene cloning and gene transfer more than two decades ago animal cell culture has undergone a dramatic transformation from a research tool and highly specialized niche technology to a central resource of innovation in pharmaceutical research and development. Many of the most successful drugs introduced to the market in recent years were discovered in cell culture supernatants, evaluated in cell culture models and produced by large scale animal cell fermentation. Best known examples include interferons, erythropoetin, various interleukins and cytokines and tissue-type plasminogen activator (t-PA).

As the build-up of blood clots in the vasculature is the most prominent cause of death in men and women above 40 years in the industrialized countries and there is currently no treatment for the underlying degenerative vascular disease, drugs that prevent clot formation or rapidly dissolve clots after they have formed in the vasculature are promising agents for the prevention and the treatment of acute myocardial infarction, pulmonary embolism and ischemic stroke. In the context of a discovery program aimed at new drugs for the treatment of cardiovascular diseases we have recently focussed on hematophageous animals as a source for compounds that interfere with blood coagulation. Animal cell technology has played a decisive role in the process of discovery and preclinical development of these drug candidates.

Vampire bat (Desmodus rotundus) salivary plasminogen activator

Plasminogen activators (PAs) are enzymes that generate the protease plasmin from its precursor plasminogen by the cleavage of a single peptide bond. Plasminogen is synthetized in the liver and circulates in high concentrations in the blood and PAs are made by most cells following mitogenic or inflammatory stimuli. Plasmin very efficiently

M.J.T. Carrondo et al. (eds.), Animal Cell Technology, 7–14.

digests fibrin, the insoluble meshwork that forms the solid support of the blood clot. Thrombolytic (=clot dissolving therapy) with PAs is routinely performed for the treatment of acute myocardial infarction. However, thrombolytic agents currently in use such as streptokinase (SK), tissue-type plasminogen activator (t-PA) and urokinase (u-PA) exhibit various shortcomings with respect to their pharmacodynamic and pharmacokinetic characteristics and may in addition cause life threatening side effects. With the aim of creating an improved fibrinolytic agent in mind, we have cloned and expressed the plasminogen activators from the saliva of the vampire bat *Desmodus rotundus*, originally described by Hawkey (1). Four *Desmodus rotundus* salivary Plasminogen Activators (DSPAs) were identified, which are composed of various conserved domains known from related proteins such as t-PA and u-PA. DSPAα1 and DSPAα2 exhibit the structural formula Finger (F), EGF (E), Kringle (K), Protease (P) and DSPAβ and γ the formulas EKP and KP respectively. Subtle sequence differences and data from Southern blot hybridization analysis indicate that the four enzymes are coded by four different genes and are not generated by differential splicing of a single primary transcript (2). A preliminary biochemical and pharmacological analysis indicated that DSPAα1 exhibited the most favorable pharmacological profile and was therefore chosen for further study. A recombinant CHO-cell line for the production of DSPA and a purification protocol were established to obtain material that fulfilled the specifications for preclinical and clinical development (3).

The most important characteristic that distinguishes DSPA from other plasminogen activators is its extraordinary fibrin specificity. In fact the activity of DSPAα1 is 105.000 times higher in the presence of fibrin than in its absence. The respective factor for t-PA is only 550. Likewise, fibrinogen which is a fairly potent cofactor of plasminogen activation by t-PA has hardly any effect on DSPAα1. Therefore the factor of fibrin selectivity expressed as the quotient of activity in the presence of fibrin vs. activity in the presence of fibrinogen is 12900 for DSPA but only 72 for t-PA. Therefore DSPAα1 could fulfill the hopes originally linked to t-PA: plasminogen activation restricted to the clot surface without the systemic activation that inexorably leads to fibrinogen consumption, "plasminogen steal" and degradation of clotting factors VIII and V (4).

This fibrin specificity and selectivity of DSPAα1 may translate into a more favorable side effect profile. Pharmacological studies using a lung embolism model in rats demonstrated that recombinant DSPAα1 is more potent and more clot specific than t-PA (5). A higher potency, clot specificity and prolonged halflife of DSPAα1 over t-PA were also verified in an arterial thrombosis model in rats. In a coronary thrombosis model of acute myocardial infarction in dogs, DSPAα1 lead to a faster recanalization and a lower incidence of reocclusion if compared to t-PA (6). On the other hand, a lower incidence of bleeding was observed in a rat mesenteric vein model. DSPAα1 was characterized pharmacokinetically in rats and cynomolgus monkeys. In comparison to t-PA a lower total clearance and a longer terminal halflife could be demonstrated. These data encourage the development of an i.v. bolus regimen in humans. A detailed toxicological study has been performed in rats and monkeys. No direct toxic effects could be observed and secondary effects appeared only at extremely high doses of DSPAα1 (7). A phase I clinical trial with healthy volunteers was performed where subtherapeutic doses were applied by intravenous bolus administration. No clinical relevant changes nor negative effects on blood coagulation were observed. In addition the pharmacokinetic predictions obtained from animal studies could be confirmed and the therapeutically effective dose of DSPAα1 was predicted to be 0.5 mg/kg of body weight (7).

In summary our data strongly support the notion that DSPAα1 may be a safer thrombolytic agent than others currently in use.

Pallidipin

Pallidipin is a 19 kDa protein isolated from the saliva of the blood sucking assassin bug *Triatoma pallidipennis,* which specifically inhibits collagen-induced platelet aggregation (8). We have produced this protein in cell culture and in *E. coli* to provide the amounts necessary for a detailed characterization of its mechanism of action (9). Recombinant pallidipin exhibited the same inhibitory activity in the platelet aggregation assay as the native protein. No effect was noted on the aggregation induced by the thromboxane mimetic U46619 or by arachidonic acid, but the collagen-mediated thromboxane formation was inhibited. Likewise, the secretion of ATP was prevented. Increasing concentrations of collagen were able to overcome the inhibitory effect. The dose-response curve of collagen-induced aggregation was shifted to the right with no concomitant depression of the maximal response. Aggregation induced by agonists known to act via a collagen receptor on platelets i.e. WGA and trimucytin was not affected. The pathway of pallidipin mediated inhibition of platelet aggregation is not yet completely understood. Nevertheless, pallidipin is an elegant tool to dissect so far unknown mechanisms of platelet activation and in consequence provides us with an understanding that will lead us to more effective and safer inhibitors of platelet aggregation.

Triabin

In the saliva of *Triatoma pallidipennis* we have also identified a new thrombin specific inhibitor. This 17 kDa protein, which we named triabin, was purified to homogeneity using gel filtration and thrombin affinity chromatography. It showed similar inhibitory activity as hirudin in the platelet aggregation and thrombin clotting time assays, but unlike hirudin, it did not inhibit the amidolytic activity of thrombin. The effect of thrombomodulin on thrombin was blocked as measured by a protein C activation assay. Triabin protected the anion-binding exosite 1 of a-thrombin from proteolysis by trypsin. Taken together, these results indicate that triabin exerts its inhibitory effect via the binding to the anion-binding exosite 1 of thrombin (10).
Using two degenerated primers deduced from the N-terminus and an internal fragment of triabin for PCR amplification, a specific triabin-coding fragment was identified and used

for the screening of a *T. pallidipennis* salivary gland cDNA library. Triabin cDNA, which was isolated and sequenced, codes for a 160 amino acid-long protein, including a 18 amino acid-long signal peptide. The sequence bears no significant homology to other proteins. Expression in SF9 insect cells using a baculovirus derived expression system yielded a recombinant protein with the same inhibitory activity as native triabin. The anion binding exosite 1 of thrombin is the binding site for fibrinogen. Hirudin blocks both this site and the active site. Therefore the therapeutic dose range of hirudin is very narrow, i.e. the therapeutically effective dose is very close to the dose where bleeding occurs. In contrast to the thrombin-hirudin complex the thrombin-triabin complex retains some coagulant activity. Triabin is - figuratively spoken - a thrombin inhibitor with a safety valve. This property puts triabin in the position of a prime candidate for the treatment of restenosis after percutane transluminal coronary angioplasty (PTCA), a very common surgical procedure for the treatment of coronary heart disease.

New surrogate systems for the pharmacodynamic evaluation of nuclear receptor ligands

Nuclear receptors are a class of related proteins that function as ligand-dependent transcription factors. Best known examples include the receptor for the classical steroid hormones estrogen, androgen, progesteron, glucocorticoids and mineralocorticoids. Also retinoids (Vitamin A) and calcitriols (Vitamin D) are known to function via this type of receptor. Many nuclear receptor ligands are known to be valuable drugs for the treatment of such diverse conditions such as rheumatoid arthritis, skin rashes, various forms of cancer, bone diseases and for the control of fertility in women. In addition many members of this family with no known ligands were identified by homology cloning. These receptors are called orphan receptors.

Pharmacodynamic evaluation of nuclear receptor ligands has traditionally depended on whole animal models. Estrogenic activity for example is measured by increase of uterus weight in ovarectomized rats after treatment with the hormone. Since the availability of cloned receptor genes and suitable reporter systems, cell culture assays have become the mainstay of first line screening efforts. The reporter genes most commonly used in animal cells are bacterial chloramphenicol transacetylase (CAT), firefly luciferase (luc), and the more recently introduced jelly fish green fluorescent protein (GFP). All three systems however are fraught with various shortcomings related to expenditure, speed, sensitivity and waste disposal. We have therefore developed an alternative system, based on secreted miniurokinase-type plasminogen activator (mu-PA) as a reporter. mu-PA specifically converts plasminogen (plg) into the protease plasmin, which is easily quantified using a chromogenic peptide substrate. The assay compares favorably with other systems in terms of convenience, speed, sensitivity and cost and has been adapted to a 96 well formate amenable to robotization. Using this system we have identified several new ligands for nuclear receptors from natural sources as well as from organochemical libraries. It appears as if this assay is particularly suitable for the identification of ligands for orphan receptors (11).

Ion channels as drug

Ion channels are the central mediators of neuronal function. Ligand dependent ion channels such as the g-aminobutyric acid (GABA) dependent chloride channel modulate anxiety and arousal behaviour. Exicatory amino acid receptors such as the glutamate receptors play an important role in learning, addictive behaviour and the induction of seizures. Such ion channels are important targets for anxiolytic, antiepileptic and cytoprotective drugs. Heterologous expression of these channels is a key technology for the identification of pharmacologically pure ligands.

The application of the techniques of molecular biology to the structural elucidation and the functional analysis of ion channels has in fact revolutionized neurobiology to an extent that would have seemed inconceivable 15 years ago. The techniques of expression cloning in combination with voltage clamp and patch clamp electrophysiology have led to the isolation of numerous channels on the basis of their function. Computer analysis of the deduced amino acid sequence allows the prediction of putative transmembrane topology, secondary structure and posttranslational modifications. Expression of the isolated channel genes in animal cell culture further allows both pharmacological and functional studies and the analysis of channel structure assembly and regulation. Structure/function can further be studied by the opportunities generated by site-directed mutagenesis and the possibilities offered by site specific antibodies as biochemical probes. We have expressed various subunit combinations of the human GABA-receptor in SF-9 insect cells using recombinant baculovirus as a vector. This system allowed us to screen for various agonists of the GABA receptor from the ß-carboline series and to identify compounds with a degree of specificity for certain subunit combinations.

Recombinant cells for the study of cytochrome P450 dependent drug metabolism

Human liver microsomal Cyt P450 enzymes catalyze the oxidation of a wide variety of structurally different compounds ranging from endogenous substrates such as steroids, fatty acids, and prostaglandins to exogenous chemicals including drugs and lipophilic xenobiotics. Oxidation can result in either bioactivation (production of reactive metabolites) or in deactivation and elimination of the compound (conjugation by phase II enzymes). Four gene families, designated CYP 1 - 4, and multiple subfamilies are currently known whose principle function is to metabolize drugs and foreign chemicals. Recombinant cell lines expressing selected major human P450 enzymes of this group might be useful for the specific identification / isolation of (steroidal) drug metabolites and for performing inhibition studies in order to reveal potential drug-drug interactions on the enzyme level. We decided to express selected human Cyt P450 enzymes in mammalian cells beginning with Cyt P450 IIIA4, a major inducible liver P450 enzyme involved in steroid metabolism, and the corresponding auxiliary Cyt P450 reductase the endogenous level of which is decreased in most cell lines.

We have demonstrated the conversion of testosteron into 6ß-OH-testosterone by this recombinant cell line. These results demonstrate that recombinant cell lines are useful in pharmacokinetic studies and that they may be particularly useful for the dissection of the complicated metabolic pathways that determine the fate of a drug within a living organism. Even though recombinant cell lines are only of minor significance in pharmacokinetic analysis today this role is likely to increase and to play a prominent role in the preclinical dossiers of the future.

Conclusion and perspectives

The introduction of gene cloning and gene transfer technology has created a shift of paradigm that has provided a new and greatly expanded base for animal cell technology. This technology is already now indispensible for the process of drug discovery, development and production and it is likely that this role will become even more prominent in the future.

References

1. Hawkey C (1966) Plasminogen activator in the saliva of the vampire bat *Desmodus rotundus*. Nature 211, 434-435.
2. Krätzschmar J, Haendler B, Langer G, Boidol W, Bringmann P, Alagon A, Donner P & Schleuning W-D (1991) The plasminogen activator family from the salivary gland of the vampire bat *Desmodus rotundus*: cloning and expression. Gene 105, 229-237.
3. Petri T, Langer G, Bringmann P, Cashion L, Shallow S, Schleuning W-D & Donner P (1995) Production of vampire bat plasminogen activator DSPAα_1 in CHO and insect cells. J. Biotechnology 39, 75-83.
4. Bringmann P, Gruber D, Liese A, Toschi L, Krätzschmar J, Schleuning W-D & Donner P (1995) Structural Features Mediating Fibrin Selectivity of Vampire Bat Plasminogen Activators. J Biol Chem 270, 25596-25603.
5. Witt W, Baldus B, Bringmann P, Cashion L, Donner P & Schleuning W-D (1992) Thrombolytic Properties of *Desmodus rotundus* (Vampire Bat) Salivary Plasminogen Activator in Experimental Pulmonary Embolism in Rats. Blood 79, 1213-1217.
6. Witt W, Maaß B, Baldus B, Hildebrand M, Donner P & Schleuning W-D (1994) Coronary Thrombolysis with Desmodus Salivary Plasminogen Activator (DSPA) in Dogs: Fast and Persistent Recanalization by Intravenous Bolus Administration. Circulation 90, 421-426.
7. Schleuning W-D, Bhargava A & Donner P (1996) DSPAa1: A new and novel vampire bat (*Desmodus rotundus*) plasminogen activator. In: A. Sasahara & J. Loscalzo, New Therapeutic Agents in Thrombosis and Thrombolysis, in press
8. Noeske-Jungblut C, Krätzschmar J, Haendler B, Alagon A, Possani L, Verhallen P, Donner P & Schleuning W-D (1994) An inhibitor of collagen-induced platelet aggregation from the saliva of *Triatoma pallidipennis*. J Biol Chem 269, 5050-5053.

9. Haendler B, Becker A, Noeske-Jungblut C, Krätzschmar J, Donner P & Schleuning W-D (1996) Expression, purification and characterisation of recombinant pallidipin, a novel platelet aggregation inhibitor from the haematophageous triatomine but *Triatoma pallidipennis*. Blood Coagulation and Fibrinolysis 7, 183-186.
10. Noeske-Jungblut C, Haendler B, Donner P, Alagon A, Possani L & Schleuning W-D (1995) Triabin, a highly potent exosite inhibitor of thrombin. J Biol Chem 270, 28629-28634.
12. Langer G, Toschi L, Dieckmann J & Schleuning W-D (1995) Low M_r urokinase-type plasminogen activator as a reporter gene. Gene 161, 287-292.

Discussion

Hauser: You mention the fusion protein of GFP and oestrogen receptor. Was this fusion protein active in the sense that it was fluorescent in the absence of oestrogen or did you need the ligand?

Schleuning: No, it is fluorescent in the absence of the ligand. It is biologically fully active.

Hauser: Was the full receptor moiety used or only the C.terminal part?

Schleuning: No, we used it on the C.terminal part.

Bernard: Is the GFP fluorescence dependent on the environmental conditions and physiological state of the cells?

Schleuning: Yes, at least the conventional GFP does not work very well in animal cells but there are mutations which are functioning and giving a better signal noise ratio. In terms of sensitivity in a single cell reporter system the sensitivity is not high enough to use the florescent protein. We believe our chicken plasminogen activator bioassay is preferable to the GFP assay.

Noe: Can you say something about the specificity of the DSPA and its half-life in blood compared to the other tPA?

Schleuning: The half-life time is considerably longer than tPA but I cannot recall the exact times. I think tPA is 5 minutes but the DSPA has a biphasic elimination, one fast phase and a slow one. In the right context the half-life is 30 minutes.

Noe: What about specificity?

Schleuning: If you compare tPA and DSPA and look at fibrin specificity then it is known that tPA activity is enhanced in the presence of fibrin 500 fold. With DSPA the activity is enhanced more than 100,000 fold. This factor becomes more relevant if you compare activation by fibrinogen. DSPA is hardly activated at all, attributed to a lack of a fibrinogen binding site located in kringle 2. The activation of tPA is believed to be due to fibrinogen in the vascular system and this does not happen with DSPA.

BIOGENESIS OF SECRETORY PROTEINS AND CELL SURFACE RECEPTORS IN MAMMALIAN CELLS:

POSTTRANSLATIONAL MODIFICATIONS AND PROTEIN FOLDING WITHIN THE ROUGH ENDOPLASMIC RETICULUM

Harvey F. Lodish, Whitehead Institute for Biomedical Research and Department of Biology, Massachusetts Institute of Technology Cambridge, MA, 02142

INTRODUCTION

In all eukaryotic cells secretory proteins and cell surface receptors are synthesized on ribosomes attached to the rough endoplasmic reticulum (ER) (15). The growing polypeptide chain is extruded from the ribosome and transported in an unfolded state through a protein- lined channel in the ER membrane into the ER lumen; it is never exposed to the cytosol. Secretory proteins are completely transported into the ER lumen; membrane proteins become anchored in the ER membrane *via* one or more ~23 amino acid hydrophobic segments that form membrane- spanning α- helixes. Once synthesis is complete, newly introduced polypeptides in the membrane and lumen of the ER must be folded, matured, sorted, and transported. Secretory or membrane proteins undergo many modifications during their transit to the cell surface:

(i) cleavage of signal sequences
(ii) formation and rearrangement of disulfide bonds
(iii) other protein folding steps, including formation of multichain proteins
(iv) addition and modification of N- and O- linked oligosaccharides
(v) other specific proteolytic cleavages.

Each modification takes place in a specific organelle through which these proteins pass. The first three of these reactions take place exclusively in the ER, and addition of N- linked oligosaccharides also occurs in this organelle. Protein catalysts within the ER lumen catalyze these and other steps in protein folding, such as isomerization of peptidyl-prolyl bonds. In general, all steps in protein folding, including formation of homo- and hetero-oligomers, are completed before the protein exists the ER *en route* to the Golgi complex.

The reactions that occur in the ER are essential for the protein to achieve its functional form. Only properly folded and assembled proteins are transported from the ER to the Golgi and to the cell surface. Unfolded, misfolded, or partly folded and assembled proteins are selectively retained in the ER, or are retrieved to the ER from the *cis*- Golgi reticulum. Frequently, misfolded proteins and unassembled subunits of multi-polypeptide complexes are degraded within the ER. Such "quality control" on transport of proteins to the Golgi insures that misfolded proteins do not reach the cell surface, where they might be recognized by the immune system as a "foreign" protein and induce an immunological response leading to the destruction of the cell. Folding within the ER is frequently the rate- limiting step in production of recombinant proteins; enhanced expression of secretory proteins frequently results in accumulation of misfolded proteins within the ER (5, 9).

M.J.T. Carrondo et al. (eds.), Animal Cell Technology, 15–26.

Here I review many of the post-translational modifications localized to the rough ER, particularly the formation of disulfide bonds. I also review evidence indicating that the rate and extent of folding of certain cell surface receptor proteins is poor, and I show how folding can be enhanced by specific mutations in the protein itself.

FORMATION OF DISULFIDE BONDS IN THE ER LUMEN SOON AFTER PROTEIN SYNTHESIS

In eukaryotic cells, formation of all disulfide bonds (Cys-S-S-Cys) occurs in the lumen of the rough ER, never in the cytosol. Disulfide bonds are confined to secretory proteins and to the exoplasmic domains of membrane proteins; such bonds are important stabilizing forces in the tertiary structure of these proteins.

The tripeptide glutathione is the major thiol (SH)-containing molecule in eukaryotic cells, and serves two important functions -- to prevent the formation of disulfide bonds in the cytosol and to catalyze their formation in the ER. Glutathione shuttles between the reduced form, abbreviated **GSH**, and the oxidized disulfide-linked dimer, **GSSG**. Formation of a disulfide bond in a protein involves two thiol-disulfide exchange reactions with glutathione. In the first, an -SH group of a cysteine residue reacts with one of the two sulfur atoms in a **GSSG** disulfide bond, forming a Cys - SS-**G** disulfide bond and one molecule of reduced **GSH**. In the second, another cysteine residue in the protein reacts with the Cys - SS-**G** bond, forming a Cys - SS- Cys disulfide bond and a second molecule of **GSH**.

To measure the redox state in the ER compartment and to establish the nature of the redox buffer (10), we synthesized the tetrapeptide N-Ac-Asn-Tyr-Thr-Cys-NH_2 (NYTC). It contains a tyrosine residue for iodination, a consensus sequence for N-linked glycosylation, and a cysteine residue for undergoing reversible SH - SS exchange reactions. This peptide can diffuse through membranes into microsomes or whole cells and become glycosylated. We showed that the glycosylated peptide becomes disulfide-bonded to glutathione, indicating that GSH is the major redox buffer within the secretory compartment. By measuring the ratio of glycosylated free NYTC and glycosylated NYTC- S- S- **G**, we could show that in the lumen of the ER the ratio of **GSH** to **GSSG** is maintained at about 5:1, sufficient to allow formation of protein disulfide bonds. Consistent with this, experiments in which reduced and denatured secretory proteins are re-folded in cell-free systems showed that a ratio of **GSH** to **GSSG** of ~5:1 indeed is optimal for disulfide bond formation (18). This observation suggests that the redox environment of the ER is optimal for disulfide bond formation.

In the cytosol of eukaryotic cells glutathione is also present at high concentration, about 0.01 M. In contrast to the ER, the ratio of **GSH** to **GSSG** in the cytosol is over 50:1. Oxidized **GSSG** in the cytosol is reduced by the enzyme glutathione reductase, using electrons from the reducing agent NADPH. Thus cytosolic proteins in bacterial and eukaryotic cells do not utilize the disulfide bond as a stabilizing force.

Thus, the ER SH/SS redox state is about ten-fold more oxidized than that in the cytosol. Using a cell-free system, we could detect transport of GSSG and cystine into microsomes. The Km's for GSSG and cystine were approximately 9 and 1 mM, respectively. Reduced glutathione or reduced cysteine were transported at less than one-tenth the rates of the oxidized species. Results from the direct measurement of microsomal thiol/disulfide (SH/SS) redox state using ^{35}S-cyst(e)ine confirmed that the microsomal redox state is at least ten-fold more oxidized than that of the surrounding cytosol. The oxidizing redox state of the ER lumen thus is maintained, in part, by selective import of oxidized GSSG from the cytosol (10).

REARRANGEMENTS OF DISULFIDE BONDS

In proteins that contain more than one disulfide bond, the proper pairing of cysteine residues is essential for normal structure and enzymatic or hormone activity. In some proteins disulfide bonds are formed while the polypeptide is still growing on the ribosome. In the case of secreted immunoglobulin (Ig) light-chain polypeptides, disulfide bonds form sequentially: the first cysteine forms a disulfide bond with the second cysteine before the third has even been translated, automatically insuring the correct pairing of cysteines (1). Similarly, the third pairs with the fourth. The two disulfide bonds stabilize the two separate domains in this protein.

Sequential formation does not occur in all disulfide-bonded proteins. For example, proinsulin has three disulfide bonds that are not sequential. In this and most other cases, the first disulfide bonds that form spontaneously by oxidation of the Cys-SH groups may not be characteristic of the proper folded conformation of the protein. Disulfide bonds often rearrange, eventually forming the proper S-S bonds and stabilizing the proper protein conformation, but studies of folding in cell-free systems of purified proteins with many S-S bonds showed that this spontaneous process is very slow, requiring minutes to hours. In the ER, this rearrangement process is accelerated by the enzyme protein disulfide isomerase (PDI), which is found in abundance in the ER of secretory tissues in such organs as the liver and pancreas. PDI acts by catalyzing the rearrangement of disulfide bonds on a broad range of protein substrates, allowing them to generate their thermodynamically most stable configurations.

A lack of understanding of the process of disulfide bonds has cost several biotechnology companies considerable amounts of money in trying to synthesize functional therapeutic proteins by recombinant DNA techniques in bacterial cells. When mammalian secretory proteins are synthesized by recombinant DNA techniques in bacterial cells, these proteins generally are not secreted (even when a bacterial signal sequence replaces the normal one) and they accumulate in the cytosol. There, such proteins often denature and precipitate, in part because disulfide bonds do not form. Indeed, chemical methods must be used to refold the denatured bacterially-produced proteins, or cultured animal cells are employed to produce such proteins in quantity.

POST- TRANSLATIONAL FORMATION OF DISULFIDE BONDS

Recently we showed that addition of the potent reducing agent dithiothreitol (DTT) to HepG2 human hepatoma cells blocks secretion of newly-made albumin and causes it to accumulate in the endoplasmic reticulum in the reduced state. Subsequent incubation of the cells in the absence of DTT allows the reduced albumin to form apparently normal disulfide bonds and to be secreted normally (17). Thus, apparently normal formation of disulfide bonds on this protein can occur posttranslationally, as has been shown by others for the influenza HA hemagglutinin (2). α_1-antitrypsin contains no disulfide bonds; in the continuous presence of DTT it acquires a normal complement of complex oligosaccharides and is secreted at an only slightly reduced rate. Thus, the secretory pathway functions efficiently even when it is reduced by DTT. ER retention of albumin and other proteins during treatment with DTT occurs because, without disulfide bonds, these proteins cannot fold properly, not because of any defect in the overall processes of vesicular transport and protein secretion (17).

CHAPERONES AND FOLDING OF NEWLY-MADE PROTEINS

Many reduced, denatured proteins will spontaneously refold into the native state in cell- free systems. In most cases refolding requires hours to reach completion, yet secretory proteins generally fold in the ER lumen within minutes after their synthesis. The endoplasmic reticulum contains several proteins that facilitate the folding of other proteins within the ER lumen. Protein disulfide isomerase is one such chaperone, or folding catalyst. Bip is another, as it transiently binds to exposed hydrophobic segments of proteins, prevents them from misfolding or forming aggregates, and thus enhances their ability to fold into the proper conformation (6).

Other examples of chaperones are peptidyl-prolyl isomerases, a family of enzymes that accelerate the rotation about peptidyl- prolyl bonds in unfolded segments of polypeptide; such isomerizations are frequently rate-limiting for folding of protein domains. Most peptidyl-prolyl isomerases will catalyze rotation of any exposed peptidyl-prolyl bond. Others have very specific substrates. One such ER peptidyl-prolyl isomerase, called ninaA, is expressed in certain eye cells of Drosophila. This chaperone is required for folding only of opsin, the membrane protein that absorbs light and that is a major protein synthesized in these cells (19). This example illustrates an important point about chaperones: they can be remarkably specific in the proteins they help fold; folding of certain polypeptides in the ER can be absolutely dependent on a particular chaperone, while folding of other proteins may not require that catalyst.

DIFFERENT CATALYSTS REQUIRED FOR FOLDING OF DIFFERENT PROTEINS

The endoplasmic reticulum is a major intracellular reservoir of Ca^{2+}. The ER contains many Ca^{2+} - binding proteins, including protein disulfide isomerase, Bip, calreticulin, and calnexin. We showed that the high lumenal concentration of Ca^{2+} is essential for proper folding of certain newly- made secretory proteins. More specifically, we showed that exit from the ER of newly- made α_1- antitrypsin in HepG2 cells was completely blocked by treatment of cells with the Ca^{2+} ionophores A23187 or ionomycin; secretion of albumin, in contrast, was only minimally affected (16). Although several indirect experiments indicated that Ca^{2+} was required for an early step in folding of α_1- antitrypsin, this protein has no disulfide bonds and there was no simple way to analyze its state of folding within the ER.

Both H1 and the related H2 subunit of the asialoglycoprotein receptor are type II membrane proteins (~ 46 to 50 kDa) with a short cytosolic domain and a single transmembrane segment that also functions as an internal signal-anchor sequence for ER insertion (14). Of the ~230 amino acids in the exoplasmic domain of H1, the carboxy-terminal 150 constitute the Ca^{2+} - dependent galactose- binding domain. This domain contains ~30 conserved amino acids, including four conserved cysteine residues, that are characteristic of the C- class of animal lectins. By resolving immunoprecipitates on non-reducing SDS gels we have detected several disulfide-bonded intermediates in folding within the endoplasmic reticulum of newly-made H1 subunits. H1 in the ER can be partially unfolded by treatment of cells with dithiothreitol, but H1 in Golgi or post-Golgi organelles is resistant to such unfolding. This defines a late step in H1 folding that occurs just prior to exit from the ER. Depletion of calcium from the endoplasmic reticulum, either by treatment with A23187 or thapsigargin, totally blocks H1 maturation from the ER. No normal ER intermediates in H1 folding are formed in cells treated with A23187 or thapsigargin, indicating that at least an early step in H1 folding requires a high Ca^{2+} concentration in the ER lumen. Calcium binds reversibly to the mature asialoglycoprotein receptor, and is essential for it to bind galactose, and our results suggest that $Ca2^{+}$ is also essential for the exoplasmic domain of the receptor polypeptides to fold in the ER (20). As noted, however, many secretory proteins fold normally even if the Ca^{2+} is depleted.

FORMATION OF OLIGOMERIC PROTEINS IN THE ER

Many important secretory and membrane proteins are oligomers: they are built of two or more polypeptides. In all cases, these oligomeric proteins are assembled in the ER. One important example is the immunoglobulins, which contain two H and two L chains, all linked by S-S bonds. A second well-studied example (5, 7) is the assembly of the influenza hemagglutinin protein that forms the spikes protruding from the surface of the virus particle. Each spike is formed within the ER from three copies of a precursor protein, termed HA_0, that has a single membrane-spanning α- helix. In the Golgi complex each HA_0 is cleaved to an HA_1 and HA_2 polypeptide. The trimer is stabilized by interactions between the exoplasmic domains of the constituent polypeptides, as well by interactions between the three cytosolic and membrane-spanning domains. A newly made HA_0 polypeptide requires approximately 7 min. to fold into a trimer. Several approaches have been used in influenza-infected cells to detect intermediates in folding of the monomers and their subsequent assembly into trimers. Certain monoclonal antibodies, for instance, bind only to a region of the HA_0 polypeptide that is exposed on the monomer surface but folded inside the rigid trimer. Others bind only to a region on the surface of the correctly folded trimer and thus immunoprecipitate trimers, not monomers. In a typical experiment, virus-infected cells are pulsed-labeled with a radioactive amino acid, and at various times during the subsequent chase period membranes are solubilized by detergent. Immediately after the pulse, all of the radioactive HA_0 protein is immunoprecipitated by the monomer-specific monoclonal antibody. These newly- made polypeptides have formed some, but not all, of the disulfide bonds characteristic of the mature protein. During the chase period, increasing amounts of radioactive HA_0 protein react only with the trimer-specific monoclonal, and lower amounts with the other. In this manner, one can follow the folding of newly-made HA_0 monomers and also the monomer - to - trimer conversion of within a living cell.

QUALITY CONTROL IN THE ENDOPLASMIC RETICULUM

After synthesis, both secretory and membrane proteins are packaged into small transport vesicles that carry them to the Golgi vesicles, but only properly folded proteins are generally allowed to enter the vesicles. Incompletely folded proteins do not reach the plasma membrane or are not secreted. Some of these polypeptides accumulate in the ER; others are degraded in the ER.

Mutations in almost any secretory or membrane protein that prevent it from folding properly also prevent the polypeptide from moving out of the ER. One example is a mutation in the secretory protein α_1- antitrypsin (α_1- antiprotease), which is secreted by hepatocytes and macrophages; this protein binds to and inhibits trypsin and also the blood protease elastase. In the absence of α_1- antitrypsin, elastase degrades the fine tissue in the lung that participates in the absorption of oxygen. A genetic inability to produce α_1- antitrypsin, widespread in Caucasians, is the major genetic cause of emphysema. The defect is due to a single mutation in α_1- antitrypsin in which $glutamate_{342}$ is replaced by lysine. As a result, the protein is produced normally in the rough ER, but the α_1- antitrypsin forms an almost crystalline aggregate that is not exported beyond the ER (12). The hepatocytes become full of rough ER that contains aggregated α_1- antitrypsin, and secretion of other important liver proteins also becomes impaired.

Finally, the most common mutation that causes cystic fibrosis is a deletion of amino acid 508 (ΔF 508) in the CFTR protein. The mutant protein like the wild type CFTR protein forms a functional chloride channel soon after its synthesis in the ER. The mutant protein, however, does not move from the ER to the plasma membrane at normal body temperature (37° C) but will at lower temperatures (25° C). Thus the ΔF 508 mutation causes a temperature-sensitive defect in protein maturation, apparently because the ΔF508 polypeptide causes the protein to assume a slightly incorrect structure at 37° C but a normal one at 25° C (4).

In addition to retention of misfolded proteins in the ER, another form of quality control is degradation of unfolded or misfolded polypeptides, particularly subunits of oligomeric proteins, while they are in the ER. Well-worked out examples include the α subunit of the T-cell receptor (11). It is normally assembled into a multi-subunit oligomeric complex while it is in the ER, and only the completely assembled oligomer is allowed to exit the ER en route to the Golgi. In contrast, if the α subunit is expressed in a cell in the absence of the other subunits, the polypeptide is retained in the ER and is selectively degraded there within one to two hours of its synthesis. Thus, the ER contains a set of proteolytic enzymes that somehow recognizes only unassembled polypeptides and degrades them. Using the H2a subunit of the asialoglycoprotein receptor as a model system, we showed that there are two distinct pathways for its degradation in the endoplasmic reticulum. Unfortunately, little is known about the putative ER proteases, except that signal peptidase participates in one of the pathways of H2a degradation (22).

INCREASED CELL SURFACE EXPRESSION AND ENHANCED FOLDING IN THE ENDOPLASMIC RETICULUM OF MUTANT FORMS OF THE ERYTHROPOIETIN AND ASIALOGLYCOPROTEIN RECEPTORS

In contrast to secretory proteins, many newly- made cell surface receptors are processed inefficiently, in that only a small fraction fold properly and appear on the plasma membrane. We have identified point mutations in two receptors - the erythropoietin receptor and one subunit of the human asialoglycoprotein receptor - that increase the rate and/or extent of receptor folding and processing. Inefficient processing of wild-type receptor proteins is one mechanism for controlling the number of plasma membrane receptors, and the yield of recombinant receptors can be increased by mutagenesis.

The erythropoietin receptor (EPO-R) is a member of the cytokine receptor superfamily. The defining features of this group of transmembrane glycoproteins are present in the extracellular domain and include four cysteine residues, the spacing of which is conserved, a series of aromatic residues, and the five amino acid motif Trp-Ser-Xaa-Trp-Ser (WSXWS). The extracellular segment of the growth hormone receptor is homologous in sequence to that of the EPO-R, and structural studies have determined that it is composed of two domains, each of which contain seven ß-strands that form a barrel. The four conserved cysteines and one conserved tryptophan are buried within the interior of the first ß-barrel and presumably stabilize this domain (3). Although the homologue of the WSXWS motif in the growth hormone receptor, YGEFS, is close to a solvent accessible surface, it is neither part of the hormone-binding nor the receptor dimerization surfaces, and its function was undefined.

Our work suggests that the WSAWS sequence of the EPO-R is a structurally important element essential for proper folding of the receptor (8, 21). We have generated and analyzed all 95 possible single amino acid substitutions of these five residues in the EPO-R. All mutants were synthesized at the same level; however, their passage from the endoplasmic reticulum (ER) to the Golgi apparatus and cell surface differed. The tolerance of the tryptophan and serine residues to substitution was narrow; most mutant proteins were retained in the ER, were not processed through the Golgi complex, and showed no cell surface expression. Although many mutants with substitutions at the middle residue of the motif reached the cell surface less efficiently or as efficiently as the wild type receptor, it was notable that one mutant, A234E, was processed more efficiently than the wild type receptor and was expressed in elevated numbers at the cell surface (8). This is the first example of a mutant receptor polypeptide that is processed more efficiently than its wild-type counterpart in the ER, and suggests that the inefficient processing of the wild-type EPO-R is one mechanism for controlling the number of plasma membrane receptors. Despite the variation in surface expression among the WSXWS mutants, all mutant receptors that reached the cell surface bound and internalized EPO normally, and were able to transmit a proliferative signal. These results suggest that the WSXWS motif plays a crucial role in generating a correctly folded receptor which is competent to exit the ER and be transported to the cell surface.

The two subtypes of the H2 subunit of the asialoglycoprotein receptor, H2a and H2b, a galactose specific C-type lectin, differ only by a pentapeptide insertion in the exoplasmic juxtamembrane region of H2a. All newly-made H2a is degraded in the endoplasmic reticulum (ER), but up to 20% of H2b matures to the cell surface (13, 22). Unfolded forms of the H2b subunit are degraded in the ER while folded forms of the protein can mature to the cell surface (20). There are eight cysteines in the exoplasmic domain of the protein, forming four disulfide bonds in the folded protein. We have constructed double cysteine to alanine mutants for each of the four disulfide bonds and examined the folding and metabolic fate of each of the mutants in transfected 3T3 fibroblasts (23).

As determined by migration on non-reducing SDS-PAGE, and by sensitivity of the proteins to unfolding by the reducing agent DTT, we found that mutation of the two cysteines nearest to the transmembrane region (C1) does not prevent proper folding of the protein. In contrast, mutations of the other three disulfides prevent proper folding of the protein and all of the mutant proteins are degraded in the ER. A normal (~20%) fraction of the C1 mutant protein exits the endoplasmic reticulum and is processed in the Golgi complex, but it does so at a ~3- fold faster rate compared to the wild-type. The C1 mutant protein is expressed on the cell surface and can form a functional receptor with the H1 subunit with similar binding affinities for natural ligands as that of the wild-type receptor. The same fraction of newly-made mutant and wild-type proteins (~80%) remain in the ER, but the mutant protein is degraded more quickly. Thus, the presence of the C1 disulfide bond in the wild-type receptor both reduces the rate of protein folding and exit to the Golgi, and also slows the rate of ER degradation of the portion (~80%) of the receptor that never folds properly. This is another example of enhanced folding of mutant forms of cell- surface receptor proteins. Inefficient or slow processing of wild-type receptor proteins is one mechanism for controlling the number of plasma membrane receptors, and, importantly, the yield of recombinant receptors can be increased by mutagenesis.

REFERENCES

1. Bergman, L. W., and W. M. Kuehl. 1979. Formation of an intrachain disulfide bond on nascent immunoglobulin chains. *J. Biol. Chem.* 254:8869-8876.

2. Braakman, I., J. Helenius, and A. Helenius. 1992. Manipulating disulfide bond formation and protein folding in the endoplasmic reticulum. *Embo J.* 11:1717-22.

3. De Vos, A. M., M. Ultsch, and A. A. Kossiakoff. 1992. Human growth hormone and extracellular domain of its receptor: crystal structure of the complex. *Science.* 255:306-311.

4. Denning, G. M., M. P. Anderson, J. F. Amara, J. Marshall, A. E. Smith, and M. J. Welsh. 1992. Processing of mutant cystic fibrosis transmembrane conductance regulator is temperature-sensitive. *Nature.* 358:761-4.

5. Doms, R. W., R. A. Lamb, J. K. Rose, and A. Helenius. 1993. Folding and assembly of viral membrane proteins. *Virology.* 193:545-62.

6. Gething, M. J., and J. Sambrook. 1992. Protein folding in the cell. *Nature.* 355:33-45.

7. Hammond, C., and A. Helenius. 1995. Quality control in the secretory pathway. *Curr Opin Cell Biol.* 7:523-9.

8. Hilton, D. J., S. S. Watowich, P. J. Murray, and H. F. Lodish. 1995. Increased cell surface expression and enhanced folding in the endoplasmic reticulum of a mutant erythropoietin receptor. *Proc. Natl. Acad. Sci. USA.* 92:190- 194.

9. Hurtley, S. M., and A. Helenius. 1989. Protein oligomerization in the endoplasmic reticulum. *Annu Rev Cell Biol.* 5:277-307.

10. Hwang, C., Sinskey, A. J., and H. F. Lodish. 1992. Oxidized redox state of glutathione in the endoplasmic reticulum. *Science.* 257:1496-1502.

11. Klausner, R. D., and R. Sitia. 1990. Protein degradation in the endoplasmic reticulum. *Cell.* 62:611-4.

12. Le, A., G. A. Ferrell, D. S. Dishon, Q. Q. Le, and R. N. Sifers. 1992. Soluble aggregates of the human PiZ alpha 1-antitrypsin variant are degraded within the endoplasmic reticulum by a mechanism sensitive to inhibitors of protein synthesis. *J Biol Chem.* 267:1072-80.

13. Lederkremer, G. Z., and H. F. Lodish. 1991. An alternatively spliced miniexon alters the subcellular fate of the human asialoglycoprotein receptor H2 subunit. Endoplasmic reticulum retention and degradation or cell surface expression. *J Biol Chem.* 266:1237-44.

14. Lodish, H. F. 1991. Recognition of complex oligosaccharides by the multi-subunit asialoglycoprotein receptor. *Trends Biochem Sci.* 16:374-7.

15. Lodish, H. F., D. Baltimore, A. Berk, L. Zipursky, P. Matsudaira, and J. D. Darnell. 1995. Molecular Cell Biology, 3rd ed. Chapter 16. Scientific American Press, New York.

16. Lodish, H. F., and N. Kong. 1990. Perturbation of cellular calcium blocks exit of secretory proteins from the rough endoplasmic reticulum. *J Biol Chem.* 265:10893-9.

17. Lodish, H. F., and N. Kong. 1993. The secretory pathway is normal in dithiothreitol- treated cells, but disulfide bonded proteins are reduced and reversibly retained in the endoplasmic reticulum. *J. Biol. Chem.* 268:20598-20605.

18. Lyles, M. M., and H. F. Gilbert. 1991. Catalysis of the oxidative folding of ribonuclease A by protein disulfide isomerase: dependence of the rate on the composition of the redox buffer. *Biochemistry.* 30:613-9.

19. Stamnes, M. A., and C. S. Zuker. 1990. Peptidyl-prolyl cis-trans isomerases, cyclophilin, FK506-binding protein, and ninaA: four of a kind. *Curr Opin Cell Biol.* 2:1104-7.

20. Wikström , L., and H. F. Lodish. 1993. Unfolded H2b asialoglycoprotein receptor subunit polypeptides are selectively degraded within the endoplasmic reticulum. *J. Biol. Chem.* 268:14412-14416.

21. Yoshimura, A., T. Zimmers, D. Neumann, G. Longmore, Y. Yoshimura, and H. F. Lodish. 1992. Mutations in the trp-ser-X-trp-ser motif of the erythropoietin receptor abolish processing, ligand binding, and activation of the receptor. *J. Biol. Chem.* 267:11619-11625.

22. Yuk, M. H., and H. F. Lodish. 1993. Two pathways for the degradation of the H2 subunit of the asialoglycoprotein receptor in the endoplasmic reticulum. *J Cell Biol.* 123:1735-49.

23. Yuk, M. H., and H. F. Lodish. 1995. Enhanced folding and processing of a disulfide mutant of the human asialoglycoprotein receptor H2b subunit. *J Biol Chem.* 270:20169-76.

Discussion

Spier What you are implying by saying the rate limiting step is sitting with these post-translational modifications in the Golgi is that we have 2 ways of proceeding to actually enhance the productivity of systems. On the one hand we can use liver cells and transform them and then actually produce recombinant proteins in liver cells where all of the enzymes sit. Alternatively, to transfect those particular genes for those particular enzymes into BHK or CHO cells. Can you advise us on which way to go?

Lodish: Many people have experimented over-producing some of these catalysts. The results are mixed. In some cases it will enhance protein secretion, presumably by preventing mis-folding. In other cases it inhibits secretion by causing enhanced retention. My instincts are rather than to engineer standard cells to have a look at myeloma cells, hepatoma cells, or some of the others - particularly if you are having trouble making a protein with a lot of disulphide bonds. My instincts are to go back to nature and try and get the cells to do the work for you rather than to try and take a fibroblast which, of course, specialises in making collagen and try to re-engineer all the proteins most of which we do not know, to get the cell to make say a disulphide bonded protein.

Bernard: A question on your receptors work. Have you done the reverse experiment of taking the ones which are naturally E in the middle and rotated them back to something else?

Lodish: The question was have we mutated other receptors with the glutamate. We are doing this now, but we do not yet have any results.

Barteling: Does the role of BIPS and other protein chaperones in the folding of the protein mean that predicting the structure of proteins by computer model becomes very difficult?

Lodish: An interesting question on whether the existence of chaperones may alter the structure. The simple answer is that I do not know but I think it might because we do not know enough about whether protein folding is thermodynamic or kinetic in nature. By preventing pathways which may be thermodynamically favoured, that is some of the misfolding events, it may in fact alter the state of the protein. It is certainly worth looking at. I guess, if I was convinced that all proteins were made in a thermodynamically most favoured state I would say there is no reason to think so.

Bailey: Is not an aggregated mass of unfolded protein the minimum free energy state?

Lodish: Yes, I think it is the prevention of that by BIP which would drive it down a more productive pathway. It is an interesting point that you might get a computer model to say that the most favoured state is an aggregate.

Hoffman: Do you think, or know, whether renaturation *in vitro* is correlated to the retention time in the ER?

Lodish: I do not know of any systematic attempt to do this. These experiments would have to be done under defined redox conditions. Many of the other experiments were simply done by air oxidation which may not be the best way. The way to do the experiment would be *in vitro* folding in a redox buffer with protein disulphide isomerase and then make the correlation. My instincts are that it would.

Handa-Corrigan: I am concerned in that we put low molecular weight surfactants in our medium. These surfactants accumulate in the cytoplasm. Whether they localise in the ER or Golgi I am not sure. I am concerned that we are changing our protein folding and output out of the cell with these intracellular polymers. Do you think something like lung surfactant which is less non-specific but still highly surface active might be an alternative for biotechnologists to use other than these man-made polymers?

Lodish: I am not an expert on surfactants so I do not really want to answer the specific question. I do think there are attempts, sometimes successful, of putting in small amounts of chaeotropic agents into cells to try to catalyse or reverse the folding of misfolded proteins. CFTR, the Cystic Fibrosis protein, has been used successfully where you can in fact generate the mutant protein on the surface. It does illustrate that they do have affects at least in model systems on protein folding.

SESSION I

Cells and Vectors for Genetic Medicine

POLYMER ENCAPSULATED XENOGENEIC CELL LINES: A NOVEL APPROACH FOR GENE THERAPY

P. Aebischer
Division of Surgical Research and Gene Therapy Center, Centre Hospitalier Universitaire Vaudois, Lausanne University Medical School, Switzerland.

In *ex-vivo* gene therapy, a gene product can be delivered through the transplantation of autologous cells genetically engineered to express the desired gene. The encapsulation technology which consists in surrounding cells by a semipermeable polymeric membrane allows cells from either an allogeneic or xenogeneic source to be transplanted across species. This technology offers interesting potential for gene therapy such as the use of cell lines. Cell lines offer several advantages including unlimited availability, the possibility of rapid *in vitro* screening for the presence of pathogens from which cell bank are established, and the suitability for stable gene transfer using non-viral based recombinant DNA techniques. Use of a cell line of xenogeneic origin provides an additional advantage since the transplanted cells will be rejected by the host immune system in the event of device breakage. To increase the safety level, the engineered cells also express a suicide gene (thymidine kinase) rendering the cells sensitive to gancyclovir mediated destruction.

The technique of transplanting polymer encapsulated cells genetically engineered to release bioactive molecules is currently investigated in our laboratory for several diseases including anemia with the release of erythropoeitin, diabetes with conditionally immortalized ß-cell lines, Parkinson's disease with the release of either dopamine or glial cell-line-derived neurotrophic factor (GDNF) and Lou Gehrig disease or ALS with the release of cocktails of neurotrophic factors. A first clinical trial based on the intrathecal delivery of ciliary neurotrophic factor (CNTF) by an encapsulated baby hamster kidney cell line genetically engineered to release CNTF has demonstrated the safety of the technique, the viability of the transgene.

M.J.T. Carrondo et al. (eds.), Animal Cell Technology, 29–31.

Discussion

Cunningham: Have you looked, and if so, have you detected any rDNA sequences from your caplets?

Aebischer: We have not looked specifically for DNA. The cut-off is 50,000 so the likelihood of having anything significant leaching out is small.

Hauser: How can your cells survive such a long time in one system? Could you also find this in *in vitro* culture?

Aebischer: I think BHK cells are known to be fairly hardy, CSF is not the best culture medium, being low in protein and oxygen. However, the cells survive both *in vivo* and *in vitro,* and not only BHK cells but the majority of cell lines we have used showed good survival in CSF.

Al-Rubeai: With a MW cut-off of 50,000 a lot of molecules, like growth factors, can be released.

Aebischer: Cut-off is a Gaussian distribution so you will have, in fact, some molecules of higher MWt that will diffuse. It is a blunting system. One cannot keep all antibodies and complement out. The dynamics are such that the system can deal with them and we would want to open the cut-off to let more protein in and out which would help cell survival.

Sinacore: It was not clear to me how you carry out the encapsulation process to create the fibres with the transfected BHK cells. Can you provide more detail of how this is done?

Aebischer: There are 2 general ways. Either you precipitate the membrane around the cell core by a phase inversion process as you would normally produce hollow fibres. The other way is to fill a preformed hollow fibre at both ends and this is the way we do it in the clinic. The most difficult part is the closure of the device because it has to be done in the wet stage.

Weissmann: Would it be reasonable to introduce Factor VIII or Factor IX producing cells in such an encapsulated form?

Aebischer: Yes. We are studying EPO but the problem is the MWt of the molecule you are delivering. We thought at the beginning it would perhaps be a problem when you have large molecules as you cannot distinguish between the immune system and the bioactive molecule. Now it is evident that with cells expressing complement inhibiting molecules that you can avoid rejection. This opens up the target molecules that can be used in this technology.

Weissmann: With the productivity you have could you make enough?

Aebischer: It depends on how many cells you can put in the device. Outside the nervous system such as you would have for Factor VIII or IX you are not confined to hollow fibre geometry and you can use disc based systems which we are using for peripheral applications. Then billions of cells can be encapsulated in a disc of 3-5 cm which should allow a sufficient dose to be delivered. A combination of high expression levels, these discs and complement-inhibitory factor, should allow treatment of haemophilia and diseases requiring high MWt molecules.

Handa-Corrigan: What cytochemical strains did you use to distinguish viable apoptotic cells in your system?

Aebischer: Just standard H&E. We measure the transient expression when we retrieve the implant and we were within 10-20% of that at implantation. So we did not put a lot of emphasis on vital dyes. The key for us is expression.

Handa-Corrigan: A similar system used for Islet cell culture has been shown to fail after 3 months because of cell necrosis.

Aebischer: That depends on clump size. With Islets people used primary cells from an extensive enzyme isolation system which gave poor viability. We control the clump size and distribution within the matrix and capsules.

Handa: Corrigan: The size of the clumps you showed appeared larger than any Islet clump.

Aebischer: This is post-mitotic. For mitotic cells we have necrosis.

GENE THERAPY AND GENETIC VACCINATION : TWO HALLMARKS OF GENETIC MEDICINE.

Dalemans Wilfried. (SmithKline Beecham Biologicals, Research and Development. Rixensart - Belgium)

The remarkable progress in the isolation and characterisation of new genes as well as in the understanding of gene function in health and disease has now led to direct application of genes in what is called genetic medicine. Two approaches of using genes, rather than their protein products, in the treatment or prophylaxis of disease are intensively explored.

Gene therapy consists in delivering, either by ex vivo genetic modification of body cells or by direct in vivo delivery, genes whose expression will allow to exert a therapeutic effect in the patient. Gene therapy is being evaluated under the form of gene replacement therapy in patients who suffer from the lack of a functional protein. It is also used to battle cancer through genetic modification of cancer or immune cells.

Genetic vaccines are currently being explored as a promising new way of vaccination. In this approach, gene expression cassettes encoding antigens are administered intramuscularly, and the subsequent in vivo expression of the antigens induces specific immune responses leading eventually to protection against pathogens.

In both these types of genetic medicine, vector development and gene delivery methods are key issues since they will determine the efficacy, safety, and long-term outcome of genetic medication.

M.J.T. Carrondo et al. (eds.), Animal Cell Technology, 33–35.

Discussion

Brown: The guinea pig model is the gold standard for protection in the HSV field. I did not see any of those data. Has that been done?

Dalemans: Yes, by other groups as well as ours and we have seen efficacy towards HSV infection. There is a need for improvement because if you then need to use the gold standard for the HSV model which is an adjuvented protein these do perform better. If you go to a bigger animal it is clear that you have to add something more.

Brown: Some of the DNA infection protocols require a membrane disturbing agent.

Dalemans: We did not check this. It is a very controversial way of doing it. It has been shown that if you pre-treat the muscle with de-stabilising agents, or which lyse the muscle, and then inject the DNA, you have a much better transfection and expression efficiency due to the regeneration of the muscle. It is also claimed that this gives a better immune response. Others claim it does not work.

Onions: We have been doing a number of DNA vaccinations of domestic animal species. One of these is a Nef IV model for HIV and we see a consistent pattern with the DNA vaccination using a defective single cycle vector. It is put in as DNA in parallel experiments with virus as a vector. We see a consistent qualitative change in the immune response between the envelope and the *gag* proteins with a shift in response towards the *gag* proteins in the DNA vaccinates as opposed to the ones given the whole vector. Do you see similar changes?

Dalemans: We have not been comparing DNA and live vectors in HIV or in any other field. A shift in immunological response has been seen by others in multiple DNA vaccinations. It shifts from a cellular response to a CH2, or antibody type response. This is dependent on the antigen being very prominent for the envelope, much less so for the *gag*.

Weissmann: There was a report last year (by Krieg) that using oxy-G rich oligonucleotides you could boost the immune response. Has anyone tried to combine this oligonucleotide treatment with DNA vector treatment?

Koell: Not yet but this technique is prominent with bacterial DNA.

PCR DETECTION OF MURINE RETROVIRUSES AND CHARACTERIZATION OF A NEW XENOTROPIC RETROVIRUS FROM MYELOMA CELLS

Manfred Wirth, Thomas Heinemeyer, Lorin Schumacher, Andreas Klingenhoff, Wiebke Hansen, Zoltan Suemeghy[1], Gabriela Huber[1], Beatrice Susanne Schläpfer[1], Dagmar Wirth and Hansjörg Hauser

GBF, Braunschweig, F.R.G. and Sandoz Pharma AG [1], Basle, Switzerland

SUMMARY

We have established a 'nested PCR' technique for the group-specific detection of murine C-type retroviruses to screen cell lines for the presence of contaminants. Analyses confirmed that a well known myeloma cell line, Sp2/0, is contaminated with a retrovirus. The genome of the retrovirus was isolated and characterized by nucleic acid sequencing. The genomic organization of this retrovirus is typical for a murine C-type retrovirus. However, numerous differences could be found in the primary structure of *gag* and the *integrase* protein. The amino acid sequence of the *envelope* region and infection experiments with cell lines of different species indicate a xenotropic host range of the virus. The *envelope* region of the Sp2/0 virus shows strong homology to the xenotropic retrovirus found in CWD mice.

INTRODUCTION AND OBJECTIVES

Contamination of mammalian cells with viruses is a well known obstacle associated with the production of pharmaceuticals in cell cultures. Despite the ability of certain protein purification processes to remove definitely contaminations, e.g. retroviruses, the screening for viral contaminations and the determination of the viral burden represent basic demands of biosafety. Exogenous mouse retroviruses, which are released as infectious particles, are found in certain cell lines derived from mouse tissue and have been observed occasionally in other rodent species. Endogenous retroviruses are widely distributed in rodents, but due to defects in their regulatory region or structural genes they form infectious particles only under certain conditions. In many cases a retrovirus infection is not accompanied with a visible cytopathic effect, and therefore remains undetected, especially in the case of slow virus replication and a low titer. For this purpose we have developed an ultrasensitive tool for detection of mouse C-type retroviruses. While screening established cell lines for their viral load, we confirmed that the Sp2/0-Ag14 cell line, a descendant of a BALB/c mouse tumor is infected with mouse C-type retroviruses. We found that some related myelomas are contaminated with these viruses, too. The Sp2/0 derived retrovirus is of particular interest because of its low replication rate and its host range. Thus, the aim of our study was the characterization of the Sp2/0 retrovirus.

M.J.T. Carrondo et al. (eds.), Animal Cell Technology, 37–42.

RESULTS AND DISCUSSION

The nested RT-PCR assay for detection of mouse retroviruses

The nested RT-PCR assay (Fig.1) applies degenerate primers from the highly conserved parts of the *gag* region. It represents a reliable, specific and sensitive test for the presence of mouse C-type retroviral RNA and/or proviral DNA. Up to one infectious particle can be detected in cell culture supernatants making the assay the most sensitive detection method for this group of retroviruses.

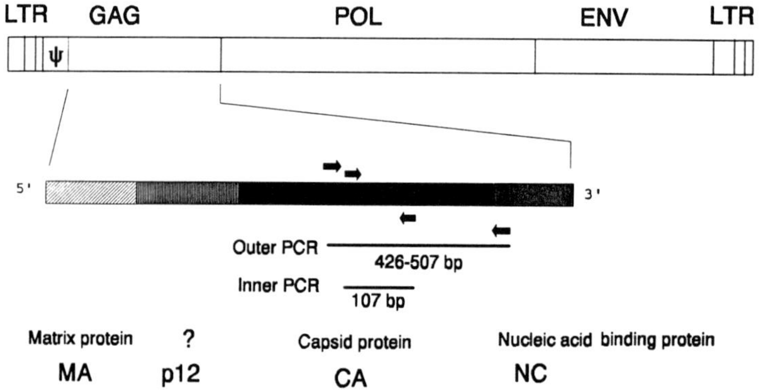

Fig. 1**: **Localization of PCR primers in the genome of murine retroviruses.
The length of the products of amplification is indicated. The details of the nested RT-PCR procedure will be published elsewhere (1).

Due to increased sensitivity compared to traditional detection methods for mouse retroviruses, the RT-PCR allowed us to discover retroviruses in cell lines previously concerned retrovirus-free using traditional methods (e.g. the radioactive version of the RT-assay). Investigation of cell lines by RT-PCR showed that several cell lines, release viral RNA containing particles (Tab.1). Some of these cell lines were previously stated as retrovirus-free based on standard detection methods. Among them, three closely related myeloma cell lines were considered virus-positive, namely Sp2/0, FOY-NY and X63-NSI/1-Ag4.1 cells. Mouse retroviruses were found independently in Sp2/0 cells originating from four different sources, which makes the adventitious infection by virus during cell handling unlikely. All cell lines are descendants of P3K, a cell line established from the BALB/c mouse tumor MOPC21 (Fig.4). These cell lines represent common fusion parents for the construction of hybridomas. Estimations of viral titers by RT-PCR with serial dilutions of supernatants revealed that the Sp2/0 cell line released 2-3 orders of magnitude less virus compared to the MoMuLV infected NIH3T3 cells. These virus replication properties may be due to unusual features in the protein coding or regulatory regions of the viral genome or the locus of integration into the hosts genome.

Table 1: Screening for murine C-type proviruses in the chromosomal DNA and virion in supernatants of cell lines

cell line	Virus released	PCR (chromosomal)	PCR (Supernatant)
NIH3T3MLV	MoMuLV	++	++
NIH3T3 SV40PRO	rec. MPSV-MoMuLV	n.d.	++
Sp2/0 ATCC CRL1581	MuRV, xenotropic	+ +	+
Sp2/0 DSM ACC 146	MuRV	n.d.	+
Sp2/0 source3	MuRV, xenotropic	++	+
Sp2/0 source4	MuRV	n.d.	+
NRK-SF+22	F-MuLV/SFFV	n.d.	++
MDTFmRec	unknown	+	+
MN7		n.d.	-
MDBK		n.d.	-
BHK		-	-
CHO		+	-
CCL64		-	-
293		-	-
NIH3T3		++	-
Karpas 299		n. d.	-
P3/NSI/1-Ag4-1		n. d.	+
X63-Ag8.653		n.d.	-
FOX-NY		n.d.	+

Characterization of the Sp2/0 retrovirus

The host range of the Sp2/0 mouse retrovirus was investigated. Mouse NIH3T3 , human 293 cells and two additional human cell lines could not be infected productively with the Sp2/0 retrovirus. However, mink lung cells (ATCC CCL64) were infected and support its replication. The ability to multiply on non-mouse cells and the exclusion of mouse cells as infectable hosts defines a xenotropic host range for the Sp2/0 mouse retrovirus. The failure to establish a productive infection in human cells until now is remarkable, as the related NZB-retrovirus has a highly homologous *env* protein and is able to infect human cells. To characterize the new retrovirus in detail, the complete nucleic acid sequence of the provirus was elucidated by PCR cloning and sequencing starting with chromosomal DNA of infected CCL64 cells. The primary structure reveals features typical for a mouse C-type retrovirus. The retroviral regulatory regions for transcription, RNA processing, translation, protein processing and the signals neccessary for replication closely resemble those in the ecotropic MoMuLV retrovirus. The *envelope* or *SU* protein exhibits 99.4% and 96.7% homology to the *envelopes* of the xenotropic retrovirus of the CWD mouse (2) or the NZB xenotropic retrovirus (3) (Fig. 2).

```
     | SP ->                    <- SP | GP70 ->
  1: MEGPAFSKPLKDKINPWGPLIVIGILVRAGASVQRDSPHQVFNVTWRVTNLMTGQnsNATSLLGTMTDTFPKLYFDLCDLVGDHWDDPEPDIGDGCRSPGGRKRtRLYDF
  1: MEGPAFSKPLKDKINPWGPLIVIGILVRAGASVQRDSPHQVFNVTWRVTNLMTGQtaNATSLLGTMTDTFPKLYFDLCDLVGDHWDDPEPDIGDGCRSPGGRKRsRLYDF
  1: *******************************************************  *********************************************** *****

111: YVCPGHTVPIGCGGPGEGYCGKWGCETTGQAYWKPSSSWDLISLKRGNTPKDQGPCYDSSVSSGVQGATPGGRCNPLVLEFTDAGKKASWDAPKVWGLRLYRSTGADPVT
111: YVCPGHTVPIGCGGPGEGYCGKWGCETTGQAYWKPSSSWDLISLKRGNTPKDQGPCYDSSVSSGVQGATPGGRCNPLVLEFTDAGKKASWDAPKVWGLRLYRSTGADPVT
111: **************************************************************************************************************

221: RFSLTRQVLNVGPRVPIGPNPVITEQLPPSQPVQIMLPRPHPPPSGAASMVPGAPPPSQQPGTGDRLLNLVKGAYQALNLTSPDRTQgCWLCLVSGPPYYEGVAVLGTY
221: RFSLTRQVLNVGPRVPIGPNPVITEQLPPSQPVQIMLPRPHPPPSGAASMVPGAPPPSQQPGTGDRLLNLVKGAYQALNLTSPDRTQeCWLCLVSGPPYYEGVAVLGTY
221: *************************************************************************************** **********************

331: SNHTSAPANCSVASQHKLTLSEVTGQGLCVGAVPKTHQALCNTTQKASDGSYYLAAPAGTIWACNTGLTPCLSTTVLNLTTDYCVLVELWPKVTYHSPGYVYDQFERKTK
331: SNHTSAPANCSVASQHKLTLSEVTGQGLCVGAVPKTHQALCNTTQKASDGSYYLAAPAGTIWACNTGLTPCLSTTVLNLTTDYCVLVELWPKVTYHSPGYVYDQFERKTK
331: **************************************************************************************************************
<- GP70 | p15e ->
441: YKREPVSLTLALLLGGLTMGGIAAGVGTGTTALVATKQFEQLQAAIHTDLGALEKSVSALEKSLTSLSEVVLQNRRGLDLLFLKEGGLCAALKEECCFYADHTGVVRDSM
441: YKREPVSLTLALLLGGLTMGGIAAGVGTGTTALVATKQFEQLQAAIHTDLGALEKSVSALEKSLTSLSEVVLQNRRGLDLLFLKEGGLCAALKEECCFYADHTGVVRDSM
441: **************************************************************************************************************
                                                                          <- p15e | p2e ->    <- p2e |
551: AKLRERLNQRQKLFESGQGWFEgLFNRSPWFTTLISTIMGPLIVLLLILLLGPCILNRLVQFVKDRISVVQALVLTQQYHQLKSIDPEEVESRE*
551: AKLRERLNQRQKLFESGQGWFEaLFNRSPWFTTLISTIMGPLIVLLLILLLGPCILNRLVQFVKDRISVVQALVLTQQYHQLKSIDPEEVESRE*
551: ********************** ***********************************************************************
```

***Fig. 2*: Comparison of xenotropic envelope proteins.** *The env proteins of the Sp2/0 derived retrovirus (top) and the CWM-S-5X retrovirus of the CWD mouse (bottom) (2) were aligned using the GENMON program package (GBF, Braunschweig). Positions marked with * are totally conserved. Amino acid substitutions are indicated in lower case letters.*

The *gagpol* region shows extensive homology to the respective region of the spleen focus forming virus (SFFV) , mink cell focus forming viruses (MCF), the radiation leukemia virus (RadLV) and endogenous mouse retroviruses, e.g. AKV (not shown). However, differences are found in the primary structures of the gag proteins if compared with mouse leukemia viruses like the MoMuLV (Fig.3). Most prominent, due to insertion of three bases a 75 nt *gag* segment is out of frame. The changes in the nucleic acid sequences are concomitant with substitutions of 21 amino acids in the *p12 / CA* boundary (Fig. 3). Strikingly, the phenylalanine / proline protease cleavage site for processing of the *gag*-precursor protein into *p12* and *CA*, which lies in the region of divergence, is conserved. The influence of the terminal alterations on *p12 / CA* function has not yet been elucidated. While the function of *p12* is still unknown, *CA* is known to form the ' inner core ' or capsid of the virus by oligomerization. Earlier investigations (4) by linker scanning mutagenesis within the *CA* region of a related retrovirus have shown, that full integrity of *CA* is a prerequisite of efficient virus particle formation. Thus, the divergence in the Sp2/0 amino-terminus can influence the efficiency of particle formation and may explain the low viral titer found in Sp2/0 cells. Amino acid substiutions were also found in the region coding for distinct retroviral enzymes. In contrast to the evolutionary well conserved *reverse transcriptase* of the *pol* region, the *integrase* shows a remarkable degree of divergence if compared with the MoMuLV protein.

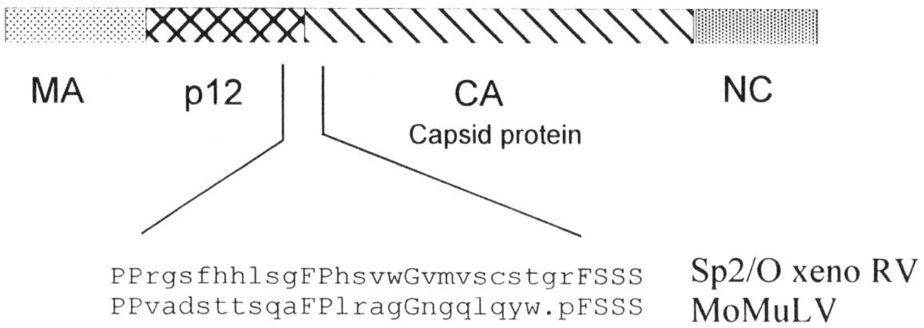

Fig. 3: Localization of the 75 b frameshift region in the gag polyprotein. *The amino acid sequences of the p12/capsid boundary of the Sp2/0 derived retrovirus and the Moloney murine leukemia virus are displayed. Remarkably, the protease cleavage site (amino acids F P) in the region of non-homology is conserved.*

The origin of the Sp2/0 retrovirus and the time of its emergence are still unclear. Recombinations of an ecotropic retrovirus with endogenous retroviral sequences may have contributed to the formation of such a xenotropic retrovirus. Interestingly, Deo et al. (5) recently discovered an ecotropic mouse retrovirus in Sp2/0 cells using a new and sensitive serological focus assay. However, more information is necessary to clarify the open questions. Investigation of the Sp2/0 precursors will be helpful to throw light on the history of Sp2/0 infection (Fig.4).

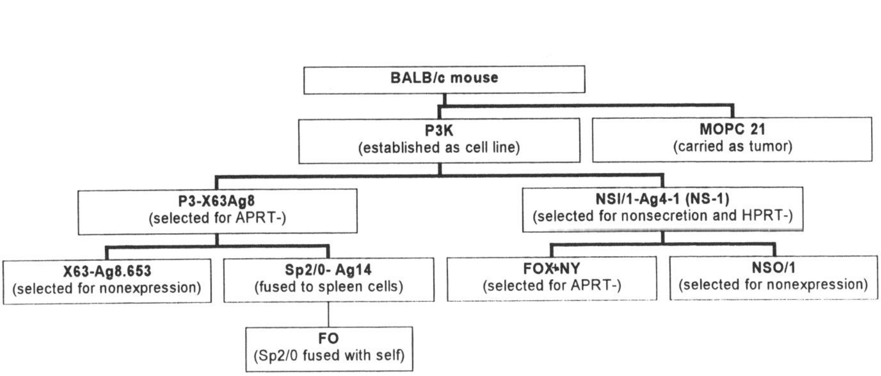

Fig. 4. Myeloma family tree

The sensibility of the virus towards detergents, heat treatment, acid etc. is comparable to other retroviruses. Downstream processing as used for antibody purification completely removes the viral contamination of Sp2/0 cells.

REFERENCES

1. Heinemeyer T, Klingenhoff A, Hansen W, Schumacher L, Hauser H and Wirth M, A sensitive method for the detection of murine C-type retroviruses, submitted.
2. Massey AC, Coppola MA and Thomas CY (1990) Origin of pathogenic determinants of recombinant murine leukemia viruses: Analysis of Bxv-1-related xenotropic viruses from CWD mice. J. Virol. 64, 5491-5499.
3. O'Neill RR, Buckler CE, Theodore TS, Martin MA and Repaske R (1985)Envelope and long terminal repeat sequences of a cloned infectious NZB xenotropic murine leukemia virus. J. Virol. 53, 100-106.
4. Hansen M, Jelinek L, Whiting S, and Barklis E (1990) Transport and assembly of *gag* proteins into Moloney murine leukemia virus. J. Virol. 64, 5306-5316.
5. Deo Y, Ghebremariam, H and Cloyd, M (1994) Detection and characterization of murine ecotropic recombinant virus in myeloma and hybridoma cells . Hybridoma 13, 69-76.

TOWARD A DNA VACCINE AGAINST LYSSAVIRUSES.

C. Bahloul, F. Fouque, N. Tordo, Y. Jacob and P. Perrin.
Laboratoire des Lyssavirus
Institut Pasteur, 25 rue du Docteur Roux, 75724 Paris Cedex 15, France

Abstract

Classical rabies vaccine strains (genotype 1) protect inefficiently against rabies-related lyssaviruses (other genotypes) and in particular against the Mokola virus (genotype 3). DNA vaccine modern approach was used to produce Mokola valency. Purified Plasmid vector was injected into each anterior tibialis muscle of BALB/c mice. It appeared that the Mokola DNA vaccine has induced a sustained level of VNAb, a T helper cell response and confered protection against Mokola virus challenge. Therefore, it seems that the DNA vaccine approach is a very promising technology, which needs further investigations for a possible application.

Introduction

The aetiologic agents of rabies encephalomyelitis are classified into six genotypes within the *Lyssavirus* genus (1). Their genome is an unsegmented negative-stranded RNA, which codes from 3' to 5' for 5 viral proteins: the N (nucleoprotein) strongly associated with the genome to form the ribonucleoprotein complex, the M1 (phosphoprotein), the M2 (membrane protein), the G (transmembrane glycoprotein) and the L (polymerase). The glycoprotein is the major immunogenic protein. Classical rabies vaccine strains (genotype 1) protect inefficiently against rabies-related lyssaviruses (other genotypes) and in particular against Mokola virus (genotype 3) which is among the most genetically and antigenically distant. To extend the spectrum of rabies vaccines towards the other genotypes, different procedures were adopted among which we have focused on: i) the use of an experimental Mokola vaccine by expressing a recombinant glycoprotein (G) in the baculovirus system (2), ii) the alternative was the use of the DNA vaccine approach (3). The DNA vaccine is based on the direct injection of naked DNA, which consists of a eukaryotic vector driving the expression of the protein of interest. The coding gene is transcribed, then translated *in vivo* to induce an immune response. This type of immunization was reported in different systems (4, 5, 6), and it was shown that it confers a full spectrum of immunological responses by stimulating both CD4+ and CD8+ T cells. Therefore, since the DNA vaccine is both simple to manufacture and efficient, it would be a promising candidate for a new generation of vaccines.

Procedure

Material and Methods

Mokola glycoprotein gene was amplified by RT-PCR, then inserted into a eukaryotic expression vector, under the control of the CMV promoter-enhancer. The plasmid was amplified and purified on a CsCl gradient and used for the immunization. 100 µg of this DNA was injected to female BALB/c mice in the anterior tibialis muscle, 4 days after a degeneration with cardiotoxin. A second set of similar mice were immunized without a prior muscle degeneration. The mice were bled at different intervals and the blood

M.J.T. Carrondo et al. (eds.), Animal Cell Technology, 43–46.

samples were assayed for Mokola virus neutralizing-antibodies (VNAb), using the Rapid Focus Fluorescent Inhibition Test (RFFIT) on BHK cell (7). CD4+ T helper cell response was tested by *in vitro* stimulation of splenocytes from immunized or non immunized BALB/c mice as previously reported (8). Interleukin-2 (IL-2) release was bioassayed for the proliferation of IL-2 dependant cells (CTLL). The challenge with Mokola virus was carried out intra-cerebrally at 72 and 190 days post-vaccination of the mice without or with muscle degeneration, respectively.

Results

As shown on figure-1A, all vaccinated mice with 100 µg of Mokola DNA vaccine after muscle degeneration, have seroconverted, with rather similar VNAb titers. On average, the Mokola VNAb titer reached 7 IU/ml at day 19, increased to 27 IU/ml at day 42 and was maintained at this level up to six months post-vaccination. The figure-1B tells us that the muscle degenerated mice gave a higher level of VNAb than mice without muscle degeneration. Nevertheless, the titers of the last group remained at an acceptable level. In figure-2, we have shown that the splenocytes of vaccinated mice with Mokola DNA vaccine induced IL-2 release when stimulated *in vitro* with Mokola glycoprotein. Hence, the Mokola DNA vaccine induced CD4+ T helper cell response. Figure-3 shows that the survival rate of the vaccinated mice with Mokola DNA was 100 percent, compared to 25 percent for control animals. The rates are relative to both sets of vaccinated mice, muscle degenerated and non degenerated ones. Therefore, it appears that this kind of vaccine confers a full level of protection against a homologous challenge.

Discussion

Classical rabies vaccines are mostly inactivated virus vaccines which confer the induction of virus neutralizing antibodies and the stimulation of CD4+ T helper cells. In normal conditions, this type of vaccines does not stimulate CD8+ T cells, therefore fails to induce T cytotoxic activity. Nevertheless, the level of protection induced against rabies by these vaccines is very high when the infection is by a virus that belongs to the genotype 1. The DNA vaccine was shown to be a means of vaccination which stimulates in addition to CD4+ (9), the CD8+ (10) T cell production. So far, we have shown that the Mokola DNA vaccine stimulates efficiently the CD4+ T cells (IL-2 release), and the induction of VNAb against Mokola. On the other hand, experiments about the T cytotoxic activity are under way. Consequently, we can evaluate the importance of this means for a further understanding of the immunological responses confered with a particular immunogen. Concerning the sustained level of VNAb, we can presume that the transfected cells trigger the release of Mokola glycoprotein at a low level in an immunocompetent form which continuously update the immune system. This might be an advantage over classical vaccines, since fewer or no boosters are needed to actualize the immune system. In addition, if it turns out that the T cytotoxic activity plays a role in the increase of the protection spectrum, it would be a further justification for the DNA vaccine application. In this respect, it was shown in rabies that the CD4+ knockout mice are poorly protected by the T cytotoxic activity (11). Nevertheless, the prospect is that this activity in addition to the VNAb may enhance the protection to other genotypes.
The eventual drawback for this new technology comes from the possible side effects. So far, the hypothesis of a possible integration of the plasmid vector into the host genome has not yet been dropped, although there are no reports of such a case. Furthermore, the problem of the induction by the organism of an immune response to the injected nucleic acid seems to be very unlikely. Therefore, if the DNA vaccine technology succeeded in

crossing the barrier of safety, it would have a major role to play on the control of rabies. For this purpose, new vaccines with a larger spectrum of protection, to target the rabies-related genotypes as well as the genotype 1, could be of great interest. Similarly, new formulations of DNA vaccine administred by other routes like the oral one and the decrease of the injected plasmid quantity, might be of further interest.

References

1. Bourhy H, Kissi B and Tordo N (1993) Molecular Diversity of the Lyssavirus genus. Virology 194, 70-81.
2. Tordo N, Bourhy H, Sather S and Ollo R (1993) Structure and Expression in Baculovirus of the Mokola Virus Glycoprotein: An Efficient Recombinant Vaccine. Virology 194, 59-69.
3. Tang D, DeVit M and Johnston SA (1992) Genetic immunization is a simple method for eliciting an immune response. Nature 356, 152-154.
4. Yankauckas MA, Morrow JE, Parker SE, Abai A, Rhodes GH, Dwarki VJ and Gromkowski SH (1993) Long-Term Anti-Nucleoprotein Cellular and Humoral Immunity Is Induced by Intramuscular Injection of Plasmid DNA Containing NP gene. DNA and Cell Biology 12, 771-776.
5. Wang B, Ugen KE, Srikantan V, Agadjanyan MG, Dang K,Refaeli Y, Sato AI, Boyer J, Williams WV and Weiner DB (1993) Gene inoculation generates immune responses against human immunodeficiency virus type 1. Proc. Natl. Acad. Sci. USA 90, 4156-4160.
6. Xiang ZQ, Spitalink S, Tran M, Wunner WH, Cheng J and Ertl HCJ (1994) Vaccination with a Plasmid Vector Carrying the Rabies Virus Glycoprotein Gene Induces Protective Immunity against Rabies Virus. Virology 199, 132-140.
7. Smith J, Yager P, Baer G (1996) A rapid fluorescent focus inhibition test (RFFIT) for determining virus-neutralizing antibody. In: Meslin FX, Kaplan M and Koprowski H, Laboratory Techniques in Rabies, pp 181-189, WHO, Geneva, Switzerland.
8. Perrin P, DeFranco M, Jallet C, Fouque F, Morgeaux S, Tordo N and Colle JH (1996) The antigen-specific cell-mediated immune response in mice is suppressed by infection with pathogenic lyssaviruses. Res. Virol. 147, In press.
9. Fynan EF, Webster RG, Fuller DH, Haynes JR, Santoro JC and Robinson HL (1993) DNA vaccines: Protective immunizations by parenteral, mucosal, and gene-gun inoculations. Proc. Natl. Acad. Sci. USA 90, 11478-11482.
10. Ulmer JB, Donnelly JJ, Parker SE, Rhodes GH, Felgner PL, Dwarki VJ, Gromkowski SH, Deck RR, DeWitt CM, Friedman A, Hawe LA, leander KR, Martinez D, Perry HC, Shiver JW, Montgomery DL and Liu MA (1993) Heterologous Protection Against Influenza by Injection of DNA Encoding a Viral Protein. Science 259, 1745-1749.
11. Xiang ZQ, Knowles BB, McCarrick JW and Ertl HCJ (1995) Immune Effector Mechanisms Required for Protection to Rabies Virus. Virology 214, 398-404.

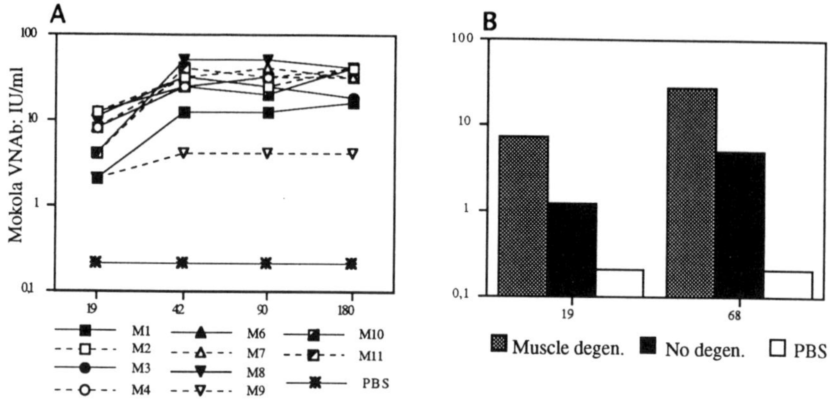

Figure-1: Mokola VNAb production in BALB/c mice immunized with 100μg of Mokola DNA vaccine. **A**: VNAb of individual mice vaccinated four days after muscle degeneration. **B**: Comparison of the means of VNAb induced by the DNA vaccine, in mice with (▩) or without (■) muscle degeneration. VNAb assays were carried out at different days post-vaccination.

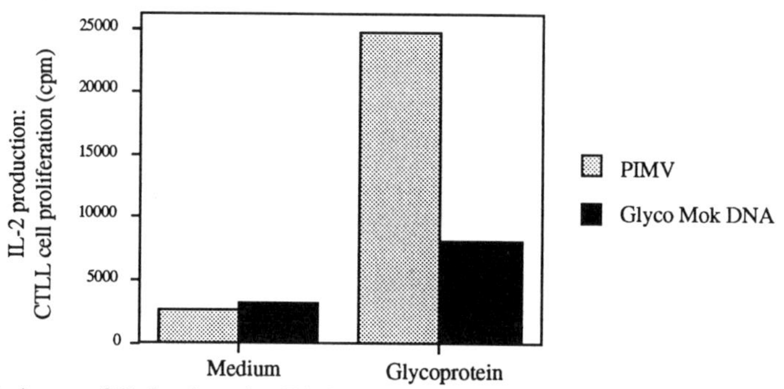

Figure-2: Assay of IL-2 release by T helper cells in BALB/c mice injected with Mokola DNA. Splenocytes were collected 42 days after the immunization of mice with purified and inactivated Mokola virus (PIMV), or with 100μg of Mokola DNA vaccine, and stimulated with Mokola glycoprotein. The IL-2 dependant cell proliferation was appraised according to ^{3}H-thymidine incorporation (cpm).

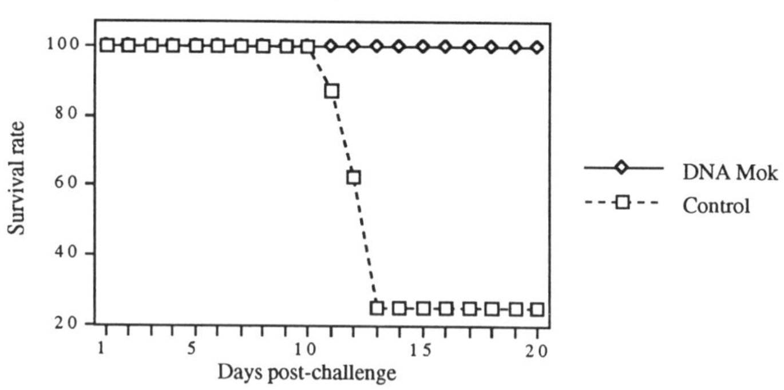

Figure-3: Protection confered by 100μg of Mokola DNA vaccine in BALB/c mice after an intra-cerebral challenge with Mokola virus.

PRINCIPLES OF A SCALEABLE TRANSIENT TRANSFECTION AND EXPRESSION TECHNOLOGY FOR MAMMALIAN CELLS

MARTIN JORDAN*, CHRISTIANE KÖHNE, FLORIAN M.WURM*

Genentech Inc., South San Francisco, California 94080, USA
** Present address: Dept. Chemistry, Swiss Federal Institute of Technology, 1015 Lausanne, Switzerland*

ABSTRACT

We report the first use of DNA-calcium phosphate co-precipitates for DNA transfer to suspended mammalian cells in bioreactors or spinners. Our goal was a scaleable transient transfection method with efficient DNA transfer and rapid product formation. In a batch process $CaPO_4$-DNA complexes were directly injected into the culture medium of human immortalized kidney cells[1] (293 HEK). Low transfection rates were observed at times, especially at pH values of the culture medium lower than 7. However, at high calcium concentrations, sensitivity to low pH was decreased. By adjusting the calcium concentration in the medium, precipitate quality as a prerequisite for efficient transfection could be addressed. Indeed, we found product titers from suspended cells to be similar to those achieved in plates: In a 2L suspension, containing 293 HEK cells transfected with 1mg of purified plasmid DNA, we obtained 2 mg of recombinant plasminogen after 5 days.

INTRODUCTION

Our recent efforts to improve and standardize the calcium phosphate mediated transfection method[2] clearly indicated that maintenance of a certain precipitate quality was a critical characteristic for high levels of expression of recombinant proteins in adherent CHO and 293 HEK cells. While optimizing this method for use in bioreactors or spinners we encountered the challenge of maintaining precipitate quality in suspension. We therefore extended our studies of the physicochemical parameters of precipitate formation and stability, resulting finally in a modified process for transient transfection in bioreactors or spinners.

METHODS

Transfecting cells in spinner flasks

Exponentially growing cells were centrifuged and resuspended in fresh medium (DMEM/F12, 2% FCS, pH = 7.4, 50 ml per 100ml spinner flask) 1 - 2 hours before the transfection. The DNA containing calcium solution and the HEPES buffered phosphate solution were prepared as published[3]. The volume of transferred precipitate suspension was reduced (as compared to standard protocol for plates) and extra calcium was added to the culture medium to achieve saturated conditions. 50 - 90% of the required calcium was given to the cells immediately before precipitate preparation. Precipitate was added to cultures 60 - 90 seconds after mixing of appropriate solutions. After 4 hours, 50 ml of

M.J.T. Carrondo et al. (eds.), Animal Cell Technology, 47–50.

fresh medium was added and the cells were cultivated under standard conditions until the product was harvested.

RESULTS AND DISCUSSION

Stability of the precipitate in the medium

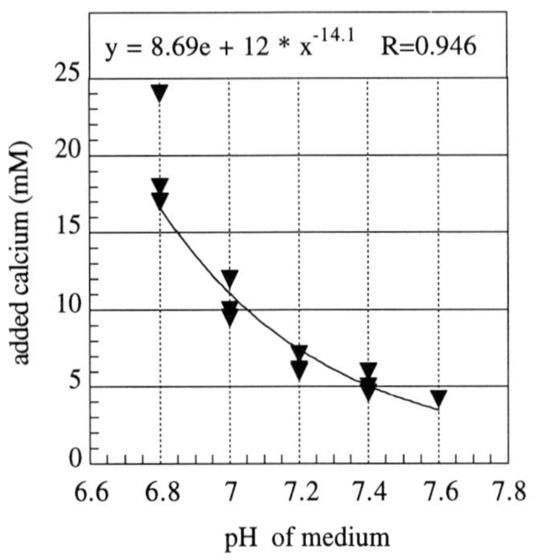

Fig.1: solubility of calcium in DMEM/F12 medium at various pH values. Data points represent minimum concentrations required for a visible precipitate.

Solubility is mainly affected by pH, calcium and phosphate concentration. When the precipitate is added to cells, the calcium concentration in the medium is increased dramatically. It was questioned what amount of total calcium would be needed in serumfree DMEM/F12 medium to create conditions for maintaining stability of the precipitate. The minimal amount of added calcium resulting in a stable precipitate, detectable with the microscope, is plotted in fig. 1. To demonstrate the influence of pH, the medium was buffered with 30 mM HEPES at 5 different pH values. The fitted curve represents the critical quantity of added calcium as a function of pH: At quantities below this value, the precipitate would dissolve, whereas at higher quantities, calcium continues to precipitate with the phosphate in the medium.

Effect of the solubility on the transfection efficiency

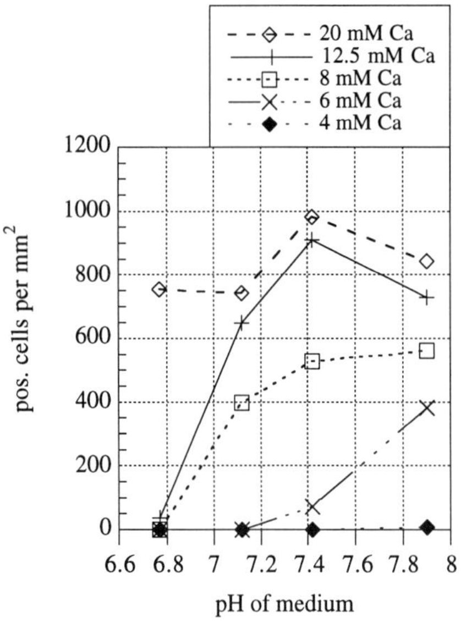

Fig. 2: Transfection efficiency as a function of pH and calcium concentration in the medium.

The next question was if the solubility per se is relevant for the transfection of cells. In a multiwell experiment one precipitate preparation was used to perform transfections in individual wells. The amount of precipitate per well was varied in order to achieve different final quantities of calcium. Due to this set-up, DNA quantity per well increases along with the varied amount of added precipitate. Different pH values were tested (buffered with HEPES). To minimize pH effects on cell growth, the medium was exchanged immediately before the precipitate was added. The indicated pH value in the graph represents the period of 3 hours post addition of the precipitate. The transfection efficiency, determined within 25 hours using X-gal for detection of beta-gal activity, was found to be strongly affected by the solubility (fig.2): only highly saturated conditions (i.e. a visible precipitate after two hours) gave positive results. Surprisingly, pH itself had only minor effects, with a minor optimum at 7.4, when calcium concentration is in the appropriate range. Also, DNA quantities had only minor effects in this experiment.

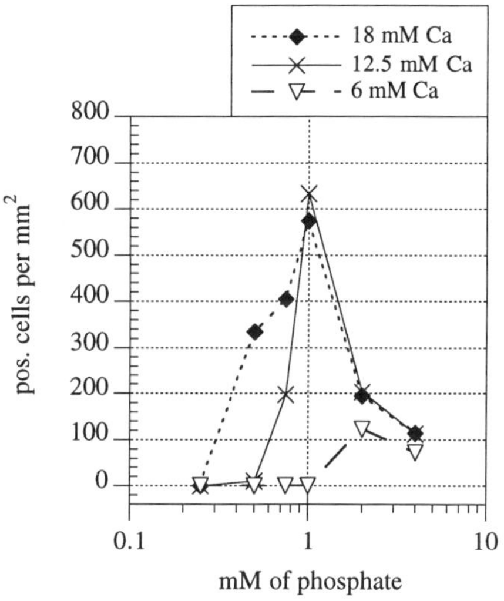

Fig. 3: Transfection efficiency as a function of phosphate concentration in the medium.

The third important parameter to be analyzed was phosphate. The transfection experiment was done the same way as in fig. 2 except that medium with different concentrations of phosphate was used during the 3 hours incubation period. As it can be seen in fig. 3 the highest transfection efficiency is achieved for phosphate concentrations of about 1mM (standard 0.95mM in DMEM/F12). Reduced efficiency at 2 or 4 mM phosphate is probably not only a function of solubility, since the efficiency is not significantly improved by adjusting the calcium concentration. Based on this data, the phosphate concentration in the medium was not changed in subsequent experiments.

Transfections in spinner flasks

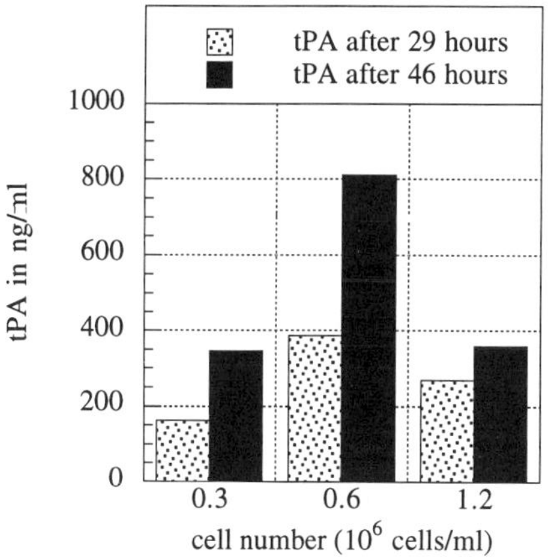

Fig. 4: transfection of 293 cells in 100 ml spinner flasks.

In the expectation that more cells in a suspension, exposed to the same quantity of DNA would increase product titers, different cell densities were tested in spinner flasks. A typical result is shown in fig. 4. As it can be seen, titers increased only up to about 0.6×10^6 cells/ml and decreased at 1.2×10^6 cells/ml. It appeared that a critical relationship between DNA quantity and the number of cells exposed to DNA exists but more work is needed to understand and optimize effects and relationships of concentration of cells and DNA.

Transfection in the bioreactor

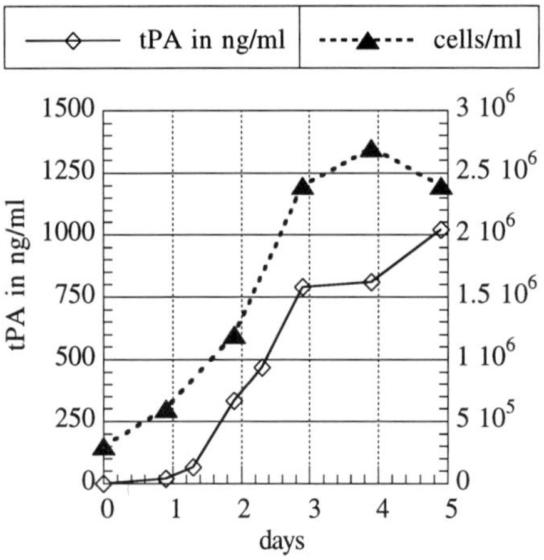

Fig. 5: Transfection with plasmid DNA and transient expression of tPA in a 2 L bioreactor.

Finally, we proved the concept for transfecting suspended cells in a stirred 2L-bioreactor. Cells were seeded at 0.3 x 10^6 cells/ml in 1.2 L of fresh medium. 30 minutes before the transfection, the pH was adjusted to 7.4. Calcium concentration in the medium was elevated to 10 mM immediately before the addition of the precipitate. 30 ml of precipitation mixture containing 1 mg of 2xCsCl gradient purified plasmid was prepared under sterile conditions, loaded into a syringe and injected into the bioreactor within 90 seconds upon mixing of the components. 4 hours later 0.8L of fresh medium was added and the pH was readjusted to 7.2. The cells grew exponentially and started to secrete tPA one day after the transfection (fig. 5).

SUMMARY AND CONCLUSION

We demonstrate here that calcium phosphate mediated DNA transfer is feasible in mammalian suspension cell cultures up to the 2L scale. We characterized the interdependence of calcium concentration and pH as critical parameters for stability and/or solubility of the precipitate in the medium. We demonstrate that precipitate quality as prerequisite for transfection efficiency is dependent on calcium concentrations in cell culture media much higher than typically used for cell growth. Higher concentrations of calcium also increase the stability of effective precipitate formulations at low pH values. By adjusting calcium concentrations prior to DNA transfer, we were able to reduce the required volume of added precipitate from one tenth to one seventieth of the working volume of the reactor. The data suggest the prospective utility of this method at much larger scales, envisioning product synthesis of gram quantities of recombinant proteins in a few days.

Acknowledgment. This work was supported by Swiss National Science Foundation Grant 5002-38003

References

[1] Graham F.L., Smiley J., Russel W.C. and Nairn R. (1977) Charactestistics of a human line transformed by DNA from human adenovirus type 5. J. gen. Virol. 36 pp. 59-72

[2] Jordan M. and Wurm F.M (1995) In: Animal cell technology: Developments towards the 21st century (ed. by Beuvery E.C., Griffiths J.B., Zeijlemaker W.P.) pp. 49-53 Kluver Academic Publishers, The Netherlands

[3] Jordan M., Schallhorn A. and Wurm F.M. (1996) Transfecting mammalian cells: optimization of critical parameters affecting calcium-phosphate precipitate formation. Nucleic Acids Res. 24 pp. 596-601

Neurological Therapy - Adenovirus Mediated Gene Therapy in Cells of the Central Nervous System

S.D. Sivasubramaniam., A.R. Fooks., J. Lee., G. Stacey and A.D. Jennings. Centre for Applied Microbiology and Research (CAMR), Salisbury SP4 0JG, Wiltshire, United Kingdom.

Key words: Neurological gene therapy; adenovirus; β-galactosidase; genetic disorder.

Introduction

Recently, the treatment of human genetic disorders has been mediated by gene transfer technology especially for monogenic disorders. Further advances in gene therapy have become possible following the first successful treatment of patients deficient of the enzyme adenosine deaminase (ADA).[1] This study explores the feasibility of using a defective non-replicating recombinant adenovirus vector to transfer a reporter gene (β-galactosidase) into neuronal cells. By their nature viruses efficiently transfer their genes to cells and promote expression of their encoded gene products in a precise and predetermined fashion. This property has been exploited to construct a replication-deficient adenoviral vector (RAd35).[2] Unlike some other expression systems, defective adenoviral vectors grow to a high titre in a helper cell line (293 cells)[3] and provide high levels of protein expression from a strong constitutive human viral promoter.[4]

Materials and Methods

Replication-deficient adenovirus vectors have become a commonly used viral vector expression system for use in vaccine design and gene therapy.[5] The recombinant adenovirus vector (RAd35) was constructed by cloning the reporter gene β-galactosidase from the *lacZ* gene of *Escherichia coli.* The β-gal gene was selected as the reporter gene because the expression of the β-gal enzyme can be readily detected, and there is no interference or competition from other enzymes. The reporter gene was placed under control of the strong human cytomegalovirus major IE promoter (hCMV IE)[6]. The expression cassette containing the CMV IE promoter, reporter gene and terminator region was transferred into the vector (pJM17) containing the adenovirus genome[7]. Human embryo kidney cells (293) were used as "the packaging"cell line[8]. The required gene (β-gal) was cloned into the adenovirus transfer vector (pMV60) to produce the plasmid, pMV35. Following co-transfection with pMV35 and the plasmid pJM17 into 293 cells, homologous recombination resulted in the insertion of the CMV IE promoter/β -galactosidase cassette. Recombination of pMV35 and pJM17 leads to the loss of the pBRX region, reducing the overall size of pJM17 allowing packaging into the adenovirus capsid, producing the recombinant adenovirus, RAd35.

M.J.T. Carrondo et al. (eds.), Animal Cell Technology, 51–56.

Results

Table 1 β-galactosidase expression in cell lines following infection with RAd35 (moi 5)

Cell line	**Description**	**ECACC No.**	**Level of β-gal expression**
MRC-5	Human foetal lung	84101801	++
1321N1	Human brain astrocytoma	86030402	++
TE671	Human medulloblastoma	89071904	++
MO-GG-CC-M	Human brain astrocytoma	86022702	++
BC3H1	Mouse brain tumour	86093001	+/–
B50	Rat neuronal tissue	85042302	+/–
NG108-15	Mouse neuroblastoma x Rat glioma hybrid	88112302	+
ND7/23	Mouse neuroblastoma x Rat neuronal hybrid	92090903	++
P/GNC	Cerebellar granule primary neurones	Not applicable	++
NTera-2	Human embryonal carcinoma (neuronal cell precursor) cells	96041736	++

Key :-

+++ = Expression of β-gal observed in >90% of cells.	**++ = >50% of cells expressing β-gal**	**+ = 10-50% of cells expressing β-gal**	**+/– = <10% of cells expressing β-gal**

1a RAd35 infected MRC-5 cells
(Human foetal lung)

Uninfected MRC5 cells

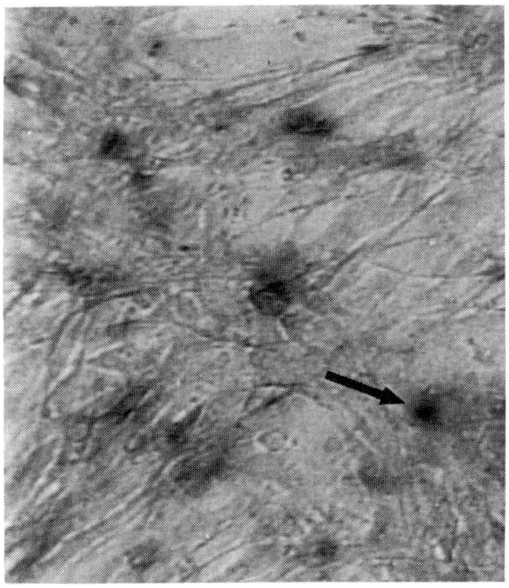

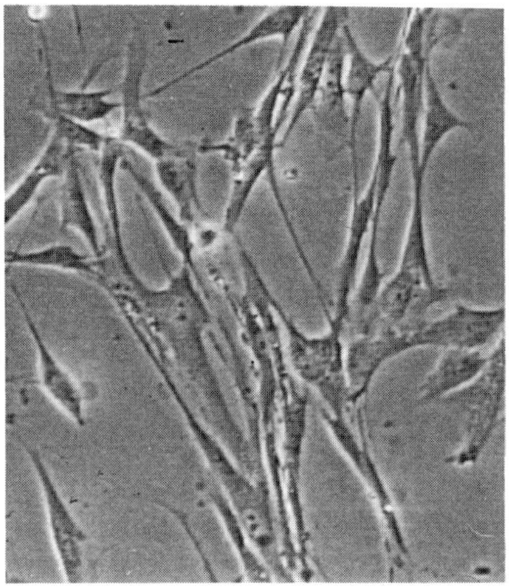

1b RAd35 infected MO-GG-CC-M cells
(Human brain astrocytoma)

Uninfected MO-GG-CC-M cells

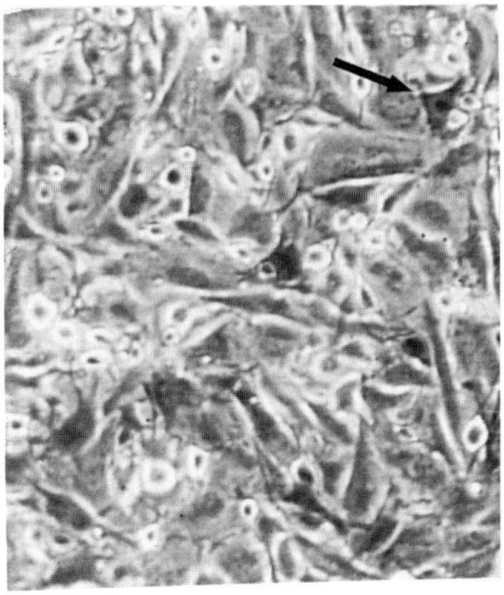

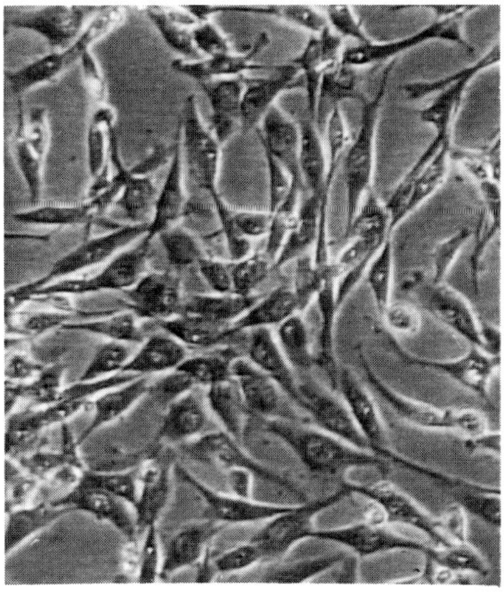

Fig 1 Expression of β-galactosidase in human foetal lung fibroblasts (MRC-5) and a human brain astrocytoma (MO-GG-CC-M) cell line following infection with RAd35 (m.o.i = 5). Uninfected MRC-5 and MO-GG-CC-M are shown as controls.

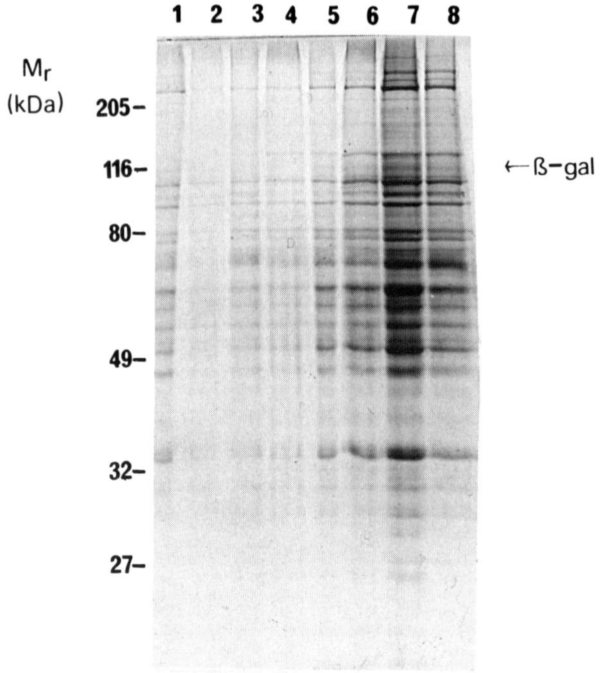

Fig 2 Time course of β-gal expression in RAd35 infected MO-GG-CC-M cells by SDS-PAGE analysis. Lanes were as follows: (1) Molecular weight marker, (2) Un-infected MO-GG-CC-M cells, (3) 0 hrs, (4) 48 hrs, (5) 48 hrs + forskolin, (6) 144 hrs, (7) 144 hrs + forskolin, (8) 168 hrs.

Discussion

Previous studies using non-replicating adenovirus vectors have concentrated on the effective stimulation of both humoral and cell-mediated immune responses to elicit a protective response in the design of novel vaccine candidates against infectious diseases[4,9]. Furthermore, replicating adenoviruses have been used successfully as vaccines by their administration to American troops to protect them against acute respiratory disease (ARD)[10]. Recently, non-replicating adenoviruses containing the cystic fibrosis transmembrane regulator (CFTR) gene were administered to the lungs of patients to treat cystic fibrosis[11]. Gene therapy has an obvious therapeutic objective by replacing a specific gene sequence into a target cell. This approach is only possible for monogenic diseases where the genetic disorder has been identified and is caused as a result of a single genetic defect. This study has been used to demonstrate the efficiency of using a replication-deficient adenovirus vector as a vehicle to transfer genes into cells of the central nervous system (CNS) i.e. "neurological gene therapy"[12].

Neurological gene therapy enables the use of defective adenovirus vectors to deliver genes to cells of the CNS. Advantages of this system for the treatment of neurological genetic disorders are that: (i) the recombinant adenoviruses are easily produced by routine cloning and transfection procedures (ii) adenoviruses replicate to extremely high titres in helper cell lines (iii) efficient gene delivery is achieved (iv) expression of the transgene can be detected in the majority of the target cell population *in vitro* (vi) a wide tissue tropism has been demonstrated including neuronal cells (v) expression can be obtained in non-replicating and terminally differentiated cells (vi) high level expression is achieved from the strong constitutive CMV IE promoter. Our results have shown that RAd35 infected non-permissive neuronal, glial, astrocytes and medulloblastoma cell lines expressed β-galactosidase *in vitro* at a low m.o.i. This suggests that the adenovirus vector RAd35 is capable of delivering the cloned foreign gene into the cells of central nervous system from different mammalian species including human neuronal cells. The expression of β-galactosidase (M_r 116kDa) in MO-GG-CC-M cells was increased by stimulating the CMV IE promoter using a chemical stimulant, forskolin. Expression increased with time up to a maximum of 168 hrs post-infection when cell lysis occurred. An estimation of β-gal expression in human brain astrocytoma cells based on the size of the protein band accounted for approximately 10-20% of the total cell protein. Protein expression from the hCMV IE promoter in human neuronal cells using defective adenovirus vectors to introduce the genes into the CNS is in this case, high level protein expression. The successful introduction of foreign genes into neuronal cells may lead to a drastic improvement in controlling monogenic genetic diseases in which only a minority of cells are required for genetic alteration in order to replace a deficient enzyme and therefore correct the disorder i.e. "enzyme replacement therapy". Further *in vivo* experiments are currently underway in our laboratory to explore the possibilities of using this vector system as a tool for replacing missing or damaged genes in cells of the CNS.

Conclusion

The versatility of replication-deficient adenoviruses has enabled this vector system to demonstrate distinct advantages in its ability to deliver genes into neuronal cells, whether *in vitro* or *in vivo*. In our laboratory, we are extending the use of this delivery system as a model for the targeted expression of single gene defect as an application of "neurological gene therapy".

Acknowledgements

The authors wish to thank Dr. J.C.S. Clegg and Dr.A. Doyle for their continual support, encouragement and constructive discussions throughout this study.

References

1. French-Anderson, W. (1992) Human Gene Therapy. *Science* **256**, 808-814.
2. Wilkinson, G.W.G., & Akrigg., A. (1992) Constitutive and enhanced expression from the CMV major IE promoter in a defective adenovirus vector. *Nucleic Acids Res.* **20**, 2233-2239.
3. Racher, A.J., Fooks, A.R. and Griffiths, J.B. (1995) Culture of 293 cells in different culture systems: Cell growth and recombinant adenovirus production. *Biotechnology Techniques* **9**, (3); 169-174.
4. Fooks, A.R., Schadeck, E., Liebert, U.G., Dowsett, A.B., Rima, B.K., Steward, M., Stephenson, J.R. & Wilkinson G.W.G. (1995) High-level expression of the measles virus nucleocapsid protein by using a replication-deficient adenovirus vector: Induction of an MHC-1-Restricted CTL response and protection in a murine model.*Virology* **210**, 456-465.
5. Warnes, A., & Fooks, A.R. (1996) Live Viral vectors: Construction of a replication-deficient recombinant adenovirus. In: Vaccine Protocols (A. Robinson, G. Farrar and C. Wiblin, Eds.) pp. 33-45, "Methods in Molecular Medicine Series" Chapman and Hall, London, UK.
6. Wilkinson, G.W.G. and Akrigg, A. (1991) The cytomegalovirus major immediate early promoter and its use in eukaryotic expression systems. *Adv.Gene.Tech.* **2**, 287-310.
7. Bett., A.J., Krougliak, V., & Graham., F.L. (1995) DNA sequence of the deletion/insertion in early region 3 of Ad5 *dl*309. *Virus Research* **39**, 75-82.
8. McGrory, W.J., Bautista, D.S. & Graham, F.L. (1988). A simple technique for the rescue of early region I mutations into infectious human adenovirus type 5. *Virology* **163**, 614-617.
9. Schindler, C., Fooks, A.R., Stephenson, J.R. & Liebert, U.G. (1994) Replication-incompetant adenoviruses as vectors for protective immunization against measles virus infection. *Behring Inst.Mitt*, **95**, 109-115.
10. Top, F.H., Buescher, E.L., Bancroft, W.H. & Russell, P.K. (1971) Immunization with live types 7 and 4 adenovirus vaccines. II. Antibody response and protective effect against acute respiratory disease due to adenovirus type 7. *J.Infect.Dis.* **124**, 155-160.
11. Crystal, R.G., McElvaney, N.G., Rosenfield, M.A., Chu, C-S., Mastrangeli, A., Fay, J.G., Brody, S.L., Jaffe, H.A., Eissa, N.T., & Danel, C. (1994) Administartion of an adenovirus containing the human CFTR cDNA to the respiratory tract of individuals with cystic fibrosis. *Nature Genetics* **8**, 42-51.
12. Lowenstein, P.R., Wilkinson, G.W.G., Castro, M.G., Shering, A.F., Fooks, A.R. & D.Bain. (1996) Non-neurotropic adenovirus: a vector for gene transfer to the brain and possible gene therapy of neurological disorders. In: Genetic manipulation of the nervous system. (edited by Prof. D.S. Latchman), pp.11-32, "Neuroscience Perspectives Series" Academic Press, London, UK.

BOVINE POLYOMA VIRUS DETECTION BY ROUTINE PCR SCREENING

Joanne Proffitt, Keri Jarvis and Carl Martin

Biotechnology Department, Corning Hazleton, Otley Road, Harrogate, North Yorkshire, HG3 1PY, UK

Abstract

Bovine polyoma virus (BPyV) has been detected with relatively high frequency in clinically normal calves indicating widespread presence of this virus in cattle. Polyoma virus contamination of a wide range of cell cultures has been linked to the supplementation of culture medium with contaminated calf serum. BPyV has been detected in approximately 70% of batches of calf serum tested (Schuurman et al., 1991). This high incidence of contamination raises safety concerns for biopharmaceuticals.

We have established a reliable and sensitive PCR assay which can detect 10 pg BPyV DNA. This was developed using a BPyV infected cell line (a gift from D. Westcott, CVL) as a source of positive control DNA, and our own stock of MDBK cells as a source of negative control DNA. This assay can be used as a routine screen for the detection of BPyV in established cell lines and continuous cell cultures.

Introduction

Bovine polyoma virus (BPyV) is a small double stranded DNA virus (4697 bp in size) which has been isolated from normal calves, indicating widespread infection in cattle (1). Polyoma virus infection detected in cell cultures of varied origin has been attributed to the use of medium supplemented with BPyV contaminated calf serum. One study detected BPyV in 70% of batches of calf serum analysed (2).

The polymerase chain reaction (PCR) is an *in vitro* DNA amplification technique which employs a thermostable DNA polymerase and oligonucleotide primers complimentary to a known DNA sequence. A predetermined number of cycles of template denaturation, primer annealing and primer extension allows the exponential accumulation of a specific DNA fragment of interest (Figure 1). The fragments can then be analysed by agarose gel electrophoresis, comparing electrophoretic migration to that of molecular weight size markers.

M.J.T. Carrondo et al. (eds.), Animal Cell Technology, 57–62.

Aims

- To develop a reproducible and reliable methodology for routine screening of cell lines for BPyV.
- To screen a range of our own cell lines commonly used in infectivity assays.

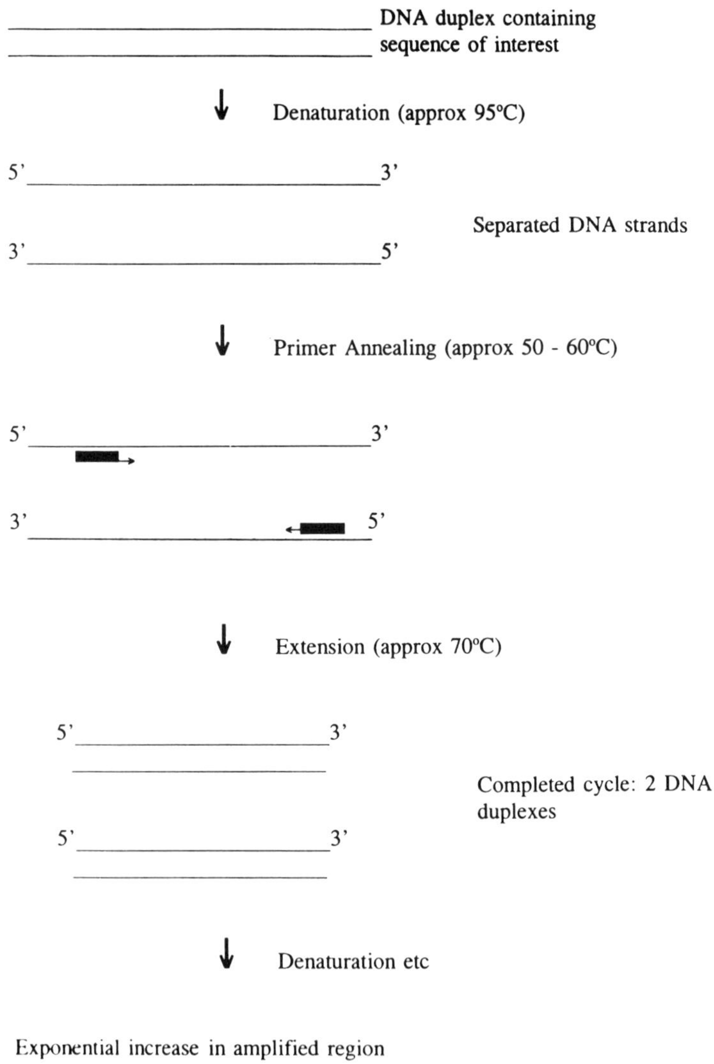

Figure 1: Schematic representation of the polymerase chain reaction (PCR)

Methodology
DNA was prepared by SDS/proteinase K lysis, followed by phenol/chloroform extraction and isopropanol precipitation, from a range of test cell lines.
Positive control DNA was prepared using the same procedure, from BPyV-infected primary calf kidney cells (a gift from D Westcott, CVL, UK).
PCR primers were synthesised to two regions of the BPyV genome; primers L complimentary to the late region of the genome and primers T to part of the large T coding sequence (Figure 2).

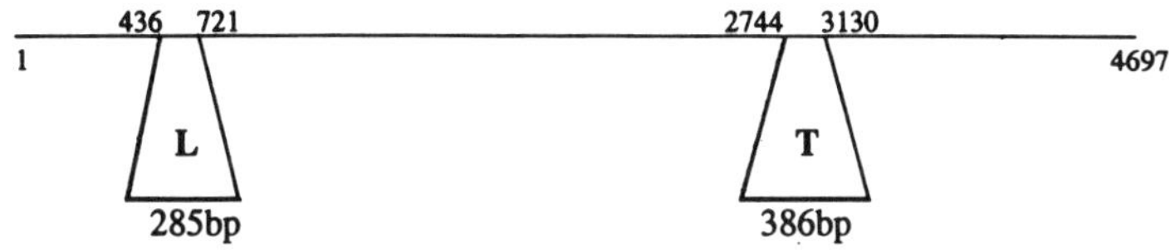

Figure 2: **Schematic representation of the location of primers designed to amplify regions of the BPyV genome.**

PCR was performed using both sets of primers with negative controls, a range of positive control DNA concentrations and 1μg each test DNA sample. Negative controls, including a water blank and sentinel control (which remained open during PCR set up to monitor air borne contaminants), were included to test for cross-contamination.
PCR reactions contained 50 mM KCl, 10 mM Tris pH 8.3, 0.01% gelatine, 0.3 mM each primer, 0.2 mM each of dATP, dGTP, dCTP, dTTP, 1.25 units *Taq* polymerase, 2.5 - 3.5 mM $MgCl_2$ and the appropriate DNA template.
Standard precautions employed in the performance of PCR at CHE were adhered to. These included, but were not limited to, the following:

a) separate work stations for the key steps of PCR
b) prevention of back flow from post-PCR work station
c) dedicated equipment and sterile consumables
d) work station-specific labwear

The following PCR profile was used: 1 cycle of 95°C 1.5 min, 55°C 1 min (58°C 0.5 min), 72°C 2 min (0.5 min); 35 cycles of 95°C 1 min, 55°C 1 min (58°C 0.5 min), 72°C 2 min (0.5 min), followed by a final extension cycle of 72°C for 8 min (varied conditions used for T primers are given in parentheses).

PCR reactions were analysed by agarose gel electrophoresis using standard procedures (Figures 3 and 4).

Results

- PCR using L primers amplified a DNA fragment of the expected size (285 bp) with BPyV positive control DNA samples in a range of 1 - 100 ng DNA (Figure 3).
- No amplified products of the correct size were detected in PCR using the L primers and 1 μg DNA isolated from MDBK, Vero, BT and XC cell lines (Figure 3).
- PCR using T primers amplified a DNA fragment of the expected size (386 bp) with BPyV positive control DNA samples in a range of 0.01 - 100 ng DNA (Figure 4).
- No amplified products of the correct size were detected in PCR using the T primers and 1 μg DNA isolated from MDBK, Vero, BT and XC cell lines. Faint background bands were identified with MDBK and Vero DNA, however, these were not of the correct size and were attributed to the use of excess template DNA resulting in the generation of non-specific PCR artefacts (Figure 4).
- No amplified products were detected in PCR using either primer pair, for the blank and sentinel control reactions (Figures 3 and 4).

Discussion

The results indicate the following:

- PCR for the detection of BPyV was successful with two primer pairs designed to different regions of the genome. The assay sensitivity was determined as 1 ng and 0.01 ng (10 pg) for primers L and T respectively. The use of two sets of primers specific to different regions of the genome enabled unequivocal confirmation of results when non-specific background was observed at high levels of test DNA.
- All cell lines examined were negative for the presence of BPyV DNA and therefore were determined as being BPyV free.
- No back flow or cross-contamination occurred during the assay, as demonstrated by the negative control results. This proves that precautions observed at CHE during the performance of PCR are appropriate and that the results generated with the positive controls are *bona fide*.

Conclusion

We have established a sensitive and accurate PCR assay for the detection of BPyV in cell lines. This assay can be used for the rapid and routine screening of established cell lines and continuous cell cultures for the presence of BPyV.

References

1. Westcott, D. G. F., Ticehurst, J., Chaplin, M., Lukey, J.R & Lucas, M. (1987) The Isolation of a Virus Resembling a Polyomavirus from Normal Calves. Veterinary Microbiology **15** 175-180.

2. Schuurman, B., van Steenis, B., van Strien, A., van der Noordaa, J. & Sol, C. Frequent Detection of Bovine Polyomavirus in Commercial Batches of Calf Serum by Using the Polymerase Chain Reaction. Journal General Virology (1991) **72** 2739-2745.

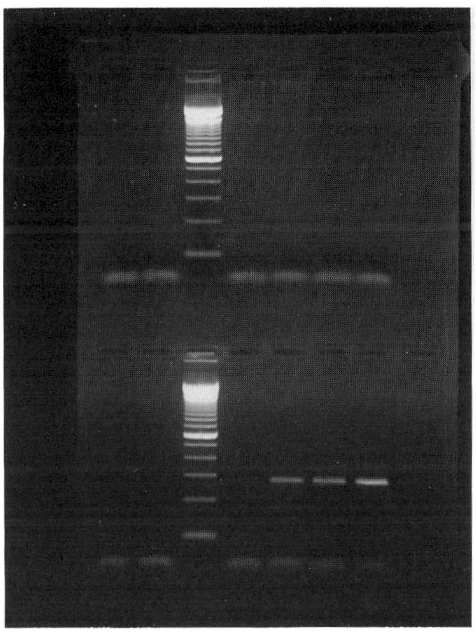

1	= Sentinel tube	7	= BPyV positive control DNA (0.01 ng)
2	= Purified water (Blank)	8	= BPyV positive control DNA (0.1 ng)
M	= DNA size marker 100 bp DNA ladder	M	= DNA size marker 100 bp DNA ladder
3	= MDBK DNA (1 μg)	9	= BPyV positive control DNA (1 ng)
4	= Vero DNA (1 μg)	10	= BPyV positive control DNA (10 ng)
5	= BT DNA (1 μg)	11	= BPyV positive control DNA (50 ng)
6	= XC DNA (1 μg)	12	= BPyV positive control DNA (100 ng)

Figure 3: PCR analysis of cell lines using L primers to the BPyV genome.

1 2 3 M 4 5 6

Molecular Weight Markers (bp) (key sizes only shown)

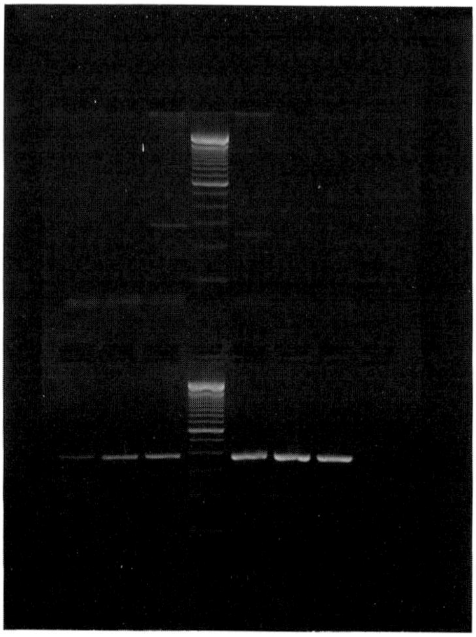

386 bp PCR product →

600 (high intensity band)
500
400
300
200
100

600 (high intensity band)
500
400
300
200
100

7 8 9 M 10 11 12

1	= Sentinel tube	7	= BPyV positive control DNA (0.01 ng)
2	= Purified water (Blank)	8	= BPyV positive control DNA (0.1 ng)
3	= MDBK DNA (1 μg)	9	= BPyV positive control DNA (1 ng)
M	= DNA size marker 100 bp DNA ladder	M	= DNA size marker 100 bp DNA ladder
4	= Vero DNA (1 μg)	10	= BPyV positive control DNA (10 ng)
5	= BT DNA (1 μg)	11	= BPyV positive control DNA (50 ng)
6	= XC DNA (1 μg)	12	= BPyV positive control DNA (100 ng)

Figure 4: PCR analysis of cell lines using T primers to the BPyV genome.

Acknowledgement

Thanks to:

David Westcott of the Central Veterinary Laboratories, Weybridge, UK, for kindly supplying us with the BPyV infected primary calf kidney cells used as a source of positive control DNA. Janine Clark for assistance with preparation of this manuscript.

SESSION II

Animal Cells as Tools for Discovery and Testing

Mammalian Cells in High Throughput Screening

William Janzen
Sphinx Pharmaceuticals, a Division of Eli Lilly and Company
Durham, North Carolina, United States of America

High Throughput Screening (HTS) is the process of testing large numbers molecules for their effect on model biological systems. These targets can be roughly divided into two classes; whole cell assays and biochemical assays. In an arena where hundreds of thousands of compounds are tested in a few months, the robustness, ease, and relevance of an assay become increasingly critical to success. A long standing question in the field has been the relative utility and merit of using mammalian cells as targets in HTS.

The process of drug discovery and development can be viewed as a funnel (figure 1). In this model, a large number of candidate compounds enter the system and are tested in primary screens. A small number (<1%) of these are classified as active and re-tested to confirm their activity and eliminate false positives. Of these, less than 1 in 10 will pass selection criteria to become true hits. If a diverse chemical library has been tested in a well designed screen, these hits will yield a small number of lead structures. These leads will be tested more rigorously in *in vitro* secondary and tertiary assays and *in vivo* for toxicity and pharmacological properties. following this, less than 1 in every 200,000 compounds tested will ever reach the clinic for testing in humans. Most of these drugs will fail in the clinic. In the final analysis, almost 2,000,000 tests must be performed to yield a single successful drug.

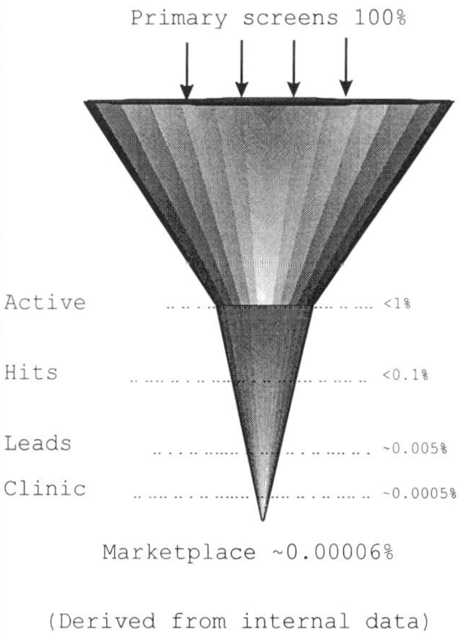

Figure 1

M.J.T. Carrondo et al. (eds.), Animal Cell Technology, 65–71.

The approach at Eli Lilly and Company is to emphasize testing a significant quantity of compounds in each screen, but once this threshold is reached, to test more targets rather than more compounds. This has led to a paradigm of focused screening in which screens average four months in duration and test approximately 125,000 compounds each. The chemical diversity set available for HTS is made up of:

- Natural product extracts
- Combinatorial chemistry compounds
- Eli Lilly organic files.

The flexibility exists to test these compound libraries in any combination within a screen. In conjunction with a commitment to proper quality control (QC) and quality assurance (QA), this model allows the maximization of lead generation from HTS.

This methodology was adopted in September, 1994 and was accompanied by a series of process changes and automation acquisitions. The process improvement which had the largest impact was a change to modular unit automation. Here assay microplates are moved to the most efficient station for processing and batch processing is applied wherever possible. The relative impact on Sphinx's screening throughput is shown in figure 2. Using either the number of targets or the combined throughput as an indicator, productivity has increased a hundred fold. It should be noted that during this period, staffing levels were not increased.

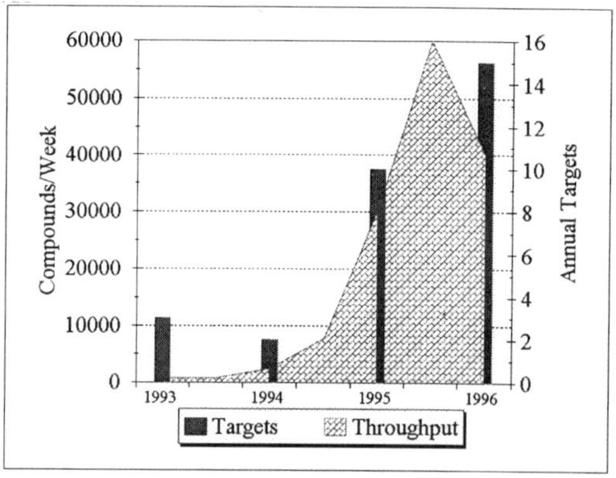

Figure 2

Throughput of Sphinx HTS facility

Bars represent the number of targets tested per calendar year. Throughput is measured as the sum of the number of compounds entered in the database for all targets.

In conjunction with these changes in the HTS facility, a new paradigm was adopted for the development of screens. Screen development was defined as a core function and divided into four basic stages. These phases of development are; Alpha - determination of the basic format and parameters of the screen, Beta - validation of the screen in 96 well plates, Gamma - scaleup and revalidation as a high throughput screen, and Delta - the actual HTS run. During this process the screen must also pass a technical assessment review for its Beta Phase validation. This review is conducted by a committee representing Lilly's Therapeutic Strategy Groups and all of the groups involved in HTS or Screen Development. Prioritization of screens is conducted by the Technology Core Strategy Group, which represents the management of the Therapeutic Areas.

Organizationally, there are several groups involved in the development and operation of a screen (table 1). The Biomolecular Screening group's primary responsibility is for screen operations in HTS. This includes compound preparation and management, data management, and the maintenance and development of the necessary automation and equipment for HTS. The Screen Interface and Screen Development groups are Screening's communication network with the Biologist's in the Therapeutic Groups. They are also responsible for developing screens and new method development. The Biomolecular Research Group is located at Sphinx Pharmaceuticals and in addition to screen development and novel research, holds responsibility for providing cell culture support to all screens. Finally, there are resources with Eli Lilly's Biotechnology Research Group which provide large scale cell expression and protein production for HTS.

Table 1: Functional Groups Involved In Screen Development And Operation

Biomolecular Screening Screen Operations Compound Preparation Data Management Automation/Equipment	**Biomolecular Research** Cell Support Screen Development New Methods Development
Screen Interface and Development Screen Development New Methods Development	**Biotechnology Research** Large Scale Cell Expression Protein Production

Automation has become synonymous with HTS and it is certainly true that automation is required to attain the testing rates needed. To this end, a fairly extensive battery of equipment has been assembled at Sphinx (Table 2). Not shown in this list are the many devices employed in semi-automated screens for processing, quantification, and dispensing. In fact, although automation is employed widely in the HTS program, the majority of the targets are tested employing unit automation (benchtop instruments capable of processing single plates or batches of plates) or workstations.

Table 2: Automated Systems Employed By Sphinx HTS Facility
1 Orca system with 2 Robotic Sample Processors for automated compound preparation
2 Orca systems with multiple peripherals for automated screening assays (Figure 3)
5 Tecan Robotic Sample Processors (Figures 4 and 5)
1 Beckman Biomek 1000
1 Beckman Biomek 2000 with Sideloader (Figure 6)
2 ICN MRD-8's (Figure 7)
1 FLIPR (Fluorescence Imaging Plate Reader, Figure 8)

Mammalian Cells have played an important role in drug discovery at Sphinx and Lilly. A summary of data accumulated since 1994 is presented in Table 3. About one half of the targets have utilized mammalian cell culture. A classification system has been established to quantify the relative difficulty of targets. In this system targets are assigned a relative difficulty on a scale of one to five, with one being the least resource intensive. These are roughly equivalent to the personnel, including cell support, required to run the screen. By this scale, cellular assays are roughly equivalent to non-

Figure 3

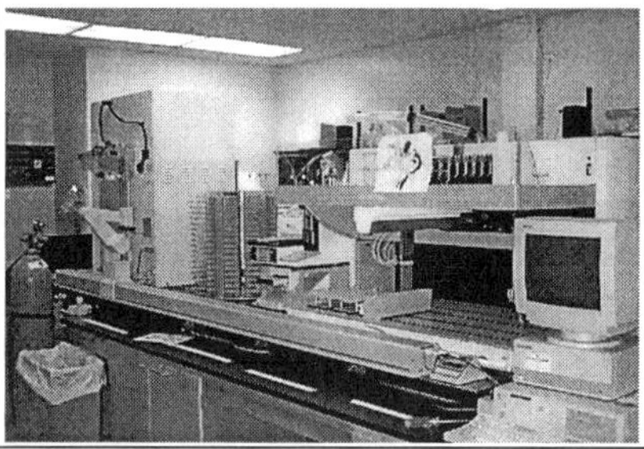

Figure 4

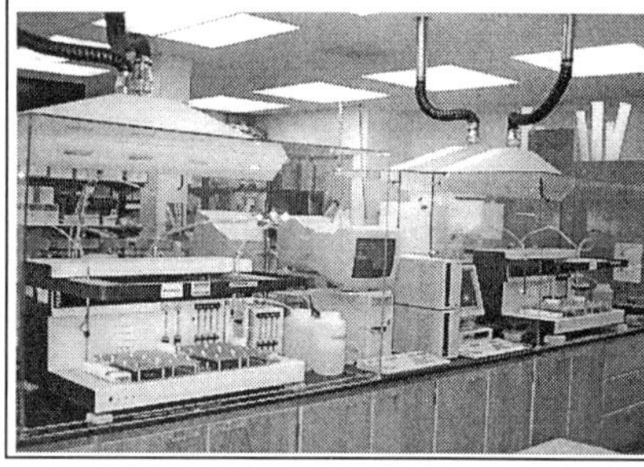

Figure 5

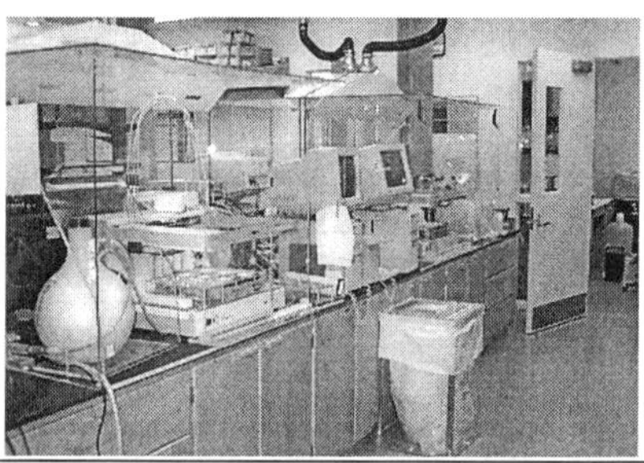

Figure 6

Figure 7

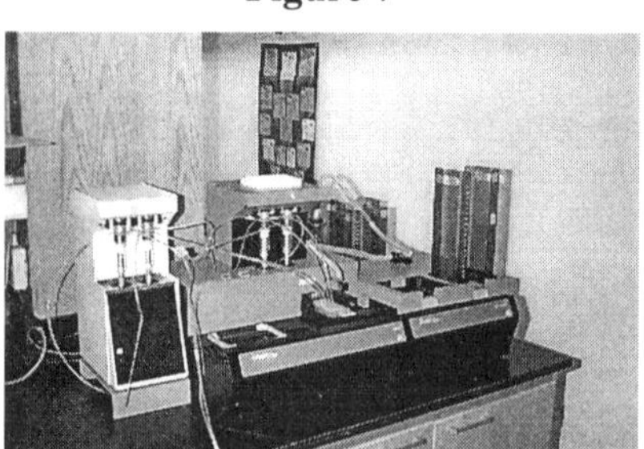

Figure 8

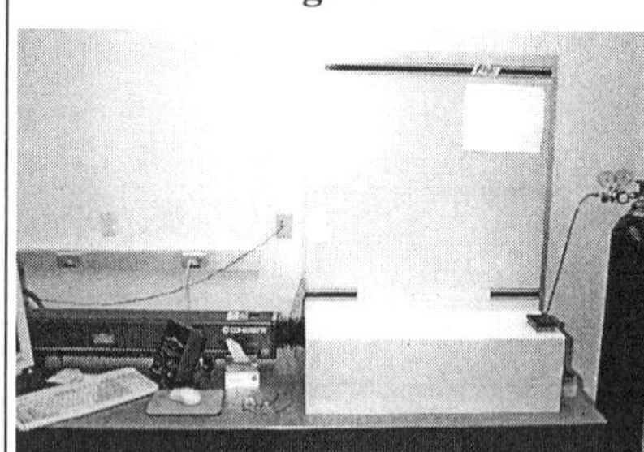

cell assays. The hit rate, or number of confirmed active from the screens, is significantly lower for cell based assays. This trend hold true through secondary and tertiary assays as well. It should be noted that 95% of these 2° and 3° tests are also cellular. In general, this is seen as a positive point, because it means that resources are not being wasted on false positives or leads which are not physiologically active.

Quality control data is collected on all targets and stored in a database. A summary of this information is also shown in Table 3. The repeat rate is the number of assay plates which have failed preset QC limits during screen operation. Once again, cellular targets are roughly equivalent to non-cellular. Another measure of the quality of a screen is the reproducibility among duplicates. Here cellular targets fair slightly better that biochemical assays, but still well within the error of the determination.

Table 3: Summary Of Data For HTS Targets

	Cellular	Other Targets
Percentage of Total	47%	53%
Difficulty (scale of 1-5)	2.2	2.4
Hit Rate	0.04%	0.14%
Lead Generation Rate	0.0014%	0.0042%
Repeat Rate	4%	3%
Bad Duplicates	1.5%	2.9%
Data represents analysis of 11,354,879 data points from all HTS screens since August, 1994.		

The methodology employed in all HTS screens at Sphinx is particularly applicable to cellular screening. To avoid variations within a screening run, single lots of reagents are purchased. Normally enough of all the materials to test over 450,000 data points must be on hand before Gamma Phase begins. This includes supplies such as microplates and tissue culture flasks. For cellular assays the cell line of interest is expanded by the Large Scale Cell Expression Lab and aliquoted. Wherever possible, aliquot sizes are determined such that a single vial can be thawed and minimally expanded for a single days testing. This frees the local cell lab in Biomolecular Research from carrying the cell line. Standardized routines and techniques must be employed during Delta Phase to minimize day to day variability.

Some techniques are associated specifically with cellular screening. Chief among these are plating methods, washing, and sterility. Most large scale cell preparation at Sphinx is done using a Multidrop (ICN or Labsystems). This instrument works well because it has a removable autoclavable head and dispenses via gentle peristaltic action. Adherent cells are plated at or near final desired density and allowed to attach overnight. Cells are never grown in 96 well plates.

In cell washing, the common knowledge is that more washing is better. Most protocols call for "wash extensively" or "wash three times". Empirical data, however, suggests that minimal washing is superior. Data from a screen validation comparing multiple washing and a single wash is shown in Figures 9 and 10. This result is typical of what has been seen from screen validations, where extensive washing is almost never profitable and sometimes detrimental.

Sterility is possibly the most misunderstood issue in HTS cellular assays. Again, common mythology is to err on the side of caution. Most cell biologists have experienced massive contamination of cultures at some time in their career and carry a natural aversion to non-sterile cell work. However, for most cellular screens, clean conditions are sufficient. If incubation times are short (<4 hours), then no precautions

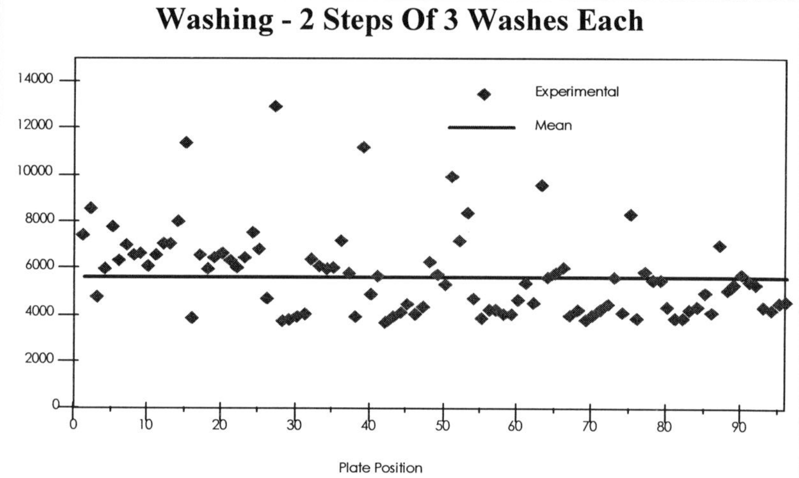

Figure 9

Validation study of a cellular target. X-axis represents well number in a 96-well microplate. Y-axis represents radioligand release in CPM.

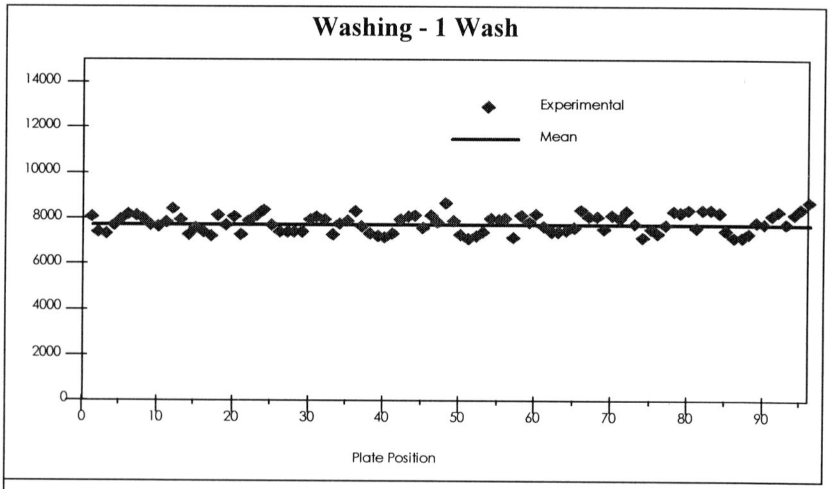

Figure 10

Validation study of a cellular target. X-axis represents well number in a 96-well microplate. Y-axis represents radioligand release in CPM

are necessary. For longer incubations, the inclusion of antibiotics and antifungals usually provides sufficient protection (data not shown). In cases where sterility is an issue, standard equipment and robotics can still be employed if care is taken. In these situations, the liquid filled systems of the liquid handling systems will be cleaned with

70% EtOH and sterilized water used for refilling. When these techniques were employed on a fully automated system (Figure 3) for a bacterial growth target no contamination (measured by overnight incubation in tryptic soy broth with phenol red) was seen in over 5,000 plates.

Finally, a major concern with cellular assays is edge effect. This is a reduction in signal and or cell growth around the edge of a microplate which usually attributed to evaporation or heat exchange. To date, all edge effects have been controllable and were eliminated during Beta Phase validation. When they do occur during Gamma or Delta Phases, they have been caused by equipment malfunctions or due to heat exchange from overcrowding of incubators.

In conclusion, whole cell and cell free assays are approximately equivalent in terms of reproducibility, difficulty, robustness, and lead generation. Cell assays offer several advantages. They offer unique functional readouts such as G-Protein coupled receptors, receptor agonist screens, and transcriptional regulation reporters. Information about a compound's ability to cross membranes as well as cytotoxicity information can be inclusive in the screen results and unknown modes of action can be explored.

Cellular screening does offer disadvantages as well. Under different circumstances many of the advantages of cellular testing may be liabilities. Cytotoxicity or the inability to cross membranes can mask activity such as binding or enzyme inhibition. Likewise, unknown modes of action can make analog generation for structure activity relationship (SAR) determination difficult.

The use of mammalian cells in HTS requires careful attention to detail in cell culture, quality control, and assay design. With these methods in place, assays based on mammalian cells are viable as a discovery tool. Sound scientific principles and judgment should be used in selecting assay methods.

Acknowledgments

The author would like to acknowledge the contributions of all the members of Biomolecular Screening, Biomolecular Research, The Screen Development and Interface Teams and Biotechnology Research. In addition, the following individuals have made particular contributions to the material presented here.

Biomolecular Screening

Mark Lister

Kiyomi Kalter

Biomolecular Research

Kate Johnston

Nancy Rankl

John Rice

Interface Development Team

Joyce Boadt

Biotechnology Research

Don McClure

Metabolism and Toxicological Studies in Immortalised Rat Hepatocyte Cell Lines

C MacDonald(1), F Reid(1), K Anderson(2), L Yin(2), E Hill(1), H Kelly(1) and M H Grant(2), (1)Department of Biological Sciences, University of Paisley, Paisley PA1 2BE, UK, (2)Bioengineering Unit, University of Strathclyde, Glasgow G4 0NW, UK.

Abstract

We have investigated the potential use of immortalised rat hepatocyte lines as an alternative to primary rat hepatocytes for drug metabolism and screening. Primary hepatocytes are widely used in the pharmaceutical industry for *in vitro* studies because established hepatoma cell lines no longer express the relevant drug-metabolising enzyme systems. Immortalised hepatocyte lines have been isolated by transfection of primary cells with SV40 early region DNA. These lines have been investigated for the retention of drug-metabolising enzymes such as glutathione-S-transferases, cytochrome P450s and bilirubin UDPGT. We have shown that immortalised lines retain a drug-metabolising capability but that these enzymes demonstrate the instability of expression with time in culture which is also a feature of primary cells.

Introduction

Primary hepatocytes are widely used in the pharmaceutical industry for *in vitro* drug metabolism studies. However, there are problems associated with the use of primary cells, because drug metabolism is not maintained in a stable fashion. They cannot be replaced by established hepatoma cell lines because the existing lines no longer express the relevant drug-metabolising enzyme systems. Immortalised hepatocyte cell lines have been isolated by transfecting primary cells with SV40 early region DNA (Yin *et al,* 1996; MacDonald and Willett 1996), but these lines behave like primary cells and instability is still a problem. Levels of drug metabolising enzymes can be induced in primary cells, eg by compounds such as benzanthracene and phenobarbitone (Jauregui *et al,* 1991) and we have investigated this approach for the immortalised cell lines. We have also modified the culture conditions for the cells and shown here and elsewhere (Goncalves *et al,* this volume) that this is another approach to increase the activity of the drug metabolising enzymes.

M.J.T. Carrondo et al. (eds.), Animal Cell Technology, 73–78.

Procedure

Materials and Methods

Cell culture. Hepatocytes were transfected with SV40 early region DNA by either calcium phosphate precipitation (SV40RH1, P9 and C2.1.2) or electroporation (LQC3 and LQC6) and cultured as previously described (MacDonald *et al*, 1994; Willett and MacDonald, 1996). Drug metabolism. Enzyme activities were measured in cell homogenates harvested with a 'rubber policeman' and prepared in 0.1M sodium phosphate buffer pH 7.6 as described previously (Grant *et al*, 1988). GST activities were measured towards the general substrate 1-chloro-2,4-dinitrobenzene (CDNB; 50µM) and towards ethacrynic acid (EA; 0.2mM). The activities were measured in 0.1mM sodium phosphate buffer, pH 7.6, in the presence of 1mM reduced glutathione (GSH) as described by Habig and Jakoby (1981). Bilirubin UDPGT activity was measured in the presence of 0.4mM bilirubin, 4mM UDP-glucuronic acid and 10mM $MgCl_2$ using the diazotisation procedure described by Dutton (1980). Total cell protein was quantified by the method of Lowry and coworkers (1951) using bovine serum albumin as standard. Cytochrome P450 dependent mixed function oxidase (MFO) activities were measured using 5µM ethoxyresorufin as substrate in the presence of 0.3mM NADPH. Reactions were carried out for 2 hours at 37^0C under yellow light and the fluorescence of the product, resorufin, detected at 580nm excitation and 600nm emission as described previously (Grant *et al*, 1988). Testosterone metabolism was used as a marker for the activity of several cytochrome P450 isoforms. Freshly isolated or primary cells were incubated for 20 minutes with 100µM testosterone and cell lines for 24 hours then the metabolic profiles were analysed by HPLC (Wortelboer *et al*, 1990).

Results and Discussion

As has been previously reported (MacDonald *et al*, 1994) the immortalised lines retain many of the characteristics of primary hepatocytes. They express bilirubin UDP glucuronosyltransferase (UDPGT) at levels which are comparable with those of fresh hepatocytes. They do not show γ-glutamyl transferase (γGT) activity (a marker found in pre-neoplastic hepatic tissue, but not in normal cells) and have a high ratio of total glutathione-S-transferase (GST) activity measured by CDNB conjugation compared to that detected by metabolism of EA. Initially this latter activity was thought to be due to the presence of the pi isoform of GST but Western blotting suggests that other isoforms capable of metabolising EA may also be present (data not shown). In addition, the immortalised cell lines retain the ability to synthesise GSH from L-methionine via the cystathionine pathway. This ability is lost during primary culture of hepatocytes and is also absent from de-differentiated hepatoma cell lines. Although these enzyme activities are present in the immortalised lines, levels of expression are not stable: for example, the ratio of CDNB to EA GST activities and GSH resynthesis decline as the cells are passaged. In contrast, however, γGT has not been detected even after continued culture.

We have adopted two approaches to increase levels of drug metabolising enzymes in the immortalised cell lines. The first has been to culture the cells in the presence of an inducing agent which is added to a confluent culture 5 days after passage for 3 days with daily renewal in fresh medium. Addition of 2mM phenobarbitone resulted in a 2.5-fold increase in UDPGT activity. It had no effect on GST activity towards CDNB but gave a 4-fold enhancement of GST activity with EA as substrate. In contrast, 25μM 1,2 benzanthracene in 0.05% (v/v) DMSO inhibited both bilirubin UDPGT and GST activities, but increased MFO activity towards ethoxyresorufin 4-fold.

A second approach to increasing the level of expression of bilirubin UDPGT was to modify the culture conditions of the cells. Two immortal lines were cultured in either DMEM with 10% foetal bovine serum or in a hormonally-defined medium based on Ham's F12 and Williams E media (Isom and Georgoff, 1984). In both lines the level of bilirubin UDPGT rose when the cells were cultured in the hormonally-defined, serum-free medium: in the case of P9 cells rising from 1.40 ± 0.25 pmol/min/mg protein to 7.72 ± 0.67 pmol/min/mg protein, a level close to that observed in hepatocytes.

The cytochrome P450-dependent mixed function oxidase system (MFO) is involved in both detoxification and activation of a wide range of chemicals. It is therefore particularly important that it is retained in a cell line which is to be used as an *in vitro* model for drug screening or toxicity testing. Although only modest levels of MFO were detected in P9 cells (9.96 ± 0.90 pmol resorufin formed/hour/mg protein), this activity was increased 4-fold by treatment with benzanthracene, 2-fold with 1μM dexamethasone and in combination the effect of these two inducers was additive.

In order to investigate the levels of expression of the isozymic forms of cytochrome P450 the cells were incubated with 100μM testosterone and the metabolites formed separated by HPLC. Eight different peaks could be identified as a consequence of the formation of seven different metabolites plus non-metabolised testosterone (Table 1). Metabolites were identified and quantified using authentic standards.

Table 1: Elution of testosterone metabolites from HPLC column and major P450 isozymes involved in their formation.

ELUTION ORDER	METABOLITE NAME	MAJOR P450 ISOZYMES INVOLVED IN METABOLISM
Peak 1	6α-hydroxytestosterone	2A1
Peak 2	7α-hydroxytestosterone	2A1, 2A2
Peak 3	6β-hydroxytestosterone	3A1, 3A2
Peak 4	16α-hydroxytestosterone	2B1, 2C11
Peak 5	16β-hydroxytestosterone	2B1, 2B2
Peak 6	2α-hydroxytestosterone	2C11
Peak 7	androstenedione	2B1, 2C11, 2B2
Peak 8	testosterone	NA

Table 2 shows the testosterone metabolism profiles of fresh hepatocytes, those cultured for 24 hours and three of the immortalised lines. A comparison of profiles of freshly isolated hepatocytes with those which had been cultured for 24 hours showed considerable differences. As a % of the total metabolites formed 16β - hydroxytestosterone dropped substantially after culture, and 2α - hydroxytestosterone and androstenedione increased. The effect of culture on the immortalised line LQC3 was studied by comparing profiles at passage 6 and 10. Once again the proportion of testosterone metabolised to 16β dropped, and metabolism to androstenedione increased. At passage 6 LQC3 also shows retention of 16α and 2α forms, but these are lost with continued culture. In order to identify unequivocally which of the P450 isozymes are expressed in these cells we are examining expression of P450 mRNAs. Preliminary data suggest that mRNAs encoding cytochromes 2B1, 2B2 and 3A1 are present. However, the PCR product amplified in some of the cell lines is different from, and longer than that observed in hepatocytes. The significance of these results is not clear and therefore further PCR analysis is required to obtain a more informed conclusion.

Table 2: Testosterone profile of hepatocytes quantified as a percentage of total metabolites present. Results are mean $\pm$ SEM, $n \geq 3$ (apart from LQC pass 6, where n=1).

Cell Name	7α	6β	16α	16β	2α	AD
Fresh hepatocytes	16.5±3.4	20.3±4.4	15.9±4.9	16.8±4.8	27.0±8.3	3.0±0.9
24h primary hepatocytes	2.2±1.6	20.4±7.7	17.2±1.6	0	34.7±12.9	9.2±2.1
C2.1.2 (pass 8)	0	0	0	5.6±2.7	0	94.3±2.7
P9 (pass 7)	0	0	0	6.4±1.3	0	93.5±1.3
LQC3 (pass 6)	0	0	10.9	9.3	56.7	23.1
LQC3(pass 10)	0	0	0	4.9±2.6	0	95.1±2.6

Conclusions

We have shown that immortalised rat hepatocyte cell lines isolated either by electroporation or calcium phosphate transfection retain some of the drug metabolising capability of normal rat hepatocytes. The lines also show an instability in culture which is comparable to that observed with freshly isolated cells. However, enzyme levels can be enhanced by exposing the cells to inducers and preliminary evidence indicates that cytochrome P450-dependent MFO levels can be stabilised by modification of culture conditions.

Acknowledgements

This work is supported by grants from the Biotechnology and Biological Sciences Research Council and the British Council.

References

Dutton G J (1980). Principles of assay of glucuronidation in biological tissues or fluids and selected practical procedures. In: Glucuronidation of Drugs and Other Compounds. Boca Raton, Fl., pp 183-204.

Goncalves L, Lopez T, MacDonald C, Grant M H and Carrondo M J T. Three dimensional cell culture systems to stabilize the differentiation of hepatocyte cell lines, this volume.

Grant M H, Duthie S J, Gray A G and Burke, M D (1988). Mixed function oxidase and UDP-glucuronyltransferase activities in the human Hep G2 hepatoma cell line. Biochemical Pharmacology, 37, 4111-4116.

Habig W H and Jakoby W B (1981). Assays for the differentiation of glutathione-S-transferases. Methods in Enzymology, 77, 398-405.

Isom H C and Georgoff I (1984). Quantitative assay for albumin-producing liver cells after SV40 transformation of rat hepatocytes maintained in chemically defined medium. Proceedings of the National Academy of Sciences USA, 81, 6378-6382.

Jauregui H O, Ng S-F, Gann K L and Waxman D J (1991). Xenobiotic induction of P450 PB-4 (IIB1) and P-450c (IA2) and associated mono oxygenase activities in primary cultures of adult rat hepatocytes. Xenobiotica, 21, 1091-1106.

Lowry O H, Rosebrough N J, Farr A L and Randall R J (1951). Protein measurement with the Folin phenol reagent. Journal of Biological Chemistry, 193, 265-275.

MacDonald C, Vass M, Willett B, Scott A and Grant M H (1994). Expression of liver functions in immortalised rat hepatocyte cell lines. Human Experimental Toxicology, 13, 439-444.

MacDonald C and Willett B (1996). The immortalisation of rat hepatocytes by transfection with SV40 sequences. Cytotechnology, in press.

Wortelboer H M, Hogberg J, De Krief C A, Van Iersel A A A, Falke H E, Noordhoek J and Blaauboer B J (1990). The isozyme pattern of cytochrome P450 in rat hepatocytes in primary culture, comparing different enzyme activities in microsomal incubations and in intact monolayers. Biochemical Pharmacology, 40, 2525-2534.

Yin L, MacDonald C and Grant H (1996). Electroporation conditions for retention of rat hepatocyte viability. The Genetic Engineer and Biotechnologist, 16, 27-34.

CELL BASED ASSAYS FOR DRUG SCREENING

<u>Elizabeth Keenan</u>, Helen Kruger, Balwinder Sambi and Dr. John S. McLean.
Department of Biological Sciences, University of Paisley, High Street, Paisley, Scotland, U.K. PA1 2BE.

ABSTRACT
A CRE-alkaline phosphatase reporter construct has been introduced into a mammalian cell line to investigate signal transduction pathways used by the calcitonin receptor. Calcitonin activated alkaline phosphatase production in the CRE-reporter cell line indicated coupling to the Gαs protein and activation could be inhibited by antagonists. These cells allow functional screening and analysis of possible novel drugs. The genetic manipulation of the cAMP/CRE/CREB pathway for cGMP sensing using the CRE-alkaline phosphatase cell based assay shall be discussed. We show that by altering the amino acid sequences of the CREB within the substrate recognition region we may be able to alter the substrate specificity of CREB for phosphorylation by PKA and PKG.

Abbreviations
cAMP:cyclic adenosine monophosphate,cGMP:cyclic guanosine monophosphate, CRE:cAMP response element, CREB:CRE binding protein, PKA/PKG: protein kinase A or G, CHO: Chinese Hamster Ovary, SPAP:secretory placental alkaline phosphatase.

INTRODUCTION
The development of cell based assays for drug screening within the pharmaceutical industry is of great importance. These assays provide a fast, economical and high throughput method of screening chemical compounds. The assays are based upon the ability of drugs to elicit a response by alteration of second messengers within the cell. and these can be sensed by use of reporter constructs within cell lines.
Alterations in the level of cGMP has been shown in a number of clinical conditions such as septic shock, hypertension and subarachnoid spasm (1). We envisage that by genetic manipulation of the cAMP assay we can develop an assay whereby the level of cGMP can be monitored which would be of great benefit to the pharmaceutical industry.

MATERIALS AND METHODS
All tissue culture medium and restriction endonucleases were obtained from Life Technologies, Paisley, Scotland. Protein kinases A and G (PKA and PKG) were obtained from Promega, Southhampton, England.Calcitonin (human), CGRP8-37(human), cAMP and cGMP were obtained from Calbiochem-Novabiochem (Nottingham, England). All other chemical agents were from Sigma(Poole, Dorset, England). CHODG44 cells were a gift from Dr. Larry Chasin (Columbia University, New York, U.S.A.). pT7CSPAP was a gift from Dr. Don Wallace (Glaxo, Greenford, England).

M.J.T. Carrondo et al. (eds.), Animal Cell Technology, 79–84.

Cell Culture and generation of CRE-SPAP cell line
Cells were routinely grown in growth media (Dulbecco's modified essential medium (DMEM) supplemented with 10% foetal calf serum (FCS), 1x non-essential amino acids (NEAA), penicillin and streptomycin). Linearized plasmids, pT7CSPAP and pCMVHyg, DNA were transfected by the calcium phosphate precipitation method (4) into Chinese hamster ovary DG44 cells (CHODG44) for 16h. Clones were selected by addition of Hygromycin (443.92μg/ml) to the medium. Clones were expanded and assayed for the presence of the CRE-SPAP construct by forskolin (20μM) induced alkaline phosphatase release assays (data not shown).

Alkaline Phosphatase Release Assay
Cells were seeded into 24 well plates at a density such that the next day they were confluent or approaching confluence. The cells were then allowed to reach quiescence for 48hr at 37°C in growth medium containing 0.5% FCS. The cells were washed with non-phenol red growth media and incubated overnight at 37°C with the test agent in 400μl in non-phenol red growth media containing 0.5%. Supernatants were removed and heat inactivated at 65°C for 45min. After heat inactivation 2 volumes of substrate [5mM p-nitrophenol phosphate in DEA buffer (1M diethanolamine, 0.28MNaCl and 0.5mM MgCl2, pH 9.85) buffer] was added to each sample. The samples were then plated out into a 96 well plate and incubated at 37°C for 24 hours and the absorbance read at 405nm.

Peptide Design and Synthesis
Peptides were designed to introduce single and multiple mutations into the substrate determinant sequence of CREB and these were synthesised by Cambridge Research Biochemicals, Cheshire, England.
ILSRRPSYRKILNDL [peptide(p) 1,CREB positive control], ILSKRPSYRKILNDL (p2), ILSRKPSYRKILNDL(p3), ILSRRPTYRKILNDL(p4),ILSRRPSKRKILNDL (p5),ILSRRPSHRKILNDL(p6), ILSRRPSYRKFLNDL (p7), ILSKKPSYRKILNDL(p8), ILSKKPTYRKILNDL (p9), ILSKKPTYRKFLNDL(p10), ILSKKPTKRKFLNDL(p11), ILSRRPAYRKILNDL (p12, negative control)

Peptide Kinase Assays
Kinase assays were carried out following the procedure of Colbran *et al* (4) using 2.2nM PKA, 0.61nM or 2.2nM PKG and ten μl of 0.1mM peptide in a 40μl reaction mixture. Ten μl of the reaction was terminated and placed in 7ml of scintillant and counted.

RESULTS AND DISCUSSION

Calcitonin induced alkaline phosphatase release from a CRE-SPAP reporter construct in CHODG44 cells.

CHODG44 cells expressed an endogenous calcitonin binding receptor as indicated by the induced release of alkaline phosphatase by calcitonin occurred in a dose dependent manner, EC50 ~6×10^{-8}M (Figure 1). There is at the present time no specific antagonist of calcitonin, however calcitonin gene related peptide (CGRP) antagonist, CGRP8-37 inhibited calcitonin induced alkaline phosphatase release in the CHODG44 CRE-SPAP cell line. The IC50 of CGRP8-37 on calcitonin induced alkaline phosphatase release was approximately 5x10-8M (Figure 1).

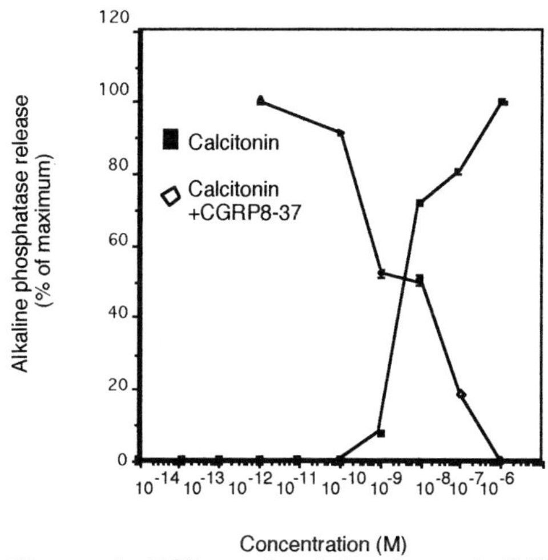

Figure 1: Effect of calcitonin and CGRP8-37 on alkaline phosphatase induced release from CHODG44 CRE-SPAP cells.
Addition of varying concentrations of calcitonin on alkaline phosphatase release (squares). Cells were stimulated with 5μM calcitonin and incubated with varying concentraions CGRP8-37 (diamonds)

To determine whether calcitonin induced alkaline phophatase transcription via the cAMP second messenger pathway we incubated calcitonin (5μM) stimulated cells with varying concentrations of the protein kinase A inhibitor H-89. As shown in Figure 2 H-89 inhibits calcitonin induced alkaline phosphatase release in a dose dependent manner (IC50 ~9x10^{-7}M).

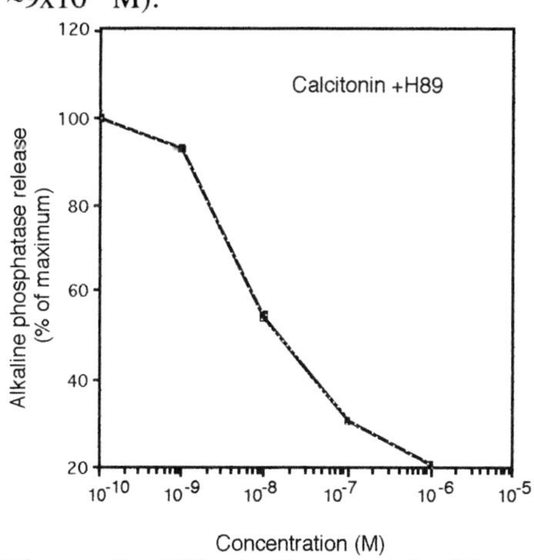

Figure 2:- Effect of a protein kinase A inhibitor (H-89) on calcitonin induced alkaline phosphatase release in CHODG44 CRE-SPAP cells.

Mutation of the substrate determinant sequence of CREB

Protein kinase A phosphorylates the serine at position 133 in the CREB sequence with a high degree of selectivity (5). Using published data we designed a number of mutant CREB peptides (see materials and methods). These contained either single, double or multiple mutations between the -3 and +4 amino acid surrounding Ser 133.

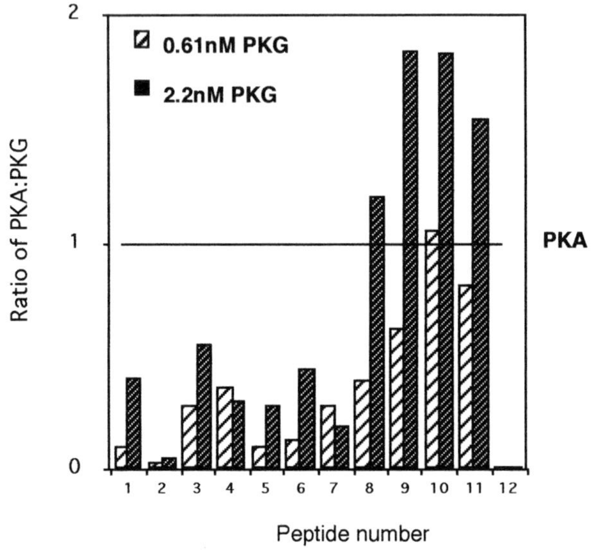

Figure 3:- Kinase assays for substate specificity of CREB and mutant CREB peptides by PKA and PKG. For peptide sequences see materials and methods.

The above data shows that the introduction of mutants into the CREB substrate determinant region can alter substrate selectivity. The greatest alteration in substrate selection was seen by mutation of the sequence RRPSYRKI to KKPTYRKF (peptide 10). We are at present introducing these mutations into the CREB DNA by site directed mutagenesis. We envisage that introduction of the mutant CREB into a mammalian expression vector and stable transfection into CHODG44 CRE-SPAP cell line will allow the monitoring of the action of chemical entities acting via the cGMP/PKG pathway as the PKG will phosphorylate the mutated CREB and allow CRE-SPAP transcription.

Crosstalk between cAMP and cGMP pathways

To ensure that there are no crosstalk pathways whereby, cGMP analogues or drugs which cause increased synthesis of guanylate cyclase, can activate transcription in the CRE-SPAP construct in CHODG44 cells, we stimulated the cells with various cAMP/cGMP agonist (Figure 4). These results show that there is very little crosstalk and that this cell line can be used to introduce a mutant CREB mammalin expression construct.

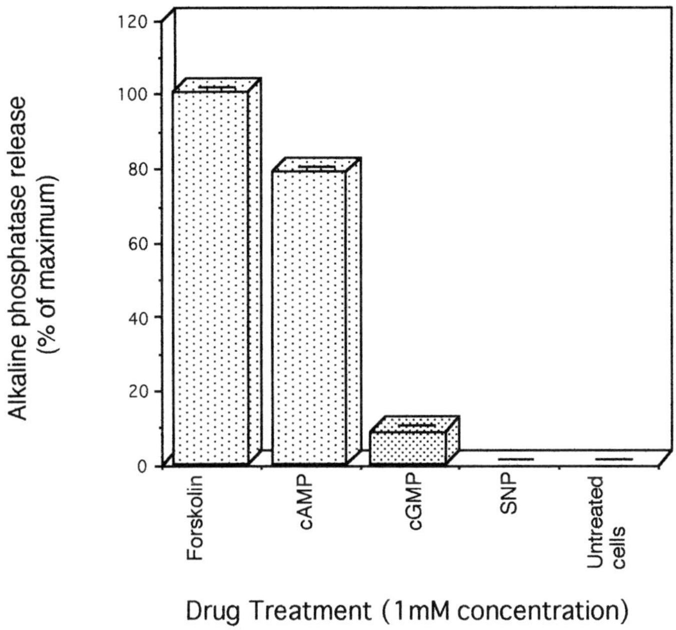

Figure 4:- Effect of cAMP/cGMP agonist on alkaline phosphatase release from a CHODG44 CRE-SPAP cell line.

In conclusion we have a cell based assay whereby the monitoring of drugs acting via the cAMP pathway can be determined in a quick, economical and functional manner. We have shown that mutation of the substrate determinant sequence within the CREB can be altered in order that it is a greater substrate for phosphorylation by protein kinase G. This mutant CREB can be introduced into the cAMP assay and the effects of drugs on the cGMP signal transduction pathway monitored.

REFERENCES

1. Walters U (1989) Rev. Physiol. Biochem. Physiol. 113, 41-88.
2. Kennelly P J & Krebs E G (1991) J. Biol. Chem. 266, 15555-15558.
3. Wigler M, Sweet R, Sim G K etal (1979) Cell 16, 777-785.
4. Colbran J L, Francis S H, Leach A B, Thomas M K, Jiang H, McAllister L M & Corbin J D (1992) J. Biol. Chem. 267 14 9589-9594.
5. Gonzalez G A & Montminny M R (1989) Cell 59, 675-680.

Discussion

Bernard: In your construct what number of repeats of the CRE elements did you have and did this have any effect on cyclic AMP response?

Keenan: There are 4 copies of the CRE element in the construct. If you take it down to 1 cyclic AMP response element you do not get very good induction of the cAMP pathway.

Bernard: Where are the specific applications of cGMP sensing in functional responses? Are there specific signal transduction pathways that are more amenable to cGMP detection rather than cAMP?

Keenan: cGMP has been shown to activate transcription via the AP1 transcription factor on a phorbol ester response element. I have carried out work with a phorbol ester response element linked to alkaline phosphatase and transfected into BHK cells. You did see an induction of alkaline phosphatase with cAMP but it was not high enough to be the basis of a functional assay.

DEVELOPMENT OF DRUG METABOLISM SIMULATOR USING THREE-DIMENSIONAL CULTURE OF HEPATOCYTES OF RATS AND DOGS

LIDOCAINE METABOLISM IN HEPATOCYTE/SPHEROID CULTURE OF RATS AND DOGS

KAZUMORI FUNATSU[1], KOHJI NAKAZAWA[1], HIROSHI MIZUMOTO[1], KENJI TAKENAKA[2] and TAKU MATSUSHITA[1]

Department of Chemical Engineering[1] and Surgery[2], Kyushu University
6-10-1Hakozaki, Higashi-ku, Fukuoka, 812-81, JAPAN

ABSTRACTS. Primary hepatocytes of rats and dogs formed spheroids in the pores of polyurethane foam (PUF) used as a culture substratum. The hepatocytes in monolayer and spheroid cultures of stationary state converted lidocaine to monoethylglycinexylidide (MEGX) which was N-deethylation of lidocaine. These metabolic activities of the hepatocytes/spheroid stationary culture system were 1.5~2.0-fold higher than those of the hepatocyte monolayer culture for 10 days. Consequently, PUF/spheroid culture maintained high drug metabolizing activity for more than 10 days and seems to be promising as a drug metabolism simulator which will be used for cytotoxicity tests and drug screening.

1. Introduction

In vitro experiments using cultured cells are increasingly popular to investigate toxic effects and metabolisms of new chemicals for novel drugs. Especially hepatocytes are frequently important cells for investigating chemically mediated toxicity of drugs, because the liver is an important organ in the metabolism and biotransformation of xenobiotics. Though, primary hepatocytes have been often used for these experiments, in the usual monolayer culture, their differentiated functions rapidly decrease in a week or so. Therefore, it is very important to develop a culture technology of primary hepatocytes which make it possible to maintain these functions stably *in vitro*.

M.J.T. Carrondo et al. (eds.), Animal Cell Technology, 85–90.

We have already reported that by three-dimensional culture of rat hepatocytes (spheroid culture) using PUF as a culture substratum, hepatocyte specific functions such as albumin production and urea synthesis can be maintained for more than 2 weeks [1]. Furthermore, we indicated that as an application of the three-dimensional culture PUF/spheroid culture was promising in development of a hybrid artificial liver support system [2].

In this study, we investigated the drug metabolizing activities of primary hepatocytes/spheroid culture of rats and dogs using PUF for the development of a simulation device of drug metabolism of liver *in vitro* that is, a drug metabolism simulator.

2. Materials and Methods

2.1. Hepatocytes preparation and their culture

Hepatocytes were isolated from adult Wistar rats (7 weeks, 180~230g) and dogs (12~20kg) by the liver perfusion method [3] with 0.05% collagenase. In this experiment, the viability of isolated hepatocytes was more than 80%.

The isolated hepatocytes were inoculated for monolayer culture at a density of 5.0×10^5 viable cells in 2ml of medium per tissue culture dish of 35 mm in diameter which was coated by Type I-collagen. For PUF/spheroid stationary culture, the isolated hepatocytes were inoculated at a density of 3.0×10^6 viable cells in 2ml of medium per dish containing a piece of PUF plate, 25 mm long, 25 mm wide and 1 mm thick. The details of hormone-defined medium (HDM) and PUF for hepatocyte culture were reported in our previous paper [1].

2.2. Metabolic reaction from lidocaine in cultured hepatocytes of the two species

Metabolic reaction from lidocaine in cultured hepatocytes of the two species in monolayer culture and PUF/spheroid stationary culture was performed in HDM containing 0.25mM lidocaine with a final concentration of 0.5% ethanol. The reaction in stationary culture was kept continuously for 24 hours with the medium exchanged at 2-day intervals on a rotary shaker at 50 rpm. Lidocaine and its metabolic products were extracted from cells and medium with toluene. *p*-Diethoxybenzene was added to each sample for the internal standard of quantitative analysis of GC/MS.

The viable cell numbers were measured by modified MTT method [4]. The calibration curves were constructed by measurement of isolated hepatocytes in a range from 1.0×10^4

to 1.0x10^6 viable cells.

3. Result

3.1. Metabolic reaction from lidocaine of primary hepatocytes of the two species in monolayer culture and spheroid stationary culture

Lidocaine metabolic activity of primary hepatocytes of the two species in monolayer culture and PUF/spheroid stationary culture was evaluated by decreasing lidocaine. The lidocaine metabolites in both cultures were identified by GC/MS mass fragment pattern of lidocaine and monoethylglycinexylidide (MEGX). The hepatocytes of the two species in both cultures converted lidocaine to MEGX, which is N-deethylation of lidocaine (Fig.1).

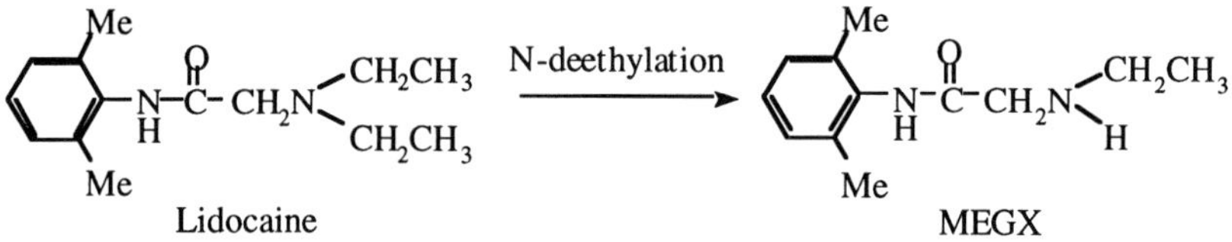

Fig.1 Metabolic pathway from lidocaine to MEGX through N-deethylation

In both hepatocytes of the two species, the lidocaine disappearance rates per viable hepatocytes in monolayer and PUF/spheroid culture were almost the same at the first day, but became different after 3 days (Table 1). The lidocaine disappearance rates continuously decreased from the initial level for 10 days in monolayer culture of primary hepatocytes of the two species, while it was maintained at almost the initial level of 30 μg/10^5viable cells/day for 10 days (Table 1) of rat hepatocytes and 5.7 μg/10^5viable cells/day of dog hepatocytes in PUF/spheroid culture. The lidocaine disappearance rate of primary hepatocytes of the two species in PUF/spheroid culture was maintained at a higher level than that in monolayer culture (Table 1). On the other hand, MEGX appearance rates in monolayer and PUF/spheroid culture were not so different (data not shown).

Table 1. Comparison of lidocaine disappearance rates between monolayer and PUF/spheroid stationary culture of rat and dog hepatocytes

Culture time [day]	Lidocaine disappearance rate [μg/10^5 viable cells/day]			
	Rat hepatocyte		Dog hepatocyte	
	Monolayer	PUF/spheroid	Monolayer	PUF/spheroid
1	27.82±1.51	32.74±0.84	7.13±0.32	5.75±0.54
3	25.00±0.47	36.60±5.27	6.95±0.26	7.56±0.84
5	17.75±1.74	30.15±6.44	4.07±0.52	10.85±0.56
7	11.57±4.67	26.89±8.61	4.06±0.51	9.13±1.29
10	8.73±4.29	29.33±11.65	3.29±0.31	5.74±0.63

4. Discussion

It is very important to be able to assess *in vitro* even approximately *in vivo* drug metabolism in the bodies of animals and humans, because, if it becomes possible, a replacement and reduction of animal testing and animal experimentation will be achieved. These achievements will make it possible to decrease the period and cost for development of new drugs. When studying the biotransformation of novel compounds *in vitro*, with the aim of predicting the *in vivo* liver-based metabolism of this drug, it would be advantageous to have a model that retains the characteristics of the *in vivo* hepatocyte.

Conventionally cultured primary hepatocytes in monolayer lose drug metabolizing activities within the first 2~3 days. On the other hand, the activity of hepatocytes/ spheroid of the two species in 0.25mM lidocaine was maintained for 10 days at 1.5~2.0 times higher levels than that of hepatocytes/monolayer as indicated in Table1. Such a stable maintenance of the activity of primary hepatocytes *in vitro* is very important for assessing drug metabolism in animals and humans.

The results described above suggest that hepatocytes/spheroid using PUF as a culture substratum maintains drug metabolism activity as *in vivo* for a comparatively longer time than conventional culture of hepatocytes. Consequently, PUF/spheroid 3-dimensional culture seems to provide a promising culture system for development of a drug metabolism simulator.

5. Acknowledgment

This study was supported in part by Grant-in-Aids for Developmental Scientific Research (B ; 06555248) and Scientific Research (B ; 06453107) from the Ministry of Education, Science and Culture of Japan, and by Terumo Life Science Foundation (No. 95Bb005).

6. Reference

[1] Matsushita T, Ijima H, Koide N and Funatsu K (1991) High albumin production by multicellular spheroids of adult rat hepatocytes formed in the pores of polyurethane foam. *Appl. Microbiol. Biotechnol.* **36**: 324-326.

[2] Ijima H, Matsushita T and Funatsu K (1994) Development of a hybrid artificial liver using multi capillary PUF/spheroid packed bed. *Japanese Journal of Artificial Organs* **23**: 463-468.

[3] Seglen P O (1976) Preparation of isolated rat liver cells. In: Prescott DM (ed.) *Methods Cell Biol.* vol. 113, Academic press, New York, pp. 29-83.

[4] A. G. E. Pearse and R. Hess (1961) Substantivity and other factors responsible for formazan patterns in dehydrogenase histochemistry. *Experimentia* **17**: 136.

Discussion

Martin: Do you think we can maintain a culture of lymphocytes with your PUF system for 1-2 weeks?

Funatsu: We have experimented with hepatocyte cells only. It may be possible for other cell types.

IMMOBILIZATION OF PRIMARY BRAIN CELLS IN POROUS MICROCARRIERS FOR *ON LINE* NMR SPECTROSCOPY

P. M. Alves, M. J. T. Carrondo and H. Santos

IBET/ITQB, Apartado 12, P- 2780 Oeiras, Portugal

Abstract

Mice primary brain cell cultures can be used as model systems to study biochemical processes in the brain. As the brain consists of a complex mixture of different cell types, using NMR (Nuclear Magnetic Resonance) spectroscopy and suitable cell culturing techniques, it is possible to study cell metabolism of each cell type separately or in defined co-cultures.

The aim of this work was the development of a suitable perfusion system to keep primary brain cells under physiological conditions for long periods in the NMR tube, without losses in viability.

Immobilization of astrocytes and neurons in three-dimensional structures is described. These immobilization techniques allowed high cell densities and protection from shear stress due to perfusion flow. Their feasibility for perfusion in the spectrometer is demonstrated using ^{31}P-NMR spectra to monitor cell energy state and thus well-being of the cells.

Key words: Brain cells, Porous microcarriers, *on line* monitoring ^{31}P NMR

Introduction

Major neurological disorders such as Parkinson's disease, Alzheimer's disease, stroke, epilepsy and trauma are poorly understood and currently applied treatment is at best ameliorative than curative. Also unknown are the metabolic mechanisms underlying cell degeneration in these disorders. As a strategy to overcome the complexity of brain and to elucidate cerebral metabolism, several models have been used in the last years. Studies using functioning brains, brain slices, cell aggregates, primary brain cell cultures and cell extracts have contributed with a great deal of information leading to a better knowledge

M.J.T. Carrondo et al. (eds.), Animal Cell Technology, 91–98.

of the CNS (central nervous system) biochemical processes [1-3].

Table I: Advantages and disadvantages of different culture systems for NMR spectroscopy

	Advantages	Disadvantages
Monolayers	• easy visualization • pure primary neurons and glial cultures can be obtained • ideal for cell extracts	• cells can not be easily moved to the NMR tube
Brain Slices	• easy to handle and to move to the NMR tube • high cell densities	• co-cultures • cell damage (due to cut of the slices and to perfusion). Slices lifetime is between 8 and 10 hours. • thickness of the slice (important to avoid necrotic areas)
Cell aggregates	• easy to handle and to move to the NMR tube • high cell densities	• co-cultures • cell damage (due to perfusion) • aggregates size (important to avoid diffusion problems inside the aggregate structure that lead to necrotic centers) • a large inoculum is required
Non Porous Microcarriers	• easy visualization • pure primary neuronal and glial cultures • no transport limitations	• shear damage due to perfusion • low cell densities • ^{13}C-NMR signals from the matrix may overlap resonances of the metabolites.
Porous Microcarriers	• pure primary neuronal and glial cultures • high cell densities • protection from shear stress due to perfusion flow • easy to handle and to move • not very expensive	• direct visualization is impossible • diffusion problems inside the pores (oxygen and nutrients limitations) • indirect quantification of cell number (protein, MTT assay, DNA) • ^{13}C-NMR signals from the matrix may overlap resonances of the metabolites.
Gel threads	• pure primary neuronal and glial cultures • easy visualization • high cell densities • protection from shear stress due to perfusion flow • easy to handle and to move	• diffusion problems inside the threads (oxygen and nutrients limitations) • indirect quantification of cell number (protein, MTT assay, DNA) • high price

NMR techniques are potentially very attractive to explore cell metabolism due to their unique non-invasive characteristics. In order to be able to profit from the advantages of *in vivo* NMR to study biochemical processes in brain cell cultures, it is essential to develop techniques for cell immobilization and adequate perfusion inside the NMR tube.

Living samples must be maintained under defined and homogeneous physiological conditions in order to obtain useful results. Various metabolic manipulations and pharmacological situations can then be simulated. *On line* monitoring of metabolic responses and the detailed characterization of cerebral energy metabolism under metabolic insults are possible. Nevertheless, this goal is hampered by the need for high cell densities due to the low sensitivity of the NMR method. Thus, an immobilization system for primary brain cells that allows high densities is needed.

Table I summarizes the main advantages and disadvantages of several approaches for NMR spectroscopic studies of brain cells metabolism.

In this work several porous microcarriers were tested for primary neurons and primary astrocytes immobilization. Their performance for real time NMR spectroscopy and their feasibility for perfusion is demonstrated by ^{31}P-NMR.

Materials and Methods

Materials

Dulbecco's modified Eagle's medium (DMEM), fetal calf serum (FCS), phosphate-buffered saline (PBS) and trypsin/EDTA were purchased from GIBCO (England). Culture dishes and flasks were obtained from NUNC (Roskilde, Denmark). Cytopore 2 microcarrier was purchased from Pharmacia (Uppsala, Sweden). ImmobaSil FS microcarriers were obtained from ASL (Leicestershire, England). Siran beads S120 (pore size <120 μm) were obtained from Schott (Mainz, Germany). CultiSpher G and GH were a generous gift from Percell Biolytica. All other chemicals were of the purest grade available from regular commercial sources.

Cell Culture

Primary neurons were prepared from cerebella from 7-day-old mice as described before [4,5]. After centrifugation cells were resuspended (2-3x10^6 cell/ml) in DMEM containing 10% FCS and immediately immobilized in different porous microcarriers (Cytopore 2, ImmobaSil FS, S120, CultiSpher G and CultiSpher GH) according to the manufacture's recommendations. Typically cells were grown in culture dishes (3 cm in diameter) using 0.5 ml of microcarriers and 2 ml of medium in a humidified atmosphere of 10% CO_2/90% air at 37°C. At 48 h after inoculation cytosine arabinoside (20 μM) was added to prevent astrocyte proliferation. Glucose content in the medium was monitored every 24 h and total protein was measured at the end of maturation period (i.e. 9 days).

Primary astrocytes were prepared from newborn (1-2 day-old) mice brains [4,6]. Cerebral hemispheres were carefully freed of the meninges and mechanically disrupted. Single cell suspensions were plated in culture flasks (2 brains/flask) in DMEM containing 10% FCS. After 7 days of preculturing, the astrocytes were removed by mild trypsinization and subcultured in culture dishes for another 15 days. Cells were kept in a humidified atmosphere of 10% CO_2/90% air at 37°C. Growth medium was changed twice per week. After this period, $3x10^6$ cells were harvested by trypsinization, pelleted by centrifugation, resuspended in DMEM containing 10% FCS and immediately immobilized in 0.5 ml of microcarriers as described above. Immobilized cells were allowed to grow in culture dishes for 5-6 days. Glucose content in the medium was monitored every 24 h and total protein was measured at day 6 after immobilization.

Control cultures (monolayers) in T-flasks were done in 3-cm dishes. For neuronal cultures a previous coating of the flask with poly-L-lysine was done for control cultures.

For NMR experiments which require a large amount of biomass, 14-cm culture dishes with 10 ml of medium and 2.5 ml of microcarriers were used. Cells were used for perfusion in the NMR apparatus approximately one week after immobilization.

Perfusion System

For NMR measurements the microcarriers were transferred to a 10 mm NMR tube fitted with a perfusion insert (Figure 1). The whole perfusion system was previously sterilized. Cells were perfused at 37° with a DMEM oxygenated with 95% O_2/5% CO_2 at 1-1.5 ml/min.

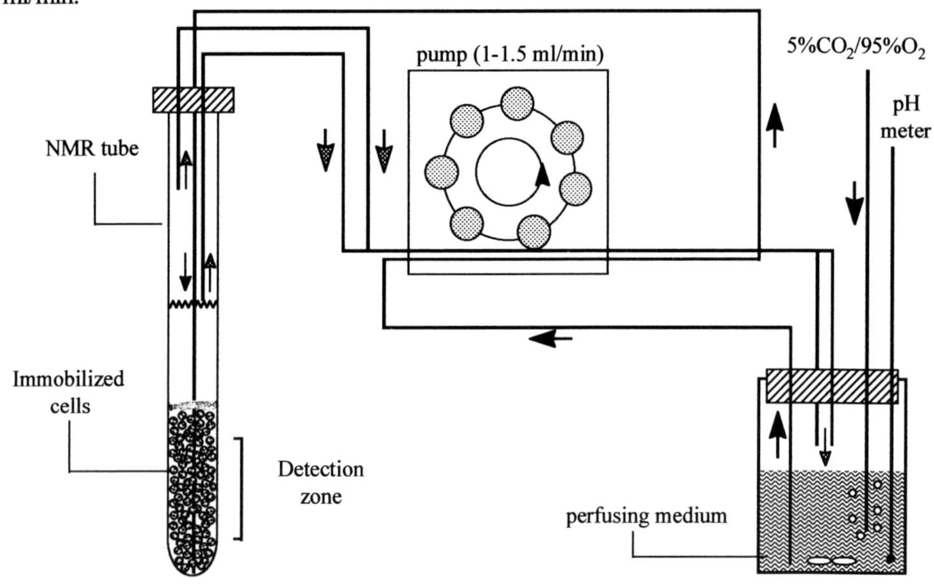

Figure 1: Diagram of the perfusion apparatus for *in vivo* NMR experiments

Analytical Determinations

Glucose was measured using standard high performance liquid chromatography techniques with a Shodex Sugar SH column from Waters (Massachusetts, USA). Protein was determined by the method using bicinchonic acid assay reagent patented by Pierce.

The content and distribution of cells immobilized in microcarriers was monitored by the MTT assay.

Spectroscopy
^{31}P NMR spectra were obtained using a 10 mm broad band probehead on a Bruker AMX-500 spectrometer operating at 202.45 MHz. Spectra were acquired using a 66° flip angle and 2.5 s repetition time; 4 K data points were collected over a spectral width of 4.5 kHz.

Results and Discussion

There are specific requirements to be taken into account to obtain viable and functional primary cultures of brain cells. Both neurons and astrocytes are strictly anchorage-dependent and, in the case of neurons, culture conditions are even more demanding because these cells do not divide (initial cell number is similar to final cell number) which makes the inoculation procedure a very critical step.

Neurons in microcarriers
Neurons are very sensitive to the anchorage surface and to culture medium changes. The life span of the culture is a very important parameter in neuronal cultures, too. Usually it is impossible to avoid significant (at least 5% of total cell number) glial contamination in the primary neurons after 9 days in culture.

Among the five different porous microcarriers tested, CultiSpher, Cytopore 2 and ImmobaSil FS were suited for maturation of cerebellar neurons as judged from the protein content measured after 9 days in culture (Figure 2a). Significantly lower biomass was obtained for Siran 120 microcarriers.

Glucose consumption *per* unit of biomass for the different systems tested is represented in Figure 2b. Except for S120, metabolism of neurons in microcarriers was similar to that of control cultures. The data

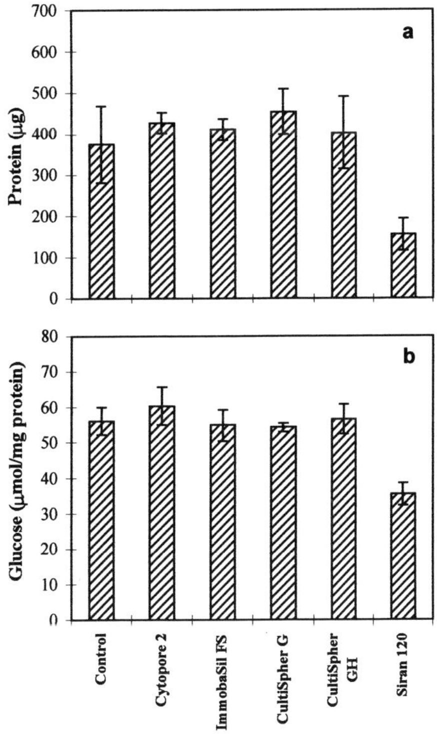

Figure 2: Comparison of various microcarriers for immobilization of neurons. **a:** Total protein determined 9 days after inoculation. **b:** Total glucose consumption per mg of protein during the first 7 days of maturation. Each bar represents the average value ± SD of five independent experiments with three replicates for each experiment.

obtained suggest that Siran carriers were not adequate for neurons maturation. This can probably be due to the physico-chemical characteristics of their surface. Glass does not seem to be a good matrix for neurons to adhere and the lower level of glucose consumed per mg of protein, as compared to control systems, may indicate that the well-being of cells is not assured when this immobilization system is used.

Astrocytes in microcarriers

Astrocytic cultures are, in general, easier to handle. These cells are more resistant to environment changes and, since they can divide (even though with very long duplication times), high cell densities can be achieved.

All the microcarriers tested were suited for astrocyte growth (Figure 3). As astrocytes can divide until they reach confluence, and the available area for growth is higher inside the three-dimensional structure of the porous microcarriers, a larger amount of biomass was obtained for the immobilized cultures (Fig. 3a).

The amount of glucose consumed *per* mg of protein was independent on the type of support (Figure 3b) which suggests that no diffusion problems inside the structures occurred.

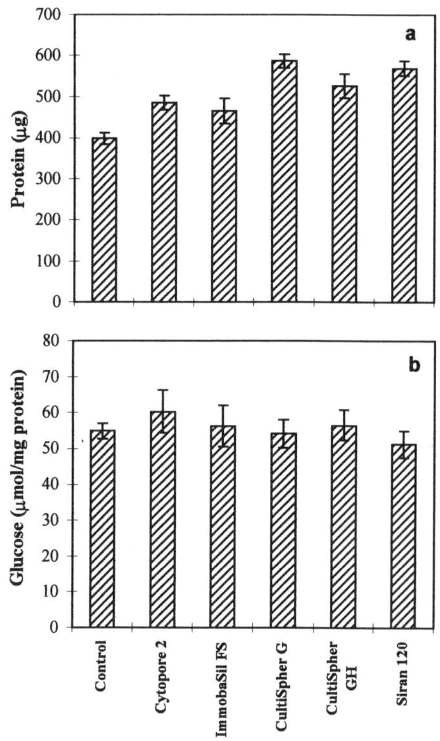

Figure 3: Comparison of various microcarriers for immobilization of astrocytes. **a:** Total protein determined 6 days after immobilization. **b:** Total glucose consumption per mg of protein during the first 5 days of growth inside the porous microcarriers. Each bar represents the average value ± SD of five independent experiments with three replicates for each experiment.

Perfusion for NMR measurements

Microcarriers performance for perfusion inside the NMR tube was tested. The interference of the matrix (due to its chemical composition) in the spectra as well as the geometry of the microcarrier (pore size, particle size and shape) are important parameters to take into account in the selection of the best immobilization system for *in vivo* NMR because they may affect the quality of the spectra.

From all the microcarriers tested, CultiSpher and Cytopore 2 were the best suited for NMR spectroscopy. The signals obtained for Siran 120 were very distorted and a good shimming was impossible to achieve. ImmobaSil FS were not good for perfusion because, even with low perfusion rates (0.5 ml/min), wash-out of the cells was observed, both for neurons and for astrocyte cultures. These microcarriers have a larger pore size (50-150 μm) than CultiSpher and Cytopore (20-30 μm), so cells are more exposed to shear damage caused by the ascensional perfusion flow.

CultiSpher G was chosen as support material for on line ^{31}P-NMR experiments because these microcarriers allowed a better resolution (line width at half height of the water resonance was 15 Hz for G, 40 Hz for GH and 25 Hz for Cytopore 2).

Figure 4 shows a ^{31}P-NMR spectrum of cortical astrocytes immobilized on CultiSpher G microcarriers perfused with DMEM at a flow rate of 1 ml/min.

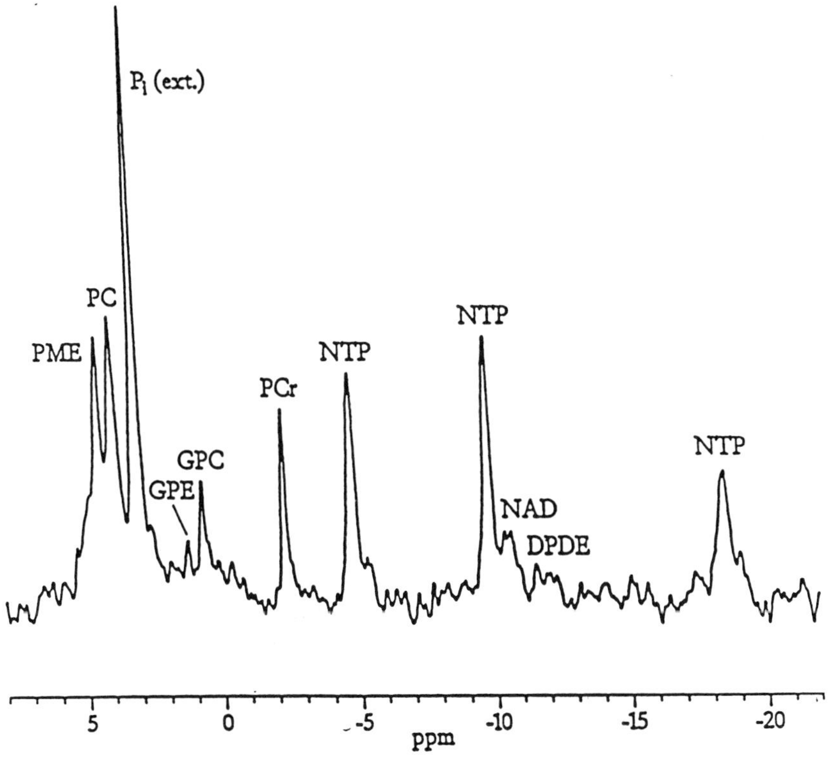

Figure 4: ^{31}P-NMR spectrum of cortical astrocytes immobilized in CultiSpher G microcarriers perfused with DMEM (30 minutes acquisition). Assignments: NTP, nucleoside triphosphate; Pi (ext.), extracellular inorganic phosphate; PCr, phosphocreatine; PC, phosphocholine; PME, phosphomonoesters; GPC, glycerophosphocholine; GPE, glycerophosphoethanolamine; NAD, nicotinamide adenine dinucleotide; DPDE, diphosphodiesters.

A very good time resolution was achieved (30 minutes) for immobilized astrocytes. The ratio PCr/NTP indicates that the cells are well protected within the support and nutrient supply was efficient at the perfusion rate used.The energetic status of astrocytes remained stable, at least for 48 hours in the spectrometer thus, no losses in cell viability were detected (data not shown).

Spectra were also acquired for immobilized neurons. These cells required longer acquisition times (4 hours) due to the lower amount of immobilized biomass *per* ml of microcarrier (data not shown).

Conclusion

This work demonstrates that primary brain cell cultures can be immobilized in porous microcarriers under conditions supporting growth.

The feasibility of these culturing techniques for perfusion in the spectrometer was demonstrated using ^{31}P-NMR spectra to monitor the cell energy status. Work is in progress in order to further optimize immobilization techniques for neuronal immobilization and consequently to decrease the acquisition times.

Acknowledgments

The authors acknowledge the technical assistance of Rosário Clemente and the financial support received from Junta Nacional de Investigação Científica e Tecnológica - Portugal (PRAXIS XXI/BD/2721/94).

References

1. Cox DWG, Morris PG, Feeney J & Bachelard HS (1983) ^{31}P-n.m.r. Studies on Cerebral Energy Metabolism under Conditions of Hypoglycaemia and Hypoxia *in vitro*. Biochem. J. 212, 365-370.

2. Brand A, Engelmann J & Leibfritz D (1992) A ^{13}C NMR Study on Fluxes into TCA Cycle of Neuronal and Glial cell line and Primary Cells. Biochimie 74: 941-948

3. Sonnewald U, Westergaard N, Schousboe A, Svendson JS, Unsgard G & Petersen SB (1993) Direct Demonstration by [^{13}C]NMRspectroscopy that Glutamine from Astrocytes is a Precursor for GABA synthesis in Neurons. Neurochem. Int. 22: 19-29

4. Alves PM, Flögel U, Brand A, Leibfritz D, Carrondo MJT, Santos H, Sonnewald U (1996) Immobilization of Primsary Astrocytes and Neurons for On-Line Monitoring of Biochemical Processes by NMR. Dev. Neurosci., in press.

5. Schousboe A, Meier E, Drejer J & Hertz L (1989) Preparation of Primary Cultures of Mouse (rat) Cerebellar Granule Cells. In: Shahar A, De Vellis J, Vernadakis A & Haber B (eds), A Dissecation and Tissue Culture Manual of the Nervous System, pp 203-206. Alan R. Liss, New York, EUA

6. Hertz L, Juurlink BHJ, Hertz EH, Fosmark H. & Schousboe A (1989) Preparation of primary cultures of mouse (rat) astrocytes. In: Shahar A, De Vellis J, Vernadakis A & Haber B (eds), A Dissecation and Tissue Culture Manual of the Nervous System, pp 105-108. Alan R. Liss, New York, EUA

Application of primary hepatocyte sandwich cultures for predictive studies of drug metabolism and precarcinogen activation

Bader A[1], Fahrig R[2], Steinkamp A[2], Frühauf N[1], Crome O [1], Zech K [3], Esselmann H [1], Gonschior AK[1] , Sewing K.-Fr[1]

[1]Institut für Allgemeine Pharmakologie, Medizinische Hochschule Hannover, [2]Fraunhofer-Institut für Toxikologie und Aerosolforschung, Hannover, [3]Byk Gulden Konstanz, Germany

Abstract

Loss of cytochrome P450 content is a common feature in conventional culture systems of primary hepatocytes. *In vivo* each hepatocyte is exposed to an extracellular matrix (space of Disse) at two opposing basolateral surfaces. This *in vivo* symmetry can be reconstructed by culturing hepatocytes within two layers of collagen, thus forming a sandwich configuration. Methods: Metabolism of various model drugs (tacrolimus, urapidil) and of the environmental pollutant dimethylbenzanthracene (DMBA) was studied in rat and human hepatocytes. Mutagenic effects of precarcinogen activation were studied in a threedimensional coculture model between sandwich hepatocytes and V79 cells using HPRT tests. Metabolites were analyzed by HPLC and LC/MS. Results: Sandwich hepatocytes generated metabolites and correctly reflected species specific metabolic patterns. Maintenance of metabolic properties in hepatocytes was dependent on extracellular matrix geometry. The amount of DMBA induced mutations tended to be higher than in standard S9 Mix assays. Conclusion: This model may be a valuable tool for predictive analysis of species dependent drug biotransformation and toxicological risk assessment of pharmaceutical and environmental compounds.

Keywords

Sandwich technique, cytochrome P450, primary hepatocyte culture, tacrolimus, FK 506, DMBA, HPRT, mutagens

Introduction

In contrast to cells engineered to express a selection of cytochrome P450 enzymes or liver microsomal preparations, viable primary hepatocytes possess in theory all the features necessary to produce a complete and integrative picture of hepatocellular drug metabolism. Unfortunately rapid time-dependent loss of drug metabolizing capacities or qualitative changes are a common feature in conventional culture systems for primary hepatocytes and parallels general loss of function. Established *in vitro* models for studies of hepatic drug biotransformation include the use of primary hepatocytes. In normal liver the space of Disse provides the possibility of bilateral attachment to extracellular matrix for each hepatocyte. This configuration is disrupted by the cell isolation procedure of normal liver tissue, which delivers suspensions of round shaped cells. In standard culture configurations this unphysiologic cell shape terminates in a morphological dedifferentiation and inability to biotransform drugs (1; 5). This study analyzes

M.J.T. Carrondo et al. (eds.), Animal Cell Technology, 99–102.

the relevance of extracellular matrix geometry in hepatocyte monolayer configurations for expression and activity of cytochrome P450 3A. This enzyme is involved in the biotransformation of a large number of pharmaceuticals including the immunosuppressant tacrolimus.

Microsomal preparations (S9 Mix) to be used in HPRT tests are prepared by Aroclor induction of the *in vivo* rat liver. This is a painful method for the animal resulting eventually in severe damage to the liver. In addition the S9 Mix used for activating chemical compounds in combination with *in vitro* mutagenicity assays can be used successfully for short term coincubations only of a few hours. Mutagenicity though is not only a function of the dose of a potentially mutagenic compound but also of its exposure time. Microsomal preparations generate metabolites in a different pattern than primary hepatocytes, which produce a more complete and *in vivo* - like pattern. The use of primary hepatocytes as an activating system could increase the predictive reliability of the available mutagenicity assays. We have previously shown that organotypical cultures of primary hepatocytes maintain cytochrome P450 dependent metabolic activity in culture by reestablishing cellular polarity (1). The purpose of this study was to develop a functionally stable coculture model between primary hepatocytes and V79 cells, which could be used as a screening model to assess mutagenicity of chemical compounds.

Methods

In vivo each hepatocyte is exposed to extracellular matrix (space of Disse) via two opposing basolateral surfaces. This normal *in vivo* symmetry was reconstructed by culturing primary hepatocytes (70000 cells/cm^2) within two layers of collagen, thus forming a double gel (sandwich) configuration. Rat and human hepatocyte cultures were exposed to the antihypertensive drug urapidil at a concentration of 1.25 µg/ml for various periods of time (1, 2, 5 or 24 hours). Metabolites were measured using HPLC. Structure of metabolites was elucidated by mass spectrometry. Synthetic metabolites were used for identification. Expression and activity of cytochrome P450 3A was studied by Western blot. The role of matrix geometry *in vitro* to stabilize cytochrome P450 3A activity was studied by use of the immunosuppressive drug tacrolimus. Metabolites were analyzed by HPLC/MS.

We selected dimethyl benzanthracene (DMBA) as a model compound in our 3-D coculture model between sandwich hepatocytes and V79 cells to study its metabolic activation. This substance is used as a positive control in standard S9 Mix assays. Cell cultures were exposed to DMBA (5 µg/ml, total volume of 4 ml) for 24 h following the 2nd, 7th and 14th day in culture. HPRT (hypoxanthine phosphoribosyl transferase) genotoxicity tests were performed to detect mutations induced by the DMBA metabolites. In conventional HPRT tests S9 Mix was used at a concentration of 15 mg/ml. DMBA was added with S9 Mix at a concentration of 10µg/ml (total volume of 50 ml). In the coculture model primary hepatocytes replaced the function of the S9 Mix.

Results

Matrix overlayed hepatocytes expressed higher biotransformation rates and generated full metabolic patterns of tacrolimus (8 metabolites). Cytochrome P450 3A expression was maintained for at least 14 days in culture in sandwich cultures. The species dependence of metabolite formation was shown for the model drug urapidil. Urapidil metabolite M1 (hydroxylated product) concentrations (mean ± SEM) in rat hepatocyte cultures were 0.002 ± 0.001 µg/ml at 1 h and increased steadily to 0.046 ± 0.016 µg/ml 24 h after exposure. Urapidil metabolites M2/M3 (O-demethylated compound and uracil-N-demethylated compound, the combined detection of M2/M3 depend on the HPLC technique used) reached 0.093 ± 0.019 µg/ml at 24 h. M1 in human hepatocytes was 0.024 ± 0.013 µg/ml at 1 h and 0.314 ± 0.065 µg/ml at 24 h. M2/M3 was 0.009 ± 0.003 µg/ml at 1 h and 0.071 ± 0.014 µg/ml at 24 h.
Cocultivated V79 cells rapidly proliferated on top of the second layer of matrix. V79 cells could be removed mechanically for further analysis using HPRT tests. The amount of DMBA induced mutations in our coculture model was higher than in standard S9 Mix assays. DMBA induced a comparable amount of mutations in 2 day and 7 day old cultures of primary hepatocytes. The spontaneous mutation rate in controls not exposed to DMBA was between 0 - $22/10^6$ cells.

Discussion

Reconstitution of a physiologic matrix geometry *in vitro* is crucial for the expression of cytochrome P450 3A and corresponding metabolic activity or other physiologic processes (2; 3; 5) *in vitro*. As shown for tacrolimus biotransformation, the model may be useful to study complex metabolic patterns. In addition this model may facilitate studies of the kinetics of hepatocellular drug biotransformation in a setting with prolonged stability (1).
The preferential production of M1 in human hepatocytes corresponded to the *in vivo* situation as measured in blood and urine of normal healthy volunteers. M1 has no intrinsic anti-hypertensive activity. The preferential production of M2/M3 in rat hepatocytes is due mainly to M2 and corresponds to the metabolite pattern found *in vivo* in rats. M2 was shown to have intrinsic anti-hypertensive activity. Knowledge of human pathways and rapid detection of possible species differences in drug metabolism already at initial stages of drug development would be a valuable tool to direct consecutive toxicological or pharmacodynamic studies in preclinical and clinical evaluation of novel compounds (4).
The reproducibility of the mutagenic effect of DMBA after 2 and 7 days in culture indicates that metabolizing enzymes are expressed in a stable manner. Using human hepatocytes in sandwich culture for bioactivation of chemical compounds could completely replace the aroclor induced liver model by a more relevant approach without having to resort to the use of animals.

Acknowlegment: We are grateful to U. Christians for instrumental support in his laboratory

References

1. Bader A, Knop E, Böker KHW, Crome O, Frühauf N, Gonschior A, Christians U, Esselman H, Winkler M, Pichlmayr R & Sewing KF (1996) Tacrolims (FK 506) metabolism in primary rat hepatocytes depends on extracellular matrix geometry. NS Arch Pharm 353, 1-13.

2. Bader A, Reimer P, Knop E, Weissleder R & Sewing K-F (1995) Receptor mediated endocytosis in matrix overlaid primary hepatocytes. J Exp Clin Cancer Res 14, 185-7.

3. Bader A, Rinkes IHB, Closs IE, Ryan CM, Toner M, Cunningham JM, Tompkins GR & Yarmush ML (1992) A stable long-term hepatocyte culture system for studies of physiologic processes: Cytokine stimulation of the acute phase response in rat and human hepatocytes. Biotechnol Prog 8, 219-225.

4. Bader A, Zech K, Crome O, Christians U, Pichlmayr R & Sewing K-F (1994) Use of organotypical cultures of primary hepatocytes to analyze drug biotransformation in man and animals. Xenobiotica 24, 623-633.

5. Knop E, Bader A, Böker K, Pichlmayr R & Sewing KF (1995) Normal ultrastructure of primary hepatocytes depends on bipolar attachment to extracellular matrix. Anatomical records 242, 337-349.

Increased cytotoxic sensibility of cultured fathead minnow fish cells by simultaneous treatment with sodium dodecyl sulfate and buthionine sulfoximine.

Paul J. Dierickx and Loeki Govers.
Instituut voor Hygiëne en Epidemiologie, Afdeling Toxikologie, Wytsmanstraat 14, B-1050 Brussel, Belgium.

The increasing number of anthropogenic chemical pollutants causes heavily contaminated fresh-, coastal- and seawater in almost all the industrialized regions of the world. Since a number of ecotoxicity tests, especially on fish, are very costly, it is not surprising that different cultured fish cell lines have thus far been used for measuring the cytotoxicity of several groups of chemicals (1). We have previously shown that a good correlation exists between the in vitro cytotoxicity to cultured fathead minnow (FHM) fish cells and the in vivo fish lethality data for a series of 50 chemicals belonging to very different groups (2). A linear relationship between chemical structure and cytotoxicity was observed when the cytotoxicity of compounds belonging to the same chemical group was considered. This was true for n-alcohols in function of their chain length, for acetic acids in function of the chloro-substitution and for amines in function of the ethanol-substitution (3). A good correlation between the cytotoxicity of inorganic metal compounds to FHM cells and the in vivo toxicity to *Daphnia magna* was also found (4).

However, the cytotoxicity values were about one order of magnitude higher than the in vivo results. Therefore, we investigated whether the simultaneous treatment of the cells with sodium dodecyl sulfate (SDS), pretreatment of the cells with L-buthionine-S,R-sulfoximine (BSO), or a combination of both, would increase the sensibility of the cytotoxicity, as measured by the neutral red (NR) uptake inhibition assay. The additive SDS was chosen as one of the multiple possible surfactants, and BSO because it is a specific inhibitor of gamma-glutamylcysteine synthetase, causing a strongly reduced endogenous glutathione (GSH) content (5). GSH has a protective effect against a very large number of xenobiotics (6).

Materials and methods

FHM cells are an established fish cell line (ATCC N° CCL42) derived from tissue posterior to the anus from fathead minnow (*Pimephales promelas*). The cells were cultured in Dulbecco's modified Eagle's medium (DMEM) supplemented with 10% fetal calf serum, 1% non-essential amino acids, 100 units/ml penicillin, and 0.1 mg/ml streptomycin (complete medium) and incubated in a humidified atmosphere of 5% CO_2 at 34 °C.

In order to know the effect of BSO (Sigma) on the endogenous GSH content in FHM cells, $9x10^6$ cells were seeded in petri dishes, i.e. the same cell density as used for the cytotoxixity experiments. After 24h the cells were treated for 24h with different BSO concentrations. GSH and proteins were then quantified as described (7).

M.J.T. Carrondo et al. (eds.), Animal Cell Technology, 103–107.

The cytotoxicity was measured by the slightly adapted (3) NR uptake inhibition assay of Borenfreund and Shopsis (8). The cells were seeded into 64 wells of a titer plate (Nunc) at $6x10^4$ cells in 0.2 ml complete medium per well. After incubation for 24h the cultures were treated with 0.2 ml aliquots of different concentrations of the freshly prepared test compounds in complete medium (8 wells/concentration). For the BSO experiments the cells were pretreated for an additional 24h with 400 μM BSO. SDS and BSO were added to all test solutions, including the controls. The cells were treated for 2h or 24h and further analyzed as previously decribed (3,8). The results were calculated in percentages compared to the control cultures. The relative toxicity of the test compounds is established by the determination of the NI50. This is the concentration of test compound required to induce a 50% inhibition in NR uptake. Standard deviations of less than 5% were found in completely independent assays.

Results and discussion

A BSO concentration dependent decrease in endogenous GSH content was observed in FHM cells (Fig. 1). After 24 h the GSH content was reduced to 28% of that

Fig. 1. The influence of BSO on the GSH content in FHM cells. The cells were treated with BSO for 24h. Control cells contained 18.1 ± 0.2 µg GSH/mg protein.

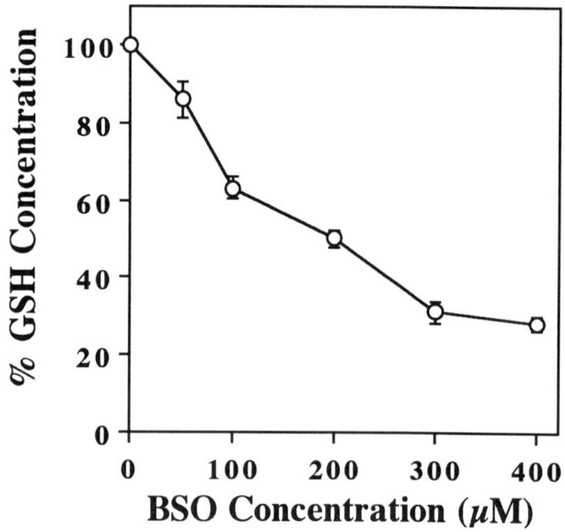

in the control cells. However, when the cells were cultured for another 24h period in complete medium without BSO, the GSH content nearly completely recovered till 95±1% of the controls. The influence of the reduced GSH content on the cytotoxicity was, therefore, investigated in the presence of 400 μM BSO (also for the control cells) after a pretreatment period of 24h. No toxic effect of BSO was observed at this concentration (results not shown).

The cytotoxicity of SDS was measured in some preliminary experiments. These are summarized in Fig. 2. On the basis of these results the following SDS concentrations were selected for simultaneous treatment with the test chemicals : 0.20 mM SDS for a treatment of 2h, 0.15 mM for a treatment of 24h, and 0.10 mM for a treatment of 24 h in the presence of 400 μM BSO. Fig. 3 shows typical results for the NR uptake inhibition after treatment of the FHM cells with $MnCl_2$, supplemented with respectively SDS alone, BSO alone, and a combination of both. In the latter case $MnCl_2$ was 5 times more toxic than in the control assay conditions.

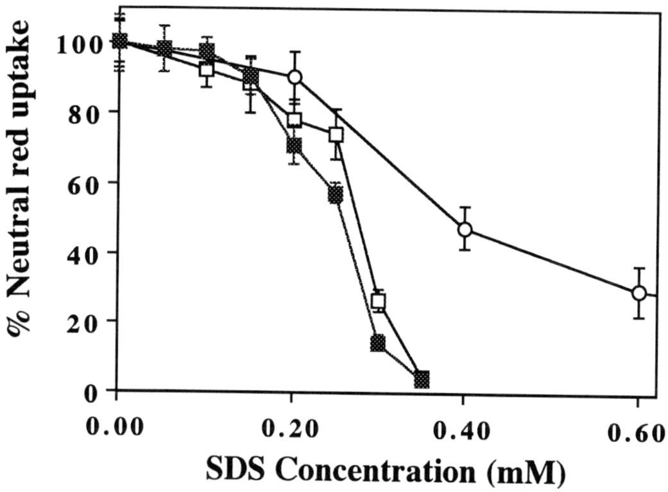

Fig. 2. Effect of various concentrations of SDS on the neutral red uptake in FHM cells, after 2h (○), after 24h (□), and after 24h in the presence of 400 μM BSO (■).

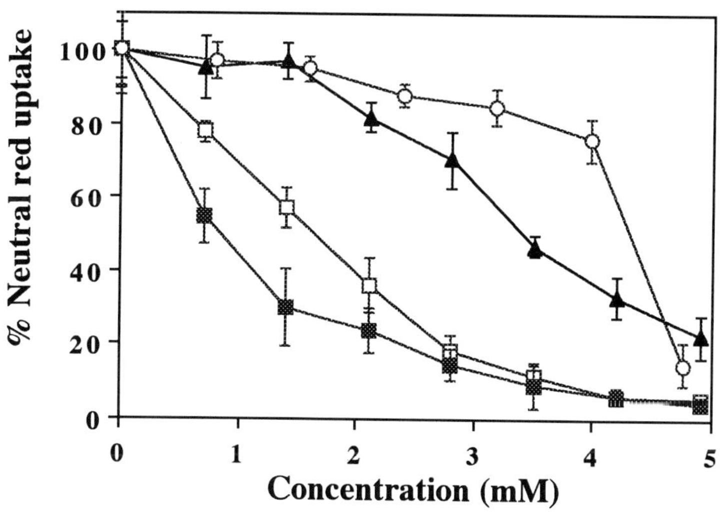

Fig. 3. Effect of a 24h treatment with various concentrations of $MnCl_2$ on the neutral red uptake in control conditions (○), and in the presence of 0.15 mM SDS (▲), 400 μM BSO (□), and 0.1 mM SDS + 400 μM BSO (■).

In a first series of experiments the FHM cells were treated for 2h with a series of 13 chemicals belonging to different chemical classes (Table 1). No significant effect of SDS and/or BSO was observed for n-butanone. For the remainder chemicals an increased cytotoxic sensibility was obtained. The increase with 0.20 mM SDS was, generally, about the same as in the presence of 400 μM BSO. The increased toxicity in the latter case is due to a reduced GSH content of the cells (Fig. 1). Simultaneous treatment with SDS and

Table 1. Neutral red uptake inhibition by compounds of several chemical groups in FHM cells after treatment for 2h. The influence of SDS, BSO and a combination of both was measured. The results are given as the NI50 values (in mM). The percentages of the NI50, as compared to control values, are given within brackets.

Test Compound	Control conditions[a]	+ 0.20 mM SDS	With 400 μM BSO - SDS	With 400 μM BSO + 0.2 mM SDS
$ZnCl_2$	0.48	0.31 (64.6)	0.26 (54.2)	0.22 (45.8)
n-Octanol	2.0	1.55 (77.5)	1.70 (85.0)	0.92 (46.0)
$NiCl_2.6H_2O$	9.2	4.4 (47.8)	5.1 (55.4)	5.2 (56.6)
Chloral hydrate	13	5.8 (44.6)	7.4 (56.9)	6.7 (51.5)
n-Pentanol	43	36 (83.7)	34 (79.1)	33 (76.7)
Acetic acid	49	42 (85.7)	36 (73.5)	26 (53.1)
Ethylenediamine	52	28 (53.9)	26 (50.0)	23 (44.2)
Potassium chlorate	142	87 (61.3)	77 (54.2)	53 (37.3)
n-Butanone	198	182 (91.9)	207 (104.5)	197 (99.5)
n-Propanol	300	178 (59.3)	226 (75.3)	225 (75.0)
Isopropanol	418	235 (56.2)	240 (57.4)	208 (49.8)
Acetone	592	338 (57.1)	297 (50.2)	156 (26.3)
Acetonitrile	903	624 (69.1)	622 (68.9)	514 (56.9)

[a]*Results from (2).*

Table 2. Neutral red uptake inhibition by several metal compounds in FHM cells after treatment for 24h. The influence of SDS, BSO and a combination of both was measured. The results are given as the NI50 values (in mM). The percentages of the NI50, as compared to control values, are given within brackets.

Test Compound	Control conditions[a]	+ 0.15 mM SDS	With 400 μM BSO - SDS	With 400 μM BSO + 0.1 mM SDS
$ZnSO_4.7H_2O$	0.43	0.28 (65.1)	0.08 (18.6)	0.05 (11.6)
$CoCl_2.6H_2O$	0.56	0.48 (85.7)	0.26 (46.4)	0.31 (55.3)
$CuSO_4.5H_2O$	0.58	0.38 (65.5)	0.06 (10.3)	0.06 (10.3)
$NiCl_2.6H_2O$	1.8	1.18 (66.7)	0.59 (32.8)	0.63 (35.0)
$FeSO_4.7H_2O$	4.3	0.50 (11.6)	0.38 (8.8)	0.44 (10.2)
$MnCl_2$	4.3	3.4 (79.0)	1.6 (37.2)	0.83 (19.3)
$SnCl_2.2H_2O$	4.8	4.5 (93.7)	0.45 (9.4)	0.52 (10.8)
$CaCl_2.2H_2O$	34	13 (38.2)	9.5 (27.9)	5.3 (15.6)
$MgSO_4.7H_2O$	52	96 (184.6)	97 (186.5)	90 (173.0)
KCl	70	65 (92.8)	90 (128.6)	98 (140.0)
NaCl	112	81 (72.3)	113 (100.9)	106 (94.6)

[a]*Results from (4).*

BSO lowered the NI50 values with ±50%. These experiments were promising, but the goal was to increase the sensibility with one order of magnitude. In a second series of experiments the influence of SDS and/or BSO on the NR uptake inhibition was, therefore, investigated for 11 metal compounds after a treatment period of 24h (Table 2). Surprisingly, $MgSO_4$ and KCl were less cytotoxic, and NaCl nearly equally cytotoxic as the controls. However, for 6 of the remainder chemicals the NI50 values in the SDS+BSO treated cells were reduced to 10-20% of the controls.

The results described in this paper show that the sensibility of cytotoxicity tests can be increased importantly. Further research in the direction outlined here will make it possible to develop very sensitive in vitro cytotoxicity tests, allowing to measure the toxicity of environmental water samples.

Acknowledgement

This text represents research of the Belgian program on Interuniversity Poles of Attraction, initiated by the Belgian State, Prime Minister's Office, Science Policy programming.

References

1. Babich H & Borenfreund E (1991) Cytotoxicity and genotoxicity assays with cultured fish cells : a review. Toxicol In Vitro 5, 91-100.
2. Brandao JC, Bohets HH, Van De Vyver IE & Dierickx PJ (1992) Correlation between the in vitro cytotoxicity to cultured fathead minnow fish cells and fish lethality data for 50 chemicals. Chemosphere 25, 553-562.
3. Dierickx PJ & Van De Vyver IE (1991) Correlation of the neutral red uptake inhibition assay of cultured fathead minnow fish cells with fish lethality tests. Bull Environ Contam Toxicol 46, 649-653.
4. Dierickx PJ & Bredael-Rozen E (1996) Correlation between the in vitro cytotoxicity of inorganic metal compounds to cultured fathead minnow fish cells and the toxicity to Daphnia magna. Bull Environ Contam Toxicol 57, in press.
5. Griffith OW (1982) Mechanism of action, metabolism, and toxicity of buthionine sulfoximine and its higher homologs, potent inhibitors of glutathione synthesis. J Biol Chem 257, 13074-13712.
6. Chasseaud LF (1979) The role of glutathione and glutathione S-transferases in the metabolism of chemical carcinogens and other electrophilic agents. Adv Cancer Res 29, 175-274.
7. Dierickx PJ (1987) Inhibition of glutathione-dependent uridine uptake in cultured human hepatoma cells. Med Sci Res 15, 1349-1350.
8. Borenfreund E & Shopsis C (1985) Toxicity monitored with a correlated set of cell culture assays. Xenobiotica 15, 705-711.

IMMORTALIZATION AND CHARACTERIZATION OF PORCINE BRAIN MICROCAPILLARY ENDOTHELIAL CELLS

Michael Teifel and Peter Friedl

Institut für Biochemie, Technische Hochschule Darmstadt

Petersenstraße 22, 64287 Darmstadt, Germany

Tel. (49) 6151-163655, Fax (49) 6151-164759, e-mail: di58@pop.th-darmstadt.de

Abstract:

Porcine brain microvascular endothelial cells (PBMEC) were transfected with pRNS-1, encoding for the small and large T-antigens of SV 40, by means of lipofection. After selection with G-418, clones were isolated and one clone, PBMEC/C1-2 was further characterized. PBMEC/C1-2 was up to now cultivated for nearly one year with 250 cumulative population doublings and showed the typical morphology of capillary endothelial cells *in vitro*. SV 40 T-antigens were present in the nuclei, the cells showed the typical granular staining of von Willebrand factor (vWF) and were positive for *Bandeiraea simplicifolia* isolectin B_4. Furthermore, PBMEC/C1-2 exhibited expression of non-muscular and smooth-muscle type myosins, and the presence of the blood-brain barrier-associated markers γ-glutamyltranspeptidase (γ-GT), the glucose transporter Glut-1 and apolipoprotein A-1 (Apo A-1). In addition, enzymatic activity of γ-GT and alkaline phosphatase (ALP) was detected in cell suspensions. In summary, PBMEC/C1-2 represent an immortalized PBMEC line exhibiting endothelial characteristics as well as typical markers of the blood-brain barrier.

Introduction:

Endothelial cells of the brain microvasculature provide a physical barrier between blood and brain, preventing an uncontrolled passage of most polar substances. The importance of the blood-brain barrier for the maintenance of a constant inner milieu of the brain, is responsible for the great efforts in the establishment of an *in vitro* model. Most groups use primary isolated cells for their studies, accepting time- and cost-effective preparation steps and common variations in the cell's physiology. Furthermore, *in vitro*, cells were subject to dedifferentiation [1] and exhibit a general plasticity according to the culture conditions [2]. To overcome this problem some groups set out to immortalize bovine and rat brain capillary endothelial cells [3,4]. Both cell lines exhibited a nontransformed phenotype and retained some characteristic markers of brain capillary endothelial cells. Porcine microvascular endothelial cells are a commonly used and well characterized *in vitro* model for studying blood-brain barrier features. To our knowledge no immortalized porcine counterpart has yet been described, so we tried to isolate an immortalized cell line in this study.

M.J.T. Carrondo et al. (eds.), Animal Cell Technology, 109–113.

Methods:

Isolation and culture of PBMEC. Porcine brain microvascular endothelial cells (PBMEC) were prepared by a modified procedure, previously described by Mischeck et al. [1]. Cells were seeded into gelatin-coated 25 cm^2 culture flasks (Greiner) at a density of 80,000 cells/cm^2 and cultivated in modified IF medium (1:1 mixture of Iscove's modified Dulbecco medium and Ham's F-12, supplemented with 7.5% (v/v) newborn calf serum (NCS, Sebak), 5 μg/ml human transferrin (Behring), 5 μg/ml heparin (Serva), 12.5 μM β-mercaptoethanol (Merck), 12 μM hypoxanthine (Sigma), 250 μM N-acetyl-cysteine (Sigma), 7 mM L-glutamine (Seromed), 350 ng/ml crude fibroblast growth factor (FGF) [5], 100 μg/ml penicillin (Boehringer Mannheim) and 100 U/ml strepto-mycin (Boehringer Mannheim)) at 37°C and 5% (v/v) CO_2 in a humidified atmosphere.

Transfection of PBMEC with pRNS-1 and isolation of clones. Transfections of PBMEC were performed in the primary cultures at 50% confluency. For transfer of the plasmid DNA, the lipofection method [6] was applied using azolectin/DDAB-liposomes [7]. Plasmid pRNS-1, containing the small and the large T-antigens of the simian virus 40 and the *neo* gene, was a kind gift of H. L. Ozer, New Jersey Medical School, Newark, and was described earlier [8]. Briefly, 100 μl azolectin/DDAB liposomes were mixed with 25 μg pRNS-1 in 1 ml serum-free medium. After 30 min incubation at ambient temperature, the suspensions were added to 25 cm^2 culture flasks, containing 4 ml of serum-free medium. Following an incubation period of 5 h at culture conditions, medium was changed to complete medium. The day after transfection the cells were passaged into 145 cm^2 culture plates. After three days of culture the medium was changed to selective medium (complete medium with 200 μg/ml G-418, Gibco). Following a selec-tion period of 3 weeks, clones of an appropriate size were isolated and gradually passaged into 12 and 6 well plates and finally into 25 cm^2 culture flasks.

Characterization of isolated clones. The cultures were characterized by microscopical examination of morphology, immunofluorescence and lectin binding. For indirect immunofluorescence confluent cultures were fixed with ethanol, blocked with NCS and incubated with primary antibodies against simian virus 40 T-antigens (Dianova), von Willebrand factor (Behring), glucose transporter 1 (Chemicon), apolipoprotein A-1 (kindly provided by B. Möckel, Institut für Biochemie, Technische Hochschule Darmstadt), γ-glutamyltranspeptidase (kindly provided by H. G. Gassen, Institut für Biochemie, Technische Hochschule Darmstadt), non-muscle (Paesel und Lorei) and smooth-muscle (kindly provided by U. Gröschel-Stewart, Institut für Zoologie, Technische Hochschule Darmstadt) myosin types at a concentration of 20-100 μg/ml. As controls, the cultures were incubated with non-immune serum at corresponding concentrations. Bound antibodies were visualized with biotinylated second antibodies (Dianova) and streptavidin-labeled Cy3™-conjugates (10 μg/ml, Dianova). For lectin binding, the cultures were fixed with ethanol and incubated with 1 μg/ml biotinylated *Bandeiraea simplicifolia* isolectin B_4 (Sigma). Bound lectins were also visualized with streptavidin-labeled Cy3™-conjugates. The cultures were then examined under a Nikon TMD epiphot microscope with attached fluorescence unit.

Results:

Isolation and transfection of PBMEC. The isolation procedure yielded approximately 4×10^7 cells per brain. The isolated cells were highly viable and reached confluency within 3-4 days. Microscopical observation indicated that the cultures persisted predominantly of microvascular endothelial cells with the typical elongated cell shape. Transfections with pRNS-1 were performed two days after cell preparation in primary cultures at approximately 50% confluency. The next day the cultures were examined microscopically, indicating that transfection with azolectin/DDAB-liposomes exhibited no toxic effect.

Morphology of the clone PBMEC/C1-2. Cells of the clone PBMEC/C1-2 retained the characteristic morphology of capillary endothelial cells. At low cell densities they grow preferentially in “islands”, reaching confluency they build a confluent monolayer (Fig. 1). PBMEC/C1-2 retained their morphology even at high passage numbers and were cultured for nearly one year with about 250 cumulative population doublings. PBMEC could only be cultured for a maximum of 12 passages with a total of 25 CPD, so PBMEC/C1-2 were regarded to be immortal.

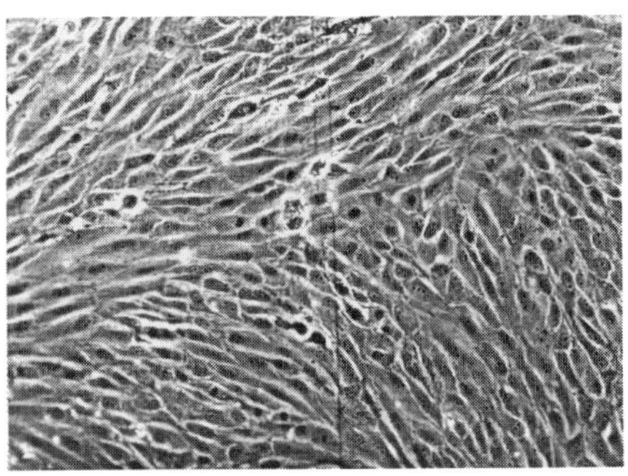

Figure 1. Confluent cultures of PBMEC/C1-2.

Immunological and lectin binding characterization of the clone PBMEC/C1-2. To verify the expression of SV 40 T-antigens, PBMEC/C1-2 were incubated with antibodies to the small and large T-antigens of SV 40. Fig. 2 shows the strong nuclear staining of SV 40 T-antigens. Staining with antibodies to von Willebrand factor showed the characteristic granular staining (Fig. 3b). Expression of markers, decided to be specific for cells constituting the blood-brain barrier, was confirmed by incubation of the cells with antibodies to Glut-1, γ-GT and Apo A-1 [9,10,11]. Whereas antibodies to Glut-1 showed a strong granular staining in some cells (Fig. 3c), γ-GT and Apo A-1 antibodies stained preferably the whole cell membrane (not shown). Non-muscle, as well as smooth-muscle myosin, was also found in PBMEC/C1-2 (Fig. 3d,e) and supported the hypothesis, that also endothelial cells, not only pericytes, exhibit contractile functions in capillaries [12]. Binding of *Bandeiraea simplicifolia* isolectin B_4 was a further proof of the endothelial origin of PBMEC/C1-2 (Fig. 3f) [13].

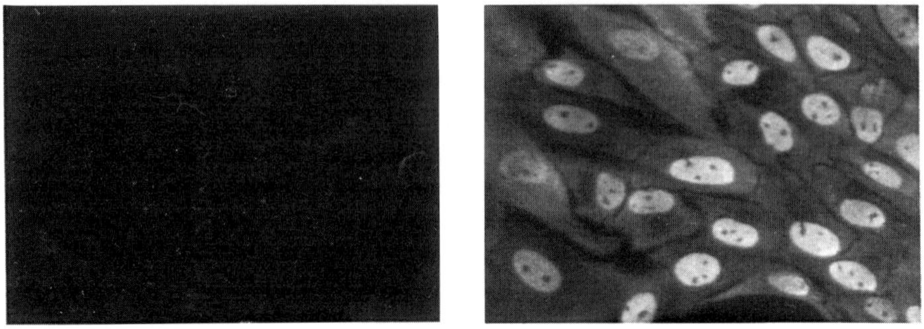

A **B**

Figure 2. Nuclear localization of SV 40 T-Ags in PBMEC/ C1-2. Incubation with non-immune mouse IgG (a). Staining of SV 40 T-Ags (b).

A **B**

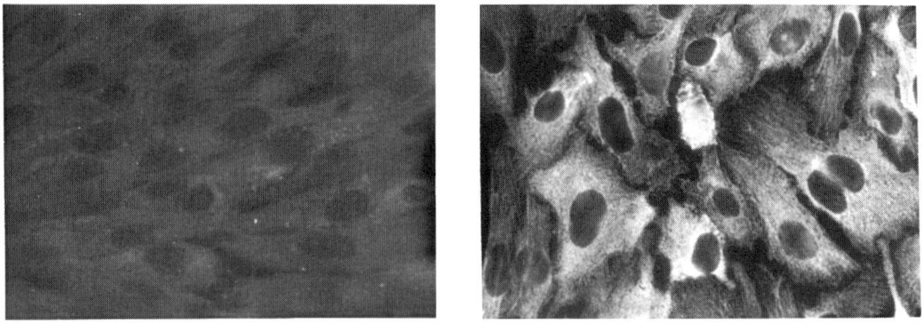

C **D**

(Legend of Fig. 3 on next page)

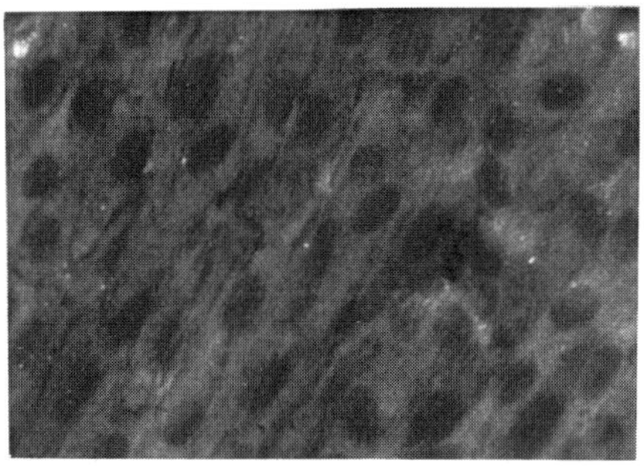

E

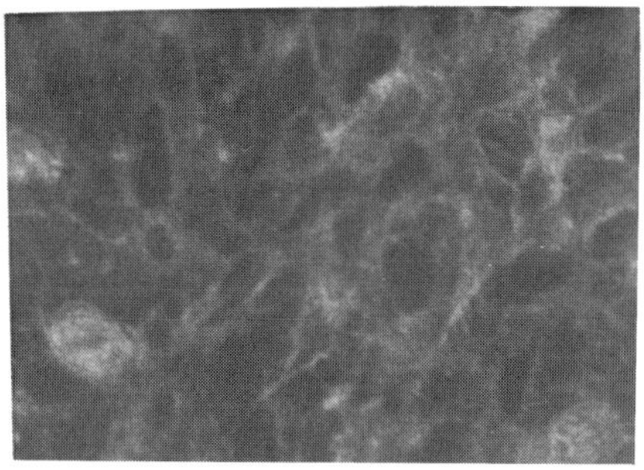

F

Figure 3. Immunofluorescence-labeling of PBMEC/C1-2. Incubation with rabbit-0-serum (a). Detection of antigens by incubation with antibodies to von Willebrand factor (b), glucose transporter Glut-1 (c), non-muscle myosin (d) and smooth-muscle myosin (e). Binding of *Bandeiraea simplicifolia* isolectin B_4 (f).

References:

1. Mischeck, U., Meyer, J., Galla, H.-J. (1989) *Cell Tissue Res.* **256**, 221-226.
2. Robinson, D. H., Kang, Y.-H., Deschner, S. H., Nielsen, T. B. (1990) *In Vitro Cell. Dev. Biol.* **26**, 169-180.
3. Durieu-Trautmann, O., Foignant-Chaverot, N., Perdomo, J., Gounon, P., Strosberg, A. D., Couraud, P. O. (1991) *In Vitro Cell. Dev. Biol.* **27A**, 771-778.
4. Roux, F., Durieu-Trautmann, O., Chaverot, N., Claire, M., Mailly, P., Bourre, J.-M., Strosberg, A. D., Couraud, P.-O. (1994) *J. Cell. Phys.* **159**, 101-113.
5. Friedl, P., Tatje, D., Czapla, R. (1989) *Cytotechnology* **2**, 171-179.
6. Felgner, P. L., Gadek, T. R., Holm, M., Roman, R., Chan, H. W., Wenz, M., Northrop, J. P., Ringolds, G. M., Danielsen, M. (1987) *Proc. Natl. Acad. Sci. USA* **84**, 7413-7417.
7. Teifel, M., Friedl, P. (1995) *BioTechniques* **19**, 79-82.
8. Litzkas, P., Jha, K. K., Ozer, H. L. (1984) *Mol. Cell. Biol.* **4**, 2549-2552.
9. Weiler-Güttler, H., Zinke, H., Möckel, B., Frey, A., Gassen, H. G. (1989) *Biol. Chem. Hoppe-Seyler* **370**, 467-473.
10. De Bault, L. E., Henriquez, E., Hart, M. N., Cancilla, P. A. (1981) *In Vitro Cell. Dev. Biol.* **17**, 480-494.
11. Möckel, B., Zinke, H., Flach, R., Weiß, B., Weiler-Güttler, H., Gassen, H. G. (1994) *J. Neurochem.* **62**, 788-798.
12. Ehler, E., Karlhuber, G., Bauer, H.-C., Draeger, A. (1995) *Cell. Tissue Res.* **279**, 393-403.
13. Tontsch, U., Bauer, H. C. (1989) *Microvasc. Res.* **37**, 148-161.

Three-dimensional Cell Culture Systems Stabilise the Differentiation of Hepatocyte Cell Lines

Gonçalves, LMD[1], T. Lopez[1], C. MacDonald[2], MH Grant[3], MJT Carrondo[1,4]
[1]IBET/ITQB, Apartado 12, Oeiras, Portugal; [2]University of Paisley, Glasgow, UK; [3]University of Strathclyde, Glasgow, UK; [4]FCT/UNL, Monte de Caparica, Portugal.

Key words: three-dimensional cell culture; extracellular matrix; long-term cytochrome P450 activities; differentiation.

Abstract

The suitability of a hepatocyte cell line for *in vitro* toxicity and drug metabolism largely depends in the expression of hepatic differentiated functions. In the present study the activity of phase I enzymes was examined in a long-term run using immobilised (Cultispher GH) and aggregates of hepatocyte cell lines derived from Sprague-Dawley male rats. With that purpose the metabolic profile of testosterone was followed as the oxidative metabolism of this hormone is catalysed, with a high degree of regio- and stereoselectivity, by different cytochrome P450 families (1A, 2A, 2B, 2C and 3A). The results revealed that the cytochrome P450 activities were maintained for higher periods of time in three-dimensional cell systems then monolayer (conventional) cell cultures. Conventional cultures showed a rapid loss of cytochrome P450, which were undetectable after few passages. In contrast, some isoenzymes assayed were measurable in three-dimensional cell cultures systems for as long as 4 months. The beneficial effects of three-dimensional systems seemed to be attributed to the cell-cell contacts that will promote their phenotypic stability or to the retention of globular cell structure.

Introduction

Liver cell cultures are becoming specially valuable as tools for investigation of hepatotoxic action and metabolic profile of drugs, as well as their interactions with other therapeutic agents or xenobiotics. However, the catalytic activities of Cytochrome P450 monooxygenases exhibit a rapid decline in primary cultured hepatocytes. This represents a serious drawback to using this experimental model for long-term *in vitro* pharmacological studies. Among others, the ways studied to overcome this limitation have been the use of selective supplemented media, cultivating hepatocytes on extracellular matrices, and co-culturing hepatocytes with other cell types, in order to reduce changes in Cytochrome P450 levels and related activities. All these parameters can be improved in three dimensional cell cultures where an appropriate microenvironment for cell differentiation is reached.

M.J.T. Carrondo et al. (eds.), Animal Cell Technology, 115–119.

Differentiated three dimensional tissue can be produced in aggregates or by immobilising cells into macroporous supports; thus, a model that resembles the structure and function of the parental tissue should be more closely representative of the live situation then any other *in vitro* model system.

Different microcarriers were considered for immobilisation of immortalised liver cells, in order to evaluate the influence of extracelluar matrix components on the phenotype stabilisation and to analyse if the maintenance of hepatocellular functions requires the preservation of cell-cell interaction that can be achieved inside the pores of those matrices.

Materials and Methods

Immortalised hepatocytes cell lines

Immortalised cell lines P9 produced by calcium phosphate precipitation and LQC-3 by electroporation were obtained from Prof. Caroline MacDonald (Paisley University, Scotland).

Culture medium, inoculum and conditions

The cells were propagated in 25-cm^2 T-flasks 37 °C in a humidified atmosphere containing 7% $C0_2$. Dulbecco's modified Eagle's Medium (DMEM) was used for both batch propagation and for immobilisation tests. The medium was supplemented with 10% FCS for propagation and 5% FCS for all other experiments. The perfused-batch studies were performed in spinner-flasks (Corning) of 150 ml maximum culture volume and stirred at 75 rpm. The microcarrier selection was done in 6 wells microplate (Griner, England). Inocula were prepared from cells in the late exponential growth phase (passage number 12 and 16 for LQC-3 and P9, respectively). Inoculum concentration was 1*10^5 cells/ml for the different conditions of culture studied. The medium was replaced when cells entered the death phase, which is followed by an accumulation of dead cells in the media.

Analytical Procedures

The metabolites (2α-OHT, 6α-OHT, 6β-OHT, 7α-OHT, 16α-OHT, 16β-OHT (OHT-hidroxitestosterone), and androstenodione) of testosterone were analysed by HPLC at 40°C on a reverse-phase C_{18} column (5mm, 250x4 mm I.D., Lichrospher RP18, Merck, Germany) preceded by a 2cm LC-18 guard column. A gradient from 70% solvent A (methanol:water:acetonitrile, 39:60:1) to 90% solvent B (methanol:water:acetonitrile, 80:18:2) was operated over 45 min at 1.0 ml/min; spectrophotometric detection was done at 254 nm. This method was adapted from (1). The number of immobilised cells was counted by using the trypan blue exclusion and also by the reduction of MTT (3-(4,5-dimethylthiazol-2-yl)-2,5-diphenyl tetrazolium bromide) that allows to observe using an inverted microscope in which extension active cells colonise the microcarriers and, after extraction, to quantify them (2).

Porous supports

The microcarriers tested for immobilisation of P9 and LQC-3 cell lines and to be used during stirred-tank reactor studies were Cytopore 2, Cytodex 3 (Pharmacia, Sweden) and Cultispher GH, Cultispher G (Percell Biolytica, Sweden)

Results and discussion

Microcarrier selection

Microcarrier selection was done in plates of six wells that were covered with microcarriers. Periodic observations were done at the inverted microscope. At the end of growth phase MTT was used to stain cells and observe their distribution into the microcarrier (results not shown). Higher cell affinity for Cultispher GH and Cytodex 3 was observed. The microcarrier Cultispher GH was chosen for the phenotype long-term stability studies as it presents advantages over Cytodex 3; it is macroporous, cells can have three-dimensional contact, and it is easy to digest by enzymes, permitting easy access for cell counting and other analysis.

Immobilised cell bioreactor: long-term stability studies

Cell attachment occurs during the first day after inoculation. The complete saturation of the support was obtained after four days. The P9 cells naturally form cell aggregates in opposition to LQC-3 cells that are only grow viably attached to the macroporous microcarrier. Cells grown immobilised in macroporous supports can colonise a t-flask surface, confirming the proliferation capacity presented by cells immobilised in three-dimensional structures.

Figure 1 and Table 1 show the results of testosterone metabolism in long-term spinner flasks with P9 and LQC-3 cells immobilised in Cultispher GH.

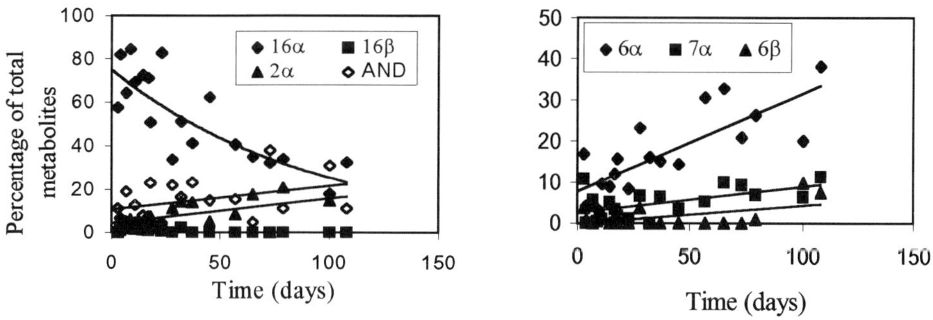

Figure 1- Evolution with time of testosterone metabolism in LQC-3 cells immobilised in Cultispher GH (Spinner flasks at 75 rpm).

LQC-3 cell line retain the ability to form 16α (Cytochrome P450 2B1 and 2C11), 6α, 7α (Cytochrome P450 2A1 and 2A2), and 2α (Cytochrome P450 2C11) metabolites. The same behaviour is observed with P9 cells. During cell culture it can be observed that the decrease of 16α and 2α metabolites is accomplished by an increase of the 6α and 7α. This reflects the increased activity of Cytochrome P450 2A1 and 2A2 and the decrease of Cytochrome P450 2B1 and 2C11. Androstenedione is the more stable metabolite present during the time of the cultivation of immobilised cells, which can be attributed to the stable activity of Cytochrome P450 B2.

Table 1-Testosterone metabolism in P9 cells immobilised in Cultispher GH and cultivated in Spinner flasks*

	Days of Culture				
Metabolite	3	15	32	65	108
6 α	10	18	23	36	26
7 α	-	13	-	19	15
6 β	-	-	-	-	-
16 α	53	20	28	27	32
16 β	-	-	-	-	-
2 α	-	17	17	-	-
Andr.	37	32	32	18	27

*Metabolites of testosterone hidroxilated in different positions are presented as the percentage of total amount of metabolites produced (6α, 7α, 6β, 16α, 16β, 2α - hidroxy-testosterone; Andr.: androstenodione).

Comparison between different culture systems

In table 2 different reactor configurations and ways of operation are compared. As can be seen from table 2 aggregates and immobilised cell culture systems retain the liver-specific phenotype functions for very long periods of time, as shown by the metabolism of testosterone, in opposition to monolayer cell culture. This result confirms that immobilised cell cultures present advantages over monolayer cell culture as has been demonstrated before for product expression by hybridoma cells (3).

Table 2 - Testosterone metabolism in hepatocytes cultured in different systems*

	LQC-3				P9				
Metabolite	Mono (P16)	Imm 2 days	Imm 108 days	Aggr 2 days	Mono (P16)	Imm 2 days	Imm 108 days	Aggr 2 days	Aggr 90 days
6 α	14	23	38	8	-	19	26	20	20
7 α	-	-	11	-	-	-	15	-	-
6 β	-	-	8	-	-	-	-	-	-
16 α	52	55	32	3	-	6	32	14	43
16 β	-	-	-	-	-	-	-	-	-
2 α	-	-	-	-	-	-	-	-	-
Andr.	34	22	11	89	-	75	27	66	42

(Mono-Monolayer; Imm- Immobilised; Aggr-Aggregates)*The immobilised cell bioreactors were inoculated with the cell lines at passage number 12 and 14 for LQC-3 and P9, respectively. Metabolites of testosterone hidroxilated in assinalated different positions are presented as the percentage of total amount of metabolites produced (6α, 7α, 6β, 16α, 16β, 2α - hidroxy-testosterone; Andr.: androstenodione).

Conclusions

The best macroporous microcarrier to immobilise LQC-3 and P9 cell lines was Cultispher GH. P9 cell line can grow in cell aggregates for long periods of time but LQC-3 cell aggregates were not very stable. This confirms that the two cell lines are also different in their behaviour in bioreactor operation.

Cells growing in immobilised macroporous supports can colonise a t-flask surface confirming their proliferative state inside of three-dimensional structures.

The use of macroporous microcarriers allowed the operation of a stirred-tank reactor for long periods, keeping the system stable. This can be attributed to the cell-to-cell and cell-to-matrix contact possible within the three-dimensional structure of pores of microcarriers.

Cell culture systems involving immobilised cells in macroporous microcarriers (LQC-3 and P9 cell lines) or cell aggregates (P9 cell line) were clearly more advantageous over monolayer cell cultures on the maintenance of cell differentiation in terms of Cytochrome P450 dependent metabolism.

References

1. Reinerink EJM, Door L, Jansen EHJM, and Van Iersel AAJ (1991) Measurement of enzyme activities of cytochrome P-450 isoenzymes by high-performance liquid chromatographic analysis of products. Journal of Chromatography, 553, 233-241.
2. Al-Rubeai, M., and R. Spier (1992) Use of the MTT assay for the study of hybridoma cells in homogeneous and heterogeneous cultures. In: Spier RE, Griffiths JB, Stephenne J and Crooy P, Advances in Animal Cell Biology and Technology for Bioprocesses, pp 143-155, Butterworths, London.
3. Gonçalves LMD, Rodrigues JM, Aunins JG, Carrondo MJT (1995) Immobilization of anchorage-independent cells to porous glass beads: mass transfer phenomena in long-term packed bed reactor. In: Beuvery EC, Griffiths JB and Zeijlemaker WP, Animal Cell Technology: Developments Towards the 21st Century, pp 723-727, Kluwer Academic Publishers, Dordrecht, The Netherlands.

Acnowledgements

The authors acknowledge JNICT, Junta Nacional de Investigação Científica e Tecnológica, by the financial support (Project Nº PBIC/C/BIO/2017/95 and Grant PRAXIS XXI/BPD/4153/94). The technical assistance by Rosário Clemente is also acknowledged.

IN-VITRO TOXICOLOGICAL EFFECTS OF PLURONIC F-68

A. Handa-Corrigan, N.P. Weeratunge, J.H. Braybrook*, G.A. Mackay* and D. Chescoe. University of Surrey and * Laboratory of the Government Chemist, U.K.

Abstract
In this study, organic contaminants present in Pluronic F-68 were extracted using Supercritical Fluid Extraction (SFE). Combined Gas Chromatography-Mass Spectrophotometry analyses of extracted components of Pluronic F-68 (obtained from two commercial suppliers) were shown to contain butylated hydroxy toluene (BHT) antioxidant and phthalate plasticizer. Both sources of Pluronic F-68 polymers and their SFE components were shown to be cytotoxic to LL24 human lung fibroblast cells. Cells which were cultivated in unpurified Pluronic F-68 were observed to have a large number of vacuoles and lipid or polymer-loaded vesicles, as evidenced by Transmission Electron Microscopy (TEM). In medium supplemented with SFE extracts of Pluronic, the LL24 cells had significantly reduced viability and cell size and increased lipid droplet formation. We conclude from these findings that contaminating molecules (such as phthalates and BHT) in Pluronic F-68 are cytotoxic and may cause impairment of lipid catabolism.

Introduction
The Pluronics are non-ionic surfactants which consist of a central, hydrophobic, polyoxypropylene block sandwiched between two chains of hydrophilic polyoxyethylenes. In recent years, Pluronic F-68 (molecular weight 8500) has been exploited increasingly in the pharmaceutical and biotechnology areas. It is used to emulsify perfluorochemicals in artificial blood substitutes,[1] it is currently being tested as an injectable drug (RheothRx) for abnormal blood rheological conditions;[2,3] other potential uses include immunomodulators,[4] anti-tumour agents,[9] anti-adhesives,[5] and treatment of myocardial reperfusion injury.[6] The use of Pluronic F-68 as an injectable therapeutic drug, or as an additive in the manufacture of therapeutic biologicals warrants further investigations due to concerns about its toxicity.[7,8,9,10] In the investigations presented in the paper we have systematically studied the in-vitro cytotoxicity effects of supercritical fluid extracted and unpurified grades of Pluronic F-68.

Materials and Methods
Supercritical Fluid Extraction (SFE): SFE is a technique which employs supercritical CO_2, for the extraction and recovery of organic analytes from solid matrices. In this

M.J.T. Carrondo et al. (eds.), Animal Cell Technology, 121–126.

work, a Lee Scientific series 600 SF chromatography (SFC) system incorporating an on-line SFE system was used. For analytical scale extraction, a $0.5m^3$ extraction cell was filled with the appropriate Pluronic powder. The cell was bolted to a heating block, maintained at 33°C. SFE grade CO_2 was passed through the cell at 300 atmos., for increments of 0.5 hr. SFE was carried out on two batches of Pluronic F-68:
(a) Pluronic F-68 (BASF), not previously cytoxicity tested.
(b) Pluronic F-68 (Sigma), cell culture tested by the distributor company.
Cell Culture: LL24 primary, human-lung, fibroblast cells were supplied by ECACC (Salisbury, U.K.). Cells were expanded and passaged twelve times for all experiments. The LL24 cells were routinely cultivated in RPMI 1640 medium supplemented with 5% new born calf serum and incubated at 37°C in a 5% CO_2 in air atmosphere.
Combined Gas Chromatography-Mass Spectrometry (GC-MS): GC-MS analyses of the polymer extracts were obtained using a Finnigan MAT 4500 quadropole instrument which was interfaced with a Finnigan INCOS data system. The temperature programme employed was 60°C for 30 minutes followed by a ramp of 40°C/min to 260°C, at which point the temperature was held until all compounds had eluted. The compounds were scanned from m/z 33 to 500 with a cycle time of 15. The source temperature was 190°C. A data library system was used for the identification of all the extracts.
Transmission Electron Microscopy (TEM): 4% glutaraldehyde (BDH) buffered with 0.1M cacodylate (BDH) was added to the culture flasks. Excess gluteraldehyde was removed by washing with cacodylate buffer. The cells were counter fixed in 2% osmic acid/0.1M cacodylate, dehydrated in ethanol and finally washed in Epon/absolute ethanol. TEM was carried out with a Phillips 400-T Transmission Electron Microscope.

Results and Discussion

SFE purification of Pluronic F-68: Unpurified Pluronic F-68 is a sticky, white powder containing aggregates of spherical particles (Fig. 1a). SFE purification of Pluronic F-68 (Sigma and BASF) resulted in fragmentation and dehydration of the Pluronic particles. (Fig. 1b). GC-MS analyses showed that both the unpurified and SF extracted samples of Pluronic F-68 contained two types of contaminants: butylated hydroxy toluene (BHT) antioxidant and phthalate plasticizer (Fig.s 2a, 2b, 3a and 3b). The Sigma Pluronic F-68 contained these materials in larger quantities. Furthermore, low molecular weight fragments were also detected (Fig.s 2b and 3c) possibly arising from low MW oligomers or as a result of SFE dehydration effects.

The addition of BHT as an antioxidant during Pluronic F-68 sterilization has been mentioned by Ingram *et al.* [11] BHT is an oxygen radical scavenger (antioxidant), commonly used to preserve food products.[12] BHT has been demonstrated to be effective in preventing certain cancers[13] and has an antiaterogenic effect in cholesterol-fed rabbits.[14] High doses (0.4%) of BHT have shown to be carcinogenic in rodents. Metabolism of BHT by liver and lung microsomes resulted in minor toxic metabolites which were involved in BHT-induced pneumotoxicity and mouse skin tumorigenesis.[15] The presence of phthalates in the Pluronic F-68 samples was surprising. Phthalate esters are commonly used as plasticisers which impart flexibility to polymers such as plasticised poly vinyl chloride (PVC). Plasticised PVC is used widely in food wrappings, biomedical devices, tubings, etc. Human exposure to phthalate esters is therefore extensive and chronic. The toxicity of phthalate esters to man has been investigated extensively and they have been shown to have a wide range of adverse effects.[16] The presence of phthalates in samples of Pluronic F-68 may have arisen from a number of sources, including the containers in which the powders were stored and transported.

In-vitro Cytotoxicity Tests: Cultivation of LL24 cells in medium supplemented with various concentrations (0.2 - 1.6% w/v) of unpurified Pluronic F-68 resulted in reduced viable cell densities and extended lag phases (Fig. 4). These growth inhibiting effects were more pronounced at the higher concentrations of Pluronic F-68. Similar cytotoxic effects were observed for both commerical grades of Pluronic F-68. Fig. 5 shows a comparison of LL24 cells cultivated in cell culture medium supplemented with 0.001% (w/v) and 0.005% (w/v) of SF extracted samples. Interestingly, these results showed that very low concentrations of SF extracts were required to cause greater than 50% reduction in viable cell densities. Therefore it is possible that the growth inhibiting effects of Pluronic F-68 were caused by the contaminating molecules (e.g. phthalate and/or BHT) extracted from the unpurified polymer.

Transmission Electron Microscopy: Incubation of LL24 cells in medium supplemented with 0.8% unpurified Pluronic F-68 (Sigma and BASF) resulted in increased cell size, larger mitochondria and sharply defined nuclear and cell membranes. In addition, the cytoplasm was seen to contain a large number of vacuoles and lipid or polymer-loaded vesicles (Fig. 7a). Markedly increased vacuolisation was observed at higher Pluronic concentrations (Fig. 7b). TEM of LL24 cells cultivated in medium supplemented with either 0.001% and 0.005% SFE extracted samples showed these cells to be significantly smaller than the control cells (Fig. 8). SF extracted contaminants from Pluronic did not cause increased vacuolization but increased lipid droplet formation was observed in the cells.

Concluding Remarks

We have demonstrated that Pluronic F-68 is contaminated with phthalates and butylated hydroxytoluene. The results presented above indicate that lipid droplet accumulation can be directly attributed to the presence of phthalate and BHT contaminants. The biochemical effects of increased lipid droplet and vacuole formation in cells is not clear, but may imply an impairment of lipid catabolism.[9,17,18]

References

1. Lowe, K.C. (1988) Biological assessment of perfluorochemical emulsions. In: *Blood Substitutes* (Lowe, K.C. ed) Ellis Horwood, Chichester, 149-172.
2. Carter, C., Fisher, T.C. Hamai, H., Johnson, C.S. Meiselman, H.J., Nash, G.B. and Stuart, J. (1995)*Clinical Heamorheology*, 12, 109-120.
3. Hunter, R. L., Bennet, B. and Check, I.J. (1990) *Fibrinolysis*, 4, 117-123.
4. Topchieva, I.N., Erokhin, V.N., Osipova, S.V., Khrutskaya, M.M., Kupriyanova, T.A. and Bykovskaya, S.N. (1991) *Biomedical Sci.* 2, 38-44.
5. Steinleitner, A., Lambert, H., Kazensky, C. and Centor, B. (1991) *Adhesion Reformation*, 77, 48-53.
6. Forman, M.B., Ingram, D.A. and Murray, J.J. (1992)*Am Heart J.,* 24, 1347-1357.

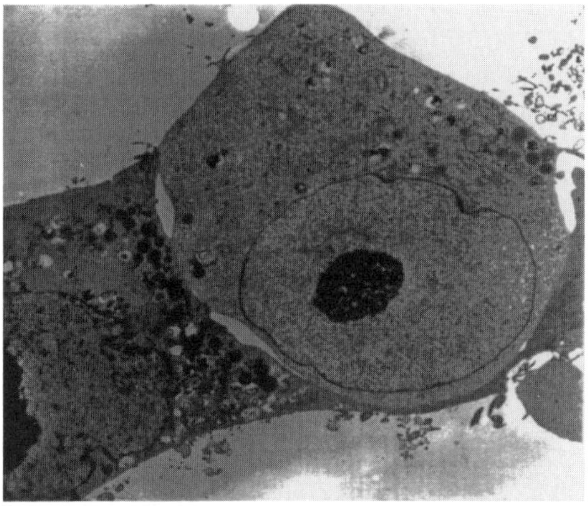

Figure 7a: TEM of LL24 lung fibroblast cells (x2.8K) cultivated in medium supplemented with 0.8% (w/v) unpurified Pluronic F-68. Cells were sampled at 120hr culture time.

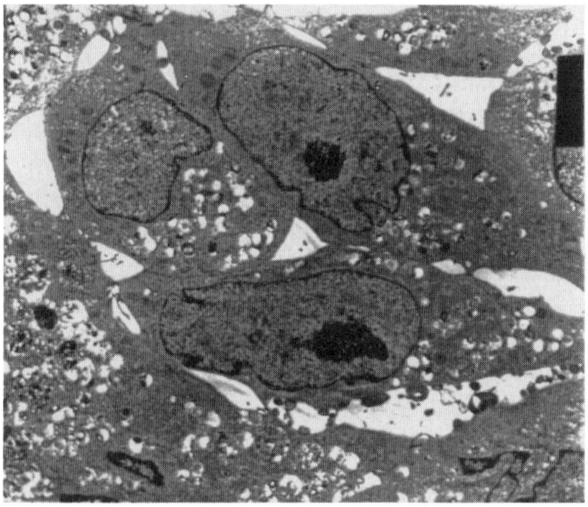

Figure 7b: TEM of LL24 lung fibroblast cells (x2.8k) cultivated in medium supplemented with 1.6% (w/v) unpurified Pluronic F-68. Cells were sampled at 120hr culture time.

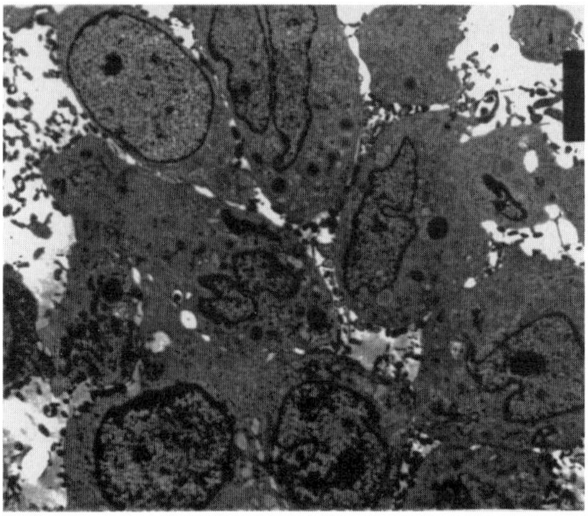

Figure 8: TEM of LL24 lung fibroblast cells (x2.8k) cultivated in medium supplemented with 0.005% (w/v) of SF extracts of Pluronic F-68.

Figure 3a - Phthalate

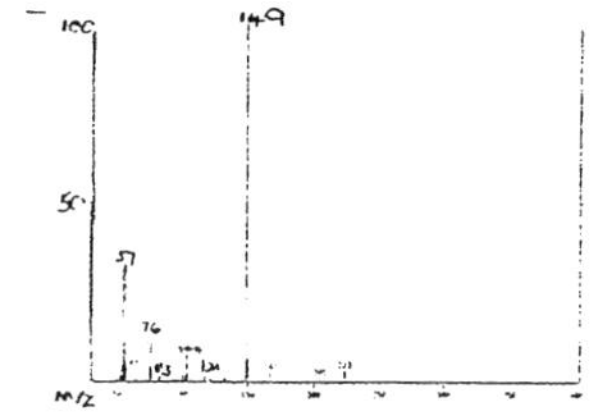

Figure 3b - Butylated hydroxytoluene (BHT)

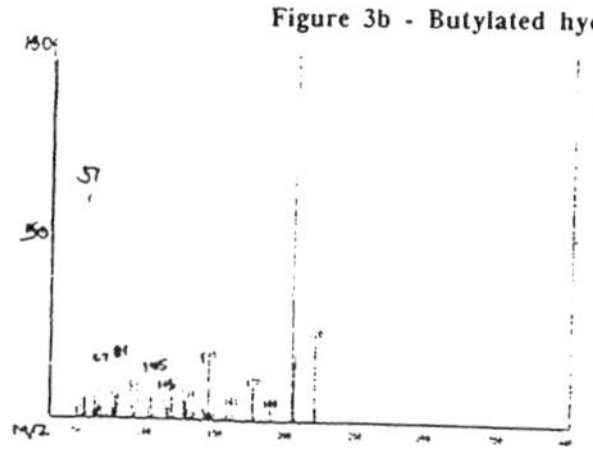

Figure 3c - Additional components of BASF and Sigma Pluronic F-68

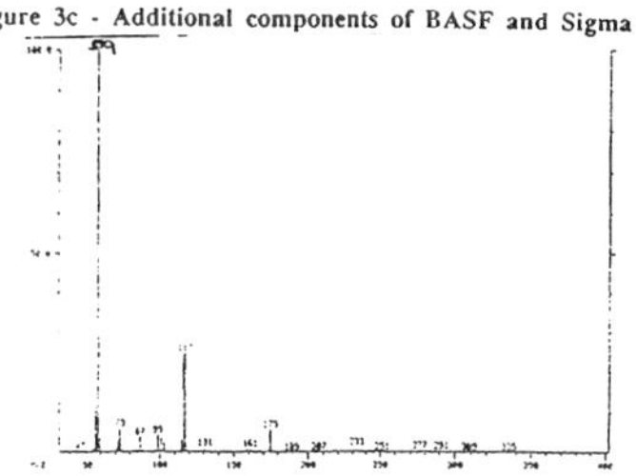

Figure 3a, b and c: Mass spectra of phthalate, butylated hydroxythouene (BHT) and additional components of BASF and Sigma Pluronic F-68s, respectively.

Fig 4 : Growth of LL24 cells in Pluronic supplemented media

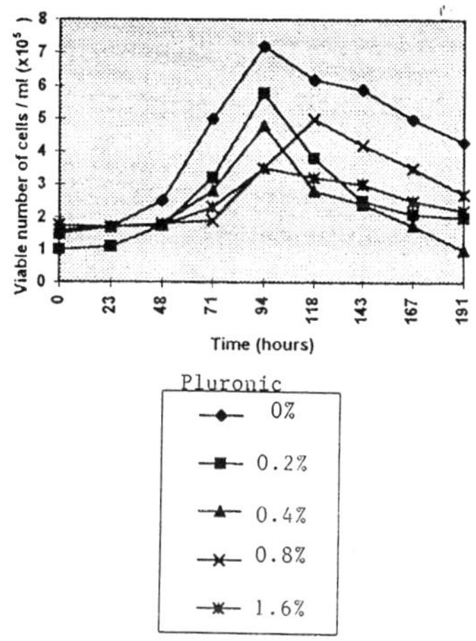

Fig.5 Effect of 0.001% & 0.005% SFE Extract Addition on Growth of LL24 Cells

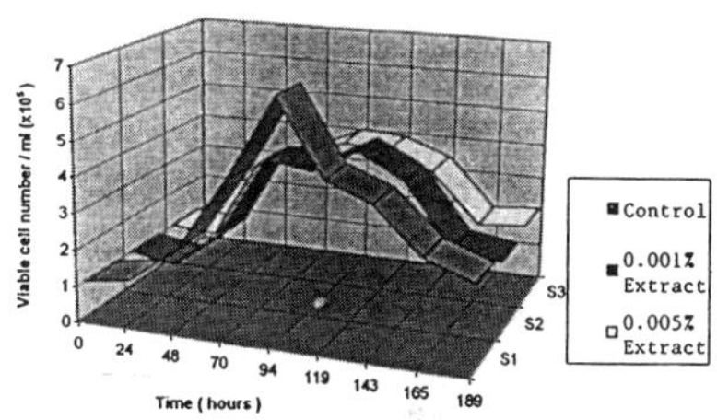

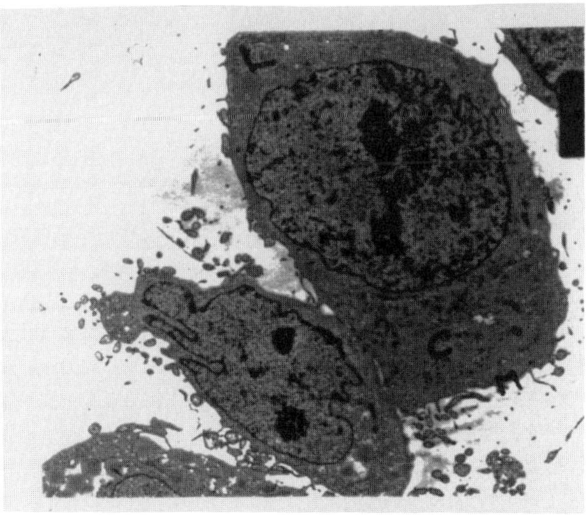

Figure 6: TEM of control LL24 lung fibroblast cells (x 2.8K) at 120hr culture time.
C - cytoplasm
M - cell membrane
N - nucleus
Nu - nucleolus
L - lipid droplet

7. Maruyama, I., Hasegawa, T., Yamamoto, T. and Mamose, K. (1989) *J. Toxicol. Sci.* 14, 153-163.
8. Davidorf, F.H., Chambers, R.B., Kwon, O.H.W., Doyle, W., Gresak, P. and Frank S.G. (1990) *Retina*, 10, 297-300.
9. Nutting, D.F. and Tso, P. *Horm. metabol. Res.* (1989) 21, 113-115.
10. Lane, T.A. and Krukonis, V. (1988) *Transfusion*, 28, 375-378.
11. Ingram, D.A., Forman, M.B. and Murray, J.J. (1992) *AJP* 140, 1081-1087.
12. Parke, D.V. and Lewis, D.F.V. (1992) *Food Additives and Contaminants*, 9, 561.
13. Wattenberg, L.W. (1978) *Adv. Cancer Research*, 26, 197-226.
14. Björkhem, I., Henriksson-Freyschuss, A., Breuer, O., Diezefalusy, V., Berglund, L and Henriksson, P. (1991). *Atherosclerosis and Thrombosis* , 11, 15-22.
15. Guyton, K.Z., Bhan, P, Kuppusamy, P., Zweier, J.L., Trush, M.A. and Kensler, T.W. (1991) *Proc. Nat. Academy Sci.*, 88, 946-950.
16. Thomas, J.A. and Thomas, M.J. (1985) *Critical Reviews in Toxicology*, 13, 283-290.
17. Nutting, D., Hall, J., Barrowman, J.A. and Tso, P. (1989) *Biochimica et Biophysica Acta*, 1004, 357-362.
18. Reabor, M.J. (1981) *Toxicology*, 20, 1-33.

Fig 1a UNPURIFIED PLURONIC F-68

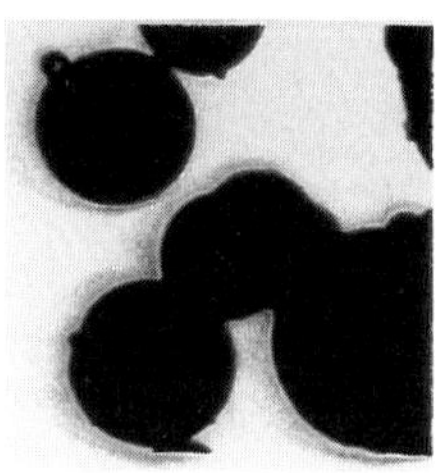

Fig 1b PURIFIED PLURONIC F-68

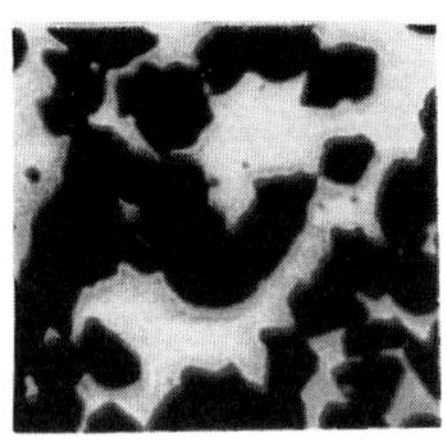

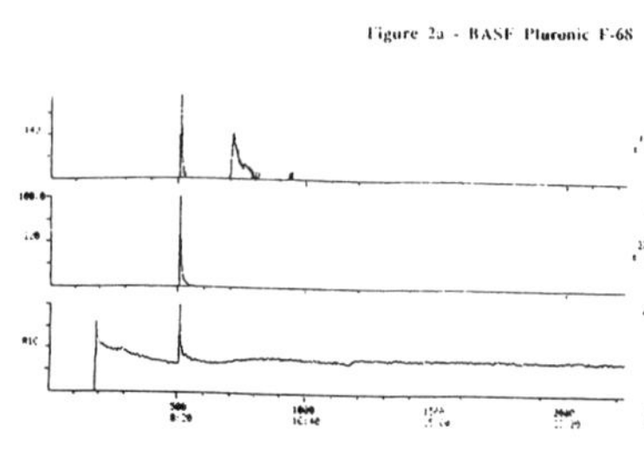

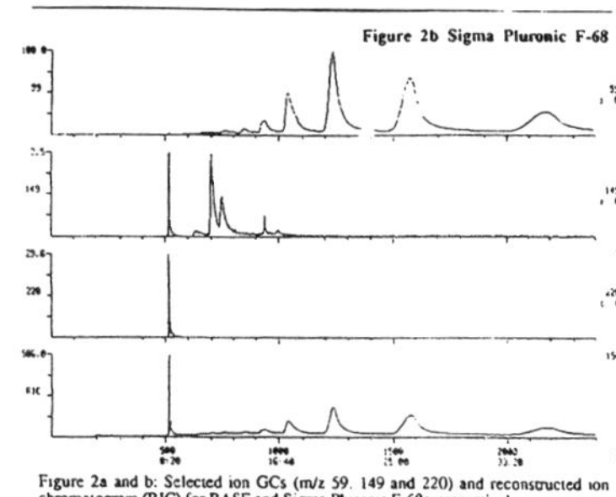

Figure 2a and b: Selected ion GCs (m/z 59, 149 and 220) and reconstructed ion chromatogram (RIC) for BASF and Sigma Pluronic F-68s, respectively

Genotoxicity of bacterial ribonuclease in mammalian cells after its application *in vitro* and *in vivo*.

Olga B.Ivanchenko, Nazira S.Karamova, Olga N.Ilinskaya
Laboratory of Genetic Toxicology, Microbiology Department, Kazan State University, Lenin str., 18, Kazan 420008, Russia

Keywords: Ribonuclease; Micronucleus; Unscheduled DNA synthesis; Genotoxicity

INTRODUCTION

Binase is an abbreviated name of guanylspecific small RNase produced and secreted by *Bacillus intermedius*. It is a single chain protein with 109 amino acids and no disulphide briges [1]. Binase shows marked specificity towards purine bases of polynucleotides and dinucleoside-phosphates and is guanyl-specific in reaction with nucleoside-2'3'-cyclophosphates [2].

Binase at microconcentrations was demonstrated to exhibit a number of biological effects, namely stimulation of microbial and plant growth [3], enhantion of mitochondrial respiration, and increasing of electrophoretic mobility of microbial cells [4]. One of the most important properties for medical application of RNases is an highly effective antitumor action of enzymes [5]

In comparison with pancreatic RNase, binase was more effective inhibitor of growth of Erlih's carcinoma and ascitic form of lympholeukemia NK/Ly [6].

The possible application of binase as a therapeutic agent is connected with the determination of all unfavourable effects, especialy genotoxic ones.

At GENTOX laboratory of Kazan State University the pure enzyme samples of *Bacillus intermedius* ribonuclease were investigated for genotoxicity in a set of microbial tests. The weak mutagenic effect of binase at high concentrations (1mg/plate) was established by induction of forward Ara^R-mutations (Ara-test) and histidine-reverse mutations (Ames test) using *Salmonella typhimurium* strains. *Bacillus intermedius* RNase appears to possess a weak DNA damaging activity in *uvr A*$^-$ and *pol A*$^-$ mutants of *Escherichia coli* strains and can cause the prophage induction in the two lysogenic microorganisms: *Bacillus subtilis* (phage φ 105 wt) and *Streptomyces lavendulae* (phage 3311) [7].

However the primary aim of mutagenicity testing is to assess the risk of heritable damage induced by chemicals for human. The data obtained using microbial tests did not reflect all possible genotoxic effects which can take place in mammalian organisms including man. So, the testing with eucaryotic cells in both modification *in vivo* and *in vitro* is a necessary step for genotoxicity estimation.

M.J.T. Carrondo et al. (eds.), Animal Cell Technology, 127–132.

The objective of this experimental study was to analyse the genotoxic effects of binase *in vitro* using human lymphocyte culture (unscheduled DNA syntesis test, UDS) and *in vivo* using mouse erythrocytes (micronucleus assay, MNA).

MATERIALS AND METHODS

Chemicals

The ribonuclease-binase (EC 3.1.4.23) was isolated from *Bacillus intermedius 7P* (strain no. B 3073 in the All-Union Collection of microorganisms, Genetics Research Institute, Moscow, Russia) culture fluid as homogenous protein with m = 12300 Da, p*I*=9.5 and maximum activity 1 000 000 U/mg at pH 8.5. (An activity unit is the amount of enzyme capable of increasing the extinction at 260 nm of acid-soluble RNA hydrolysis products to 1 U/min at 37° C, pH 8.5). The ribonuclease is free of contamination with DNase, phosphomonoesterase and phosphodiesterase. Chemical homogenicity of the RNase was checked by gradient chromatography of a commercial sample on phosphocellulose. Binase showed marked specificity towards purine bases of the RNA molecule and was guanine-specific in the synthesis of oligoribonucleotide phosphodiester linkages. The enzymatic, physical and chemical properties was described by Golubenko *et al.*[8] and Balaban *et al.* [9].

The compounds used for positive mutagenicity control were obtained from Experimental Biochemistry Factory, Kurgan, Russia (thiophosphamide), and from Sigma, FRG (colchicine). All other compounds were purchased from Sigma, FRG, excepting Giemsa solution was from Koch-Light Laboratories, Berks, England.

Unscheduled DNA synthesis test

Lymphocytes were isolated from samples of peripheral blood of healthy donors according to Boym [10].Immediately after isolation lymphocytes were cultured 2h in Hank's balanced salt solution (HBSS) supplemented with RNase at different concentrations in a humidified incubator with 5% CO_2.

Viable cells were determined by trypan blue exclusion.

DNA replicative synthesis inhibited by adding of 2,5 mM hydroxyurea. After 20 min of incubation with hydroxyurea the tritiated thymidine (6 μCi/ml) was added to the cells and incubation continuated 2h at 37° C. Then the exposure mixture was removed, cells were rinsed and fixed with methanol on the glass slide. Incorporated radioactivity was measured by autoradiography. The slides with lymphocytes were dipped in autoradiographic emulsion and exposed at 4° C in the dark for 14 days before the emulsion is developed. The cells were stained with Giemsa solution and grains in the emulsion over the nuclei and in the three background regions adjacent to the nuclei was counted visually.

The increase of grains amount in treated cells in comparison with control cells without treatment reflect the intensity of repair synthesis induced by RNase.

Physiological solt solution (0,9 % NaCl) was used as the solvent control.
Thiophosphamide (10 μg/ml) was used as knowh inductor of UDS.

Micronucleus assay

The male mice (line CBAxC5713 L/G) between 10-16 weeks of age weigthing 25-30 g were used for all studies.

RNase was dissolved in physiological solt solution and injected once intraperitoneally (25 mg/kg). Control animals received the equivalent volumes of salt solution. Colchicine (125 μg/kg) was used as known inductor of micronucleus formation.

Peripheral blood from a tail vein (5μl) was collected 72 h after treatment and placed without anticoagulants on the glass slide.

Smears were prepared according to Schmid [11]. Smears were air dried, fixed in metanol and stained with Giemsa (Giemsa/H_2O, 1/3, 30 min).

Micronucleated cell frequencies were determined by scoring a minimum of 1000 - 2000 polychromatic erythrocytes for each slide.

The data were analyzed for statistical significance with Student's t-test. The significance level for all tested doses of compounds yielding a positive response ranged from $P < 0.05$ to 0.001.

RESULTS

Induced damage to the nucleic acids of mammalian cells generally results in a valiant effort by the cell's enzymatic systems to repair the detect, resulting in the unscheduled synthesis of DNA [12,13].

The effect of RNase on UDS was examined by measuring the uptake of ^3H-TdR in lymphocytes. The doses of RNase tested were 200 μg/ml and 1000 μg/ml. In Fig.1 we report the autoradiographic data reflecting the grain numbers on nuclei of lymphocytes treated by RNase at different doses. RNase at 200 μg/ml did not result in significant increasing of radioactive thymidine incorporation into DNA and the corresponding numbers of silver grains on nuclei in comparision with control level.The high concentration of the tested enzyme (1 mg/ml) induced the grain number approximately 1.5-fold higher as in variant without RNase. So, data obtained can be taken to indicate that in human lymphocytes treatments with high dose of RNase produce DNA damage and stimulate DNA excision repair.

UDS assay are measure the repair of DNA damage over the entire genome, while micronuclei are thought to arise from both clastogenic (chromosome breakage) and aneugenic (chromosome lagging and effects on spindle) effects. The micronucleus assay is widely used as a screening method for detecting clastogenicity of substance *in vivo* [14].

Our experimental results indicate that RNase at 25 mg/kg (0.1 LD50) was found very weak cytogenetic activity. For injection this dose was choosen because it overcovers all therapeutic doses reccomended for antitumor and antiviral action of enzyme [6,15]. As shown in Fig.2, no appreciable changes in the number of micronuclei in polychromatic erythrocytes of peripheral blood of mice subjected to RNase was detected if compared with the control groups. A significant increase in micronuclei was observed after three days of treatment by colchicine, a well-known inductor of micronuclei. After the same time the RNase induced micronucleus

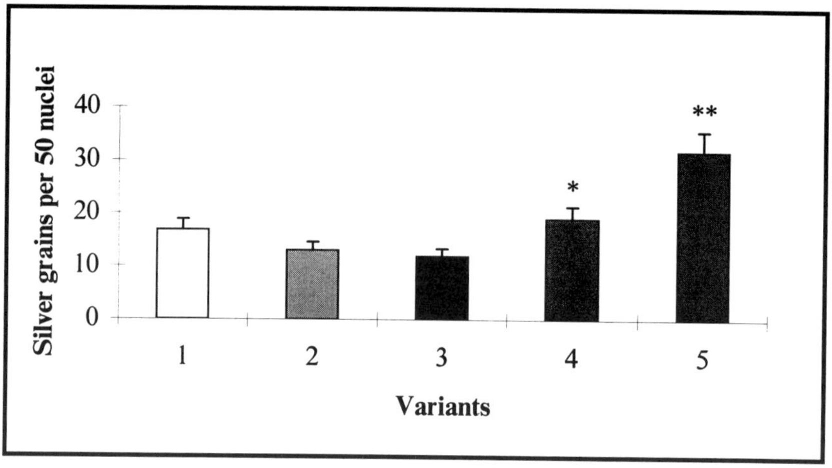

Fig. 1 3H-Tdr incorporation into DNA in lymphocytes treated with RNase. 1-DNA replicative and repair synthesis; 2-DNA repair synthesis; 3-RNase, 200 μg/ml; 4- RNase, 1000 μg/ml; 5-thiophosphamide 10 μg/ml. Data represent mean of 3-5 independent experiments. * indicates significant difference from control P< 0.05, **- P< 0.001.

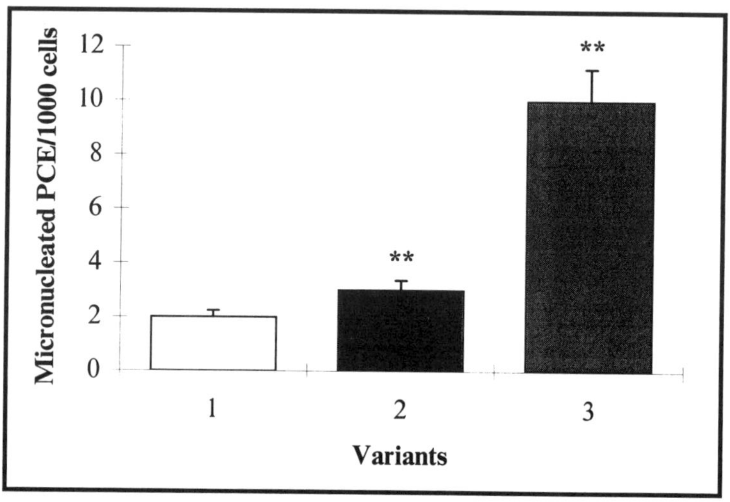

Fig.2 Clastogenicity of RNase in an *in vivo* micro-nucleus test using the mouse peripheral blood erythrocytes.

1-Spontaneous level; 2-RNase, 25 mg/kg; 3-colchicine, 125 μg/kg.

** - P<0.001.

number was lower. The failure of micronucleus to increase after i.p. application of RNase suggests that RNase can be involved with the low probability in the nuclear transformation process of animal blood cells.

DISCUSSION

The data presented in this report indicate that RNase at the high dose (1000 μg/ml) induced unscheduled DNA synthesis and resulted in the incorporation of radiolabeled thymidine into the nucleic acid of lymphocytes *in vitro*. Investigation of RNase action in mouse erythrocytes *in vivo* was found a weak positive genotoxic effect too.

The data of UDS test and micronucleus test are in argeement with results of our previous investigations of genotoxic activity in microbial tests. Binase exhibited a weak mutagenic activity inducing different types of mutations (both frameshift mutations and base-pair substitutions) and displaes a prophage induction effect [7]. The enhanced phage induction proves that binase had activated RecA protein, the main regulator of cell SOS-functions. However, binase propably does not act as a direct genotoxic agent interacting with DNA, but could provoke nucleotide imbalance through its catalytic action with membrane-associated RNA, which results in alteration of DNA replication and, as a consequence, in *recA*-dependent mutagenesis.

It is probably that the biochemical nature of process leading to the genotoxic effects induced by RNase in eucariotic cells is similar to that in bacterial cells. At the present time the data have been accumulated which reflect the SOS-response exists not only in bacterial, but in mammalian cells too [16,17].

On the basis of our experiments with the eucariotic cells *in vitro* and *in vivo* we can suppose the ability of binase high concentrations to induce the genotoxic effects in mammalian organisms including man.

REFERENCES

1. Aphanasenko GA, Dudkin SM, Kaminir LB, Leschinskaya IB, Severin EB (1979) Primary structure of ribonuclease from Bacillus intermedius. FEBS Letters 97, 77-79.
2. Yakovlev GI, Chepurnova NK, Moiseev GP, Bocharov AL, Lopatnev SV (1987) Specificity of the Bacillus intermedius 7P ribonuclease in reactions of olyribonucleotide cleavage. Bioorg.Khimia 13, 338-343.
3. Kolpakov AI & Kuprijanova FG (1992) Influence of exogenous ribonuclease on division of Candida tropicalis yeast. Microbiologiya 61, 969-974.
4. Ilinskaya ON & Krylova NI (1993) Effect of exogenous Rnase on the cell of lower eukaryotes. Prikladnaya biokhimiya i mikrobiologiya 29, 280-285.
5. Kapetsky TR & Levy CC (1981) Human ribonucleases. In Holcenberg S, Roberts J (Ed) Enzymes as Drugs, New York, Chichester, Brisbane, Toronto, 103-166.

6. Kurinenko BM, Sobchuk LI, Haibullina SA, Karpova SI (1988) Antitumor activity of ribonuclease Bacillus intermedius 7P. Experimentalnaya oncologiya 6, 54-57.
7. Ilinskaya ON, Ivanchenko OB, Karamova NS (1995) Bacterial ribonuclease: mutagenic effect in microbial test-systems. Mutagenesis 10, 165-170.
8. Golubenko IA et al. (1979) Ribonuclease of Bacillus intermedius 7P: purification by chromatography on phosphocellulose and some properties of the homogenous enzyme. Biokhimiya 44, 640-648.
9.Balaban NP, Sharipova FR, Golubenko IA, Aphanasenko GA, Volkova TI, Vershinina VI, Znamenskaya LV & Leschinskaya IB (1980) Physico-chemical characteristics of ribonuclease Bacillus intermedius. Proceedings of the fist International Meeting "Structure and chemistry of ribonuclease". Moscow, 28 Nov-2 Dec.1988, 349-352.
10. Boym A (1977) Separation of lymphocytes, lymphocyte subgroups and monocytes: a review. Lymphology 10, 71-76.
11. Schmid W (1975) The micronucleus test. Mutation Res. 31, 9-15.
12. Williams GM (1977) Detection of chemical carcinogens by unscheduled DNA synthesis in rat liver primary cell cultures. Cancer.Res. 37, 1845-1851.
13. Kornburst D & Barfknecht T (1985) Testing of 24 food, drug and cosmetic and fabric dyes in the in vivo and the in vivo/in vitro rat hepatocyte primary culture DNA repair assay. Environ. Mutagen. 7, 101-120.
14. Schmid W (1977) The micronucleus test. In: Kilbey BJ, Legator M, Nichols & Ramel (Eds), Handbook of Mutagenicity test procedures, Elsevier, Amsterdam, 235-243.
15. Kurinenko BM et al. (1989) Study of toxicity of Bacillus intermedius RNase in vitro and in vivo. Antibiotiki i chemiotherapia 34, 266-270.
16. Sarasin A (1985) SOS response in mammalian cells. Cancer.Invest. 3, 163-174.
17. Elespuru R (1987) Inducible responses to DNA damage in bacteria and in mammalian cells. Environ. and molecular mutagenesis 10, 97-116.

USE OF CELL-BASED ASSAYS FOR IN VITRO STUDIES OF DRUGS

S.Riva, C.Brunati, W.Pollini, C.Quarta and M.L.Nolli
Lepetit Research Center, via R.Lepetit 34, 21040 Gerenzano (VA)
Italy

ABSTRACT

One of the most recent and interesting applications of cell cultures is their use for set up of in vitro tests to be used for different purposes: screening of new drugs, cytotoxicity and monitoring of certain attributes during drug development. Few but important factors have influenced this new development: one is the possibility to cultivate cells that maintain their functions in microtiter plates and therefore "validate" them as reagents; another one is the achievement in cells handling automation and robotization that makes possible to easily run a lot of samples in a short time. A typical cell-based assay consists of two phases:1) The interaction of the cells with the drug of interest 2) The measurement of the cell affected function (target of the assay) by fluorimetric, ELISA or enzymatic test. In addition, on the same cells, it is possible to do viability test (e.g. MTT test) to have an indication on the "cytotoxic behaviour" of the drug of interest. We report here two examples of cell-based tests used to study the in vitro behaviour of drugs. One is a Caco-2 test for characterization of intestinal absorption of antibiotics. Another one is a macrophage test for studying the immunomodulation of TNF-α.

KEY WORDS: Drug discovery, cell-based assay, automation, Caco2, Macrophage.

INTRODUCTION

The use of cultured animal cells for the development of cell-based assays has become more and more important in the last few years. This type of assays can be utilized for different purposes: 1) discovery of new drugs 2) toxicology testing 3) monitoring of certain attributes during drug development. The advantage of cell-based assays either used in the selection of new drugs or in the testing of molecules at different stages of

M.J.T. Carrondo et al. (eds.), Animal Cell Technology, 133–138.

their development is the possibility to study the behaviour of a molecule that can: 1) aspecifically alter the cell functionality 2) selectively modify a cell specific function 3) use the cell to be transported through the tissues. This is done in a situation quite similar to physiological conditions.The big progress in DNA cloning and expression techniques and cell culture media has allowed the availability of many transformed and recombinant cell lines with a large variety of different phenotypes that can be easily maintained in culture. Moreover the possibility to grow the cells in microtiter plates without loosing their functions, in reproducible conditions was a fundamental achievement for cell handling automation and robotization, a needed prerequisite for dealing with large number of samples in a short time. We report here the complete process for the introduction of two cell-based assays in a screening program for new drugs.

MATERIALS AND METHODS

Caco2 cell line, a human colon adenocarcinoma, and the related assay has been described elsewhere (1). Macrophage assay was performed with the stabilized MT2 macrophage cell line (2). The Elisa and MTT tests have been described elsewhere (3).

DEVELOPMENT AND OPTIMIZATION OF A CELL BASED ASSAY

- Cell culture standardization
  - weekly flasks preparation
  - standard splitting method
- Cell plating reliability
  - cells at the same level of confluence or differentiation
  - same number of cells in each well (dispensing precision)
- Kinetic studies for evaluating time effect
  - on cell response
  - on reagents stability
  - on chemical reactions
  - on cytotoxicity
- Selection of drug concentration to reduce cytotoxic effects
- Optimization of the signal/noise ratio
- Assay optimization
  - number of cells
  - reagents concentrations
  - time
  - temperature
  - detection
  - recovery
- Validation with standard molecules
  - according with EN45000 or ISO9000

AUTOMATION

PARAMETERS

- Productivity (number of samples tested in time units)
- Handling of a large numbers of reagents
- Simultaneous detection of multiple tests

AUTOMATION

- Automatic data reduction and data management

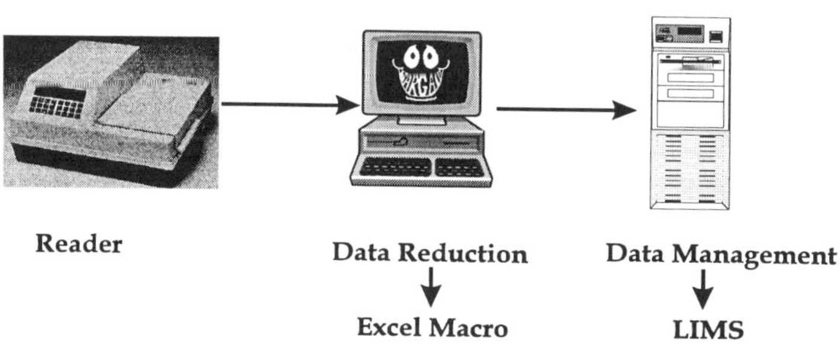

The CACO2 ASSAY DESIGN

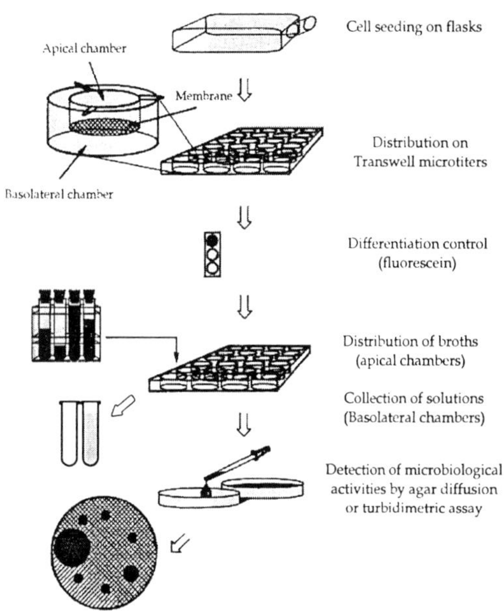

VALIDATION WITH STANDARD ANTIBIOTICS

Antibiotics	In vivo	Caco2	% of passage
Rifampin	yes	yes	1-10
Rifapentine	yes	yes	>10
Ampicillin	yes	yes	1-10
Tetracycline	yes	yes	1-10
Erytromycin	yes	yes	1-10
Novobiocin	yes	yes	1-10
Cycloserine	yes	yes	>10
Trimethoprim	yes	yes	>10
Teicoplanin	no	no	<1
Vancomycin	no	no	<1
Neomycin	no	no	<1
Kanamycin	no	no	<1
Cephaloridin	no	no	<1
Rifamycin SV	no	no	<1

USE OF CACO2 ASSAY IN SCREENING PROCESS (Activities distribution)

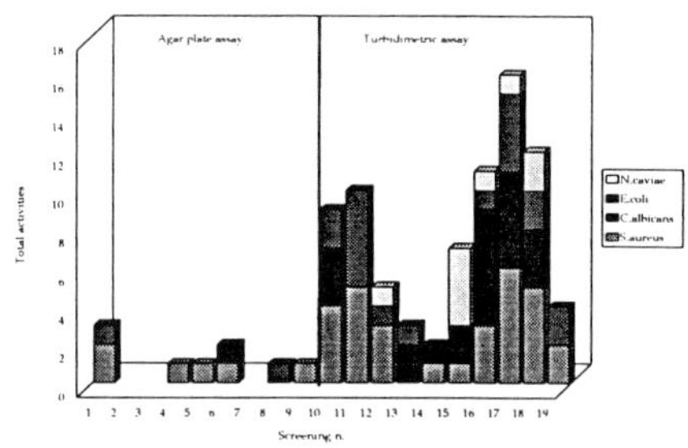

MACROPHAGE ASSAY DESIGN

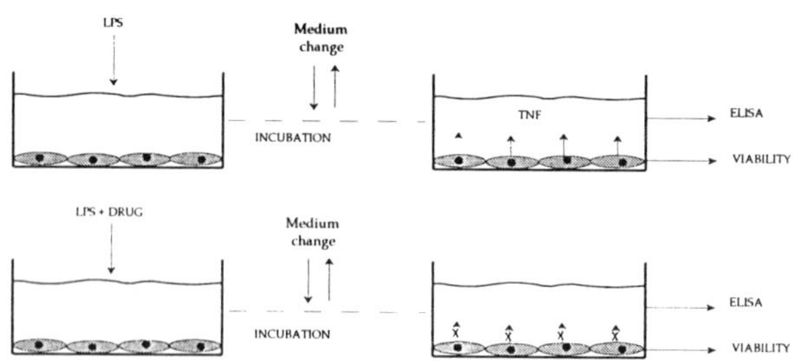

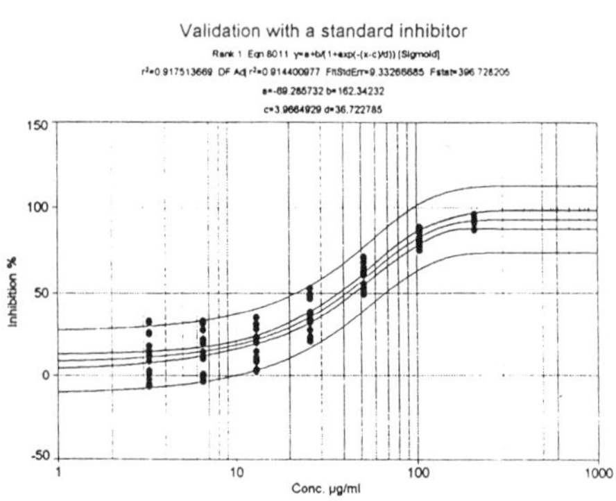

DISCUSSION

Cell-based assays play an important role in the screening for new drugs. Since they are close to a physiological environment, they offer a great variety of information very useful in the early stages of a discovery program. Usually they are not selected as primary assays for their intrinsic complexity but as we have shown in this paper it is possible to design and use them in a routine screening program. During the set up of these assays many parameters must be studied; it is better to assess them before the beginning of a screening program to control the variation that is a peculiarity of biological systems. In particular the validation and automation of cell-based assays is a crucial and fundamental phase. The two assays we have described give completely different answers, they are used for different targets and are structurally different. In the first case (CaCO2) the assay is used for the selection of potential orally absorbable molecules, while the other (macrophage) has been set up to select molecules which modulate or inhibit the release of a cytokine important in inflammation processes. Even though the two assays are conceptually simple to do, some important parameters have to be checked to set up the assay and to implement the scaling up to HTS. At the beginning is important to standardize cell culture conditions and cell plating reliability to enhance the reproducibility of the response. Kinetic studies of the assay (incubation times, reagents stability etc.) play an important role for the automation. In fact in this phase there are some constrains that should be taken into consideration to obtain a maximum of efficiency from the robotic systems. Other parameter to check are the cell integrity and functionality. In the reported cases two different approaches have been used. In one case (macrophage), since cell functionality was the critical point, the MTT viability test was used to monitor cell status while in the other case (Caco2) it was relevant for the test to demonstrate the integrity of cell monolayer using the fluorescein passage test (manuscript in preparation). Moreover the ability to select interesting molecules has to be checked. For this purpose it is necessary to ensure the reproducibility and reliability of a test through a validation program. This can be done running standards (when available) through the process, collecting data and estimating the related statistical parameters after data reduction . As in the final process (HTS) large volume of data are involved, a good data base and a good data analysis package are needed to support an efficient drug discovery process.

CONCLUSIONS

- Cell-based assays can be applied to the screening of molecules with different activities (e.g. absorbable antibiotics, immunomodulators).
- The validation and optimization of these assays allow the use of cells as biochemical reagents for the improvement of test confidence in a screening program.
- The automation and robotization of cell-based assays allow handling and management of a large number of samples and data.
- Cell-based assays offer the possibility of monitoring simultaneously specific activities and aspecific effects as cytotoxicity.

REFERENCES

1. Ranaldi G, Islam K and Sambuy Y (1992). Epithelial Cells in Culture as a Model for the Intestinal Transport of Antimicrobial Agents. Antimicrob. Agents Chemother. 36:1374-1381.
2. Sassano M, Granucci F, Seveso M, Marconi G, Foti M, Ricciardi-Castagnoli P (1994). Molecular Cloning of a Recombinant Retrovirus Carrying a Mutated env^{AKR}-myc^{MH2} Fusion Gene Immortalizing Cells of the Monocytic-Macrophage Lineage. Oncogene 9:1473-1477.
3. Riva S, Nolli ML, Lutz MB, Citterio S, Girolomoni G, Winzler C and Ricciardi-Castagnoli P (1996). Bacteria and Bacterial Cell Wall Constituents Induce the Production of Regulatory Cytokines in Dendritic Cell Clones. Journal of Inflammation 46:98-105.

Evaluation of Cytotoxicity with Fish Cell Lines: A Case Study for α-Oligothiophenes (1-5 Rings).

Lopes, A.[1,2], L.M.D. Gonçalves[1,2], M.J.T. Carrondo[2,3]
[1] Instituto de Tecnologia Química e Biológica, Ap 127, P-2780 Oeiras, Portugal; [2] Instituto de Biologia Experimental e Tecnológia, Oeiras, Portugal; [3] Lab. Engª Bioquímica, FCT/UNL, Monte da Caparica, Portugal.

Abstract

This study was designed to compare two continuous fish cell lines in culture, BB and RTG-2, for their ability to determine cytotoxicity based upon the phototoxicological action of a set of α-oligothiophenes with 1-5 rings (α1,..,α5). Cells were incubated in optimal growing conditions for 1, 2, and 6 days after inoculated with thiophenes and irradiated 20 minutes with Phillips TL D lamps (15W, spectral distribution 300-500nm). Three end-points were measured as markers of phototoxicity (compared to "dark" experiments): *i)* mitochondrial activity (MTT reduction), *ii)* membrane integrity (LDH released activity), and *iii)* cell adherence. Inhibitory concentrations 50% (IC50) were obtained from concentration effect curves.

The concentrations of thiophenes causing IC50 in the MTT reduction. and the pattern of phototoxicity, α3>α4≅α5≅α2>α1, was identical for both cell lines). These results agree very well with the predicted phototoxicity taking into account both the partition coefficients in micro-heterogeneous systems that mimic the environment of the molecules *in vivo*, and the "killing factor", the efficiency of singlet oxygen production. The maximum of the predicted as well as the experimental values occur at α3, the member of the family which occur in larger amount in the nature.

Keywords: *in vitro* toxicology, non-target organisms, pesticides, phototoxicolgy, singlet oxygen, partition coefficients.

Introduction

Oligothiophenes are one of the recently most diffused and studied compounds covering research areas from photobiology (with the particular case of α-terthienyl, n=3) to non-linear optics and electronic devices (oligothiophenes with n ≥ 5).

M.J.T. Carrondo et al. (eds.), Animal Cell Technology, 139–143.

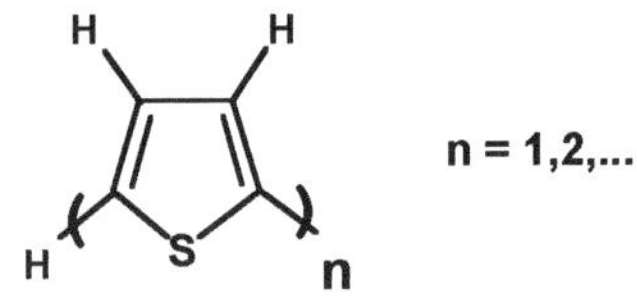

Since it is accepted that the main photodamage mechanism of this family of compounds occurs *via* $^1O_2$ production [1], we need to take into account physico-chemical parameters, in order to complete rationalize its phototoxicological behavior, as: i)the preferential localization in the target areas, namely lipidic reservoirs - K_p; ii)the amount of absorbed light (UVA, visible) by the compound; and iii)the singlet oxygen quantum yield formation, Φ_Δ.

Moreover, it is interesting to see if the predicted trends of cytotoxicity gives a highly phototoxic α-terthienyl (α-3) - the member of the family which occur in larger amount in the nature. It was recently shown [2] that both the triplet lifetime, τ_T, and the O_2 ($^1\Delta_g$) quantum yield formation, Φ_Δ, decreases continuously from n=2 to 7. By other hand, preliminary results with fungi cultures [3] have shown that the phototoxicity increases from α-2 to α-3, and then decreases with the increase of n.

In order to understand why the two sets of data are not mutually exclusive we have tried to rationalize the photobiological action of α-1 to α-5 taking into account the amount of absorbed light, the characteristics of the triplet state (τ_T, Φ_Δ), and the extension of the retention of these compounds in the lipophilic media, evaluated from the usual bioconcentration and bioaccumulation predictors, the distribution between 1-octanol and water, $\boldsymbol{K_{ow}}$, and the partition coefficients, $\boldsymbol{K_p}$, in micro-heterogeneous systems, *e.g.* micelles, phospholipid vesicles, that mimic the biologic target areas -the membranes.

Methods

BB and RTG-2 fish cell lines consist of fibroblast cells initiated from the posterior trunk tissue of brown bullhead catfish, and from pooled gonadal tissues of rainbow trout respectively (ECACC, UK). Cells were grown in humidified atmosphere with 5% CO2/air mixture at 22°C. Dulbecco´s Modified Eagle´s Medium (DMEM) supplemented with fetal bovine serum (FBS) was used for both batch propagation and toxicity tests.

For the phototests individual wells of a tissue culture (96-well microtiter plate) inoculated with 0.2 ml of medium containing 2×10^4 cells were used. After 2 days of incubation the medium was replaced with the exposure medium containing different concentrations of oligothiophenes dissolved in DMSO (final concentrations of DMSO < toxic levels).

The plates were irradiated for 20 min with a PHILIPS TLK 40W/09N lamp. Control plates which had not been irradiated aloud to distinguish between dark toxicity and phototoxicity [3]. The end-points used for the toxicity evaluation were: i) lactate dehydrogenase activity, LDH; ii) total soluble protein, TSP; iii) reduction of MTT.

When applicable, for each of the end-points analyzed, LD50 values were obtained from the fits of the experimental point with a GRANPROBIT routine. The normalized values of the photobiological action of each α-*n* were determined assuming an effective concentration of excited species proportional to the overlap integral between the emission of the lamp and the absorption of each compound (corrected for the scattered light by the culture medium/plate).

The $^1O_2$ formation efficiency was taken from literature [2] and the extension of the retention of these compounds in the lipophilic media was obtained for the entire family by two distinct methods:

i) the usual bioconcentration and bioaccumulation predictors - the distribution between 1-octanol and water, K_{ow}, (homogeneous system);

ii) the partition coefficients, K_p, in amphiphilic micro-heterogeneous systems that mimic the biological membranes - neutral ($C_{12}E_{10}$) and charged (SDS) micelles.

The distribution coefficients between 1-octanol and water were obtained using the *shake flask method* [4]. Localization and partition coefficients, $\boldsymbol{K}_p$, in neutral ($C_{12}E_{10}$) and charged (SDS) micelles were inferred from the *fluorescence anisotropy method* [5] by fitting the *fluorescence anisotropy*, r, as a function of the *volume fraction* of the *lipophilic medium*, γ, (ξ is a proportionality constant, and W and L refer to water and lipidic phase, respectively).

$$r = \frac{r_W + \gamma \xi K_p r_L}{1 + \gamma \xi K_p} \quad (1)$$

The outcome variables, K_p, and r_L, are related to the hydrophobicity and the intrinsic micro-viscosity in the restricted environment where the molecule stands, respectively.

Results and Discussion

Fig. 1 shows the phototoxicological behavior of α-oligothiophene family measured as the growth inhibition of the cells culture (●) (compared to dark experiments), and the predicted value (■) obtained by the product of the amount of compound in the lipophilic medium (derived from $\boldsymbol{K}_p$), by the "*killing power factor*" - considered to be Φ_Δ. The maximum of the predicted (■) as well as the experimental (●) values occur at α-3, the member of the family which occur in larger amount in the nature.

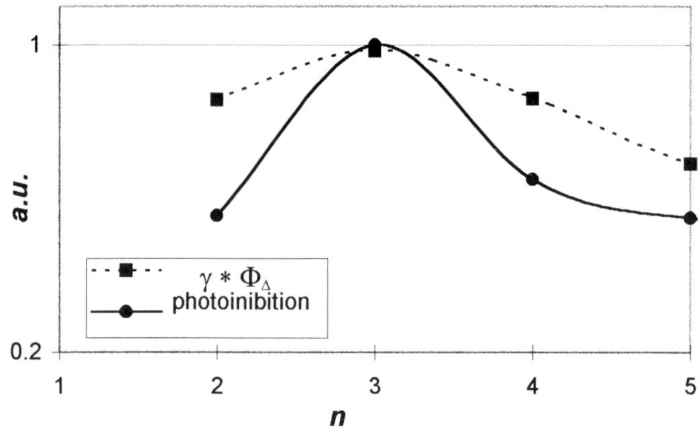

Fig. 1 - Photobiological action, predicted (■) and experimental (●) values, of α-1 to α-5 towards RTG-2 cells (BB cells gives identical results).

It is concluded that the photobiological action of the thiophene family for these animal cells can be described in a simple fashion - only the molecules that stands in a lipophilic environment are able to (after irradiation) generate the O_2 ($^1\Delta_g$) which in turns act as the *killing agent.* These results agree with the ones we find with fungi [3].

Moreover, accordingly to the three end-points used in this work, MTT reduction, LDH activity, and cell adherence the two continuous fish cell lines analyzed, BB and RTG-2, shown the same ability to determine cytotoxicity based upon the phototoxicological action of the α-oligothiophene family [6].

Future experiments will be carry out with other organisms, *e.g.* plants, plant cells, bacteria, in order to investigate whether or not α-oligothiophenes show the same kind of behavior, and in order to search the preferential localization of these molecules *in vivo* (membrane, mitochondria, etc.).

References

1. Towers G (1986), J.Chem. Ecology 12, 813.
2. Becker R, Melo J, Maçanita A, Elisei F (1995), Pure and Appl Chem 67, 9.
3. Lopes, A, Borges M, Matos O, Pais I, Melo J, Becker R (1995), XVIIth International Conference on Photochemistry, London, pp 4P38.
4. Connell D (1990), In Bioaccumulation of Xenobiotic Compounds, Connell D. Ed., CRC Press.
5. Lopes A, Maçanita A, Pina F, Melo E, Wahmoff H (1992), Environ Sci Technol 26, 448.

6. Gonçalves, LMD, A Lopes, MJT Carrondo (1996) Development of Cell Model for Phototoxicyty: Mechanism of Toxicity and Structure Activity Relationship. In Vitro Cell Dev Biol 32 (3) part II, pp 64A.

Acnowledgements

The authors acknowledge JNICT, Junta Nacional de Investigação Científica e Tecnológica, by the financial support (Project Nº PBIC/C/BIO/2017/95 and Grant PRAXIS XXI/BPD/4153/94).

The Effect of Cell Culture System on the Phenotypic Stability in Primary Hepatocyte Cells

Lopez, T.[1] , M.J.T. Carrondo[1,2], L.M.D. Gonçalves[1]
[1] IBET/ITQB, Oeiras, Portugal
[2] FCT/UNL, Monte de Caparica, Portugal

Abstract

Liver cell culture become specially valuable as a tool for investigation of hepatotoxic action and metabolic profile of drugs, as well as their interactions with other therapeutic agents or xenobiotics. However, the catalytic activities of cytochrome P450 monooxygenases exhibit a rapid decline in primary cultured hepatocytes. This represents a serious drawback to using this experimental model for long-term *in vitro* pharmacological studies. Among others, the strategies used to overcome this limitation include the use of selective supplemented media, cultivating hepatocytes on extracellular matrices, and co-culturing hepatocytes with other cell types. In the present work are compared monolayer cell cultures with three-dimensional systems; where the phenotype stability of primary hepatocytes is evaluated. Cells were immobilized in different microcarriers (Cultispher G, Cytopore 2 and Cytodex 3) and were tested for their ability to metabolise xenobiotics using the metabolism of testosterone as a marker for phase I reactions of biotransformation.

Key words: cell culture; three dimensional; testosterone; matriz; aggregates

Introduction

The liver plays a key role in the metabolism of most endogenous and exogenous substances. The reactions are mainly catalysed by cytochrome P450, a family of haemoproteins, which are found at high concentrations in hepatocytes. Some forms of this enzymes are present constitutively, whereas others are induced to high levels of expression upon exposure to certain foreign compounds.

Primary cultures of rat hepatocytes have been used extensively to study xenobiotic metabolism and toxicity *in vitro* but their usefulness is limited because the cells rapidly undergo de-differentiation in culture and some metabolic functions, particularly oxidative and postoxidative reactions are not well preserved in culture. So, this represents a serious drawback to using this primary cultures in long term *in vitro* experiments. Different strategies have been used to improve maintenance of the biotransformation system, introduction of hormonally defined media, supplemented media, ligands, heme

M.J.T. Carrondo et al. (eds.), Animal Cell Technology, 145–149.

precursors co-culture systems and culture of hepatocytes on extracellular matrices (1). In this work, was considered the effect of extracelular matrix where cells presents a three dimensional intercellular interaction. Different microcarriers were considered for immobilization of fresh isolated hepatocytes, in order to evaluate the influence of a extracelular matrix on the phenotype stabilisation and maintenance of the biotransformation capacity.

Material and Methods

Isolation of hepatocytes

Hepatocytes were isolated from male rats Wistar (250-350 g) using the perfusion technique described by Moldéus (2). The technique is based on liver perfusion with collagenase after removal of Ca^{2+} by perfusion with a chelator. All the solutions were bubbled with carbogen gas (95% O_2, 5% CO_2) during perfusion of the liver. The rats were anaesthetised with diethyl ether and the peritoneal cavity was opened by a midventral incision.

Culture medium, inoculum and conditions

Cells were suspended in DMEM/Ham's F12 (1:1) supplemented with 5% new-born bovine serum, 50 U/mL penicillin, 50 mg/mL streptomycin, 0.2% bovine serum albumin and 10^{-8} M insulin. Spinner-flasks (Corning) 50 mL with different microcarriers (0.1 g of Cytopore 2 (Pharmacia, Sweden) and Cultispher G (Percell Biolytica,Sweden) and 0.25 g of Cytodex 3 (Pharmacia, Sweden) and one without microcarrier were inoculated with 1.8×10^5 cells. Two 24 well plates were inoculated with 3.6×10^5 cells. Initially only half of the spinners were filled with the supplemented medium. After 2 h the medium was renewed. Spinners were filled until its maximum culture volume and stirred at 100 rpm. The medium was renewed daily.

Analytical Procedures

The cell number was counted by using the trypan blue exclusion in a Neubauer chamber. Reduction of MTT (3-(4,5-dimethylthiazol-2-yl)-2,5-diphenyl tetrazolium bromide) was used to evaluate the extension of the microcarriers colonisation by active cells (3). The metabolites (2α-OHT, 6α-OHT, 6β-OHT, 7α-OHT, 16α-OHT, 16β-OHT (OHT-hidroxitestosterone), and androstenodione) of testosterone were analysed by HPLC at 40°C on a reverse phase C_{18} column (5 mm, 250×4 mm I.D., Lichrospher RP18, Merck, Germany) preceded by a 2 cm LC-18 gard column. A gradient from 70% solvent A (methanol:water:acetonitrile, 39:60:1) to 90% solvent B (methanol:water:acetonitrile, 80:18:2) was operated over 45 min at 1.0 mL/min. The detection was at 254 nm. This method was adapted from (4). The standards were purchased from Steraloids (UK).

Results and Discussion

The metabolism of testosterone can be used as a marker for the activities of several cytochrome P450 subfamilies (2A, 2B, 2C and 3A). The comparison of the monolayer cell culture with three dimensional systems was done by following the testosterone metabolism profile of hepatocytes during several days. Table 1 shows the preliminary results of the testosterone metabolism.

Table 1- Testosterone metabolism profile of hepatocytes in different types of culture

Type of Culture	Days of culture			
	0.5	1	4	17
	6α-hydroxytestosterone			
Cytodex 3	11.2	-	N.A.	N.D.
Aggregates	25.8	N.A.	2.9	-
Cytopore 2	16.8	-	N.A.	N.D.
Cultispher G	13.9	2.6	N.A.	0.8
Plate	-	25.6	19.6	N.D.
	7α-hydroxytestosterone			
Cytodex 3	0.6	9.2	N.A.	N.D.
Aggregates	0.6	N.A.	1.4	-
Cytopore 2	0.7	1.3	N.A.	N.D.
Cultispher G	1.6	7.7	N.A.	23.0
Plate	0.5	1.5	2.8	N.D.
	16α-hydroxytestosterone			
Cytodex 3	-	-	N.A.	N.D.
Aggregates	53.9	N.A.	-	0.5
Cytopore 2	-	7.8	N.A.	N.D.
Cultispher G	-	2.1	N.A.	1.8
Plate	0.2	2.8	-	N.D.
	6β-hydroxytestosterone			
Cytodex 3	82.2	80.1	N.A.	N.D.
Aggregates	1.5	N.A.	87.7	13.7
Cytopore 2	75.8	74.5	N.A.	N.D.
Cultispher G	75.6	79.9	N.A.	-
Plate	21.3	53.3	58.7	N.D.
	16β-hydroxytestosterone			
Cytodex 3	1.2	2.0	N.A.	N.D.
Aggregates	2.4	N.A.	3.0	0.4
Cytopore 2	1.2	3.6	N.A.	N.D.
Cultispher G	1.7	0.6	N.A.	1.0
Plate	0.17	2.7	2.8	N.D.
	2α-hydroxytestosterone			
Cytodex 3	3.8	5.8	N.A.	N.D.
Aggregates	8.3	N.A.	-	2.7
Cytopore 2	4.4	8.2	N.A.	N.D.
Cultispher G	4.7	9.7	N.A.	1.4
Plate	3.5	11.2	14.0	N.D.
	Androstenedione			
Cytodex 3	1.0	2.8	N.A.	N.D.
Aggregates	7.5	N.A.	5.0	82.7
Cytopore 2	1.2	3.2	N.A.	N.D.
Cultispher G	2.5	-	N.A.	72.0
Plate	74.3	2.9	2.2	N.D.

Results are quantified as the percentage of total metabolites present
(N.D.- Not determined; N.A. Not analysed.)

As can be seen, after hepatocytes isolation the major P450 isoenzymes involved in the testosterone metabolism are still active. After 0.5 days of isolation, the hepatocytes predominantly form 6α and 6β-OHT reflecting the activities of cytochromes P450 2A1 and 3A1, respectively. After several days of isolation the major metabolite detected is 6β-OHT, and increases in monolayer cultures in opposition to the percentage of 6α that is decreasing. The 6α-OHT is still detected after 17 days of isolation in cells immobilised in Cultispher G and the 6β-OHT in aggregate cells.
From table 1 it can be observed that the metabolism to 7α and 16β-OHT, which reflects the activities of cytochromes P450 2A1, and 2B1/2B2, respectively, is very low, in isolated hepatocytes, these results are in agreement with those obtained by Donato *et al* (5). The same behaviour is observed with androstenedione that is the most stable activity, increasing his percentage with the time of culture.

Conclusions

The P450 isoenzymes involved in the metabolism of testosterone are still active after 17 days of hepatocyte isolation. The metabolites which reflect the activity of P450 isoenzyme 2B are detected in low quantities in all types of cell culture studied, whereas the metabolism to 6β -OHT is maintained in all types of culture.
With this preliminary results is not possible to choose the best type three dimensional cell culture system. However, the results indicate that this type of cell culture stabilise the phenotypic differentiation of isolated hepatocytes as in co-cultures (5).

References

1. MacDonald C, Vass M, Willet B, Scott A and Grant H (1994) Expression of liver functions in immortalised rat hepatocyte cell lines. Human Exp. Toxicol. 13, 439-444.
2. Moldéus P, Hogberg J and Orrenius S (1978) Isolation and use of liver cells. Methods in Enzymology, 5260-71.
3. Al-Rubeai M and Spier R (1992) Use of the MTT assay for the study of hybridoma cells in homogeneous and heterogeneous cultures. In: Spier RE, Griffiths JB, Stephenne J and Crooy P, Advances in Animal Cell Biology and Technology for Bioprocesses, pp 143-155, butterworths, London.
4. Reinerink EJM, Door L, Jansen EHJM and Van Iersel AAJ (1991) Measurements of enzyme activities of cytochrome P-450 isoenzymes by high-performance liquid chromatographic analysis of products. Journal of Chromatography, 553, 233-241.
5. Donato MT, Castell JV, and Gómez-Lechón MJ (1994) Cytochrome P450 activities in pure and co-culture rat heaptocytes. Effects of model inducers. In Vitro Cell Dev Biol 30A, 825-832.

Acnowledgements
The authors acknowledge JNICT, Junta Nacional de Investigação Científica e Tecnológica, by the financial support (Project Nº PBIC/C/BIO/2017/95 and Grant PRAXIS XXI/BPD/4153/94).

SESSION III

Animal Cell Vaccines - Present and Future

NEW EMERGING VIRUSES :
POTENTIAL RISK OF CONTAMINATION OF BIOLOGICAL PRODUCTS

J.F. SALUZZO
Pasteur Mérieux Sérums & Vaccins, Virology Division,
69280 Marcy l'Etoile, France

ABSTRACT

Following the AIDS pandemic, a conference sponsored by NIH took place, in 1989, in Washington D.C. at which the concept of emerging viruses was presented.
We analysed an example of the virus diversity in a tropical area. A long term study performed by the Institut Pasteur in Central African Republic revealed an extraordinary diversity of arboviruses and zoonotic viruses. Almost 30 new viruses (i.e described scientifically for the first time) were isolated during this epidemiological survey. The viral evolutionary mechanisms and factors of emergence will be assessed.
Equally, the potential for viral contamination of biological products for human use and particulary for vaccines will be discussed. Sources of viral contamination for vaccines produced on cell lines include : cell substrate, environmental factors, laboratory personnel, raw materials, cross contamination during campaigning. Possible solutions which limit the risk of viral contamination will be presented : origin of raw material, γ-irradiation, thermal and chemical virus inactivation, campaigning strategy.

INTRODUCTION

Following the AIDS pandemic, a conference sponsored by Rockefeller University, the NIH and the Forgaty International took place in May 1989 in Washington from which the concept of emerging viruses was presented.
The world-wide epidemic of HIV infection and the acquired immune deficiency syndrome is, of course, the major emerging viral threat. Other viruses may be introduced to humans in various ways, including by mosquitos and other arthropods, or from contact with rodents or other vertebrates. Arbovirologists have a large experience in the description of newly recognized viruses. The 1985 International Catalogue of arboviruses identified 504 viruses world wide, 124 of which have been associated with a disease in man. For this reason this scientist group became leaders in the study of emerging viruses.

VIRUS IN TROPICAL AREA

To assess the role of tropical area in the emergence of new viruses, I would like to summarize a virological investigation that we conducted at the Pasteur Institute in Bangui (Central African Republic CAR) from 1962 to 1993. The goal of the study was to perform an inventory of virus circulating in arthropods, birds and rodents collected in different areas. This ecological study was associated with extensive surveillance of arbovirus disease in man.

M.J.T. Carrondo et al. (eds.), Animal Cell Technology, 153–157.

A total of 919 viruses were isolated of which 84 different species belonged to 12 genera. Of those 36 were "new" viruses. The term "new" means "newly described". From man, 18 viruses were isolated mainly from febrile infection or exanthematic fever. Seven viruses isolated from man were "orphan" (no known human disease associated).

In the field of tropical virology, a group of viruses associated with hemorrhagic fever (HF) gained a particular interest. A total of 14 viruses were isolated world-wide. Some viruses like Ebola and Hantaan have become movie material. HF syndrome therefore is extremely well-defined. This can be attributed to two reasons : firstly the pioneering work of Dr Karl JOHNSON (CDC) and his collegues ; secondly HF is a syndrome which relatively easy to diagnose in tropical areas. The CAR programme concentrated on another syndrome, exanthematic fever (EF), which is more difficult to analyse but in fact very common. A total of 13 viruses were associated with EF. It is probable that a similar investigation, in any tropical area, conducted for example, on neurological syndromes, would lead to the isolation and idendification of numerous "new" viruses.

The experience gained in the CAR revealed :

- A huge diversity of viruses.
- Newly described viruses are very common.
- Some viruses are associated with human disease, but some are still orphan viruses.
- Intensive research on a syndrome will permit the isolation of new virus.
- And finally it is likely that emerging viruses are not newly evolved. Rather it appears that they already exist in nature.

I should like to make the point that, taking HIV as an example, research work in tropical areas has the tendency to concentrate on viruses associated with acute infection. The facility of diagnosis is probably the main reason for this emphasis. For HIV, probably of African origin, it is clear that the nature of the research during the virus' emergence (1970-1980) did not allow for identification of a virus associated with a chronic disease. It is noteworthy that international organisations continue to focus on viral-associated acute syndromes such as those caused by Ebola rather than those linked to chronic disease.

MECHANISM OF VIRUS EMERGENCE

Given such virus diversity as has been described in tropical areas, we need to take in consideration the potential variation of RNA viruses which operate in the process of emergence. In fact, it is remarkable that so few new diseases emerge. Apparently, genetic structured and environmental factors are important in virus evolution. Interestingly, the discovery of Hantavirus world-wide correlates with its recent isolation and development of adapted technology (ELISA, PCR) and is not related to the spread of "new" virus. Like other rodent-associated viruses, Hantaan is probably an old virus coevolving with its host, and the distribution of virus is strictly limited to the rodent activity. This explains the emergence of a huge diversity of Hantavirus topotype.

Taking in consideration the AIDS experience, the possibility for a virus to spread as a pandemic disease from a natural focus is correlated mainly with the biological properties of the virus. Viruses inducing an acute severe syndrome like Ebola and

Lassa, are easy to diagnose and sufficiently debilitating to limit the spread of the disease. In contrast viruses inducing a chronic infection, with a long incubation period will escape detection. In such cases, as for HIV, the main source of viral contamination remains blood.
In the past, viral transmission due to a vaccine product, have been attributed to viral contamination of cell substrates : yellow fever vaccine contaminated with avian leucosis virus and polio vaccine contaminated with SV40. These contaminations fortunately did not have adverse consequences. The use of human sera added as stabilizer as been a source of Hepatitis B transmission by yellow fever vaccine. It should be recognized that in the last three decades incidents linked to viral vaccines have become rare events.
The use of cell banks and their extensive characterisation offers a great security. In contrast the use of primary cells, particulary in the case of primate cells, such as monkey kidney, can represents a higher risk of viral contamination.
Raw materials including media components should also be taken into consideration when examining potential sources of viral contamination.
The main sources of viral contamination for vaccine is summarized in table 1.

Table 1 : Source of viral contamination

- Source material may be contamined with a indigeneous virus.
    - Cell bank
    - Virus seed lot
- Adventitious viruses may be introduced by the use of contaminated animal products in the production process.
    - Bovine sera
    - Media
    - Trypsin
- Failure of GMP.
    - Multi-use manufacturing facilities, with inadequate decontamination between campaigns.
    - Introduction of adventitious viruses during production.

PREVENTION OF VIRAL CONTAMINATION DURING PRODUCTION

Selecting and extensive screening of mammalian cells and viral seed with *in vivo* and *in vitro* assays for adventitious agents and endogenous viruses is the principal goal for minimizing the risk of contamination (table 2 summarizes classical tests to perform). This list is not exhaustive, industrial scientists should have the freedom to choose appropriate tests.

Table 2 : Prevention of viral contamination

Typical tests used to confirm absence of viral contamination.

- Antibody production : MAP, RAT, HAP and by extension, inoculation into monkey for the research of antibody against primate viruses.
- In vivo virus screen : inoculation into cells, eggs, animals.
- EM.
- Reverse transcriptase.
- Retrovirus infectivity.
- PCR : HIV, Hepatitis viruses, HTLV, HAV 6, EBV, CMV.

As stated above, table 3 summarizes the strategy to prevent viral contamination. We believe that strict selection, and vigorous supplier validation γ-irradiation with spiked validation of sera and more recently powder media constitute a comprehensive approach to reducing the potential risk of viral contamination, without adversed affection on the quality of the raw material itself.

Table 3 : Prevention of viral contamination : raw material

	Origin	Country	Treatment
Bovine serum	- Donor	- USA, New Zealand, Australia, Canada,	γ irradiation*
Trypsin	- Porcine	- USA	γ irradiation*
Powder media (MEM, BME, ISCOVE)	NS	NS	γ irradiation*

* 2.5 Mrad
NS : Non Specified

Finally we have also to take in consideration a viral contamination during the production following GMP failure. Particular attention must be paid during campaigning of different viruses to prevent cross contamination.
Table 4 summarizes critical success factors for the multi-use manufacturing facilities. And table 5 summarizes the viral decontamination obtained with polio and rabies virus. The importance of dessication in viral decontamination is noteworthy. Based on such figures a minimal time internal between campaigns in a given building can be determined.

Table 4 : Multi-use manufacturing facilities
(Prevention of cross contamination)

Critical success factors :

- Engineering control (closed system - fermentors)
- Material flow
- Personnal flow
- Dedicated equipment
- Environmental monitoring program
- Cleaning validation
- Training of personnal
- Changeover procedure

Table 5 : Viral decontamination

Step	Virus titer reduction (log_{10})	
	Polio virus	**Rabies**
Dessication (70 h)	4	6
Dessication + formolisation (formaldehyde)	≥ 7 (no virus detectable)	≥ 8 (no virus detectable)

CONCLUSION

The world's tropical zones represent the harbour for virus diversity. Virus emergence from these areas requires major ecological modification as was seen in the case of the HIV pandemic.

In my view in order to prevent such emergence, it is imperative that we study the potential modifications. In particular it is clear that tourism, particularly in the olderly population will logically represent a source of viral spread in the future.

For future prevention of pandemics our attention must turn to viruses associated with viral chronic infection with a long incubation period.

From a vaccine production viewpoint ours energies must concentrate on ensuring "viral-free" raw materials, cell banks and seed lots as a function of technology advances.

Key words : Emerging viruses, Central African Republic, Prevention, viral decontamination

DEVELOPMENT AND PRODUCTION OF NEW LIVE MEASLES VACCINE BASED ON HUMAN EMBRYO LUNG DIPLOID CELL CULTURE L-68

Nechaeva E.A., Kolokoltsova T.D., Jurchenko N.D., Kits I.V., Jilina N.., Sandakhchiev L.S.

Institute of Cell Cultures, State Research Center of Virology and Biotechnology "VECTOR",633159 Koltsovo, Novosibirsk Region, Russia

ABSTRACT

Measles dominates among child droplet infections in many countries. This disease is still an important health problem and is still the major killer in children. Immunization of all humans sensitive to this infection is the only radical measure in the struggle against measles. The quality of a vaccine is primarily determined by the properties of the virus strains, technology of production and the cell substrate. Live measles vaccine is produced at present in 15 countries of the world. Hen eggs from SPF farm and MRC-5 diploid cells of human embryo lung are used as a substrate. Production of live measles vaccine in primary cell cultures has a number of drawbacks caused by nonstandard character of the substrate and probability of contamination. The use of certified human diploid cells deposited in liquid nitrogen in sufficient quantities shows promise most of all. We have developed a new technology of live measles vaccine production by using the Leningrad -16 virus strain (L-16) on the basis of the L-68 diploid cell culture of human embryo lung.

INTRODUCTION

Measles dominates among child droplet infections in many countries. This disease is still an important health problem and is still the major killer in children. In several developing countries the mortality level of this infection reaches 10-32% in certain years. Now the immunization of all populations sensitive to this infection is the only radical measure in the struggle against measles.

As the quality of vaccines is first of all determined by the properties of the virus strain and the used substrate as well as the technology of production, the studies on evaluation of the possible use of a cell substrate with stable properties are rather topical.

M.J.T. Carrondo et al. (eds.), Animal Cell Technology, 159–164.

At present live measles vaccine (LMV) is produced in Russia on the basis of Japanese quail embryo fibroblasts (JQF) used as a cell substrate. At the same time difficulties associated with transportation and work with quail eggs, nonstandard character and genetical heterogeneity of the produced material are prerequisitives for the substrate substitution. In this case the use of a diploid cell culture having standard biological and genetical properties gives doubtless advantages. According to the WHO's classification, the diploid cells L-68, belong to the group of substrates of the least risk.

The main purpose of this work was development of a new live measles vaccine based on a human embryo lung diploid cell culture L-68, obtaining of experimental and production series of the vaccine and their certification according to the WHO requirements.

MATERIALS AND METHODS

Measles virus. In our country, the strain Leningrad-16 (or L-16) is used for production of measles vaccine. The strain was isolated in 1960 in the Leningrad Pasteur Institute of Epidemiology and Microbiology from the blood of measles-infected child and adapted on the primary cell culture of guinea pig kidneys. This strain is highly immune, causes seroconversion in 98-100 % of primarily vaccinated persons, and is less reactogenic [1].

Cell culture. The diploid cell culture L-68 was obtained from the lung of an 11-day-old human embryo [2].

Certification of the vaccine samples was carried out according to the physical and chemical properties, anaphylactogenicity, safety in laboratory animals and foreign microorganisms content according to the State Pharmacopoeia (3). Biological activity of measles virus was determined by cytopathic effect in Vero cell culture, the content of outside viruses, mycoplasmas, and mycobacteria was evaluated by the methods described in Pharmacopoeial Item (4). The control of residual neurovirulence and antigenicity was carried out in green monkeys after intracerebral inoculation of the vaccine at 2000-16000 TCE50 in accordance with (4); 10 experimental and 4 control monkeys were used in the experiment for each vaccine series. Humoral immune response to the vaccine inoculation at the dose of 2000-4000 TCE50 in guinea pigs and monkeys was evaluated in reaction of inhibition of hemagglutination (RIHA) with 4 AU of measles virus prepared according to Norrby method (5).

RESULTS AND DISCUSSION

In 1990 the World Health Organization carried out the comparative trials of measles vaccines made on the basis of the strains Edmonston-Zagreb, L-16, Schwarz, AIK-C strain in volunteers. The results prove that L-16 strain is comparable with the foreign ones in its antigenic activity [6]. In 1991 in our Institute the production strain of measles virus on the basis of L-16 vaccinal strain was obtained and certified . The official permission of the National Institute of Standardization and Control (NISC) for its use in production of measles vaccine was received. By now 4 working banks of measles virus L-16 strain have been thoroughly certified, including such parameter as neurovirulence in trials in primates.

Cell culture L-68 does not contain mycoplasmas, viruses and bacteria, is not tumorigenic and is highly receptive to the measles virus. Morphologically the cell culture is represented by typical fibroblast-like cells of stretched form with appendices on the poles which can be seen in Fig. 1. There were no differences in the ultrastructure or karyologic characteristics of the L-68 cells at passages 21 up to 45; cells had normal human karyotype.

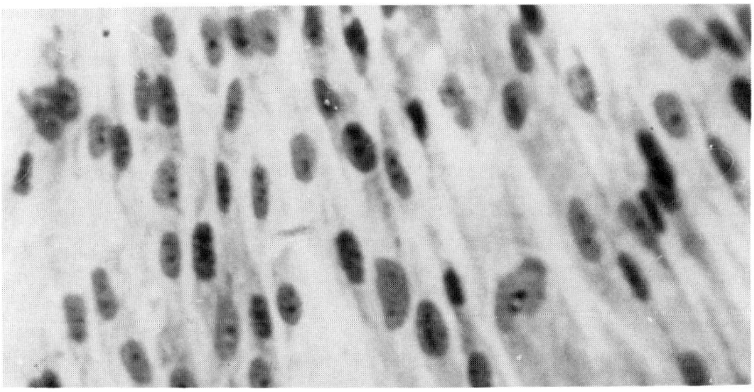

Fig.1 Morphology of human embryo lung diploid cell culture L-68. Hematoxylin-eosin stain.

According to the WHO's classification, L-68 diploid cells belong to the group of substrates of the least risk. By now the main and 3 working banks of L-68 cells have been obtained by our Center. Banks of cell cultures were certified and officially approved by the National Institute of Standardization and Control (NISC) for production of measles vaccine.

When L-68 cell culture is infected by the measles virus L-16, an active virogenesis accompanied by the development of the typical cytopathic effect takes place. Titer of the virus reaches the level of 5 to 5.7 lg of TCE 50 per half ml, which is consistent with the data of Mirchamsy et.al. on cultivation of different measles virus in MRC-5 cell culture [7].

Comparative study of antigenic properties of measles virus passaged in different cell substrates (L-68 cell culture and Japanese quail fibroblast) was carried out in reaction of inhibition of hemagglutination with guinea pig sera preimmunized with culture fluid containing L-16 measles virus at the doses of 2.68-6.2 lg TCE50/0.5 ml. 10 animals were used in each experiment. Table 1 presents the obtained results.

Table 1

Control of antigenic properties of L-16 measles virus passaged in different cell substrates

Viral material (passages in cell cultures)	Titer of initial preparation lgТЦД50/0,5мл	Serocon-version %	Titers in RIHA (inverse value)
1 passage in JQF	2,84	11,1	4
1 passage in L-68	4,2	22,2	8-16
5 passages in JQF	5,2-6,2	100,0	16-32
5 passages in L-68	4,3-5.5	100,0	16
10 passages in JQF	4,1	28,5	4-16
10 passages in L-68	4,5	62,5	8

Results on titers in RIHA demonstrate a high antigenic activity of new vaccine obtained after cultivation virus L-16 during 5 passages in L-68 cell culture.

To produce the vaccine the suspension of the certified L-68 cell culture was infected with L-16 vaccine strain of measles virus with the multiplicity of infection of 0,1-0,001 TCE50/cell and cultivated in roller bottles at 33-35^{0}C. From the 3-4th day post infection when the virus cytopathic effect on the cells began to occur, collection of virus containing fluid was performed each 24 hours followed by replacement by fresh Eagle MEM nutrient medium containing

10% of fetal calf serum and 0.3 g/l of glutamine. Sterile viral decantates containing not more than 15mkg/ml of protein and showing measles virus biological activity not less than 5.0 lg TCE50/0.5 ml were collected, sorbitol-gelatose stabilizer was added at the final concentration of 5% each, the mixture was poured into ampoules and liophilized for 48 hours.

Using L-68 cell culture we obtained the experimental series of the measles vaccine. These series passed the production control both in the Department of Biological Control of our Institute and in the National Institute of Standardization and Control in Moscow. Table 2 presents the obtained data on certification of the vaccine experimental series.

Table 2

Characteristics of experimental series of live measles vaccine prepared in L- 68 cell culture

Vaccine control parameter	Vaccine series number				
	1	2	3	4	5
Biological activity (lg TCE50/0,5 ml)	3,7	3,6	3,2	3,7	3,03
Protein content, mkg/ml	3,3	10,1	3,1	8,6	< 1.0
Residual moisture, %	1,5	1,0	1,1	1,8	1,7
Specificity	correspond				
Sterility	sterile				
Mycoplasmas present	not detected				
Mycobacteria present	not detected				
Foreign viruses present	-"-				
Anaphylactogenicity	nonanaphylactogenic				
Safety for guinea pigs and mice	safe				

The produced experimental series of a new vaccine do not contain mycoplasmas, outside viruses and bacteria, and are innocuous and non-anaphylactogenic for animals. Biological activity of the preparation amounts 3.0 to 3.7 lg of TCE 50 per half ml, the protein level does not exceed 2 micrograms per dose, and the quantity of the antibiotic canamycine monosulfate does not exceed 15 units per dose. The produced experimental series of live measles vaccine were specific, nonanaphylactogenic, safe, with normal level of residual moisture.

Control of residual neurovirulence of measles virus vaccine strain in the vaccine experimental series was performed in green monkeys by inoculation of the vaccine finished form at the dose of 4000 TCE50 and the liquid semifinished form at the dose of 16000 TCE50 to the talamic region of the brain. During the experiment the monkeys were under clinical observation, none of the animals displayed symptoms of affection of central nervous system. Histologic investigation of different regions of the brain and spinal cord (total of 1000 sections) showed the presence of feebly marked morphological changes in the inoculation zone a months after the vaccine inoculation. No pathological reactions were found in all the other studied regions of the brain (marrow).Thus, experimental series of the vaccine at the dose of 4000 TCE and liquid semifinished product at the dose of 16000 TCE possess no neurovirulent properties. Control of the monkeys' sera in RIHA with measles hemagglutinating antigen at the dose of 4 AU detected the titer of antibodies of 1/16-1/64.

CONCLUSION

1. A technology for production of live measles vaccine with standard biological and genetical properties for parenteral administration based on L-68 diploid cell culture and virus strain L-16 has been developed.

2. Experimental and production series of the vaccine produced by the present technology correspond to the national requirements and are not inferior to the commercial analogue in the studied parameters.

REFERENCE

1. Taros L.Ju. (1967) Experience of isolation of measles virus on the monolayer guinea pigs kidney cell cultures and on the chiken fibroblasts. Voprosi virusologii 4,399-402,The Russia.
2. Stepanova, L.G., Alekseev, S.B., Zgursky, A.A.(1986) Isolation and characterization of a new strain of diploid cells from human embrionic lung tissue.Tsitologiya 28(12),1373-1376. The Russia.
3. State Pharmacopeia of the USSR (1987), 11 ed. The Russia.
4. TPC 42-179 VS-94 (1994) "Live dry culture measles vaccine".The Russia.
5. Norrby E.,Gollmar Y.(1975) Identification of measles virus-specific hemolysis-inhibiting antibodies separate from hemagglutinin-inhibiting antibodies.Infection and Immunity 11,231-239.
6. Markowitz, L.E., Sepulwede, J., Diaz-Ortega, et al.(1990) New Engl. J.Med. 322(2),580-582.
7. Mirchamsy H.,Shafyi A.,Bahrami S. et al.(1974) A comparative field trial of five measles vaccines produced in human diploid cell MRC-5.J.Biol.Standardization 5(1),1-182.

Development of influenza subunit vaccine produced using mammalian cell culture technology

R. Brands, G.J.M. van Scharrenburg, A.M. Palache
Solvay Duphar, P.O.Box 900, Weesp, The Netherlands

We developed MDCK-derived Influvac, a tissue culture based influenza inactivated subunit vaccine using serum free culture conditions. Extensive safety testing of the cell banks and the production process was performed by Microbiological Associates Inc. The medium EpiSerf supports both cell and virus propagation, thereby increasing overall efficiency, and compares to culture performances with FCS-containing medium at all scales applied up to production scale fermenters. All recommended WHO influenza A (H_1N_1 and H_3N_2) and B viruses from the last 4 years, deriving from egg-adapted production strains, thus were propagated successfully on MDCK cells. Since human field strains thus far can be cultured directly on MDCK cells, we expect that the tissue culture based technology reported here will be applicable as well in the case the WHO enables direct production of selected field strains, thereby reducing overall production time per dose. The production system involves expansion of cells anchored on microcarriers using stirred fermenters followed by virus infection. Total cycle time is about 5 days per batch. The down stream processing compares to that used for egg-based Influvac and involves concentration-, ion exchange chromatography-, inactivation-, ultrafiltration-, solubilisation-, (ultra)centrifugation-, homogenisation- and sterile filtration steps. The final product obtained after down stream processing is similar to the currently produced egg-derived Influvac both with regard to immunogenicity and reactogenicity and lack of toxicity, as has been confirmed by data from several animal and human clinical studies.

Methods and Validation

Host cell line: MDCK
The subclone from the ATCC-derived MDCK host cell line (ATCC CCL34) was found to be suitable for the production of all influenza viruses that were in production over the last five years. Master, working and extended cell banks were generated by Microbiological Associates (USA) and Solvay Duphar, in which the extended cell bank

M.J.T. Carrondo et al. (eds.), Animal Cell Technology, 165–167.

was generated on production-like conditions using bioreactors and microcarriers during 40 continuous passages. Characterisation and validation of the MDCK clone was carried out by MA (Stirling, Scotland) under Eur Pharmacopeae and FDA regulations. Thus tests were performed establishing the lack of mycoplasmal, bacterial or fungal contamination and the identity of the cell line by isoenzyme and genetic analysis to be of canine origin. General safety testing was performed establishing the absence of endogenous and adventitious virus and the absence of retroviral inclusions using reverse transcriptase assays and electron microscopy. Tumorigenicity testing was performed on nude immunodeficient mice and resulted in palpable nodular growth at the site of inoculation only. When cells were injected into immunocompetent animals no growth was evident as expected. Therefore it was concluded that the investigated Duphar-MDCK cell line is to be regarded as being safe.

Culture Medium

The anchorage dependent host cell is cultured on serum free medium, based on the commercially available EpiSerf medium (Gibco Life Technologies, Scotland). The medium supports the whole range of upscaling from TC-flask to bioreactor.

Seed Virus

Influenza working seed virus is obtained by propagation of yearly regular production strains on MDCK. To date all virus strains will propagate on MDCK with minimal adaptation.

Down Stream Process and validation

The DSP compares to large extent with the current egg-derived influenza vaccine and includes concentration-, ion exchange chromatography-, inactivation-, ultrafiltration-, solubilisation-, (ultra)centrifugation-, homogenisation- and sterile filtration steps.
Validation of the DSP for viral clearance was carried out at MA (Scotland) and demonstrated efficient clearance of a wide range of small and large viruses of RNA and DNA nature.

Clinical trials

The final product, an inactivated influenza subunit vaccine, has proven to be safe and efficuous in animal (tox) studies. The data presented enabled permission for clinical trials in France and the UK from the respective authorities. The human clinical phase I studies in which the regular egg-derived vaccine was compared to the MDCK derived vaccine, confirmed the safety of the MDCK-derived vaccine. In the phase II clinical trials and in a similar comparative set-up, the vaccine was tested on elderly subjects and again proves to be safe and similar to the egg-derived product with respect to immunogenic parameters.

Conclusion

The method used enables us to produce at large scale and in a cost effective manner the MDCK-derived Influvac. Clinical data in both phase I and in phase II studies indicate that the MDCK-derived vaccine is safe, non reactogenic and can be compared in effectiveness to the regular egg-derived inactivated subunit vaccine.

References

Palache AM, Brands R, Van Scharrenburg GJM, (1996) Safety, immunogenicity and reactogenicity of MDCK-derived influenza subunit vaccine in healthy volunteers. Abstract. Options for the control of influenza III, p 71

Brands R, Palache AM, Van Scharrenburg GJM, Immunogenicity and reactogenicity of influenza subunit vaccines produced either in Madin Darby Canine Kidney (MDCK) - cells or in fertilised chicken eggs. (1996, submitted for publication)

Monitoring and control of large - scale vaccine production via dynamic oxygen uptake rate measurement

ANJA PREISSMANN, ROLF G. WERNER, WOLFGANG NOÉ
Dr. Karl Thomae GmbH, Biberach / Riss, Germany

Abstract

For large - scale production of BVD (Bovine Viral Diarrhea) antigen on adherent MDBK target cells we have established a 300 L airlift reactor with stainless steel wire springs as carrier material.

To improve economy, reliability and reproducibility of large-scale virus production, relevant information on cell number and physiological status of the cells is essential. Since direct growth monitoring of adherent cells in bioreactors by sampling is impossible, effective and simple *indirect* methods such as enzymatic assays were tested. Although they showed good results in small scale experiments, the applicability on 300 L scale was limited.

Dynamic measurement of the Oxygen Uptake Rate (OUR) has several advantages: it is simple, reliable and inexpensive. The method proved feasible for small scale cultivation of MDBK target cells: hemocytometer counts correlated very closely with calculated cell numbers. Thus, OUR measurement is an helpful tool for growth monitoring of adherent cells on 300 L scale. Moreover, characterizing the increased oxygen demand and metabolic activity during the virus phase of the production process should be possible.

Introduction

Since economy, reliability and reproducibility of large-scale virus production are crucial to the animal health business, an efficient vaccine production process is to meet three requirements:

- optimal conditions for a fast and effective proliferation of the target cell line during the growth phase: a sufficient cell density is an important prerequisite for successful virus seeding and high product yield.
- optimal virus seed timing and
- optimal virus harvest strategy

M.J.T. Carrondo et al. (eds.), Animal Cell Technology, 169–174.

To investigate each of these aspects, relevant information on the cell number and the physiological status of the cells is essential. Without this information, it is impossible to assess the performance of the process and to control the culture parameters.
The main problem involved in such large-scale production systems is gaining access to the cells *inside* the bioreactor. Since most of the vaccine processes deal with adherent cell lines, a simple cell count by sampling the cell culture fluid is not possible.

Material and Methods

Bioreactor setup and analytical methods

For the large scale production of BVD antigen on adherent MDBK target cells we have successfully constructed a 300 l airlift reactor with stainless steel wire springs (6 x 6 x 0.6 mm) as carrier material (Fig. 1).

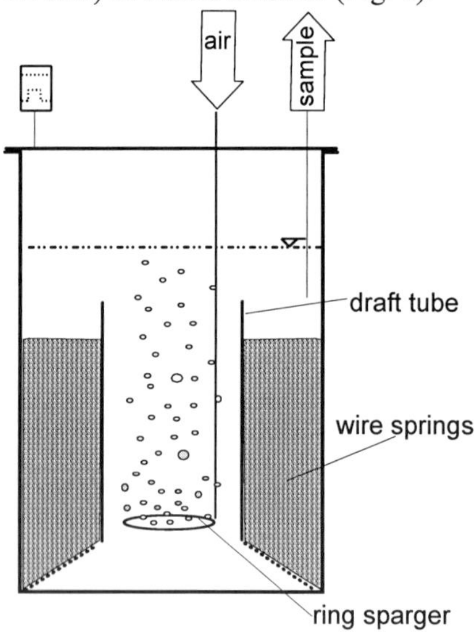

Fig. 1: Schematic drawing of the 300 l airlift reactor for large scale vaccine production

We developed an internal loop system equipped with a concentric draft tube. Located in the annulus are 150 L of carrier. These carriers are the supporting matrix for the growth of the MDBK cells. For off-line analysis we sampled the cell - free supernatant at least once a day.
Glucose and lactate concentrations were automatically monitored with a YSI 2700 system (Yellow Springs Instruments, Yellow Springs, USA). Ammonia, lactate dehydrogenase (LDH), aspartate aminotransferase (AST), acidic phosphatase (AP) and glutamine were determined by enzymatic test-kits (Boehringer Mannheim, Mannheim, Germany). Aminoacid analysis was performed by means of HPLC (OPA method).
For the investigation of Specific OUR (SOUR), a 250 mL mini airlift reactor - an exact scale-down model of the 300 L scale - was used.
To determine the cell number via hemocytometer, the medium in the small scale airlift reactor (250 mL) was replaced by the equivalent volume of trypsine solution (2.5 % in PBS, 37°C). During an incubation of 1 - 6 h the solution was agitated vigorously to suspend the cells. Samples were counted periodically ensuring a quantitative cell detection. After staining with trypan blue the samples were counted optically with a hemocytometer.

Dynamic Oxygen Uptake Rate measurements

There are various methods to monitor the OUR of a culture. A balance of gas phase or fluid phase is possible (1). The gas balance requires a very sensitive exhaust analysis and is therefore quite complicated (2).
The dynamic measurement only requires a standard oxygen probe. The gas flow is switched off and the linear decrease in DO is monitored. This decrease is caused by the oxygen consumption of the cells. The measurement was stopped at a DO concentration of 20 % in order to prevent an oxygen limitation of the cells. The critical concentration of DO was determined at 8 % for MDBK cells.
During measurement the airlift reactor was properly mixed via a marine impeller. The stirrer speed was set accordingly to maintain adequate mixing characteristics. Measurements were performed at least once a day at an interval up to 2 h.

Results and discussion

Enzymatic Assays

To determine the lethal cell number, we measured the concentration of certain enzymes in the supernatant. We investigated lactate dehydrogenase (LDH), aspartate transferase (AST) and acidic phosphatase (AP), all of which are located in the cytoplasm and the organellae (3). The concentrations of these enzymes correlate well with the number of lethal cells as shown in Figure 2.

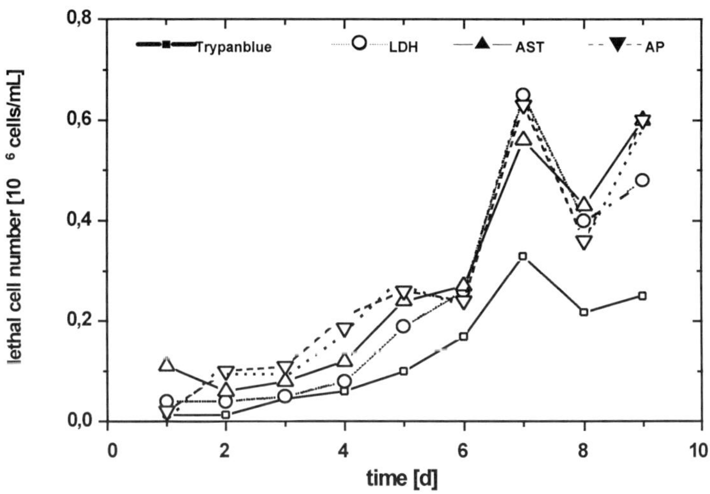

Fig. 2: Comparison of lethal cell numbers determined by enzymatic assays and trypan blue count

With the onset of cell deterioration, its membrane disintegrates and releases intracellular enzymes (4, 5). The trypan blue method, in comparison, results in lower values because optical counting does not detect cells which have already completely disintegrated.

Although good results for all assays were achieved in small scale experiments, in the large-scale airlift system only LDH measurement was successful. The signal for AST and AP was too low for reliable results. The intracellular concentration of these enzymes is not as high as that of the LDH.
However, with the help of the LDH assay we are able to determine the lethal cell number. Knowing this number is important, but does not permit any conclusion on the actual vital biomass concentration. Therefore, an additional method was needed.

Oxygen Uptake Rate monitoring

In case of the vital cell number we established the Oxygen Uptake Rate monitoring. The oxygen consumption correlates well with the cell number (6) and is a sensitive signal for process control (7, 8). We chose the dynamic measurement because it is simple, inexpensive and reliable. Its results correspond well with more complicated methods such as mass spectrometric analysis of the exhaust (9). The dynamic method is very easy to carry out and needs little additional equipment since it works with the standard oxygen probe. Of course, the airlift reactor requires a stirrer to maintain proper mixing while the gas flow is switched off.
Another advantage of the method is, that a direct assessment of cells is superfluous. This is extremely helpful after virus infection for reasons of containment. Furthermore, this method is easily transferrable to other cell lines and bioreactor configurations.

Essential for correlating the OUR signal and the vital cell number is the specific OUR of the cells. In order to ensure the applicability of the method, we closely examined the kinetics of the SOUR for the MDBK cells.

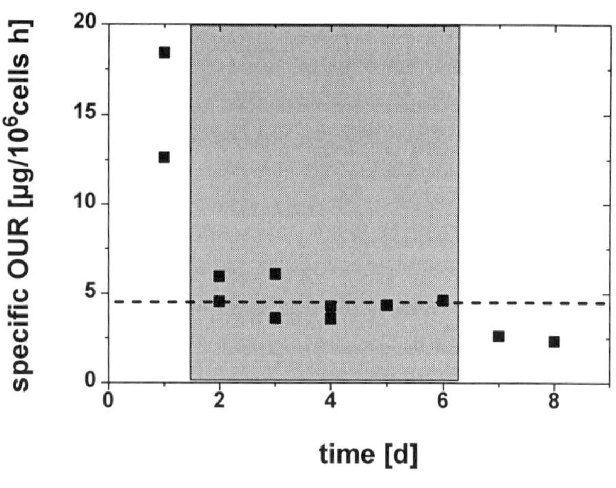

Fig. 3: Kinetics of specific OUR for MDBK cells

Vital for a *linear* correlation of OUR and cell number is a constant specific oxygen demand of the cells. The data corroborated this for most of the cultivation period. As seen in Figure 3, the oxygen demand during the adaptation phase is very high after inoculation. Afterwards the values remain constant.
A medium limitation on day 6 caused a decrease in the SOUR. Since there is no medium limitation during the growth phase of the large-scale process, we can very well asume a

constant SOUR value and a reliable correlation between OUR signal and cell number. The cell metabolism does not change significantly since there is no variation in the glutamine/glucose ratio. The medium composition is simple and there is no feed involved. Miller et al reported similar results (10). Small scale experiments proved the applicability of the method for the cultivation of MDBK target cells. Figure 4 shows the very close linear correlation of cell numbers determined by hemocytometer count and OUR measurement.

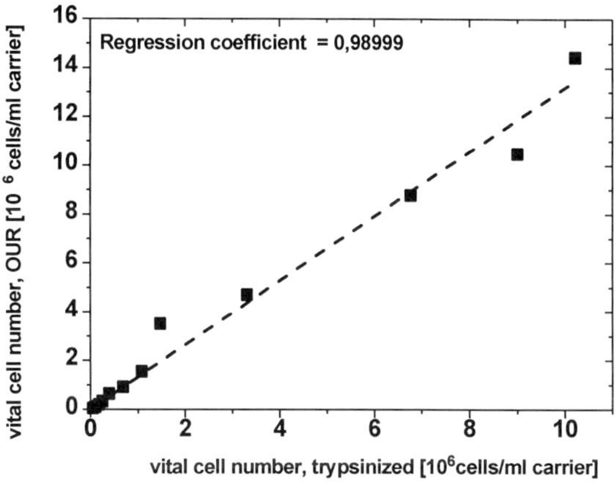

Fig. 4: Linear correlation of cell numbers determined by hemocytometer count and OUR measurement

Thus, the OUR measurement is an extremely helpful tool for growth monitoring of adherent target cells on the 300 l scale. It helped solve many questions concerning process optimization during process development.

We were able to assess the effect of medium formulation as well as the impact of different inoculation procedures on the growth rate of cells. Figure 5 shows that different inoculation procedures result in different growth rates. A reduced medium formulation has the same effect. In addition, the LDH assay enabled us to control the viability

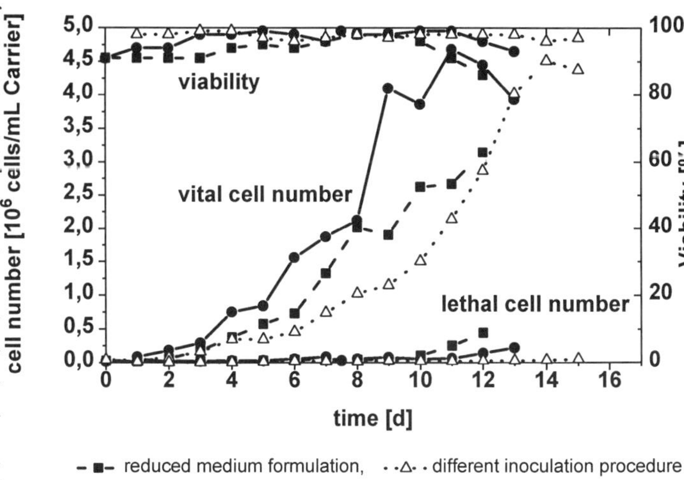

Fig. 5: Monitoring of different 300 L fermentations via dynamic OUR measurement

of the culture.
Such information makes it possible to optimize the virus seed timing and is also useful for characterizing the virus phase of the production process. This, of course, will lead to the improvement of the harvest procedure and the product yield. First results support this assumption.

Conclusion

The dynamic OUR method is an *effective and reliable indirect method* for determining the vital cell number in large-scale processes using adherent cell lines.
With this information, it is possible to optimize three of the main targets in large-scale vaccine production:
1. effective proliferation of the target cell line
2. optimal virus seed timing
3. appropriate schedule for harvesting procedures

In other words, the dynamic OUR measurement helps us to understand and control the process parameters vital for high product yield of reliable and reproducible quality.

References

1. Yoon, S.; Konstantinov, K.B. (1994) Continuous, real-time monitoring of the oxygen uptake rate (OUR) in animal cell bioreactors. *Biotech. & Bioeng.* **44** 983 - 990.
2. Oeggerli, A.; Eyer, K.; Heinzle, E. (1995) On-line gas analysis in animal cell cultivation: I. Control of dissolved oxygen and pH. *Biotech. & Bioeng.* **45** 42 - 53.
3. Bergmeyer, H.U. (1974) Methoden der enzymatischen Analyse" Band 1 3.Auflage, Verlag Chemie Weinheim
4. Renner, W.A.; Jordan, M.; Eppenberger, H.M.; Leist, C. (1993) Cell-cell adhesion and aggregation: influence on the growth behaviour of CHO cells. *Biotechnol. & Bioeng.* **41** 188 - 193.
5. Wagner, A.; Marc, A.; Engasser, J.M.; Einsele, A. (1991) The use of Lactate Dehydrogenase (LDH) release kinetics for the evaluation of death and growth of mammalian cells in perfusion reactors. *Biotechnol. & Bioeng.* **39** 320 - 326.
6. Bonarius, H.P.J.; Gooijer, C.D. de; Tramper, J.; Schmid, G. (1995) Determination of the respiration quotient in mammalian cell culture in bicarbonate buffered media. *Biotech. & Bioeng.* **45** 524 - 535.
7. Behrendt, U.; Koch, S.; Gooch, D.; Steegmans, U.; Comer, M. (1994) Mass spectrometry: A tool for on-line monitoring of animal cell cultures. *Cytotechnology* **14** 157 - 165.
8. Lovrecz, G.; Gray, P. (1994) Use of on-line gas analysis to monitor recombinant mammalian cell cultures. *Cytotechnology* **14** 167 - 175.
9. Eyer, K.; Oeggerli, A.; Heinzle, E. (1995) On-line gas analysis in animal cell cultivation: II. Methods for oxygen uptake rate estimation and its application to controlled feeding of glutamine. *Biotech. & Bioeng.* **45** 54 - 62.
10. Miller, W.M.; Wilke, C.R.; Blanch, H.W. (1987) Effects of dissolved oxygen concentration on Hybridoma growth and metabolism in continuous culture. *J. of cellular physiology* **132** 524 - 530.

Acknowledgements

The authors would like to thank all members of the Cell-Technology Group of the Department of Biotechnical Production at Dr. Karl Thomae GmbH for their assistance.

Reactor Development for the Hepatitis A Vaccine VAQTA™

* J.G. Aunins, *T.A. Bibila, †S. Gatchalian, †G.R. Hunt, †B.H. Junker, *J.A. Lewis, †P. Licari, §K. Ramasubramanyan, †C.S. Ranucci, †T.C. Seamans, *D.B. Seifert, †W. Zhou, *J. Waterbury, †B.C. Buckland

Merck Research Labs, *West Point, PA, 19486 and †Rahway, NJ, 07065
§Merck Manufacturing Division, West Point, PA, 19486

Keywords: Hepatitis A virus, bioreactor, static mixer, microcarrier, nunc cell factory, cell cube.

Abstract

Reactor choice for virus propagation is dependent not only on technological merits but also on manufacturing convenience, regulatory considerations, and process development organization culture. The different reactors selected for Hepatitis A Virus (HAV) cultivation and the issues which drove the choices over the course of development for VAQTA™, an attenuated inactivated HAV vaccine, are discussed.

Introduction

Hepatitis A virus (HAV) is a picornavirus: 28 nm dia, non-enveloped, with a single positive stranded RNA genome. Although the three major capsid proteins are not particularly complex - for example, they are not glycosylated - the immunodominant epitope on the capsid surface involves either domains from multiple capsid proteins or capsid protein conformations specific to the assembled virus. Accordingly, current vaccines to HAV are made of whole virus particles that have been formaldehyde inactivated and adsorbed to an adjuvant (usually an aluminum salt). Since the capsid is constructed of multiple proteins, recombinant DNA approaches to obtain virus-like particles are currently unattractive and viral particles are most conveniently obtained by infection of animal cell cultures with live virus.

The CR326F strain of HAV (1) produces a persistent rather than lytic infection in MRC-5 cells (human male embryonic lung diploid fibroblast) under the cultivation conditions used for the attenuated variant employed in VAQTA™ manufacture. HAV strains in general are non-lytic in *in vitro* culture, probably due to the fact that the viral 2A protease is genetically distinct from other picornavirus genera and is apparently lacking similar activity in HAV (2). In other cytolytic picornaviruses, this protease interferes with the host cell mRNA cap binding complex, inhibiting translation and diverting the host protein synthesis machinery to translation of viral proteins. HAV thus has to compete for cellular resources and possibly as a consequence of this requires long cultivation times *in vitro* to accumulate significant amounts of antigen. Since the infection is not cytolytic, the majority of viral antigen is found intracellularly and must be released from the cells to construct a vaccine.

The relatively low productivity of HAV CR326 in MRC-5 cells creates challenges for the process designer in terms of reactor choice. Large growth surfaces must be provided for the large numbers of the strictly anchorage dependent MRC-5 cells which must be cultivated. The surface must be amenable to cell attachment at temperatures below 37° C that are used to grow attenuated strains. Since the infection occurs over the course of weeks, the cells must be nourished appropriately during that time, and the growth surface will preferably be arranged to discourage 'biofouling' or overgrowth of

M.J.T. Carrondo et al. (eds.), Animal Cell Technology, 175–183.

cells in regions of the reactor. Other considerations are ease of manufacturing use and minimization of opportunities for contamination by adventitious agents.

Phase I/II Reactor Development

Reactor choice for Phase I/IIa clinical development of a vaccine or therapeutic is driven by the need to rapidly conduct proof-of-principle studies. The reactor technology chosen should at a minimum allow rapid production of enough virus to support initial clinical studies . Changes in reactors and processes between Phase I and IIa studies are not so critical from a regulatory perspective. However, if possible the technology should be translatable, that is it should resemble the reactors envisioned for full scale production of the vaccine, or allow translation of critical process features to these reactors to minimize changes in product characteristics.

For VAQTA™, roller bottles were our choice to start a clinical development program. They are an off-the-shelf technology used for a number of other successful diploid cell derived vaccines such as MERUVAX® (Rubella vaccine, live) and VARIVAX® (Varicella vaccine, live), and they allow access to the cell sheet for virus harvesting, in this case by scraping the cells from the surface. Despite their early utility, roller bottles were not optimal for large clinical programs since the number of doses cultivatable per bottle was small and large amounts of labor were required to support expanded clinical studies.

For Phase IIa and IIb studies, a more economical alternative was sought, preferably one that fulfilled the goals for manufacturing translatability. Nunc Cell Factories (NCFs) were found to be a convenient technology for this purpose. In order to employ NCFs, the obstacle of cell sheet inaccessibility to mechanical cell harvest needed to be overcome. This was addressed by developing a TRITON X-100 detergent cell lysis method. The Triton detergent solubilizes the cell membranes and cytoplasmic protein fraction to release HAV (which is not affected by the detergent) but leaves the cytoskeleton and nuclei intact, resulting in a 'cleaner' antigen stream entering purification relative to the scraper harvested and freeze-thaw lysed Phase I material. Since NCFs are much larger than roller bottles, 6000 or 24000 cm^2 versus 850 cm^2, far fewer vessels are required to generate the requisite quantity of vaccine. In principle, the technology is also translatable to manufacturing. Thus, NCF technology was employed until fairly late in the development cycle when it became apparent that alternative reactor technologies should be examined.

Phase III/Manufacturing

The considerations for Phase III and manufacturing reactor technologies differ from Phase I/II. While expedience is still important, it must be balanced against the need for manufacturing economy, consistency, monitoring capability and controllability. It is also essential to employ technology which will best meet current and future GMP requirements. As the requirements for HAV antigen mass became more apparent through VAQTA™ clinical development, it was recognized that there were alternatives to NCFs that might prove more desirable in a manufacturing setting due to the above considerations. Several reactor types - NCFs, microcarriers, static mixer reactors (fixed bed) and CellCubes (fixed bed, Corning Costar Corp.) were investigated and are described below.

Nunc Cell Factories

NCFs were thoroughly scrutinized for their utility in manufacturing as they were potentially advantageous from several standpoints. They had been shown highly useful for HAV propagation in our own clinical efforts and elsewhere (3), and they were known to be scaleable and had been used for other licensed biologicals. In addition, highly successful Phase IIb efficacy studies (4) had been conducted using NCFs, providing an impetus to remain with this technology to avoid clinical bridging study requirements.

Balancing these positive aspects, development work had shown that modifying the cell microenvironment and nutrient feeding protocols produced two to three-fold increases in antigen productivity. These process changes could not be conveniently incorporated into the existing NCF protocol in a manufacturing setting as they would have increased manipulation of the factories significantly. At manufacturing scale, hundreds of aseptic manipulations would be required, and this was seen as undesirable from manufacturing sterility risk and regulatory perspectives. The trend for biologicals is towards closed system processing which minimizes opportunities for GMP breakdowns which could compromise product safety, *e.g.* - extraneous agent contamination.

In addition to the above considerations, VAQTA™ employs a robust purification process that removes host cell protein and nucleic acid contaminants to extremely low levels (5). As a consequence of the ability of the purification train to accept different detergent-lysed feedstocks, the reactor technology was not deemed critical to maintaining final product similarity (although this was examined extensively) provided the culture environment encouraged healthy cell growth and similar virus propagation. These factors moderated the desire for constancy between Phase IIb and Phase III cultivation processes and led to investigation of other reactor options that could readily incorporate productivity improvements.

Microcarrier Culture

Like NCFs, microcarriers had been reported useful for HAV propagation (6) and are employed at very large scale for the production of poliomyelitis virus (7), another picornavirus grown on MRC-5 cells for vaccine use. The potential technical virtues of microcarriers - scaleability, a homogeneous cell environment, the ability to be used in closed clean-in-place and sterilize-in-place stirred tanks, the ability to sample cells and monitor culture behavior - have been extolled for years.

MRC-5 cells grow well on conventional Cytodex microcarriers (Pharmacia Upjohn) at low microcarrier concentrations (1 - 2 g/l) and for short times leading up to microcarrier confluence. However, as surface confluence was reached in long term cultures, cell-cell adhesion between microcarriers caused microcarrier aggregates to form, often up to several millimeters in size (8). Agitation adjustment was not found useful to prevent this phenomenon. As aggregates became larger, the cell monolayer retracted from the outermost microcarriers and expelled the bare microcarriers. This process of cell sheet contraction continued through the long infection at 32° C so that at the end of culture, microscopic observation with fluorescent dyes (fluorescein diacetate and propidium iodide) showed a mixture of non-viable cell spheroids detached from the micocarriers and microcarrier clusters with the innermost cells necrotic, presumably due to nutrient starvation. By screening a large number of microcarriers, glass coated microcarriers were identified (9) which allowed the formation of stable aggregates that did not disintegrate during infection and which allowed HAV production. However, within the scope of our development, we could not maintain a monodisperse microcarrier suspension for the duration of the culture, and the productivity of the glass aggregate microcarrier process was approximately half that of static cultures under similar conditions. Although the cultures responded to process improvements, relative productivity was always less than static cultures.

As a route to rapid and economical manufacture, microcarrier culture of HAV CR326F did not possess some of the attributes which are present for other virus cultivations. Technically, cell growth was difficult to achieve and environmental homogeneity was clearly not present. Moreover, microcarrier aggregation lead to concerns about sensitivity to scale-up and robustness in manufacturing. From a project management standpoint, considerable development effort would have had to be invested in process scale up and validation activities when the same resources could have been better spent improving culture productivity. This decreased the attractiveness of microcarriers relative to simpler reactor technologies. This situation was unique to the vaccine and to Merck's development and manufacturing infrastructure; a simpler system, a longer opportunity for development, or a process development organization possessing a long history of microcarrier use might have already accomplished many of the time and resource consuming scale-up and technology transfer tasks that posed a barrier to this project. Nevertheless, corporate experience or culture is a factor which must be considered when examining reactor technology for any vaccine project.

Static Mixer Reactors

Static (or motionless) mixers are devices used in the chemical process industries to mix viscous fluids or rapidly reacting streams within pipes in a plug flow manner. Although mixing was the original application of these devices, static mixers can be constructed with surface areas per volume greater than 5 - 10 g/L Cytodex microcarrier cultures. Static Mixer Reactor (SMR) use for animal cell and virus culture was invented at Merck in the mid-1970's (10), although the reactor has not been used commercially due to patent protection. SMR's are essentially fixed bed reactors which can be used in an internal or external loop reactor configuration.

The SMR fixed bed reactor possesses two major advantages over other fixed bed reactors due to the mixer construction. First, a uniform packed bed of ridged plates is created that disallows the fluid channeling which can potentially occur with random packed bed devices (11, 12, 13). Second, the growth surface is essentially flat on the scale of a cell, and the plates only touch each other at a few small points where the ridges meet. These features dramatically reduce the opportunity for biofouling and subsequent starvation of the cell sheet that can occur with other fixed bed reactors (13). With any packed bed reactor, fluid flows through parallel paths in the bed, and if a path of least resistance exists channeling will occur. This extra flow of nutrients encourages vigorous cell growth which in turn can choke off the flow in a bio-fouling reactor type, leading to starvation of the innermost cells and sloughing of the entire cell sheet from the attachment surface. Having a flat surface with no opportunity for cell bridging allows control of the cell sheet thickness and hence assurance that the cells proximal to the attachment surface remain healthy. When operated at circulation loop superficial velocities appropriate to supply oxygen to the culture (0.02 - 0.2 cm/s for the scales under consideration here), flow is laminar and uniform throughout the SMR, and approximate calculations of the surface shear stresses show they are substantially lower ($\tau_{wall} \approx 0.01 - 0.10$ dyne/cm^2) than those known to remove cells from growth surfaces or alter cell behavior ($\tau_{wall} \approx 1 - 10$ dyne/cm^2).

HAV culture on static mixers was found to be equivalent to NCF reactors under similar conditions (14), and the bioreactor configuration allowed the productivity-improving process modifications mentioned above to be practically implemented in manufacturing. Operation of fixed bed reactors is simple, the main process issues being inocculating the reactor growth surface uniformly and ensuring nutrient supply by appropriate superficial velocity (14). These issues were easily addressed, and the SMR was thus a very technically attractive reactor for manufacturing use. In contrast to microcarrier culture, the reactor was successful at matching static culture performance in virtually the first experiment, and required much less development manpower.

As with the other reactors above, SMRs possessed situation-specific drawbacks that also made them undesirable for rapid factory implementation for HAV. The main factor was the need to document cleanability of the mixer elements. Although some static mixer types are used in sanitary applications in the food industry, the Koch/Sulzer "SMX" design that enables high surface area per volume is not amenable to (non-destructive) surface swab testing on the innermost plates, and such testing would need to be performed for cleaning validation. Documenting cleaning for the reusable 316L stainless steel or titanium mixer elements required destructive testing and hence expensive and long-lead custom element fabrication. In addition to this constraint, the process was to be retrofitted into an existing facility which lacked clean steam and other utilities required for a skid mounted reactor, such that this option was capital intensive and required long lead times for fabrication, installation, qualification and validation.

CellCube Reactors

At the time that SMRs were being considered for use with VAQTA™, the first generation Corning-Costar CellCube (CCC) reactors were being introduced. The reactor growth surface is composed of polystyrene plates spaced *ca.* 1 mm apart (Fig. 1). Medium flow enters the front of the reactor at one corner into a header, continues through flow-controlling orifices and across the plates, and exits the reactor through another header at the corner opposite the entrance. Despite an odd diverging-converging geometry for the flow, this device possessed many of the attractive attributes of the above reactors. A simple fixed bed loop reactor, it was amenable to manufacturing operability and process flexibility, monitoring and control. Although some aseptic connections are necessary to assemble the reactor, these were reduced to *ca.* twenty versus the hundreds needed for NCF operation. The disposable reactor element did not require validation other than auditing Corning-Costar's manufacturing and sterilization facility and protocols. The aerator vessel was autoclavable and did not require extensive utilities in the production facility, and the circulation and perfusion flows were driven by simple peristaltic pumps. The entire reactor was compact enough so that the unit could be housed in conventional incubators for temperature control.

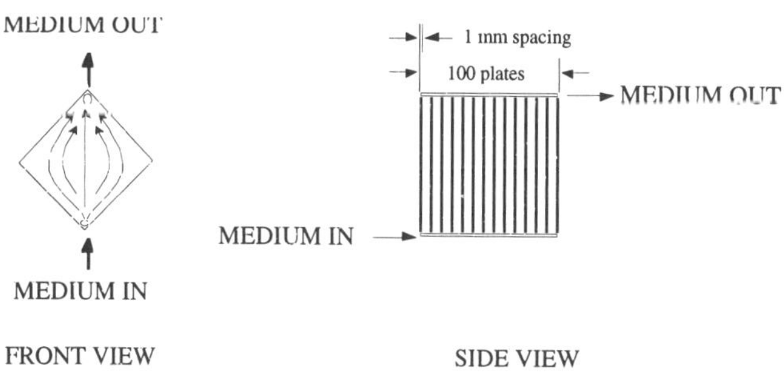

Figure 1. Schematic of Corning-Costar CellCube construction and flow pattern.

Like the SMR, the first experiments with the CCC were very successful and demonstrated equivalent productivity to static and SMR cultures. There was initial concern about the reactor geometry and whether fluid flow and hence the cell growth would be uniform and whether the cells would be well-nourished. However, simple residence time distribution information coupled with mixing cup dissolved oxygen measurements on the outlet flow from the Cubes allowed calculation of the distribution of oxygen tension in the reactor, and a circulation flow rate which ensured adequate oxygen supply to all points could be specified. Manufacturing issues revolved around pumping and controller technology, with Watson-Marlow and B.Braun Biotech equipment selected to replace the CCC units. This was due to the relative lack of experience with the CCC units and the known durability of the others. Watson-Marlow offered multichannel pump

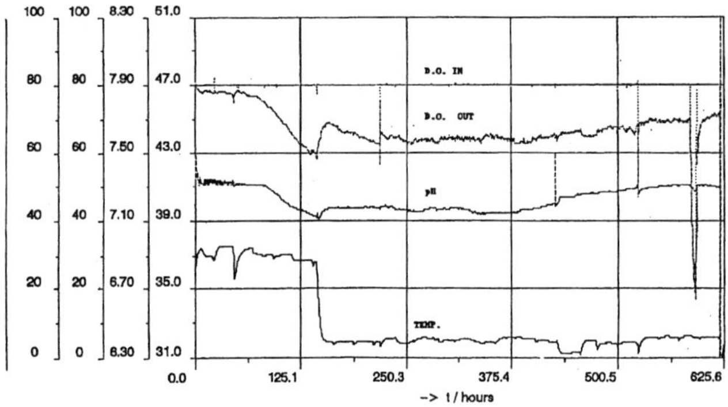

Figure 2. On-line DO_{inlet}, DO_{outlet}, pH and temperature data collected during HAV cultivation.

heads that ensured equal fluid flow to the four Cubes comprising a reactor. The Braun controllers offered supervisory computer control and data logging capabilities, an example of which is shown in Figure 2. The CCC falling film aerator mass transfer properties were examined under various flow rates and found to support the cell growth anticipated in the reactor. The CCC aerator was thus retained for its simplicity.

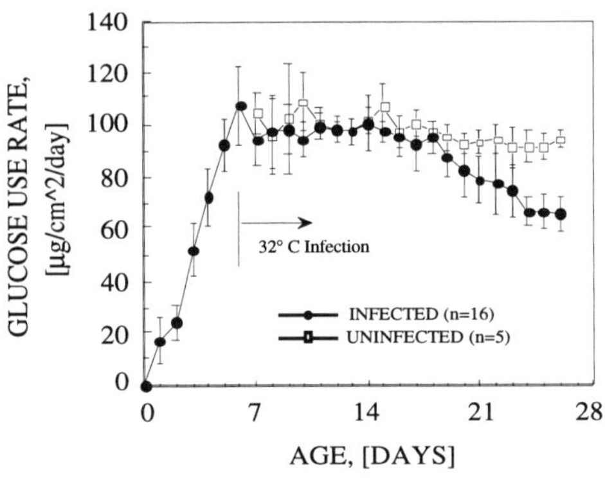

Figure 3. Glucose use rate versus time for infected and uninfected CellCube reactors in manufacturing.

Based on its technical success, ease and inexpensiveness of manufacturing introduction, and acceptability as a platform to meet future regulatory demands, the Costar-Corning CellCube was selected for use in VAQTA™. Since its introduction it has performed remarkably consistently for over a hundred production lots. Although the cell sheet is not directly accessible to sampling for cell concentration during manufacturing, cell growth has been correlated with metabolic behavior and with the lysate protein content. Cell number during 37° C growth correlates with the glucose use rate as confirmed by experiments where reactors have been trypsin harvested and the cells counted by hemacytometry. At infection, the lot-to-lot variation in glucose use rate (and hence cell number) is only *ca.* 10% RSD as shown in Figure 3. Such a correlation is not

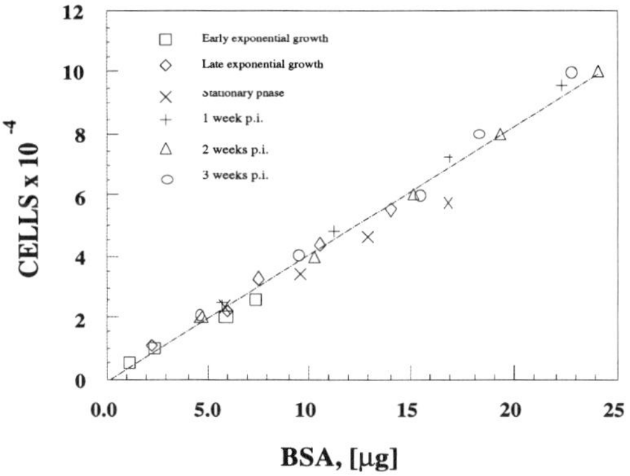

Figure 4. Cell number versus Triton X-100 detergent lysate protein content (Biorad assay calibrated with BSA) at varying stages of growth and infection in T-flasks.

possible for the 32° C infection phase since HAV infection leads to decreased glucose metabolism, also seen in Figure 3. Cell number also correlates with protein concentration in a detergent lysate of MRC-5 over a range of growth and infection conditions, as shown in Figure 4. This T-flask correlation has also been confirmed for reactors by taking some Cubes in the reactor and lysing them, and taking the other Cubes and trypsinizing them. Like the glucose use rate, lysate protein variance is small, *ca* . 20% RSD from lot-to-lot or within the variability of the protein assay. The result of this consistent cell propagation is that the HAV antigen yield from a batch also varies by only *ca.* 20% lot-to-lot RSD, again within the inherent variability of the antigen EIA. One interesting point is that since the protein per cell does not change over the course of infection as shown in Fig. 4, and since the protein contents of the uninfected and infected reactors in Fig. 3 were the same at harvest, the decrease in the glucose use rate during infection is not due to cell lysis, but due to an effect of HAV replication on cell metabolism. Such information on viral replication cannot be gained from simpler uninstrumented reactors such as NCFs.

Conclusions

Development of bioreactors for the hepatitis A vaccine VAQTA™ was influenced by multiple factors over and above the technological feasiblity and economics of each reactor process. The simplest Nunc Cell Factory reactors were not employed due to the impracticality of improved feeding strategies to boost productivity, the lack of process control and monitoring, and the many aseptic connections necessary to execute a batch.

Microcarrier reactors suffered from poor productivity even under enhanced feeding, possibly due to aggregation and partial cell starvation, and lack of the ability to control this aggregation discouraged further development. Although Static Mixer Reactors were found technically exciting in all ways, the late adoption of this technology and the necessary tasks and cost to implement an automated reactor placed it at a disadvantage to the less sophisticated but very pragmatic CellCube reactors.

General lessons which can be drawn from this experience are several. First and foremost, technologies chosen for investigation have to be appropriate for the process and the manufacturing scale, and of course must be economical. The technologies chosen for investigation also need to be suitable for the process development and manufacturing infrastructure present. Beyond this however, the reactor technology must also address the concerns of the regulatory liaison who will justify its safety, and of the manufacturing production personnel who will be responsible for validation and operation. Early collaboration among the different groups can assist the focus of efforts towards one or two universally acceptable technologies.

Bibliography

1. Provost P, Hilleman M (1979) Propagation of human hepatitis A virus in cell culture *in vitro*. Proc. Soc. Exptl. Biol. Medicine 160, 213.

2. Holinger FB, Ticehurst J (1990) Hepatitis A Virus. in "Virology" 2nd ed., Fields BN, Knipe DM *et al.* eds., Raven Press, NY, 631.

3. Siegl G, de Chastonay J, Kronauer G (1984) Propagation and assay of hepatitis A virus *in vitro*. J. Virol. Methods 9, 53.

4. Werzberger A, Mensch B, Kuter B, Brown L, Lewis J, Sitrin R, Miller W, Shouval D, Wiens B, Calandra G, Ryan J, Provost P, Nalin D (1992) A controlled trial of a formalin-inactivated hepatitis a vaccine in healthy children. New England J. Medicine 327, 453.

5. Aboud R, Aunins J, Buckland B, Hagen A, Hennessey J, Junker B, Lewis J, Oliver C, Orella C, Sitrin R (1994) Hepatitis A Virus Vaccine. International Patent Application, publication number WO 94/03589, Feb. 17, 1994.

6. Widell A, Hansson BG, Nordenfelt E (1984) A microcarrier cell culture system for large-scale production of hepatitis a virus. J. Virol. Methods 8, 63.

7. Montagnon B, Vincent-Falquet JC, Fanget B (1984) Thousand litre scale microcarrier culture of Vero cells for killed polio virus vaccine. Promising results. Developments in Biological Standardization 55, 37.

8. Junker BH, Wu F, Wang S, Waterbury J, Hunt G, Hennessey J, Aunins J, Lewis J, Silberklang M, Buckland B (1992) Evaluation of a microcarrier process for large-scale cultivation of attenuated hepatitis A. Cytotechnology 9, 173.

9. Seifert D, Ram K, Lewis D, Phillips J, Sheridan C (1994) Characterization of aggregated microcarrier cultures for the production of viral vaccines. Abstr. Pap. Amer. Chem. Soc., 207th Mtg., Pt.1, BIOT18.

10. Grabner R, Paul E (1981) Use of motionless mixer as cell culture propagator. US Patent 4,296,204.

11. Chiou TW, Murakami S, Wang DIC, Wu WT (1991) A fiber bed bioreactor for anchorage dependent animal cell cultures: Part 1. Bioreactor design and operations. Biotechnology & Bioengineering 37, 762.

12. Noe W, Bux R, Berthold W, Werz W (1994) Optimization of vaccine production for animal health. Cytotechnology 15, 169.

13. Brown PC, Figueroa C, Costello MAC, Oakley R, Maciukas SM (1988) Protein production from mammalian cells grown on glass beads. in "Animal Cell Biotechnology" vol. 3, Spier RE, Griffiths JB eds., Academic Press, San Diego, 251.

14. Junker B, Seamans TC, Ramasubramanyan K, Aunins J, Paul E, Buckland B (1994) Cultivation of attenuated hepatitis A virus antigen in a titanium static mixer reactor. Biotechnology & Bioengineering 44, 1315.

PRODUCTION OF LIVE ATTENUATED VARICELLA-ZOSTER VIRUS IN HUMAN EMBRYONIC LUNG CELLS USING MICROCARRIER

B-H Chun*, G-J Woo, H-S Cheung, H-M Moon
MOGAM Biotechnology Research Institute,
341, Pojung-Ri, Kusung-Myon, Yongin-Kun, Kyunggi-Do, Korea

Key words: Varicella-zoster virus production, cell attachment, microcarrier culture

INTRODUCTION

Varicella-zoster virus (VZV), one of the human herpes viruses, is a well known virus that can infect and induce both chicken pox (varicella) and shingles (zoster). General vaccination against VZV is now seriously considered mainly for economical reasons. A live attenuated varicella vaccine was developed using Oka strain by a cell culture method (1). This varicella vaccine was approved for use in Japan, Korea, and USA (Biken), and use for healthy children is under consideration in some countries in Europe (2). However, so far it has been difficult to obtain high yield of infectious cell-free VZV for vaccine from cell culture. As an attempt to develop a cell-free VZV production process, we established a human embryonic lung cell culture, and VZV strains were isolated from patients and attenuated.

So far tissue culture flasks have not been adequate to produce cell-free VZV on a large scale because of complexity in handling, high risk of contamination, and high time and labour costs. In previous studies, human diploid cells were cultured successfully in microcarriers (3, 4) and production of virus using microcarriers was also reported (5). To devise strategies which can improve production of VZV in microcarrier culture, an understanding of the attachment kinetics of both normal cells on microcarriers and virus-infected cells on cell-laden microcarriers are necessary.

Major objective in this study is to evaluated a microcarrier culture system for the production of VZV. Other objective is to identify the attachment characteristics of both normal human embryonic lung cells on microcarriers and virus-infected cells including normal cells onto cell-grown microcarriers at various culture conditions and to improve the VZV production by understanding of these attachment kinetics of both normal cells and virus-infected cells and by determining CPE level and MOI of virus inoculum.

MATERIALS AND METHODS

Cell Line: Human embryonic lung (LuMA) cells (PDL: 20-25th)
Virus Strain: MAV/06 strain isolated from VZV-infected patients.
Virus Culture: The VZV was inoculated onto monolayered LuMA cells and after 4 - 6 day incubation, the typical CPE of VZV infection was observed under the microscope. The VZV-infected cells showing 40-50% CPE were harvested with 0.25% trypsin and passed into T-flasks containing confluent monolayered LuMA cells at 0.1 MOI. The seed-lot system was adopted for maintenance of VZV infectivity.
Assay of Cell-Associated and Cell-Free Virus : plaque assay (6)
The stability buffer for virus assay: 5% sucrose, 2.0% gelatin hydrolysate, and 0.1% sodium glutamate in PBS. The virus-infected cell suspension was diluted with growth medium and assayed for cell-associated VZV. The cell-free VZV was prepared from VZV-infected cells by sonication method.
Cell Attachment to Microcarriers and microcarrier culture (6)

M.J.T. Carrondo et al. (eds.), Animal Cell Technology, 185–189.

RESULTS

Growth Characteristics of Cells and Virus-Infected Cells

Determination of optimum microcarrier concentration and medium exchange schedule are necessary for high-density cell culture and high product yield, and understanding the attachment kinetics of both cells and virus-infected cells on microcarriers is also needed (Fig. 1). The cell concentration per unit surface area of substrate in microcarrier culture was lower than that in T-flask culture. Decreasing microcarrier concentration resulted in increasing in the extent of final cell growth per unit carriers. The total cell concentrations after virus infection were decreased with culture time. The desirable microcarrier concentrations were shown to be 3 or 5 g/l. Glucose was not limited at this replenishment of medium, and the cells in initial growth phase require a little higher glucose concentration in comparison to late growth phase and virus propagation phase. Efficiency of the cell attachment was increased with the increase of microcarrier concentration. Normal LuMA cell attachment on microcarriers was somewhat higher than that of VZV-infected cells on cell-laden microcarriers.

Effect of Microcarrier Concentration on VZV Production

The maximum cell-associated virus titer was increased with microcarrier concentration, and the specifc VZV titer was decreased with increase of microcarrier concentration. The higher the microcarrier concentration was, the lower the yield of VZV from cells was. Although cell-associated virus was detected at low levels in the culture medium, no infectious cell-free virus was found.

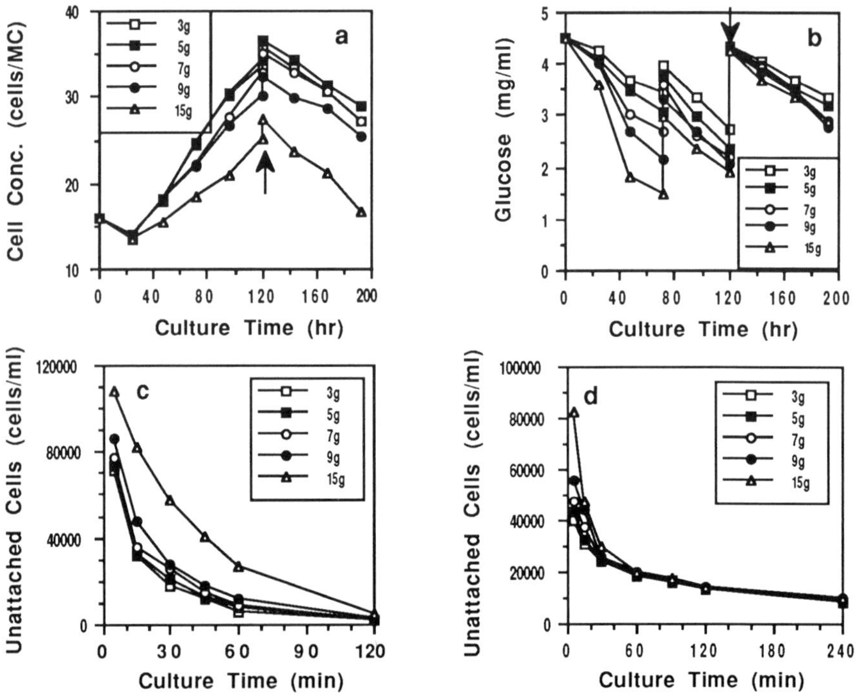

Fig.1. a) profiles of cell growth, b) profiles of glucose consumption, c) kinetics of cell attachment on microcarriers, d) kinetics of virus-infected cell attachment on the cell-laden microcarriers. Arrow is a virus infection time with VZV-infected cells.

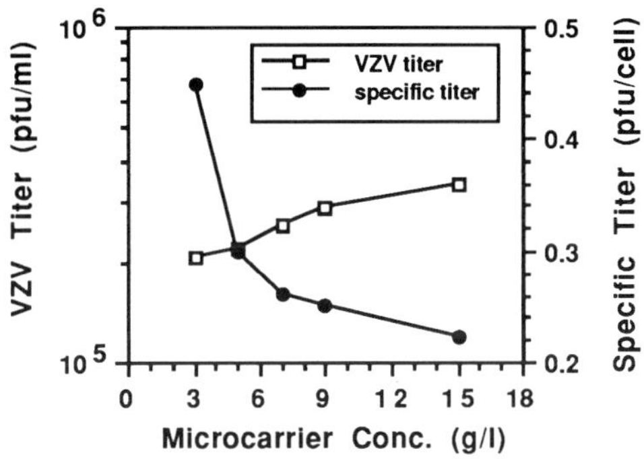

Fig. 2. Comparison of cell-associated virus production at different concentrations of microcarrier.

Effect of CPE Level

Effect of CPE on attachment: The lower the CPE level of virus inoculum, the higher the attachment rate of VZV-infected cells on cell-laden microcarriers (Fig. 3). The attachment of VZV-infected cells to cell-laden microcarriers were affected by CPE level of virus inoculum.

Effect of CPE on VZV yield: The maximum yield of VZV released into culture medium were increased with the increase of CPE level of virus inoculum (Fig. 4a). When the cell layer confluented on microcarriers was infected with VZV-infected cells, the cells had a tendency to slough off the microcarriers with culture time, because of cell disrupture by virus replication. The maximum yields of CAV and CFV was a little increased with the increase of CPE level (Fig. 4b). Based on cell densities, the use of virus inoculum showing 40% CPE led to about 2- and 1.2-fold increase in the cell-specific VZV (CAV & CFV) as compared to the culture in the use of 10% CPE. This is indicating an enhanced virus production from cell-laden microcarriers by virus-infected cells shown in 40% CPE.

Effect of MOI

Effect of MOI on attachment: In cell attachment period for virus infection, a portion of the unattached normal cells in culture medium was increased with the decrease of unattached VZV-infected cells regardless of infection ratio (Fig. 5a & 5b).

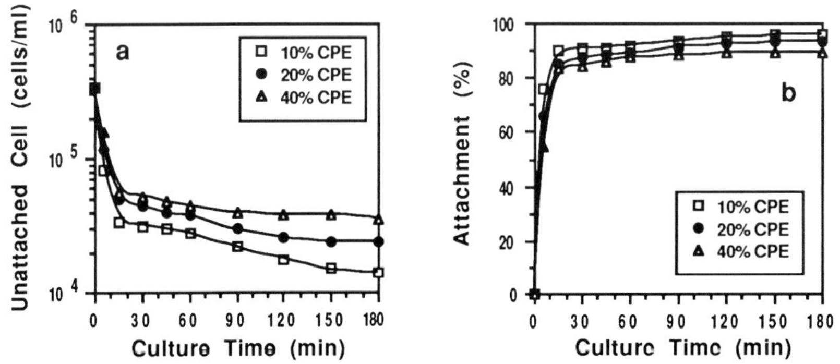

Fig. Kinetics of virus-infected LuMA cell attachment on cell-grown microcarriers at different CPE levels: 10%, 20%, and 40%.

Effect of MOI on VZV yield: VZV yield was enhanced with increase of initial virus-infected cells onto cell-laden microcarriers (Fig. 5c & 5d). By understanding relationship between the attachment kinetics of cells and virus titers and by determining virus infection ratio (MOI), it might be possible to propagate the VZV efficiently using microcarrier culture system.

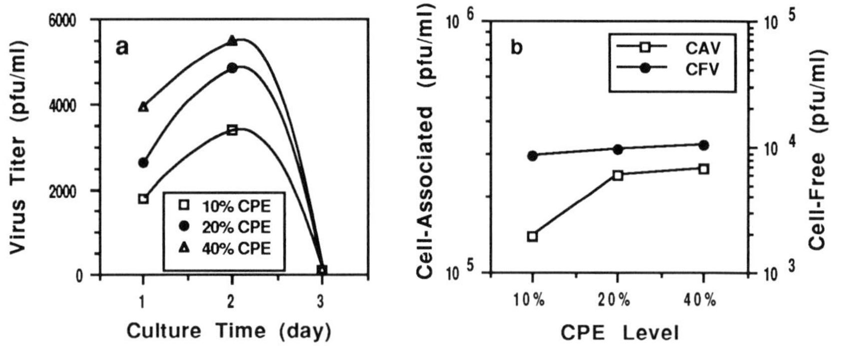

Fig. 4 Comparison of virus production at different CPE level of 10%, 20%, and 40%: a) profiles of cell-associated VZV titer in culture medium, b) comparison of the maximum extents of cell-associated and cell-free VZV production.

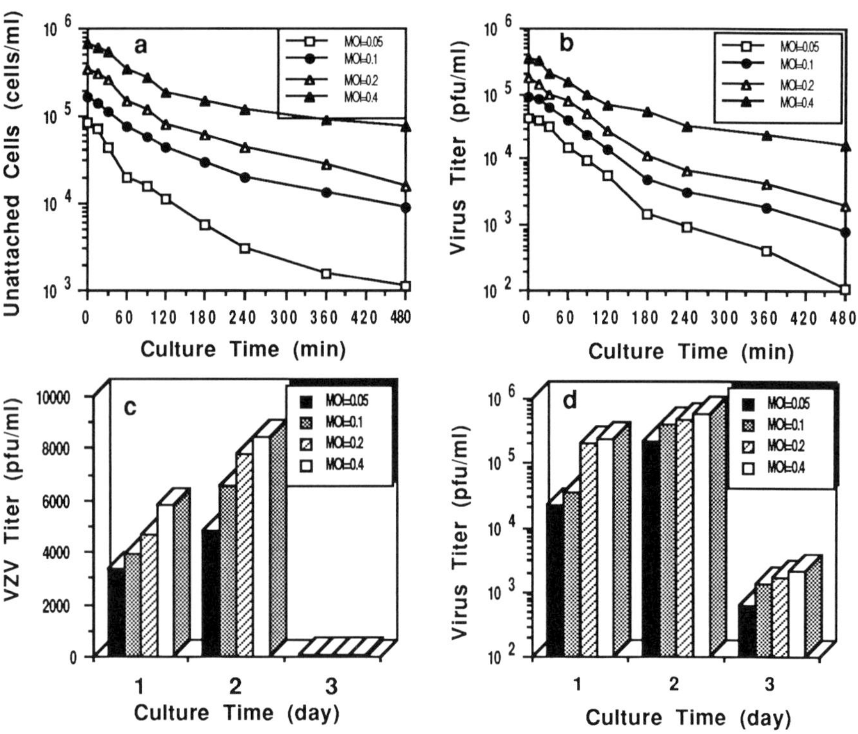

Fig. 5. a) Kinetics of VZV-infected LuMA cell attachment on cell-laden microcarriers at different MOIs; b) profiles of cell-associated virus titer at culture supernatant; c) profiles of cell-free virus production secreted into culture media during virus production; d) profiles of cell-associated virus harvested from cell-laden microcarriers.

Comparison of Virus Production at Various Cultures

Cell-free VZV production in roller bottle and multitray culture was lower than that obtained in T-flask. Although the maximum titers of cell-free VZV in microcarrier culture infected with 0.05 & 0.1 MOI were a little lower than those in T-flask culture, the maximum titer of cell-free VZV at 0.2 MOI was comparable to that in T-flask culture. Substitution of tissue culture system by microcarrier culture is recommended for the large scale production of cell-associated and cell-free VZV. The highest cell-free virus was obtained by harvesting before CPE was extensive. When the confluent cell layer was infected with VZV-infected cells, the cells had a tendency to slough off the microcarriers with culture time, because of cell disrupture by virus replication. Therefore, determination of harvest time in microcarrier culture system is very important for obtaining high-titer of cell-free virus before sloughing off.

The ratio of cell-free virus to cell-associated virus is important for efficient virus production, especially in a large scale production. The ratio obtained in microcarrier culture was lower than that obtained in T-flask. This suggests that the high MOI is required for the high production of cell-associated and cell-free VZV in microcarrier cultures. The reason for this difference in specific VZV production at different MOI may be due to a high cell affinity of varicella-zoster virus. Since infectious VZV remains cell associated, it is necessary to optimize cell growth condition to obtain the high titers of cell-associated and cell-free virus.

Table 1. Comparison of VZV production in different cell cultures

Culture Method	Growth Area (cm^2)	Relative Growth Area	Culture Volume (ml)	MOI	Maximum CAV [a] (pfu/cm^2)	Maximum CFV[b] (pfu/cm^2)	CFV/CAV	Relative CFV Yield
T-flask	175	1	35	0.2	9,140	800	0.088	1
Roller bottle	850	4.9	150	0.2	6,060	590	0.097	0.74
Multitray[c]	6,320	36.1	1,300	0.2	8,120	730	0.090	0.91
Microcarrier	2,200	12.6	100	0.05	7,730	470	0.061	0.59
	2,200	12.6	100	0.1	9,540	720	0.075	0.90
	2,200	12.6	100	0.2	12,540	860	0.069	1.08
	44,000	251.4	1,000	0.2	9,630	730	0.076	0.91

a, CAV: cell-associated virus; b, CFV: cell-free virus; c, Cell-Factory 10 (Nunc)

SUMMARY

- Varicella-zoster virus was propagated in human fetal embryonic lung (LuMA) cells grown on microcarriers to produce live attenuated varicella vaccine.
- The cell-associated and cell-free virus yields in Cytodex 1 microcarriers were comparable to those in the stationary tissue cultures in T-flask, roller bottle, and multitray.
- By understanding relationship between the attachment kinetics of cells and virus titers in the medium and by determining virus infection ratio and cytopathic effect level of virus-infected cells, it was possible to propagate effectively the VZV in microcarrier culture system.

REFERENCES

1. Takahashi M, Otsuka T, Okuno Y, Yazaki T & Isomura S(1974)Lancet ii,1288-1290.
2. Takahashi M (1992) Vaccine 10,1007-1014.
3. Morandi M & Valeri A (1982) Biotech. Letters, 4,465-468.
4. Forestell SP, Kalogerakis N, & Behie LA (1992) Appl Microbiol Biotechnol 38, 165-172.
5. Griffiths B & Thornton B (1982) J. Chem. Tech. Biotechnol. 32,324-329.
6. Chun BH, Park SY, & Moon HM (1996) Biotechnol. Techniques 10,151-156.

PRODUCTION OF LYSSAVIRUS PROTEINS IN THE BACULOVIRUS EXPRESSION SYSTEM AND THEIR USE AS POTENTIAL VACCINE

A. Drings, B. Chambert, C. Jallet, N. Tordo, P. Perrin and Y. Jacob
Laboratoire des Lyssavirus, Institut Pasteur, 25 rue du Docteur Roux, 75724 Paris Cedex 15 , France

Abstract

The aetiologic agents of rabies encephalomyelitis are classified into six genotypes within the Lyssavirus genus (1). It is generally accepted that the classical rabies vaccine strains (genotype 1) are inefficient in protecting against distantly rabies-related Lyssaviruses (genotypes 2 to 6). In an attempt to enlarge the protective activity of rabies vaccines, we have expressed the two principal viral antigens as recombinant proteins using a baculovirus expression system: 1° the glycoprotein (G) which constitutes viral spikes and induces virus-neutralizing antibody (VNAb); 2° the nucleoprotein (N) which is the major constituant of the ribonucleoprotein complex (RNP) involved in the T helper cell response. This work was performed with the Pasteur virus strain (PV: genotype 1).

Recombinant proteins were expressed in insect cells (Sf9 for N-PV; BTI-TN-5B1-4 for G-PV) using recombinant baculoviruses. After expression and partial purification, Lyssavirus antigens were tested for their immunological and protective properties alone or in association.

Recombinant proteins display their classical immunogenic properties: G induces VNAb and N protein is efficient in inducing T helper cells even administered orally. When N is used for priming, it enhances the production of both VNAb and anti-RNP antibodies when animals are boosted respectively with the inactivated and purified rabies virus (IPRV) or the RNP. Results are discussed in the context of extending the spectrum of rabies vaccines.

Introduction

Rabies virus is a member of the *Lyssavirus* genus of the *Rhabdoviridae* family. It is constituted of five proteins: the surface glycoprotein (G) and the matrix protein (M) which surrounds the ribonucleoprotein complex (RNP) comprising the nucleoprotein (N), the phosphoprotein (NS) and the polymerase (L). Two of them, G and N, are of major importance concerning the immune response versus a rabies virus infection. G is the only rabies antigen capable of inducing VNAb and protection against both a peripheral and an intra-cerebral challenge (2, 3). N does not induce VNAb, but induces a potent T helper cell response and a protection against a peripheral challenge with rabies or some rabies-related Lyssaviruses (4, 5). Whereas the impact of G in the anti-rabies protection is clear, the role of N is still a subject of controversy (6). In some experiments, RNP is considered as an adjuvant capable of priming and enhancing VNAb when animals are boosted with rabies vaccine (7).

In order to verify if the N protein must be included in a new generation rabies vaccine for enhancing its protective activity or enlarging its spectrum, both recombinant rabies glycoprotein (RRG) and nucleoprotein (RRN) were obtained with the baculovirus expression system and tested for their immunological and protective properties alone in association.

Materials and methods

Recombinant viruses and proteins

In order to transfer Glycoprotein and Nucleoprotein gene (from PV rabies virus) into the AcNPV genome, RT-PCR products were subcloned and inserted under the control of polyhedrin promoter into Baculovirus transfer vector pVL1393 (Pharmingen). The

M.J.T. Carrondo et al. (eds.), Animal Cell Technology, 191–195.

distance between the start site of polyhedrin mRNA transcription and the initiation codon of both gene was reduced to a few nucleotides by using degenerated PCR primers with artificial restriction sites. Recombinant baculoviruses were isolated after cotransfection of *Spodoptora frugiperda* cells (Sf9) with linear, non infectious, AcNPV DNA (BaculoGold TM) and recombinant transfer vector. Clonal selection of pure recombinant baculovirus were carried out by plaque-assay purification. Amplification of baculoviruses on Sf9 cells provided a viral stock with an infectious titer of 5-10 x 10^8 PFU/ml. For mass production, optimised Trichoplusia ni cells (line BTI-TN-5B1-4, High Five TM) were infected with recombinant AcNPV at a multiplicity of 5 PFU/cell. Cells and supernatant were harvested after infection. The supernatant was clarified. Recombinant protein was solubilised from the cell pellet by Octlyglycopyranoside 2%.

Antigenicity and immunogenicity

The quantities of rabies glycoprotein or nucleoprotein expressed in cells or released from cells was estimated by ELISA using specific purified-antibody as immunosorbent (8,9).

Immunogenicity of recombinant proteins was tested according to T helper cell response and antibody production. T helper response was studied by stimulating *in vitro* splenocytes from immunized and non-immunized BALB/c mice as previously reported (10). Interleukin-2 (IL-2) produced in splenocyte supernatants was assayed with proliferation of IL-2 dependent cells: CTLL (10). VNAb was titrated using the rapid fluorescent focus inhibition test (RFFIT) (11). Anti-RNP antibody was titrated by ELISA as previously described (12).

Protective activity

Protective activity of antigen preparation containing the glycoprotein was determined according the NIH test (13). The protective activity of RRN was appraised by immunizing and challenging swiss mice by the peripheral route with the CVS strain.

Results

Immunogenicity of recombinant proteins

Whatever the source of antigen (cell extract or cell supernatant) RRG induced in mice production of VNAb against the CVS virus (genotype 1: Fig 1).

When administrated either by the intraperinoneal or oral route, RRN induced the production of T helper cells which can be specifically stimulated *in vitro* at a level similar to that induced by RNP (Fig.2). The potentiality of RRN to induce antibody production is similar to that of RNP (not shown).

Protective activity

RRG was capable of protecting mice against an intracerebral challenge with the CVS strain whereas RRN was not capable of inducing protection against a peripheral challenge with the same strain (not shown).

Influence of a priming with RRN on the immunogenicity of IPRV and RNP.

When mice were primed with RRN, the immunogenicity of both IPRV and RNP was enhanced (Fig. 3): 1) anti-RNP antibodies as well as VNAb induced by IPRV increased; 2) protection against a peripheral challenge induced by RNP increased.

Influence of the association of RRN with RRG.

Association of RRN to RRG preparations did not enhance VNAb production against either CVS (genotype 1) or EBL (European Bat Lyssavirus: genotype 5) (Fig. 4). Moreover, no protection was obtained against Mokola virus (genotype 3) even when the Pasteur RRG was associated with RRN (not shown).

Conclusion

As the rabies nucleoprotein, an internal antigen, was considered to play a role as an adjuvant and/or "cross help" effector, two types of studies were performed using recombinant proteins: 1) influence of a priming with the RRN on the immunogenicity of IPRV, RNP and RRG; 2) influence of the presence of RRN on VNAb induction by RRG and on protection against rabies-related lyssaviruses. Even though we have shown that a priming of IPRV with the RRN of the same genotype enhances VNAb production against this genotype, no production (against Mokola virus) or enhancement (against EBL) was observed despite the potent immunological activity of both antigens.

Our results indicate that while the glycoprotein seems to be a major constituent of rabies vaccine, its association with the nucleoprotein does not extend the protection induced across the genotypes despite the priming properties of the nucleoprotein.

References

1. Bourhy, H., Kissi, B., and Tordo, N. (1993) Molecular diversity of Lyssavirus genes. Virol. 194. 70-81
2. Wiktor, T., E. Gyorgy, et al. (1973). Antigenic properties of rabies virus components. J. Immunol. **110**: 269-276.
3. Perrin, P., P. Sureau, et al. (1985). Structural and immunogenic characteristics of rabies immunosomes. Develop. Biol. Standard. **60**: 483-491.
4. Ertl, H. (1989). Dominant T-helper cell epitopes of the rabies virus nucleoprotein." In Rabies (Eds O. Thraenhart et al. Wells Medical, Limited, England): 238-242.
5. Dietzschold, B., M. Tollis, et al. (1987). Antigenic variation in rabies and rabies-related viruses: cross-protection independent of the glycoprotein-mediated virus neutralizing antibody. J. Infect. Dis. **156**: 815-822.
6. Ertl, H. and Z. Xiang (1996). Novel vaccine approaches. J. Immunol. **156**: 3579-3582.
7. Tollis, M., B. Dietzschold, et al. (1991). Immunization of monkeys with rabies ibonucleoprotein (RNP) confers protective immunity against rabies. Vaccine **9**: 134-136.
8. Perrin, P., S. Morgeaux, et al. (1990). In vitro rabies vaccine potency appraisal by ELISA : Advantages of the immunocapture method with a neutralizing anti-glycoprotein monoclonal antibody. Biologicals **18**: 321-330.
9. Perrin, P., P. Rollin, et al. (1986). A rapid rabies enzyme immuno-diagnosis (RREID): a useful and simple technique for the routine diagnosis of rabies. J. Biol. Standard. **14**: 217-222.
10. Perrin P., De Franco, M., Jallet, C., Fouque, F., Morgeaux, S., Tordo, N., and Colle, J-H. (1996). The antigen-specific cell-mediated immune response in mice is uppressed by infection with pathogenic lyssaviruses. Res. Virol. **147:** (in press)
11. Smith, J., P. Yager, et al. (1996). A rapid fluorescent focus inhibition test (RFFIT) for determining virus-neutralizing antibody. In "Laboratory techniques in rabies", Fourth edition (Eds Meslin, F-X; Kaplan, M and Koprowski, H) WHO, Geneva.: 181-189.
12. Perrin, P., P. Versmisse, et al. (1986). The influence of the type of immunosorbent on rabies antibody EIA ; advantages of purified glycoprotein over whole virus. J. Biol. Standard. **14**: 95.
13. Seligmann, E. (1973). The NIH test for potency. In "Laboratory techniques in rabies", Third edition (Eds Kaplan, M and Koprowski, H) WHO, Geneva.: 279-286.

Acknowledgments: A. Drings was a recipient of a fellowship by the Gottlieb Daimler - and Karl Benz - Foundation (Germany).

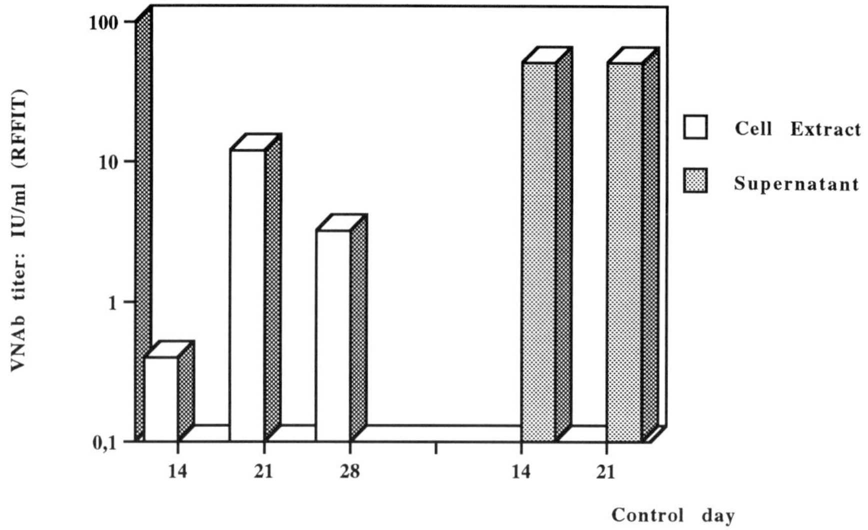

Figure 1: Induction of VNAb by RRG. BALB/c mice were immunized (days 0 and 7) with 1µg of RRG extracted from infected cells (cell extract: solubilization by 2% octyl-glucopyranoside and dialysis) or clarified from infected cell supernatant. Sera were assayed at different times.

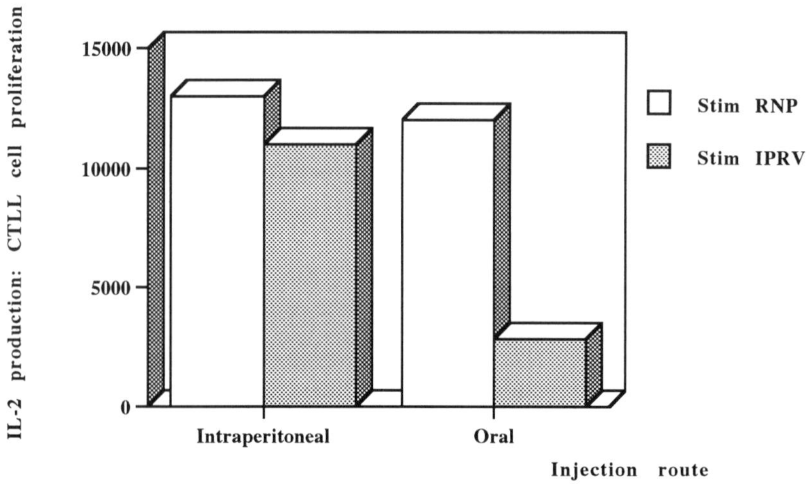

Figure 2: Induction of T helper cell response (IL-2 production) by RRN. RRN was intra-peritonealy (1µg) or orally (5µg) adminitrated to BALB/c mice. Their splenocytes were stimulated in vitro with RNP or IPRV. IL-2 produced by T helper cells was tested of IL-2 dependent cells (CTLL).

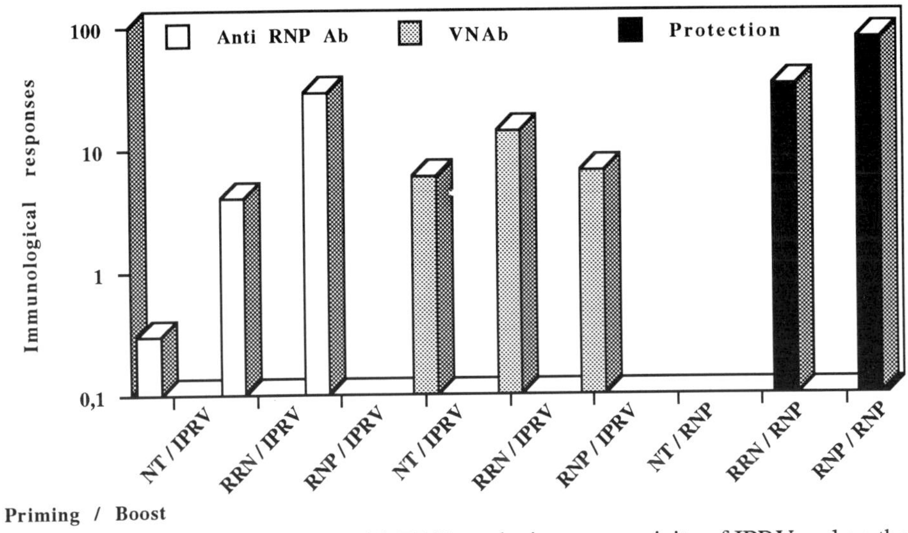

Figure 3: Influence of a priming with RRN on the immunogenicity of IPRV and on the protective activity of RNP. C3H mice were primed with NT buffer (NT), RRN (1.5µg) or RNP (0.5µg). Mice boosted with IPRV were tested for antibody responses (anti RNP and VNAb) and those boosted with RNP were challenged by the peripheral route.

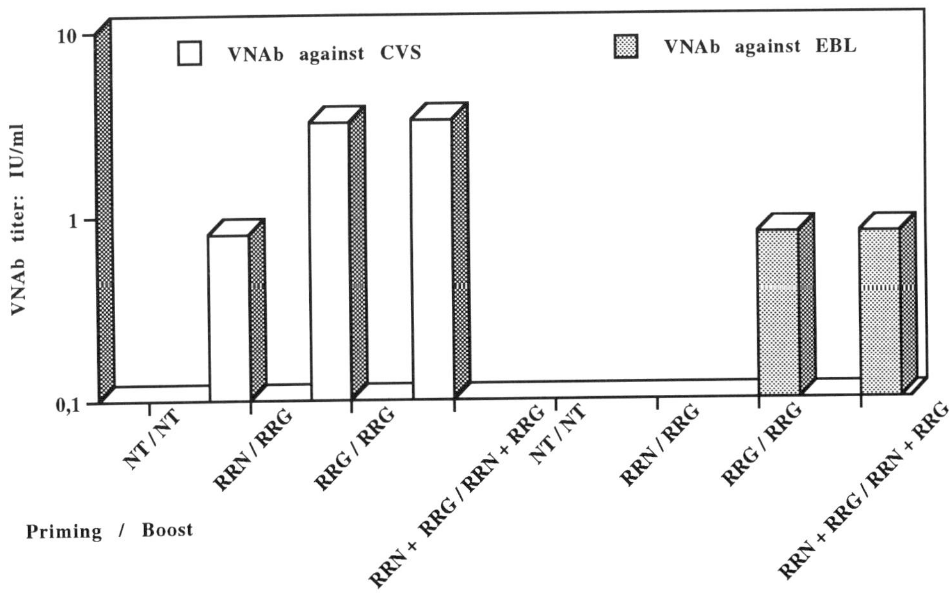

Figure 4: VNAb production by RRG associated with RRN. BALB/c mice were primed (day 0) with NT buffer (NT), RRN (1µg) or RRG (0.4µg) and boosted (day 10) with NT, RRN (1µg) or RRG (0.4µg). Sera were assayed (day 21) for VNAb against CVS (genotype 1) or EBL (genotype 5).

DEVELOPMENT OF PRODUCTION TECHNOLOGY OF LIVE MEASLES VACCINE FOR PERORAL ADMINISTRATION

Nechaeva E.A., Varaksin N.A., Ryabicheva T.A., Kolokoltsova T.D., Ignatyev I.M., Agafonov A.P., Kolokoltsov A.A., Yurchenko N.D., Sandakhchiev L.S.

Institute of Cell Cultures, State Research Center of Virology and Biotechnology VECTOR, RUSSIA

ABSTRACT

According to modern views, an ideal vaccine must provide a life-long immunization effect in 100% of the vaccinated persons, should be safe, thermostable, cheap, and should not be administered parenterally. As known, the most important feature of live vaccines against respiratory infections is rapid development of local secretory immunity in the portal of respiratory and intestinal tracts. We have developed a new technology for production of peroral live measles vaccine where a special attention is paid to optimization of biological stabilizers. The vaccine trials in rabbits, the tablet and capsule forms of the preparation being administered intraduodenally, revealed a stimulation of humoral and cellular immunity and demonstrated the absence of side effects.

INTRODUCTION

Immunization with live measles vaccine is presently an obligatory and the only radical measure in the struggle against measles. Owing to the vaccination, the measles morbidity in our country decreased by a factor of 50. However, the administration of parenteral vaccines may lead to infection with AIDS-causing and hepatites viruses in the course of vaccination. Besides, the parenteral way of administration of measles vaccine not always results in the formation of specific protective factors (1). Thus, the development of production technology of live measles vaccine for per os administration is a topical problem.

Moreover, according to the modern views, an ideal vaccine must provide a life-long immunity in 100% of the vaccinated persons, should be innocuous, thermostable, cheap, and should not be administered parenterally (2).

The characteristic feature of oral immunization is penetration of the antigen in the lymphoid tissue of the alimentary tract and subsequent forming of the specific sec-

M.J.T. Carrondo et al. (eds.), Animal Cell Technology, 197–202.

retory immunoglobulins. The latter play the key role in the functioning of the unitary secretory immunity system, including that of the mucosa of the upper respiratory system (3).

In connection with the above stated, the main objective of the present work was production of the tablet form of live measles vaccine and evaluation of prospects associated with the use of this vaccine in experiments involving animals.

MATERIAL AND METHODS

Measles virus. The strain Leningrad-16 (or L-16) of measles vaccine was isolated in 1960 in the Leningrad Pasteur Institute of Epidemiology and Microbiology from the blood of a measles-infected child (4).
Cell culture. *The diploid cell culture L-68 was obtained from the lung of an 11-day-old human embryo* and has a normal karyotype (5). By now the main and 3 working banks of L-68 cells have been obtained by our Center. Banks of cell cultures were certified and officially approved by the National Institute of Standardization and Control (NISC) for production of measles vaccine. Certification of experimental batches of live measles vaccine (LMV) in tablets was carried out according to the physical and chemical properties, anaphylactogenicity, safety in laboratory animals and foreign microorganisms content according to the State Pharmacopoeia (6). Biological activity of measles virus was determined by cytopathic effect in Vero cell culture, the content of foreign viruses, micoplasmas, and mycobacteria was evaluated by the methods described in Pharmacopoeial Item (7). Control of immunogenicity of the LMV experimental batches in tablets for oral administration was performed in rabbits by intraduodenal preparation administration. Blood samples were taken from the vaccinated animals on days 0, 3, 9, 14, 21 post vaccination to check the changes of local (secretory) and general immunity. The intensity of humoral immunity was estimated by the increase of the titers of antimeasles antibodies in the blood sera of immunized animals in reaction of passive hemagglutination (RPHA), reaction of inhibition hemagglutination (RIHA) and by immmunoenzyme assay (ELISA) (7,8). Control of cellular immunity was performed in reaction of blast-transformation of lymphocytes (RBTL) by purified measles virus, concanavalin A, and lipopolysaccharide of E.coli (9).

RESULTS AND DISCUSSION

Concept of the unity of mucosal immune system within the human organism can be applied in the development and use of the new nonconventional expedients of vaccination against various human and animal infectious diseases. Comprehensive investigations aimed at the development of oral vaccines have been carried out in our country since 60s. Positive results have been obtained in immunization with live measles vaccines against plague, poliomyelitis, rabbit fever, Q-fever, smallpox, influenza, whooping cough, mumps, tick-borne encephalitis, brucellosis, cholera, and other infections . A significant increase in secretory IgA was recorded after oral vaccination with vaccines against influenza, polioviral infection, German measles, poliomyelitis (10,11.12).

The data on intranasal and aerosol vaccination of men with the live measles vaccine of Edmonstone-Zagreb and L-16 strains also testifies to the possible increase in the protective properties of the vaccines and reduction of the risk of side effects (13,14).

Production of quantities of the measles virus and collection of the virus-containing overflows were performed in L-68 and JQF cell cultures. The sterile viral overflows with a biological activity of not less than 4 lg of TCE50/0.5 ml were used for production of LMV. The material was mixed with stabilizers, frozen in trays or glass vials for 2 days at -50^0 C and subsequently subjected to lyophilization drying for 48 hours. Dry virus-containing material was mixed with a filler and tabletted.

Nine laboratory experimental batches of the LMV tablet form (500 pieces in each batch) were produced on the laboratory unit using the optimum composition of stabilizers and fillers. Additionally, an experimental pilot batch of the tablet LMV (400 tablets) was manufactured. Lactose (49%) and calcium stearate (1%) were used as fillers.

The tablets were coated with an acid-resistant layer on the basis of acetylphtalylcellulose. All tablet batches produced were characterized in their physical-chemical properties, biological activity, occurrence of alien microflora and meet the requirements of the Russian Pharmacopeial Committee for tablet forms of the preparation. Biological activity of the virus in this batch amounted 3.4- 4.2 lg of TCE50 per tablet (see Tabl.1).

Table 1

Data on certification of experimental specimens of the live measles vaccine in tablet form

Quality characteristics	Requirements to quality characteristics (State Pharmacopea)	Match between the requirements and characteristics of the experimental vaccine batches (batch number)			
		1	2	3	4
1. Mass	0.1-1.0 g	0.22	0.245	0.23	0.207
2. Dissociative characteristics of the tablet with acid-resistant coating	May not dissolve during 1 hour in 0.1M HCl; dissociates during 1 hour in sodium hydrocarbonate solution.	match	match	match	match
3. Moisture mass fraction	Not over 3%	1.44	1.83	1.56	1.61
4. Biological activity	Not less than 3 lg of TCE50 per tablet	3.4	3.52	4.15	3.45
5. Acute toxicity on animals	Not toxic	-	-	match	-
6. Alien microflora	Not over 1,000 live bacteria, not over 100 fungi	match	match	match	match

Control of immunogenicity of the LMV experimental batches of tablets for oral administration was performed in rabbits by intraduodenal preparation administration. Determination of viremia, changes in humoral and cellular immunity in blood sera of rabbits immunized with the live measles vaccine was performed on days 0, 7, 14, 21 post immunization. Table 2 presents the obtained results.

The performed studies revealed that the LMV for oral administration causes the stimulation of humoral and cellular immunity in a major part of laboratory animals. This stimulation manifests itself as an induction of antibodies in the blood serum and formation of the clone of memory cells which acquire the ability to transform into the blastic forms when encountering in vitro with the antigen for the second time. The lack of immune response in a part of animals may be associated with the inadequacy of the laboratory model for estimation of the live measles vaccine.

Table 2

Determination of viremia, changes in humoral and cellular immunity in blood sera of rabbits immunized with the live measles vaccine

Preparation, dose per animal	Biologycal activity (lg of TCE50 /tablet)	Rabbit's number	Viremia (day)	RHAR (inverse titer)		IEA (inverse titer)		RBTL (proliferation index)
				0	*21*	*0*	*21*	
Placebo, 2 tablets	0	1	-	<2	<2	<8	<8	1.22
		2	-	4	4	8	8	1.22
LMV for injections, 1 dose	4	3	-	8	16	16	32	2.2-1.6
		4	-	<2	<2	<8	<8	1.07
		5	7	4	16	16	32	1.12
		6	-	4	4	8	8	1.02
LMV (tablet form), 2 tablets	3.98	7	-	<2	4	<8	16	0.8
	3.98	8	-	<2	4	<8	32	1.0
	3.98	9	7-9	<2	16	8	64	1.64
	4.15	10	-	<2	2	<8	8	1.68
	4.15	11	-	<2	8	8	16	1.76
	4.15	12	-	<2	<2	<8	<8	1.26

Immunization of green monkey with the dose of 2000-4000 TCE_{50} also proves the antibody induction in some animals.

Control of innocuousness of the experimental batch of the LMV produced in L-68 cell culture was performed in white mice and guinea pigs, 1, 2, and 5 doses of the vaccine being administered orally. The toxic effect was estimated according to the general condition and behavior of the animals, changes in the blood formula and biochemical tests of the liver (protein content, content of protein fractions, alanine and aspartate-aminotransferases, urea content in blood serum). The recorded deviations were of reversible character and laid within the range of permissible values.

Thus, studies carried out to test the peroral form in vivo and in vitro showed the possibility of creation of LMV tablet form, nontoxicity for animals and causes the stimulation of humoral and cellular immunity in the laboratory animals.

REFERENCE

1. Krasnova VN (1989) Abstracts of Candidate thesis,Moskow,The Russia.
2. Bektimirow TA (1990) Actual problems of global measles liquidation.ZhMEI 9,104-108,The Russia.
3. Husband AJ (1993) Novel vaccination.Strategies for the control of mucosal infection.Vaccine 11(2),107-112.
4. Taros LJu (1967) Experience of isolation of measles virus on the monolayer guinea pigs kidney cell cultures and on the chiken fibroblasts. Voprosi virusologii 4,399-402,The Russia.
5. Stepanova LG, Alekseev SB,Zgursky AA (1986) Isolation and characterization of a new strain of diploid cells from human embrionic lung tissue.Tsitologiya 28(12),1373-1376.The Russia.
6. State Pharmacopeia of the USSR (1987), 11 ed.,The Russia.
7. TPC 42-179 VS-94 (1994) "Live dry culture measles vaccine",The Russia.
8. Voller A, Bidwell D,Bartlett D (1976) Microplate enzyme immunoassays for the immunodiagnosis of virus infections:In Rose NR,Ed. Friedman H.Manual of Clinical Immunology,Washington, DC, American Society for Microbiology,pp 506-512.
9. Nait S (1988) Analysis of Lymphocyte proliferation.In: Klaus GGB (Ed.) Lymphocytes A practical approach,IRL,Press, Oxford,Washington DC,pp 286-311.
10. Andries K,Rombaunt B,Dewidt B (1994) Discrepancy bettwen infectivity and antigenicity stabilization of oral poliovirus vaccine a capsid -linding compound.J.Virol 68(5),3397-4000.
11. Grillner L (1976) Immunity of intranasal challenge with rubella virus two years after vaccination comparison of three vaccines J. Infec.Diseases 133(6), 637-642.
12. Freestone D, Kelly A,Ferris S,Simmons R (1980) Comparison of antibody responses and virus sheadding following administration on of trivalent oral polyomyolitis vaccines prepared either on monkey kidney or human diploid substrates.J.Hug 84(1),17-28.
13. Sabin A,Arechigo A, Fernandez de Castro (1983) Succeful immunization of infants with and without material antobody by aerosolized measles vaccine.JAMA 249, 2651-2662.
14. Yuminova NV,Krasnova VP,Ljashenko VA (1993) Specific activity and immunologic sabety of intranasal revaccination with live measles vaccine of Leningrad-16 strain.Voprosii Virusologii 3,111-113,The Russia.

ACKNOWLEDGEMENTS.

This work was supported by Grant from the International Scientific Tecnical Center, Project 133.

SESSION IV

Animal Cell Process Engineering

ON-LINE MONITORING FOR CONSISTENT AND OPTIMAL PRODUCTION OF BIOLOGICALS FROM MAMMALIAN CELL CULTURES

S. S. OZTURK, J. BLACKIE, P. WU, R. TATICEK, K. KONSTANTINOV, C. MATANGUIHAN, Y. TSAI, S. YOON, J. THRIFT, and D. NAVEH
Bayer Corporation, Biotechnology, Berkeley, CA 94701

1. Abstract

Despite all efforts, process monitoring and control of mammalian cell cultures are still limited to physical variables such as pH, DO, and temperature. Although these are important parameters, cellular activities are equally governed by the biochemical composition of the media in the fermenter. If the cellular environment is not controlled, then cells alter their physiological state leading to batch-to-batch variations in production. To overcome some of these problems, we have developed a new process control system that allows on-line continuous determination of cell density, respiration activity, and metabolite and product concentrations. Cell densities are estimated by using on-line optical sensors. The respiration activities are determined using DO control output coupled with a mass transfer model. Measurement of glucose, lactate, and monoclonal antibody concentrations for a hybridoma cell culture was accomplished by using a YSI analyzer and a BioCad/RPM system operated in series. The application of this highly instrumented fermenter system is discussed in batch and continuous fermentation. Both direct and indirect (metabolic rate based) cell density measurements allowed adjustment of perfusion rate proportional to cell density. Using this "cell specific perfusion rate" control we could establish a constant environment for the cells and minimize variations in production.

2. Introduction

Successful operation of bioreactors for mammalian cell culture often requires a close monitoring of the cellular environment. Cell physiology is affected by environmental conditions such as temperature, pH, dissolved oxygen (DO), and metabolite concentrations (Ozturk & Palsson, 1991; Miller et al, 1988). Although some process variables such as temperature, pH, and DO are monitored and controlled continuously, metabolite concentrations normally require off-line sampling and analysis. Traditionally samples are drawn once a day and the analysis is performed afterwards. Rapid assessment of metabolite concentrations is critical in more sophisticated mammalian cell bioreactors such as fed-batch and high density perfusion cultures (Zhou & Hu, 1994; Ozturk, 1996). In fed-batch culture, limiting nutrient(s) are added continually to the reactor to maintain a certain level of nutrient and/or waste products. In perfusion reactors, the feed rate to the reactor is adjusted constantly in response to cellular activity so that a relatively constant cellular environment can be accomplished. In both cases, the success of the feeding strategy is dependent on the frequency of sampling and the process control algorithm used for the adjustment of feed rate.

M.J.T. Carrondo et al. (eds.), Animal Cell Technology, 205–214.

Since the inception of mammalian cell culture technology, a significant research effort has been dedicated to on-line monitoring and control of cell culture fermentation (Ozturk, 1996). The ultimate goal of these studies is to control the cell physiology by manipulating the culture conditions (Figure 1). On-line monitoring of cellular environment (pH, DO, metabolite and product concentrations) can be combined with mass balances to calculate several rates, including metabolite consumption and productivity in the fermenter. Estimation of cell density allows direct assessment of cellular activities. A simple feed-back algorithm can then be established to control the cell physiology based on several control strategies, including knowledge-based and adaptive controllers. The control of cell physiology in the fermenter by directly monitoring the cellular activities minimizes batch-to-batch variations in manufacturing and allow for operation of the fermenters at an optimal level.

In this paper we share our experience with on-line monitoring and control of mammalian cell culture bioreactors. Methods were developed for on-line monitoring of cell density, respiration rates, metabolite and product concentrations, and finally real time monitoring of cell physiology. On-line monitoring of cell density and cellular activities allowed the development of a new method for the control of perfusion cultures. This method is based on a feeding strategy proportional to cell density (cell specific perfusion rate control). This control system is also discussed in this paper.

3. On-line Monitoring of Cell Culture

Several techniques have been developed for monitoring cell density, respiration rates, metabolite and product concentration. These methods can also allow the calculation of cellular activities, such as specific growth, oxygen and metabolite consumption, and specific production rates.

3.1. Monitoring of Cell Density

Cell density is a very important parameter for the monitoring and control of fermentation processes. In the past two decades several probes were developed for the monitoring of cell density. Several in-situ probes have proven to be applicable for cell culture fermentation (Merten et al, 1985; Reardon & Scheper,1991; Yamane, 1993; Wu et al, 1995; Ozturk et al, 1995b).

Figure 2 presents the performance of optical (Aquasant, Switzerland) and capacitance (Aber Instruments, Ltd., England) cell density probes for perfusion cultures. Although optical density probes could not be recommended for low viability cultures (decline phase in batch culture), the probe worked extremely well in perfusion cultures. The calibration curve was linear up to 50 million cells and probe fouling was not a problem in most of the runs. The capacitance probe measures directly the viable cell density and it's performance is not affected by culture viability (data not shown). The problem we had with the capacitance probe was related to the necessity of regular cleaning of the probe tips, which is done externally. Once we established a good cleaning cycle, the probe reading stabilized (Figure 2).

3.2. Monitoring of Respiration Rates

Several methods have been developed for the measurement of oxygen consumption rates (OCR) in mammalian cell cultures. Although it can be automated, the dynamic method (Merten et al, 1986; Kyung et al, 1994) cannot provide continuous signal, and the measurement cannot be performed if the DO levels are low. Gas phase mass balances also have limitations for cell culture as the difference in the gas composition between the inlet and outlet is minimal for cell culture fermentation (Bonarius et al, 1995; Eyer et al, 1995). Alternatively, the OCR can be measured using gas phase oxygen composition (measured by a gas analyzer) and dissolved oxygen (measured by DO probe) using a mass transfer method (Ramirez and Mutharasan, 1990; Eyer et al, 1995).

We have developed an alternative method for the estimation of OCR for cell culture fermentation (Ozturk et al, 1995a). The need for a gas analyzer was eliminated by the calculation of inlet gas composition from mass flow meter outputs controlled by a DO controller. By the use of the DO probe reading and a mathematical model, we could estimate the OCR and outlet gas composition (no additional hardware). Figure 3 presents some of the data obtained for the OCR estimations for a perfusion culture. The OCR data could then be used for the estimation of cell density as presented in Figure 3.

3.3. Monitoring of Metabolite and Product Concentrations

Despite all efforts, no reliable in-situ probe is available for metabolite and product concentrations. However, several investigators have developed FIA or HPLC techniques for analyzing the fermenter samples using off-line analyzers (Huang et al, 1991; Kurokawa et al, 1994, Ozturk et al, 1996; Fenge et al, 1991; Freitag et al, 1991; Middendorf et al, 1993; Polzius et al, 1993; Gebbert et al, 1994; Ozturk et al, 1995c). If the sampling and analysis is fast enough, then "real-time" data can be generated and used for process control. The key to these developments is the use of an aseptic sampling device that allows the sterility of the operation and the generation of samples for a long period of time.

For the measurement of metabolites such as glucose, lactate, and glutamine a commercial YSI analyzer can be coupled to the fermenter (Figure 4). Cell free samples are generated by the use of on-line or off-line filtration devices and aseptic sampling is achieved by alternating the flow of sample and cleaning solution (NaOH) at the end of the sampling loop. The performance of the on-line YSI analyzer for glucose measurements is presented in Figure 4 for a perfusion culture. Although operational problems such as fouling of the filtration devices and changing the membranes of the YSI analyzer have been encountered, the performance of YSI analyzer was satisfactory. Figure 4 also indicates glucose control at desired set points. Glucose was monitored continuously and the perfusion rate was altered dynamically to maintain the glucose set points..

A new analyzer from YSI can measure glutamine and glutamate. The performance of this analyzer for glutamine measurements is presented in Figure 5. The YSI analyzer performed well against the more laborious HPLC technique. The YSI glutamine analyzer was also coupled to the fermenter for on-line monitoring. The time profile of glutamine concentrations are presented in Figure 5 for a perfusion culture. Despite the variations in cell density, the continual adjustment of perfusion rate resulted in relatively constant glutamine levels.

The on-line measurement of metabolite concentrations (glucose, lactate, glutamine) allow the calculation of metabolic rates. These metabolic rates, in turn, can be used for the estimation of cell densities (Ozturk et al, 1995b; Zhou and Hu, 1994). We have successfully used, for instance, the glucose consumption rates for cell density estimation (data not shown). These estimates of cell density are useful for process control purposes as we will discuss later.

The use of fast affinity chromatography allowed monitoring of product concentrations in "real-time". Figure 6 demonstrates the system used for on-line monitoring of proteins in a cell culture fermentation. We have previously demonstrated the utilization of BioCad/RPM system (Perseptive Biosystems) for the monitoring of IgGs in batch and perfusion cultures (Ozturk et al, 1995c). The same methodology can be applied to a variety of products. When a specific antibody for the product is conjugated to the columns, then affinity chromatography can be utilized for the determination of the product concentration. Figure 6 presents the data obtained for t-ICAM production in perfusion cultures. After the selection of a proper filtration device, we could obtain continuous data on product concentrations, and these data compared well with those obtained from off-line measurements.

4. Development of Cell Specific Perfusion Rate (CSPR) Control

Our main goal for dealing with on-line monitoring systems was the development of a control system for perfusion cultures. These systems operate under inherently unstable conditions with dynamically changing cell densities and cellular activities (Ozturk et al, 1996). The fermenters run for long periods of time and slight variations in cellular environment during the startup of the cultures can affect the long term performance. As a result, batch-to-batch variations are more pronounced for perfusion cultures.

We have developed a simple yet powerful strategy for the control of perfusion reactors (Ozturk, 1996). This method is based on a feeding strategy proportional to cell density (cell specific perfusion rate, CSPR, control) which can be measured directly or can be estimated from metabolic rates. This algorithm was implemented on a Macintosh computer.

Figure 8 illustrates the performance of CSPR control for establishing a constant environment for the cells. Cell densities were estimated using a combination of on-line cell density probe readings and oxygen consumption rates. The controller then adjusted the perfusion rate proportional to cell density to maintain a constant CSPR. The net result was maintenance of constant metabolite and product concentrations (only glucose and lactate is shown).

The stability of *CSPR* control is also presented in Figure 8. The system was stable for 16 days. Then a disturbance was applied to the system by purging the cells from the reactor. The control system could react to the disturbance quickly and continued to maintain glucose and lactate concentrations after the incident. Figure 6B also presents the product (IgG) concentration also determined on-line (Ozturk et al, 1995c). The product levels also remained constant after the disturbance.

5. Conclusions

The following conclusions can be made from the present study:

1. Cell density sensors work reasonably well for estimating the cell number in the reactor as long as viability remains high.
2. Monitoring of metabolite and product concentrations needs development of in-situ probes. In the mean time, however, on-line measurements can be performed using special aseptic sampling devices.
3. Cell density in the reactor can be estimated using metabolic rates
4. A constant cellular environment for the cells can be created by continually adjusting the perfusion rate proportional to cell density (CSPR Control).
5. Cell specific perfusion rate controls cell metabolism and production rates, thus can be used as a "lumped" control parameter for optimal and consistent production.

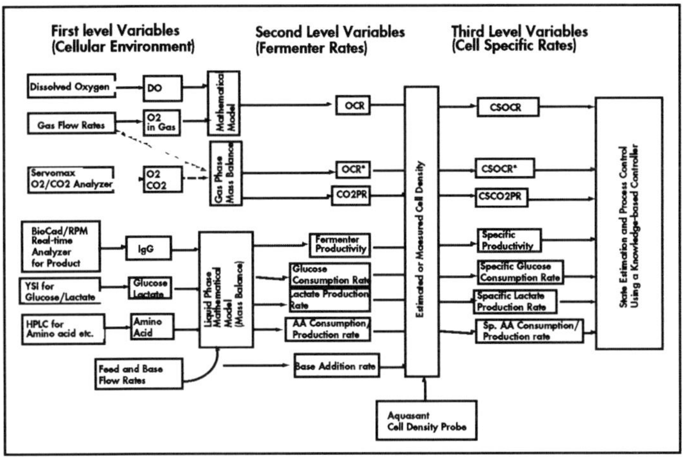

Figure 1. Continuous On-Line Measurement of Cell Density and Cellular Activities

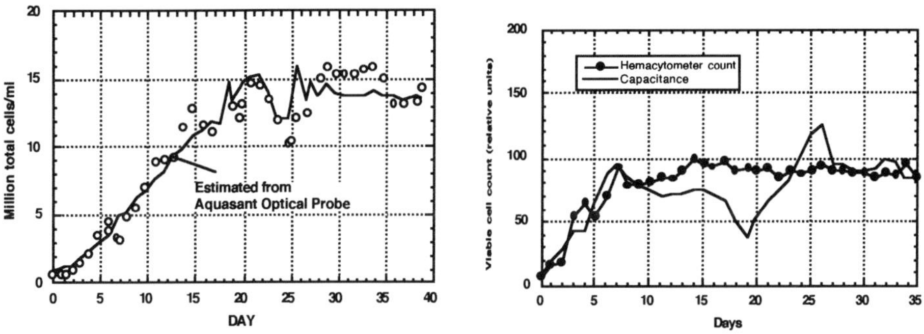

Figure 2. On-line estimation of cell densities using cell density probes

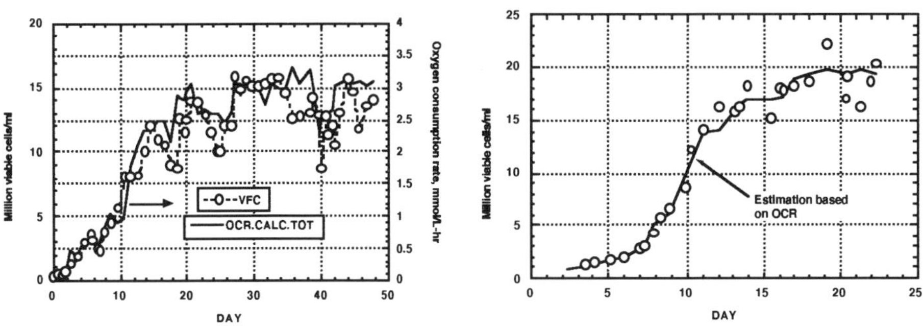

Figure 3. Continuous on-line estimation of oxygen consumption rate (OCR) and estimation of cell density based on OCR (different fermenters and different cell lines are used in these plots).

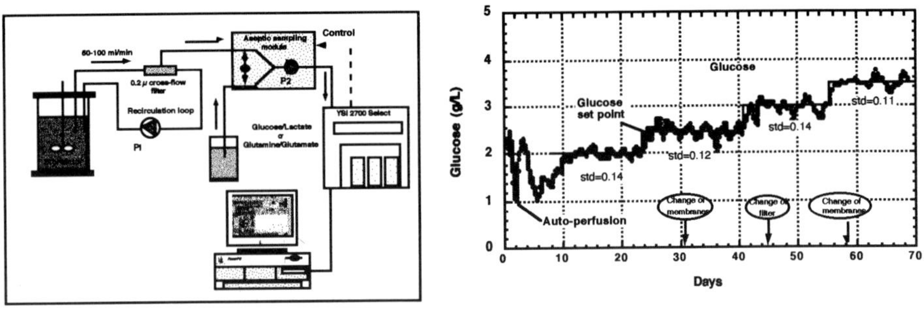

Figure 4. Left panel: System for on-line monitoring of glucose, lactate, glutamine, and glutamate using the YSI analyzer. Right panel: Monitoring and control of glucose.

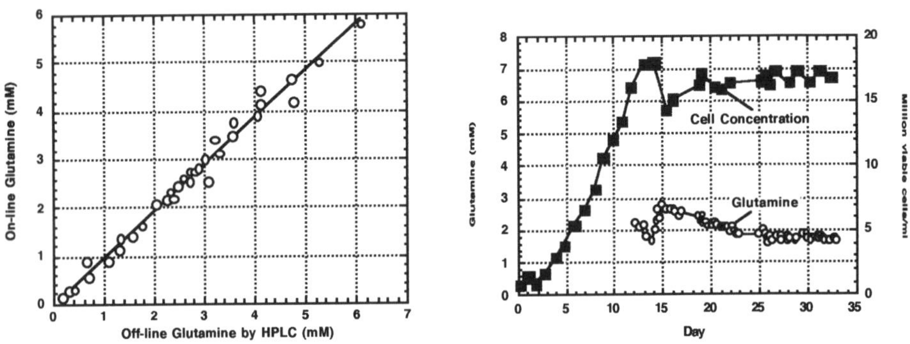

Figure 5. On-Line Monitoring of Glutamine Using the YSI Analyzer

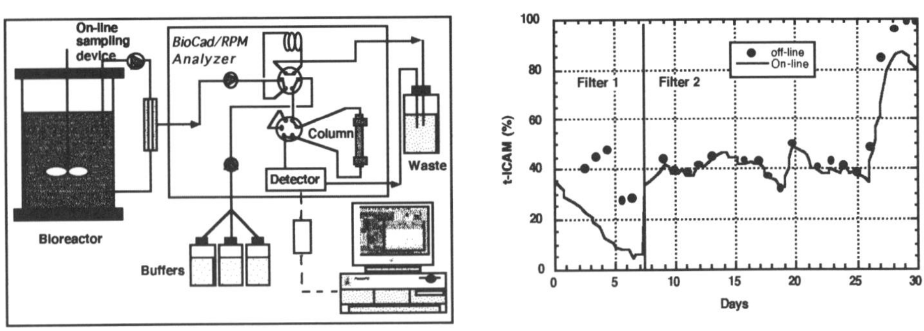

Figure 6. On-Line Monitoring of proteins from cell culture using fast affinity chromatography.

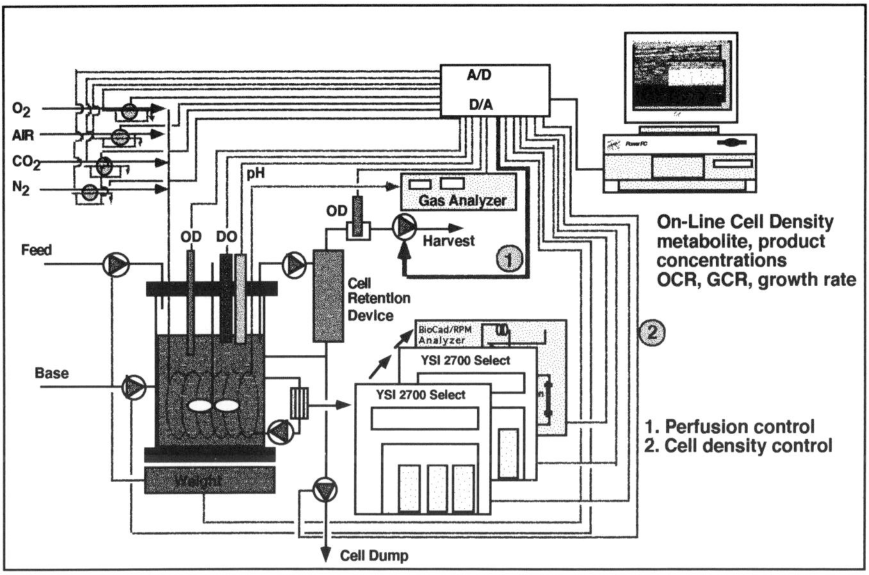

Figure 7. Reactor system and control system for perfusion rate and cell density control.

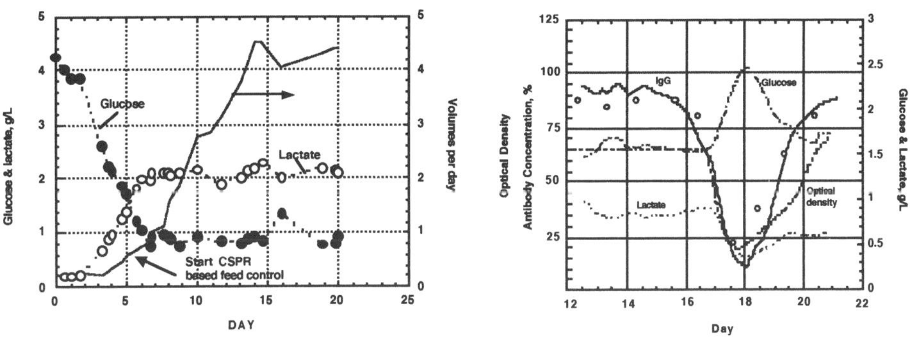

Figure 8. Cell specific perfusion rate control. Left panel: Establishment of Constant Environment for the cells (constant glucose and lactate), Right panel: Stability of CSPR control. After the major disturbance, the controller brings the conditions back to the previous set points.

6. References

Bonarius, H.P., de Gooijer, C.D., Tramper, J. Schmid, G. (1995) in Animal Cell Culture Technology, Ed. E.C. Beuvery et al, Kluwer Academic Publishers, Netherlands, pp. 677-681

Eyer, K., Oeggerli, A., Heinzle, E. (1995) On-line gas analysis in animal cell cultivation: II. Methods for OUR measurements and its application for control, Biotech. Bioeng. **45**: 54-65

Fenge C. et al (1991) On-line monitoring of monoclonal antibody formation in high density perfusion culture using FIA, Cytotechnology, **6**: 55-63

Freitag, R.., Fenge, C., Scheper, T., Schugerl, K. (1991) Immunological on-line detection of specific proteins during fermentation processes, Anal. Chim. Acta, **249**: 113-122

Gebbert, A., Alvarez-Icaza, M., Peters, H., Jager, V., Schmid, R.D. (1994) On-line monitoring of antibody production with regerable flow injection immuno system. J. Biotechnology **32**: 382-388

Huang, Y., Li, S., Drmel, B., Bilitewski, U., Schmid, R. (1991) J. of Biotechnology **18**: 161-172.

Kurokawa, H., Park, Y. S. , Iijima, S. , Kobayashi, T. (1994). Growth characteristics in fed-batch culture of hybridoma cells with control of glucose and glutamine concentrations. Biotech. Bioeng., **44**: 95-103.

Kyung, Y.-S., Peshwa, M.V., Gryte, D.M., Hu, W.-S. (1994) High density culture of mammalian cells with dynamic perfusion based on on-line oxygen uptake measurements, Cytotechnology, **14**: 183-190

Miller, W. M.; Blanch, H. W.; Wilke, C. R. (1988) Biotech. Bioeng., **32**: 947-965.

Merten, O., Palfi, G., Steiner, J. (1986) On-line determination of biochemical/physiological parameters in the fermentation of animal cells in a continuous and discontinuous mode. Adv. Biotechnol. **6**: 111-178

Merten, O. W., Palfi, G. E., Staheli, J., Steiner, J. (1985) Invasive infrared sensor for the determination of the cell number in a continuous fermentation of hybridomas. Develop. Biol. Standard. **66:** 357-360.

Middendorf, C., Schulze, B., Freitag, R., Scheper, T., Howaldt, M., Hoffmann, H. (1993) On-Line immunoanalysis for bioprocess control. J. Biotechnology **31**: 395-403.

Ozturk, S.S., Palsson, B.O. (1991). Growth, metabolic, and antibody production kinetics of hybridoma cell culture: 2. Effects of serum concentration, dissolved oxygen concentration, and medium pH in a batch reactor. Biotechnol. Prog, **7**, 481-494.

Ozturk, S. S., Blackie, J., Thrift, J., Naveh, D. (1995a) Perfusion rate control based on on-line monitoring of oxygen consumption rates in mammalian cell fermentation, ACS National Meeting, San Diego, CA

Ozturk, S. S., Blackie, J. D., Wu, P., Figueroa, C., Thrift, J. C., Naveh, D. (1995b) in Animal Cell Culture Technology, Ed. E.C. Beuvery et al, Kluwer Academic Publishers, Netherlands, pp. 301-305

Ozturk, S. S., Blackie, J., Thrift, J., Naveh, D. (1995c) Real-time monitoring of mammalian cell fermentation for monoclonal antibodies using a computer controlled immunodetection (BioCad/RPM) system, Biotech. and Bioeng., in press

Ozturk, S. S., Blackie, J., Thrift, J., Naveh, D. (1996) Real-time monitoring and control of glucose and lactate in mammalian cell perfusion reactor, Biotech. and Bioeng., in press

Ozturk, S.S. (1996) Engineering challenges in high density bioreactors, Cytotechnology, to appear

Polzius, R., Bier, F.F., Bilitewski, U., Jager, V., Schmid, R.D. (1993) On line monitoring of monoclonal antibodies in animal cell culture using a grating coupler. Biotech. and Bioeng. **42**: 1287-1292.

Ramirez, O.T., Mutharasan, R. (1990) Cell cycle and growth phase dependent variations in size distribution, antibody productivity and oxygen demand in hybridoma cultures, Biotech. and Bioeng., **36**: 839-848

Reardon, K. F., Scheper, T. H. (1991) Determination of cell concentration and characterization of cells. p. 179-223 In: Shugerl, K. (ed.) Biotechnology, vol. 4. VCH, New York.

Wu, P., Ozturk, S.S., Blackie, J., Thrift, J.., Figueroa, C., Naveh, D. (1995) Evaluation and applications of optical cell density probes in mammalian cell bioreactors. Biotech. Bioeng. **42**: 312-323

Yamane, T. (1993) Application of an on-line turbidimeter for the automation of fed-batch cultures. Biotech. Prog. **9:** 81-85.

Zhou, W., Hu, W-S. (1994) On-line characterization of a hybridoma cell culture process, Biotech. & Bioeng. **44**: 170-177.

Acknowledgment

Experimental help from Gary Johnson, Susanti Martin, and Leonard McDaniels are greatly appreciated. Andy Johnson implemented the process control algorithm on Macintosh computers using LabView programming language.

Discussion

Bernard: Can you tell us what the effect of your strategy of maintaining your constant CSTR is on your product quality?

Ozturk: You want to be sure you are getting the same product, not only the concentration but consistency. You keep a constant environment for the cells - constant glucose, lactate, growth rate etc - then you will get a better quality product with good consistency.

Bonarius: I have a question on CO_2 production rates.. You said it was difficult to measure CO_2 production rates because of the sensitivity of the mass spectrometer. Are there other reasons for that like accumulation kinetics as mass spectrometry is very sensitive?

Ozturk: I talked about O_2 consumption rate, not CO_2 accumulation rate. For CO_2, as the fermenter does not use it at the inlet, the sensitivity will be high. I agree if you want to calculate the CO_2 production rate you have to look at accumulation in terms of bicarbonate and other components.

Zeng: If you have a number of sensors for on-line monitoring, how can you use this for control eg you use a YSI analyser for glucose and lactate but you need to control glucose at a very low level for optimum control. I am not sure that the YSI analyser can do that. Also specific perfusion rate. You showed a ratio of glucose/lactate of about 1•5. If you reduced the glucose concentration below 1 mM/litre you would get a much lower ratio. The ratio would be interesting for process control. Can you comment on how you could use this for an optimum control process?

Ozturk: You can use the YSI separately to control the glucose level but then we gave up on it. If you use a cell specific perfusion rate and you have a relationship between this and glucose then you can control the glucose by the perfusion rate. This is a better control than using off-line measuring systems. You can apply it for scale-up.

Fenge: What kind of product quality parameters are you looking at when you do your on-line product quality measurements?

Ozturk: I said it was possible that when you had a sample coming from the fermenter then you can also do tests for product quality using digestive methods etc.

On the Use of Electric Fields to Prevent Membrane Fouling of Process Integrated Cross Flow Filtration at Hybridoma Cell Cultures

P. v. Zumbusch*, R. Pörtner, B. Brun**, G. Brunner*, H. Märkl****

*** Arbeitsbereich Verfahrenstechnik II, TU Hamburg-Harburg, 21071 Hamburg**
**** Arbeitsbereich Biotechnologie I, TU Hamburg-Harburg, 21071 Hamburg**

Abstract

This work describes the use of alternating electric fields to prevent membrane fouling during process integrated filtration at low Reynolds-numbers. An experimental setup for perfusion cultur of suspended mouse/mouse hybridoma was realized. The filtration was performed in an external crossflow module which allows to generate an electric field over the membrane.

The first part of the study concerned the influence of shear stress on mammalian cells. The results imply that shear stresses in backpressure regulators severely damage the cells. Hence, backpressure regulators are not suitable to generate the transmembrane pressure difference during crossflow filtration of mammalian cells. Furthermore, wall shear rates in the filtration module above 0.97 N m^{-2} are detrimental to the hybridoma cells. That limits conventional crossflow filtration to Reynolds-numbers below 648.

Filtration experiments showed, that the alternating field diminishes the fouling layer on the membrane. By this action the permeate flux is enhanced. The electric field can only increase the life time of the membrane if it is combined with a constant flux mode of the filtration.

The alternating electric field does not harm the sensitive animal cells as long as parameters of the electric fields are chosen, where no electrolysis products are set free from the electrodes.

Introduction

Bioseparation on the basis of crossflow-filtration (CFF) is strongly limited by biofouling. Conventional anti-fouling techniques mostly involve the introduction of high shear stress to the membrane surface to sweep away fouling layers. Animal cells are highly sensitive to shear stress and thus most conventional anti-fouling techniques face severe problems in cell culture technology.

M.J.T. Carrondo et al. (eds.), Animal Cell Technology, 215–220.

The use of an electric ac field, generated perpendicular to the membrane to break the fouling layer by electrophoretic and electroosmotic movement is a new approach to prevent fouling of micro- and ultrafiltration membranes in biotechnology (see fig.1).

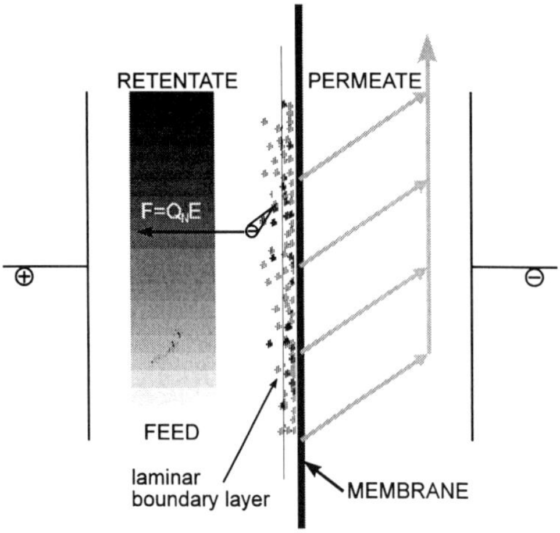

Fig. 1: Principle of crossflow-electrofiltration.

Material and Methods

The experimental setup is shown in fig. 2. The fermentation was performed in a stirred reactor with a working volume of 450 ml. The filtration device is a custom made plate and frame modul with an active membrane area of 9,17 cm^2. The transmembrane pressure difference was generated on the permeate side by a vacuum pump. The electrodes are rinsed separately with sterile fermentation medium. A polyether sulfone membrane (Hoechst, UF-PES 50H) with a cut off of 50 000 Dalton was used.

The m/m-hybridoma (IV F 19.23) was cultivated in complex medium (1:1 IMDM/HAMS F12), supplemented with 3% HS and other additives.

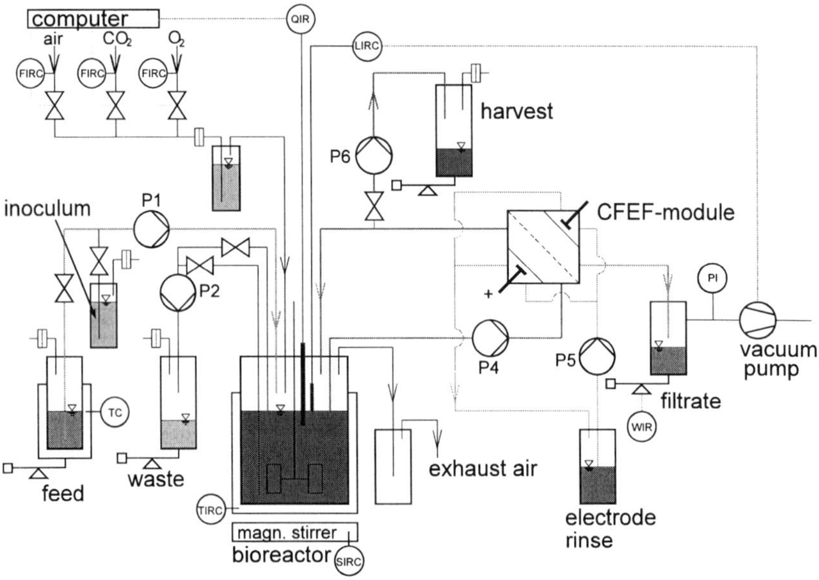

Fig. 2: Flow scheme of the fermentation.

Results and Discussion

The experiments to investigate the influence of shear stress in back pressure regulators on hybridoma cells were performed with a simplified experimental setup (see fig. 3). The cells are not damaged without pressure difference. That shows that a peristaltic pump is suitable for pumping animal cells. If the backpressure regulator generates a pressure difference, the number of cells decreases rapidly. An increasing pump rate does not increase the deterioration, whereas a rising pressure difference does. This implies that backpressure regulators shall not be used in the retentate stream of a filtration setup to filter hybridoma cells.

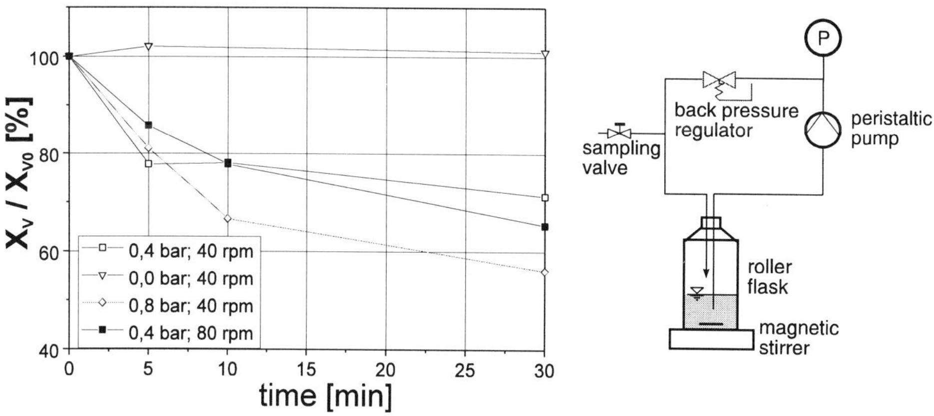

Fig. 3: Influence of shear stress in a backpressure regulator on the viability of hybridoma.

Fig. 4 shows the influence of the Reynolds-number in the filtration module on filtrate flux and cell viability. The filtrate flux in fig. 4 is the arithmetic mean of the filtrate flux during filtration experiments from the 20th to the 60th minute divided by the pure water flux of the clean membrane. The numbers of viable cells after 4 hours of a filtration experiment are related to the cell number at the beginning of each experiment.

The filtrate flux decreases with decreasing Reynolds-numbers. Even in the laminar region an increase of N_{Re} from 650 to 1200 can significantly ameliorate the filtrate flux. The filtration experiments were performed with sterile medium without cells and without serum. The filtration behavior with cells and serum is even worse.

At the same time, the number of viable cells after 4 hours at a Reynolds-number above 648 decreases. This Reynolds-number corresponds in our experimental setup to a wall shear rate of $\tau_w = 0.97 \text{ N m}^{-2}$.

The sensibility of animal cells against shear stress limits the conventional crossflow filtration to the region of very low Reynolds-numbers. The strong fouling tendency under that flow regime limits the life time of the membranes and yields a low transmission of macromolecular products. Ultrafiltration under those circumstances is hardly an economic separation process. Those results are a good motivation to search for efficient anti-fouling techniques that do not harm animal cells.

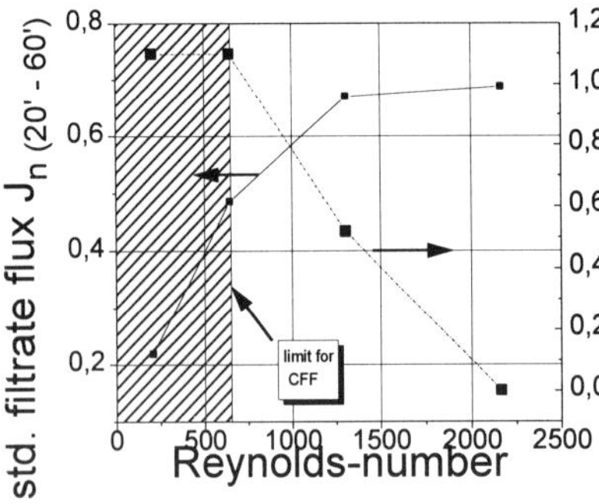

Fig. 4: Limitation of cross-flow filtration of hybridoma by shear stress.

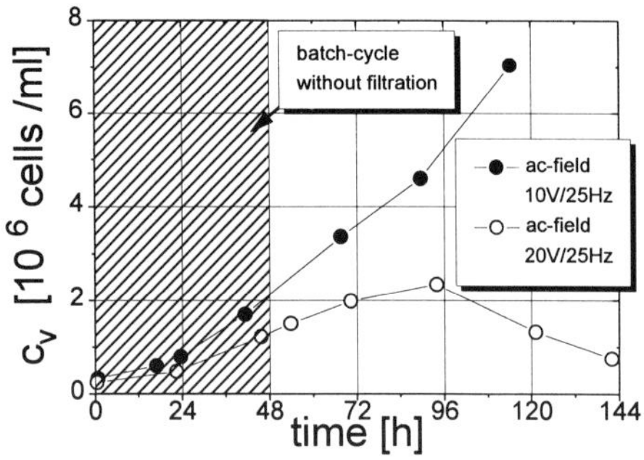

Fig. 5: cell density during fermentations with CFEF at different voltages.

The use of alternating electric fields does not affect the growth of animal cells, if the voltage is kept below a certain level. The critical voltage can be determined by observation of the electrode rinsing solution. If little bubbles are visible in the rinsing solution, the cells were allways affected. Those bubbles are electrolysis products that are set free at the electrodes. Whether or not electrolysis products are set free depends on the frequency, the voltage and the conductivity of the medium. A systematic study yielded that 15 V are the critical voltage at 25 Hz with fermentation medium in the filtration module. Fig. 5 displays that a fermentation accompanied by crossflow electrofiltration (CFEF) with a voltage (10 V) below that critical voltage is growing over several days. The filtration results of that experiment are shown in fig. 6. If a voltage of 20 V is applied, the growth is slower in the beginning of the experiment and finally the cell number is decreasing.

Furthermore the use of alternating electric fields does not lead to a shift of the pH-value or other negative symptoms that accompany the use of direct current [1, 2].

Fig. 6 shows the filtration results of the afore mentioned experiment. The electric field was not continously operated. It was switched manually as cleaning cycle. The application of the electric field yields a sharp rise of the permeate flux. This rise is followed by a decrease of the permeate flux, which is due to the fact that the high permeate flux carries an elevated amount of fouling matter towards the membrane. This not only counteracts the beneficial effect of the electric field. Sometimes the fouling after application of the electric field can even be worse than without anti-fouling techniques. That might be explained by the fact, that the fouling matter is transported convectively to the membrane with a higher speed than usually and thus it penetrates deeper into the membrane or it forms a fouling layer with higher density. This behavior was also observed for the use of direct current electric fields [2].

Those zig-zag shaped graphs for the permeate flux as a function of time are commonly reported [3] if cyclic anti-fouling techniques are combined with a filtration setup, that generates a constant transmembrane pressure (TMP).

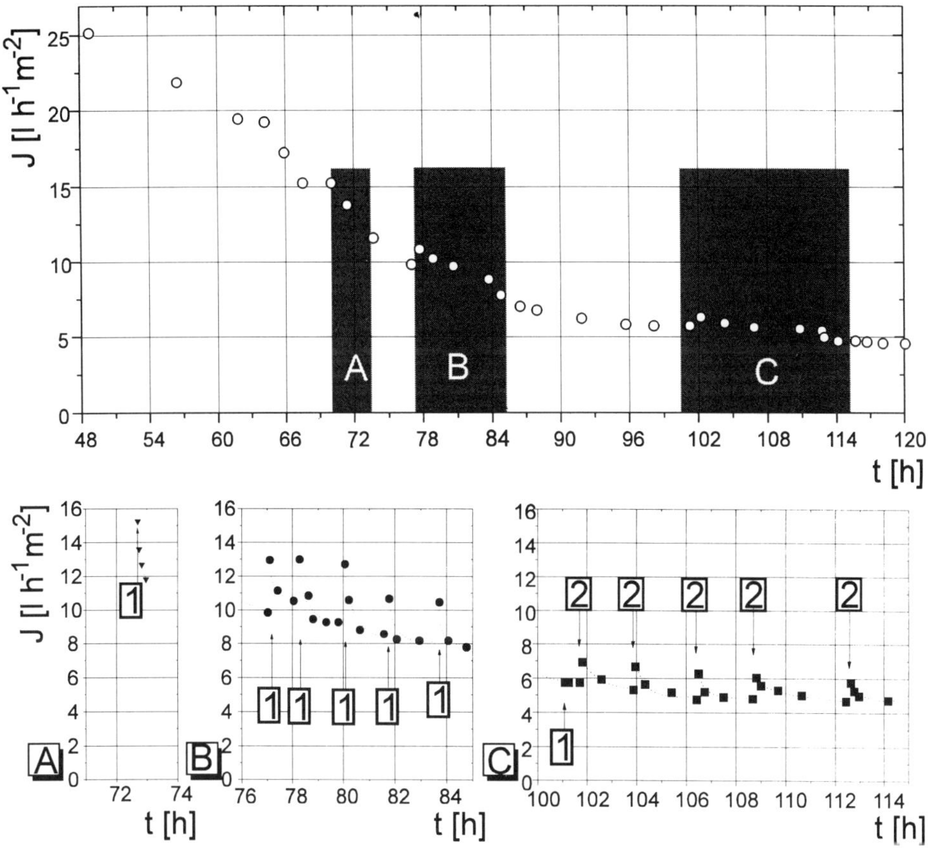

Fig. 6: filtrate flux during a perfusion fermentation with CFEF. TMP = 0,25 bar, Re = 648, Membran: PES 50H. <u>Top:</u> overview with average values; <u>Bottom:</u> during periods A,B and C the electric field was switched manually:
[1] = 5 min/±10V/25Hz; [2] = 5min/±20V/25Hz.

One possible solution to handle that problem might be the combination of anti-fouling techniques with a constant flow mode. The principle of that idea is presented in fig. 7. If a pump sucks a constant permeate flow over the membrane the TMP depends on the hydraulic resistance of membrane and fouling layer. If an anti-fouling technique diminishes the fouling layer the transmembrane pressure is decreased, but the filtrate flow stays constant.

The constant flux mode yields a better life time of the membrane (fig. 7). Combined with a constant permeate flux, the electric field diminishes the membrane fouling and the beneficial effect of the electric field is not counteracted by peaks of the filtrate velocity.

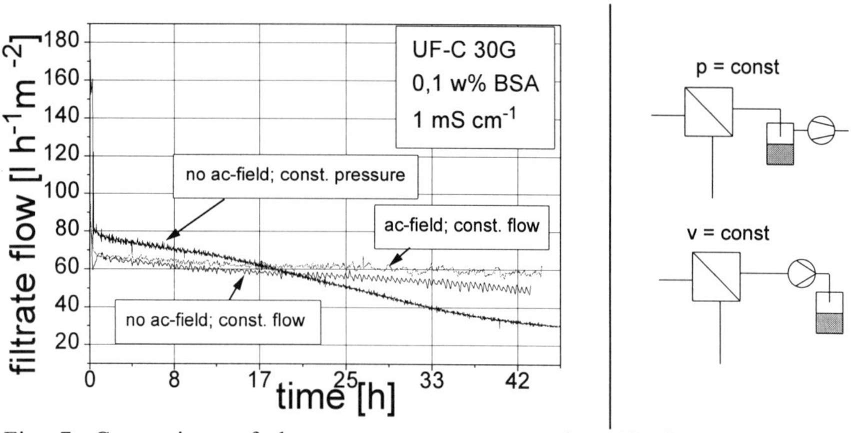

Fig. 7: Comparison of the constant pressure mode with the constant flux mode. Filtration of aqueous bovine serum albumin (BSA) solution.

References

1. Robinson CW, Siegel MH, Condemine A, Fee C, Fahidy TZ, Glick BR, *Pulsed-Electric-Field Crossflow Ultrafiltration of Bovine Serum Albumin*, J. Membr. Sci., 80 (1993) 209 - 220.

2. Okoro E, *Ultrafiltration proteinhaltiger Lösungen im überlagerten elektrischen Gleichstromfeld*, Fortschr.-Ber. VDI, Reihe 3 Nr 232, VDI-Verlag (1991).

3. Bowen WR, Sabuni HAM, *Pulsed Electrokinetic Cleaning of Cellulose Nitrate Microfiltration Membranes*, Ind. Eng. Chem. Res., 31 (1992b) 515 - 523.

Acknowledgements

The authors gratefully acknowledge financial support of this work by the Deutsche Forschungsgemeinschaft as part of the Sonderforschungsbereich 238 (Teilprojekt C3).

Simplified screening of monoclonal antibodies for immunoaffinity chromatography - A case study on anti-hIL-2-antibodies

J. H. Vogel, I. Randolph and K.-H. Kroner

Bioseparation Techniques Department
Gesellschaft für Biotechnologische Forschung mbH; Mascheroder Weg 1
38124 Braunschweig, Germany

1. Introduction

Immunoaffinity chromatography is an interesting feature for the direct capture of pharmaceutical proteins out of cell culture broths, which can lead to improved production schemes. A prerequisite for the development of such a capture step is the choice of the appropriate antibody ligand. However, testing and selection of antibodies by means of chromatographic procedures is tedious and needs larger amounts of antibodies to be applied. Therefore, ELISA techniques have been used for the estimation of binding and elution properties, leading to a pre-selection of monoclonal antibodies [1]. Nevertheless, the pre-selected antibodies had still to be produced in larger amounts in order to test the effects of covalent immobilization on antigen binding and long-term stability using standard column chromatography procedures. To circumvent this problem, a more simplified screening procedure (fig.1) has been developed based on ELISA techniques and covalent coupling to magnetic beads. The applicability of this novel strategy was successfully tested for the selection of an appropriate antibody for the immunoaffinity chromatography of recombinant human IL-2.

2. Materials and Methods

All tested anti-hIL-2 antibodies were mouse IgG1 or IgG2b, respectively. The antibodies 3C2, 3D2, 14D3, 6F3, 13C2, 1E7 and 1F10 were kindly provided as 1 ml ascites fluid by Prof. Knapp, University of Vienna. The E10 antibody was yielded out of cultivations of the

M.J.T. Carrondo et al. (eds.), Animal Cell Technology, 221–226.

E10 hybridoma cell-line (kindly provided by Prof. Schimpl, University of Würzburg) in a special medium developed in-house (DIF 1000 with supplements and 5% FCS). For comparison, a commercial anti-hIL-2 mab (Genzyme) was used.

An overview of the main steps of the screening procedure is given by fig.1.

2.1 Direct ELISA for quick estimation of binding/elution properties

ELISA plates (Maxisorp, Nunc) are coated overnight with purified hIL-2 (1 µg/ml in coating buffer: carbonate buffer pH 9.4). 10% FCS (Gibco) in phosphate buffer is used to block free binding sites (30 minutes) followed by washing with washing buffer (0.05 % Tween in deion. water). The monoclonal anti-hIL-2 antibodies are pre-diluted to equivalent concentrations after which the hIL-2 coated plates are incubated with dilution series of the different antibodies (mab concentrations c_0 from 1 µg/ml to 0.005 µg/ml) for 2 hours. Half of each plate is washed with washing buffer (method 1a), the other half with elution buffer (0.1 M glycine pH 2.5; method 1b). As can be shown in preliminary experiments, the adsorbed hIL-2 is not eluted using this procedure. After immunostaining with peroxidase-conjugated anti-mouse antibodies and o-phenylenediamine as substrate, the optical densities (ODs) are measured in an ELISA reader at 492 nm. The ODs yielded by method 1a and method 1b are plotted against the mab concentrations c_0. The OD values corresponding to a c_0 in the linear range of the curves can be used for a qualitative comparison of binding and elution properties. The elution yield is calculated by equation 1, with $OD_{c0,binding}$ and $OD_{c0,elution}$ as the OD values yielded by method 1a and 1b, respectively, for the same mab-concentration c_0.

$$elution\ yield\ \% = \frac{OD_{c0,binding} - OD_{c0,elution}}{OD_{c0,binding}} * 100\ \% \qquad (1)$$

1. Direct ELISA for quick estimation of binding/elution properties

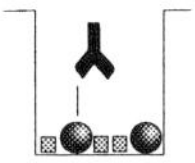

1. Binding of antibody to antigen-coated ELISA plate
2. Washing with washing or elution buffer, respectively
3. Detection of bound antibody

2. Combination of ELISA techniques for precise measurement of affinity constants

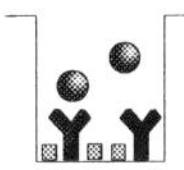

1. Coating of ELISA plate with antibody; determination of unbound antibody by sandwich ELISA
2. Incubation with antigen
3. Quantification of unbound antigen by ELISA assay

3. Covalent immobilization of selected antibodies on magnetic beads for testing of stability in repeated binding/elution cycles

1. Immobilization of antibody on magnetic beads with selected coupling chemistry
2. Testing of influence of covalent coupling on antigen binding
3. Repeated binding/elution cycles in microtubes

Fig. 1: Screening scheme

2.2 Measurement of affinity constants

ELISA plates (Maxisorp, Nunc) are incubated overnight at 4°C with the different mabs (3 µg/ml in coating buffer: carbonate buffer pH 9.4) The unbound mab concentration is measured by using an ELISA assay, so that the amount of antibodies adsorbed to each well is known. After blocking with FCS (as described in 2.1), the plates are incubated with dilution series of purified hIL-2. The amount of unbound hIL-2 in each well is quantified by using an ELISA assay. For each well the original concentration of the antibodies adsorbed in the well [AB] is known, as well as the starting antigene concentration [AG] and the unbound antigene concentration in equilbrium $[AG_{unbound}]$. The concentration of the antibody-antigene complex in each well is calculated from $[AG\text{-}AB] = [AG]-[AG_{unbound}]$ assuming monovalent binding. The affinity constant is calculated using the law of mass action. Arithmetic averaging leads to the K_A values given in results and discussion.

2.3 Immobilization on magnetic beads

In this study, magnetic beads precoated with anti-mouse antibodies (No. 14-104-4344, Paesel and Lorrei) were used for the coupling of the preselected monoclonal antibodies. 5 mg beads are incubated with 1 ml prediluted antibody solution (100 µg mab/ml) in a 1.5 ml microtube (Eppendorf). A powerful magnet (Magnetic Particle Concentrator, Dynal) is used for the separation of magnetic beads and supernatant. The beads are washed with PBS. Covalent cross-linking of the monoclonals and the anti-mouse antibodies (which functioned as spacer molecules) was performed according to [2]. The antibody concentrations in the supernatants of all steps of the coupling and cross-linking procedures are measured in order to calculate the achieved coupling density. The magnetic immunoaffinity beads are used in repeated binding/elution cycles as well as for measurements of the adsorption isotherm. All steps are carried out in 1.5 ml microtubes. In all cases hIL-2 concentrations are measured by an ELISA assay.

3. Results and Discussion

The procedure has been successfully tested for the selection of an appropriate antibody against hIL-2 (fig. 2; fig. 3). Studies with several mabs revealed that an antibody which

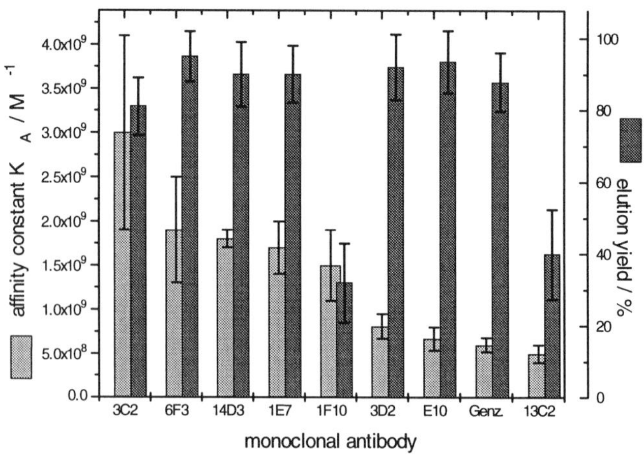

Fig.2: Determination of affinity constants (fig .1, method 2) and elution yields (fig. 1, method 1); elution buffer: 0.1 M glycine pH 2.5;

shows comparatively low affinity under binding conditions must not necessarily allow easy elution (fig.2). Thus the antibodies 1F10 and 13C2 showed medium or low affinity, respectively, but also low elution yield at pH-shift. On the contrary, it was possible to select

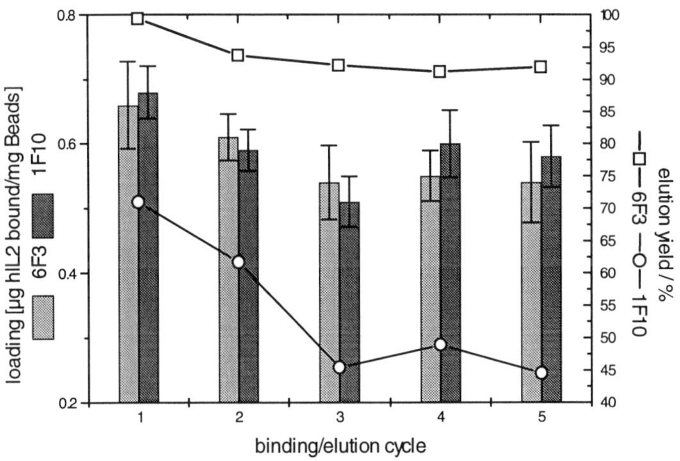

Fig. 3: Comparison of covalently immobilized mabs in repeated binding/elution cycles (fig. 1, method 3); coupling density: approx. 30 µg mab/mg beads for both mabs; elution buffer: 0.1 M glycine pH 2.5

an antibody (6F3, see fig.2) which shows comparatively high affinity under binding conditions, but simultaneously good elution efficiency at pH-shift. This antibody was coupled covalently to magnetic beads. Measurements of the adsorption isotherm showed an affinity constant K_A of $1.8*10^9$ M^{-1} compared to $1.9*10^9$ M^{-1} measured in the ELISA procedure, thus no negative influence of the immobilization procedure has been found. In repeated binding/elution cycles, the hIL-2 loading capacity decreased at first due to leakage of not covalently bound antibody, but stabilized after the second cycle (fig. 3).

The leakage of antibodies (and associated IL-2) also led to the very high apparent elution efficiency observed in the first two elution cycles; but as expected from ELISA results, the elution yield generally remained above 90 %.

Negative control experiments with the antibody 1F10, which was also covalently coupled to magnetic beads, showed much lower yields (fig. 3), which is also consistent with ELISA measurements (compare fig. 2). These results clearly demonstrate the usefulness and efficiency of the suggested screening procedure.

Conclusion

Applying the novel strategy, it is possible to evaluate not only affinity constants and elution efficiency, but also effects of covalent immobilization on antigen binding, as well as stability in repeated binding/elution cycles. The procedure is capable of testing many mabs in short time at micro-scale.

References

1. Bonde, M., Frokier, H. and Pepper, D. S. (1991): Selection of monoclonal antibodies for immunoaffinity chromatography: model studies with antibodies against soy bean trypsin inhibitor. Journal of Biochemical and Biophysical Methods, 23; 73-82.

2. Schneider, C., Newman, R. A., Sutherland, D. R., Asser, U., Greaves, M.F. (1982): A one-step purification of membrane proteins using a high efficiency immunomatrix. Journal of Biological Chemistry 257, 10766

Acknowledgements

The authors wish to thank Prof. Schimpl, University of Würzburg for kindly providing the E10 cell line and Prof. Knapp, University of Vienna for providing antibody samples.

INTEGRATIVE ANTIBODY PURIFICATION FROM WHOLE BROTH BY FLUIDIZED BED ADSORPTION

Jörg Thömmes, Andrea Bader, Andreas Karau, Maria-Regina Kula

Institut für Enzymtechnologie der Heinrich Heine Universität Düsseldorf, P.O. Box 2050, D-52404 Jülich, Germany

Introduction

The first operation during purification of proteins produced by mammalian cells usually is a solid-liquid separation of the cell containing culture broth, which is supposed to deliver a cell free supernatant. The protein concentration then is enhanced by a concentration step prior to further processing. This initial sequence of purification steps is referred to as primary recovery. Clarification of mammalian cell broths is performed by centrifugation (1) or crossflow filtration (2) successfully on a very large scale, but both methods may have certain disadvantages. Shear sensitive cells may be damaged in the centrifugal field of a large scale separator, thus contaminating the cell free supernatant with intracellular proteins or nucleic acids. Conventional crossflow filtration suffers from low flux rates as well as from fouling of the membrane by lipids, cell debris, and protein precipitates. Additionally the protein of interest may be adsorbed irreversibly to the membrane leading to a reduction in product yield during primary recovery. Advanced methods of dynamic filtration indicate solutions to circumvent these problems (3). In addition to some drawbacks of conventional technologies for primary recovery the number of sequential steps in a protein purification correlates with the overall time requirements of the process and thus with the total costs of protein production. Furthermore each additional purification step reduces the overall yield of product from the isolation procedure due to the inherent product loss of the individual operation. As the downstream process may contribute up to 80 % of the total cost of a protein production (4), a reduction of the number of steps using an integrative purification technology especially in the primary recovery phase, may prove to be beneficial to process economics.

M.J.T. Carrondo et al. (eds.), Animal Cell Technology, 227–233.

Fluidized Bed Adsorption

One possible approach to integrative primary recovery is the use of a fluidized adsorbent (5). A suitable chromatographic matrix is fluidized by a liquid flow directed upwards, in a steady state the particle supporting forces are balanced by the particle weight, so a stable fluidized bed is formed. The central property of the fluidized bed is the increased interstitial volume within the matrix, allowing the introduction of a particle containing feedstock. Choosing a fluidized adsorbent which is able to bind the protein of interest, a chromatographic separation starting from cell containing culture broth is possible. Thus a combination of clarification and initial purification in the sense of an integrative primary recovery is achieved. The situation is depicted schematically in figure 1.

Figure 1: Principle of fluidized bed adsorption

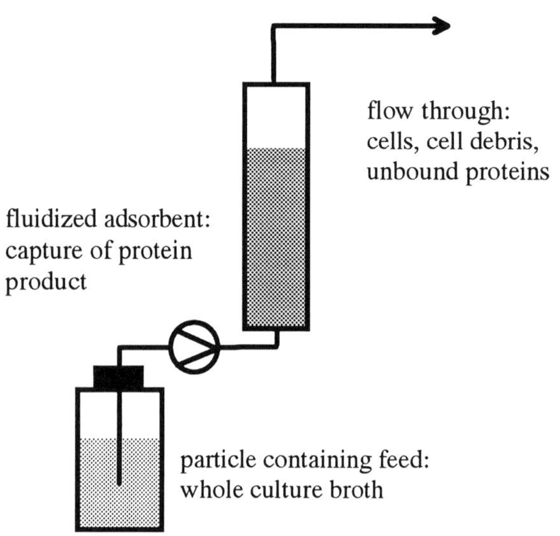

An important criterion for the success of a fluidized bed adsorption is the stability of the fluidization. A situation of reduced axial mixing in the solid and liquid phase is required for sufficient capacity and resolution of the capture process analogous to a standard purification in a packed bed, which is hampered by axial mixing as well. This is achieved by the development of a classified fluidized bed, where a distribution in adsorbent particle size and/or density causes a stratification of particles within the bed, leading to a reduced local mobility of the adsorbent particles and thus to the desired reduction in axial mixing. Using matrices with suitable properties, adsorption characteristics comparable to a conventional packed bed may be found (6).

Several products from mammalian cells have been purified successfully using fluidized adsorbents starting from hybridoma as well as from CHO whole cell broth (7-11). In the following, a procedure for the purification of different antibodies with a cation exchange as well as with an affinity adsorbent will be shown.

Cation Exchange Process

A commercially available cation exchange fluidized bed adsorbent (Streamline SP, Pharmacia, Uppsala, Sweden) was used in a laboratory scale column of 5 cm diameter. 300 mL adsorbent volume were fluidized with 10 mM potassium phosphate buffer pH 5.5 at 5 cm/min linear flowrate until a stable classified bed was formed. 10 L of cell containing hybridoma culture broth (approx. $1*10^6$ cells/mL) were adjusted to pH 5.5 and were subsequently applied to the stabilised bed at 5 cm/min. After sample application unbound proteins and residual cells were washed out of the column with equilibration buffer, finally the bound antibody was eluted with 10 mM potassium phosphate buffer containing 1 M NaCl at pH 8.0. The matrix was regenerated with 0.5 M NaOH and 1 M NaCl.

This procedure was performed with three antibodies produced by different hybridoma cell lines. The results are summarised in table 1. The capacity was calculated from the amount of antibody eluted per unit volume of adsorbent. The purification factor was determined from the ratio of MAb concentration to total protein concentration in the feed solution to the ratio in the eluate. The concentration factor was calculated from the ratio of the MAb concentration in the eluate to the MAb concentration in the feed.

As can be taken from table 1, the purification of the antibodies may be performed with reasonable purification and concentration of the protein of interest. The particle concentration in the eluate was reduced more than 100 fold compared to the particle load of the feed, so a successful combination of clarification, concentration and initial purification in the sense of an integrative primary recovery was achieved. A drawback of the method are the comparatively low capacities of the cation exchange matrix for capture of the different antibodies. This could be attributed to the high ionic strength of the feed solution. As the only manipulation of the broth was pH adjustion to 5.5, the high conductivity of the culture medium (13 mS/cm) remained unchanged during sample application. This high conductivity is detrimental to an ion exchange process, the concomitant reduction in adsorbent capacity is similar if the cell free culture supernatant is applied to a packed bed of a commercial cation exchanger (S Sepharose FastFlow, Pharmacia, Uppsala, Sweden), so the performance of the matrix in this case was not compromised by the fluidization process or the presence of cells but by the high conductivity of the original cell culture broth. Dilution increases the capacity of the matrix but necessitates additional manipulation and increases the sample volume, so a compromise has to be found between the elegance of original broth processing and performance of the ion exchange adsorbent.

Table 1: MAb purification from whole hybridoma culture broth with a fluidized cation exchange adsorbent

Antibody	Streamline SP Capacity (mg MAb/mL of adsorbent)	Purification (fold)	Concentration (fold)
IgG2a	0.58	4	4
IgG2b	0.6	4	13
IgG1	1.5	4	8

Protein A Affinity Process

As the performance of the ion exchange process was compromised by the high conductivity of the medium, a charge insensitive affinity process based on the ProteinA-antibody interaction was tested. A pre-commercial sample of Streamline rProteinA (Pharmacia, Uppsala, Sweden), was used. The material consists of weighted agarose, the development of a classified fluidized bed is attained by a density and size distribution within a batch of adsorbent particles. In principle the experimental procedure was performed analogous to the ion exchange process, the specific conditions, however, were adapted to the affinity interaction. The equilibration and wash buffer consisted of 50 mM potassium phosphate buffer pH 7.0, the elution was performed with 100 mM sodium citrate at pH 4.0. 60 L of cell containing broth were applied without prior conditioning at 6 cm/min linear flow rate. After elution the particle concentration in the feed was reduced more than 200 fold, a 30-50 fold concentrated antibody solution could be obtained at very high purity. In figure 2 the application of two different batches of 60 L cell containing hybridoma culture broth to 150 mL of fluidized Streamline rProteinA is shown. As can be seen from the effluent concentration, only small amounts of MAb were lost during sample application demonstrating the highly efficient MAb capture. The overall yield of the process was 85 % at 14 mg/mL capacity of the adsorbent.

Comparing the results of the affinity recovery in the fluidized bed to the ion exchange procedure, the increased efficiency of an affinity process is evident. The culture broth can be applied without pre-treatment, the capacity of the adsorbent is increased tenfold compared to the ion exchange operation due to the insensitivity of the affinity step to the high ionic strength in the broth.

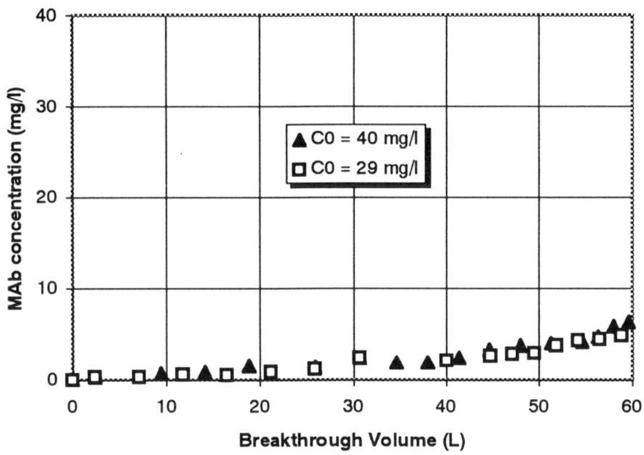

Figure 2: Application of two 60 L batches of cell containing hybridoma culture broth to 150 mL of rProteinA fluidized bed adsorbent (linear flow rate 6 cm/min)

There are, however, also some drawbacks of the protein A step. Cation exchange adsorbents are very robust and can be cleaned and sanitised with NaOH. This well established procedure cannot be applied to a protein A coated adsorbent. Additionally, in some cases adsorption of cells and cell debris to the matrix occurred, leading to the formation of cell/adsorbent aggregates. These aggregates may severely hamper the chromatographic procedure, in the worst case they may lead to the collapse of the fluidized bed. A possible solution is avoiding the formation of channels and dead zones within the bed from the very beginning of the process, as in these zones the formation of aggregates is initiated. A thorough cleaning of the adsorbent after each purification cycle is strongly recommended, as residual cells or cell debris still bound to the adsorbent after elution and regeneration may catalyse the formation of aggregates in the subsequent cycles. In our case matrix regeneration was performed using 2 M urea and 1 M acetic acid. During 15 purification cycles no deterioration of the performance of the adsorbent could be detected.

Conclusions

Fluidized bed adsorption may be an interesting alternative to conventional primary recovery of mammalian cell products. By fluidizing suitable adsorbents, cell containing broths can be applied to a chromatographic matrix, thus integrating the initial solid-liquid separation step into a chromatographic purification procedure. Cation exchange adsorbents can be used for application of whole broth at pH 5.5, their performance, however, is compromised by the high ionic strength in the culture medium. Nevertheless, a reliable process can be established, delivering a fourfold purified and up to 13 fold concentrated antibody preparation at more than 100 fold reduced particle load in the eluate. Protein A coated affinity adsorbents may be used as well, the affinity purification delivers an up to 50 fold concentrated product of very high purity at more than 200 fold reduced particle concentration in the eluate. The capacity of the affinity sorbent was 10 fold increased compared to the ion exchange matrix. Problems caused by adsorption of cells or cell debris to the matrix must be avoided by careful control of the fluidization process and thorough cleaning of the adsorbent after the purification cycles.

References

(1) Kempken R, Preissmann A & Berthold W (1995) Assessment of a disk-stack centrifuge for use in mammalian cell separation. Biotechnol. Bioeng. 46, 132-138.

(2) van Reis R, Leonard LC, Hsu CC & Builder SE (1991) Industrial scale harvest of proteins from mammalian cell culture by tangential flow filtration. Biotechnol. Bioeng. 38, 413-422.

(3) Berndt R, Ernst H & Kracht D (1993) Dynamische Filtration in Scherspaltfiltern - apparative Gestaltung, Anwendungsperspektiven und Einsatzvorbereitung. Chemische Technik 45, 145-151.

(4) Spalding BJ (1991) Downstream processing: Key to slashing production costs 100 fold. BIO/TECHNOLOGY 9, 229-233.

(5) Chase HA (1994) Purification of proteins by adsorption chromatography in expanded beds. TIBTECH 12, 296-303.

(6) Hjorth R, Kämpe S & Carlsson M (1995) Analysis of some operating parameters of novel adsorbents for recovery of proteins in expanded beds. Bioseparation 5, 217-223.

(7) Spence C, Schaffer CA, Kessler S & Bailon P (1994) Fluidized bed receptor affinity chromatography. Biomed. Chromatogr. 8, 236-2411.

(8) Erickson JC, Finch JD & Greene DC (1994) Direct capture of recombinant proteins from animal cell culture media using a fluidized bed adsorber. In: Griffiths B, Spier RE & Berthold W, Animal Cell Technology: Products for today, prospects for tomorrow, p.p. 557-560, Butterworth & Heinemann, Oxford.

(9) Thömmes J, Halfar M, Lenz S & Kula M-R (1995) Purification of monoclonal antibodies from whole hybridoma fermentation broth by fluidized bed adsorption. Biotechnol.Bioeng. 45, 203-211.

(10) Batt BC, Yabannavar VM & Singh V (1995) Expanded bed adsorption process for protein recovery from whole mammalian cell culture broth. Bioseparation 5, 41-52.

(11) Born C, Thömmes J, Biselli M, Wandrey C & Kula M-R (1996) An approach to integrated antibody production: coupling of fluidized bed production to fluidized bed adsorption. Bioproc. Eng. 15, 21-29.

Discussion

Lehmann: Your eluate is not totally particulate-free. Did you measure the particle size distribution to make sure no cells appear?

Thommes: Yes. The particle size distribution was measured with a Coulter counter with the orifice set at 50 microns, the detection limit is 1 micron, and the distribution was towards the 1 micron size. There are no cells.

Al-Rubeai: How did you sterilise this material? I understand protein A is not sterilisable.

Thommes: No, we do not sterilise protein A. The cationic exchanger can be autoclaved. The integrated process I showed used the cationic exchanger because you then have a totally autoclavable system.

Al-Rubeai: Have you tried to use your system with suspension culture?

Thommes: These are hybridoma cells.

Al-Rubeai: They are in microcarriers.

Thommes: With this fluidised bed process cell concentration in the outlet is 1 to $1 \cdot 5 \times 10^6$ cells/ml ie equivalent to a suspension culture.

Al-Rubeai: Do cells come in contact with your affinity legend?

Thommes: Yes.

Al-Rubeai: Are the cells damaged?

Thommes: No, the cells are viable. The residence time is only about 4-5 minutes. There is a large particle distance in the bed. The Reynolds number is $0 \cdot 1$ so you are in the laminar region.

Large-Scale Production of Humanized Monoclonal Antibody Expressed in a GS-NSO Cell Line

Nitya G. Ray, Roberto Rivera, Rajeew Gupta and Dale Mueller

Hoffmann-La Roche, Biopharmaceuticals Department, 340 Kingsland Street, Nutley,New Jersey 07110, USA

INTRODUCTION

Monoclonal antibodies intended for use as therapeutic agents must be produced in large quantities due to their high *in vivo* dosages.[1] Therefore, development of a cost-effective process that can be scaled-up without affecting product quality, is critical. It is also essential that the process be capable of producing consistent batches under CGMP regulations. In developing such a production strategy, the various process elements, from cell line and media to bioreactor design and control, must be critically examined and integrated using systems-level approach. Figure 1 illustrates a schematic for such an approach.

Final product titer is determined by culture longevity (integral cell density) and specific productivity, both of which can be significantly influenced by manipulating the environmental conditions. Fed-batch systems have been used successfully to enhance product titer, primarily by improving longevity of the cultures.[2] In this paper we present data from a 2000L bioreactor culture operated in fed-batch suspension mode to produce a humanized monoclonal antibody. The antibody is expressed in a recombinant mouse myeloma NSO cell line with the Glutamine Synthetase (GS) expression system.[3] In developing the cell culture process, particular emphasis was placed on predictive scale-up studies to identify the critical process parameters under operating conditions expected in the large scale bioreactors. For example, gas mass transfer to remove carbon dioxide, a metabolic byproduct, is an important factor in cell culture process optimization and scale-up.[4] Transfer of oxygen into and removal of carbon dioxide from the culture liquid are enhanced by bioreactor agitation and sparge rates. However, high agitation and sparge rates can be detrimental to cell growth and viability.[5] A significant difference in cellular metabolism, particularly in specific lactate production rates, was observed under different air sparge rates in a 2000L bioreactor culture.

Often, a process initially used to produce a biopharmaceutical product for clinical studies is modified to improve productivity in order to meet the market needs and to reduce the cost of goods. This is more pronounced for therapeutic antibodies where dosage requirements are generally significantly greater than those for cytokines, hormones, etc. The most dramatic productivity improvement can be achieved by changing the expression system to enhance the antibody secretion rate. However,this strategy often calls for a change in the host cell line and the expression vector.

M.J.T. Carrondo et al. (eds.), Animal Cell Technology, 235–241.

Regulatory acceptance of the modified process is subject to assesment of product equivalency. The humanized antibody presented in this paper was previously expressed in a recombinant SP 2/0 myeloma cell line. Biochemical characterization of the antibodies produced by both these cell lines at the 2000L bioreactor scale demonstrated that the products were equivalent.

Figure 1. Cell culture process optimization for production

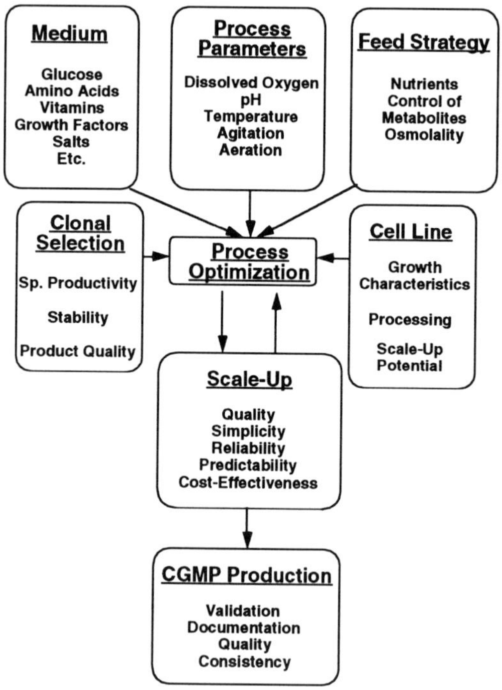

MATERIALS AND METHODS

Bioreactor Culture

The scaled-up process for the GS-NSO cell line consists of a train of seed bioreactors and a 2000L production bioreactor. The culture in the bioreactor is kept in suspension and is well mixed by agitation. Microprocessor based automation is provided to each bioreactor for control of temperature, pressure, dissolved oxygen, pH, agitation and gas flow rates. The inoculum culture for the first seed bioreactor is propagated using tissue culture flasks and spinners. Earlier, a 2000L airlift bioreactor at a different facility was used to produce the same antibody expressed in a recombinant SP 2/O myeloma cell line. Both the bioreactor cultures were operated in fed-batch mode and used serum-free media.

Analytics

The following off-line analytical measurements were performed on the bioreactor samples from the GS-NSO culture:

Total and viable cell counts were determined by dye (trypan blue) exclusion technique using a hemacytometer. Glucose and lactate were measured by a dual channel analyzer (YSI, model 2700 Select).

Antibody was quantified using an affinity chromatography column.

Antibodies produced by the GS-NSO and SP 2/O cell lines were compared by physicochemical and biochemical examination. The following analytical tests were performed:

SDS-PAGE, size exclusion chromatography, Western analysis, isoelectric focusing, peptide mapping, N-terminal analysis and carbohydrate analysis. A cell based bioassay was also used to compare bioactivity of the antibodies.

RESULTS AND DISCUSSION

Figure 2 represents typical GS-NSO cell growth and antibody production kinetics in the 2000L bioreactor culture. A concentrated mixture of supplemental feed, containing glucose and amino acids, was added at different time intervals. The antibody titer vs. integral viable cell density plot indicates that the specific antibody secretion rate remains relatively constant throughout the duration of the culture. The slope of the plot is a measure of specific antibody secretion rate.

Figure 2. Fed-batch culture of GS-NSO cells in a 2000L bioreactor: cell growth and antibody secretion kinetics

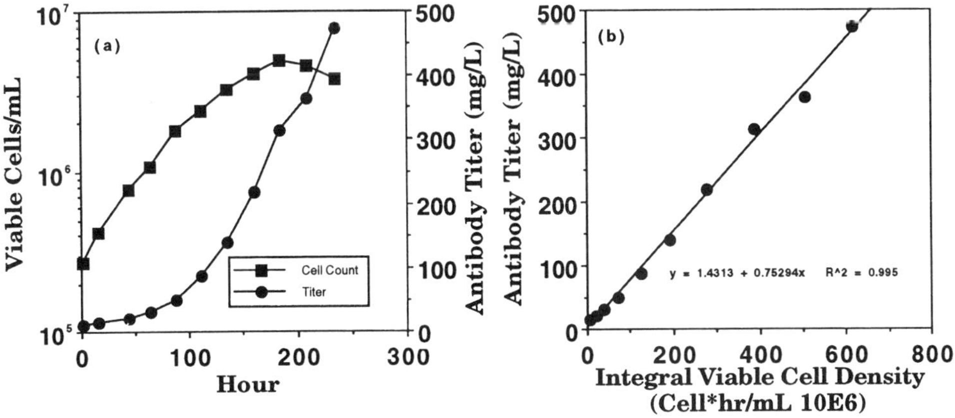

The effect of air sparge on cellular metabolism in the 2000L bioreactor culture is shown in figure 3. Figure 3(a) shows air sparge profiles for two different runs. In both the runs a 5µm sintered stainless steel sparger was used, dissolved oxygen levels were controlled and agitation rates were kept constant. Lactate concentrations are plotted against integral cell density in figure 3(b). The slopes of the lines represent specific lactate production rates (q_{lac}). q_{lac}'s were similar in both the runs until integral cell density reached ~ 200 10E6cell X hour/mL, which corresponds to approximately 120 hours of culture time in the bioreactor. Following this period, the q_{lac} for lower sparged condition increased significantly, while it remained essentially constant at higher sparged condition. The increase in q_{lac} at lower sparge rate was also accompanied by an increase in specific glucose consumption rate (data not shown). The observed differences in metabolism could be attributed to levels of carbon dioxide, a metabolic byproduct, in the culture. At the lower sparge rates, less gas-liquid interfacial area is available for removing carbon dioxide, which, at an elevated level is presumed to inhibit oxidation of glucose carbon and favor formation of lactic acid. The GS-NSO medium does not contain any glutamine.

Figure 3. Effect of air sparge rate on lactate production in a 2000L stirred-tank bioreactor culture producing a humanized monoclonal antibody expressed in a GS-NSO cell line

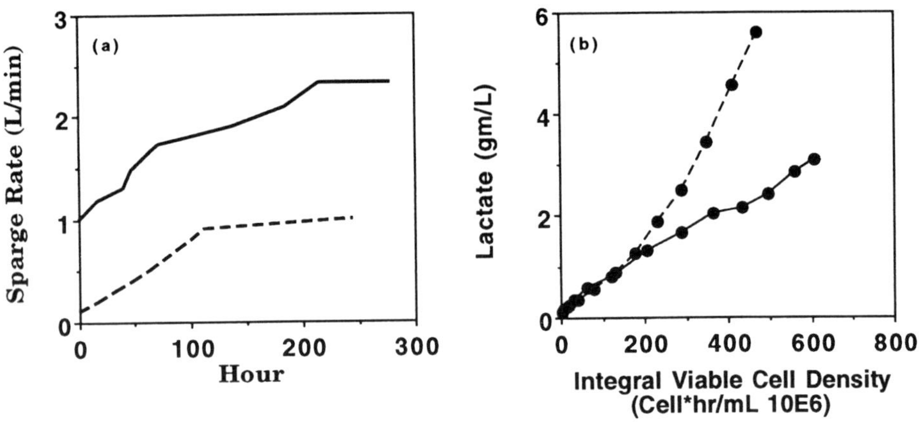

Consistency of the process at the 2000L bioreactor scale is demonstrated in Figure 4. The results from these four runs are summerized in Table 1.

Figure 4. Consistency of a cell culture process producing a humanized monoclonal antibody expressed in a GS-NSO cell line at 2000L bioreactor scale

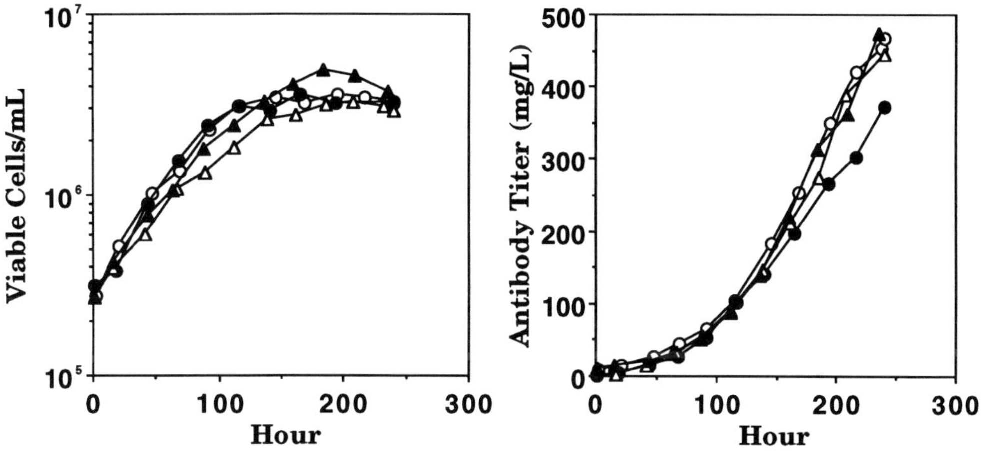

Table 1: Performance summary of the GS-NSO batches in 2000L bioreactor

Batch	Culture Duration (Hours)	Mab Titer (mg/L)	Integral Viable Cell (10E6CellXhr/mL)	Sp.Productivity (μg/10E6 cell/hr)	Viability at Harvest (%)
1	240.0	468	576	0.81	76
2	240.0	372	563	0.66	70
3	240.0	444	472	0.94	80
4	235.2	473	616	0.77	80
Average		439±11%	557±11%	0.80±14%	77±6%

As indicated in Figure 4 and Table 1, the viable cell density and culture viability remained high at the time of harvest. It has been observed that a 50% improvement in product titer can be achieved by extending the culture duration another three days (data not shown).

CONCLUSION

Development of a predictable cell culture process for high-yield large-scale production of monoclonal antibodies requires that critical issues, like expression systems, media, culture stability, scale-up parameters, product quality, etc. are addressed in the context of a systems-level approach. Of particular importance in process scale-up is gas mass transfer, specifically, removal of carbon dioxide. It is demonstrated that air sparge rate in a large-scale bioreactor is an important factor influencing cellular metabolism and nutritional requirements.

Consistency of a large-scale (2000L) fed-batch process using a GS-NSO cell line for the production of a humanized monoclonal antibody has been demonstrated. The same antibody was also expressed in a SP 2/O myeloma cell line. Although the antibody was expressed in two different host cell lines and different bioreactor systems were used for large-scale production, an extensive physicochemical and biochemical characterization of the protein demonstrated that both products are equivalent.

REFERENCES

1. Aulitzky,W., Schulz,T., Tilg,H., Niederwieser,D., Larcher,K., Ostberg,L., Scriba,M., Martindale,J., Stern,A., Grass,P., Mach,M., Dierich,M. and Huber,C. (1991) J. Infec. Dis. 163, pp 1344-1347.
2. Bibila,T.A. and Robinson,D.K. (1995) Biotechnol. Prog. 11, pp 1-13.
3. Brown,M.E., Renner,G., Field,R.P. and Hassell,T. (1992) Cytotechnology 9, pp 231-236.
4. Kimura,R. and Miller,W.M. (1994) Extended Abstract, AIChE Annual Meeting, November 13-18,1994, San Francisco.
5. Michaels,J.D., Kunas,K.T. and Papoutsakis,E.T. (1992) Chem. Eng. Commun. 118, pp 341-360.

Discussion

Zeng: You showed lactate production increased at lower sparging rates. Do you know the reasons for this? In scale-up do you use different or the same sparging rate to small scale?

Ray: You have a variable surface: volume ratio at small scale. The mass transfer interface is thus variable. At large scale it is not so variable. You can increase Kl_a by increasing the sparge rate but this will affect cell damage/viability. Using Pluronic F68 one can increase sparging but there is a limit. At large scale if CO_2 is not removed efficiently, it inhibits glucose oxidation and more carbon is channelled into lactic acid. At a higher sparge rate you have a higher Kl_a, better mass transfer and better removal of CO_2, so you get less lactic acid production.

2D PAGE examination of the proteins extracted from high density perfused HeLa cell cultures controlled by an on-line capacitance-based biomass probe

V. Degouys[1], C. Gatot[2], S. Sonderhoff[3], F. Trampler[3] and A.O.A. Miller [1*]
*1 Biochimie-Technologie Cellulaire - Université de Mons-Hainaut, 24 Avenue du Champ de Mars B-7000 Mons, Belgium; 2 Computer Cell Culture Center (4C) s.a., Belgium; 3 SONOSEP BIOTECH Richmond B.C., Canada; * Corresponding author.*

Keywords: Acoustic resonance, animal cells, biomass, biotechnology, capacitance, culture, perfusion.

Abstract

Capacitance measurement performed at 1MHz have been used to on-line, non invasively monitor and possibly control the biomass of long term high density perfused HeLa cells suspension cultures. Perfusion is realized by the combined action of a spin filter and an acoustic resonance retention device coupled together.

Introduction

Perfusion allows cells to be grown at high concentrations. Densities around 10^7 cells/ml - one order of magnitude higher than today's values - are not uncommon with this technique[(1)].
Validation of such high density, long term (4-6 months), perfused cultures and product consistency ultimately depend on the measurement and the close control of the cultivation process.
In this paper, cell consistency has been evaluated by comparing the 2D polyacrylamide gel electropherograms of total cell proteins extracted from HeLa cells collected at low ($1,8\times10^6$ cells/ml), middle (6×10^6 cells/ml) and high (12×10^6 cells/ml) densities from perfused cultures using an acoustic cell retention device and an on-line non invasive capacitance-based probe to evaluate the biomass.

Materials and Methods

HeLa cells strain Oxford have been grown in MEMS in a nominal 5L capacity glass jacketed APPLIKON bioreactor equipped with the Agitation Filtration Module (AFM), a spinfilter, built in house with improved performances. Perfusion occurs at one culture volume/day - 1.500 ml/24 h - and 120 rpm). Five percent CO_2 in air mixture under a pressure of 100 mm Hg at 10 ml/min flows in semi permeable silicone tubing wound inside the bioreactor. The gas mixture then flows above the surface of the culture and exits to atmosphere through a condenser cooled by tap water.

M.J.T. Carrondo et al. (eds.), Animal Cell Technology, 243–249.

Concentration of dissolved oxygen is kept at 20 % saturation using an Ingold Oxygen polarographic probe coupled to the APPLIKON's ADI 1020 controller. Capacitance measurements have been made using the Biomass monitor model 214 from Aber[2]. Cells have been acoustically retained in the 10 ml ultrasonic resonator linked to the UCCS-05 ultrasonic separation control system from SONOSEP. All parameters such as speed and length of forward and backward flows, the elapsed time between start of recirculation and interruption of forward flow and the slope of the ramp, can be individually adjusted by means of the Acoustic Retention Timer (ART) an electronic package developed by 4C which is connected to the various pumps as well as to the UCCS-5 unit. Cell concentration has been determined by counting an appropriately diluted cell culture sample in a Bürke counting chamber.

High resolution two-dimensional polyacrylamide gel electrophoresis (2D PAGE) is performed using the Millipore Investigator 2D Electrophoresis System (Millipore Corporation, Bedford, MA).

The gels are analysed on a Sun Sparc Station with the BioImage 2D Analyser Software 6.1 (B.I. Systems Corporation, Ann Arbor, MI).

Cell retention is realized by backflushing into the bioreactor the cell aggregates which accumulate in the resonator during forward flow as a result of the energy dissipated by the resonating acoustic wave. This is accomplished by reversing the pump providing the forward flow, progressively using a linear ramp to minimize "hydraulic shock". Sometime before this step, recirculation of cell free growth medium, pumped from inside of the AFM, is initiated. The stream of cell-free liquid thus obtained is injected obliquely in the lower conical part of the resonator, contributing by a kind of Venturi effect, to the emptying of the resonator. The simultaneous action of the reverse flow and of the recirculation last long enough to clean the resonator of the cells it contains. Recirculation then stops and forward flow resumes.

Careful quantitative and qualitative analysis of normalized electrophoretic patterns obtained from low-, middle- and high concentration HeLa cells cultures have been made and compared amongst themselves using the software to hand. To reduce complexity, the gels have been stained with Coomassie Blue only. Figure 4 is one such electropherogram.

2D PAGE is performed according to the method of O'Farrell with modifications as described by Garrels[3]. The proteins are purified from nuclear extracts corresponding to $1.25x10^8$ cells (according to the Millipore procedure [3]). The gels are analysed according to the B.I. Systems Corporation procedure [4].

Results

1) Evaluation and control of the biomass by capacitance measurements

In batch cultures, small changes of Gmedium, the conductance of the growth medium in which cells propagate, severely and unpredictably affect the medium capacitance Cmedium, so much so that using the relation Ccells = Csuspension - Cmedium, determination of Ccells, the capacitance of the cells to which biomass is proportional, can no longer be inferred from a measurement by the Biomass monitor of the total capacitance of the cell suspension Csuspension[2].

Continued removal by perfusion of wastes produced as cells multiply, dampens the effect that small changes in conductance have on capacitance. The almost complete suppression of the C-G change dependency observed around 1MHz with all the growth media (MEMS, DMEM, F12) most commonly used to propagate animal cells (Figure 1), itself adds to the above effect, the dip in the sensitivity 1MHz to 0,25 MHz being offset by the higher cell concentration.

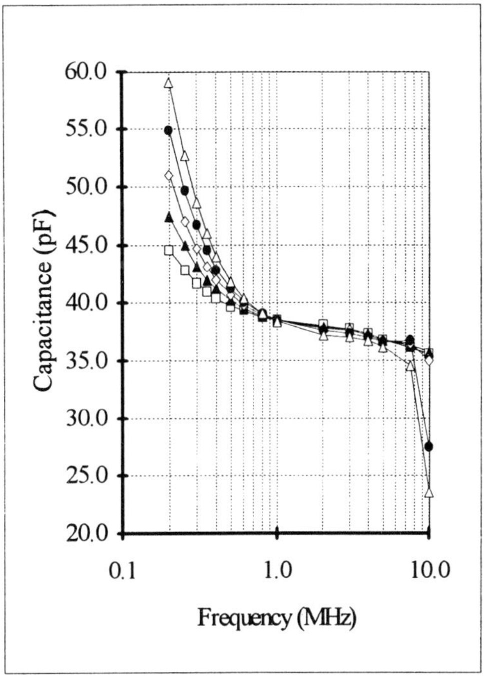

Figure 1 : Dielectric dispersion of complete MEMS growth medium.
The dielectric spectrum of complete MEMS growth medium is determined by the progressive, controlled addition of five times concentrated complete MEMS to water. The curves correspond to dilution whose conductance are 7 mS (□), 9 mS (▲), 11 mS (◇), 13 mS (●) and 15 mS (△).

As a result of Gmedium remaining constant, it now becomes possible for the first time, using the capacitance probe permanently inserted into the bioreactor and the Biomass monitor set at 1MHz, to non invasively monitor on-line the changes of biomass of HeLa cells suspension cultures kept between 10^6 and $10x10^6$ cells/ml (Figure 2). Uninterrupted monitoring periods can be obtained extending for as long as one month provided the Biomass monitor is shut off between successive measurements to avoid fouling.

Figure 2 also shows that the slope as well as the abscissal intercepts of the plots of Csuspension versus cell concentration change unpredictably after each autoclaving, requiring this relationship to be established anew during the first two to three days which follow inoculation of a culture.

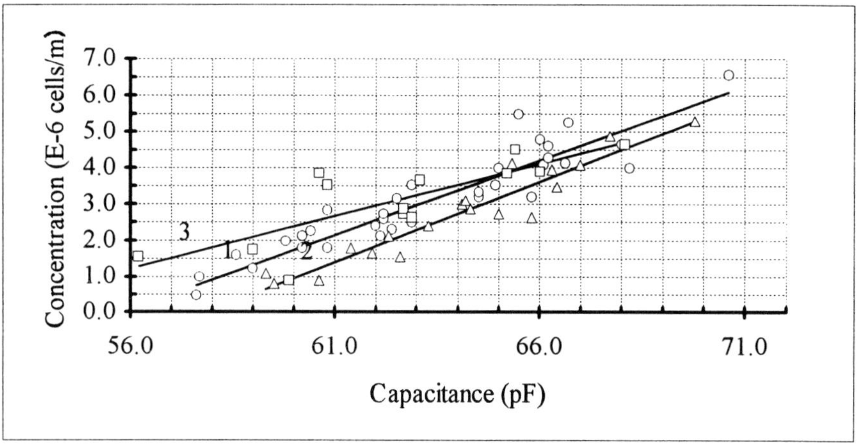

Figure 2 : Effect of successive autoclavings on the suspension capacitance (Csuspension) versus HeLa cells concentration lines : use of the capacitance probe for monitoring purposes.
The capacitance of 3 successive independent perfused cultures of HeLa cells performed in october 94 (O), march(△) and april (□) 1995 have been measured on-line, non invasively at 1MHz and the concentration simultaneously determined. For each culture, the capacitance probe is autoclaved anew. Each set of data is computer-linearized (linear regression) to give lines 1 (october), 2 (march) and 3 (april).

As shown in Figure 3, attempts to use the capacitance signals to control cell density of perfused cytostat cultures i.e. cultures where cell concentration is kept constant at a preset value by daily cell removal, met with little success.

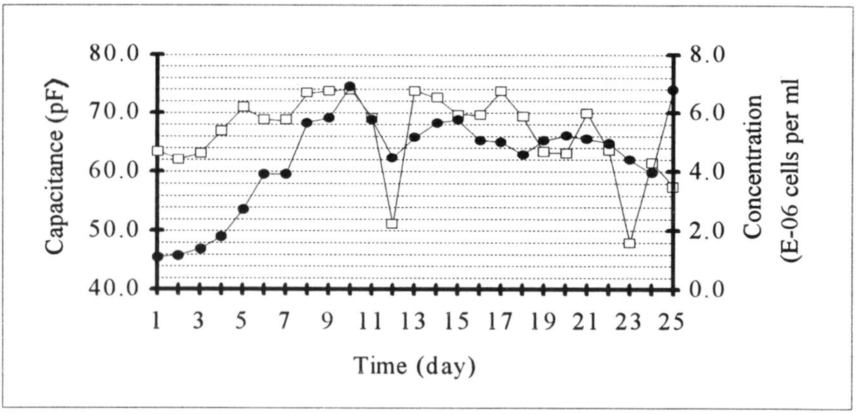

Figure 3 : Use of the capacitance (□) probe (1MHz) to control the concentration (●) of CHO cells suspension run in perfused cytostat mode. The capacitance (1MHz) and concentration of a perfused CHO cells suspension culture kept at $5x10^6$ cells/ml by daily elimination of cells and addition of fresh medium have been recorded during one month.

2) 2D PAGE evaluation of total cell proteins

For the sake of simplicity, only the electropherogram obtained from low density cell culture is given (Figure 4). Of the 29 dark spots which correspond to polypeptides present in all three electropherograms, ten (anchorpoints) are linked together to form the anchoring network. When superimposed, all three anchoring networks coincide, an indication that no gel distorsion occurred. The other eighteen white spots in Figure 4 correspond to polypeptides present in cells grown at low concentration but not detected in cells propagated at middle or high concentration. The results of the analysis of all three electropherograms numbered 1 to 3 are summarized in Table 1.

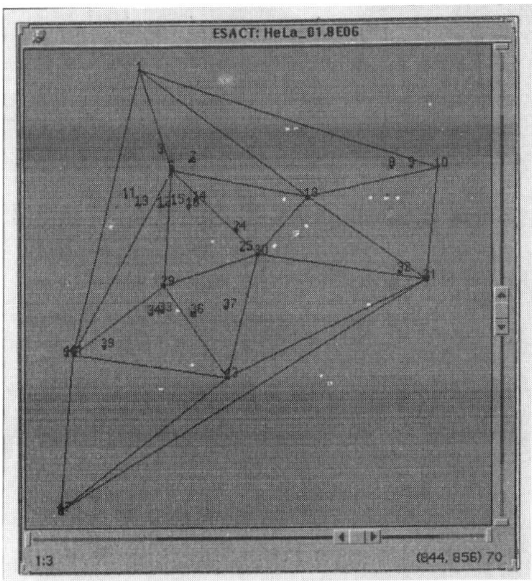

Figure 4 : High resolution 2D Polyacrylamide gel electropherogram of total proteins extracted from low density HeLa cell culture.

Table 1. Match results of the electropherograms from cultures at low- ($1,8x10^6$ cells/ml -image 1), middle- ($6,0x10^6$ cells/ml - image 2) and high ($12x10^6$ cells/ml - image 3) cell concentration (a)

(column A)	(column B)	Match (b)	Unmatched (c)	Unmatched (d)
Image 1	Image 2	29	18	13
Image 1	Image 3	35	12	13
Image 2	Image 3	30	12	18

(a) Total matches in all 3 images : 29 ; Total unmatched images (spots absent in either one or two images but not all three images) : 36.
(b) Spots present simultaneously in images from columns A and B.
(c) Spots present in image column A but not in image column B.
(d) Spots present in image column B but not in image column A.

Conclusions

Continuous perfusion using the ART-controlled acoustic cell retention system coupled to a mechanical cell filtration device (AFM), allows high density cultures of HeLa (CHO) cells to be reached and kept for as long as one month without visible signs of malfunctioning.

On-line non invasive monitoring of cell growth can be achieved using a capacitance-based probe and the corresponding monitor set at 1MHz. However, using the same experimental set up, control of cultures operated in perfused cytostat mode, does not seem realizable.

As inferred from high resolution 2D PAGE, under the experimental conditions described, quality of HeLa cells collected from high density perfused cultures does not seem to depart much from that of cells cultivated at low densities.

Acknowledgement

We thank J. Harfield for the extensive revision of the english.

References

1. Ozturk SS, Blackie J, Wu P, Figueroa C, Thrift J & Naveh D (1995) Investigation of high density cell culture kinetics using on-line cell density and metabolic rate measurements. In: Beuvery EC, Griffiths JB & Zylemaker WP (eds), Animal Cell Technology : Developments Towards the 21st Century, pp 301-305. Kluwer Academic Publishers, Dordrecht, The Netherlands.
2. Cerckel I, Garcia A, Degouys V, Dubois D, Fabry L & Miller AOA (1993) Dielectric spectroscopy of mammalian cells 1. Evaluation of the biomass of HeLa- and CHO cells in suspension by low-frequency dielectric spectroscopy. Cytotechnology 13, 185-193.
3. Investigator 2D Electrophoresis System : Operating and Maintenance Manual (1993). Part Number P 17957, Rev. B. Millipore Corporation, Bedford, MA.
4. BioImage 2D Analyser : User's Manual (1995). Version 6.1 B.I. Systems Corporation Ann Arbor, MI.

Discussion

Aunins: HeLa cells aggregate naturally. Why are you using an acoustic aggregating device?

Miller: HeLa cells do not aggregate. It depends upon the cell line. My HeLa cells came from Oxford and I have used them for 20 years forcing them to grow in suspension. CHO can aggregate depending upon the serum concentration.

Experimental and theoretical analysis of cell separations by ultrasonic forces

Steven M. Woodside[1,2], Bruce D. Bowen[2], James M. Piret[1,2]
[1]Biotechnology Laboratory and [2]Department of Chemical Engineering, University of British Columbia, Vancouver, BC, Canada V6T 1Z3

Introduction

The ultrasonic separation of mammalian cells has been successfully used for cell retention in long-term perfusion processes [1,2]. Ultrasonic separators are free of the problems typically encountered with conventional cell separation systems such as filter fouling, long sedimentation times or complex centrifuge operation. Acoustic separation technology relies on the aggregation of cells by ultrasonic forces in a standing wave field. In this work, a versatile method is presented for the experimental measurement of the forces on single particles as a function of position in the standing wave field. A detailed understanding of the forces should accelerate the rational design of acoustic separation devices.

Background

A piezo-electric ceramic transducer oscillating in response to an applied high frequency alternating voltage signal generates a standing wave field in an ultrasonic resonator. Compressional waves travel axially from the transducer to the reflector. The reflected waves superimpose with the incoming waves to create the standing wave. The differing acoustic properties (i.e. density and compressibility) of the particles and suspending fluid generate the axial primary radiation force (PRF) and the transverse Bernoulli force (BF). The PRF rapidly concentrates cells at the pressure node planes, where the pressure amplitude is zero. The BF then aggregates the cells at local maxima of the acoustic energy density (Fig. 1).

Yosioka and Kawasima [3] derived a relation for the PRF on compressible spheres of radius $r<1/k$ in an ideal fluid, where the wave number $k=2\pi/\lambda$. In a plane standing wave field the PRF (F_P) is a sinusoidal function of the axial cell position (z) and is proportional to the cell volume (V_0), the acoustic energy density of the field (E_{ac}), the wave number and the acoustic contrast factor (G):

$$F_P = V_0 E_{ac} k G \sin(2kz)\,, \qquad (1a)$$

where

M.J.T. Carrondo et al. (eds.), Animal Cell Technology, 251–256.

$$G = \left(\frac{\beta_f - \beta_p}{\beta_f} + \frac{3(\rho_p - \rho_f)}{\rho_f + 2\rho_p} \right), \tag{1b}$$

ρ is the density and β is the compressibility of the (*p*)article and (*f*)luid.

A standing wave of non-uniform amplitude distribution will generate the transverse BF [4,5]. This force ($\vec{F}_B$) depends on the gradient in the energy density ($\vec{\nabla} E_{ac}$). At a pressure node it may be expressed as:

$$\vec{F}_B = V_0 \vec{\nabla} E_{ac} \left(\frac{3(\rho_p - \rho_f)}{\rho_f + 2\rho_p} \right). \tag{2}$$

where $\vec{\nabla}$ is the gradient operator in the transverse plane.

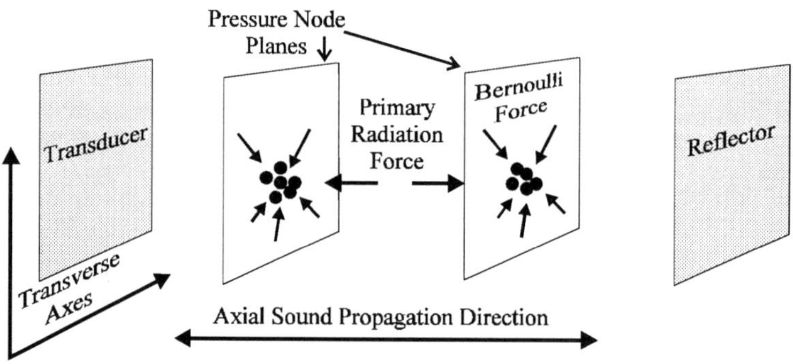

Figure 1 - Aggregation of suspended cells by the axial primary radiation force and transverse Bernoulli force at local acoustic energy density maxima on the pressure node planes. The distance between pressure nodes at a resonance frequency of 2.23 MHz is ≈350 μm.

Ultrasonic Force Measurement System

Polystyrene particles of 10.2 μm diameter (Seradyn, Indianapolis IN) suspended in distilled water were employed as model particles because they have acoustic properties similar to those of mammalian cells, and are highly uniform in diameter (s.d.=0.1 μm). Particle interactions were neglected due to the low particle concentration employed (<0.01% v/v) which resulted in an average interparticle distance greater than 20 particle diameters, and a 10 particle diameter separation when uniformly dispersed on the pressure node plane. The maximum observed particle velocity was of the order of 1 mm/s (Re_p<0.01). Thus, the drag force on each particle is given by Stokes' law. The particles are sufficiently small that inertial effects can be neglected and, consequently, the PRF can be equated to the drag force, i.e.:

$$F_p = 6\pi \mu r v \, , \tag{3}$$

where μ is the fluid viscosity and v is the particle velocity.

To permit image acquisition, the ultrasonic resonance chamber was constructed with two side windows and a glass reflector. An inverted microscope (Wild Heerbrugg) equipped with a CCD camera (COHU, San Diego, CA, USA) and a computer controlled video recorder (Sony SLV-380) was used to acquire sequential images using a frame grabber (Coreco Oculus TCX, St. Laurent, PQ, Canada). Image analysis (Bioscan Optimas 4.0, Bothell, WA, USA) was employed to determined particle displacements from frame to frame using a modified nearest neighbours algorithm [6]. The single particle velocities were calculated from the displacement and the time interval between frames.

Experimental Results

The measured particle velocities and calculated PRF values (normalized with respect to the net force of gravity on the particles) are shown in Fig. 2 as a function of the axial particle position in the field. The sequential data for each particle are represented by different symbols. A positive axial velocity indicates displacement towards the

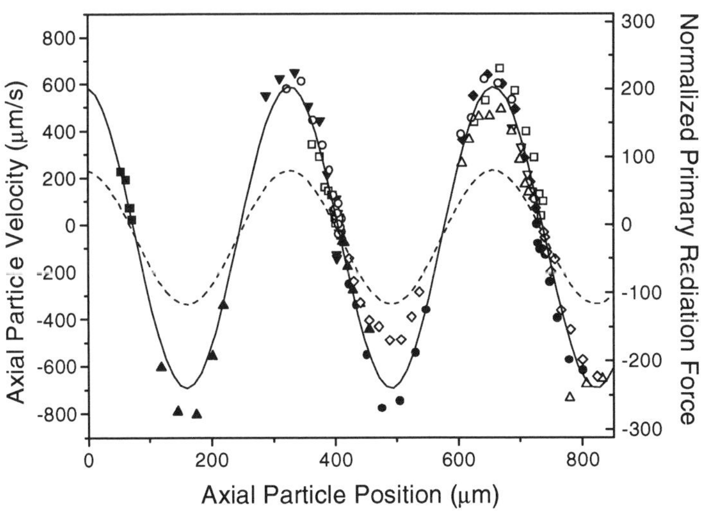

Figure 2 - Particle velocity (left) and PRF normalized with respect to the net force of gravity on the particles (right) as a function of axial position. The line of best fit (solid line) has a velocity amplitude of 640 μm/s. The expected velocity profile (dashed line) has an amplitude of 285 μm/s (f=2.23 MHz, P=24 W/L).

transducer. As expected the PRF fell to zero at the pressure nodes (i.e. at 70, 410, 750 μm). The calculated forces on the particles followed the sinusoidal variation predicted by the PRF theory (Eq. 1).

A non-linear least squares fit of the PRF data was performed for variables E_{ac} and k using Eq. 1 modified to include a phase shift and offset. The value of the fitted wavelength was 680 μm, while the corresponding calculated value was 671 μm (c=1497 m/s and f=2.23 MHz). The fitted PRF amplitude corresponded to an apparent E_{ac} = 20.5 J/m^3. Based on the electronically measured P=24 W/L, an E_{ac}= 9.2 J/m^3 was expected. This significant discrepancy was principally due to the non-uniform distribution of E_{ac} in the resonator volume, which is responsible for the observed BF.

The substantial variation in the maximum velocity of different particles is also apparent in Fig. 2. PRF measurements in the central regions of the resonator revealed a two-fold variation in E_{ac}. At the side walls the E_{ac} fell to zero. An examination of the E_{ac} variation near a hot-spot revealed that the PRF amplitude on individual particles decreased by up to 30% within one image area over a transverse distance of 600 μm. Based on this gradient information the BF was estimated to be more than two orders of magnitude lower than the PRF (Table 1 - P=44 W/L).

To determine the BF directly, the microscope was focused on a pressure node plane and images where acquired at 1 s intervals of the particles converging within this plane (Fig. 3). Again, it was assumed that the acoustic force could be equated with Stokes' drag. The effect of gravity was included because it was significant compared to the BF.

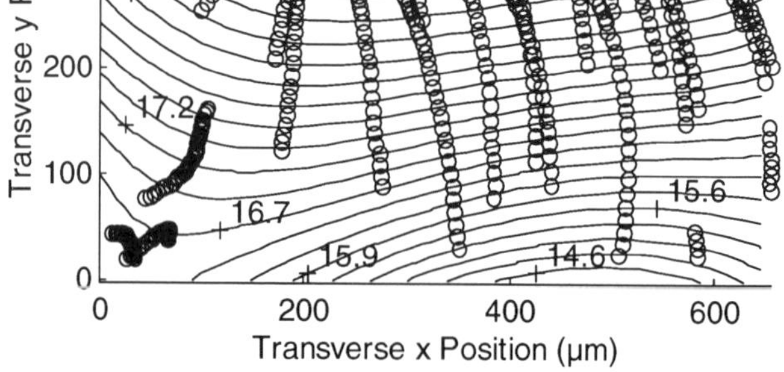

Figure 3 - Particle displacement in a transverse pressure node plane due to the Bernoulli force. The E_{ac} distribution is plotted as solid contour lines with values in J/m^3 obtained by fitting the particle displacement data to Eq. 2.

The forces and the corresponding E_{ac} gradients were calculated for each particle as a function of position. The E_{ac} distribution in the measurement area of the pressure node plane was determined by fitting a 5th-order polynomial in transverse coordinates x and y to the calculated gradient data. The E_{ac} is plotted as solid countour lines in Fig. 3, with values marked in J/m^3. The particles collected slightly below the region of highest E_{ac} due to the effect of gravity. The maximum BF calculated from this distribution is in good agreement with the estimate obtained from the PRF data when normalized for P (Table 1 - P=62 W/L).

Extrapolating from these data to a power level typical of operating conditions for the ultrasonic filter (i.e. 750 W/L), the BF would be an order of magnitude greater than the net gravitational force on a single particle (Table 1). The approximate flow rate against which a single particle could be held by the BF at these power levels is 0.06 mm/s, which is too low for most practical applications. In contrast, it had already been shown that cells are retained by the ultrasonic field at greater than 95% efficiency for superficial velocities over 1 mm/s in a chamber with flow parallel to the nodal planes [1]. It appears that multiparticle aggregates forming at the node were held against higher flow velocities by the total BF on the many particles in the aggregate volume, while the drag on the aggregate only increased approximately as a linear function of the aggregate width.

Table 1 - Comparison of ultrasonic forces normalized to the net force of gravity on these particles.

Input Power (W/L)	Bernoulli Force (F_g)	Average Peak PRF (F_g)
44	1.1	210
62	1.7	
750	19	1300

Conclusions

In most of the previously proposed acoustic particle separation systems the stronger PRF was the principal force used. There were likely two important limitations to this approach. First of all, the PRF acted only on single particles and therefore did not exploit the apparent cumulative force on aggregates. Secondly, as nodal planes became filled by particles they would obstruct flow perpendicular to the planes. In the device developed by Trampler et al. [1] with the flow parallel to the nodal planes, the PRF acts over the short inter-node distance and clears cells into the nodal planes. The BF and gravity retain the particles while the fluid flows freely between the pressure node planes.

Notation

c	sound speed	V_0	volume of particle
E_{ac}	acoustic energy density	x, y	transverse particle position
f	frequency	z	axial particle position
F_B	Bernoulli force	β	compressibility
F_P	primary radiation force	λ	wavelength
G	acoustic contrast factor (Eq. 1)	μ	viscosity
k	wave number ($2\pi f/c$)	ρ	density
P	power		
r	particle radius	Subscripts	
Re_p	Reynold's number ($=2\rho vr/\mu$)	f	fluid
v	velocity of particle	p	particle

Acknowledgments

This project was supported by the Natural Sciences and Engineering Research Council of Canada and by SonoSep Biotech Inc. (Richmond B.C.). SMW holds fellowships from NSERC and the Science Council of B.C.

References

1. Trampler F., Sonderhoff S.A., Pui P.W.S., Kilburn D.G., Piret J.M. (1994) Acoustic cell filter for high density perfusion culture of hybridoma cells, *Bio/Technol.,* **12**, 281.
2. Doblhoff-Dier O., Gaida Th., Katinger H. (1994) A novel ultrasonic resonance field device for the retention of animal cells, *Biotechnol. Prog.,* **10**, 428.
3. Yosioka K., Kawasima Y. (1955) Acoustic radiation pressure on a compressible sphere, *Acustica,* **5,** 167.
4. Coakley W.T., Whitworth G., Grundy M.A., Gould R.K., Allman R. (1994) Ultrasonic manipulation of particles and cells, *Bioseparation,* **4**, 73.
5. Hager F., Benes E. (1991) A summary of all forces acting on spherical particles in an ideal fluid, *Proc. Ultrasonics Int. Conf.,* 283.
6. Noble P.B., Levine M.D. (1986) Computer-assisted analysis of chemotaxis, CRC Press:Boca Raton, Fla., USA.

Factors Affecting the Production and Quality of IgA Antibodies Active Against *V. cholerae*

Ian Marison, Thibaud S. Stoll[1], Elke Lüllau and Urs von Stockar, Institute of Chemical Engineering, Swiss Federal Institute of Technology, CH-1015 Lausanne, Switzerland. [1] present address: Cell Culture and fermentation R&D, Genentech Inc., 460 Point San Bruno Blvd, South San Francisco CA 94080, USA.

Abstract

Anti- *V.cholerae* monoclonal IgA antibodies have been produced using a hybridoma cell line (Zac3) grown in RPMI medium containing 10% Foetal calf serum (FCS) and a completely defined protein- free medium. The latter has been optimised with respect to amino acid, pluronic, glucose and glutamine content to give batch growth, total IgA concentration, IgA productivity and dimeric IgA fraction which is even higher than that obtained during growth on the complex medium. The hybridoma has been grown in batch and continuous modes in a range of bioreactor configurations and the quantity and quality of the IgA produced determined. The results show that high cell density systems (hollow fibre reactors) produce the highest concentrations and productivities. However, the high concentration can result in the formation of high levels of polymeric forms which creates difficulties for the downstream processing. The bioreactor configuration had an important effect on IgA productivity, ratio of mono-, di- and poly- meric forms as well as antigen binding activity.

Keywords: Monoclonal antibodies, IgA, cell culture, hybridoma, product quality

Introduction

Of all of the immunoglobulins produced in mammalian systems, the daily quantity of sIgA antibodies produced exceeds that of all other forms combined. The main role of these antibodies is the protection of mucosal surfaces, such as the lungs, gastrointestinal and urogenital tracts from pathogen attack and entry. As a result these antibodies have been identified as having considerable potential as a prophylactic agent for the passive mucosal protection of immuno- compromised (those receiving treatments with immunosuppressive drugs or as a result of infection) or immuno- inexperienced individuals (e.g. neonates), or indeed for the treatment of infections for which no efficient vaccine is yet available. In order to establish this potential it is necessary to produce high quality (s)IgA of defined activity in quantities sufficient to allow extensive testing with animal model systems.

A major difficulty is the complex structure of sIgA class 2 which naturally occur as dimeric immunoglobulins, with covalent linkages between the heavy chains (α), the light chains (κ) but not between the α- and κ- chains. In addition the monomeric IgA units are covalently attached to a J- chain peptide to form the dimeric molecule, which in turn is covalently bound to a 5-domain peptide termed the secretory component (SC) (1). The latter is believed to protect the IgA molecule from proteolytic attack. Until now the only methods for producing the intact sIgA molecule involve purification from colostrum, milk, faeces or other mucosal secretions (2, 3). In order to overcome the difficulty of supply a two- step process has been developed in which dimeric IgA with associated J-chain is

M.J.T. Carrondo et al. (eds.), Animal Cell Technology, 257–264.

produced from hybridoma cells while SC is produced from transfected CHO cells, re-association of the two being undertaken in vitro. A recent publication showed how the total concentration of an anti- *V. cholerae* IgA secreted by a murine hybridoma varied as a function of the bioreactor system used for growing the cells (4). However the complexity of the IgA structure is such that optimum conditions for the production of active, dimeric antibodies must be carefully established. Furthermore the different bioreactor conditions and media composition had a significant effect on the procedure used for subsequent purification. The latter problems form the basis of the work to be presented here.

Procedure

Cell line and growth conditions. A murine hybridoma cell line (Zac3) made by fusion of the mouse myeloma cell P3X63/AgU.1 with a lymphocyte from Peyer's patches of a BALB/c mouse were kindly provided by Prof. J.-P. Kraehenbuhl (Swiss Institute for Experimental Cancer Research, Lausanne, CH) and was used for all experiments. This cell line secretes a monoclonal IgA antibody of allotype 2, directed against the LPS of the envelop of *Vibrio cholerae*. Cells were maintained at 36.5°C under a 95% air/5% CO_2 atmosphere in a humidified incubator. The cells were routinely grown in RPMI medium containing 10% FCS but were also adapted to stable growth in a completely defined serum- and protein- free medium (FMX- Turbodoma, Cell Culture Technologies, Glattbrugg, CH). This basal defined medium (FMX-0) was supplemented with glucose, glutamine, pluronic acid F-68 and a balanced cocktail of amino acids to yield FMX-3 as previously described (5).

Bioreactors. All conditions for the growth of the hybridoma in T- flasks, batch stirred tank reactor (STR) and hollow fibre reactors (HFR) have been described in detail elsewhere (4).

Analytical methods. Culture samples were removed at intervals (thrice daily) and cells counted using a haemocytometer. Viable cell number was determined using the Trypan blue exclusion method. Samples were centrifuged and the supernatant stored at -20°C until required for further analysis. The analysis of glucose, glutamine and amino acids was performed as described earlier (4,6). The sandwich ELISA method for the determination of total IgA, the indirect ELISA method using LPS of *V. cholerae*, for the determination of active IgA, SDS- PAGE and Western blotting have been described elsewhere (5).

Purification methods. A CSTR culture of Zac3 growing on FMX-3 medium and operating at a dilution rate of 0.3 h^{-1} was harvested to yield 10 litres of culture and subjected to microfiltration (0.65 μm membranes, Sartoclean, Sartorius, Germany) to remove cells and cellular debris. The filtrate was concentrated by ultrafiltration (Omega 100 kD mol. wt. cut-off membranes 2.46 m^2 surface area, Skan, Switzerland) and diafiltered to yield approximately 10 fold concentrated filtrate in 10 mM phosphate buffer pH 7.3 containing 100 mM NaCl. This was loaded onto a 500 ml DEAE- Sepharose FF (Pharmacia) column (5 cm x 25 cm) and eluted batchwise with elution buffer containing 200 mM NaCl. The eluate was diluted two- fold and applied to a 50 ml DEAE- Sepharose FF column (2.6 cm x 10 cm) and eluted batchwise with 10 mM phosphate buffer pH 7.3 containing 200 mM NaCl. The concentrated IgA fractions were applied directly to a Sephacryl S300 column (Pharmacia, 2.6 cm x 200 cm) equilibrated and eluted with phosphate- buffered saline (PBS). Fractions containing IgA were pooled and stored at 4°C until required. Details of the purification methods have been described elsewhere (7).

Results and Discussion

Medium optimisation. All hybridoma cell cultures were routinely grown in a batch STR with RPMI 1640+ 10% FCS medium and subsequently adapted to growth on a completely defined serum- and protein- free medium, FMX- Turbodoma. The latter medium was used to identify potential limiting nutrients, and supplemented with glucose, glutamine and other amino acids to obtain optimum cell growth and antibody productivity.

Table 1 Kinetic parameters for Zac3 hybridoma grown in a batch STR on different media

Parameter	RPMI+10%FCS[1]	FMX-0[2]	FMX-3[3]
μ (h^{-1})	0.048 ± 0.004	0.047 ± 0.002	0.047 ± 0.002
$X_{v,max}$ (cell ml^{-1})	(9.5 ± 0.2) x 10^5	(1.1 ± 0.2) x 10^6	(2.5 ± 0.2) x 10^6
$Y_{lact/glc}$ (mol mol^{-1})	2.0 ± 0.1	1.8 ± 0.1	1.9 ± 0.1
$q_{glc,max}$ (pmol (cell h)$^{-1}$)	0.6 ± 0.1	0.7 ± 0.1	0.8 ± 0.1
$q_{lact,max}$ (pmol (cell h)$^{-1}$)	1.3 ± 0.2	1.2 ± 0.2	1.3 ± 0.2
$Y_{NH3/gln}$ (mol mol^{-1})	0.77 ± 0.04	0.80 ± 0.04	0.78 ± 0.04
$q_{gln,max}$ (pmol (cell h)$^{-1}$)	0.12 ± 0.02	0.08 ± 0.02	0.07 ± 0.02
$q_{NH3,max}$ (pmol (cell h)$^{-1}$)	0.10 ± 0.02	0.07 ± 0.02	0.06 ± 0.02
q_{IgA} (pg (cell h)$^{-1}$)	1.9 ± 0.3	0.7 ± 0.1	0.8 ± 0.1
$[IgA]_{max}$ (mg l^{-1})	47 ± 5	14 ± 3	114 ± 10
Active fraction (%)	58 ± 6	81 ± 5	85 ± 5

[1] Medium contains glucose (11.1 mM) and glutamine (2 mM); [2] FMX-0 medium (5) contains glucose (13.9 mM) and glutamine (2.71 mM); [3] FMX-0 medium supplemented with glucose (25 mM), glutamine (5 mM), Pluronic acid F-68 (0.1%) and a balanced combination of amino acids (5).

The results (Table 1) show that the specific growth rate (μ), maximum viable cell concentration attained ($X_{v,max}$), lactic acid yield with respect to glucose consumed ($Y_{lact/glc}$), maximum specific glucose consumption rate ($q_{glc,max}$), maximum specific lactate production rate ($q_{lact,max}$), ammonia yield with respect to glutamine consumed ($Y_{NH3/gln}$), maximum specific glutamine consumption rate ($q_{gln,max}$) and maximum specific ammonia production rate ($q_{NH3,max}$) were not significantly influenced by the composition of the medium. However the specific IgA production rate (q_{IgA}) was found to be approximately 2.4 fold higher in the complex medium compared with the defined optimised medium. Furthermore the maximum IgA concentration ($[IgA]_{max}$) was initially 3.4 fold lower in FMX-0 compared with the complex medium, although this was dramatically increased by over 8 fold by subsequent addition of a balanced amino acid supplement to yield 114 mg IgA l^{-1}. This difference was almost entirely due to an extended IgA production period once the maximum cell viability had been attained, since in the complex medium a rapid decrease in viability occurred immediately after the maximum, whereas in the defined medium viability could be maintained for extended periods.

The composition of the medium was also shown to have an important effect on the active fraction of the IgA formed with the defined media typically having 81-85% of the total IgA secreted in the active form (meaning the ability to bind to LPS antigen in an ELISA assay) while only 58% of the total IgA produced in the complex medium was active.

Comparison of growth and IgA production in different bioreactor configurations. The Zac3 hybridoma cell line has been grown in a range of bioreactor configurations in batch mode including T-flasks (175 cm^3, 80 ml medium) and STR (1.6 litre working volume) and in continuous mode in an STR, 300 cm^2 (7ml extracapillary volume) hollow fibre reactor (Tecnomouse, Integra Biosciences) and a 1.1m^2 (380 ml extracapillary volume) hollow fibre reactor (Endotronics Maximizer) in both complex and defined media. The results are presented in Table 2.

Table 2 Comparison of IgA production for different bioreactor configurations and media

Bioreactor	**Medium**	**[IgA] (mg l^{-1})**	**$Q_{IgA,max}$ (mg day^{-1})**	**Q_{IgA} (mg day^{-1})**	**Active fraction (%)**
ascitic fluid		2200	n.d.	0.4	85
T- flask	Complex[1]	42	18	18	60
	Defined[2]	120	23	23	78
STR (batch)	Complex[1]	47	19	10	60
	Defined[2]	115	24	17	81
Tecnomouse	Complex[1]	1000-13'000	2-13	4	58-38
	Defined[3]	1000-7000	1-14	4.6	76-42
Maximizer	Complex[1]	1000-14'000	400	200	57-30
	Defined[3]	500-5000	245	130	77-35

[1] RPMI + 10% FCS; [2] FMX-3; [3] FMX-0. [IgA], Maximum concentration of IgA attained (mg l^{-1}), $Q_{IgA,max}$, maximum IgA productivity (mg day^{-1}), Q_{IgA} average IgA productivity (mg day^{-1}).

The results clearly show differences in both total IgA concentration, productivity and the fraction of the IgA which actively binds to antigen, both with respect to the medium and bioreactor configuration. Thus, in general the maximum concentration of total IgA formed and productivity are higher in the defined protein- free medium compared with the complex serum- containing. The higher values of concentration and productivity for the complex medium in the hollow fibre reactors result from the use of a defined medium which was not optimised (FMX-0 from Table 1) with respect to amino acids. More importantly the fraction of active IgA is always higher for a particular reactor with the defined medium. This is in part a reflection of the relative amounts of monomer, dimeric and polymeric forms of IgA formed in the different reactor configurations (Table 3) which in turn is directly linked to the concentration of the IgA formed. Thus very high IgA concentrations can be attained in the hollow fibre reactors, especially in defined medium, yet the polymeric forms represent more than 45% of the total IgA with only 44-46% in the dimeric form. Since the active fraction is between 38 and 83% of the total IgA, this suggests that at least a fraction of the polymeric IgA is also capable of binding to the LPS antigen of *V. cholerae* in the activity assay. This has been confirmed by purification of monomeric, dimeric and polymeric IgA followed by testing of antigen binding activity (results not shown). Monomeric IgA showed very weak binding whereas dimeric and polymeric IgA bound strongly.

Figure 3 Data from scanning of Western blots of IgA produced by the hybridoma Zac3 in different media and different bioreactor configurations

Bioreactor	[IgA] (mg l^{-1})	Active fraction (%)	Polymer (%)	Dimer (%)	Dimer* (%)	Monomer (%)
STR[1]	11	59	8 ± 4	61 ± 10	8 ± 2	23 ± 6
STR[1]	147	59	16 ± 4	55 ± 10	6 ± 2	23 ± 6
STR[2]	3	88	20 ± 5	52 ± 10	4 ± 2	24 ± 5
STR[2]	114	79	25 ± 5	61 ± 10	3 ± 2	12 ± 4
Tecnomouse[2]	5950	38	45 ± 10	46 ± 10	3 ± 2	6 ± 2
Maximizer[2]	990	83	48 ± 10	44 ± 10	4 ± 2	5 ± 2

* Incomplete dimer; [1] RPMI+10% FCS medium; [2] FMX-3 medium

Purification of IgA. Culture harvests from protein- free medium (FMX-3) contained a mixture of monomeric, dimeric, polymeric and aggregate forms of IgA in addition to contaminating proteins. By a combination of microfiltration, diafiltration, ion- exchange and size- exclusion chromatography it was possible to purify IgA to homogeneity and to completely separate the different molecular forms of IgA (Table 4).

Table 4 Data from the final step of the purification procedure to show the separation of the different molecular forms of IgA, purity (as determined by reduced SDS- PAGE) and yield.

Procedure	Volume [l]	Protein[1] [gl^{-1}]	IgA [mgl^{-1}]	Purity	Yield [%]
Culture harvest	10	0.226	26.4	0.12	100
Sephacryl S300	2.016				45
Aggregate/Polymer	0.126		9.5		0.5
Polymer	0.315	0.985	97.5	0.99	11.6
Polymer/Dimer	0.378	0.585	57.9	0.99	8.3
Dimer	0.315	0.136	134.6	0.99	16.3
Dimer/Monomer	0.252		46.4		4.4
Monomer	0.630	0.021	19.0	0.90	4.5

[1] Total protein determined with the Bio-Rad protein assay using IgA as standard.

Of the 10 litres of culture harvest a total of 315 ml of dimeric IgA solution of greater than 99% purity were obtained representing a yield of 16.3% with respect to total IgA in the original harvest. However if it is considered that only 30% of the original harvest was of the dimeric form the true yield is about 53.7% of which greater than 95% was in the active form and contained covalently bound J- chain. Considerable difficulty was observed to completely separate the different molecular forms of IgA which eluted from the size exclusion column in peaks which overlapped. The results presented in Table 4 were obtained by carefully taking a large number of small eluate fractions and pooling only those which could be determined from ELISA and PAGE/Western blotting to be free of the other molecular forms. A similar purification procedure was applied to culture harvests from other bioreactor configurations and using different culture media, however separation was considerably reduced and significant modifications to the basic purification procedure would be required to improve separation. This is in part due to the lowered dimeric IgA content with respect to other molecular forms observed with harvests from hollow fibre reactors and to the considerable amount of albumin and other proteins in media containing FCS. The latter has been shown to associate with IgA and affect binding to ion- exchange columns and mobility in size exclusion columns (results not shown).

Conclusions

The results of this work show that pure dimeric IgA, in a form which is capable of binding to antigen, can be produced using hybridoma cells in a variety of bioreactor configurations. However the absence of serum and other proteins from culture media, as well as the bioreactor configuration, significantly influence the quantity and quality of the antibodies formed and the ease with which dimeric IgA may be purified. It would appear that high cell density hollow fibre reactors using protein- free media could provide optimal productivity and quality providing that some method to determine the concentration of IgA on- line can be developed and used to define the optimum harvesting regime to avoid the formation of polymeric and aggregate forms. This work further shows the importance of co-ordination of production optimisation and protein purification since small changes in the former may exert significant changes on the latter.

References

1. Kerr M A (1990) The structure and function of human IgA. Biochem. J. 271, 285-296.
2. Bouige P, Iscaki S and Pillot J (1990) Immune complexes as immunizing agents to increase the number of monoclonal antibody producing hybrids and to deviate the response to poorly immunogenic epitopes. Hybridoma 9, 519- 526.
3. Parr E L, Bozzola J J and Parr M B (1995) Purification and measurement of secretory IgA in mouse milk. J. Immunol. Methods 180, 147-157.
4. Stoll T, Perregaux C, von Stockar U and Marison I W (1995) Production of immunoglobulin A in different bioreactor configurations. Cytotechnology 17, 53-63.
5. Stoll T, S Mühlethaler K, von Stockar U and Marison I W (1996) Systematic improvement of a chemically- defined protein- free medium for hybridoma growth and monoclonal antibody production. J. Biotechnol. 45, 111-123.
6. Stoll T, Pugeaud P, von Stockar U and Marison I W (1994) A simple HPLC technique for accurate monitoring of mammalian cell metabolism. Cytotechnology 14, 123-128.
7. Lüllau E, Heyse S, Vogel H, Marison I, von Stockar U, Kraehenbuhl J-P and Corthésy B. (1996) Antigen binding properties of purified IgA and reconstituted secretory IgA antibodies. J. Biol. Chem. (accepted for publication).

Acknowledgements

This work was undertaken with funding by the Swiss National Science Foundation (FNRS) as part of the Swiss Priority Program in Biotechnology Module 1. We are grateful to Mr. P. Pugeaud for help with HPLC analysis.

Discussion

Bidawid: How reproducible are the yields between the different bioreactors?

Morrison: Within 12%.

Butler: You have a higher quality product in defined medium. Do you know what the active parameters are? Was it due to something in the defined media, or due to the absence of serum?

Morrison: I do not know. We tested a whole range of additions and I cannot say why it is more active except for effect of the concentration of course. The defined medium gives much higher concentration. The cell number increases by 2•5, the decline phase is much lower so we can maintain the cells for a much longer productive time and much greater yields. It depends where you harvest. Ideally one wants semi-continuous harvesting.

Ryll: Have you tried to characterise your antibodies, checking lack of glycosylation, different types of glycosylation or other factors which lead to the low productivity?

Morrison: No, the effect of protease's with high cell density systems where you get 10-15 g/l gives the probability that there is an accumulation of protease's. We did test for that but it did not seem to be the problem. We have not looked at glycosylation. This project is part of a Swiss National Programme and a group in Ciba Geigy are doing a detailed glycosylation study to find out the effect of different media and bioreactor configurations. Also the effect of ammonia on glycosylation. We know these probably have an effect.

Bernard: It looked as if running the hollow-fibre bioreactor at sub-maximal IgA concentration would get you to the same point. Was there any reason for choosing the CSTR versus the hollow fibre?

Morrison: There are the obvious reasons. The hollow fibre system is only scaleable linearly. We wanted a scaleable system to produce kilos per year. Ideally one would like to have a parameter to measure IgA concentration at any time, especially the IgA active fraction. As we do not have an on-line method for this it is hit or miss how you harvest as you do not have a steady state system in a hollow fibre. We tried to relate it to other parameters eg oxygen consumption, amino acids. Certain amino acids may be a good parameter.

Miller: Why do you use in succession 2 anionic exchange chromatography columns?

Morrison: Simply, the second one is a concentration step. The membranes in a hollow fibre system have a high affinity for the antibodies.

CERAMIC HYDROXYAPATITE: A NEW TOOL FOR SEPARATION AND ANALYSIS OF IGA MONOCLONAL ANTIBODIES

E. Lüllau, I. W. Marison and U. von Stockar
Swiss Federal Institute of Technology, 1015 Lausanne, Switzerland

Abstract

A novel downstream process is reported for preparative scale production of dimeric Immunoglobulin A (IgA), a protein with potential pharmaceutical applications. Time-consuming size exclusion chromatography was replaced by ceramic hydroxyapatite chromatography (CHT). A clear baseline separation of the dimeric IgA from the polymeric forms using a step-gradient-step protein elution program is described. Application of this method for IgA analysis during production is suggested.

Introduction

Immunoglobulin A (IgA) plays a key role in host protection at mucosal surfaces. The hybridoma cell line *ZAC3* secretes specific mouse IgA molecules against *Vibrio cholerae* lipopolysaccharide as a heterogeneous population of monomeric (IgA_m), dimeric (IgA_d) and polymeric (IgA_p) forms. Until today size-exclusion chromatography was the method of choice for the separation of the different forms of IgA [1, 2]. We used this method in a purification process in preparative scale [3]. But this technique is accompanied by some disadvantages: A column length of 200 cm was necessary for the separation of dimers and polymers. One separation run took 36 h with a maximal load of 25-30 mg protein. For 1 g dimeric IgA at least 200 runs would be needed. Hydroxyapatite chromatography, introduced by Tiselius et al. in 1956 [4], has been used for column chromatography separations for many years, although its drawbacks deriving from difficulties of constant quality of the calcium phosphate (hydroxyapatite) led to a declining acceptance of this method.

The secretion of various molecular forms of IgA antibodies by hybridomas is well known [1, 2]. The aim of this study was therefore to establish a method for the separation of large amounts of IgA. 1993 Aoyama and Chiba [5] used spherical hydroxyapatite beads for this purpose. Thus, a ceramic type of hydroxyapatite seemed to be a possible alternative to reach our goal.

M.J.T. Carrondo et al. (eds.), Animal Cell Technology, 265–269.

Materials and Methods

Figure 1: Bioprocess Flowchart

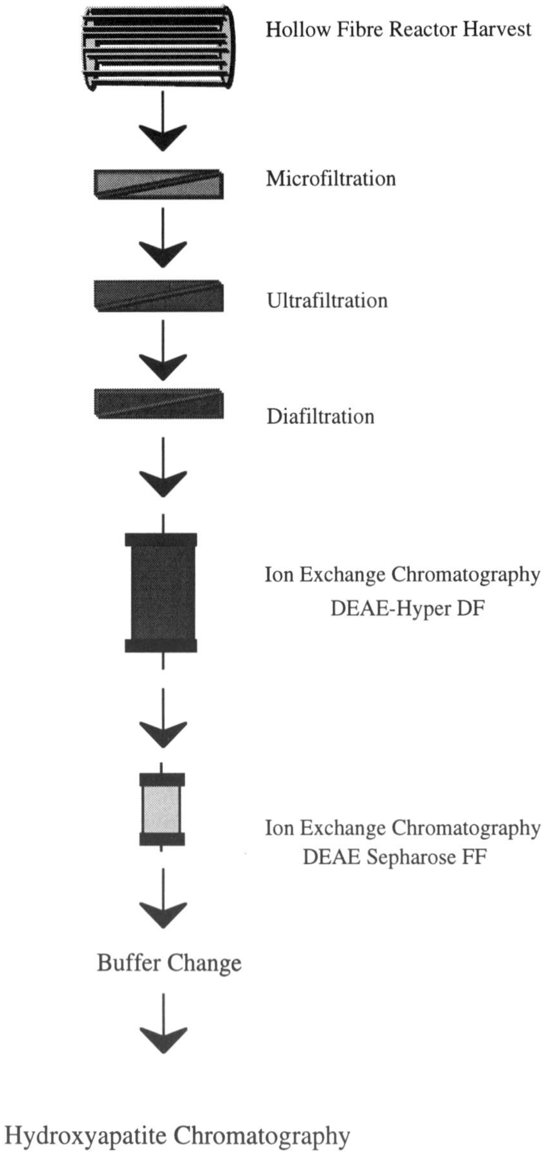

Cell culture

The IgA antibody corresponding to the human allotype A2(m1) [1] is produced by the murine hybridoma cell line ZAC3 (provided by Prof. J.-P. Kraehenbuhl from the Swiss Institute for Experimental Cancer Research). The culture was grown in a protein-free Turbodoma FMX standard medium (F. Messi Cell Culture Technologies, Glattbrugg, CH) in a hollow-fiber reactor (Acusyst Maximizer™ 500, Endotronics, Minneapolis, MN, USA) Stoll et al. 1995 [6]. The cell free harvest was stored under sterile conditions at 4°C for further processing.

Down-stream processing

Figure 1 shows a flowchart of the IgA bioprocess. Detailed information about chromatography and product assessment has been given elsewhere [3].

Hydroxyapatite chromatography was performed with pre-packed CHT-I columns on a Bio-Logic FPLC system (Bio-Rad Laboratories AG, Glattbrugg, CH). A 2 ml CHT-I column was used for analytical and a 20 ml CHT-I column for preparative IgA separation. Potassium Phosphate was used as buffer system (buffer A: 5 mM potassium phosphate, pH 6.8; buffer B: 500 mM potassium phosphate, pH 6.8). The chromatography was performed with a combined step-gradient elution profile represented by the conductivity profiles in figures 2 and 4. Samples were equilibrated in loading buffer (buffer A) using PD-10 desalting columns or a 1000 ml Sephadex G25 column (5 cm x 50 cm) (Pharmacia Biotech, Dübendorf, CH).

Results and Discussion

A ceramic type of hydroxyapatite (CHT I) was selected for the separation of the IgA multimers. A combined step-gradient-step elution program was developed giving a clear baseline separation between IgA dimers and polymers (Fig. 2).

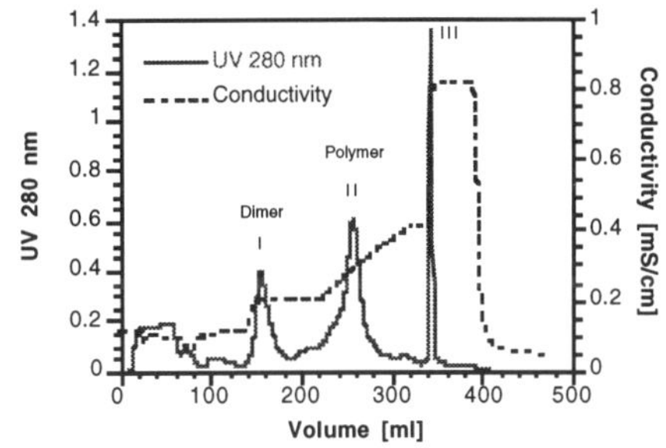

Figure 2: *Elution profile; preparative CHT*

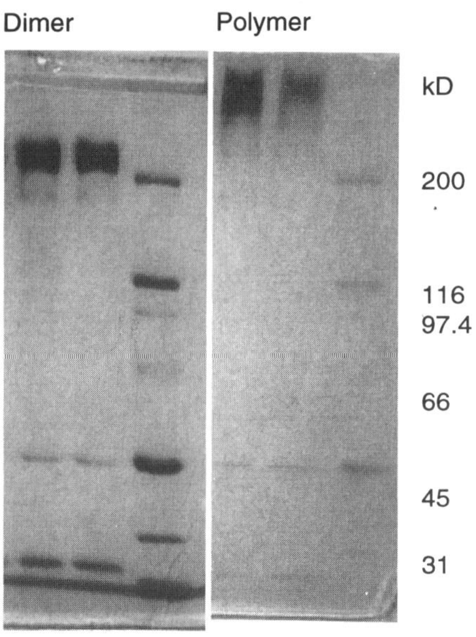

Figure 3: *SDS-PAGE non reduced conditions, silver staining*

Its high resolution was obtained by loading maximal 100 mg of IgA mixture at a flow rate of 2.5 ml/min onto a 20 ml preparative hydroxyapatite column. This procedure allowed the recovery of purified dimer in ca. 3 h. IgA monomers were shown to be eliminated during the purification procedure and not to bind to hydroxyapatite under the given conditions. The IgA dimers (peak I) and polymers (peak II) were analysed by SDS-PAGE and found to be free of any protein and IgA impurities (Figure 3)

The biological activity of the separated immunoglobulins proteins was assessed as the binding of IgA to lipopolysaccharide antigen from the outer membrane of *V. cholerae* . Both polymers and dimers showed similar binding to the LPS-antigen in a range from 70 to 95 %. This results were identical with those results determined for IgA produced in a continuous stirred tank reactor and separated by size exclusion chromatography [6]. Finally the reproducibility of this method was demonstrated with pre-purified IgA mixtures produced in a fluidized bed reactor and a continuous stirred tank reactor (data not shown). The elution peak III, observed regularly in every chromatogram, was shown to contain mainly DNA fragments in the range from 500 to 1000 bp and small amounts of protein.

Using a linear gradient of phosphate ion concentration for protein elution as described in [5] failed in IgA separation. This was referred to the nature of this IgA proteins and differences in the HA matrix used.

The average IgAd concentration in the protein mixture varied from 20 to 40%.

Using a 20 ml CHT I column under optimal conditions 100 h would be needed for the separation of 1 g dimeric IgA, which is 72 times faster than size exclusion chromatography.
These results implicated a further application for HA-chromatography in IgA analysis during production. The determination of the proportion of IgAd and IgAp would be helpful in optimizing the production towards the desired dimeric IgA.

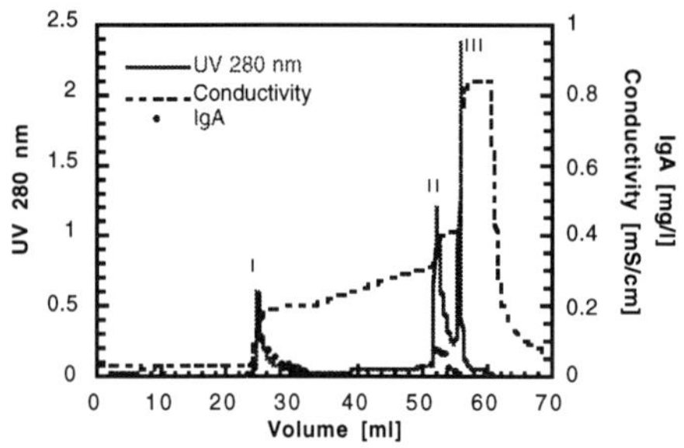

Figure 4: *Elution profile; analytical CHT*

Figure 4 shows an analytical run of a hollow fibre sample passed over a 2 ml analytical CHT-I column. The phosphate concentration necessary for the IgA elution changed when unpurified HFR samples were applied. Comparing UV absorption and the measured IgA concentrations dimeric IgA (peak I) seemed less accompanied by foreign proteins than IgAp (peak II). Further investigation for possible monitoring are currently being undertaken in our laboratory.

Conclusion

We established a novel approach to achieve separation of large amounts of pure dimeric and polymeric IgA in a preparative scale necessary for prospective future large scale operations. A traditional method was replaced by a faster and more efficient technique which has been tailored substantially for our purpose. The use of the ceramic type of hydroxyapatite speeded up IgA purification. We achieved a clear baseline separation of IgA dimers from polymers by introduction of a step-gradient-step protein elution program rather than a traditional linear gradient as assessed by immunodetection, SDS-PAGE and biological activity.
Finally reproducibility of this method has been demonstrated with IgA mixtures from a FBR and a CSTR. The pattern of the elution peaks could be used for the monitoring of the different polymeric forms of IgA during production.

References

1. Mestecky J & Kilian M (1985) Immunoglobulin A (IgA). Methods in Enzymology 116, 37 - 75.
2. Vaerman JP, Langendries A & Vander Maelen C (1995) Homogenous IgA, Monomers, Dimers, Trimers and Tetramers from the same IgA Myeloma Serum. Immunol. Investigations 24(4), 631 - 641.
3. Lüllau E, Heyse S, Vogel H, Marison I, von Stockar U, Kraehenbuhl JP & Corthésy B (1996) Antigen Binding Properties of Purified IgA and Reconstituted Secretory IgA Antibodies. J. Biol. Chem. (in press)
4. Tiselius A, Hierten S & Levin O (1956) Protein Chromatography on Calcium Phosphate Colunms. Arch. Biochem. Biophys. 65, 132 - 155.

5. Aoyama K & Chiba J (1993) Separation of Different Molecular Forms of Mouse IgA and IgM Monoclonal Antibodies by High-Performance Liquid Chromatography on Spherical Hydroxyapatite Beads. J. Immunol. Methods 162, 201 - 210.
6. Stoll TS, Ruffieux PA, Lüllau E, von Stockar U & Marison IW (1995) Characterization of Monoclonal IgA Production and Activity in Hollow-Fibre and Fluidized-Bed Reactors. In: Immobilized Cell: Basics and Applications, Noordwijlerhout, The Netherlands (accepted for publication)

Acknowledgements

We are grateful to P. Pugeaud for assistence during purification. This work was supported by research funds from the Swiss National Sience Foundation, Biotechnology Priority Program

"Bionebulisation": an alternative to Dounce homogenization for the preparation of mammalian cells nuclei in bulk

Degouys V.[1], Montoisy C.[2], Wanderpepen M.-P.[3], Togasaki R.K.[4], Surzycki S.J.[4], Ségault V.[5] and Miller A.O.A.[1*]

*1 Biochimie-Technologie Cellulaire - Université de Mons-Hainaut, 24 Avenue du Champ de Mars B-7000 Mons, Belgium, 2 Institut Provincial d'Enseignement Supérieur Agricole et Technique (IPESAT) Ath, Belgium, 3 Computer Cell Culture Center S.A. (4C) Mons, Belgium, 4 Department of Biology - Indiana University Bloomington IN, USA, 5 Institut fur Molekularbiologie und Tumorforschung Marburg, Germany, * To whom requests for reprint must be addressed*

Keywords: Animal cells, splicing, nebulization, Dounce.

Abstract

HeLa cell nuclei, the reference source for the isolation of factors involved in reactions such as initiation of transcription, splicing, polyadenylation, are obtained following Dounce homogenization of hypotonically fragilized cells(1, 2, 3).
Replacing the homogenization step by nebulization, using the purpose-built BioNeb™, makes it possible to process much larger number of cells than the 2 E+10 which constitute the maximum daily production.
Based on polyacrylamide gel electrophoresis of snRNPs and the evaluation of the splicing activity of nuclear extracts, the quality of nuclei prepared with this instrument compares favourably with the results obtained by douncing.

Introduction

Perfusion allows the routine production of quality HeLa cells, so much so that the problems associated with the isolation of molecules present either in the cytoplasm or in the nucleus are now shifted from cell growth to cell fractionation. Up to now, douncing is the method of choice to prepare homogenates from which nuclear extracts are obtained(1, 2, 3). However, because of the physical strain involved in douncing large numbers of cells, this procedure constitutes a stumbling block if not a nightmare for the persons in charge of this operation. Developed a few years ago by Togasaki and Surzycki, the BioNeb™, is an instrument which generates shearing forces capable to successfully break plant cells in a controlled way(4).

M.J.T. Carrondo et al. (eds.), Animal Cell Technology, 271–275.

In order to replace the Dounce by the BioNeb^{TM}, it is imperative to validate the quality of the nuclear extracts obtained with the latter instrument using the Dounce as a reference. This work aims at presenting a comparative evaluation of the splicing activity displayed by HeLa nuclear extracts obtained by using both instruments and the adaptations brought to the experimental protocol to improve the procedure and bring it in line with the standard protocol of Dignam.

Materials and Methods

HeLa cells strain Oxford are grown in perfused 3 liter Applikon glass jacketed bioreactor and nuclear extract are prepared from exponentionally growing cells.

The cells from one lot are homogenized either by ten strokes of a Dounce homogenizer (B type pestle) or by nebulization with the BioNeb^{TM} disruption system using He at 1 bar(5). The nuclei are enumerated in a Bürke counting chamber following a 60 min. incubation at 37 °C in 0.1 % crystal violet in 0.1 M citrate solution.

U-snRNPs are isolated from the nuclear extract by anti-m^{3}G immunoaffinity chromatography(6). The corresponding proteins are obtained by Phenol : Choroform : Isoamyl Alcohol extraction and ultimately analysed by SDS-PAGE.

Splicing is performed as described earlier(7) by incubating for 90 min. at 30 °C 1 ng of ^{32}P-labelled MINX pre-mRNA in a 25 µl reaction containing 7.5 µl HeLa nuclear extract (± 20 µg proteins), 2.7 µl buffer D (20 mM Hepes/KOH pH 7.9, 100 mM KCl, 20 % glycerol, 1.5 mM $MgCl_2$, 0.2 mM EDTA, 0.5 mM DTE, 0.5 PMSF), 1 µl 120 mM $MgCl_2$, 0.5 µl 25 mM ATP, 0.5 µl 1 M creatine phosphate, 1 µl 28 U/ml RNasin, 0.5 µl 2 µg/µl *E.coli* tRNA.
The splicing products are purified and analysed on denaturing 13 % polyacrylamide-urea gels. Radiolabelled MINX pre-mRNA is prepared by *in vitro* transcription of BamHI linearized pMINX with SP6 polymerase.

Results

Examination of immunoaffinity purified snRNP proteins electropherograms does not allow us to distinguish any qualitative difference between Dounce homogenization and nebulization, neither for the core proteins (present in all U-snRNPs), nor for the other proteins (specific to some U-snRNPs).
The separation on polyacrylamide-urea gel of the different components (pre-mRNA, mRNA, splicing intermediates) of the splicing reaction following the incubation of the MINX pre-mRNA with HeLa nuclear extracts prepared from cells collected at different concentrations (results not shown) shows that the splicing assay realized with the nuclear extract prepared by the Dignam's procedure following Dounce homogenization compares unfavourably with the same assay prepared using an extract obtained by nebulization.

Figure 1 shows the result obtained for nuclear extract prepared from HeLa cells collected at 6.1 E+06 cells per ml. It can be observed that the percentage of mRNA spliced from MINX pre-mRNA is higher with nebulization than with douncing. Extracts prepared from cells collected at other concentrations (1.8 E+06 and 12 E+06 cells per ml, results not shown) always show a higher increase of the splicing activity following BioNeb^{TM} utilization.

Figure 2 shows the time necessary to break 5 E+09 cells when, preceeding the homogenization, different treatments are applied. In comparison with Dounce homogenization performed with the "low viscosity" buffer recommended by Dignam, using the same buffer but the BioNeb^{TM}, longer times are needed to homogenize 5 E+09 HeLa cells. Using the same "low viscosity" buffer but doubling the cell concentration results in a "high viscosity" suspension. Then the time taken to homogenize 5 E+09 cells shortens considerably.

Figure 3 shows that the yield of nuclei obtained by Dounce homogenization is around 76 %. Comparison between the different treatments shows a decrease of this yield following nebulization whith "low viscosity" hypotonic buffer (± 60 %). This value comes back to normal (± 80 %) with "high viscosity" buffer.

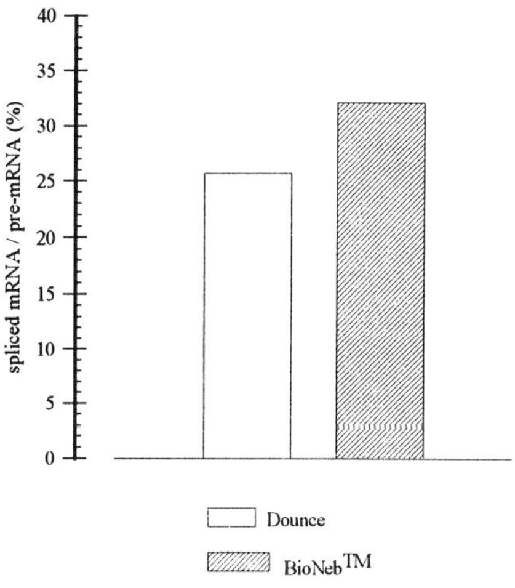

Figure 1 : Comparison of the splicing activity of HeLa nuclear extracts prepared by Dounce homogenization or nebulization.

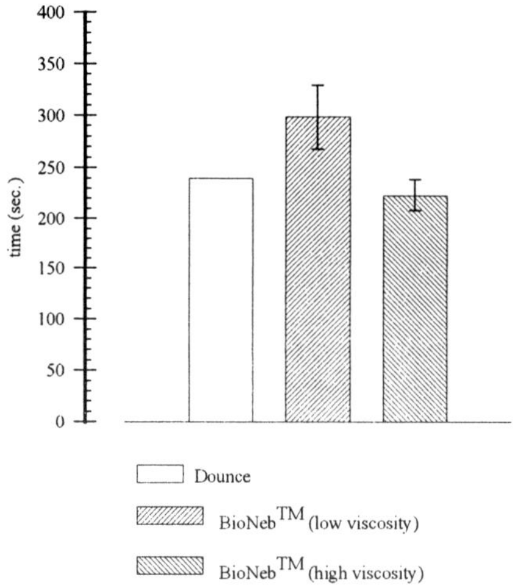

Figure 2 : Comparison of the time necessary to break 5 E+09 HeLa cells respectively with a Dounce homogenizer, the BioNeb™ and a "low viscosity" hypotonic buffer or the BioNeb™ and a "high viscosity" hypotonic buffer.

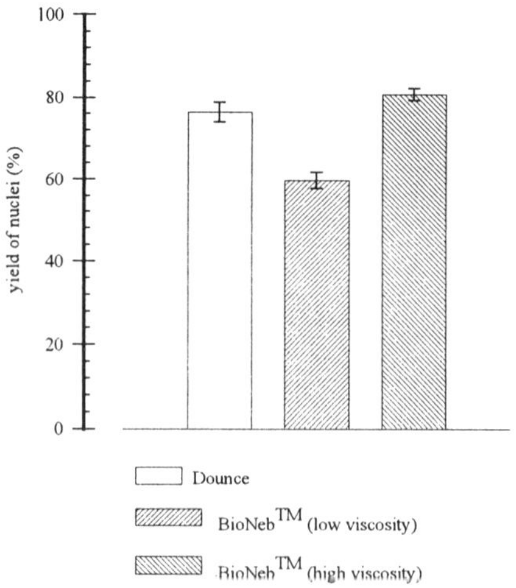

Figure 3 : Comparison of the yields of nuclei recovered respectively after Dounce homogenization, nebulization in a "low viscosity" hypotonic buffer or nebulization in a "high viscosity" hypotonic buffer.

Conclusions

In order to prepare HeLa cells homogenates, the BioNeb™ can advantageously replace the Dounce homogenizer. Using He at 1 bar and "high viscosity" suspension, the yields obtained by nebulization and the time necessary to break the cells are comparable to those obtained by "Douncing". The U-snRNPs protein electropherograms show no differences between the two techniques. Whatever the concentration of the culture at which the cells are collected, extracts prepared with the BioNeb™, show superior splicing activity.

References

1. Dignam J.D., Lebovitz R.M. and Roeder R.G. (1983) Accurate transcription initiation by RNA polymerase II in a soluble extract from isolated mammalian nuclei. Nucleic Acids Research 11, 1475-1489
2. Krämer A., Frick M. and Keller W. (1987) Separation of multiple components of HeLa cell nuclear extracts required for pre-messenger RNA splicing. J. Biol. Chem. 262, 17630-17640
3. Will C.L., Kastner B. and Lührmann R. (1994) Analysis of ribonucleoprotein interactions. In: Higgins S.J. and Hames B.D., RNA processing (vol.I): A practical approach, pp 141-177, Oxford University Press, Oxford, United Kingdom.
4. Okpodu C.M., Robertson D., Boss W.F., Togasaki R.K. and Surzycki S.J. (1994) Rapid isolation of nuclei from carrot suspension culture cells using a BioNebulizer. BioTechniques 16, 154-158
5. Degouys V., Crohain S., Togasaki R.K., Surzycki S.J. and Miller A.O.A. Mass preparation of HeLa cell nuclei using the BioNebulizer. Cytotechnology, in press
6. Bringmann P. and Lürhmann R. (1986) Purification of the individual snRNPs U1, U2, U5 and U4,U6 from HeLa cells and characterization of their protein constituents. EMBO J. 5, 3509-3516
7. Wolff T. and Bindereif A. (1992) Reconstituted mammalian U4/U6 snRNP complements splicing: a mutational analysis. EMBO J. 11, 345-359

EPITOPE DETERMINATION FOR ANTIBODIES RAISED AGAINST RECOMBINANT HUMAN INTERFERON-GAMMA

Andrew D. Hooker[1], Nicola H. Green[1], David C. James,[1] Philip G. Strange[1], Anthony J. Baines[1], Alan T. Bull[1] and Nigel Jenkins[2].
[1]*Dept. of Biosciences, University of Kent, Canterbury, U.K.*
[2]*Dept. of Biological Sciences, De Montfort University, Leicester, U.K.*

1. ABSTRACT

Oligosaccharides, like peptides, make excellent antigens due to their hydrophilicity and complexity and can also mask potential antigenic sites on the peptide backbone. Human interferon-gamma (IFN-γ) has been purified by immunoaffinity chromatography from the supernatant of batch cultures of recombinant Chinese hamster ovary cells to determine its glycosylation profile. To validate this purification procedure and the quantification of IFN-γ by enzyme-linked immunosorbent assay (ELISA) with mouse monoclonal antibody 20B8, the effects of glycosylation upon antibody recognition were investigated. Preliminary experiments with glycosylated and deglycosylated IFN-γ demonstrated that there was no difference in antibody binding by ELISA. To confirm this result, IFN-γ was digested with trypsin and the resulting peptides separated by reverse-phase high-performance liquid chromatography, the isolated peptides being identified by matrix-assisted laser desorption/ionisation mass spectrometry. Glycopeptides containing the Asn_{25} and Asn_{97} glycosylation sites were not recognised by anti-IFN-γ 20B8 in ELISA studies. Its epitopes were located on non-glycosylated regions of IFN-γ, 20B8 recognising protein epitopes rather than the carbohydrate moieties. The purification and quantification of IFN-γ by ELISA with this antibody are therefore independent of the glycosylation status and are not selecting a sub-population of IFN-γ molecules.

2. INTRODUCTION

N-glycosylation is a diverse post-translational modification which can lead to extensive heterogeneity of recombinant glycoproteins produced by eukaryotic cells. Recombinant human IFN-γ is used as a model glycoprotein and has been expressed and characterised in Chinese hamster ovary (CHO) cells, baculovirus-infected insect cells and the mammary glands of transgenic mice (1). This protein contains two N-glycosylation sites at Asn_{25} and Asn_{97}, the Asn_{97} site showing more variable site occupancy and a higher percentage of truncated and oligomannose structures. Matrix-assisted laser desorption/ionisation mass spectrometry (MALDI-MS) techniques have been developed to monitor site-specific changes in glycosylation for IFN-γ throughout batch cultures of CHO cells (2). The quantification by ELISA and immunoaffinity purification of IFN-γ from the cell culture supernatant are a pre-requisite to glycosylation analysis. Highly specific antibodies are therefore required which recognise protein epitopes and do not distinguish between glycoform variants.

M.J.T. Carrondo et al. (eds.), Animal Cell Technology, 277–282.

A variety of methods have been described for the study of antibody-antigen interactions. Recently, the use of MALDI-MS has been reported for the rapid mapping of linear protein epitopes that are bound by monoclonal antibodies (3,4).
In this study, we have investigated the localisation of the IFN-γ epitope for mouse monoclonal antibody 20B8 to determine whether the ELISA quantification and immunoaffinity purification for recombinant human IFN-γ are independent of the glycosylation status of the protein.

3. MATERIALS AND METHODS

3.1 Cell line and culture conditions

A mutant CHO-Duk-B11 cell line lacking dihydrofolate reductase ($DHFR^-$) has been co-transfected with the genes for IFN-γ and DHFR. This cell line was adapted for growth in RPMI-1640 based serum-free medium supplemented with BSA, insulin, transferrin and trace elements and the genes for IFN-γ and DHFR co-amplified by methotrexate selection (5).

3.2 Purification and assay of IFN-γ

IFN-γ was purified from the cell culture supernatant by immunoaffinity chromatography on Sepharose linked to monoclonal antibody 20B8 raised against *E. coli* derived recombinant IFN-γ. IFN-γ concentrations were determined by a monoclonal antibody based sandwich ELISA. Microtitre plates were coated with rabbit polyclonal antibody R1PA. The plates were blocked with casein, then with sequential additions of serial dilutions of IFN-γ standards and samples, monoclonal antibody 20B8 and with anti-mouse polyvalent immunoglobulins, peroxidase conjugate, with intermittent washes as described previously (5). Binding was determined at 490nm following an incubation with o-phenylenediamine dihydrochloride (OPD). Deglycosylated samples of IFN-γ were prepared by digestion with PNGase F from *Flavobacterium meningosepticum* (Oxford Glycosystems, UK) according to the manufacturers instructions and the deglycosylation confirmed by sodium dodecyl sulphate polyacrylamide gel electrophoresis (SDS-PAGE).

3.3 Trypsin digestion and separation of tryptic peptides

Purified IFN-γ was digested with trypsin (1.5μg:50μg) for 24h at 30^0C and the resulting peptides separated and isolated by reverse-phase high-performance liquid chromatography (rp-HPLC). Peptides were lyophilised and stored at -20°C.

3.4 Peptide-binding ELISA

The binding of rabbit polyclonal and mouse monoclonal antibodies to the IFN-γ tryptic peptides was determined by ELISA. Microtitre plates were coated with the isolated peptides. The plates were blocked with casein (0.5%, w/v), then with sequential additions of the antibody of interest (1:100, v/v) and anti-mouse or anti-rabbit polyvalent immunoglobulin, peroxidase conjugates (1:1000, v/v) with intermittent washes. Binding was determined at 490nm, following incubation with OPD.

3.5 MALDI-MS

Peptide sample (0.5μL) was mixed with a saturated solution of alpha-cyano-4-hydroxy cinnaminic acid in 60% (v/v) aqueous CH_3CN (0.5μL) and allowed to co-crystallize on stainless steel sample discs. Mass spectrometry was performed with a VG Tof Spec mass spectrometer using a N_2 laser at 337nm. Mass spectra were calibrated with an external peptide standard, vasoactive intestinal peptide (fragment 1-12, Sigma). Digestion of glycopeptides with PNGase F prior to analysis by mass spectrometry confirmed the mass of the core peptides to be 2252.5 for tryptic glycopeptide 4 (T4, containing Asn_{25}) and 1522.7 for tryptic glycopeptide 19 (T19, containing Asn_{97}).

4. RESULTS AND DISCUSSION

4.1 Initial observations: glycosylated and non-glycosylated IFN-γ

IFN-γ produced by *E. coli* is non-glycosylated and migrates by SDS-PAGE as a single protein band (18 kDa). It is recognised by mouse monoclonal antibody 20B8, suggesting that this antibody recognises a protein epitope. In contrast, IFN-γ from CHO cells can exist in three forms: doubly glycosylated (2N - occupied at both Asn_{25} and Asn_{97}), singly glycosylated (1N - occupied at Asn_{25} only) and non-glycosylated (0N).

To confirm that monoclonal antibody 20B8 recognises a protein epitope, we compared its ability to recognise glycosylated and deglycosylated IFN-γ from CHO cells by ELISA (Figure 1). 20μg of purified IFN-γ was incubated at 37°C for 18h ± PNGase F and the deglycosylation of the IFN-γ samples confirmed by SDS-PAGE. Both glycosylated and deglycosylated IFN-γ were recognised by ELISA with monoclonal antibody 20B8. However, the presence of the carbohydrate moiety does appear to slightly reduce the binding sensitivity of monoclonal antibody 20B8, possibly due to steric hindrance.

4.2 IFN-γ epitope mapping by ELISA

Rabbit polyclonal (R1PA and R2PA) and mouse monoclonal (20D7 and 20B8) antibodies raised against human IFN-γ were investigated for their ability to bind protein or carbohydrate epitopes. Purified IFN-γ was digested with trypsin and the peptides separated and isolated by rp-HPLC (Figure 2). Individual peptide fractions were coated onto ELISA plates and antibodies screened for their ability to bind to the different peptides (Figure 3). Those peptides strongly recognised by the antibodies using the modified ELISA protocol were identified by MALDI-MS (Table 1). The monoclonal antibodies 20D7 and 20B8 bound to peptides which were non-glycosylated and which appeared to possess a linear epitope, whereas the polyclonal antibodies R1PA and R2PA bound both peptides and glycopeptides and appeared to possess non-linear epitopes.

The glycopeptides T4 and T19 were digested with PNGase F, re-purified by rp-HPLC and the mass of the deglycosylated peptides confirmed by MALDI-MS. Polyclonal antibodies R1PA and R2PA bound to both glycosylated and deglycosylated T4 and T19 to a similar extent by ELISA, suggesting that they recognise only protein epitopes.

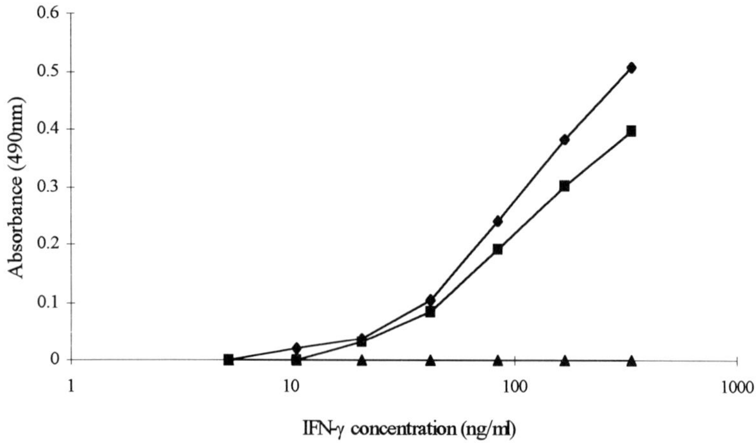

Figure 1 **Comparison of glycosylated and deglycosylated IFN-γ by ELISA with monoclonal antibody 20B8.**

◆ = Deglycosylated IFN-γ (following PNGase F digestion),

■ = Glycosylated IFN-γ; ▲ = PNGase F.

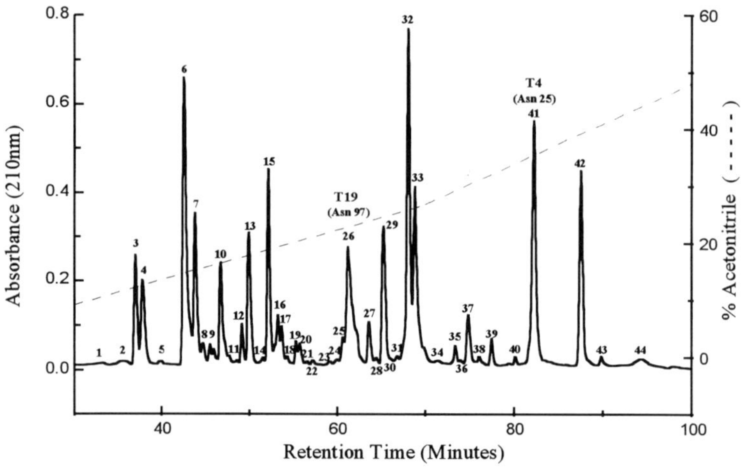

Figure 2 **Separation and isolation of the tryptic fragments of IFN-γ by rp-HPLC.**

The Asn_{25} - containing glycopeptide (T4) was isolated at a retention time of 81 minutes and the Asn_{97} - containing glycopeptide (T19) was isolated at a retention time of 61 minutes.

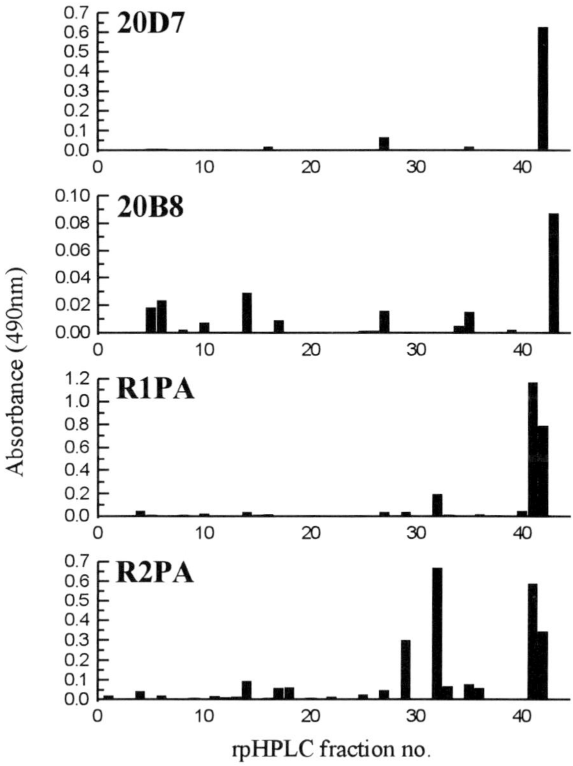

Figure 3 **IFN-γ epitope mapping of tryptic fragments by ELISA**

HPLC Fraction Number	Antibody Binding	Mass (M/Z)	Peptide Identification
29	R2PA	1525.1	$T19_{(98\text{-}110)}$
32	R1PA ; R2PA	1413.5	$T1 + T2_{(1\text{-}15)}$
41	R1PA ; R2PA	*	$T4_{(17\text{-}37)}$
42	20D7 ; R1PA ; R2PA	1820.3	$T21_{(112\text{-}128)}$
43	20B8	1420.6 1512.2	$T12 + T13_{(72\text{-}83)}$ $T11 + T12_{(65\text{-}77)}$

Table 1 **MALDI-MS identification of major epitope-containing peptides using monoclonal and polyclonal antibodies**

* = Heterogeneous signal suggesting a glycopeptide

The numbers in brackets indicate the amino acid sequence

5. SUMMARY

Antibodies can bind either a linear epitope containing a stretch of contiguous amino acids or a non-linear epitope which is composed of residues that can be distant in the primary sequence but close in space in the folded protein. Techniques for antigenic site mapping of linear epitopes include binding assays of synthetic peptides that span the protein and the use of bacteriophage peptide libraries. However, these approaches are costly and time consuming, whereas the technique described in this paper is particularly suitable for the rapid mapping of linear protein epitopes and takes advantage of the high accuracy (>0.1%), speed of analysis (<1 min) and sensitivity (<1 pmol) of MALDI-MS.

The data presented here strongly indicates that the rabbit polyclonal (R1PA and R2PA) and mouse monoclonal (20D7 and 20B8) antibodies to human IFN-γ used in our laboratory recognise protein epitopes, as opposed to carbohydrate structures associated with the molecule. The ELISA and immunoaffinity purification procedures used regularly for human IFN-γ are therefore not influenced by the changes in glycosylation that occur throughout batch culture of CHO-320 cells (2).

6. ACKNOWLEDEMENTS

This work was supported by the Biotechnology and Biological Sciences Research Council (BBSRC). We thank Glaxo-Wellcome for the gift of recombinant CHO 320 cells and Celltech Biologics Ltd. for monoclonal 20B8 and 20D7.

7. REFERENCES

1. James, D.C., Freedman, R.B., Hoare, M., Ogonah, O.W., Rooney, B.C., Larionov, O.A., Dobrovolsky, V.N., Lagutin, O.V. and Jenkins, N. 1995. N-glycosylation of recombinant human interferon-gamma produced in different animal expression systems. Bio/technology 13(6): 592-596

2. Hooker A.D., Goldman M.H., Markham N.H., James D.C., Ison A.P., Bull, A.T., Strange, P.G., Salmon I., Baines, A.J. and Jenkins, N. 1995. N-glycans of recombinant human interferon-γ change during batch culture of Chinese hamster ovary cells. Biotechnology and Bioengineering 44: 639-648.

3. Zhao Y. and Chalt B.T. (1994). Protein epitope mapping by mass spectrometry. Analytical Chemistry 66: 3723-3726.

4. Papac D.I., Hoyes J. and Tomer K.B. (1994). Epitope mapping of the gastrin-releasing peptide/anti-bombesin monoclonal antibody complex by proteolysis followed by matrix-assisted laser desorption mass spectrometry. Protein Science 3:1485-1492.

5. Hayter, P.M., Curling, E.M., Baines, A.J., Jenkins, N., Salmon, I., Strange, P.G. and Bull, A.T. 1991. Chinese hamster ovary growth and interferon production kinetics in stirred batch culture. Appl. Microbiol. Biotechnol. 34: 559-564.

A NEW SPINFILTER DESIGN

E. Fraune[1] S. Meichsner[1], M. N. Kamal[2]
[1] B. Braun Biotech International, 34212 Melsungen, Germany
[2] Bioprocess Engineering Dep., University of Technology Malaysia, 54100 Kuala Lumpur, Malaysia

Abstract

A design of a flexible industrial scale bioreactor, for perfusion of mammalian cells, must meet process specific requirements and general design criteria for a production plant. A new spinfilter design has been developed for perfusion via an external loop of a bioreactor. The filter is operating in a glass cylinder with stainless steel top and bottom plates. Clogging by cell particles and proteins is minimized by the formation of taylor vortices at the surface of the 10 μm filter. During a 1400 hour perfusion experiment with hybridoma cells, various operating parameters were investigated with respect to retention of viable and dead cells. The rotational speed of the spinfilter, the recirculation rate of the cultivation fluid, and the dilution rate were varied. Optimum operating conditions were found under which the retention of viable cells is significantly higher than that of dead cells.

Key words: external spinfilter, perfusion, taylor vortices, hybridoma cells

Introduction

The current trend in mammalian cell culture focuses on the development of effective and reliable cell retention systems. The use of vortex flow filtration has turned out to be highly effective for filtration of cell culture fluids /1/.

The filtration of cell culture media is more difficult to perform due to high protein amounts in the medium and shear sensitivity of mammalian and insect cells. Small cell particles and cell proteins have a negative effect on the filtration process. Filtration equipment based on membranes has some major disadvantages such as filter fouling, which make special cleaning and validation procedures necessary. This problems can be avoided and a prolongation of the operation time can be obtained by employing stainless steel as a filter material due to its cleanability. Spinfilters make use of the fact that dead cells are smaller and their cell wall is less rigid as compared to intact viable cells so that they can easily pass a filter mesh.

M.J.T. Carrondo et al. (eds.), Animal Cell Technology, 283–287.

Here, we applied a 10 μm stainless steel spinfilter which is operated in an external loop of the bioreactor /2/. The external set-up of the spinfilter has some major advantages as compared to conventional built-in spinfilters /3/:

- clogging by cells and cell particles is kept at a minimum by the formation of taylor vortices at the surface of the filter,
- the spinfilter is easily scalable regardless of the overall bioreactor design,
- the spinning velocity of the filter can be chosen independently from the stirring velocity of the bioreactor,
- the working volume of the bioreactor can be varied during operation,
- the working volume of the bioreactor is not reduced by the filter volume,
- the filter can be exchanged during operation of the bioreactor,
- a spinfilter of given size can be connected to different sizes of bioreactors.

Materials and Methods

Bioreactor

The bioreactor is designed for mammalian cells and equipped with membrane aeration and agitators for shear sensitive cells (Biostat MCD, B.Braun Biotech International, Germany). Precise control of T, pH, DO and level is managed by the DCU (digital control unit, B.Braun Biotech International, Germany).

External Rotating Filter

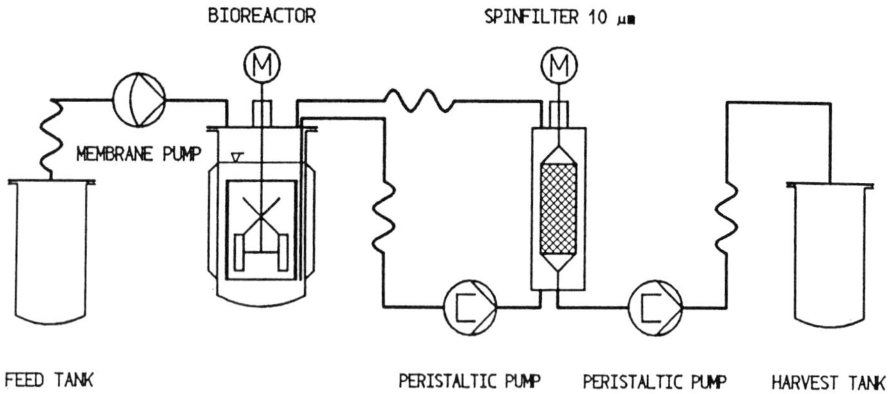

Figure 1: Experimental set-up of external spinfilter

The filter is operated in a glass cylinder with stainless steel top and bottom plates. The inlet of the retentate is located at the bottom plate, the outlet at the top plate. The filtrate is collected through the hollow shaft via 4 symmetric openings. Medium is recirculated and harvested by means of peristaltic pumps (FE 411, B.Braun Biotech International, Germany). An adjustable stirrer (IKA, Germany) is used for rotating the filter. The stirrer can be adjusted from 60 to 2000 rpm. The rotating filter is mechanically sealed. The filter mesh is made of stainless steel with a mesh size of 10 μm. The surface of the spinfilter amounts to 300 cm^2, the annular gap between the filter and the filter housing is 0.5 cm **(figure 1)**.

Cells and Medium

The HB 124 hybridoma cell (DB9G8, ATCC) is cultivated in DME-medium with FCS in decreasing concentrations of 10 to 2.5 %. Samples were taken from the bioreactor vessel and the spinfilter filtrate and the cell number was counted microscopically by trypan blue exclusion.

Results and Discussion

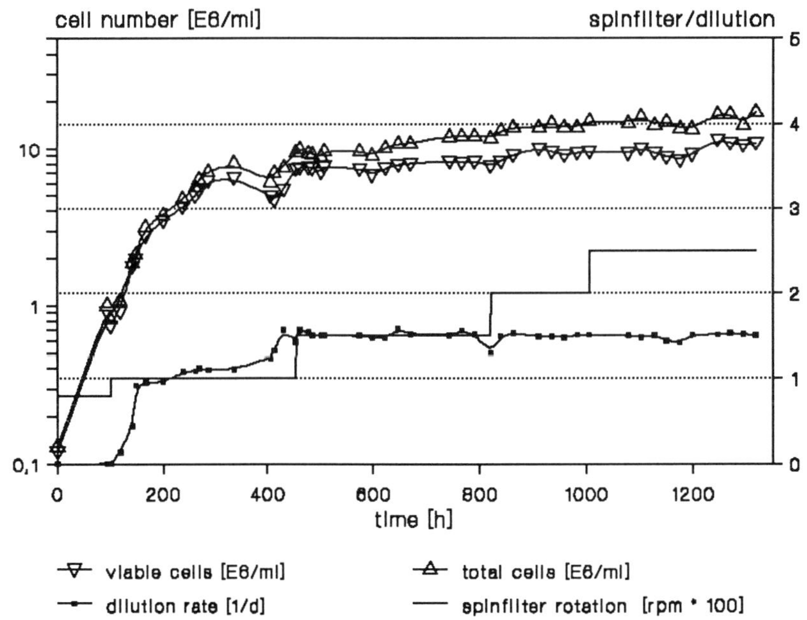

Figure 2: growth curve of HB 124 cells

During the 1400 hour perfusion experiment with hybridoma cells a maximum cell density of 1.9 * 10^7 cells/ml was reached at the end of the experiment. As a remarkable feature of the fermentation experiment, a constantly increasing cell number was obtained **(figure 2)** and so steady-state conditions could be adjusted for various parameter settings. Thus, we were able to investigate the effect different parameters had on the cell retention. During 55 days of perfusion with DME-medium containing 2.5 % to 10 % fetal calf serum the medium exchange was varied from D = 0.1 d^{-1} to D = 1.5 d^{-1}. The retention rate of viable cells was in the range from 75 % to 95 %, whereas dead cell removal ranged from 30 % to 70 % depending on the viability of the culture and the spinfilter operation mode.

The **recirculation rate** of the culture medium through the external spinfilter has an important influence on cell retention. Increasing the recirculation rate results in a critical pressure whithin the filter device since more cells are pressed through the filter and cell dilution can reach high values as shown in **figure 3**.

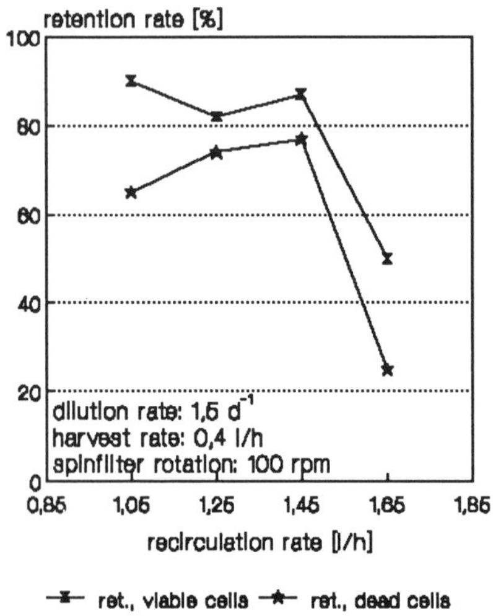

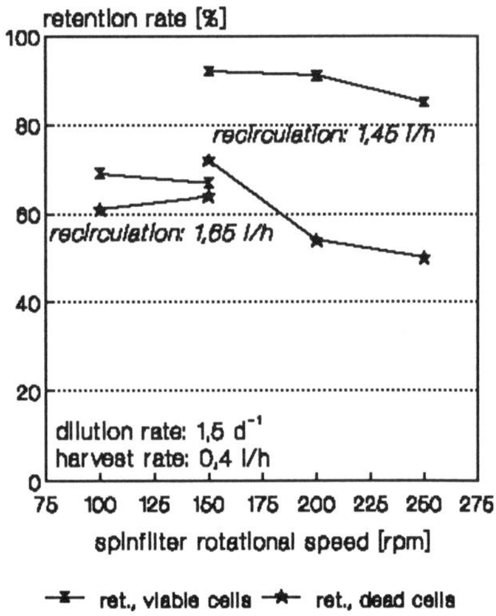

Figure 3: correlation between recirculation rate and cell retention

Figure 4: correlation between spinfilter velocity and cell retention

The variation of the **rotational speed of the spinfilter** influences the fluid flow pattern in the spinfilter device. A laminar regime in the annular gap between filter and filter housing improves the retention of cells. Turbulences, e.g. Taylor vortices formed in the annular gap, occur at a rotational speed of 250 rpm using this type of spinfilter geometry. Turbulences enhance cleaning effects at the surface of the filter and prolongate the operation time of the spinfilter. Under turbulent conditions, the retention of viable and dead cells differ significantly. Dead cells are washed out **(figure 4)**.

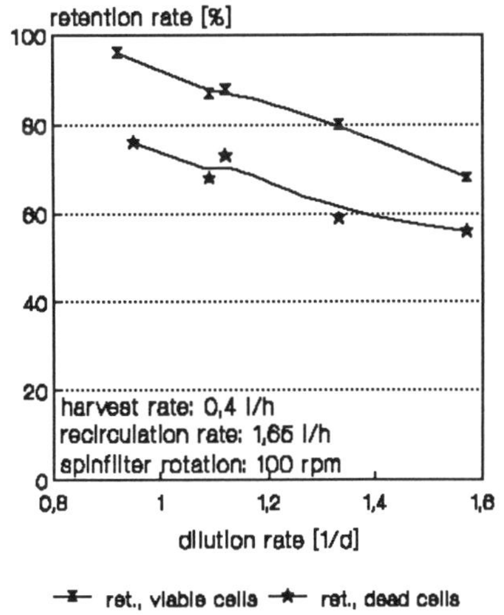

Figure 5: correlation between dilution rate and cell retention

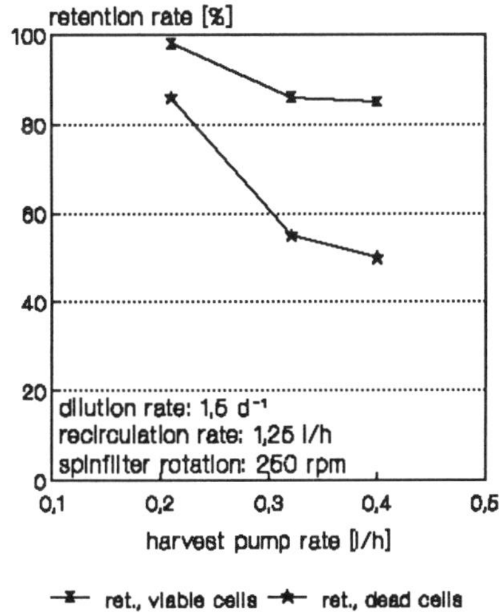

Figure 6: correlation between harvest rate and cell retention

During perfusion cultivation the medium exchange was varied from D = 0.1 to 1.5 d^{-1}. With increasing **dilution rate D** and an improved nutrient supply the number of cells increased. As a consequence, the flux through the spinfilter increases and the retention of both viable and dead cells is decreasing by the same order of magnitude **(figure 5)**.

Varying the **harvest rate**, e. g. by means of increasing the flux through the filter at constant dilution rates, results in a reduction of dead cell retention **(figure 6)**. The applied fluxes are 6.6 to 13.3 l/m^2*h.

Conclusion

We presented a new spinfilter design and tested its performance during a long-term cultivation of the hybridoma cell line HB 124. The presented data clearly demonstrate that various operating conditions e. g. rotational speed of the spinfilter, recirculation rate of the cultivation fluid through the external spinfilter, harvest rate and flux are an appropriate tool for enhancing the viability of the culture. The external spinfilter design provides a flexible easy-to-use perfusion system for mammalian and insect cells.

References

/1/ Hawrylik, S. J.; Wasilko, D. J.; Pillar, J. S.; Cheng, J. B.; Lee, S.E. (1994)
Vortex Flow Filtration Of Mammalian And Insect Cells
Cytotechnology 15: 253-258

/2/ Kamal, M. N.; Aziz, Z.; Greenfield, P. E. (1994)
High Density Hybridoma Cell Culture In A Novel Perfusion Bioreactor
3rd Asia - Pacific Biochemical Engineering Conference, Singapore

/3/ Siegel, U.; Fenge, C.; Fraune,E. (1991)
Spinfilter For Continuous Perfusion Of Suspension Cells
Animal Cell Technology: Dev., Proc. And Prod., Butterworth - Heinemann, Oxford

On-line Biomass Monitoring of Immobilized Hybridoma Cells by Dielectrical Measurements

T. Noll, M. Biselli, C. Wandrey
Institute of Biotechnology
Forschungszentrum Jülich GmbH, D-52425 Jülich, Germany

Abstract:

To prove the aptitude of dielectric spectroscopy for on-line biomass monitoring and an automated process control, this non-invasive method has been succcessfully used in the cultivation of hybridoma cells, immobilized on microcarriers, in a fluidized-bed bioreactor.

Introduction:

The estimation of parameters like cell concentration, viability, physiological states or consumption rates belongs to the most important demands in the cultivation of animal cells, because this will lead to a better understanding of the biological background and enables an optimisation of the bioprocess. Therefore many methods especially for determination of cell concentration have been developed. One of the newer methods for determination of Biomass is based on dielectric spectroscopy or also called capacitance measurement. This technique has some advantages compared with other methods. It enables an on-line biomass monitoring independent from simultaneous present solid particles or bubbles and can therefore be used also in systems with immobilized cells. We used a fluidized bed bioreactor with hybridoma cells immobilized on chemically modified open-porous microcarriers to investigate the usefulness of dielectric spectroscopy for on-line biomass monitoring and an automated process control.

Materials und Methods:

Cell line: mouse-mouse hybridoma, producing an IgG monoclonal antibody
Medium: DMEM / Ham's F12 (3:1), supplemented with amino acids, fatty acids, insulin, transferrin, buffers and 1 % FCS
Dielectrical measurements: Biomass Monitor BM 214A (Aber Instruments Ltd., Aberystwyth, UK) set on 0.6 Mhz
Cell size determination: directly with Flow Cytometry (Coulter XL, Coulter, FRG) indirectly by cellular protein content, with BCA assay (Pierce, NL)

M.J.T. Carrondo et al. (eds.), Animal Cell Technology, 289–294.

Bioreactor:

The fluidized bed bioreactor consists of a thermostated reactor tube with a conical bottom inlet, leading to a homogeneous, gentle fluidization of the carriers. The reactor was inserted in a recycle loop containing the recirculation pump, DO and pH probes and external oxygenator for bubblefree aeration by silicon tubes[1] (5m). The probe for dielectrical measurements was integrated vertically into the fluidized bed (Fig. 1).

Volume: total volume 225 ml, containing 60 ml carrier
Productivity: approximately 1 g Mab / ($l_{carrier}$*d)[2]
Carrier: open porous glas carriers, modified with gelatine[3], diameter 400-700 μm, porosity 50 %

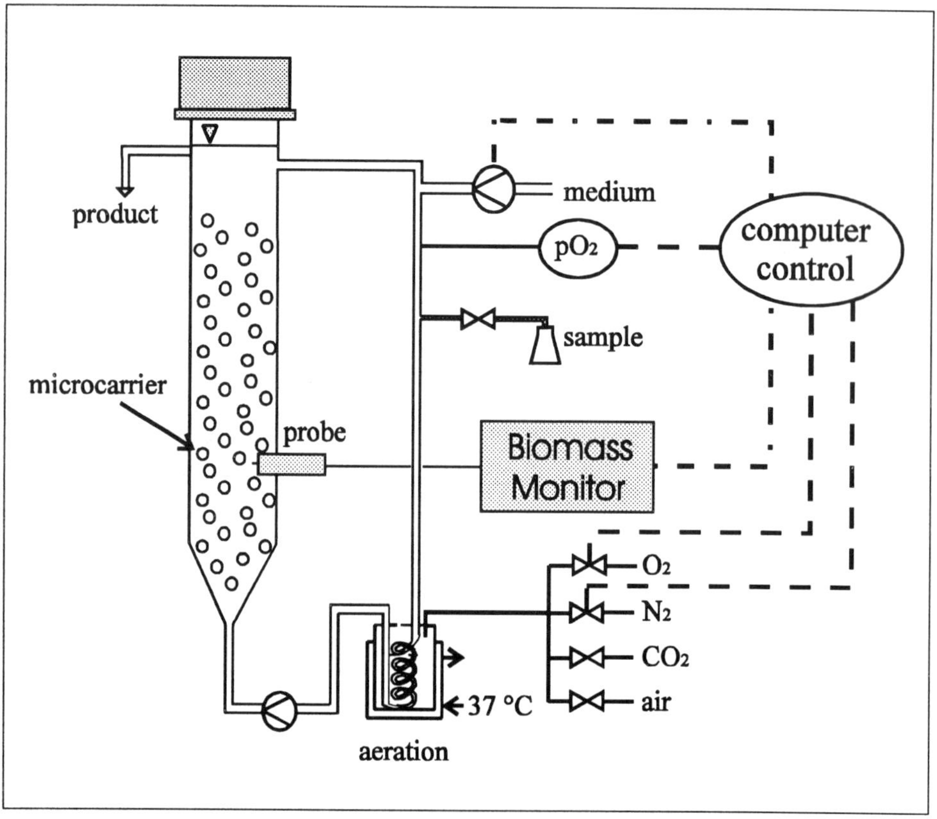

Fig. 1: Flow-sheet of the fluidized-bed bioreactor

Theoretical background:

The method of dielectric spectroscopy is based on the measurement of the capacitance and the conductance, at low frequencies (0.2 - 10 MHz). The principle may be illustrated by a suspension of mammalian cells in a ionic solution (Fig. 2). From an electrical sight biological cells consist of a non-conducting plasma membrane that separates the conducting cytoplasma from the also conducting extra-cellular environment. In an electrical field the ions in the cytoplasma as well as in the surrounding medium move towards the contrary charged electrode. The ions in the cytoplasma bump into the plasma membrane and are stopped. This leads to a polarization of the cell membrane, that can be measured as an increase of the capacitance. Solid particles or lysed cells don't get polarized, so dielectric spectroscopy detects only living cells[4].
The capacitance (C) of a suspension of ideal spherical cells with the cell density N, the radius r and the capacitance per membrane area C_m has been calculated by Schwan[5] as shown in the following equation.

$$C = 3 * \pi * r^4 * N * C_m * d/A \qquad (1)$$

This is applicable for plane-parallel electrodes of area A separated by the distance d. The measured capacitance is therefore a function of three variables, namely cell density, cell size and capacitance per membrane area.

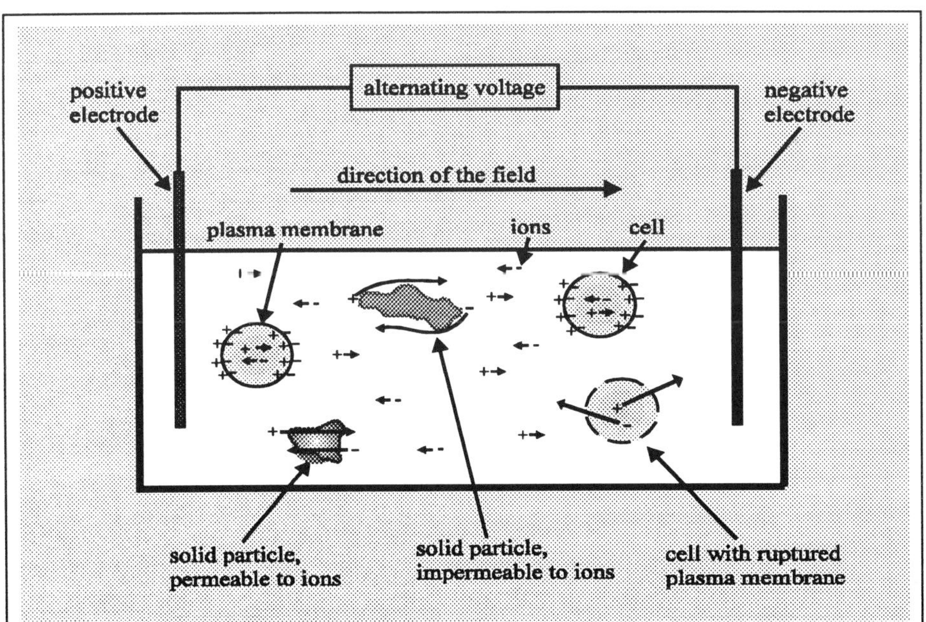

Fig. 2: Principle of dielectrical measurements

Results and Discussion:

For a constant cell size and capacitance per membran area (performed by rapid dilution and measurement of a high concentrated suspension of hybridoma cells) the capacitance of suspended cells showed a highly linear correlation to the cell density over a wide concentration range, as predicted in equation 1 (Fig. 3).

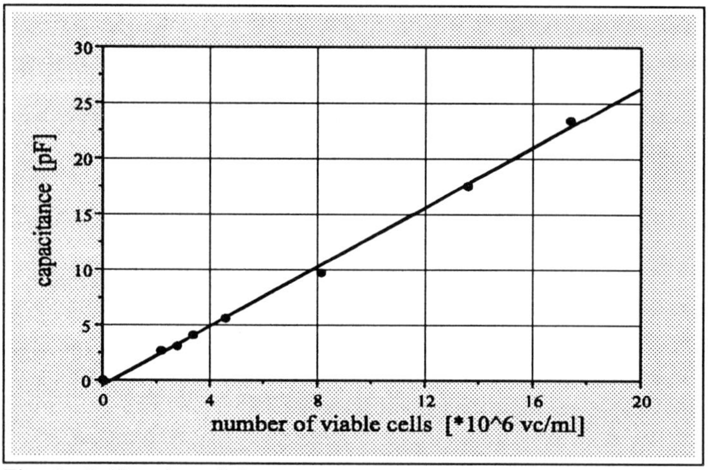

Fig. 3: Calibration of the sensor

Cell immobilization on microcarriers (after inoculum with suspended cells) could be observed in real-time by capacitance measurement (Fig. 4).

=> possibility to optimize the microcarriers (diameter, size of the pores, kind of modification)

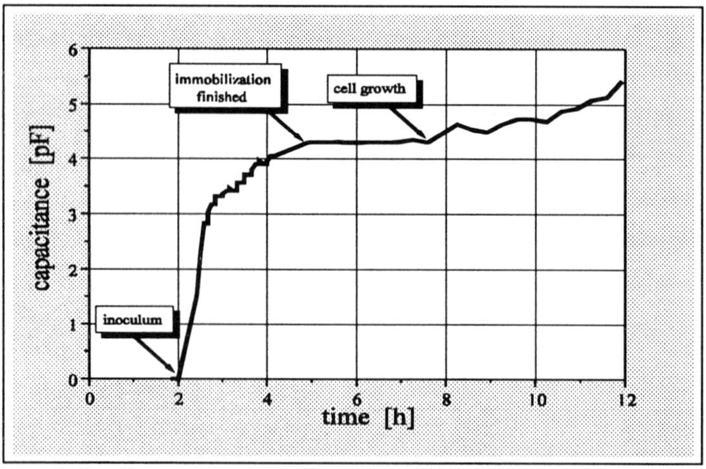

Fig. 4: On-line monitoring of cell immobilization on microcarriers

In cultivation of immobilized cells the cell size decreased significantly over the fermentation (data not shown). Therefore, in combination with the fact that the capacitance per membran area C_m varied with the physiology of the cells, an on-line estimation of the cell density is impossible. The measured capacitance and the number of immobilized cells (off-line determined by crystal violet nuclei staining) showed only a fair correlation (Fig. 5).

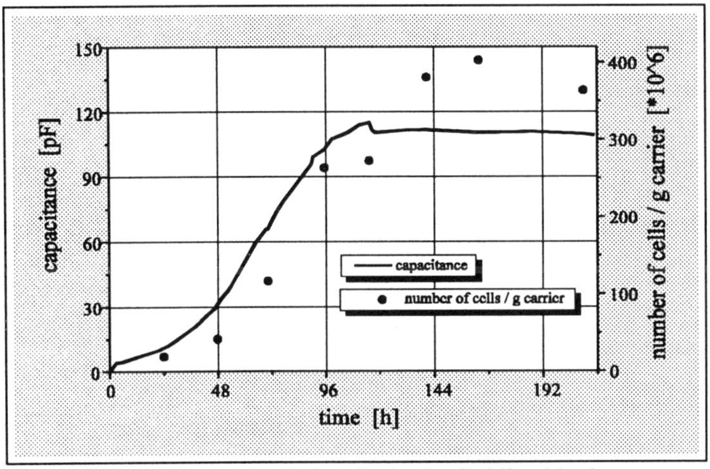

Fig. 5: Capacitance and cell density in the fluidized bed

In contrast to this the capacitance showed an exact and reproducible correlation with the actual metabolic rates in the bioreactor, especially with the glutamine consumption rate (Fig. 6).

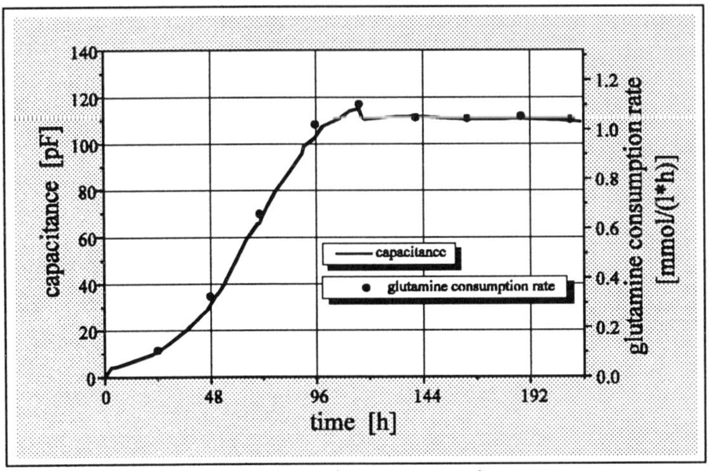

Fig. 6: Capacitance and glutamine consumption rate

=> actual glutamine consumption rate is on-line available by the measured capacitance
=> permits an automated and optimized medium dosing depending on the actual metabolic activity of the cells
=> manual process control by off-line analysing of samples, calculation of consumption rates and manual adjustment of the medium dosing is no longer necessary

Summary:

- Dielectric spectroscopy detects living cells, independent from solids
  => also possible for immobilized cells
- It permits on-line monitoring of cell immobilization.
- Capacitance is a function of cell density, cell size and physiology.
- It correlates exactly and reproducibly with the actual metabolic rates
  => on-line available
- Dielectric spectroscopy was successfully used for an automated medium dosing.

References:

[1] Thömmes J, et al. (1993) 'The influence of dissolved oxygen tension on the metabolic activity of an immobilized hybridoma population', Cytotechnology 13, 29-39
[2] Rolef G, et al. (1993) 'Optimization of antibody production in a fluidized-bed reactor', in Spier RE, Griffiths JB, McDonald C (eds.): Animal cell technology, Butterworth-Heinemann, Oxford, 481-484
[3] Lüllau E, Dreisbach C, Grogg A, Biselli M, Wandrey C (1992) 'Immobilization of animal cells on chemically modified carrier', in Spier, Griffith (eds.): Animal cell technology, Butterworth-Heinemann, 469-475
[4] Davey CL, Markx GH, Kell DB (1993) 'On the dielectric method of monitoring cellular viability', Pure&Appl. Chem. 65, 1921-1926
[5] Schwan HP (1957) 'Electric properties of tissues and cell suspensions', Adv. Biol. Med. Phys. 5, 147-208

EVALUATION OF THE PROTEOLYTIC ACTIVITY PRESENT IN CHO CELL CULTURE SUPERNATANTS

C. Tans, S. Wattiaux-De Coninck, M.-M. Gonze* and L. Fabry*

Department of Physiological Chemistry, Faculty of Medicine
Facultés Universitaires Notre-Dame de la Paix, 5000 Namur - Belgium
(*)R&D Preclinical Department, Cell Culture Technologies
SmithKline Beecham Biologicals SA, 1330 Rixensart - Belgium

ABSTRACT

The heterogeneity of recombinant glycoproteins expressed by CHO cells may partly result from a proteolytic activity present in cell culture supernatants. This study investigated the presence of proteases in CHO cell culture supernatants and their effects on the quality of a recombinant secreted viral envelope glycoprotein used as a model.

The glycoprotein characteristics were assessed by isoelectric focusing and western blot. Some of the isoforms of the glycoprotein so observed were found to be modified forms in particular at the NH_2-terminal domain of the glycoprotein. Moreover, incubation of the intact glycoprotein with parental (untransfected) CHO cell culture supernatants resulted in similarly modified isoforms. Preliminary trials with protease inhibitors suggested that a serine protease could be involved in the glycoprotein modifications.

The presence of proteases in CHO cell culture supernatants might thus influence the quality of recombinant secreted glycoprotein and has to be considered in the optimisation of the cell culture conditions in large scale process development.

INTRODUCTION

Recombinant glycoproteins secreted by animal cells in culture are not always recovered as active molecules as suggested recently for interferon-γ (1).
One possible explanation for the heterogeneity of the secreted recombinant glycoproteins is the modification by protease(s) present in the cell culture supernatants.
In the present study, we have evaluated the impact of the proteolytic activity present in CHO cell culture supernatants on the quality of a recombinant viral envelope glycoprotein secreted by CHO cells. Samples of the purified recombinant glycoprotein have been incubated in parental (untransfected) CHO cell culture supernatants. After the incubation period, the quality of the recombinant glycoprotein, especially in the NH_2 terminal domain, was analysed by isolectric focusing and western blot to assess the effects of the proteases.

MATERIAL AND METHODS

Cell culture

Cell-free supernatants (SN) were obtained from parental (untransfected) CHO cells cultures in shake flasks (150 ml) in serum-free culture medium.

M.J.T. Carrondo et al. (eds.), Animal Cell Technology, 295–300.

Incubation conditions

A purified glycoprotein sample (with a low percentage of NH_2-terminal modifications assessed by NH_2-terminal sequencing) was incubated in the presence of parental CHO cell culture supernatants (not concentrated or concentrated by ultrafiltration). Prior to incubation, the pH was adjusted to pH 7.2 with 5N NaOH. Incubation time was 44 hours at 37 °C.

Isoelectric focusing and western blotting (IEF/WB)

The glycoprotein samples were concentrated 15 fold by ultrafiltration prior to isoelectric focusing. Gel electrophoresis was performed under denaturating conditions as reported by O'Farrel (2) and Lognonné (3) in a vertical slab gel unit; a pH gradient between 3 and 6 was used; migration was stopped after 30 000 Vxhours. After several washes in a fixing solution, gels were blotted to PVDF membranes in the presence of SDS using a semidry transblot system. The blot was incubated with a polyclonal rabbit anti-recombinant glycoprotein antibody and then with an anti-rabbit horseradish peroxidase conjugated antibody; the blot was developed with 3,3' diaminobenzamidine.
The IEF/WB were analysed with a densitometer/image analyser (Bio-Rad).

RESULTS AND DISCUSSION

1. Analysis of purified recombinant glycoprotein samples by IEF/WB (figure 1)

IEF/WB of the purified recombinant glycoprotein sample with a low percentage of NH_2-terminal modications revealed six major bands at an isoelectric point (pI) between 4.91 and 5.28 (bands 2', 4, 6, 7, 8, 9).
For the sample with a high percentage of NH_2-terminal modifications, two bands called " basic bands" (bands 11, 12) as well as four "acidic bands" (bands 1, 2, 3, 5) were found in addition.
The NH_2-terminal sequencing of bands 1, 2, 2', 3, 5 revealed that they contained more than 70 % NH_2-terminal modified forms of the recombinant glycoprotein. From this observation, we have assumed that the NH_2-terminal quality of the glycoprotein can be assessed by the relative amount of the bands 1, 2, 2', 3, 5 in the IEF/WB profile as obtained from the densitometric analysis.

2. Analysis of purified recombinant glycoprotein after incubation with CHO supernatants (figure 2)

After incubation of the glycoprotein sample with cell-free supernatants (3 fold concentrated), IEF/WB analysis showed the five bands 1, 2, 2', 3, 5 (lane b) corresponding mainly to NH_2-terminal modified forms of the glycoprotein as compared with the control (lane a). When the glycoprotein was incubated in fresh culture medium, the five bands 1, 2, 2', 3, 5 did not appear (data not shown). It was also observed that a pH increase resulted in larger relative amount of the bands 1, 2, 2', 3, 5 in the IEF/WB profile.

3. Analysis of purified recombinant glycoprotein samples after incubation in the presence of protease inhibitors (figure 3)

When the purified glycoprotein sample was incubated with parental CHO cell culture supernatant (5 fold concentrated) in presence of the serine protease inhibitor aprotinin, the relative amount of the bands 1, 2, 2', 3, 5 in the IEF/WB profile decreased significantly (lane c compared with lane b).
These results suggest that the protease(s) responsible for the NH_2-terminal modifications of the recombinant glycoprotein might be a serine protease.

CONCLUSIONS

The study demonstrates that CHO cell culture supernatant contains protease(s) responsible for modifications of a secreted recombinant glycoprotein, especially in the NH_2-terminal domain (isoforms addition).
The optimal activity of the protease(s) is found to be around pH 8 and preliminary results using protease inhibitors suggest that the protease(s) is a serine protease(s).
In conclusion, protease(s) present in CHO cell culture supernatants might impact on the quality of recombinant secreted glycoproteins. Proteolytic activity should be taken into account for the development of cell culture conditions for an optimized recombinant glycoprotein production.

ACKNOWLEDGEMENTS

We thank C. Capiau, D. Plainchamps and M. Loiseau for valuable discussions and for glycoprotein samples sequencing.
This work was supported by a FIRST grant nr 2369 from the Walloon Region of Belgium.

REFERENCES

(1) James, D.C., Goldman, M.H., Hoare, M., Oliver, R.W.A., Green, B.N. and Freedman. R.B. (1996) Postranslational processing of recombinant human interferon-γ in animal expression systems. Protein Science 5, 331-340.
(2) O'Farrell, P.H. (1975) High resolution two-dimensional electrophoresis of proteins. J. Biol. Chem. 250 (10), 4007-4021
(3) Lognonné, J.-L. (1994) 2-D analysis : a practical guide to principle critical parameters. Cell. Mol. Biol. 40(1), 41-55.

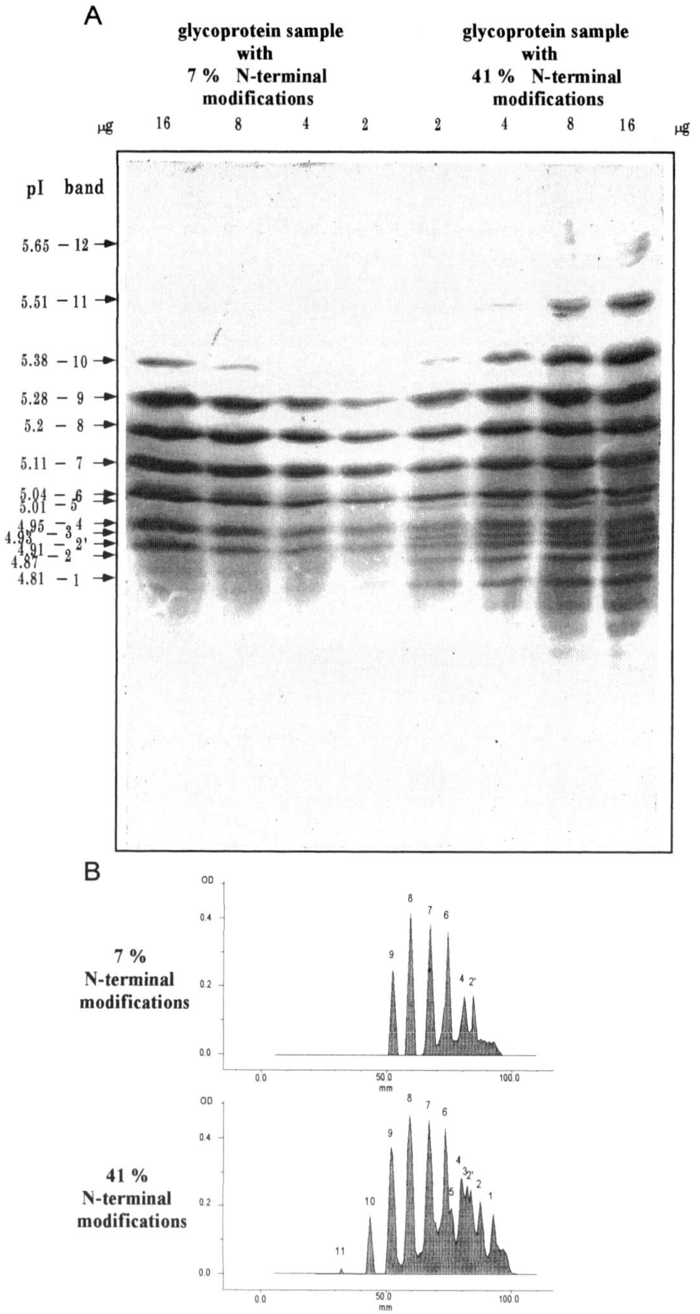

Figure 1 :
A : blot after IEF
B : glycoprotein profile after IEF/WB analysis by densitometry
Two samples of the recombinant purified glycoprotein have been analysed by IEF/WB :

- a sample with a low percentage of N-terminal modifications (7%)
- a sample with a high percentage of N-terminal modifications (41 %)

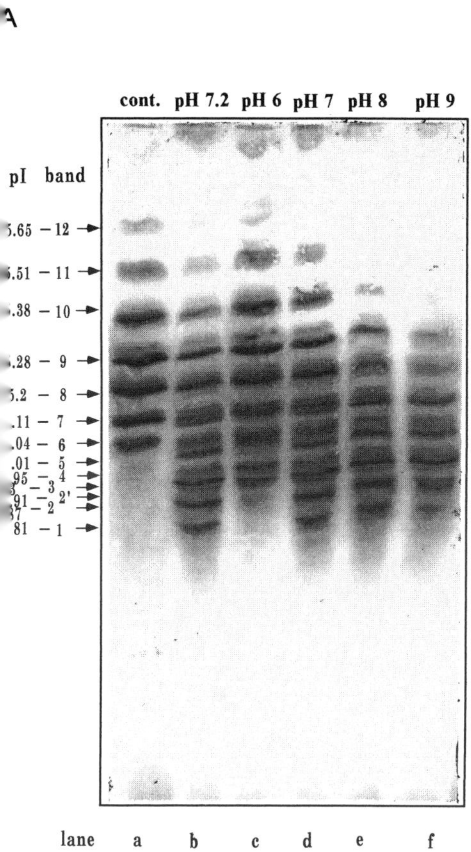

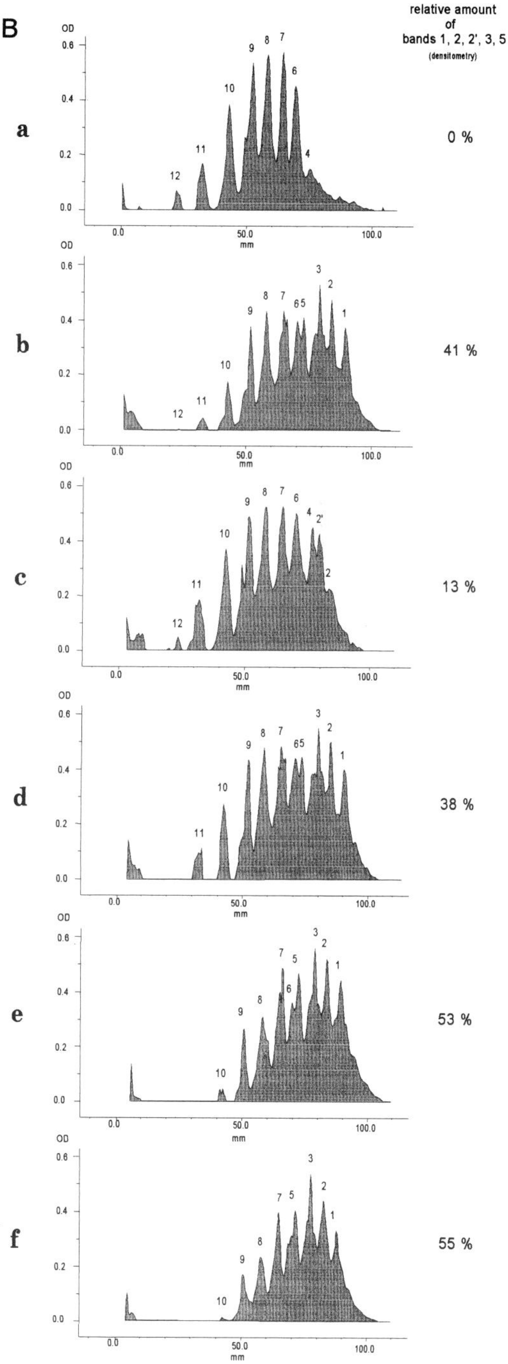

gure 2 :

: blot after IEF
glycoprotein profile after IEF/WB analysis by densitometry

ne a :
ntrol : purified glycoprotein sample with a low percentage
N-terminal modifications (4 %) without incubation
ne b, c, d, e, f :
e same glycoprotein sample was incubated with a SN obtained
er 72 hours of cell culture, 3 fold concentrated and adjusted to
ferent pH (pH 7.2, pH6, pH7, pH8, pH9)

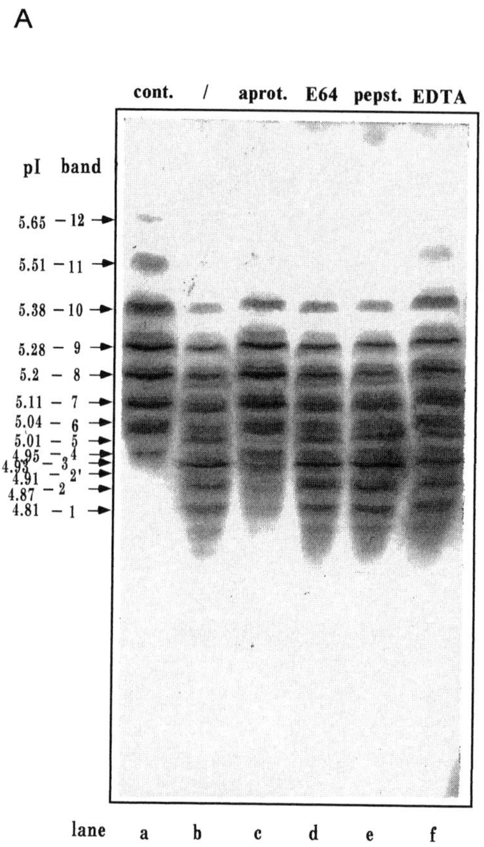

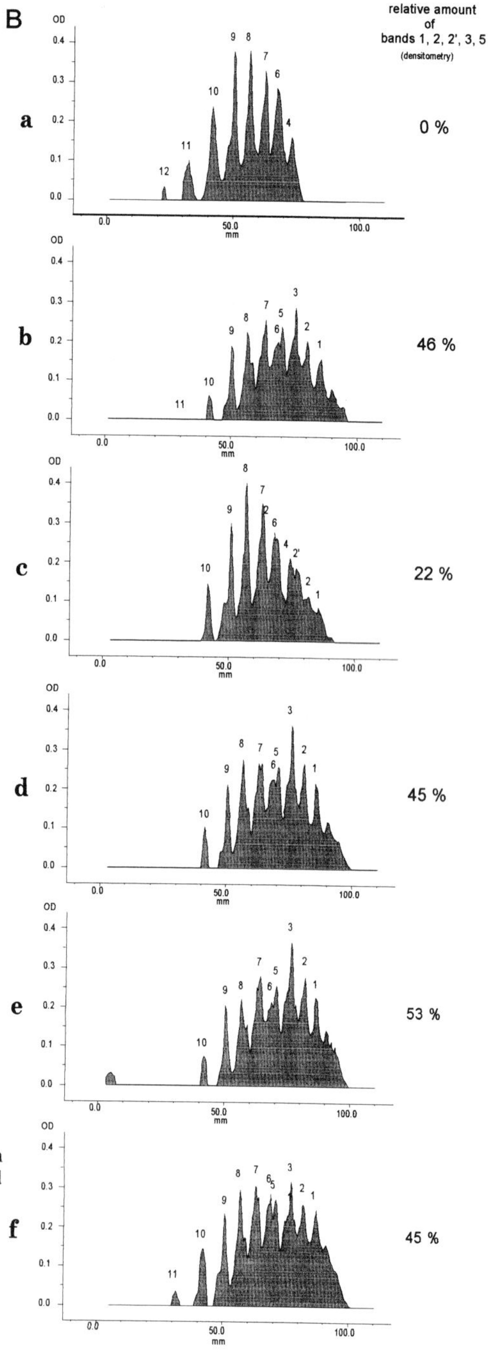

Figure 3:

A : blot after IEF
B : glycoprotein profile after IEF/WB analysis by densitometry

Lane a :
Negative control : purified glycoprotein sample with a low percentage of N-terminal modifications (4 %) without incubation

Lane b :
Positive control : the same glycoprotein sample was incubated with a SN obtained after 72 hours of cell culture and 5 fold concentrated

Lane c, d, e, f :
The same glycoprotein sample was incubated with a SN from a culture 72 hours old, 5 fold concentrated in presence of different protease inhibitors :

- lane c : 50 μM aprotinin : serine protease inhibitor
- lane d : 300 μM E64 : cystein protease inhibitor
- lane e : 50 μM pepstatin A : acid protease inhibitor
- lane f : 5 mM EDTA : metalloprotease inhibitor

CeDEX: Automated Cell Density Determination

F. Gudermann, P. Ziemeck[*] and J. Lehmann

University of Bielefeld, Institute of Cell Culture Technology,

[*] *Institute of Neuroinformatics*

P.O. Box 100131, 33501 Bielefeld, Germany

1. Abstract

The **C**ell **D**ensity **EX**amination system *CeDEX* has been developed to perform automatic and reproducable measurements of cell densities and standardized differentiation between viable and dead cells.
A large variety of different methods for cell density determination has been developed. Devices exist using e.g. light scattering, resistance, capacitance or turbidity measurements to determine the number of cells. These systems can not distinguish between dead and viable cells in an unambiguous way.
CeDEX uses the optical information of a sample of animal cell cultures. The direct and automated optical analysis with modern pattern recognition methods of the cells themselves allows cell identification and standardized differentiation between viable and dead cells. This guarantees user independent results with high accuracy.

2. Introduction

The production of complex proteins used for diagnostic or therapeutic purposes is frequently carried out in cell cultures. The production processes of these pharmaceuticals have to be monitored accuratly. Therefore, not only the concentrations of substrates and metabolites have to be measured, but also the cell densities of viable and dead cells have to be determined.
Usually the cell densities are manually determined by trypan blue dye exclusion. This method is very time intensive and leads to more or less userdependent results.
Several attempts have been accomplished to automate the cell density determination by measuring indirect parameters like turbidity, resistance or capacitance. Changes of these

M.J.T. Carrondo et al. (eds.), Animal Cell Technology, 301–305.

parameters are related to variations of the cell density. Practically, the relevant parameter is heavily influenced by other physical quantities like temperature, pH value or suspension velocity. Additionally, the state of the probes themselves is possibly altered during the cultivation or sterilisation process. Because of these influences and side-effects none of the cell density determination methods using indirect parameters lead to results qualitatively comparable to those obtained by manual determination using a microscopic image of the sample.

Automation of the manual measurement procedure is the simple and basic idea of the *CeDEX* system. The usage of the full two dimensional image information instead of indirect measurement of single parameters yields to reliable results and additional information like eccentricity and diameter spectra or debris percentage. The whole measurement process ranging from sample handling and image capture to particle detection and result computation is carried out automatically without user interaction.

3. The *CeDEX* System

The *CeDEX* system consists of three functional units; the liquid handling, the image

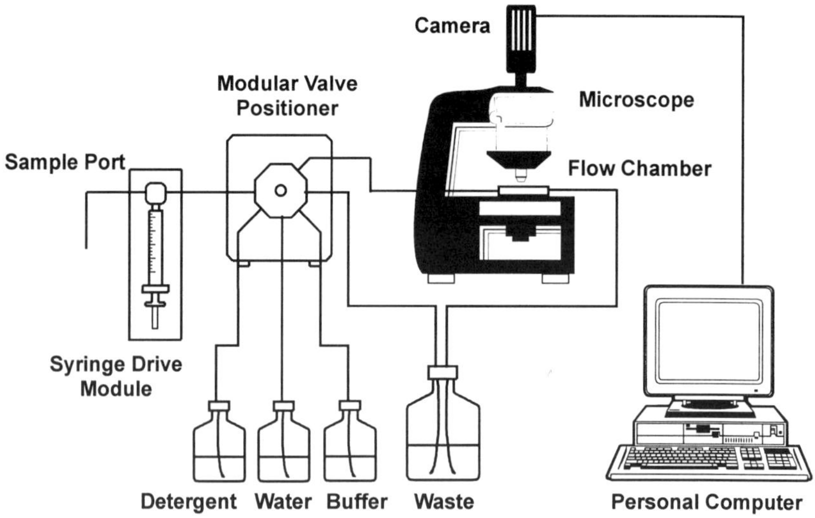

Fig. 1: The structural principle of *CeDEX*

capture optics and the data processing system containing the user interface.

The main parts of the liquid handling system are a syringe drive module, a modular valve positioner and the flow chamber (Fig.1). The trypan blue stained sample is picked up from the sample port and pumped gradually through the flow chamber to take the images. Afterwards the system cleans and prepares itself automatically for the next run.

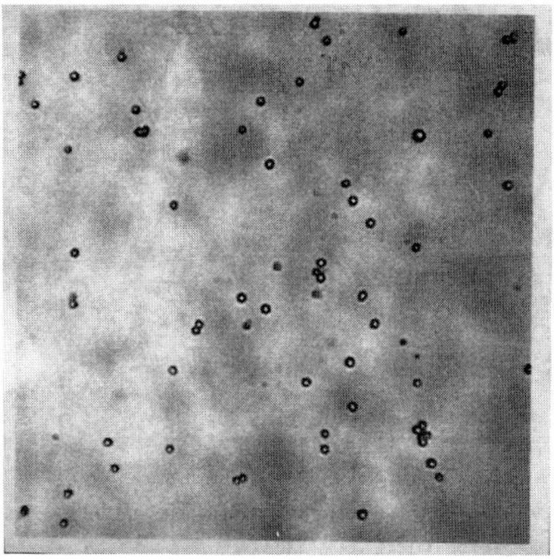

Fig. 2: The two dimensional information of the sample image allows the determination of cell densities and geometric parameters.

The frame capture is performed via CCD camera mounted on a microscope. The sample volume is constant and determined by the depth of the standadised flow chamber and the the image region being processed. The sample image (Fig. 2), also presented in the user interface, is similar to that obtained by manual sample analysis. A framegrabber takes the data of the CCD camera and pre-processes the information. Detection and differentiation of viable and dead cells is carried out by the integrated personal computer. First, the algorithm detects the cells and distinguishes them from other objects like cell debris or dirt. Once the cells are identified, they are differentiated into viable and dead cells according to their grade of staining. On user request the diameter spectrum and eccentricity histogram are presented. In addition to the cell densities, these parameters are helpful to assess the condition of the culture.

The *CeDEX* user interface (Fig. 3) is subdivided into several areas. All functions are accessible without complex user interaction. After user identification and selection of the data source, every sample is identified by bioreactor ID, cultivation ID and sample number. All further actions, sample images and analysis results are automatically saved.

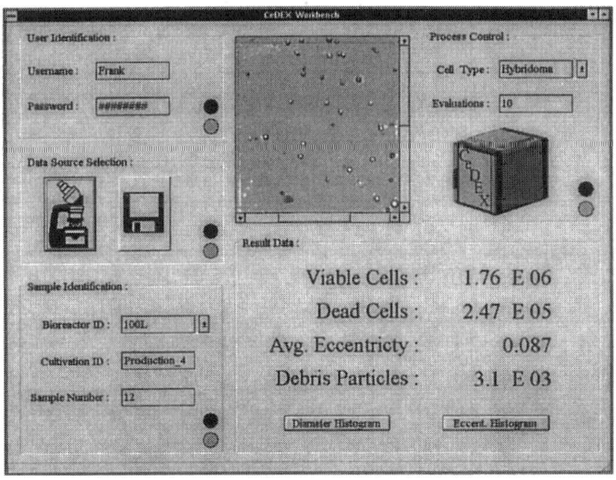

Fig. 3: The *CeDEX* user interface is subdivided into several areas. All functions are accessible without complex user interaction.

4. Results

Fig. 4 shows the time course of the cell densities of viable and dead cells during a hybridoma cultivation in 100 liter scale. The cell densities were determined manually by trypan blue dye exclusion, particle size distribution measurement by the Casy system and by the new *CeDEX* system. There exists an excellent correspondence between the manually determined data and the measurements performed by the *CeDEX* system. The sample variance of the results obtained by *CeDEX* is lower than the manual determination because of the greater sample size. Especially the variations of the results measuring the dead cell density manually are higher than those measured by *CeDEX*.

Using the Casy system one gets diameter spectra which are sometimes difficult to interpret. Cultures with low viability show two peaks in the diameter spectrum, one can be related to the dead cells the other to viable cells but when cell cultures with high viability are measured only one peak occures. The differentiation of viable and dead cells is troublesome in this case.

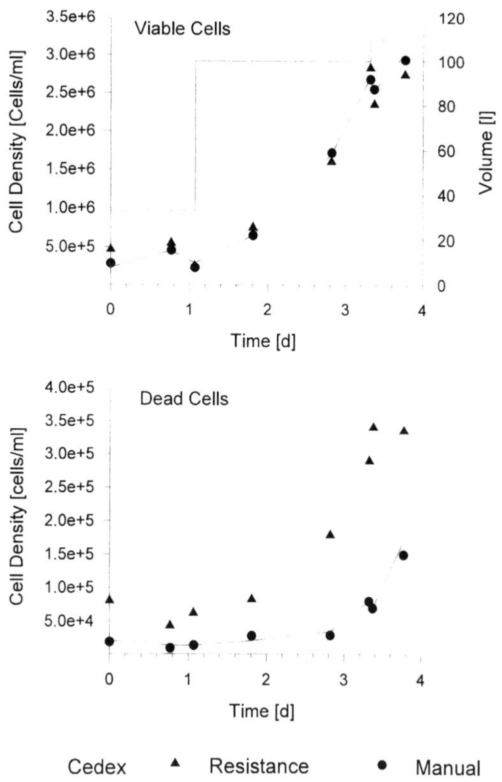

Fig. 4: Comparison of different methods of cell density determination during a hybridoma cultivation in 100 liter scale.

5. Conclusion

The *CeDEX* system is an integrated analysis tool for precise cell density determination. The liquid handling system together with the simple and effective user interface guarantees easy handling. User independ results are achieved by the high degree of automation. The analysis process is based directly on the two dimensional information of the cells and therefore much more precise and reliable then other methods using one single, indirect parameter. Because of these advantages the *CeDEX* system will be commercially available at the end of this year.

DESIGN AND PERFORMANCE OF A PERFUSION BASED PACKED BED BIOREACTOR FOR THE PRODUCTION OF SECRETED PROTEIN

Avinoam Kadouri[1], Guozheng Wang[2], Wenying Zhang[2], Corey Jacklin[2], David Freedman[2]
[1]Department of Membrane Research and Biophysics, The Weizmann Institute of Science, Rehovot, 76100 Israel,
[2]New Brunswick Scientific Co. , 44 Talmadge Road, Edison, NJ 08818-4005, USA

Abstract

A bioreactor employing a packed-bed system for the production of secreted proteins has been developed. The bioreactor is able to achieve volumetric productivities, with most cell lines, that are five to ten times higher than standard stirred tank perfusion systems. The stationary basket uses polyester fiber disks (Fibra-Cel™ - Sterilin UK & New Brunswick USA) as a support matrix for immobilized cell propagation and is capable of cultivating either anchorage dependent or anchorage independent cells. Typical cell densities achieved are about 10^8 cells per cm^3 of Fibra-Cel in serum or serum-free medium. A big advantage of this system is the fact that it can be scaled linearly. Similar productivities were obtained in the benchtop bioreactor and in the larger pilot scale system.

Key words: Cell culture Matrix, packed bed reactor, Fibra-Cel carrier

Introduction

Biologically active macromolecules are being produced today by normal, transformed and genetically engineered cells. The majority of these cells are anchorage dependent and required a solid surface to attach to in order to proliferate, metabolize and produce biomolecules. There are two key elements that affects the cell behavior and the success of every large scale cell culture process: a) the properties of the substrate on which he cells are attached, which affects the ability of the cells to proliferate and produce a desired product; b) the bioreactor configuration. The desired biotechnological system is the one that has large surface for cell adhesion and small volume.

M.J.T. Carrondo et al. (eds.), Animal Cell Technology, 307–312.

Various attempts has been taken to develop cell culture reactor with large surface area to volume. One of the most promising systems is the stationary packed bed reactor. This reactor consist of a mass of randomly orientated particulate material held in a container. The major advantages of the packed bed system are: a) Low shear, hence fragile cells can be used even in serum free medium. b) Low backmixing, hence little product feed-back inhibition. c) Potential for high density of immobilized cells. d) The system is easily washed and allows media change with continuous product absorption on a downstream affinity column. The choice of material, the size and shape of the particles are very critical parameters in the performance of these systems.

Results and discussions

Out of various materials tested in stationary packed bed reactor, the best performing one was the Fibra-Cel matrix. This carrier found to be ideal for use as a matrix for cell growth and production of biomolecules in a laboratory and industrial scale operations. This matrix was also found suitable for use in a column type reactor or in a modified stirred tank reactor. Intensive study of this carrier presents a list of advantageous which makes it superior as a packed material for stationary packed bed reactors: a) it has a large surface to volume ratio to enable high cell density. b) cell growth potential is similar to that achieved on petri dishes. c) it allows seeding at low density as low as 1% of the final density (data not presented). d) it allows high flow rate of media. e) retains the cell on changing to protein free medium. f) the carrier has no effect on the biosynthesis of cellular product (tested so far). g) have no affinity for the secreted products (tested so far).

The number of cells that can be achieved on this carrier ranged from $5X10^7$ per/ml for normal human fibroblasts up to 10^8 cells per/ml for hybridoma cells.

In all scaling up experiments we found that this carrier allows linear scaling up from small banch top reactor to a large pilot scale reactor with no difficulties.

Fibra-Cel carrier physical parameters

The Fibra-Cel carrier is made from a non-woven polyester fabric laminated to a polypropylene screen. The physical parameters of this matrix in a disk form is presented in Table 1.

Table 1 - PHYSICAL PARAMETERS OF FIBRA-CEL™ DISC CARRIER

Material	Nonwoven fabric of polyester laminated to polypropylene screen
Partial Diameter (mm)	6
Specific Gravity (g/cm^3)	~1.03
Total Available Growth Area (cm^2/g)	400-500
Projected Area (cm^2/g)	50
Weight/Packed Volume (g/cm^3)	0.085
One Disc Carrier Weight (g)	0.00572
Number of Disc Carriers in 1 gr	175
Autoclavable	Yes

Production of Biomolecules on Fibra-Cel

The carrier was originally developed (by A. Kadouri, Weizmann Institute of Science, Israel) to be used in a column type reactor with a separate medium reservoir(1). In this configuration three types of biomolecules were produced: a) ß-galactosidase from insect cells(1); b) tissue plasminogen activator (t-PA) from normal human fibroblasts(2). (Fig. 1) and an anti Leukemic factor(3) from mouse bone marrow stroma cells (Fig. 2).

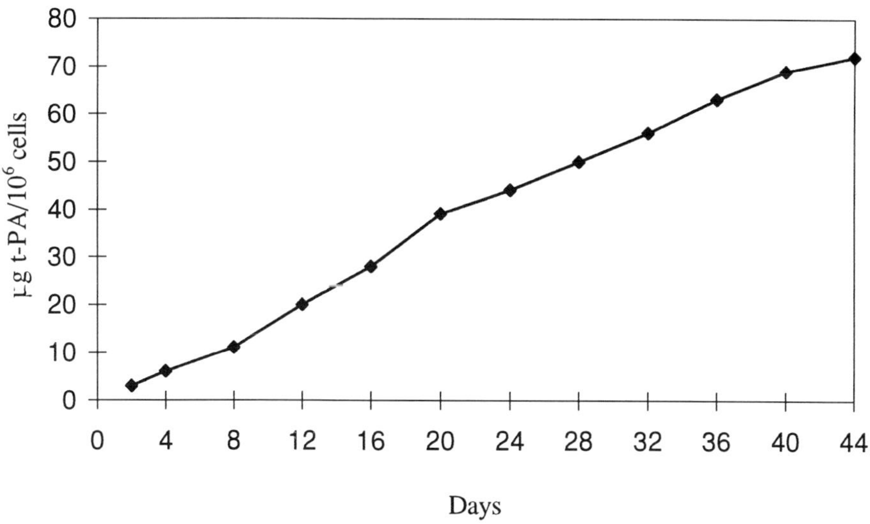

Fig. 1 - Production of t-PA from normal human Lung Fibroblasts, on Fibra-Cel in a column type bioreactor.

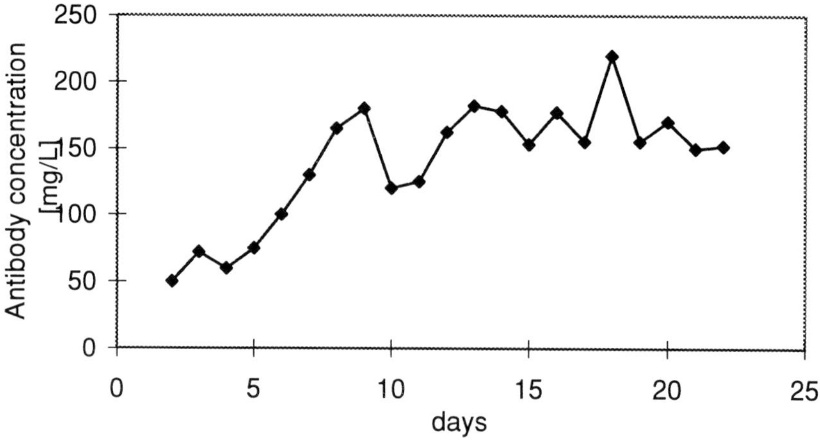

Fig. 2 - Production of antileukemic factor from mouse bone marrow stroma cells, on Fibra-cel in a column type bioreactor.

These products were produced in a protein free medium which composed of basic medium with a supplement of protein hydrolyzate[4]. These experiments show that the cells can be maintained in a productive state on this carrier for prolonged periods of time - over 200 days (Fig-2).

The stationary packed bed system with the Fibra-Cel was scaled up in a modified stirred tank reactor - CelliGen® type bioreactor (new Brunswick Scientific USA). In this case, the Fibra-Cel carrier was packed into a stainless steel perforated basket placed inside the CelliGen vessel. When modified CelliGen is used, there was no need for separate cell compartment and reservoir, since these two functions were combined together in the bioreactor vessel which simplifies the mode of operation. The modified CelliGen system (7.5 liter scale) was used for growth of hybridoma cells and for continuous production of monoclonal antibody in a medium with reduced serum or in a serum free medium[5]. This system was further scaled up successfully to 30L bioreactor (New Brunswick Scientific USA) and used for monoclonal antibody's production (Fig. 3).

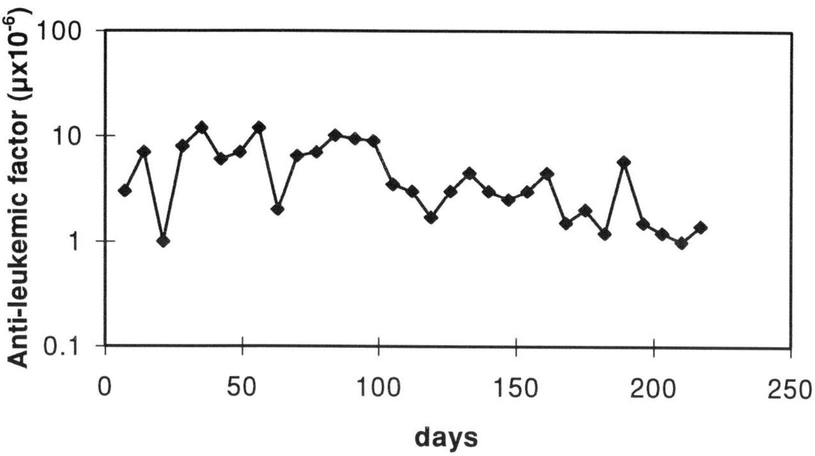

Fig. 3 - Production of monoclonal antibodies from mouse-mouse hybridoma cells in 30L modified stirred tank bioreactor equipped with basket (see Fig. 3). 900 gr of Fibra-Cel (10 liters of packed volume) was perfused with 30 liter medium per day. The product harvest was about 5.0 grams per day.

A comparison mode between the results previously obtained with 7.5L CelliGen[(5)] and the data presented in Fig. 3, proves the linear scalability of this system.

Conclusions

The stirred tank suspension bioreactors is the most commonly used technology. But, it becomes clear lately that this technology is not suitable for some large scale cell culture process, particularly for cells which are sensitive to shear forces foaming or their production capability affected by negative feedback inhibition. In these cases the other promising technology is the stationary packed bed reactors. The major drawback in this technology was the packing elements. The development of the Fibra-Cel opens a new era for this technology. The unique properties of this carrier and the fact that it can be used in a column type reactor and in a modified stirred tank reactor, makes it very compatible in the field of large scale cell culture. So far Fibra-Cel carrier has been tested and used in a verity of cell culture for research and industrial production processes. Out of the data collected from these studies one can summarize the advantages of the Fibra-Cel:

a) The Fibra-Cel matrix supports the growth of anchorage dependent and anchorage independent cells.
b) Cells grow on this carrier in three dimensional manner and they can reach a very high cell density~10^8 cell/cm^3.

c) The cell growth rate on this carrier is similar or better than on flat surface.
d) It allows high flow rate of medium through the packedbed.
e) The cells are retained on the carrier upon changing to protein free media.
f) The cells on this carrier maintained production for very long period of time.
g) In all cell lines tested so far, the productivity on Fibra-cel is always higher than on any other conventional carrier.
h) It allows linear scaling up from small laboratory system to a large pilot scale reactors.

All this makes the Fibra-Cel a carrier of choice for the production of biomolecules from mammalian cell culture.

References

1. Kompier, R., N. Kislev, I. Segal and A. Kadouri. 1991. The use of a stationary bed reactor and serum free medium for the production of recombinant proteins in insect cells. Enz. & Microb. Tech. 13, 822-827.

2. Kadouri, A. 1993. Cultivation of anchorage-dependent mammalian cells and production of various metabolites, Colloids an surfaces. B: Biointerfaces 2, 265-272.

3. Kadouri, A. R. Kompier, J. Honigwachs-Shaanani, J. Toledo, N. Brosh, D. Sternberg, A. Levy, E. Tzehoval and D. Zipori. 1992. Dynamic changes in cytokine secretion by stromal cells during prolonged maintenance under protein-free conditions. Int. J. Cell Cloning 10, (5), 299-308.

4. Kadouri, A. 1991. Use of serum free medium in large scale production of biomolecules. In: Biologicals from recombinant microorganisms and animal cells production and recovery. (M.D. White et al (eds) VCH, Weinheim - New York. 187-197.

5. Wang, G., W. Zhang, D. Freedman, L. Epstein and A. Kadouri. 1992. Modified CelliGen - packed bed bioreactor for hybridoma cell cultures. Cytotechnology 9, 41-49.

LOW- AND HIGH-PRODUCING ANIMAL CELL CULTURES IN PERFUSED, PACKED BEDS OF NON-WOVEN FABRIC

J.C. Treat, K.K. Sandelin, B.C. Beiderman, and R.J. Kirschner
Bioprocess Research Preparations
Pharmacia & Upjohn, Kalamazoo, MI, USA 49002

INTRODUCTION

Perfused packed beds containing disks of non-woven polyester fabric (Fibracel™) as the macroporous support have been used in a Celligen Plus (1.75 l basket reactor) to produce proteins from CHO and hybridoma cell lines to support research and development programs; 25 grams of a humanized anti-P-selectin mAb were produced in 21 days from a recombinant CHO line. A scale-down version (100 ml column) was developed to investigate process parameters. It was observed that cells quickly were entrapped (or bound) following inoculation of the packed beds, and then grew exponentially (at rates similar to suspension cultures) until the disks were populated; the cells remained healthy at the saturation density (5-10 x 10^7 cells/ml of packed bed) until experiments were terminated (4-9 weeks). Perfusion rates were set to meet nutritional needs.

MATERIALS AND METHODS

Cell Lines and their culture - CC3, a recombinant CHO cell line, expresses a humanized mAb (U-141,567) against P-selectin; CC1 is a hybridoma secreting the P23 murine monoclonal; and CC4 is a CHO line producing a soluble form of ICAM-1. The cell-specific expression rates for these cultures (in suspension mode) are 20-30, 2-4, and 0.1-0.3 pg/cell/day, respectively. Media were maintained at 4°C until used. The medium contained no Pluronic F68 or similar surfactant. Perfusion rates were set daily so that nutrients (i.e., glucose and glutamine) were in excess, and metabolite (i.e., lactate and ammonia) levels were non-toxic; thus, the perfusion rate was increased rapidly during the period of exponential growth following inoculation, and was changed little once the packed beds were saturated with cells.

Bioreactors - The CelliGen Plus™ Basket Bioreactor (New Brunswick Scientific) included a 5 liter vessel and 1.75 l basket; its design and general operation have been described previously (1). The basket was loaded with 175 g of Fibracel™ disks, a non-woven polyester attached to polypropylene threads (1), and the system was filled with phosphate buffered saline and sterilized in an autoclave. Medium was pumped into the bioreactor and cell-free medium was removed (harvest) to maintain a working volume of 3.5 liters. The nutrient-rich medium was circulated down through the basket, while "depleted" medium returned up through a central draft tube; DO and pH were controlled based on probes located above the basket and a second DO probe below the basket allowed the oxygen consumption to be monitored.

M.J.T. Carrondo et al. (eds.), Animal Cell Technology, 313–318.

Setpoints for control generally were agitation @ 175 rpm, temperature @ 37°C; pH @ 6.95 and dissolved oxygen @ 60%. Perfusion was initiated after the cells had attached to the disks and was monitored by the weight change of the harvest vessel. The perfusion rate was increased to meet the needs of the cells and ranged from 0.4-50 liters/day.

The scale-down (100 ml) Column Bioreactor is illustrated in Figure 1 and is functionally analogous to the CelliGen (1). In this system a 5 cm diameter x 5 cm high jacketed glass column (containing 11 g of FibraCel disks, volume about 100 ml) was connected to a 500 ml glass jar (containing 200 ml media) capped with a stainless steel headplate and containing pH, DO, and temperature probes. Metered gases were blended and added to the head space of this reservoir to control pH and DO; a wire-wound silicone blanket controlled the temperature. At the standard recycle rate of 100 ml/min, the dissolved oxygen dropped from 50-55% to 28-33% saturation during passage through the fully populated column. Fresh medium was added and cell-free medium was removed at 0.2-3.0 l/day. Sensors were monitored, parameters were controlled, and data was logged using a proprietary software, FMCS3.

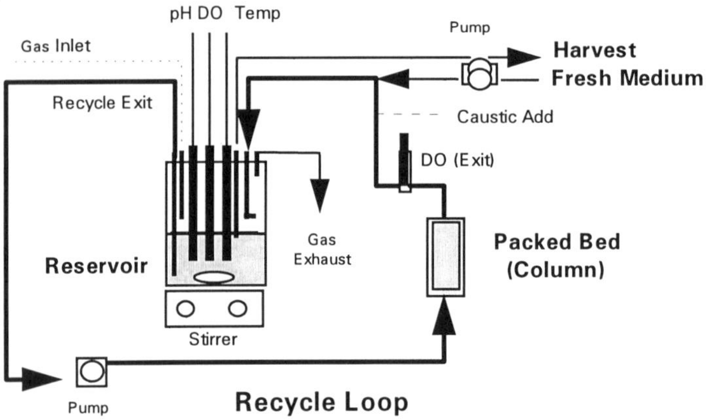

Figure 1. The perfused packed bed bioreactor.

Analysis - The harvest medium was sampled daily and assayed. A Kodak Biolyzer was used to assay glucose, lactate, ammonia, and lactate dehydrogenase levels. Glutamine and other amino acids were quantitated as phenylthiocarbamyl (PTC) derivatives. A bead-based fluoro-immuno assay was used to measure antibody levels.

Cell Number - The number of "viable cells" in the packed bed was estimated from measures of cellular metabolism, such as glucose utilization (based on suspension culture consumption of ~1 ng/cell/day). Cells in suspension were counted using a Coulter Counter, with viability estimated using Trypan Blue.

RESULTS AND DISCUSSION

Inoculation Conditions - Cells (from spinner cultures) were inoculated into the top of the packed bed. Agitation rate in the CelliGen was set to 50 rpm during inoculation of the basket; the initial turbidity due to the cells in the upper and lower medium compartments was observed to clear within 15 minutes as cells became entrapped in the disks. The column reactor, which allowed binding parameters to be more directly examined, was loaded either by gravity flow or peristaltic pumping. Figure 2 shows the fraction of cells that remained in the column after a single pass; the cells not entrapped in Fibracels eluted from the column, were collected in a bottle, and counted. Association of cells with the macroporous support is shown to be a function of flow rate during inoculation. The eluted cells demonstrated high viability and grew similarly in spinner culture to cells that had not passed through the column (not shown).

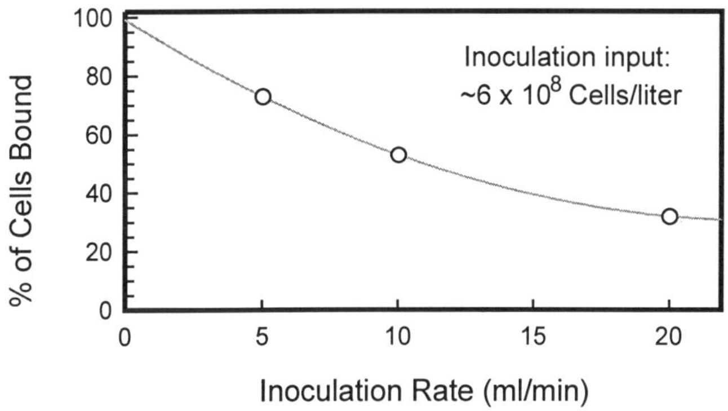

Figure 2. Inoculation flow rate affects the binding of cells to Fibracels in column.

Behavior of Three Cultures in the CelliGen Basket Bioreactor - Figure 3 illustrates that two CHO lines and a hybridoma line populated the 1.75 liter basket to similar extents: 1-2 x 10^{11} cells/basket, 5-10 x 10^{7} cells/ml, or 0.5-1 x 10^{9} cells/gram of Fibracels. While the hybridoma required almost twice the time to saturate the bed as did the CHO lines, this was expected because the hybridoma grows more slowly in suspension culture than either CHO line. The three cell lines have quite different cell-specific expression capacities in suspension culture, and this was reflected in the productivity data observed in the CelliGen Plus Basket (Figure 3, lower panel): the maximum production in 24 hours was 2150 mg (CC3), 280 mg (CC1), and 24 mg (CC4). The apparent cell-specific expression of CC3 characteristically decreased with time, however, neither CC1 nor CC4 show similar decreases. The physiology of this phenomenon is unknown.

The loss of cells into the medium during the 24 hour harvest periods was examined using the column bioreactor. Cells in the harvest bottle, which would have passed through the peristaltic pump of the recycle loop many times before being sucked into the harvest line, were counted. During the exponential growth phase, CC3 cells present in the harvest [86% viability] represented 1.5-3% of those estimated to be in the column; once the Fibracel disks were maximally populated, only 0.1% of the cells [5% viability] were in the harvest.

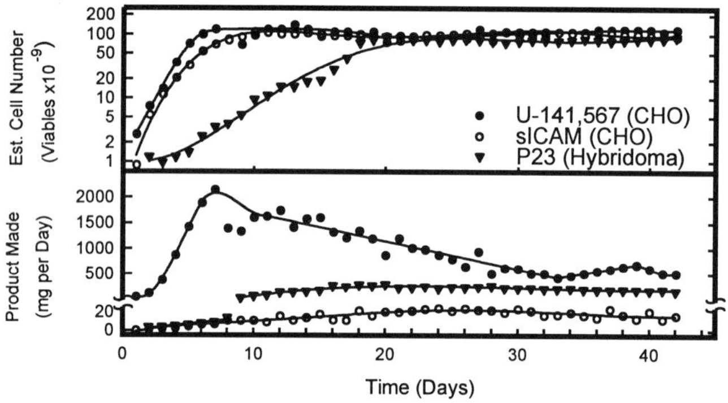

Figure 3. Growth and Expression for three cell lines in the CelliGen Basket Bioreactor. Cell numbers are estimated from glucose consumption.

Results from the basket and column bioreactors have been quite similar (Figure 4). In each case, the apparent cell number populating the bioreactor rapidly increased until the saturation point, and, thereafter, remained relatively constant for as long as the bioreactor was continued (maximum of 75 days). The dip in cell number observed with the basket run (day 18) resulted from a medium perturbation; typically, the system recovered from significant process perturbations within 1-4 days. Although the profiles of product and metabolite production were quite similar, future investigations may allow closer agreement. Among the variables that might be explored are the differences in the density of packing the beds with Fibracels (0.1 *vs* 0.11 g/ml), and in perfusion rates (see inset of Figure 4, upper panel). The latter affects the metabolite concentrations, which could alter the stress on the cells, and, consequently, mAb secretion rates.

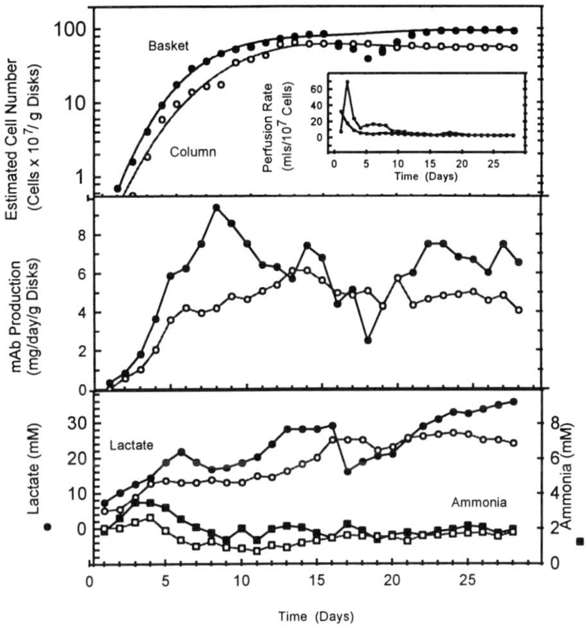

Figure 4. Growth, expression, and metabolism compared in Basket and Column modes of the packed bed bioreactor. Cell numbers are estimated from glucose consumption.

One of the major challenges in studies such as these is the inability to directly measure cell numbers. Estimation of cell numbers based on glucose utilization makes the assumption of consistent metabolism; this clearly is an error during the first day or two after inoculation where the apparent cell number is much lower than the level of cells that attached to the disks during inoculation. An increase in apparent cell number after a pH increase also may reflect a metabolic change. Other measures to estimate cell number are being investigated, such as the q_{ATP}, which is based on oxygen consumed and lactate produced (2).

CONCLUSIONS

CHO and hybridoma cells grew to, and were maintained at, about 0.5-1 x 10^9 cells/gram of non-woven fabric disk or about 5-10 x 10^7cells/ml under the packing conditions used. The cells could be maintained at these levels in a healthy, productive state for long periods. Similar results have been found for a variety of cell lines other than those shown here. Hence, relatively large quantities of proteins can be produced by this technology.

The scale-down version functions reliably, uses relatively low perfusion volumes, and requires little attention once the maximum cell capacity has been reached. It has been convenient for investigating parameters such as pH, dissolved oxygen, temperature, and rates of recycle and perfusion.

ACKNOWLEDGMENTS
The experiments reported here represented a team effort that included contributions of Joe Santek, Doug Grevenstuk, Larry Riegle, and Jim Ullrey for growth, media preparation, and harvesting; John Hoogerheide, Annette Hooker, Carla Campbell, and Marty Sears for sample analysis; Dave Fischer of the Biotechnology Group at P&U has been an important advisor, and Corey Jacklin of New Brunswick Scientific Co, has been very supportive.

REFERENCES

1. Wang G, Zhang W, Jacklin C, Freedman D, Eppstein L, and Kadouri A (1992) Modified CelliGen-Packed Bed Bioreactors for Hybridoma Cell Cultures. Cytotechnology 9, 41-49.
2. Eyer K, Elmar H (1996) On-line Estimation of Viable Cells in a Hybridoma Culture at Various DO Levels Using ATP Balancing and Redox Potential Measurement. Biotechnol. Bioeng. 49, 227-283.

TEMPERATURE
A FACTOR INFLUENCING CELL BEHAVIOUR

G. Kretzmer, B. Rössler, M. Hüners, H. Lübben, T. Reich

Institut für Technische Chemie der Universität Hannover
Callinstr. 3, 30167 Hannover, F.R.G.

Abstract

Temperature as a cultivation parameter was investigated. It was shown that the temperature has a negative effect on the growth rate but not on the maximum cell density. Lower temperature supports the resistance of the cells against shear stress. Cellular primary metabolism was reduced with lower temperature, whereas the product productivity was not influenced. The sialidase activity in the supernatant was lower at reduced temperature. For optimization the performance of the process with temperature shift was beneficial.

Introduction

The easiest parameter for process optimisation in animal cell culture is the control of nutrient and/or oxygen supply and the reactor configuration. The aim of this research was the characterisation of the influence of the temperature on the cells, especially on the metabolic activity and productivity of a recombinant blood protein. Previous results showed a decreased shear stress sensitivity of mammalian cells at lower temperatures, which allowed for example the increase of the stirrer speed. Studies were performed with rBHK and rCHO cells. The results obtained with CHO cells will be presented. The experiments with the rBHK cells showed a similiar tendency.

Cells , culture conditions and reactors

- Suspension Chinese Hamster Ovary Cells (CHO-SS0-A2) producing Antithrombin III (ATIII, secreted)
- Excell 301 (JRH Bioscience) with 25 % Iscoves Modified Eagle`s Medium (IMDM) and 25% Ham`s F12

M.J.T. Carrondo et al. (eds.), Animal Cell Technology, 319–323.

- Stock cultures and precultures were done in spinner flasks
- Batch, fed-batch and perfusion cultivations were performed in a 5-litre bioreactor (Biostat B, B. Braun, Melsungen)
- For shear stress experiments a viscosimeter with double gap system was used

Results

Cell Growth*:*
In order to study the influence of temperature on cell growth cultivation were performed as batch cultures at 37, 35, 33, and 31 °C. In all cases the growth behaviour was similar, but the growth rate decreased. The maximum cell density was the same in the range between 37 and 31 °C (37 °C: 1.34×10^6; 35 °C: 1.36×10^6; 33 °C: 1.47×10^6; 31 °C: 1.48×10^6). The adaptation phase was slightly longer at lower temperature, whereas the exponential growth phase was significantly prolonged (37 °C: 72 h; 35 °C: 108 h; 33 °C: 108 h; 31 °C: 134 h). The viability increased with reduced temperature (Fig. 1) and a stationary growth phase could be observed (37 °C: 0 h; 35 °C: 36 h; 33 °C: 48 h; 31 °C: 60 h).

Shear stress sensitivity
The shear stress sensitivity of the CHO cell line was investigated using a double gap viscosimeter. The exposure level studied were 0.30; 1.25; 1.79; 2.56 Nm^{-2}; and the exposure duration was set between 0.5 to 24 h.
Table 1 shows the results for the different exposure levels, durations and temperatures. At 37 °C cell viability dropped significantly within 2 hours while exposed to 1.79 and 2.56 Nm^{-2}. Stress levels lower than 1.25 Nm^{-2} caused no or a positive effect on cell viability.
At lower temperature, 33 and 31 °C, stress levels increased with increasing viscosity. Even with these higher levels no negative effect on cell viability was seen after 5 hours of exposure

Temperature [°C]	Stress level [Nm-2]	survival [%] 1 h	survival [%] 1.5 h	survival [%] 2 h	survival [%] 5 h	survival [%] 8 h	survival [%] 24 h
37	0.30	n.m.	n.m.	88	92	n.m.	97
37	1.25	n.m.	n.m.	97	128	106	97
37	1.79	83	82	52	n.m.		
37	2.56	74	46	25	≈0		
37	1.79	83	82	52	n.m.		
33	1.94	154	140	114	n.m.		
31	2.02	105	134	111	n.m.		
37	2.56	74	46	25	≈0		
33	2.78	116	95	99	102		
31	2.89	124	n.m.	148	135		

Table 1: Influence of temperature on shear stress sensitivity

Metabolic activity and AT III-productivity

The cultivations shown in Figure 2 and 3 were performed as batch cultures at 37 and 33 °C and with a temperature shift at 37/33 °C. This temperature shift was carried out at a viable cell count of 1.2 - 1.5×10^6 cells per ml. For an easy comparison of the data the rates were calculated before (q1) and after (q2) the time of the temperature shift for all cultivations. Reducing the temperature leads to reduced glucose consumption rates (37°C: q1= 4.38 and q2= 1.78 ; 33°C: q1= 2.65 and q2= 0.94 µmol $(10^6 \text{ cells})^{-1} d^{-1}$). Concomittandly the lactate production rates were decreased (37°C: q1= 6.8 and q2= 2.15 ; 33°C: q1= 3.12 and q2= 0.99 µmol $(10^6 \text{ cells})^{-1} d^{-1}$). The cultivation with temperature shift performs in the first part like a cultivation at 37°C and in the second part like one at 33°C (Glucose: q1=4.25; q2=0.94; Lactate: q1=7.22; q2=0.99 µmol $(10^6 \text{ cells})^{-1} d^{-1}$). The AT III-productivity was similar at all temperatures (37°C: 9.4 µg $(10^6\text{cellsd})^{-1}$; 33 °C: 9.2 µg $(10^6 \text{ cells d})^{-1}$; 37/33 °C: 10,3 µg $(10^6 \text{ cells d})^{-1}$). As a consequence of the higher viability and the prolonged growth phase the productive time span of the cultivation was enhanced and the total product amount at reduced temperature was significantly higher (37 °C: 57.8 mg l^{-1}; 33 °C: 113.3 mg l^{-1}; 37/33 °C: 103.2 mg l^{-1}).

Glycosidase activity (Fig. 4)

During the cultivation Glycosidases and Proteases are released into the cultivation medium. These Enzymes may cause severe damage to the desired protein. Therefore, the amount and the mechanism of the release is of great interest. It was found that the extracellular Sialidase activity is correlated to the viability of the culture. The extracellular sialidase activity is connected to the intracellular activity. At lower cultivation temperature (33°C) the extracellular and specific sialidase activities are lower than at 37°C. Also the intracelluar sialidase activity is less at lower temperature. Furthermore the amount of sialidase produced is higher at 37°C than at lower temperature. The total protein concentration is higher at lower temperature, although the AT III concentration is the same.

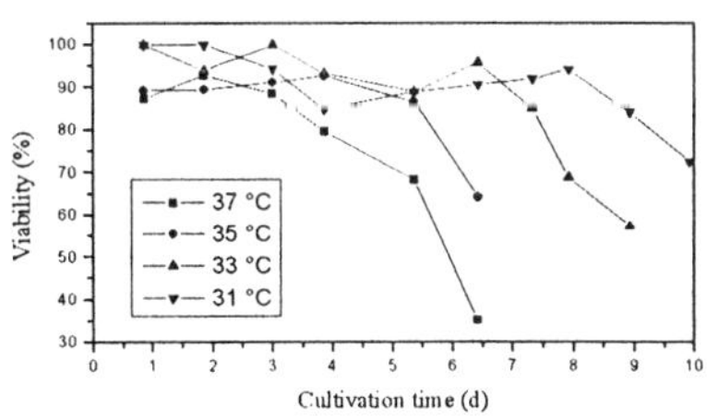

Fig.1: Cell viability of rCHO cells at different cultivation conditions

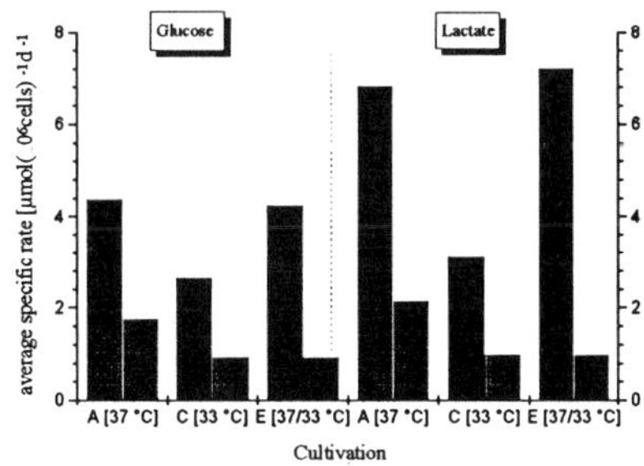

Fig. 2: Specific metabolic rate of rCHO cells at different cultivation conditions

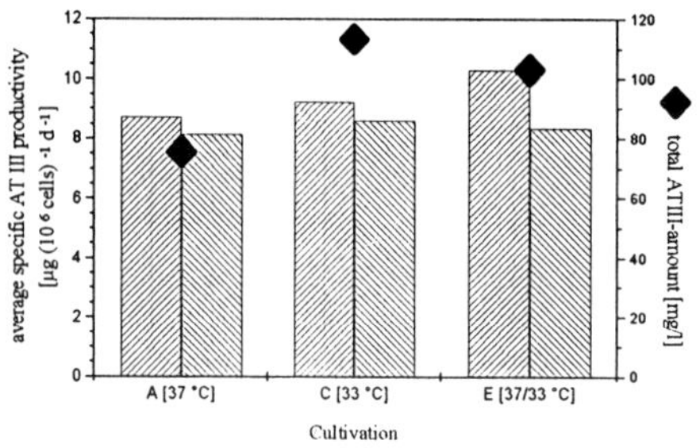

Fig. 3: AT III productivity of rCHO cells and total product amount at different cultivation conditions

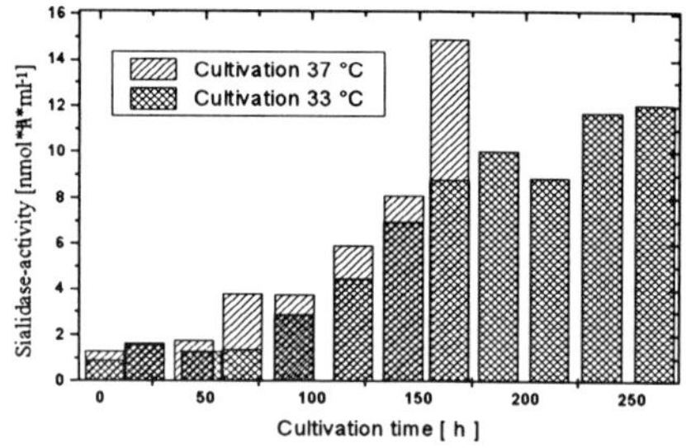

Fig. 4: Extracellular sialidase activity during rCHO cultivation

Process optimization (Table 2)

For process optimization a comparison of batch, fed-batch and perfusion cultures was carried out. The viable cell count increases significantly from batch to perfusion culture because of the better nutrient supply. Fed-batch cultures were performed at two different temperatures and with temperature shift. With all cultivations the viable cell count was the same independent from the temperature. The maximum growth rate was influenced. It decreased with decreasing temperature but with temperature shift the decrease was only slight. The metabolism of the cells was effected by the temperature as expected.
The cell specific glucose consumption rate decreased with fed-batch and perfusion cultivations at 37°C, but it was even lesser when cultivated at 33°C or with temperature shift. The opposite was true for the cell specific lactate production rate. A higher AT III-titer was found with fed-batch cultures, but the highest titer was at 33 °C.
Space-time-yield increased with fed-batch and perfusion cultivation. For the fed-batch cultures the highest space-yield was found in cultivations with temperature shift. In case of the perfusion cultures, the perfusion rate could be lowered because of the low lactate productivity. Therefore, the product titer in the harvest was high.

	Batch	Fed-Batch			Perfusion
Temperature [°C]	37	37	33	37/33	33
Viable cell count [10^6*ml^{-1}]	2,37	3,33	3,38	3,28	13,66
μ_{max} [d^{-1}]	0,93	1,35	0,52	1,03	0,43
$q_{Glucose}$	0,70	0,56	0,33	0,47	0,12
max. AT III Titer [mg*l^{-1}]	55,2	85,4	113,2	103,2	163,5
Space-Time-Yield [mg*l^{-1}*h^{-1}]	9,08	12,96	10,51	15,43	61,2

Table 2: Influence of Process mode and Temperature

Conclusion

The results show, that lowering the temperature is a good parameter for process optimisation, because the productive life-span could be prolonged. This is due to the prolongation of the exponential and stationary phase but also to the high viability for a long period of time. The higher resistancy against shear stress is one parameter supporting the high viability; another factor is the reduced metabolism of glucose and lactate. So the yield concerning the substrate was significantly better at lower temperatures and the lactate concentration during the cultivation did not rise so fast. This enables low perfusion rates, that means higher product titer. Surprisingly, the cell specific ATIII productivity was not influenced at lower temperatures, so the total ATIII amount could be increased when cultivating at 33°C. Performing the process with temperature shift prevents the drawback of the long adaption phase and uses only the benefits of the reduced metabolism and shear stress sensitivity and the improved viability. The reduced shear stress sensitivity allows higher stirrer speed without cell damage.
This claims: Temperature is a valuable parameter for process optimization

Acknowledgements

The authors wish to thank the Beringwerke AG. Marburg, (Germany) for suppling the recombinant cell line and the Bundesministerium für Forschung und Technologie (BMFT) for financial support.

References

Ludwig A., J. Tomeczkowski, G. **Kretzmer** Influence of the temperature on the shear stress sensibility of adherent BHK 21 cells Appl. Microbiol. Biotechnol. 38 (1992) 323-327

Weidemann R., A. Ludwig and G. **Kretzmer** Low Temperature Cultivation - A Step Towards Process Optimisation Cytotechnology 12 (1995) 1-6

Rössler B., H. Lübben, G. **Kretzmer** Temperature - A simple parameter for the process-optimization in Fed-batch cultures of recombinant Chinese Hamster Ovary Cells Enzyme and Microbial Technology 18 (1996) 423-427

Direct Capture of Monoclonal Antibodies using High Capacity Membrane Ion Exchangers in Pilot Scale

D. Lütkemeyer, N. Ameskamp, H. Tebbe, K. Bracht* and J. Lehmann
Institute of Cell Culture Technology, University of Bielefeld, 33501 Bielefeld, Germany
*Sartorius AG, 37070 Göttingen, Germany

Abstract

A new generation of high capacity ion exchange membrane adsorbers (MA) from Sartorius AG (Göttingen, Germany) were tested in pilot scale for the concentration and purification of monoclonal antibodies (Mabs) from 100 l batch culture. The new adsorber is based on crosslinked regenerated cellulose with a pore size of 3-5 µm. For the purification of MAbs we used a 293 mm stainless steel filter holder containing one or two membrane modules each consisting of 10 sealed sheets. The available effective adsorption area amounts to 5,500 cm^2 per stack which corresponds to 110 ml of gel matrix.

We succeeded in extracting the MAbs directly from the cell culture supernatant, rendering unnecessary any preliminary dilution or ultrafiltration. Flow rates of up to 2 l/min did not cause a significant additional pressure during the application. The dynamic binding capacity ranged from 0.05 mg/cm^2 to >0.23 mg/cm^2 (2.5 / >11.5 mg/ml gel matrix) depending on the number of membrane sheets and amount of HSA added to the culture. Using two adsorber units it took less than 2 hours to process a 100 l batch (throughput 6.4 g/h). The concentration factor of the extracted MAbs could approach 66.

Introduction

Synthetic membranes are well-known tools for biotechnology procedures [1]. Only recently have they gained growing attention as chromatographic adsorbents for protein purification [2,3]. The main advantage of a membrane-based stationary phase is the reduction of mass transfer limitations which usually result from the diffusion of the biomolecules to the binding sites located on the inner surface of the particles of conventional porous gel adsorbents. In this case a convective mass transport through the pores of the membranes brings the proteins into direct contact with the binding sites which offers an accelerated throughput without influencing the resolution. In addition low back pressure, incompressibility, and the avoidance of channelling enable high flow rates and favour rapid separation. The reusable membrane adsorbers are easy to handle, partly because cleaning requires less effort. Scale ups may be realized by increasing the diameter of the membranes or the number of sheets [4].

M.J.T. Carrondo et al. (eds.), Animal Cell Technology, 325–329.

Material and Methods

Product: Mouse IgG_1; pI 6.3

Cultivation: Mouse/mouse hybridoma cells were cultivated in serum-free standard medium containing human transferrin (10 mg/l), bovine insulin (10 mg/l) and human albumin (500 or 20 mg/l).

Bioreactor: 100 l stirred tank reactor (B. Braun International)

Cell and debris removal: Two-stage centrifugation (s. paper of Tebbe et al.)

MA-System: Stainless steel filter holder ((293 mm ∅, Sartorius AG

MA-Material: Regenerated cellulose (s. Tab. 1)

Buffer system I: A: 20 mM Acetate + 10 mM NaCl, pH 4.5 (5.0); B: Buffer A + 1 M NaCl

Buffer system II: A: 50 mM Lactate + 10 mM NaCl, pH 3 B: Buffer A + 1 M NaCl

Table 1: Characteristics of the membrane adsorbers
([1] 50 cm^2 = 1 ml membrane volume)

Characteristic	Value
Membrane material	crosslinked regenerated cellulose
Filtration area[1]	5500 cm^2, 10 sheets
Type	S = strongly acidic (sulphonic acid)
Membrane thickness	280 µm
Pore size	3-5 µm
Maximum operation pressure	5 bar
pH stability	pH 1-14

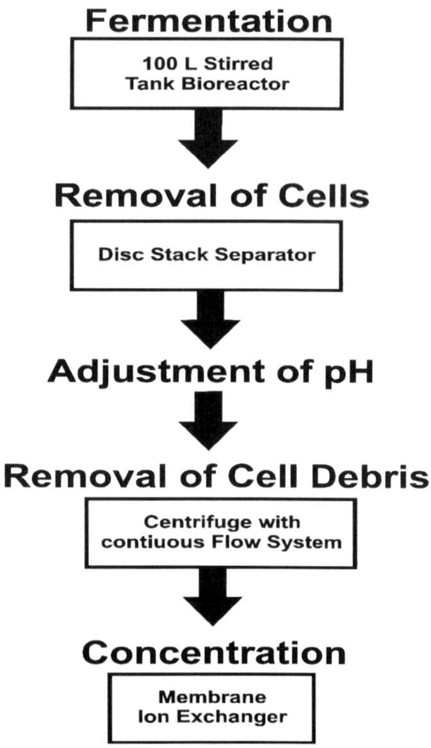

Fig. 1: Schematic diagram of the production process

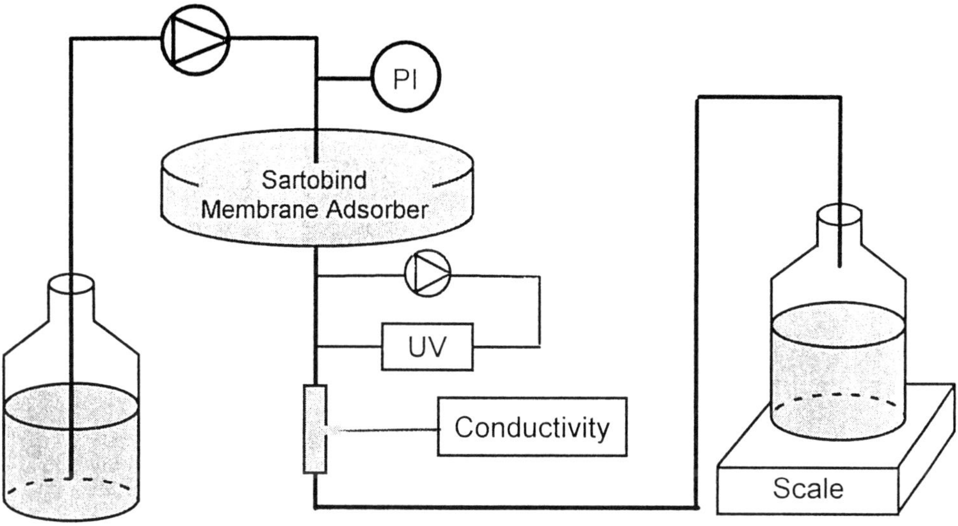

Fig. 2: Pilot system consisting of a gear wheel pump, an UV monitor, a pressure transmitter, a conductivity monitor and the stainless steel filter holder (293 mm ∅). All signals were recorded via PC and recorder. During operation the pressure of the pump was set to 1.5 bar. The flow was kept at 1.5-2.0 l/min during application and at 0.1-0.4 l/min during elution

Results

Monoclonal antibodies were purified from 100 l batch cultures. Cells were removed by gentle centrifugation using a disk stack separator (Westfalia, Fig. 1). After adjusting the pH to 3 - 5 the debris was removed by continuous centrifugation (Contifuge 300, Heraeus). For details of the two stage centrifugation step see the paper of Tebbe et al.

We succeeded in extracting the MAbs directly from the cell culture supernatant, rendering unnecessary any preliminary dilution or ultrafiltration. Flow rates of 1.5 to 2 l/min (135 - 180 cm/h) did not cause a significant additional pressure during the application. Between 7.6 and 41 l supernatant were applied (s. Tab. 2).

In the first experiments (I) the HSA concentration in the supernatant was 500 mg/l. The static capacity for one module was determined as only 920 mg IgG (0.17 mg/cm^2; 8.4 mg/ml gel matrix) due to co-chromatography of Albumin under this binding conditions. The binding rates were poor caused by overloading the membrane module. The static capacity increased to 1.4 g IgG after reducing the HSA concentration to 20 mg/ml (II) in the culture medium. The binding rates improved but there was still an overloading of the membranes by applying more than 44 l of supernatant. Using two modules with 11000 cm^2 and a total hight of approx. 6 mm, it was possible to purify more than 2.5 g of antibodies in 45 min. This correspond to an application of 41 l of supernatant with

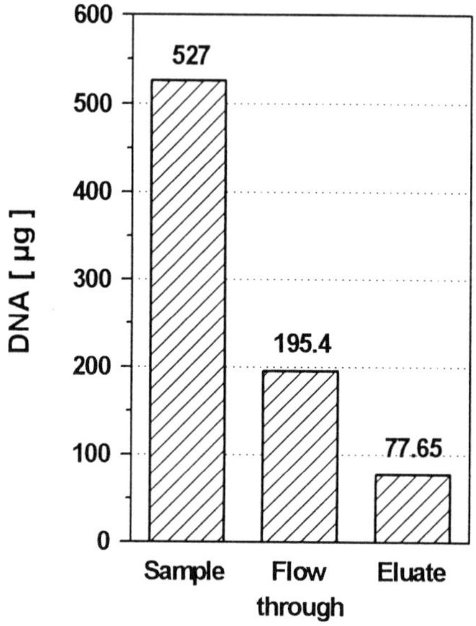

excellent binding rates. The static capacity increased drastically. Under these conditions one batch of the 100 l cultivation was processed by membrane chromatography in 2 h. Looking for the dynamic capacities (c/c_0 = 10%), there was an increase from method I to III from 0.05 mg/cm^2 to >>0.23 mg/cm^2 (2.5 / >>11.5 mg/ml gel matrix). In all experiments the recovery rates were comparable to common ion exchanger materials. The DNA depletion was 6.8 during the membrane chromatography (Fig. 3) and correspondent to other results with ion exchangers.

Fig. 3: Example of depletion of DNA during the concentration process. The DNA concentration was measured using an intercalating fluorescence dye. Values are given for the total DNA amount.

Table 2: Purification experiments using one or two membrane stacks.

Method	IgG [g]	Volume [l]	Flow rate [l/min]	IgG bound [%]	Recovery rate [%]	Concen-tration factor	Process time [min]	Through-put [g/h]
I. 1 module 500 mg HSA/l	1.613	18.0	1.6	57	75	20.9	48	(8.6)
	0.658	9.64	2.0	56	100	18.2	31	(8.2)
	0.522	7.64	2.0	72	83	13.4	23	(8.2)
II. 1 module 20 mg HSA/l	0.705	24.0	3.0	77	n.d.	22.9	36	5.3
	0.763	26.0	1.5	86	n.d.	36.1	58	2.6
	1.639	43.7	2.0	83	94	66.2	65	4.5
III. 2 modules 20 mg HSA/l	2.567	41.1	1.7	97	100	22.8	45	6.4
	2.317	39.5	1.6	98	100	43.9	54	5.6

Discussion

The direct capture of the MAbs from the cell culture supernatant without ultrafiltration or dilution offers the opportunity of a fast processing of the supernatant. The high flow rates of the MA-system resulted in throughputs of more than 6.4 g/h. The binding rates were comparable to other commonly used ion exchangers. The concentration factor is not satisfactory because of too much dead space in the filter holder. The filter holder and the applicator have to be optimised. For higher dynamic capacities it seems to be necessary to increase the number of stacks in the filter holder not only to create a bigger adsorption area but also to improve the conditions of the liquid flow. The results proved that the membrane modules can be applied on industrial levels.

References

1 Langlotz, P., Kroner, K.H. (1992)
Surface-modified membranes as a matrix for protein purification.
Chromatogr. 591, 107-113

2 Lütkemeyer, D., Bretschneider, M., Büntemeyer, H. Lehmann, J. (1993)
Membrane Chromatography for Rapid Purification of Recombinant Antithrombin III and Monoclonal Antibodies from Cell Culture Supernatant. Chromatogr. 639, 57-66

3 Finger, U.B., Thömmes, J., Kinzelt, D., Kula, M.R. (1995)
Application of Thiophilic Membranes for Purification of Monoclonal Antibodies from Cell Culture Media. Chromatogr. 664, 69-78

4 Wang, W., Lei, S., Monbouquette, H., McGregor, W.C. (1995)
Membrane Adsorber Process Development for the Isolation of a Recombinant Immunofusion Protein. BioPharm, 8, 5, 52-59

INFLUENCE OF MASS CULTURE OPERATING CONDITIONS ON THE APOPTOSIS OF OKT3 HYBRIDOMA CELLS

Hervé Pauly[1], Patricia Franck[2], Pierre Nabet[2], Marc Cherlet[1] & Annie Marc[1]

1 - Laboratoire des Sciences du Génie Chimique, CNRS-INPL, BP 451, 54001 Nancy Cedex, France
2 -Laboratoire de Biochimie Médicale, Université Henri Poincaré, BP 184, 54505 Vandoeuvre Cedex, France

Abstract

The apoptosis is one of the two cell death mechanisms which occur during animal cell culture in bioreactors. It was investigated during mass cultures of OKT3 hybridoma through combined methods which monitor morphological and biochemical characteristics of apoptotic cells. These methods were validated by using cultures in which apoptosis and necrosis were respectively induced by camptothecin and sodium azide. Then they were used to monitor death kinetics in cultures performed in standard and glutamine-free media.

Introduction

Cell death is often a critical factor limiting the productivity and the product quality during long term animal cell processes. Apoptosis is one of the two cell death mechanisms involved in large scale bioreactors (1). While apoptosis is a natural phenomenon (2), it can also result from the stimuli of the cell environment. More often in bioreactors, this environment is altered by mild stresses such as slow decrease of nutrient levels or concentration gradients (3). For these reasons, a better knowledge of the apoptosis kinetics is a prerequisiste for the optimisation of animal cell processes.

Material and Methods

Cell cultures : Cultures of OKT3 hybridoma cells were performed in 200 mL spinner flasks. The standard medium was DMEM containing 4.5 g/l glucose and 5 mM glutamine supplemented with 7% FCS, 2% essential amino-acids, 1% non-essential amino-acids. For one culture, 1μM of camptothecin in DMSO were added to 10^6 cells/mL in exponential growth phase. In another one, 0.1 % sodium azide (0.3 g/L) in PBS was added to 3.10^5 cells/mL. The last culture was performed in standard glutamine-free medium.
Cell analyses : Microscopy counting of Trypan-Blue staining cells leads to total non-lysed cells. The Coulter Counter allows the mean cell volume monitoring. The Flow Cytometer Epics C gives the evolution of the cell DNA content and the cell cycle. Apoptosis is confirmed by the observation of DNA fragmentation after DNA extraction and electrophoresis (results not shown) (4).

M.J.T. Carrondo et al. (eds.), Animal Cell Technology, 331–336.

Results

Culture with camptothecin

The camptothecin was used to induce the cell apoptosis. It has been added to the culture medium during the exponential growth phase when the cell density was about 10^6 cells/mL. The comparison for 25 h of cultures performed with or without camptothecin is presented in Figure 1.

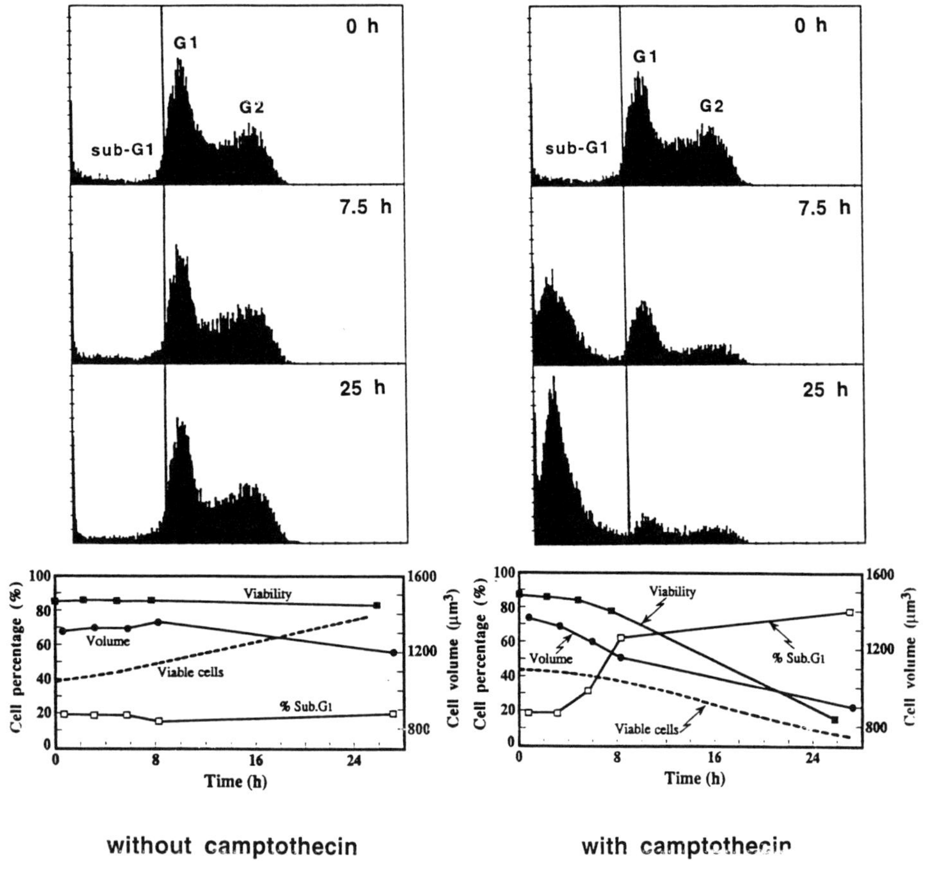

Figure 1: *Evolution of cell DNA content, cell viability (■), cell volume (●) and fraction of sub-G1 cells (□) during cultures w or w/o camptothecin.*

During the first 5 h we noticed a decrease of cell volume, a stable viability and a low sub-G1 fraction which indicate the degradation of the cytoskeleton. From 5 to 8 h, it was observed an increase of sub-G1 cell fraction, a slow decrease of the viability with a good membrane integrity and nucleosomes formation due to the DNA degradation. The viability then decreased fastly explaining the disruption of apoptotic cells called the secondary necrosis.

Culture with sodium azide

The sodium azide was used to induce the cell necrosis. This culture was realized to reach a better comparison between necrotic and apoptotic cells. The monitoring of the evolution of cell characteristics is given in the Figure 2.

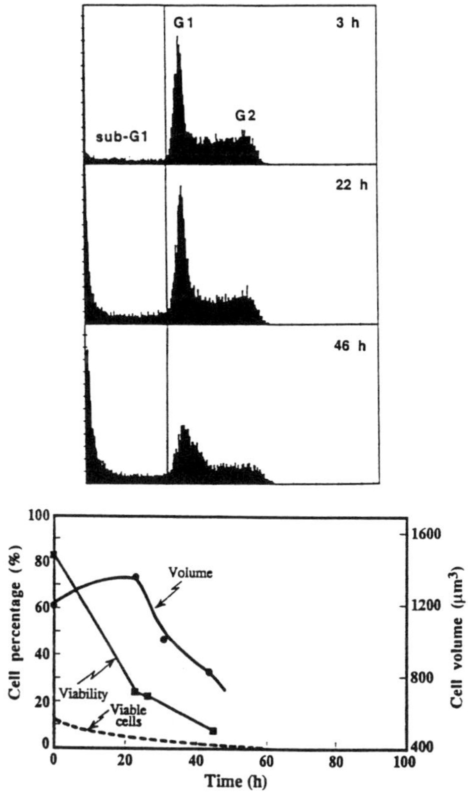

Figure 2 *Evolution of cell DNA content, cell viability (■), cell volume (●) and fraction of sub-G1 cells (□) during cultures with sodium azide.*

During 20 h the membrane permeation and the cell swelling appeared through a slow increase of the cell volume and a fast decrease of the viability. Thereafter the volume decrease and the absence of sub-G1 cells or DNA fragmentation confirmed the cell disruption without apoptosis.

Culture in standard medium

The same evolutions were monitored for cells cultivated in the standard medium. Results are shown in Figure 3. During the exponential growth phase until 50 h, viability and cell volume were remained constant, the low sub-G1 fraction was low and no DNA cleavage was noticed. The death mechanisms are not yet observed. Between 50 and 90 h, the decrease of viability and volume occured with the increase of sub-G1 cell fraction. No specific DNA cleavage appeared but smears could correspond to high DNA fragments resulting from apoptosis initiation. After 90 h the specific cleavage of DNA in nucleosomes appeared clearly confirming the late apoptosis.

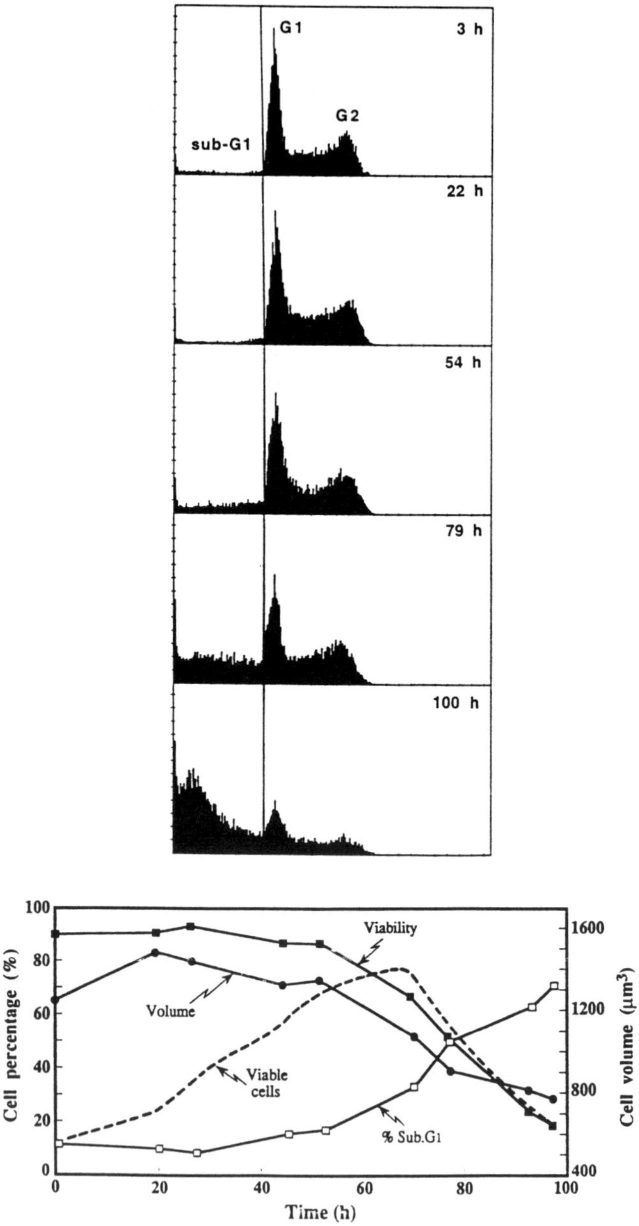

Figure 3 *Evolution of cell DNA content, cell viability (■), cell volume (●) and fraction of sub-G1 cells (□) during cultures in standard medium.*

Culture in glutamine-free medium

In the last culture, the glutamine-free medium was used as a model of nutrient deprivation which occurs at the end of batch cultures or due to concentration gradients inside large scale bioreactors. The cell evolutions are shown in Figure 4.

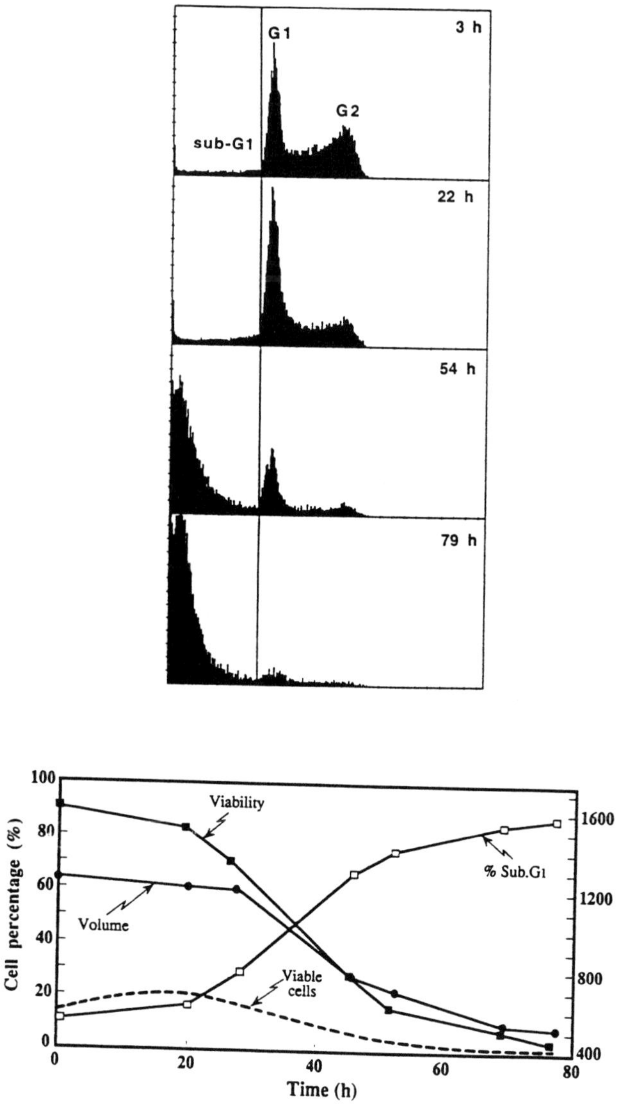

Figure 4 *Evolution of cell DNA content, cell viability (■), cell volume (●) and fraction of sub-G1 cells (□) during cultures in glutamine-free medium.*

During the first 30 h, slow induction of death occured as noticed with a constant volume, a low sub-G1 fraction, no DNA cleavage and a slow viability decrease. Between 30 and 60 h, cell apoptosis was clearly identified with the increase of sub-G1 cell fraction combined to nucleosomes formation. At the end of culture, the decrease of viability at about 5%, the increase of sub-G1 fraction at about 90% and the presence of non specific smears could be the result of the secondary necrosis with nucleosome release.

Conclusions

The monitoring of cell evolution with both sub-G1 DNA content and specific DNA cleavage in nucleosomes is required to well identify the apoptosis of OKT3 hybridoma. Indeed loss or alteration of DNA can occur during both apoptosis or necrosis and results in low DNA cells only. Non specific cleavage can also appear during necrosis or at the end of apoptosis phase by nucleosome release and results in DNA cleavage alone. In standard conditions, the apoptosis is the prevalent mode of cell death for OKT3 hybridoma. It slowly and spontaneously starts before reaching the maximal viable cell density. In glutamine-free medium, apoptosis is earlier induced. Glutamine starvation could then favors apoptosis in large scale reactors with concentration gradients. The role of both glutamine and other amino-acids as repressor of apoptosis is still to be identified (5). For all cultures, except the one with sodium azide, cell lysis is negligible. This can be confirmed by both the released LDH (results not shown) and the strong correlation between sub-G1 cell fraction and viability.

References

1. Singh RP, Al-Rubeai M, Gregory CD, Emery AN (1994) Cell death in bioreactors : a role for apoptosis. Biotechnol. Bioeng. 44, 720-726
2. Solis-Recendez MG, Perani A, D'Habit B, Stacey GN, Maugras M (1995) Hybridoma cell cultures continuously undergo apoptosis and reveal a novel 100bp DNA fragment. J. Biotechnol. 38, 117-127
3. Mercille S, Massie B (1994) Induction of apoptosis in nutrient-deprived cultures of hybridoma and myeloma cells. Biotechnol. Bioeng. 44, 1140-1154
4. Franek F, Vomastek T, Dolnikova J (1992) Fragmented DNA and apoptotic bodies document the programmed way of cell death in hybridoma cultures. Cytotechnology 9, 117-123
5. Franek F, Chladkova-Sramkova K (1995) Apoptosis and nutrition : Involvment of amino acid transport system in repression of hybridoma cell death. Cytotechnology 18, 113-117

INCREASED PRODUCTION OF RECOMBINANT hIGFBP-1 IN SELF-FUSED CHINESE HAMSTER OVARY (CHO) CELLS

CHARLOTTE DYRING
Pharmacia & Upjohn, S-112 87 Stockholm, Sweden and the Technical University of Denmark, DK-2800 Lyngby, Denmark

ABSTRACT

A recombinant chinese hamster ovary (CHO) cell clone, S1, stably expressing human insulin-like growth factor binding protein-1 (hIGFBP-1), was treated with polyethylene glycol (PEG), resulting in cell fusion, in order to further enhance the protein expression by increasing the gene copy number and/or the amount of organelles important to the protein expression/-secretion. Both the fused cell line, Peg1, and its mother cell line, S1, were adapted to serum-free growth in suspension and was characterized with respect to growth and productivity. Peg1 was easier to adapt to the serum-free suspension conditions and had a higher viability during the adaptation period than S1. Furthermore, Peg1 showed a stable productivity of hIGFBP-1 that was twice as high as that for S1 under both adherent and suspension conditions. A considerable difference in the productivity (up to 3-4 times) was noticed during the growth phase. PEG fusion experiments have earlier been studied in our laboratory with CHO cells producing recombinant factor VIII and our results correlates very well with the reults obtained with the factor VIII producing cells. Surprisingly, it was possible to obtain high producing recombinant cell lines, which were stable for more than 4 months.

INTRODUCTION:

Large scale production of recombinant proteins in mammalian cells requires high expression levels for a successful process. A common approach to obtain high expression is to increase the gene copy number e.g. by gene amplification (dihydrofolate reductase, glutamine synthetase etc.) (1). The cell productivity increases at higher gene copy number but it does not increase linearly (2, 3). The cells seem to reach a limit at which they cannot make use of any extra gene copies. However, when using the dihydrofolate reductase (DHFR) gene amplification system, increased expression can be achieved in cells cultured over a very long period of time in MTX containing medium. This does not increase the gene copy number. A likely explanation is, however, that the expression

M.J.T. Carrondo et al. (eds.), Animal Cell Technology, 337–342.

and/or secretion rate increases (3). Fusion of two different cell lines is a well-known method to achieve a new cell line with characteristics from both parent cells. Somatic cell fusion is also used to enhance specific properties of one of the parent cells. The most common use is in the establishment of hybridoma cells for the production of monoclonal antibodies.

Human IGFBP-1 is a 26 kDa IGF binding protein with a total of 234 amino acids (4). It acts as an acute modulator of the metabolic activities of IGF. To understand the regulation of IGF bioactivity at the cellular level, it is necessary to study the IGFBPs. In this paper we present a study on the self-fusion of cells from a CHO line with cells from the same line leading to an increased production of recombinant hIGFBP-1.

MATERIALS AND METHODS:

A hIGFBP-1 producing CHO cell clone, S1, was used for these experiments. Cells were seeded in tissue culture flasks at a concentration of $0.2 \cdot 10^6$ cells ml^{-1} in growth medium (DMEM/F12 (1:1) + 5% BCS (Hyclone, USA)). When the cell culture reached around 90% confluence, the cells were detached from the surface by trypsination and washed twice with PBS in a sterile plastic tube. The fusion protocol developed by Adamson and Spira (5) was followed.

In order to determine the cell size, cells were cultured in tissue culture flasks to around 70% confluence. They were detached from the surface by trypsination, washed twice with PBS and resuspended in PBS to a concentration of $1 \cdot 10^4$ cells ml^{-1}. The cell size of about 10,000 cells was measured on the cell analyzer system, CASY 1 (Schärfe System, Germany). The number and viability of the cells in suspension were determined by trypan-blue exclusion. The growth rate was determined as mean $[(t_2 - t_1)^{-1} \cdot \ln(C_{V,2} \cdot C_{V,1}^{-1})]$, where C_V is the concentration of viable cells and t is the time interval between time 2 and time 1.

S1 and Peg1 were seeded in tissue culture flasks in growth medium. When the cell culture reached around 90% confluence, the cells were detached from the surface by trypsination and transferred to 50 ml defined serum-free medium, named XDYC4 (table 1), with 8 g l^{-1} casein peptone in shake flasks (Corning, UK). The medium was fully replaced every second day. As soon as the cells began to grow as single cells, the medium was changed to XDYC4 with 4 g l^{-1} casein peptone for eight days and then changed to XDYC4 without casein peptone.

Real-time biospecific interaction analysis (BIA) has been used to quantify the amount of hIGFBP-1 produced by the cells as described in (3).

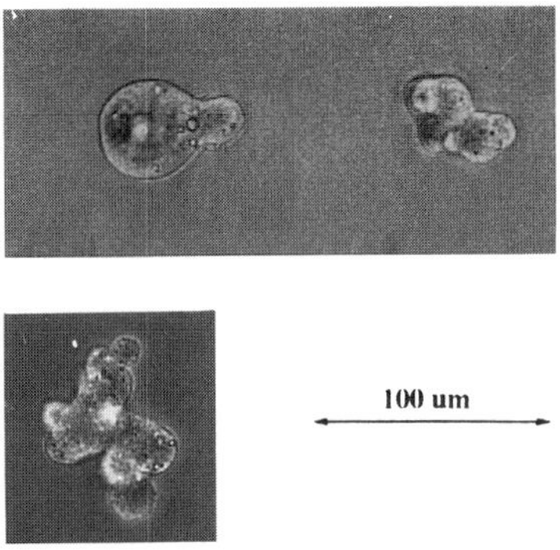

Figure 1

Different numbers of Peg2 cells are fusing shortly after the PEG treatment of the cells

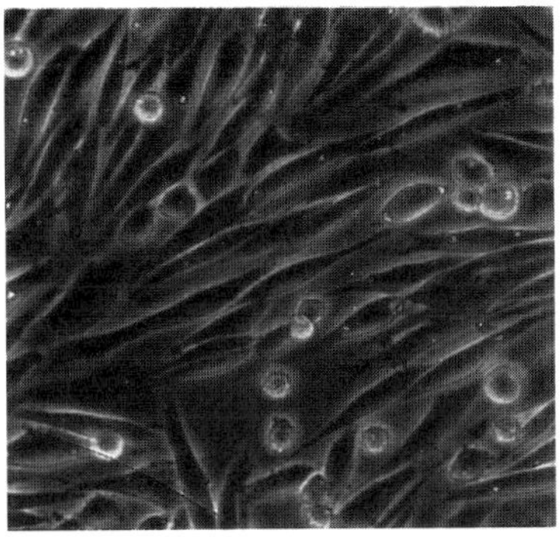

Figure 2a

Figure 2b

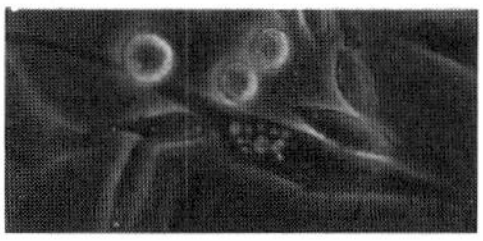

Figure 2c

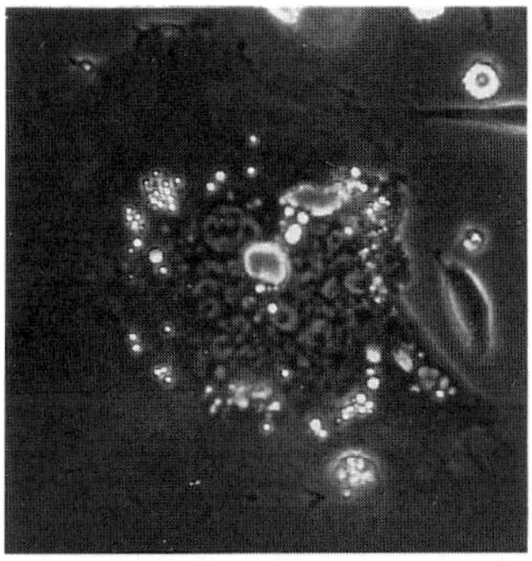

Figure 2b

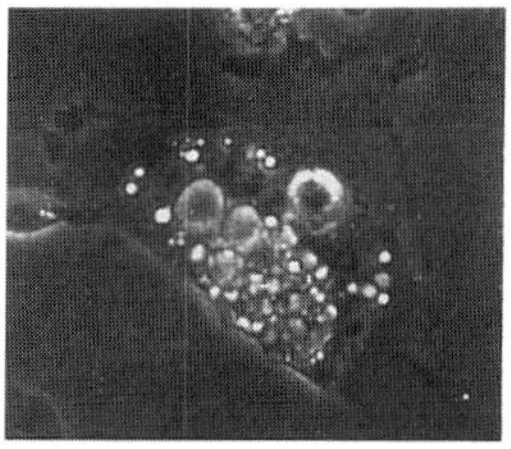

Figure 2d

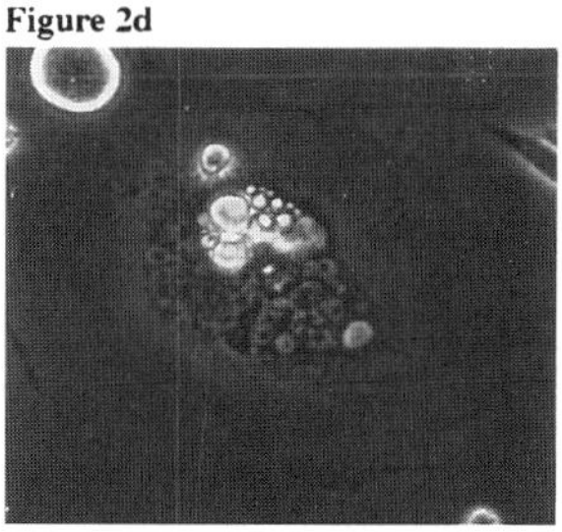

Figure 2d

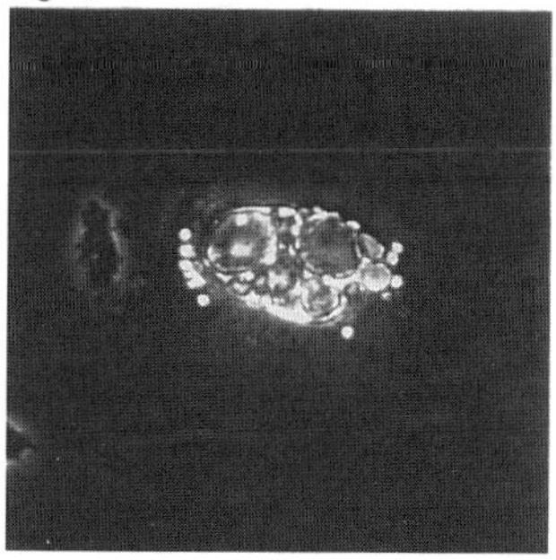

Figure 2

S1 cells before the PEG treatment (a). The cells were cultured adherent in tissue culture flasks in DMEM/F12 (1:1) + 5% BCS (Hyclone, USA). Appearance of the fused cells: (b) Peg1, (c) Peg2 and (d) Peg3. With each PEG treatment, larger and more apart looking cells were obtained

RESULTS:

After the treatment with PEG, cells were seeded in a tissue culture flask and immediately, the cells began to fuse as shown on figure 1. The picture is taken five minutes after the PEG treatment of Peg2 cells. A fusion between different number of cells can be seen. The cells rapidly attached to and spread on the surface of the tissue culture flask. The three obtained fusion cell populations, Peg1-3, appeared as heterogenous populations of non-fused and fused cells, where the latter seemed to be derived from different numbers of cells. The untreated S1 cells cultured adherently had a morhology characterized by a fibroblast-like appearance typically for CHO-DG44 cells cultured adherently (figure 2a). After the cell fusions, the appearance of the cells changed. Examples are given in figure 2b-d. The fused cells were larger than their mother cells, S1, and attached to the surface of the tissue culture flasks, cells with a diameter as big as 200 μm could be seen. The plasma membrane was very diffuse and the cells appeared extremely thin except from a more defined area in the middle of the cell. This area contained a large amount of shining spots in different sizes, which were most probably different organelles.

Determination of the cell size showed, that the mean diameter of the cell population increased with each fusion step. Analysis of the gene copy number by southern blotting one month after the PEG treatment of the cells showed that Peg1 had twice as many copies of the hIGFBP-1 gene as S1, while Peg2 and Peg3 had 1.5 and 1.2 times as many copies of the gene, respectively, as S1 (results not shown).

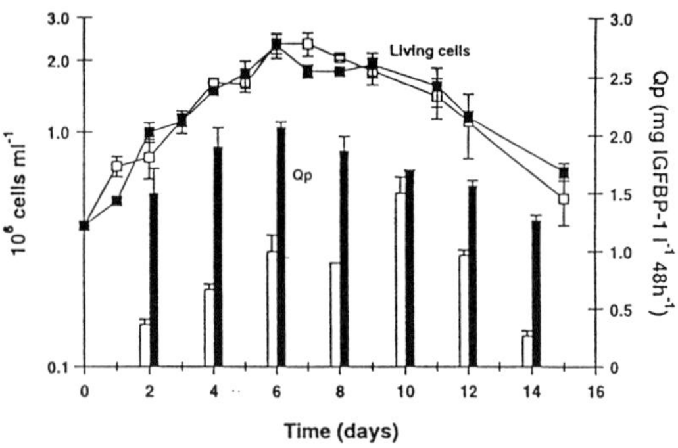

Figure 3. Growth and hIGFBP-1 production of S1 (□) and Peg1 (■) in 50 ml XDYC4 in 250 ml spinnerflasks. The results shown are a mean of a triplicate experiment performed simultaneously with each cell line. No difference in growth was observed but Peg1 produced increased amounts of IGFBP-1.

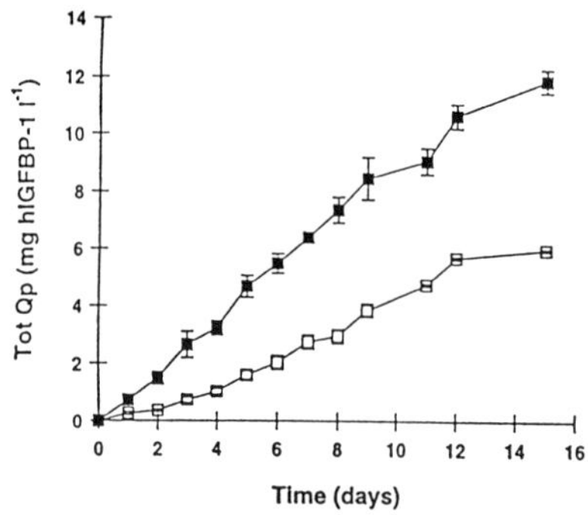

Figure 4. Accumulation of harvested hIGFBP-1 during the 15 days cultivation period in spinner flasks; S1 □, Peg1 ■. The results shown are a mean of a triplicate experiment, see figure 4, performed simultaneously with each cell line.

S1 and Peg1 were adapted to serum-free suspension cultivation and figure 3 shows the growth of S1 and Peg1 in 50 ml XDYC4 in 250 ml spinnerflasks. The results shown are a mean of three experiments performed simultaneously with each cell population. There seemed to be no difference in cell growth between the two cell populations. The viability of Peg1 was more than 90% during the adaptation to serum-free suspension growth while the viability of S1 was below 90% most of the adaptation period. Figure 3 also shows the amount of IGFBP-1 that was produced during 48 hours periods consequtively. The total amount of hIGFBP-1 which was produced during the 15 days cultivation period was twice as high for Peg1 as for S1 as shown in figure 4. According to immunoblotting, there was no apparent difference in the hIGFBP-1 produced from S1 and Peg1 (results not shown).

DISCUSSION:

The described results show, that self-fusion of the stable recombinant cell line, S1, lead to a stable cell population, Peg1, with an increased hIGFBP-1 production and with cells that had an altered appearance. It was observed, that the Peg1 population contained cells that were larger than the S1 cells three weeks after the PEG-treatment. This indicates, that at least a part of the fused cells had survived and that the average cell size had increased. After months of cultivation, the mean Peg1 cells were still larger than the non-fused S1 cells (data not shown) and were stably producing increased amounts of hIGFBP-1. These results correlates very well with the results obtained by Adamson and Spira (5). They found that, by treating stable, recombinant factor VIII producing CHO cells with PEG_{4000}, the mean diameter of the cells as well as the productivity increased. The increased size of the cells and the increased productivity of recombinant factor VIII were stable for more than five months in cultivation.

The increased productivity of hIGFBP-1 in Peg1 could either be a direct result of the increased hIGFBP-1 gene copy number and/or a result of an increased number of organelles, in particular ER, mitochondria and even Golgi as described by Adamson and Spira (5). They studied the PEG-treated recombinant factor VIII producing cells with Transmission Electron Microscopy (TEM) and detected a larger number of high energetic mitochondrias in the PEG cells, which indicated that these cells probably had a larger energy pool than their mother cells. The TEM studies were confirmed by FCM (flow cytometry) analyses of mitochondrias in living cells after incubation with Rhodamin 123 and in fixed cells after incubation with antibodies against mitochondria inner membrane. The PEG cells showed a much stronger fluorescense than the mother cell population. A further support for this theory was energy level measurement performed by Adamson and Spira (5), that showed, that the amount of ATP/cell was higher for the PEG cells. An increased requirement for ATP in the larger, fused cells may lead to an increase in metabolic capacity and production of ATP and thereby an increase in protein expression rate.

The self-fusion also lead to cells which were less sensitive to adaptation to serum-free suspension growth, observed by the higher viability during the adaptation period. Based on the results obtained by Stoicheva and Hui (6), it is likely to think, that less than 20% of the S1 cells fused, which means that the fused cells in average produced more than 6 times as much hIGFBP-1 as their mother cells. Sorting out the 20-30% largest cells in the Peg1 population by e.g. FACS could most probably lead to a population of cells with a very high production of hIGFBP-1.

Taken together, these results show that cell fusion by polyethylene glycol treatment induced a stable change in the expression capability of the hIGFBP-1 producing cells. Furthermore, cells which were easier to adapt to serum-free suspension culture and which were less sensitive to mechanical stress, were obtained.

ACKNOWLEDGEMENT

We gratefully acknowledge Jessica Heidrich for her assistance with the PEG-fusions. We also thank Mats Lake, Erik Walum, Jan Hedander and Claus Emborg for critical comments on the manuscript. Partial support was provided by the Nordic Industrial Foundation (NIFU 039).

REFERENCES

1. Sharma RC & Schimke RT (1994) The propensity for gene amplification: a comparison of protocols, cell lines and selection agents. Mutat. Res. 304(2):243-260.

2. Gu MB, Kern JA, Todd P & Kompala DS (1992) Effect of amplification of *dhfr* and lac Z genes on growth and beta-galactosidase expression in suspension cultures of recombinant CHO cells. Cytotechnology 9:237-245.

3. Dyring C & Mellström K (1996) Recombinant expression of human insulin-like growth factor binding protein-1 (hIGFBP-1) in chinese hamster ovary (CHO) cells. Submitted to Cytotechnology

4. Luthman H, Söderling-Barros J, Persson B, Engberg C, Stern I, Lake M, Franzén S, Israelsson M, Rådén B, Lindgren B, Hjelmqvist L, Enerbäck S, Carlsson P, Bjursell G, Póvoa G, Hall K & Jörnvall H (1989) Human insulin-like growth factor binding protein. Low molecular mass form: protein sequence and cDNA cloning. Eur. J. Biochem. 180:259-265

5. Adamson L & Spira J (1994) Increasing protein expression by self fusion of producer cells to form stable cell lines providing higher yields of therapeutic protein. Patent no. WO 94/19459.

6. Stoicheva NG & Hui SW (1994) Electrically induced fusion of mammalian cells in the presence of polyethylene glycol. J. Membrane. Biol. 140:177-182

NEW PROCESS FOR CELL DETACHMENT: USE OF HEPARIN

Merten, O-W, Dante, J*, Noguiez-Hellin, P*, Laune, S*, Klatzmann, D**, Salzmann, J-L**. Institut Pasteur, 25, rue du Docteur Roux, F-75724 Paris Cedex 15; * Génopoïétic, 22, rue Esquirol, F-75013 Paris; ** CERVI-Hôpital de la Pitié-Salpétrière, 83, boulevard de l'Hôpital, F-75651 Paris Cedex 13

ABSTRACT

The use of animal derived products like trypsine or serum presents a risk due to the potential transmission of adventitious agents. Their replacement by non animal derived or pharmaceutical grade substances is the best choice for avoiding this risk. Trypsin, a widely used substance for the detachment of anchorage dependent cells, can be replaced by heparin in most of the cases. At a concentration of 100IU/ml in the presence of EDTA at 0.02% in PBS, 75% (9/12) of all cell lines tested could be detached from polystyrene based growth supports (T-flasks, roller bottles, the Costar CellCube) without the need of optimizing the detachment conditions. These so detached cells could easily be subcultivated whereby a washing step for eliminating the detachment solution was preferable. The comparison of the growth rates of these subcultures revealed only insignificant differences for cells detached by trypsin/EDTA, heparin/EDTA or EDTA alone (e.g. for BHK21 C13, recombinant CHO and ψCRIP M11 cells grown on polystyrene surfaces).
The detachment of anchorage dependent cells from DEAE dextrane microcarriers was possible for 50% (2/4) of the cell lines tested, however, more difficult than for the cell detachment from polystyrene. In order to get an optimal and fast detachment the microcarriers had to be prewashed with PBS and the concentration of heparin had to be increased to 400-1600 IU/ml (tested for ψCRIP M11 and BHK21 C13 cells).
Concluding, the use of heparin instead of trypsin for detaching anchorage dependent cells is feasible and might have an important impact on cell detachment at an industrial scale.

INTRODUCTION

In general, anchorage dependent animal cells are treated with trypsin ±EDTA for detachment and subsequent passaging, in laboratory as well as in industrial scale. Although this treatment is relatively efficient, there are a lot of drawbacks, like cell, receptor or virus damage or the use of an animal derived protein (trypsin originates from porcine pancreas) demanding rigorous testing of the product for the absence of adventitious agents. These and other drawbacks have lead us to search for other means for cell detachment. Heparin, a sulfated glycosaminoglycan, is known to interact with various components of the extracellular matrix. San Antonio et al. [1] could show that heparin inhibits the attachment and growth of Balb/c-3T3 fibroblasts on collagen substrata due to the interaction of heparin - collagen leading to a disruption of the cell collagen attachment. This fact has led us to investigate the use of heparin±EDTA for detaching anchorage dependent animal cells as

M.J.T. Carrondo et al. (eds.), Animal Cell Technology, 343–348.

replacement for trypsin±EDTA. This study was conducted by using 12 cell lines of different origine (fibroblast and epithelial) cultivated in T-flasks, the CellCube or in microcarrier cultures whereby details are presented for some cell lines of industrial interest (recombined CHO, BHK21 C13, 2 ψCRIP derived cell lines). The impact of this detachment method for detaching anchorage dependent cell lines will be discussed.

MATERIALS AND METHODS

Cell lines and media: The following cell lines were cultivated in the media indicated: BHK21 C13 obtained from P. Perrin/Institut Pasteur (DMEM (AXCELL) + 5%FCS (fetal calf serum)). CHO-IC5 (producing a soluble form of H2Kd of the mouse) obtained from P. Kourilsky/Institut Pasteur (αMEM (GIBCO) + 5%SVF), M11 (ψCRIP M11, producing a retrovirus coding for TK, used in clinical studies by Génopoïétic)(DMEM + 10%NCS (new born calf serum)), ψCRIP ID15 (producing a retrovirus coding for human α-L-iduronidase) obtained from J.M. Heard/Institut Pasteur (DMEM + 10%NCS). Culture and detachment conditions: Cultures in T-flasks (TPP) were started with 10^6 c/25cm² by using 10ml of the respective medium. For cell detachment of confluent cultures, the supernatant was eliminated, the cells were once washed with 5ml of PBS (pH 7.3) and detached by the addition of 2ml of detachment solution (trypsin (0.125%) + EDTA (0.01%) in PBS (pH 7.8); 0, 100 or 400IU heparin/ml + EDTA (0.02%) in PBS (pH 7.4-8.0)). After detachment, the cells were once washed with 12ml of medium and plated at a cell density of 10^6 c/25cm². Cultures in the CellCube were performed as recommended by Costar. Generally, they were started at 14-18*10^3 c/cm² and stopped after having reached confluency at 160-190h. Only the culture of ψCRIP ID15 was performed as a long-term perfusion culture and was stopped at 1144h. For collecting and counting the cells, the CellCube was emptied and filled up with the detachment solution (trypsin (0.125%) + EDTA (0.01%) in PBS (pH 7.8); 100 IU heparin/ml + EDTA (0.02%) in PBS (pH 8.0) or 800 IU heparin/ml in DMEM). The particular conditions are presented in the results section. For starting microcarrier cultures, 4 or 5 g/l of Cytodex 1 carriers (Pharmacia) were inoculated with 3*10^5 c/ml. Cultures were performed in 50ml tubes (TPP, working volume: 25ml). The tubes were incubated at 37°C on a Thermolyne Vari-Mix agitator at 8rpm. After having obtained confluency, samples of microcarriers (0.5ml) were taken and treated with 0.5ml of detachment solution after one washing step with PBS. Cell counting was done in a Malassez chamber after dilution of the samples (1:1) with trypan blue (0.2% in PBS) or with crystal violet (0.1% in 0.1M citric acid, 1% Triton X100; incubation at 37°C for 1h) where specially indicated.

RESULTS

Detachment from polystyrene surfaces: The modification of the actual or the use of a new detachment method for cells needs a rigorous evaluation with respect to cell detachment and subcultivation performance of the detached cells. Therefore, 12 cell lines were used, of which detailed results for CHO-IC5 and BHK21 C13 are shown. The detachment of recombinant CHO cells (CHO-IC5) was comparable when trypsin/EDTA (0.125%/

0.01%), heparin/EDTA (100 IU/ml/0.02%), or EDTA (0.02%) in PBS were used. At semiconfluent cultures (about 2.5*10^6c/25cm²) the detachment took 1', 15', or 20', respectively; at confluent cultures (about 9*10^6 c/25cm²) this time was 3' for all conditions used. In the case of BHK21 C13 cells, the duration for detaching all cells from the surfaces was 2'-5' for trypsinisation, and 2'-10' or 5'-15' for the other detachment conditions, when confluent (10^7 c/25cm²) or semiconfluent (3.4*10^7 c/25cm²) cultures, respectively, were treated. The subcultivation of these detached cells (CHO-IC5, BHK21 C13) revealed that the growth was almost not influenced by the detachment conditions, when the cells were once washed with culture medium before subcultivation (Fig. 1, bars 1-6). In the case that the detached cells were not washed, the cells detached by heparin/EDTA grew better than those detached by trypsin/EDTA or by heparin alone (Fig. 1, bars 7-9). From all cell lines tested, 9 could easily be detached by using heparin/EDTA (100 IU/ml/0.02%) in PBS (HeLa R19, HT1080, ψCRIP and derivatives, Sf9, 143B), for one the conditions had to be optimized (GPenvAM12), and for two others the detachment took more than 1h (Vero) or was practically impossible (MDCK).

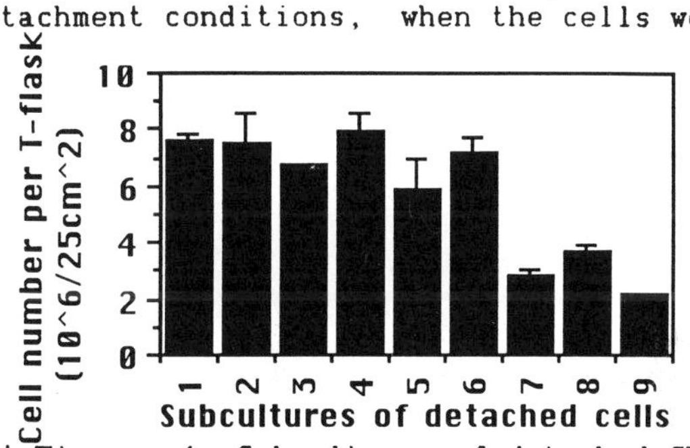

Figure 1. Subcultures of detached CHO-IC5 (bars 1-3) and BHK21 C13 (bars 4-9) cells. Conditions: the detached cells were once washed in culture medium for CHO-IC5 (bars 1-3) and BHK21 C13 (bars 4-6) or were directly subcultivated without washing after detachment (bars 7-9). The cells were plated at 10^6 c/ T-flask of 25cm². Time of counting:56h, 54.5h, and 70.5h for cultures 1-3, 4-6, and 7-9, respectively. The detachment solutions were: trypsine/EDTA (0.125%/ 0.01%) for cultures 1,4,7; 100 IU/ml heparin, 0.02% EDTA for cultures 2,5,8; 0.02% EDTA for cultures 3,6; 100 IU/ml heparin for culture 9.

The large scale production of animal cells for gene therapy is often done in the CellCube. The comparison of different detachment methods of ψCRIP cells grown in the CellCube revealed that heparin/EDTA detached faster and more efficiently without the formation of aggregates than trypsin/EDTA did. Generally, confluent ψCRIP cultures reach a maximal cell density at confluency of 330000 c/cm² at about 190 h when the cultures were started with 15-18*10^3 c/cm². The detachment of confluent ψCRIP cells by trypsin/EDTA led to a cell recovery of only 16.26% (53647 c/cm²) due to the formation of enormeous aggregates which were retained by the spacers between the plates of the CellCube (Table I.). The addition of heparin directly to the supernatant at higher concentrations (e.g. 800 IU/ml) in DMEM did not improve the detachment very much with respect to trypsin/ EDTA. About 30% (99330 c/cm²) of the cells could be recovered. Only the use of heparin/EDTA (100 IU/ml, 0.02%) in PBS allowed a rather fast (within 20') detachment of the cells and a complete recuperation of these cells. The cell number recovered depended only on the state of

Table I. Detachment of confluent cultures of ψCRIP cells (ID15. M11) from the Costar CellCube:

Cell line. detachment method	Startina cell densitv: c/cm^2	Collected cells per cm^2	% of maximallv obtainable cells	Collection time h
ID15*. 1 PBS washina. trvpsin/ EDTA in PBS for 21′	9647	53647	16.26	at 1144
M11**. 100 IU/ml heparin. 0.02% EDTA in PBS for 20′	17131.5	251435 (330000)***	76.19 (100)	at 170.6 (at 190)
M11****. 800 IU/ml heparin in DMEM for 22′	16830	70902 (99330)	21.48 (30.1)	at 168 (at 190)

Notes: *) For startina this culture the cells were thawed from Nunc Crvotubes and expanded in roller bottles. This culture was a perfusion culture for the production of retrovirus for aene therapv. therefore the verv late stop of the culture. **) Averaae of several cultures. the cells were thawed from Nunc Crvotubes and expanded in roller bottles or the cells were thawed from Baxter Crvocvte Freezina Containers (cells originated from a previous culture in a CellCube) and directly inoculated into the CellCube. ***) The values in parenthesis indicate the values which would have been obtained in the case of stopping the cultures at 190 h. The specific growth rate used for this calculaion was 0.01533/h (average growth rate in the CellCube). ****) The cells were thawed from Nunc Cryotubes and expanded in roller bottles.

---

confluencv of the culture. At 190 h about 330000 c/cm^2 (the theoretical maximal cell density obtainable with ψCRIP cells) could be recovered from the 21250 cm^2 CellCube (Table I) as well as from the 85000 cm^2 CellCube (not shown). The so detached cells could be subcultivated without any problems (Table I). All cultures done with the cell lines ψCRIP in the CellCube were precultivated in roller bottles or CellCubes. detached by heparin/EDTA and directly used for inoculation or after freezing and thawing in Baxter Cryocyte Freezing Containers.

Detachment of anchorage dependent cells from microcarriers: Because microcarrier cultures are an important mean for the large scale production of various animal cell derived products, we tested also the detachment of cells by heparin. The detachment of ψCRIP M11 cells from confluent microcarrier cultures revealed that trypsin/EDTA was rather efficient for cell dissociation. however. the cell count was significantly lower than that obtained by crystal violet. Using 100 IU/ml heparin + 0.02% EDTA or without EDTA cell detachment was feasible (detachment with respect to trypsinisation: 71.4% and 61.8%. respectively), however, the action was more due to the presence of EDTA than to the presence of heparin. At a heparin concentration of 400 IU/ml in the presence or absence of EDTA 88.5% and 81.8%. respectively, of the cells could be recovered with respect to trypsinisation (Table II). These results

Table II. Detachment of ψCRIP M11 cells from Cytodex 1 by trypsin and heparin:

Detachment solution	c/ml	% of cells detached by trypsinisation
1. Crystal violet	2.15*10^6	
Trypsin (0.125%), EDTA (0.01%)	1.75*10^6	100
100 IU/ml heparin, EDTA (0.02%)	1.25*10^6	71.4
0 IU/ml heparin, EDTA (0.02%)	1.43*10^6	82.1
100 IU/ml heparin	1.08*10^6	61.8
2. Crystal violet	2.42*10^6	
Trypsin (0.125%), EDTA (0.01%)	1.69*10^6	100
400 IU/ml heparin, EDTA (0.02%)	1.49*10^6	88.5
400 IU/ml heparin	1.38*10^6	81.8

Notes: The cells have been washed once with PBS before treatment with detachment solution, the duration for dissociation was 10'-15'.

signify that an increased heparin concentration is important for a more complete cell detachment. Similar results have been obtained with BHK21 C13 cells. The experiments done with Vero and CHO-IC5 cells were not very successful and for both cell lines, trypsinisation still rests the best method for dissociation these cells from microcarriers.

DISCUSSION

Cell detachment is a very important step in animal cell culture. Classically this is done by using trypsin/EDTA, acting in cleaving protein liaisions and in dissolving Ca-bridges between the cells and the support. Although this detachment is relatively efficient, there are major drawbacks associated with it. The action of trypsin damages cells, receptors, viruses even at relatively short incubation periods. A major risk associated with the use of trypsin is the fact that it is derived from porcine pancreas and may, therefore, contain animal viruses and other adventitious agents. Though trypsin used by the pharmaceutical industry is rigorously tested for the absence of viruses and bacterial contaminants, the appearance of new unknown viruses is possible, not easily detected and may be introduced into the biological product.
Heparin is extensively used as injectable in human therapeutics for years without any viral contamination. It is known to interact with various compounds of the extracellular matrix. San Antonio et al. [1] could show that heparin inhibits the attachment and growth of Balb/c-3T3 fibroblasts on collagen substrata. So the idea was that heparin might also provoque a detachment of various types of anchorage dependent cells from the growth support. The results showed that about 75% of all tested cells could easily be detached from polystyrene. The other 25% could be detached after optimisation of the detachment conditions or were more or less resistant to this treatment. Actually, we have no explanation for the different behaviour observed because these differences do not cor-

respond to the morphological classification (fibroblast. epithelial....) of the cells used. The differences observed between the detachment from polvstvrene (rather easv) and from DEAE-dextrane based microcarriers (relativelv difficult) are probablv due to different mechanisms of the cell attachment. However. more experiments are necessarv to understand the reasons for these differences.

Other trypsin-free detachment methods have been proposed. however. thev are still in a state of development. The use of other proteases is possible but not all are inactivated by serum (e.g. dispase [2]) leading to the need of thorough washing of the harvested cells. Other detachment methods free of any animal derived product have been described for small scale and particular applications. however. they are not relevant for the large scale application [2. 3]. The only large scale detachment method devoid of any animal derived product is the use of polystyrene surfaces or carriers which have been coated with thermally reversible copolymers [4. 5]. In these cases the cells are detached when the temperature is decreased below a critical temperature at which the surface characteristics switch from hydrophobicity to hydrophilicity leading to a detachment of the cells. Although this method is very elegant and straightforward. several problems have still to be resolved. First. the surfaces or carriers have to be coated with these polymers. and second. the cells have a tendency to detach in 'sheets' [5]. After having overcome these problems the detachment of cells using thermally reversible polymers will be of high interest.

In conclusion. for the moment. the use of heparin±EDTA is the best choice for the non-proteolytic detachment of anchorage dependent cells. Without extensive optimisation of the detachment conditions 75% of anchorage dependent cells could easily be detached from polystyrene by the use of heparin± EDTA. whereas 50% of the cell lines tested could be detached from DEAE-dextrane based microcarriers. however. only after optimisation of the detachment method. Although this method cannot be applied for the detachment of all anchorage dependent cell lines, it is the best choice when cells have to be detached without using proteases. in laboratory as well as in industry.

REFERENCES

1. San Antonio JD. Lander. AD, Wright TC & Karnovsky MJ (1992) Heparin inhibits the attachment and growth of Balb/c3T3 fibroblasts on collagen substrata. J Cell Physiol 150, 8-16.
2. Anonymus (1981) Microcarrier cell culture. Principles & Methods. Pharmacia.
3. Griffiths JB (1985) Cell Biology: Experimental Aspects. In: Animal Cell Biotechnology, eds. Spier, RE & Griffiths JB. pp. 49-83. Academic Press. London.
4. Okano T. Yamada N, Sakai H & Sakuai Y (1993) A novel recovery system for cultured cells using plasma-treated polystyrene dishes grafted with poly(N-isopropylacrylamide. J Biomed Mater Res 27. 1243-1251.
5. Rollason G. Davies JE & Sefton MV (1993) Preliminary report on cell culture on a thermally reversible copolymer. Biomaterials 14. 153-155.

DEVELOPMENT OF A HIGH SENSITIVITY RETROVIRUS SPECIFIC REVERSE TRANSCRIPTASE ASSAY

Audrey Chang, Martin Wisher, Robert Bird, and Jeffrey M. Ostrove
Microbiological Associates, 9900 Blackwell Road, Rockville MD and Stirling, Scotland

Introduction:
The safety testing of cell substrates used to produce biological products is an important issue in the Biotechnology industry and is required by regulatory authorities worldwide prior to human clinical trials (1). The most frequent viral contaminant detected in these cells are retroviruses (2,3). The Retroviridae comprise a large number of viruses known to infect many species, including humans. They have been associated with leukemias and lymphomas (4), immunodeficiencies (5), and other diseases (6,7).

The enzyme reverse transcriptase (RT) is found in all retroviral particles. Assays that detect RT activity have been used to establish the presence of retroviruses in biological materials. Recently, several PCR based RT detection assays have been developed that are several orders of magnitude more sensitive than conventional methods that measure the incorporation of radiolabelled nucleoside triphosphates (8,9,10,11). These assays are fundamentally similar in that they detect RT activity by amplifying the newly synthesized cDNA from an RNA template via PCR. It has been shown that these RT-PCR detection assays can detect RT's from various viral sources and from cell supernatants. The drawback to many assays with increased sensitivity is the elevated incidence of false positive results generated by cellular polymerases that can reverse transcribe. Although some steps can be taken to eliminate the contaminating DNA polymerases, the presence of polymerases from lysed cells and cellular debris often make interpretation of the RT assay impossible. We present a modification to a PCR based RT detection assay, PERT (8), that easily allows discrimination between contaminating DNA polymerases and true RT activity.

Materials and Methods

Reagents: Molecular biology grade chemicals were purchased from Sigma and protease inhibitors were purchased from Boehringer Mannheim. Primers used were those described by Pyra et al. (8).

Cell Cultures: All cell lines were maintained by Microbiological Associates in either DMEM or EMEM with serum (fetal bovine or newborn calf) with or without gentamycin.

M.J.T. Carrondo et al. (eds.), Animal Cell Technology, 349–354.

Sources of RT: Squirrel Monkey Retrovirus (SMRV), Equine Infectious Anemia Virus (EIAV), Rauscher Murine Leukemia Virus (R-MLV), Moloney Murine Leukemia Virus (MMLV), Gibbon Ape Leukemia Virus (GaLV), and Adenovirus virus particles were all purified in house. Raji cells persistently infected with Simian Retrovirus (SRV-2), H9 cells infected with Simian Immunodeficiency Virus (SIV), and LLCMK2 cells infected with Foamyvirus were also maintained. The Human T-Cell Leukemia Virus (HTLV-1) producing MT-2 cells were the source for HTLV-1 virus. A GaLV producing gibbon ape T-cell lymphoma was used as a GaLV source. Purified RT were purchased as follows: AMV (Amersham), HIV (Worthington), MMLV (Gibco/BRL), Superscript II (Gibco/BRL).

Sample Preparation: *Cell lysates*: Adherent cell lines were rinsed with ice cold PBS and scraped in 5 ml of ice cold PBS. Cells were centrifuged at 1000 x g and the resulting pellet was resuspended in PBS. Suspension cells were pelleted at 1000 x g and washed once with ice cold PBS, centrifuged again and the resulting pellet was resuspended in PBS. 100 μl of the resuspended cell pellet was lysed in 0.04% Triton X-100 by freezing and thawing twice in the presence of the protease inhibitors aprotinin, leupeptin, and pepstatin. Total protein concentration was determined using the BioRad Protein assay kit.

Cell supernatants: The conditioned media from actively growing cells were harvested to test for RT activity in cell supernatants. The conditioned media were on the cells for a minimum of 72 hours. One ml of medium was centrifuged at 1000 x g for 5 minutes to clear cellular debris. An equal volume of cell supernatant and buffer containing the protease inhibitors was lysed in 0.04% Triton X-100 by freezing and thawing twice.

Virus: Purified virus was diluted in buffer containing protease inhibitors and containing 0.04% Triton X-100 and the sample was freeze/thawed as described above. Subsequent 10 fold dilutions were made in the same solution.

Modified PERT Assay: The assay was similar to that previously described (8) with the following modifications: Bacteriophage MS2 genomic RNA (Boehringer Mannheim) was annealed with RT-1 in an amount equivalent to 50 reactions, stored at -20°C, and used in one reaction aliquots. Two separate RT buffers were freshly prepared as previously described (8) except Hepes was substituted for Tris at varying pH's. The RT incubation was reduced from 5 hours to 1 hour. To each RT reaction, 75 μl of PCR mix with 0.1 μg RNase A and 7.5 units of Taq polymerase is added and the reaction amplified in a PE2400. The PCR product was directly visualized on a horizontal 2.5 % Metaphor agarose (FMC BioProducts) gel. The DNA bands were visualized by placing the gel on a UV transilluminator and photographed.

Results:

Titration of purified RT: The product-enhanced RT assay (PERT) described by Pyra et al. (8), used either Southern blot or enzyme immunoassay as an endpoint. We have modified the PERT assay in several steps, including changing the endpoint. Other modifications involve changing the ingredients of the RT buffer mix, reducing the RT reaction incubation time, eliminating the separate Rnase A digestion step, and running the PCR product on a high resolution agarose gel to visualize the amplification product directly. As for any PCR procedure, it is essential to include negative controls to ensure that no contaminating amplifying DNA from reagents is present. We tested the modified PERT assay sensitivity by serially diluting the recombinant M-MLV reverse transcriptase (Superscript II) ten fold by unit activity. As shown in figure 1, the 112 bp band is generated by the process.

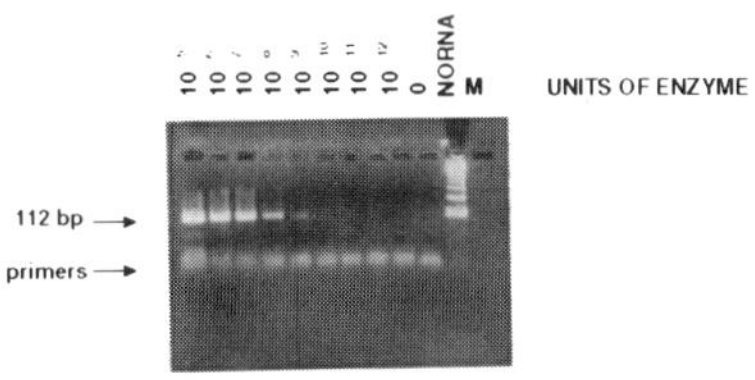

Fig 1: Titration of purified RT

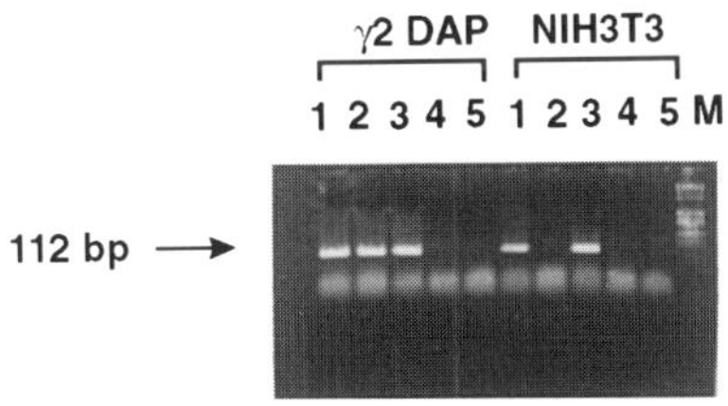

Fig 2: Cell lysate and cell supernatant of uninfected NIH/3T3 cells and γ2DAP cells. Lanes (1) cell lysate (2) cell supernatant (3) cell supernatant with RT (4) cell lysate without RNA (5) cell supernatant without RNA

The detection limit for the 112 bp PCR band is 10^{-9} units of RT. The conventional RT detection limit at best is between 10^{-3} and 10^{-4} units (unpublished observation). The modified PERT represents approximately a 10^{5} increase in sensitivity. The negative controls (No RT, and No MS2 RNA template) do not have the 112 bp band, indicating that no contaminating DNA or contaminating RT activity from Taq or Rnasin was present. The sensitivity obtained with the high resolution agarose gel is comparable to the ELISA and Southern blot methodology. Given that the specific activity of the enzyme is 350,000 units/mg protein (Gibco-BRL Superscript II Lot No. EPF401) and the molecular weight equals 78,000 daltons, 10^{-9} units corresponds to 22 molecules of RT. Other purified RT enzymes, HIV RT, AMV RT, M-MLV RT, had similar limits of detection of 10^{-9} and 10^{-10} units (data not shown).

Retrovirus infected and non-infected cell culture supernatants and cell lysates: Although the mitochondria DNA polymerase γ has reduced activity with the heteropolymer RNA template, cellular DNA polymerases that reverse transcribe can still generate false positive results. Therefore, PERT, along with other PCR based RT detection assays have been limited to detecting retrovirus in culture supernatants. This point is illustrated in Figure 2. The γ2 DAP cell line is derived from an NIH/3T3 cell line and produces recombinant retrovirus. Both the cell supernatant and cell lysate when tested for RT activity were positive. The uninfected NIH/3T3 cell supernatant tested negative for RT in the assay whereas the NIH/3T3 cell lysate generated a positive signal. Since the NIH/3T3 cell line does not harbor infectious retrovirus, and expression of full length endogenous virus IAPs is low (12), the "RT" activity seen in the NIH/3T3 cells was the result of cellular DNA polymerases.

We investigated methods that would allow discrimination between authentic RT and cellular polymerase activities. Figure 3 demonstrates that by lowering the pH of the RT reaction buffer, only the cell lysates that harbor infectious retrovirus produce the PCR 112 bp band whereas the uninfected cells do not.

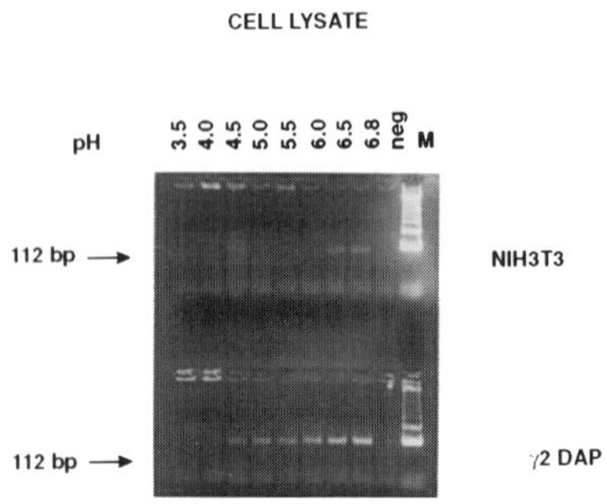

Fig 3: Effect of decreasing pH in RT reaction mix

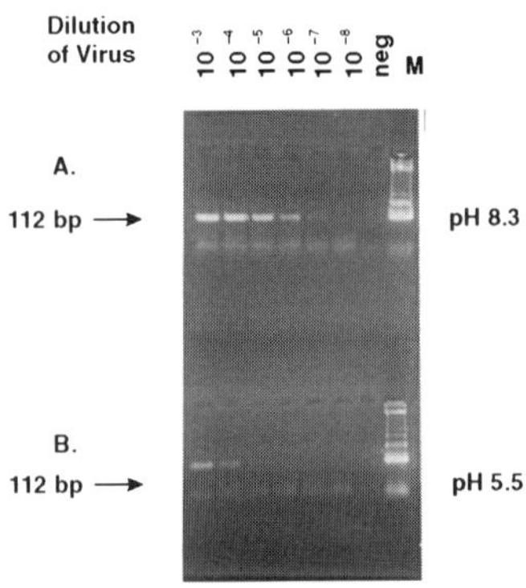

Fig 4: Titration of R-MLV with RT reaction mix at (A) pH 8.3 or (B) pH 5.5

Table 1 summarizes the results of additional uninfected cell lysates from various sources that were prepared and tested. All were negative for RT activity at pH 5.5. The negative result is not a outcome of inhibition of RT since samples spiked with known amounts of RT resulted in a positive signal.

Table 2 lists the different retroviruses that we have detected in our assay. Foamyvirus, HTLV-1, SMRV, EIAV, AMV, HIV, GaLV, represent the different retrovirus subfamilies (Oncovirus, Lentivirus, and the Spumavirus). Purified adenovirus viral particles which do not have RT, are negative for RT by the assay.

Testing Uninfected Cell Lysate			
Cell Line	pH 8.3	pH 5.5	pH 5.5 + spike
293	+	-	n/d
NIH 3T3	+	-	+
Raji	+	-	+
MRC-5	+	-	+
HeLa	+	-	+
CHO	+	-	+
BHK	+	-	+
LLC-MK2	+*	-	+
H9	+*	-	+
Mus dunni	+*	-	+

* Weak Positive

Table 1

Detecting Viral RT		
Virus or RT Species	Source	Results
Foamyvirus	infected cells	+
HTLV I	infected cells	+
GaLV	infected cells	+
SIV	infected cell supe	+
SRV 2	infected cell supe	+
SMRV	purified virus	+
EIAV	purified virus	+
R-MLV	purified virus	+
MMLV	purified virus	+
AMV	purified RT	+
HIV	purified RT	+
MMLV	purified RT	+
Adenovirus	purified virus (non-retrovirus)	-

Table 2

Rauscher MLV: A purified preparation of R-MLV viral particles was serially diluted and tested for RT at pH of 8.3 and 5.5 (Figure 4). At the higher pH, the limit of detection was at 10^{-6}. At the lower pH, the limit of detection was at 10^{-4}. The two log difference in sensitivity (data not shown) between the two pH's was also observed for purified RTs (AMV, HIV, and MMLV) but a minimum of 10^3 fold increase in sensitivity between the conventional assay and the modified PERT assay still exists.

Discussion: The successful analysis of the retroviral status in biologicals depends upon the interpretation of data produced from several assays (infectivity, EM, RT assays). The PERT assay incorporates the extremely sensitive method of PCR to detect retroviruses by the presence of viral reverse transcriptase activity. Cellular DNA polymerases can also utilize the RNA template and mask authentic RT activity producing invalid results. In this study, we have modified and improved the assay at several steps (changing the endpoint, lowering the pH of the RT reaction reducing RT incubation, adding of protease inhibitors). With these conditions it is possible to assay for retrovirus RT activity at a sensitivity level several logs greater than the conventional assays while eliminating the false positives generated by cellular polymerases. The ease of preparing samples (one template, one divalent cation, no radioactive/organic reagents, no ultracentrifugation) and the wide range of retrovirus RT that can be assayed gives our assay tremendous potential for use as a screening procedure for retroviruses.

References:

1. CBER (1993) Points to Consider in the Characterization of Cell Lines Used to Produce Biologicals
2. Moore, A. (1992), Develop. biol. Standard, 76:51-56.
3. Losikoff AM, Poiley A, Raineri R, Nelson RE, Hillesund T (1992) Develop. biol. Standard 76:187-200
4. Cann AJ & Chen ISY (1990) HTLV I and II. In: Fields BN & Knipe DH: Fields Virology, pp 1501-1528, Raven Press Ltd, New York.
5. Hirsch MS & Curran J. (1990) HIV. In: Fields BN & Knipe DH: Fields Virology, pp 1545-1570, Raven Press Ltd, New York.
6. Reddy EP, Sandberg-Wolheim M, Mettus RV, Ray PE, DeFreitas E and Koprowski H (1989) Science, 243:529-533
7. Lisby G. (1993) Acta Neurol. Scand. 88:385-387
8. Pyra H, Boni J, & Schupbach J (1994) Proc. Natl. Acad. Sci. U.S.A. 91:1544-1548
9. Heneine W, Yamamoto S, Switzer WM, Spria TJ, and Folks TM (1995) J. Infect. Dis. 151:1210-1216
10. Busso M & Resnick L (1994) J. Virol. Meth. 470:129-140.
11. Silver J, Maudru J, Fujita K, & Repaske R (1993) Nucl. Acids Research 21:3593-3594
12. Grigoryan MS, Kramerov DA, Tulchinsky E, Revasova ES & Lukanidin EM (1985) EMBO J. 4:2209-2215

KINETICS OF GLUCOSE METABOLISM AND UTILIZATION OF LACTATE IN MAMMALIAN CELL CULTURES

S. S. OZTURK, P. JORJANI, R. TATICEK, B. LOWE, S. SHACKLEFORD, D. LADEHOFF-GUILES, J. THRIFT, J. BLACKIE, and D. NAVEH
Bayer Corporation, Biotechnology, Berkeley, CA 94701

1. Abstract

Glucose is a major energy source for mammalian cells grown in culture. Although conditions are aerobic, in most cases, glucose is not completely oxidized. Depending on the culture conditions and the cell type, about 50-100% of glucose is converted to lactate. In this study, the glucose metabolism was studied in batch, chemostat, and perfusion cultures of CHO and NS0 cells. The glucose utilization rate slowed down when glucose became limiting. This decrease could also be attributed to slow growth rates at low glucose concentrations. Lactate production rate also decreased at low growth rates. When the media was enriched with amino acids, glucose was completely metabolized in batch cultures. The lactate concentration increased until the glucose is depleted. Then lactate was utilized along with other amino acids. The utilization of lactate was also observed in chemostat and perfusion cultures. Net lactate production from glucose dropped to zero at low glucose levels. A thorough analysis on lactate yield on glucose demonstrated a constant flux of glucose to cell mass. This flux was observed to be independent of lactate formation, which was proportional to glucose utilization.

2. Introduction

Mammalian cells utilize glucose for energy production and produce lactate as a metabolite by-product. In most cases, glucose is the limiting nutrient in the media and glucose concentration can affect the cell growth and reactor productivity (Miller et al, 1988a; Hu et al, 1987; Lavering et al, 1992; Racher et al, 1993; Kurokawa et al, 1993). Lactate, produced mainly as a metabolic product of glucose metabolism is also important for cell growth and metabolic activity. In cell culture fermentation, typically about 50-100% of the glucose is converted to lactate (Miller et al, 1988a; Hu et al, 1987; Lavering et al, 1992; Racher et al, 1993; Kurokawa et al, 1993; Ozturk et al, 1992). Lactate can also be produced from glutamine metabolism (Zielke, et al, 1980). High levels of lactate (in the order of 2 g/L) were shown to be inhibitory for mammalian cell growth (Miller et al, 1988b; Ozturk et al, 1992). Lactate production also contributes significantly to base addition. In poorly mixed reactors, base addition can create problems due to localized cell lysis (Ozturk, 1996).

A simplified description of mammalian cell metabolism in culture is presented in Figure 1. Cells utilize both glucose and glutamine for the production of metabolic energy. Glucose is utilized primarily through anaerobic glycolysis, with a flux diversion through the pentose pathway to give the pentose building blocks for nucleic acids. Most of the pyruvate generated from glycolysis is not utilized in the TCA cycle, but converted to

M.J.T. Carrondo et al. (eds.), Animal Cell Technology, 355–360.

lactate. The reasons for significant lactate formation from glucose under aerobic conditions is not fully understood. Loss of control in the regulation of glycolysis, presence of other energy sources (i.e., glutamine), and faulty shuttle systems to mitochondria have been suggested as possible reasons (Eigenbrodt et al, 1985).

Culture conditions can influence glucose metabolism and lactate production. Although lactate yield from glucose can be lowered by reducing the glucose levels in the medium (Glacken et al, 1988; Hu et al, 1987), complete utilization needs careful metabolic rate analysis and medium formulations. In this work we have investigated the kinetics of glucose metabolism and lactate formation in well controlled reactors. We focused our attention on very low concentrations of glucose in batch and perfusion cultures in order to define conditions for complete glucose utilization. We also examined the possibility of lactate utilization by mammalian cells.

3. Experimental Method

Reactor System:

Experiments were performed in a water-jacketed stirred tank reactor (Applikon) with 2 L working volume. Agitation was provided by 2 marine type impellers. Aeration was achieved by surface and membrane aeration. The agitation was maintained at 200 rpm. The temperature and DO were kept constant for both fermenters at 37°C and 50% air saturation, respectively. The pH was maintained at 7.2 by the use of CO_2 and 0.3 M NaOH fed to the medium line.

For perfusion operation, the fermenter was coupled with an external cell retention device. Cell densities were estimated on-line using oxygen consumption rate evaluated from mass transfer coefficients and the gas composition to the fermenters (Ozturk et al, 1995). The harvest rate was adjusted proportional to estimated cell density (cell specific perfusion rate control). The feed pump was controlled by the computer so that the weight of the reactor remained constant. Cell density was controlled at the desired set points by purging the cells from the cell separation device. All the controls were achieved using an object oriented programming language (LabView) on a Macintosh computer.

Cell lines, media, reactor operation:

Two different cell lines (CHO and myeloma) were used in this study. Serum-free media (a DMEM:F12 based amino acid fortified medium) were used for both cell lines. The glucose level in both media was 4.5 g/L. For the CHO cell line, the glutamine level was 7.5 mM. The myeloma cell line (NS0) could synthesize glutamine (GS gene incorporated) so no glutamine was added to the medium used for myeloma cultivation.

Measurement of cell counts, metabolite concentrations, and specific rates:

Daily sampling was performed off line for determination of cell and metabolite concentrations. Cell counts and viabilities were determined using a hemacytometer and trypan blue dye exclusion method. Samples were analyzed for glucose and lactate using a YSI analyzer. The cell specific rates were obtained from the measurements of cell counts and metabolite concentrations (Ozturk and Palsson, 1991).

4. Results and Discussion

Figure 2 illustrates the batch CHO cell culture time profile of cell, glucose, and lactate concentrations for the amino acid fortified medium. During the sirst 5 days, cells continued to grow until glucose was depleted while lactate accumulated in the medium. Then the cells started utilizing more amino acids (data not shown) and more importantly, the lactate. This utilization of lactate was not observed for regular DMEM:F12 medium. Apparently, the fortification of amino acids maintained active cell metabolism during the decline phase. Thus the cells could utilize lactate for energy metabolism and cell mass. This important piece of data demonstrates that the mammalian cells can indeed utilize lactate under appropriate conditions. Similar data on the lactate utilization in batch culture was observed for myeloma cells (data not shown).

Although we may have demonstrated the utilization of lactate for the cell culture, for the first time, utilization of lactate is not surprising for mammalian cells metabolism *in vivo*. The lactate can be consumed by mammalian cells after being converted to pyruvate by pyruvate dehydrogenase. The metabolic reaction between lactate and pyruvate is close to equilibrium (Eigenbrodt, et al, 1988) and higher lactate and lower pyruvate concentration favor the utilization of lactate.

The kinetics of glucose and lactate metabolism was also investigated in perfusion cultures for CHO and myeloma cell lines. In perfusion cultures we have varied the "cell specific perfusion rate (CSPR)" to obtain various steady states (Ozturk et al, 1995). Alteration of CSPR resulted in differences in the cellular environment (i.e., glucose concentration) and allowed us to manipulate the cell physiology (i.e., glucose metabolism).

Figure 3 presents the time profiles of cell concentration, % viability, glucose and lactate concentrations for perfusion culture of CHO cells. Cell concentration was maintained around 15-20 million cells/ml and the CSPR was varied in the order of 0.25->0.18>0.09->0.06->0.04->0.035->0.025->0.11 nl/cell/day. Each CSPR is presented as intervals in the x-axis and sufficient time (7-10 days) between the CSPR's was allowed for reaching a new steady state. The run lasted for 103 days and good viabilities (more than 80%, except very low perfusion rates) were obtained during the run.

As expected, lowering the perfusion rate (CSPR) resulted in low residual glucose concentrations. After reaching a CSPR of 0.04 nl/cell/day, the residual glucose level dropped to zero. The lactate concentration remained high (more than 2 g/L) until a CSPR of 0.06 nl/cell/day. Further decrease in CSPR decreased lactate levels and no lactate formation was observed below a CSPR of 0.035 nl/cell/day. In other words glucose was completely consumed without any lactate secretion at these CSPR values.

Figure 4 presents data from the perfusion culture of a myeloma cell line. The cell concentration was maintained around 10-15 million cells/ml and the CSPR was altered to obtain different glucose levels. Decrease in glucose levels resulted in an initial increase in lactate levels. However, after a CSPR of 0.06 nl/cell/day, the lactate also decreased. At a CSPR of 0.3, the viability decreased in the culture. Addition of 1.2 g/L lactate to the feed resulted in the recovery of viability and established a residual lactate of 0.8 g/L, again indicating the utilization of lactate.

Figure 5 presents the lactate production rates and lactate yields on glucose for both cell lines. In both cases, the lactate production was directly proportional to glucose consumption with a yield coefficient of 0.75 g/g. However, there was observed to be a minimal glucose flux not related to lactate formation. The data in Figure 5 can be fitted to the following equation:

$$q_{lac} = q_{Glu} - m = Y_{app} \cdot q_{Glu} = Y \cdot q_{Glu} - m$$

Note that although the apparent yield of lactate on glucose (Y_{app}) changes, the actual yield (Y) remains to be constant. The above analysis also suggests that when the glucose flux was maintained at a certain value (m=0.12 ng/cell/day), then glucose is completely metabolized with no lactate formation. Thus manipulating the glucose flux, and maintaining it at critical conditions, one can minimize or eliminate lactate formation.

5. Conclusions

The following conclusions can be made from the present study:

1. In batch culture, lactate can be utilized by myeloma and CHO cells used in this study. The utilization of lactate occurs when glucose levels drop to zero, and when the medium is enriched by amino acids so that the cells maintain active metabolism.

2. In perfusion cultures the glucose utilization and lactate production rates slow down at low perfusion rates. The decrease in metabolic rates is possibly due to nutrient limitations (in our case, glucose), and low growth rates.

3. Lactate production is correlated to glucose consumption rate with a constant yield coefficient. However, there appears to be a certain glucose flux not utilized in lactate metabolism.

4. The lactate formation and apparent lactate yield from glucose become zero at a certain glucose utilization rate. This critical glucose consumption rate is about 0.12 nmol/cell/day, and seems to be independent of cell line used in this study.

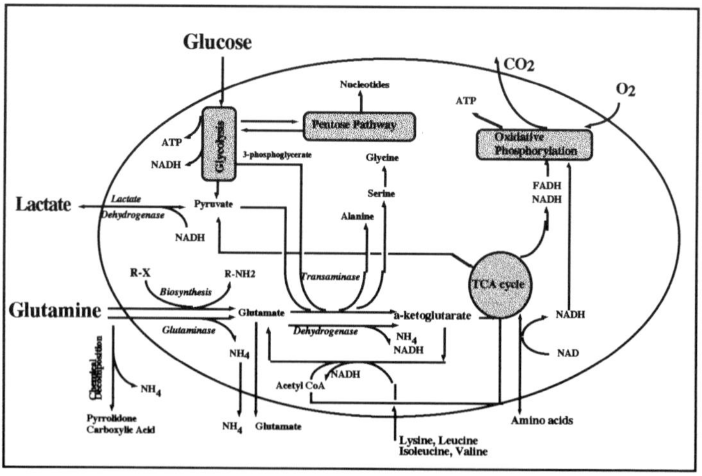

Figure 1. Simplified mammalian cell metabolism.

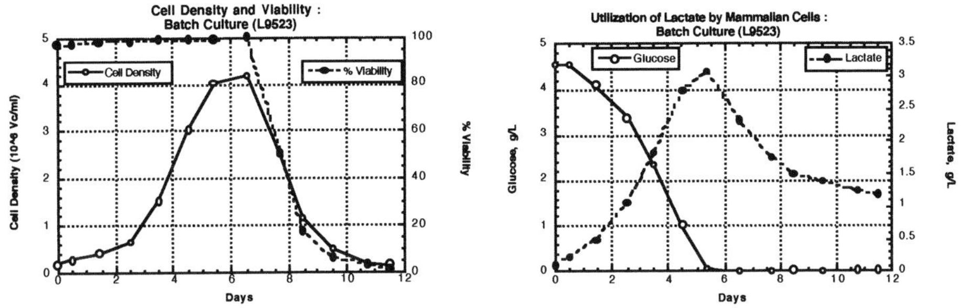

Figure 2. Mammalian cell growth and glucose/lactate metabolism in batch culture of CHO cell: Utilization of lactate after glucose depletion

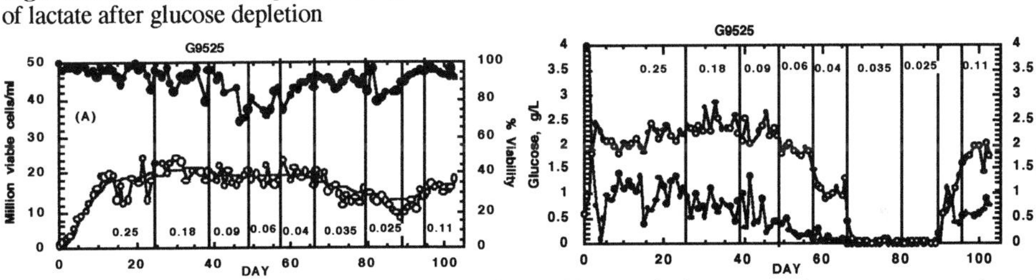

Figure 3. Time profiles of cell concentration, % viability, and glucose and lactate concentrations for perfusion culture of CHO cells: Utilization of Lactate

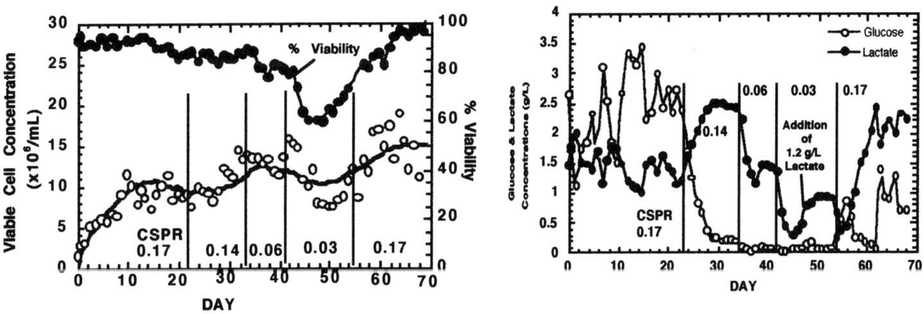

Figure 4. Time profiles of cell concentration, % viability, glucose and lactate concentrations for perfusion culture of myeloma cells: Utilization of Lactate

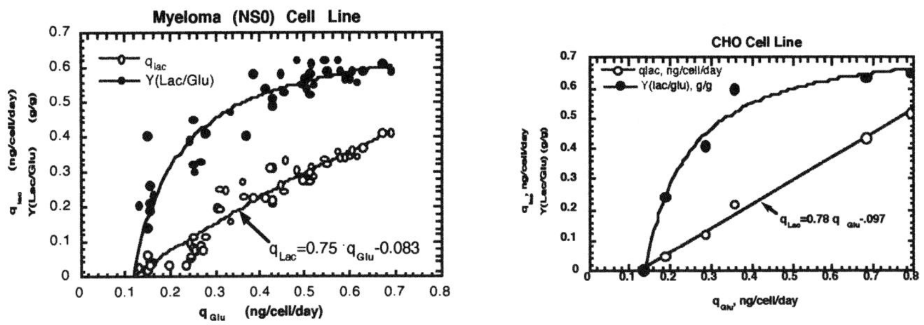

Figure 5. Specific lactate production rate and apparent yield of lactate from glucose as a function of specific glucose consumption rate for a myeloma and a CHO cell line.

6. References

1. Miller, W. M.;,Blanch, H. W., Wilke, C. R. (1988a) Kinetic analysis of hybridoma growth and metabolism in batch and continuous culture: effect of nutrient concentrations, dilution rate, and pH, Biotech. Bioeng., **32**: 947-965.

2. Hu, W.-S.; Dodge, T. C.; Frame, K. K.; Nimes, V. B. (1987) Effect of glucose on the cultivation of mammalian cells. Dev. Biol. Stand. **66**: 279-290.

3. Lavering, P., van Heijst, J., Sunnen, C. (1992) Physiology of myeloma cell grown in glucose-limited chemostat cultures, Cytotechnology, **9**: 125-130

4. Kurokawa, H., Park, Y. S. , Iijima, S. , Kobayashi, T. (1994). Growth characteristics in fed-batch culture of hybridoma cells with control of glucose and glutamine concentrations. Biotech. Bioeng., **44**: 95-103.

5. Racher, A. J., Looby, D., and Griffiths, J.B. (1993). Influence of ammonium ion and glucose on mAb production in suspension and fixed bed hybridoma cultures. J. of Biotechnology, **29**: 145-156.

6. Miller, W. M.; Blanch, H. W.; Wilke, C. R. (1988b) Transient responses of hybridoma cells to lactate and ammonia pulse and step changes in continuous culture, Bioprocess Engineering, **3**: 113-122.

7. Ozturk, S. S., Riley, M. R., Palsson, B. O. (1992). Effects of ammonia and lactate on hybridoma growth, metabolism, and antibody production. Biotech. Bioeng., **39**, 418-431.

8. Zielke, H. R., Sumbilla, C. M., Sevdalian, D. A., Hawkins, R. L., Ozand, P. T. (1980) Lactate: a major product of glutamine metabolism by human diploid fibroblasts, J. Cell Physiol. **104**: 433-441.

9. Ozturk, S.S. (1996) Engineering challenges in high density bioreactors, Cytotechnology, to appear

10. Eigenbrodt, E.; Fister, P.; Reinacher, M. (1985) New perspectives on carbohydrate metabolism in tumor cells, in regulation of carbohydrate metabolism, Vol. II., R. Beitner,, ed. CRC Press: Boca Raton, FL, pp.141-179.

11. Glacken, M. W.; Fleischaker, R. J.; Sinskey, A. J. (1988) Reduction of waste product excretion via nutrient control: possible strategies for maximizing product and cell yields on serum in cultures of mammalian cells, Biotech. Bioeng., **28**: 1376-1389.

12. Ozturk, S. S., Blackie, J., Thrift, J., and Naveh, D. (1995) Perfusion rate control based on on-line monitoring of oxygen consumption rates in mammalian cell fermentation, ACS National Meeting, San Diego, CA

13. Ozturk, S.S., Palsson, B.O. (1991) Growth, metabolic, and antibody production kinetics of hybridoma cell culture: 1. Analysis of data from controlled batch reactors." Biotechnol. Prog. 7: 471-480.

14. Ozturk, S. S., Blackie, J. D., Wu, P., Figueroa, C., Thrift, J. C., Naveh, D. (1995). Investigation of high density cell culture kinetics using on-line cell density and metabolic rate measurements. in Animal Cell Culture Technology: Developments towards the 21th Century, Ed. E.C. Beuvery et al, Kluwer Academic Publishers, Netherlands, pp. 301-305

Acknowledgment

Experimental help from Gary Johnson, Susanti Martin, Sara Stokes and Leonard McDaniels are greatly appreciated. Andy Johnson implemented the process control algorithm on Macintosh computers using LabView programming language.

"NUTRIENT-SPLIT" FEEDING STRATEGY FOR DIALYSIS CULTURES OF IMMOBILIZED HYBRIDOMA AND TRANSFECTOMA CELLS

Ines Lüdemann, Ralf Pörtner, Reno Neumann and Herbert Märkl

Technische Universität Hamburg-Harburg, Bioprozeß- und Bioverfahrenstechnik, Denickestraße 15, 21071 Hamburg, Germany

Abstract

A novel "nutrient-split" feeding strategy for cultivation of hybridoma and transfectoma cells was applied to a fixed bed reactor coupled with a dialysis membrane. Concentrated medium was fed directly to the fixed bed unit, whereas a buffer solution was used as dialysate. With both cell lines a steady state product concentration (monoclonal antibodies and Fab fragments, respectively) appr. 10 times higher compared to batch cultivation in T-flask was reached. A comparision with results of Bohmann et al. (1995), who employed the same experimental set-up and same hybridoma cell line, but used standard medium at both sides of the dialysis membrane, shows a much higher efficiency of the "nutrient-split" feeding strategy in respect to the extent of nutrient utilization.

Introduction

Bohmann et al. (1995) investigated a fixed bed system coupled with a dialysis membrane. Low molecular nutrients were supplied and low molecular toxic metabolites removed over the membrane simultaniously, monoclonal antibodies (MAb) were accumulated. They obtained MAb concentrations up to 383 mg l^{-1} during continuous cultivation, ten times more compared to chemostat cultures with suspended cells. But the efficiency of medium utilization was low, because a high concentration difference of nutrients over the membrane had to be maintained for sufficient supply.

To solve this problem a novel "nutrient-split" feeding strategy was applied. Concentrated medium was fed directly to the fixed bed unit, whereas a buffer solution was used as dialysing fluid.

Material and Methods

Cell lines

A murine hybridoma cell line IV F 19.23 producing monoclonal antibodies against penicillin-G-amidase was used to compare the "nutrient-split" feeding strategy with results of Bohmann et al. (1995). A further experiment was carried out with the human/mouse transfectoma Fab#10, which secrets monovalent chimeric Fab-antibody fragments specific for the human carcinoembryonic antigen (CEA).

M.J.T. Carrondo et al. (eds.), Animal Cell Technology, 361–366.

"Nutrient-Split" medium

The standard medium IMDM/Ham's F12 was split into a nutrient concentrate (10 x) and a buffer solution containing all inorganic components (Fig. 1). In respect to the fermenter geometry (see fig. 2) the buffer was concentrated 1.24 x to obtain 1x salt concentration in the whole fermenter. To prevent precipitation of the medium concentrate (purchased by BioConcept, Umkirch, Germany) the pH had been adjusted to rather basic conditions, which enhances glutamine decomposition. Therefore the glutamine was fed as an extra solution. Both nutrient concentrate and buffer solution were supplemented with an iron-rich, protein-free (IR-) supplement (Franěk & Dolnìková 1991).

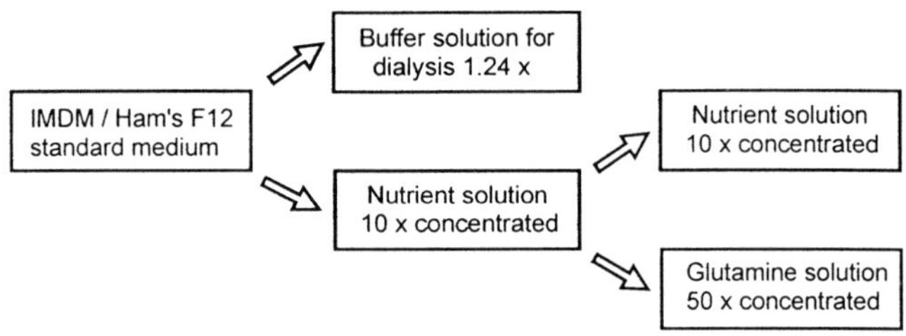

Fig. 1: "Nutrient-split" medium

Experimental set up

A membrane dialysis reactor (MDR) consisting of an inner fermentation chamber with integrated radial-flow fixed bed and an outer dialysis chamber, separated by a dialysis membrane was used. In table 1 the experimental set-ups of Bohmann et al. (1995) and of the "nutrient-split" feeding strategy are compared.

Tab. 1: Comparision of the experimental set-ups of Bohmann et al. (1995) and the "nutrient-split" feeding strategy.

Experimental set-up	MDR Bohmann et al. (1995)	MDR "nutrient-split"
Inner chamber		
Medium	Standard medium 1 x + 3 % FCS	Nutrient solution 10 x + IR-supplement
Dilution rate [d^{-1}]	0.2	0.12
Outer chamber		
Medium	Standard medium 1 x without FCS	Dialysis buffer 1.24 x + IR-supplement
Dilution rate [d^{-1}]	0.6	0.15

Applying the "nutrient-split" feeding strategy, the nutrient concentrate was fed directly to the inner chamber, while the outer chamber was perfused with buffer solution (Fig. 2).

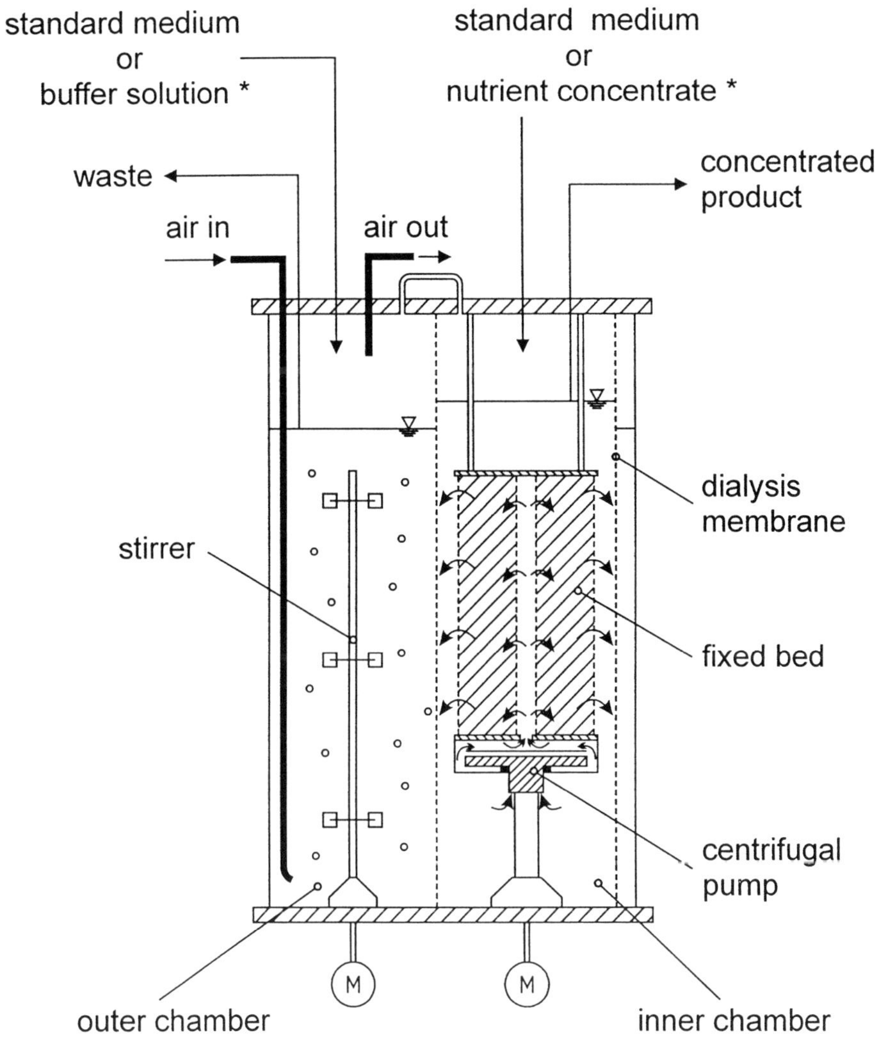

Fig.2: Membrane dialysis reactor (MDR) with integrated radial-flow fixed bed.

* : Nutrient-Split feeding strategy.

Dialysis membrane : Cuprophan (Akzo, Wuppertal, Germany);

Carrier: SIRAN® (Schott, Mainz;Germany).

Results

Figure 3 shows the time course of an experiment with the "nutrient-split" feeding strategy using the hybridoma cell line IV F 19.23.

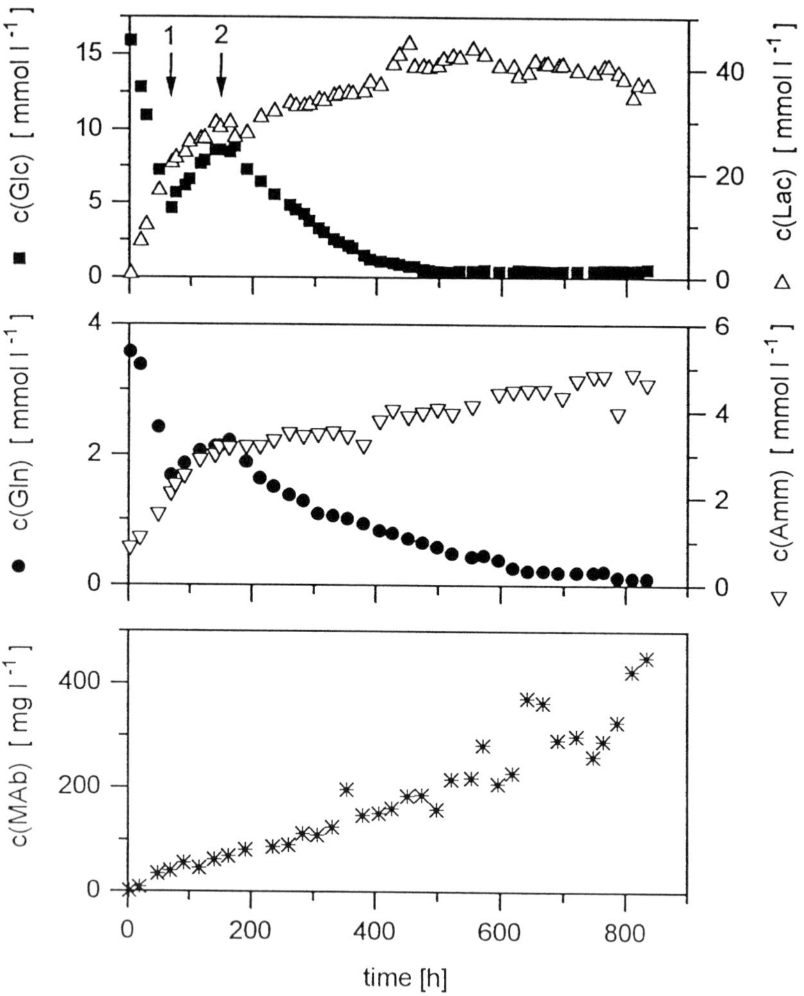

Fig 2: Time course of an experiment employing the "nutrient-split" feeding strategy in the membrane dialysis reactor (MDR) with integrated radial flow fixed bed. The shown values were obtained from the inner chamber.

1: start of perfusion of the inner chamber with medium concentrate ($D = 0.12\ d^{-1}$).

2: start of perfusion of the outer chamber with dialysis buffer ($D = 0.15\ d^{-1}$).

The experiment was started with both chambers containing complete medium. After 70 h of batch fermentation continuous perfusion of the inner chamber with nutrient concentrate was started ($D = 0.12\ d^{-1}$). To keep the ammonia concentration below toxic level perfusion of the outer chamber with dialysis buffer was started at t = 148 h with a dilution rate of $0.15\ d^{-1}$. Steady state levels for glucose and glutamine concentration were reached after 450 h of continuous fermentation (t = 600 h). Until the experiment was finished after 850 h cultivation time appr. 400 mg l^{-1} MAb were accumulated. This coincides rather well with the steady state level calculated with a volume specific MAb production rate of 4 mg $l_{FB}\ h^{-1}$.

In an experiment with the transfectoma cell line Fab#10 the same experimental set-up as decribed above (Tab.1) was used. The culture could be maintained for atleast 860 h. A steady state Fab fragment concentration of 6 mg l^{-1}, appr. nine times higher compared to batch cultivation in T25 flasks, was reached.

Discussion

The comparision of specific process parameters of the "nutrient-split" experiment with data of Bohmann et al. (1995) (Tab. 2) shows clearly the high efficiency of the new feeding strategy.

Tab. 2: Comparision of specific process parameters of a chemostat experiment in a stirred tank reactor, data of Bohmann et al. (1995) and the "nutrient-split" experiment.
i: inner chamber   o: outer chamber

	Fermentation system		
Criterion	Stirred suspension (chemostat)	MDR Bohmann et al. (1995)	MDR "nutrient-split"
Glucose concentration [mmol l^{-1}]	4	c(i) = 4.25 c(a) = 7.0	c(i) = 0.45 c(a) = 0.42
Glucose utilization [%]	76	57	98
Glutamine concentration [mmol l^{-1}]	0.5	c(i) = 2.4 c(a) = 3.2	c(i) = 0.21 c(a) = 0.20
Glutamine utilization [%]	86	26	98
MAb concentration [mg l^{-1}]	35	383	400
Medium demand [l $g(MAb)^{-1}$]	28.6	35.4	15.5 (2.5 nutr. conc.13 buffer)

Because of a lower dilution rate for the inner chamber higher antibody concentrations can accumulate at similar volume specific MAb production rates. The extent of medium utilization is much higher using the "nutrient-split" feeding strategy. The high efficiency of the "nutrient-split" feeding strategy can also be expressed in terms of medium demand per g(MAb): in the experiment shown in figure 3 at steady state 15.5 l medium per g(MAb) were required which is half of the the medium demand of a stirred chemostat culture; Bohmann et al. (1995) needed 35.4 l medium per g(MAb).

The "nutrient-split" feeding strategy is suitable to increase the potential of fixed bed reactors coupled with dialysis membranes and can be easily applied to the cultivation of other cell lines.

References

Bohmann A, Pörtner R and Märkl H (1995) Performance of a membrane dialysis bioreactor with a radial-flow fixed bed for the cultivation of a hybridoma cell line. Appl. Microbiol. Biotechnol. 43: 772-780.

Franěk F & Dolnìková J (1991) Hybridoma growth and monoclonal antibody production in an iron-rich protein-free medium: Effect of nutrient concentration. Cytotechnology 7: 33-38.

Acknowledgement

The hybridoma cell line was kindly donated by Prof. fil. dr. Volker Kasche, TU Hamburg-Harburg, Germany, the SIRAN® carrier by Schott, Mainz, Germany.

IMMOBILIZATION CAN IMPROVE THE CULTURE STABILITY OF NON-ANCHORAGE-DEPENDENT ANIMAL CELLS IN SERUM-FREE MEDIA

Ines Lüdemann*, Ralf Pörtner*, Claudia Schaefer*, Katrin Schick*, Karolina Šrámková**, Kerstin Reher***, František Franěk**, Michael Neumaier***, Herbert Märkl*

* Technische Universität Hamburg-Harburg, Bioprozess- und Bioverfahrenstechnik, Denickestr. 15, D-21071 Hamburg.

** Institute of Molecular Genetics, Czech Academy of Sciences, Vídenská 1083/270, CS-14220 Prague.

*** Universitätskrankenhaus Eppendorf, Abteilung für Klinische Chemie, Martinistr. 52, D-20246 Hamburg.

Abstract

A murine hybridoma cell line and a human/mouse transfectoma cell line were cultivated in three different media (serum-containing; low protein, serum-free; iron-rich, protein-free) in flask cultures, stirred reactors and a fixed bed reactor. In the fixed bed reactor the cells were immobilized within macroporous carriers. Stirred suspension cultures of the hybridoma cells were only successfull in serum-containing medium. Using the serum-free media a lot of problems arose which are mainly due to the level of shear stress exposed to the cells. The transfectoma cell line did not grow in stirred suspension in any of the three media. For fixed bed cultivation of both cell lines, the serum-free media, which were optimized in flask cultures, could be used successfully as well as serum-containing medium.

Introduction

Because of the disadvantages of using serum as medium supplement for animal cell culture (high costs, difficulties in product purification, contamination risks) serum-free media have been developed. But several problems arise in serum-free cultivation. The inoculation density has to be significantly higher, the cells are more sensitive to shear stress and the antibody productivity might drop or get lost completely.

In the following we will demonstrate that immobilization can improve the stability and reproducibility of cultures with non-adherent animal cells in serum-free medium.

Material and Methods

Cell lines

The mouse/mouse hybridoma cell line IV F19.23 producing monoclonal antibodies against penicillin-G-amidase /1/ was kindly donated by Prof. fil. dr. V. Kasche, TU Hamburg-Harburg, Germany.

M.J.T. Carrondo et al. (eds.), Animal Cell Technology, 367–371.

The cell line FAB#10 is a chimeric human/mouse transfectoma producing monovalent Fab-fragments against the human carcinoembryonic antigen (CEA) /2/.

Media

As basal medium a 1:1 mixture of IMDM and Ham's F12 supplemented with 2 mmol l^{-1} glutamine, 2 g l^{-1} $NaHCO_3$, 50 µmol l^{-1} ethanolamine and 0.001 % (w/v) PEG was used. To complete the medium the following supplements were added:

Tab. 1: Supplements to the basic medium to receive either serum containing (HS), low protein, serum-free (SFM3) or iron-rich, protein-free (IR) medium.

HS - serum containing -	SFM3 - low protein, serum-free -	IR - protein-free, iron-rich -
	1 mg l^{-1} transferrin	
3 % horse serum	60 nmol l^{-1} Na-selenite	200 µmol l^{-1} ferric citrate
	50 µmol l^{-1} ethanolamine	100 µmol l^{-1} ethanolamine
	(0.1 % pluronic F 68 in the 2 l suspension fermenter)	trace elements as described in /3/

In static batch cultures in flasks both cell lines showed similar good growth in all three media.

Stirred reactors

a) A 500 ml glass reactor (300 ml working volume) was agitated with a flat-blade impeller operating at 35 rpm. An estimation of the level of shear stress exposed to the cells by calculating the mean volumetric power input leads to 0.0013 W l^{-1}.

b) A 2 l reactor (working volume 1 l, Bioengineering, Wald, Switzerland) was equipped with two buffles and two rushton turbines, operated at a rotation speed of 110 rpm. The meanvolumetric power input was estimated to be 0.011 W l^{-1}.

Fixed bed reactor

An axial-flow fixed bed was filled with macroporous Siran®-carriers (Schott, Mainz, Germany) with a diameter of 3-5 mm to immobilize the cells.The volume of the fixed bed was 176 ml with a length of 90 mm. The fixed bed was connected with a conditioning vessel (working volume 1 l) for aeration and medium exchange.

Results

The results are summarized in table 2. In serum-containing medium the hybridoma cell line could be cultivated satisfactory in both stirred reactors. In serum-free media continuous cultivation was only successfull in the 500 ml reactor. In the 2l fermenter, which has a higher volumetric power input, Pluronic F68 had to be added to the SFM3 medium to protect the cells against shear stress. But even with this supplement only batch, but not chemostat mode was possible. Experiments with the IR-medium were not reproducible and at least most of them were not successfull. The transfectoma cell line did not grow in stirred reactors in any of the three media.

Tab. 2: Comparison of experimental results cultivating the hybridoma and transfectoma cell line in different cultivation systems and media.

Cell line	Hybridoma IV F 19.23			Transfectoma FAB#10		
medium	HS	SFM3	IR	HS	SFM3	IR
cultivation system						
T25 flasks	++	++	++	++	++	++
500 ml glass reactor	++	++	++	- -	- -	n.d.
2 l reactor	++	- -	(+) -	n.d.	n.d.	n.d.
fixed bed reactor	++	++	++	n.d.	n.d.	++

Media: HS - serum containing; SFM3 - low protein, serum-free; IR - protein-free, iron-rich.

++ : cultivation satisfactory; (+) - : cultivation possible, but not reproducible
- - : cultivation not possible; n.d.: not done

Good results were obtained in fixed bed experiments. The serum-free media could be used for fixed bed cultivation of both cell lines as well as serum containing medium. The volume specific glucose uptake rate as an indicator for the activity of the immobilized cells was similar in all three media.

As an example the time course of a fixed bed experiment with the transfectoma cell line FAB#10 is shown in Fig. 1. The fixed bed was inoculated with $2.7*10^6$ cells ml_{FB}^{-1}. After 70 h of batch fermentation half of the medium volume in the conditioning vessel was exchanged with fresh medium. At 90 h fermentation time continuous perfusion of the conditioning vessel with a dilution rate of 0.2 d^{-1} was started. 5 days later (t = 210 h) the glucose concentration reached a steady steady state value of 3 mmol l^{-1}. The volume specific glucose uptake rate during steady state was appr. 1 mmol l_{FB}^{-1} h^{-1}. The FAB#10 concentration reached appr. 550 µg l^{-1}. The culture could be maintained for atleast 400 h.

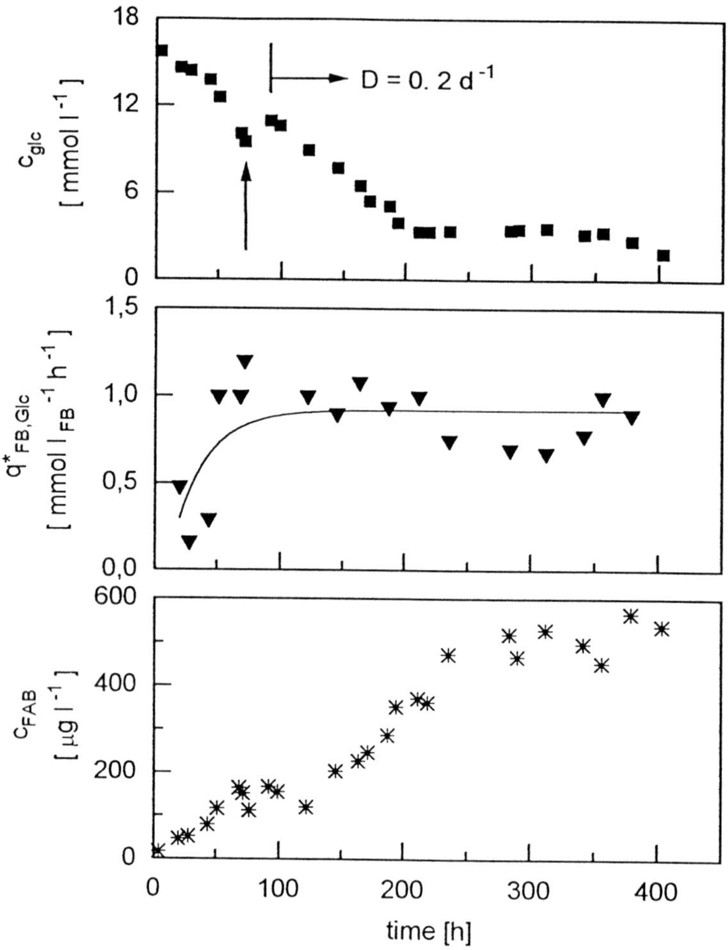

Fig.1: Fixed bed culture of the transfectoma cell line Fab#10 using iron-rich, protein-free medium.

Glucose concentration c_{Glc}, volume specific glucose uptake rate $q^*_{FB,Glc}$ and concentration of Fab#10-antibody fragments c_{FAB} vs. culture time.

Continuous exchange of medium in the conditioning vessel was started at t = 90 h.

↑ : exchange of medium in the conditioning vessel

Conclusions

Serum-free media, which were optimized in flask cultures, can be used without any further adaptation in the fixed bed culture of non-anchorage-dependent animal cells. For immobilization of the cells within the fixed bed macroporous carrier are suitable. In stirred suspension cultures a lot of problems may arise which are mainly due to the level of shear stress exposed to the cells.

References

/1/ Niebuhr-Redder CH & Kasche V (1990) Antibodies to *Escherichia coli* Penicillin-G-Amidase for Biochemical Studies and the Control of Fermentation Processes. In: Kataoka H & Märkl H, Proc. German-Japanese Workshop on Animal Cell Culture Technology, Hamburg, Germany, pp 57-60.

/2/ Lüdemann I et al. (1996) Improvement of the culture stability of non-anchorage-dependent animal cells grown in serum-free media through immobilization. Cytotechnology (in press).

/3/ Franěk F & Dolnìková J (1991) Hybridoma growth and monoclonal antibody production in an iron-rich protein-free medium: Effect of nutrient concentration. Cytotechnology 7: 33-38.

INFLUENCE OF PLURONIC F-68 ON THE ULTRAFILTRATION OF CELL CULTURE SUPERNATANTS

C. SCHULZ, J. H. VOGEL[1], K. SCHARFENBERG
GBF-Gesellschaft für Biotechnologische Forschung mbH
Cell Culture Techniques Department
[1]Bioseparation Techniques Department
Mascheroder Weg 1
38124 Braunschweig
Germany

1. Abstract

In order to characterize the influence of the cell culture medium supplement Pluronic F-68 on the ultrafiltration of CHO cell culture supernatants two preselected ultrafiltration membranes were tested and compared in a stirring cell system. It was demonstrated that Pluronic F-68 could be nearly completely removed by ultrafiltration of cell culture harvest. However, Pluronic caused an additional membrane fouling and thus an increase of the process time. Moreover, the initial product retention was decreased in the presence of Pluronic F-68 which makes initial recycling of the filtrate necessary until a steady state is reached.

2. Introduction

Ultrafiltration (UF) is one of the most efficient processes recommended as a first step in downstream procedures for the recovery of proteins from mammalian cell cultivation. Cell culture supernatants consist of a broad spectrum of compounds, which influence the performance of UF. These include supplements of the culture medium (e.g. Pluronic F-68, silicone oil) or compounds that are secreted from the cells or released after cell lysis (e.g. proteins, lipids). The nonionic block copolymer Pluronic F-68 is often supplemented to cell culture media in order to protect mammalian cells from shear stress caused by sparging [1,

M.J.T. Carrondo et al. (eds.), Animal Cell Technology, 373–378.

2]. However, it further complicates downstream processing of pharmaceutical proteins.

The overall performance of ultrafiltration processes strongly depends on solute-membrane interactions, since these different substances can adsorb onto the external or internal membrane surface and influence permeability and retention.
Additionally, information about the fouling tendency as well as the cleaning property of the membranes is important for an optimal design of the UF process [3].
In this study, two preselected membrane types of 100 kD cut-off were tested and compared under defined values of UF parameters comparable to crossflow-conditions, and the influence of the non-nutritional medium supplement Pluronic F-68 was characterized.

3. Materials and Methods

Experimental set-up:

Fig. 1 shows schematically the experimental set-up. The ultrafiltration was performed in a stirring cell with optional filtrate recycling (minimal working volume of the system: 450 ml; volume of the stirring cell: 240 ml; diameter of the stirrer: 63 mm; stirring rate: 900 rpm; transmembrane pressure: 0,9 bar).
During concentration mode and diafiltration mode the filtrate was not recycled.

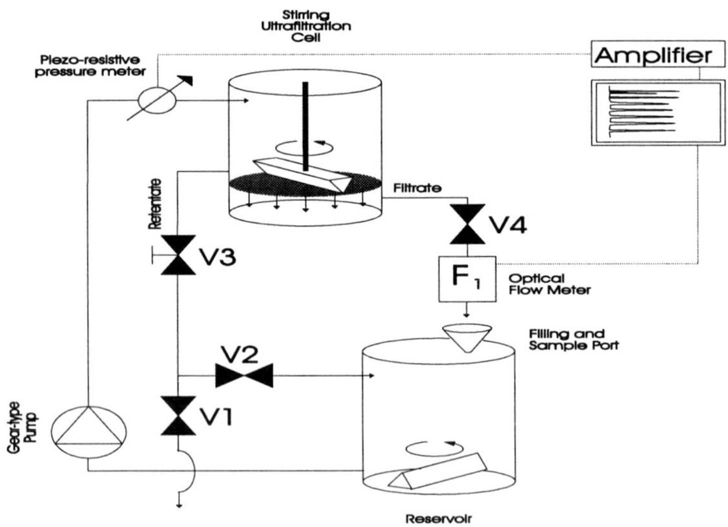

Fig. 1: Stirring cell

Membranes:

Two types of round ultrafiltration membranes (diameter: 76 mm) were compared.

membrane type	catalogue number	NMWCO	material	manufacturer
OMEGA	OS100	100 kD	hydrophilized polyethersulfone	Filtron
YM	YM100	100 kD	regenerated cellulose	Amicon

Filtrated solutions:

SMIF6 cell culture medium with and without spiking

The protein-free cell culture medium SMIF6 [4] was ultrafiltrated in order to examine the influence of its components on the ultrafiltration. In spiking experiments Pluronic F-68 was added before ultrafiltration (0.05 % w/w and 0.3 % w/w, respectively).

CHO cell culture harvest with and without spiking

Microfiltrated culture supernatant from CHO cells cultivated in SMIF6 medium (aliquots of a pool, stored frozen) was ultrafiltrated. In spiking experiments Pluronic F-68 was added to concentrations of 0.05 % and 0.3 %, respectively.

Product:

The product used as model for the UF studies was recombinant gp220/350 Epstein-Barr-Virus surface antigen. Due to the removal of the anchor sequence by gene engineering and RNA-processing within the cell, the highly glycosylated product is secreted in two protein forms with a molecular weight of 220000 and 350000 Dalton [4].

Fouling indices:

The consequence of membrane fouling can be shown by the fouling index I_F [3].

$$I_F = \frac{water\ flux\ of\ used\ membrane\ after\ filtration}{water\ flux\ of\ clean\ membrane}$$

The influence of membrane fouling and concentration polarization can be demonstrated by the index I_{K+F} [3].

$$I_{K+F} = \frac{steady\ state\ filtrate\ flux}{water\ flux\ of\ clean\ membrane}$$

I_F and I_{K+F} values of 1 would characterize ideal filtration performance (no fouling, no concentration polarization).

4. Results and Discussion

SMIF6 cell culture medium with and without spiking

Fig. 2a presents the permeate flux during the ultrafiltration of Pluronic F-68 in SMIF6 at a concentration of 0.3 %. As can be seen, Pluronic F-68 caused a fast decrease in filtrate flux comparable to protein-containing solution (see Fig. 3a) and increased the process time drastically. The flux decrease correlates with membrane fouling (see Fig. 2b). As can be seen, the OMEGA 100 showed significantly higher membrane fouling due to adsorption of Pluronic F-68 than the YM 100.

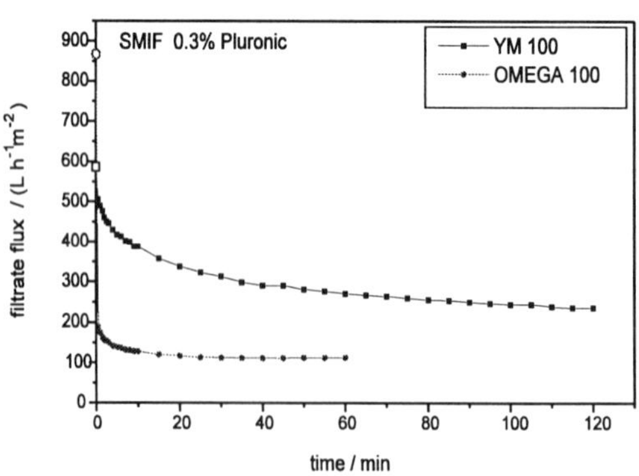

Fig. 2a

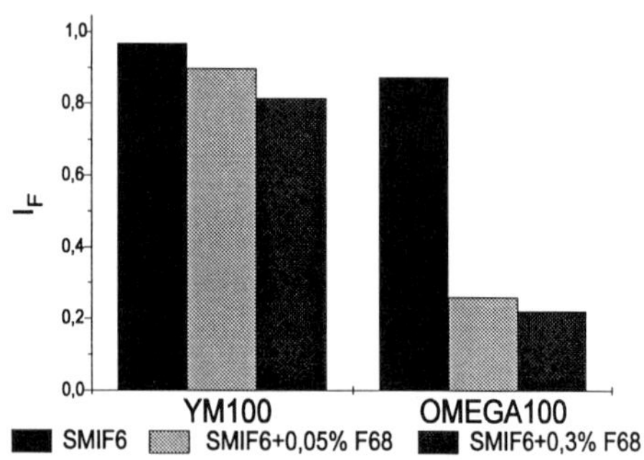

Fig. 2b

Fig. 2: Ultrafiltration of Pluronic F-68 in SMIF6

CHO cell culture harvest with and without spiking

Fig. 3a demonstrates the permeate flux during the ultrafiltration of CHO harvest with and without spiking of Pluronic F-68. In all cases steady state values were reached only after some minutes of filtration. Harvest containing Pluronic F-68 showed an additional decrease of filtrate flux. As can be seen from Fig. 3b, concentration polarization and membrane fouling were more significant in case of the OMEGA 100 membrane (see also Fig. 2b).

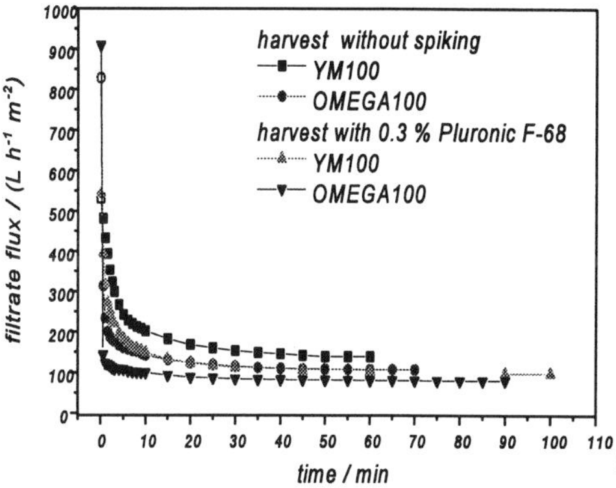

Fig. 3a

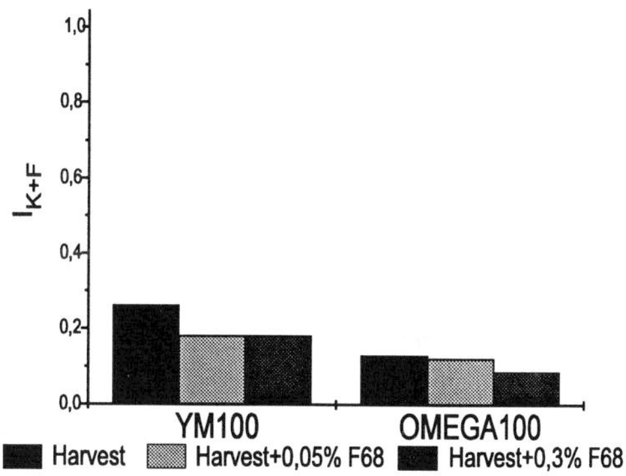

Fig. 3b

Fig. 3: Ultrafiltration of cell culture harvest with and without spiking of Pluronic F-68

Influence of Pluronic F-68 on product retention

From Fig. 4 it can be seen that product retention was delayed in the presence of Pluronic F-68 compared to the ultrafiltration of harvest without Pluronic. In conclusion, the filtrate has to be recycled until the steady state is reached in order to achieve high product recovery.

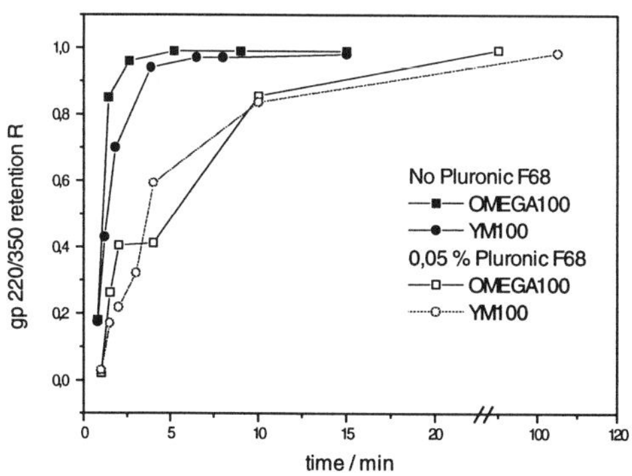

Fig. 4: Product retention with/without Pluronic F-68

Removal of Pluronic F-68 by diafiltration

Fig. 5 shows an ultrafiltration experiment with 0.30 % w/w Pluronic F-68 in MilliQ water. At the time markers concentration 1, 2 and 3 the emptied reservoir was filled with 0.3 % Pluronic F-68 solution and at the time markers diafiltration 1 and 2 with MilliQ water up to the double minimal working volume (900 ml). During volume reduction (concentration mode), no retention of Pluronic F-68 was found. Therefore the Pluronic F-68 concentration was significantly decreased during subsequent diafiltration, demonstrating that this additive can in principle be completely removed. Nevertheless, the strong adsorption of Pluronic F-68 to the membrane hindered the achievement of an increased filtrate flux during reduction of the Pluronic F-68 concentration.

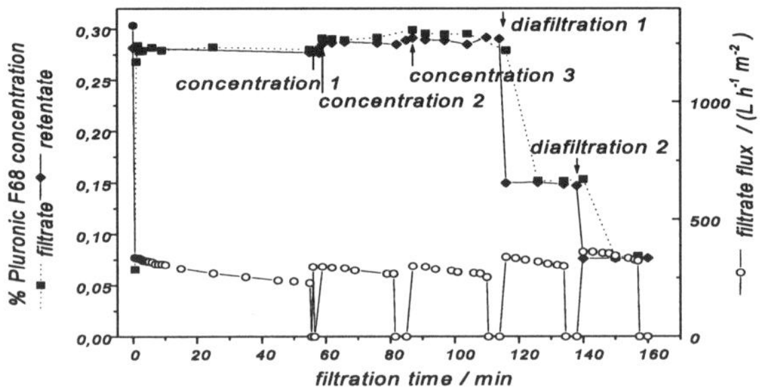

Fig. 5: Concentration and diafiltration of Pluronic F-68

5. References

[1] Murhammer, D.W.; Goochee, C.F.: (1990) Structural Features of Nonionic Polyglycol Polymer Molecules Responsible for the Protective Effect in Sparged Animal Cell Bioreactors, Biotechnology Progress; 142-148

[2] Jordan, M.; Sucker, H.; Einsele, A.; Widmer, F.; Eppenberger, H.M.: Interactions Between Animal Cells and Gas Bubbles: The Influence of Serum and Pluronic F-68 on the Physical Properties of the Bubble Surface (1994), Biotechnology and Bioengineering, Vol. 43

[3] Vogel, J.H.: Charakterisierung und Auslegung der Ultrafiltration im Verfahren zur Produktion von rekombinantem Epstein-Barr-Virus-Antigen (1994), Diploma Thesis, Technische Universität Braunschweig, Germany

[4] Wolf, H.; Scharfenberg, K.; Wagner, R.: Verfahren zur Herstellung von Aktivimpfstoff gegen Epstein-Barr-Virus-Infektionen (1993), Deutsches Patentamt P 43 13 620.6; 26.04.1993

Dynamic Membrane Filtration in Mammalian Cell Culture Harvest

Kempken, R.*; Rechtsteiner, H.*; Schäfer, J.*; Katz, U.*; Dick, O.**; Weidemeier, R.***; Sellick, I.****

* Dr. Karl Thomae GmbH, Biberach, Germany
** PharmaPlan GmbH, Bad Homberg, Germany
*** Pall GmbH, Dreieich, Germany
****Pall Europe Ltd, Havant Street, Portsmouth, PO1 3PD, United Kingdom (to whom reprint requests should be addressed)

Introduction

Many valuable biopharmaceuticals, such as those for parenteral applications, are produced by the cultivation of recombinant mammalian cells. Batch or fed batch fermentations, as well as continuous processes, are applied. After fermentation, cell harvesting is commonly done by cross-flow filtration (tangential flow filtration) or dead-end (flow-through) filtration. In most cases, the latter is too impractical and expensive for production scale applications. Therefore, continuous tangential flow filtration using hollow fibre modules or cassette modules is widely used.

In all cases, the primary recovery of the product (i.e. the removal of cells) is an important and demanding process. On one hand, product losses due to moderate concentration, long processing times, or incomplete transition of the product across the membrane are of especial importance. On the other hand, unwanted by-products, such as DNA, proteins and lipids may be released which can disturb the subsequent downstream process, or can affect (as proteases may do) the integrity, or the effectiveness of the product.

Thomae and Pall investigated the suitability of the Dynamic Membrane Filter (**DMF**) system for this separation. The main interest was focused on the performance of the system (as demonstrated by the filtrate flow rate), and on the occurrence of unwanted cell disruption.

Materials and Methods

The cell suspension used originated from 80 litre fermentations of a CHO cell line, expressing a recombinant glycoprotein of 57kD MW. The GLP fermentation process was reproducible but not optimised for product generation. Concentration of the product (a cytokine) in the pre-harvest suspension was *circa* 5 - 7 mg/l. Approximately 16 litres of

M.J.T. Carrondo et al. (eds.), Animal Cell Technology, 379–384.

Figure 1. **Cross sectional arrangement of Pall DMF unit.**

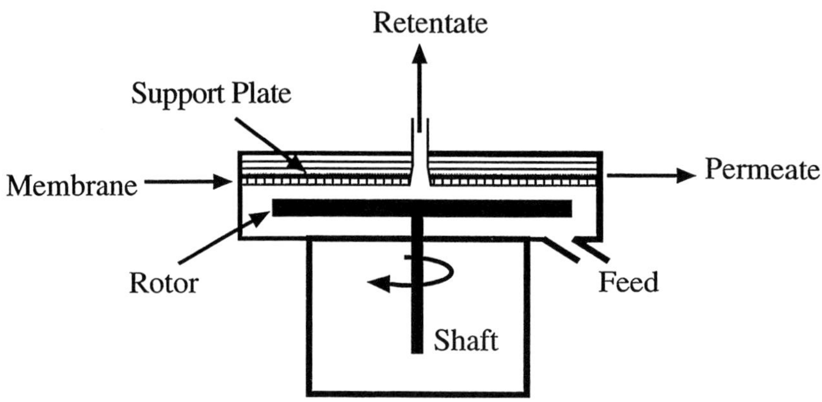

cell suspension with *circa* 14 million cells per ml were used for each separation experiment. A laboratory scale **Pall DMF** separation system was used, equipped with a single membrane of 0.0137m² and a single rotating disc of 150 mm in diameter (see Figures 1, 2).

Figure 2. **Pall DMF unit installation arrangement.**

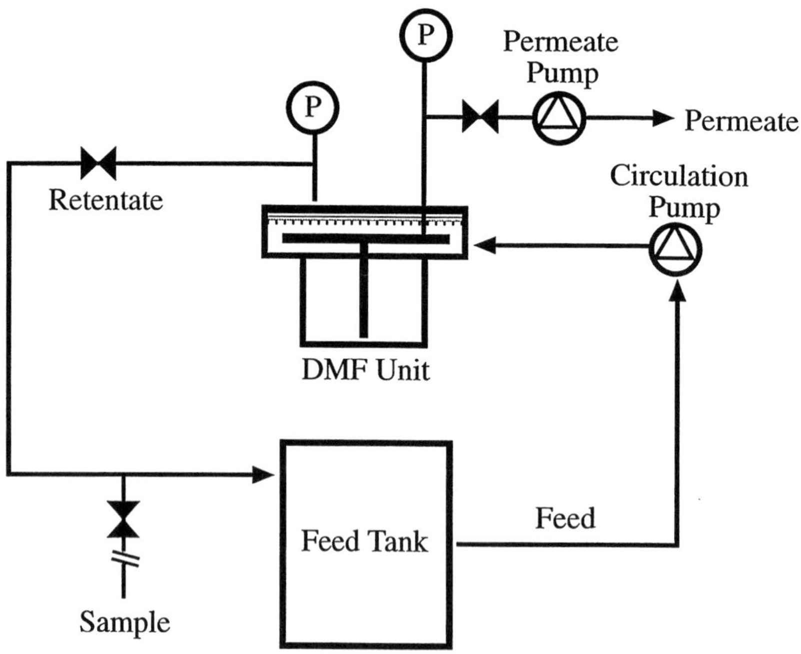

Different **Pall** membranes were used, as well as various circumferential rotor speeds, and various transmembrane pressure differences.

Either fresh cultures with high viability, or aged cultures with a large portion of lysed cells were utilised in the experiments. Accurate monitoring of the process and extensive analytical work was performed to record the concentration of the cells and of the product versus time during the separation.

Results

The process conditions used in the **DMF** system are reported in Table 1. In both experiments, the suspension was first concentrated (mode: CONC). Then a diafiltration (mode: DIA) was done by adding PBS to the remaining concentrate.

Table 1

Experiment	Phase	Membrane	Mode	Circumferential rotor speed (m/s)	Trans-membrane pressure (mbar)
1	I	**FRY** - PTFE 0.2μm	CONC	15	700
	II	**NTG** - N_{66} 0.1μm	CONC	15	700
	III	**NRG** - N_{66} 0.2μm	CONC	15	500
	IV	**NRG** - N_{66} 0.2μm	DIA	15	500
	V	**NRG** - N_{66} 0.2μm	CONC	15	800
2	I	**NRG** - N_{66} 0.2μm	CONC	10	500
	II	**NRG** - N_{66} 0.2μm	CONC	15	700
	III	**NRG** - N_{66} 0.2μm	DIA	15	700
	IV	**NRG** - N_{66} 0.2μm	CONC	15	700

The results of the experiments are shown in Figure 3 and Figure 4. A hydrophobic PTFE membrane of 0.2μm removal rating was used in Phase I of Figure 3. The initial filtrate flow rate of *circa* 200 l/(m²h) went down to 160 l/(m²h) at 20 minutes. The separation was ceased and an NTG membrane of 0.1μm removal rating was installed.

An initial flow rate of 230 l/(m²h) and a stabilisation at *circa* 80 l/(m²h) can be seen from Phase II of Figure 3. After 140 minutes separation time the process was interrupted again. An NRG membrane of 0.2μm removal rating was installed and used during the period shown in Phases III to V of Figure 3.When starting again with the 0.2μm N_{66} membrane, the cells had been concentrated by the factor 1.3, resulting in a total cell density of 20.8×10^6 cells/ml. The initial filtrate flow rate was 470 l/(m²h). After a sharp decrease the flux was stabilised at *circa* 230 l/(m²h). Due to the increasing cell density in the retentate, a further decrease in flux was observed during the extended concentration period of Phase III of Figure 3. At the end of Phase III (i.e. after a total separation time of 360 minutes), the cell suspension had been concentrated from 16 litres to 0.9 litres with *circa* 170x10^6 cells/ml. The filtrate flow rate was still at 120 l/(m^{2}h).

Figure 3. (See text for details.)

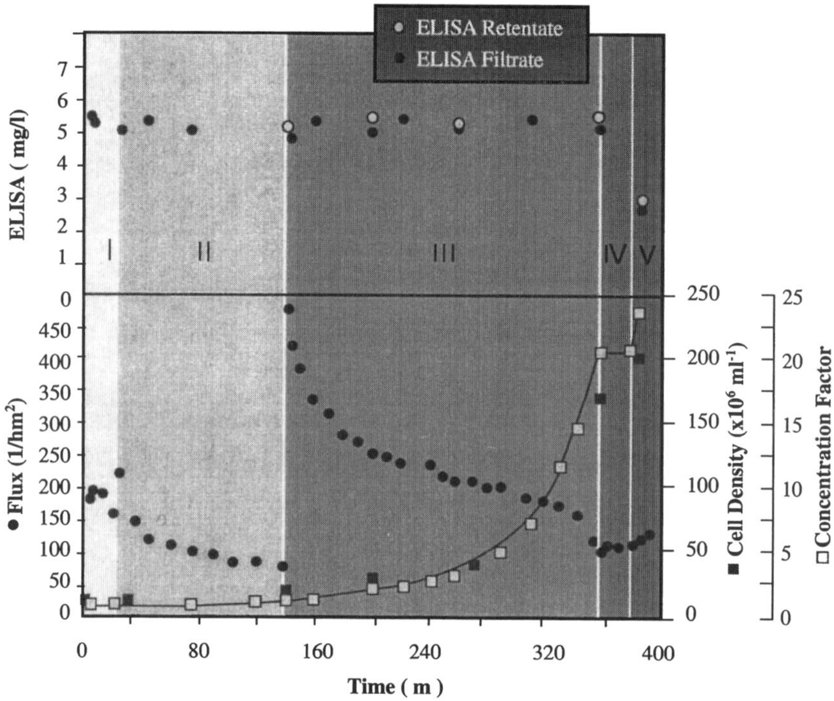

In Phase IV of Figure 3, a diafiltration was done with PBS and followed by a final concentration of the cell suspension in Phase V to the void volume of the **DMF** unit. At the end of the experiment, a concentration factor of 23.6 and a total cell density of 202.5x10^6 cells/ml were obtained. The filtrate flow rate was 120 l/(m^{2}h) at the end of the separation.

During the entire experiment, the same product concentrations were found for filtrate and retentate, i.e. the product had passed through the membrane without any loss. The product yield of the separation was 97%.

In Figure 4, separation of an analogous cell suspension was performed using an NTG 0.2µm membrane during Phases I to IV. The initial flux of *circa* 390 l/(m²h) was slightly lower than that obtained in Phase III of Figure 3, probably due to the lower circumferential rotor speed (see Table 1). After two hours process time, the filtrate flow rate stabilised at *circa* 130 l/(m²h).

At 200 minutes process time, rotor speed and transmembrane pressure were increased, which resulted in a higher flux. The latter decreased again due to higher concentration of the cells during Phase II of Figure 4. At 340 minutes separation time, the flux was *circa* 140 l/(m²h) at a total cell density of 100x10⁶ cells/ml and a viability of 12%. Since the initial cell viability was 37%, disruption of the cells is assumed to be an inevitable consequence of high cell concentrations (see Figure 4).

Figure 4. (See text for details.)

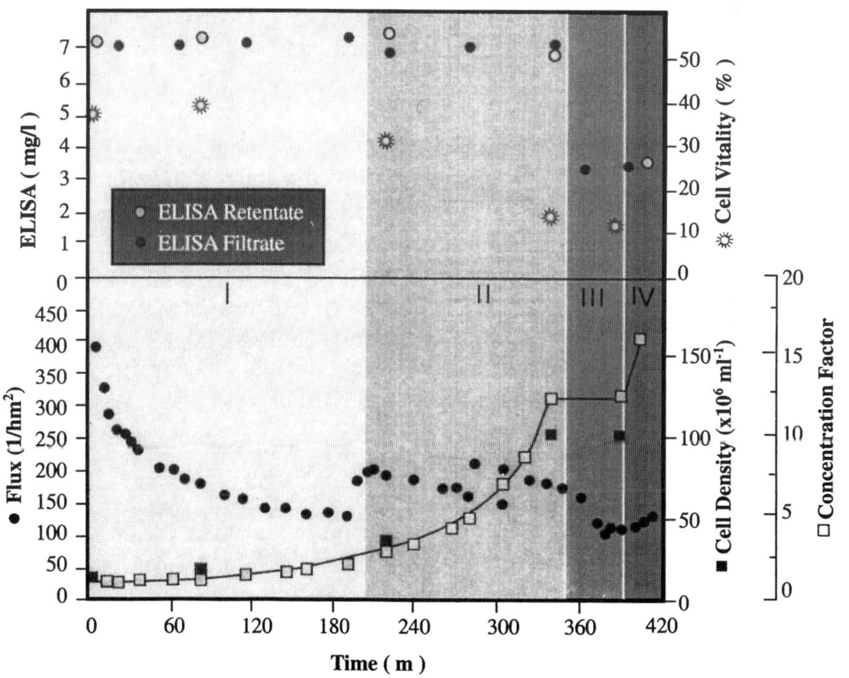

In Phase III of Figure 4, the remaining cell concentrate was diafiltered with PBS. Filtrate flow rate, total cell density and viability of the culture was not affected during this process period.

Finally, the cell suspension was concentrated during Phase IV of Figure 4 to the void volume of the **DMF** unit (*circa* 700 ml). At the end of the experiment, a concentration factor of 17.6 had been obtained, with a flux of more than 100 $l/(m^2h)$ at the end of the separation.

With respect to product concentration, the results of the first experiment were reproduced. The transition of product across the membrane was excellent. The product yield of the separation was 98%.

Discussion

The above experiments demonstrated the capability and effectiveness of the **DMF** principle to separate the product containing culture fluid from mammalian cells. Though the applicability of the method and the parameters to other cell lines has to be evaluated, the results reported here are very promising.

The **DMF** principle allows an effective decoupling of liquid velocity from system pressure. Thus, the **DMF** unit may be operated at low transmembrane pressure combined with high wall shear rates, leading to high filtrate flow rates, high product yields and minimum gel-layer formation at the membrane surface.

The best type of membrane material was nylon N_{66}. In both experiments, the filtrate flow rate was above 200 $l/(m^2h)$ on average, and was stabilised above 120 $l/(m^2h)$. The maximum final cell concentration was more than 20 times the initial cell concentration, but was affected by the high void volume compared with membrane area of the **DMF** laboratory unit. Much higher concentrations should be able to be obtained in production scale units where void volume to membrane area ratios are optimised.

Almost no loss of product was found during filtration. The **DMF** principle and system seem to offer a powerful alternative to conventional cross-flow microfiltration systems.

CULTURE OF HUMAN INTESTINAL CELLS IN SIMULATED MICROGRAVITY.

Latigo JR, Battle TP, Hall G, Cook R, Freemantle P & Stacey GN.
CAMR, Porton-Down Salisbury SP4 OJG, Wiltshire UK.

Abstract:

We have used the rotating wall vessel (RWV) technique to culture standard cell lines as cell aggregates under conditions of very low fluid shear stress in aerobic and microaerophilic conditions (90% N_2, 5% O_2, and 5% CO_2). The human intestinal adenocarcinoma cells lines Caco-2 and HT-29 in conventional two dimensional culture have been used for a wide range of research purposes. However, HT-29 cells when cultured in the RWV (Synthecon), formed large aggregates of up to 1 cm in their longest dimension. Haematoxylin and eosin stained sections of HT-29 aggregates showed large structures of cells with a small central zone of cell death compared to spheroids of HT-29 cells produced under standard two-dimensional conditions. Histochemical studies of the RWV aggregates (HT-29) showed populations of acid and basic mucin-secreting cells with basic mucin secretions predominant which encapsulated the aggregates. Caco-2 cells as expected did not show mucin secretion, but did show large vacuoles within RWV aggregates of about 30 cells which formed after 9 days in culture. Our work with the RWV system has demonstrated the value of this low shear stress system in enabling novel cell aggregate configurations with useful applications in the fields of pathogenesis of infection and tissue engineering.
Key words: Simulated microgravity, Rotating wall vessel, Caco-2,HT-29, Intestinal adenocarcinoma cells.

Introduction:

Epithelial cell lines derived from the human intestine have been developed for the study of normal and pathophysiological events *in vivo* (1). Two human colon carcinoma cell lines Caco-2 and HT-29 have proven to be particularly useful models since they undergo enterocyte differentiation *in vitro* (2). They have been used widely for studies of the barrier properties of human intestine for drug transport (3), and screening cytoprotective activity of potential drugs and their pharmaceutical formulations (4).
A new technology developed by NASA at the Johnson space centre in the USA has produced a device known as the Rotating-Wall Vessel (RWV). This device is a horizontally rotated, transparent vessel which has no head space (i.e. there is no direct contact between atmosphere and medium) and the gas liquid interface is instead mediated through a silicone membrane. The rotation of the vessel mimics microgravity, reducing shear forces which are normally associated with impeller-driven stirred bioreactors to a minimum. This device has been shown to encourage cell to cell contacts (notably between

M.J.T. Carrondo et al. (eds.), Animal Cell Technology, 385–389.

cells of different density), good mass transfer characteristics and three dimensional growth (5).
We have used this device to grow intestinal cell lines and to investigate their performance in different atmospheres *in vitro*.

Materials and Methods:

Two-dimensional cultures: Caco-2 cells were cultured in T-Flasks in MEM + non-essential amino acid (Gibco), and HT-29 in DMEM (Gibco). Both media were supplemented with 10% FCS and 1% Glutamine (2mM), in an atmosphere of 5% CO_2 in air (HT-29) or 90% N_2, 5% CO_2, and 5% O_2 (Caco-2) with saturated humidity.
Rotating Wall Vessel (Synthecon): Cells cultured in T-flasks were trypsinized to obtain a cell suspension of $2x10^5$ cell/ml for seeding after having equilibrated the medium inside the vessel overnight according to the manufacturer's recommendations. Solid-body rotation was accomplished as per manufacturer's instructions and incubated in 5% CO_2 and 95% air (HT-29) or microaerophilic condition (Caco-2). The vessel was observed each day post seeding and the rotation speed was adjusted (to maintain synchronous rotation of cells) as cell aggregate particles increased in size.
Sampling Procedure: Each day after cell seeding, samples were taken, centrifuged, and analysed for glucose and lactate concentrations (YSI 2300 Stat Analyser) and aggregates were removed after different times and fixed in 1% agar with 4% formalin for preparation as histological sections.
Medium Renewal: Medium was renewed according to levels of glucose and lactate (i.e. maintained at >1mM glucose and ≤ 20mM lactate). Great care was taken in handling the RWV cell aggregates which were allowed to settle at the bottom of the vessel before any manipulation.

Results:

Monolayer Culture: In the T-flask monolayer culture (seeded at $1x10^5$ cells/ml), Caco-2 cells were observed to proliferate after a lag period of about 2 days and confluency was reached by day 5 (Fig. 1a). After 5 days of culture, confluency of Caco-2 cells was accompanied by formation of domes in the monolayer which increased in size with time in culture. Glucose concentration decreased while lactate concentration increased with time in culture. Glucose metabolism progressed steadily at a rate of about 0.84mmol/L/day to a resulting level of 0.6mmol/L at day 5 and lactate production was also steady at a rate of about 1.80mmol/L/day with a final concentration of 8.9mmol/L at 4 days. In the T-flask monolayer culture of HT-29 cells there was a lag period of about 2 days and confluency was reached by day 4. No dome formation was observed in HT-29 monolayer after confluency was reached, and they had the appearance of undifferentiated epithelial cells (Fig. 1c). Glucose consumption was on average 2.63mmol/L/day and lactate production of 4.37mmol/L/day without media changing, glucose level fell to approximately 6mmol/L at 6 days whilst lactate production rose to about 29mmol/L in the same time.

<u>3-D Culture in simulated microgravity:</u> Caco-2 cell aggregates formed at day 2 of cell seeding when the aggregate size was roughly 1mm in diameter (Fig. 1b). The size and number of cell aggregates increased each day and cell aggregates increased in size by about 1mm a day up to day 10. The average glucose consumption per day by Caco-2 cells in 6 days of culture before media change was 0.76mmol/L/day (falling to 0.6mmol/L at 5 days) and average lactate production was 1.56mmol/L/day peak value 11.5mmol/L at 5 days. HT-29 cell aggregates formed at day 1 of cell seeding (Fig. 1d) and grew up to 1cm in their longest dimension. Large structures of cells with small central zone of cell death were observed in paraffin sections of the aggregates stained with heamatoxylin and eosin (see arrow in Fig. 1d). Populations of acid and basic mucin-secreting cells were identified by PAS and Schiff staining sections which also showed basic mucin secretions encapsulating the aggregates. In glucose containing medium, glucose consumption was 0.68mmol/L/day. Lactate production was of 3.25mmol/L/day.

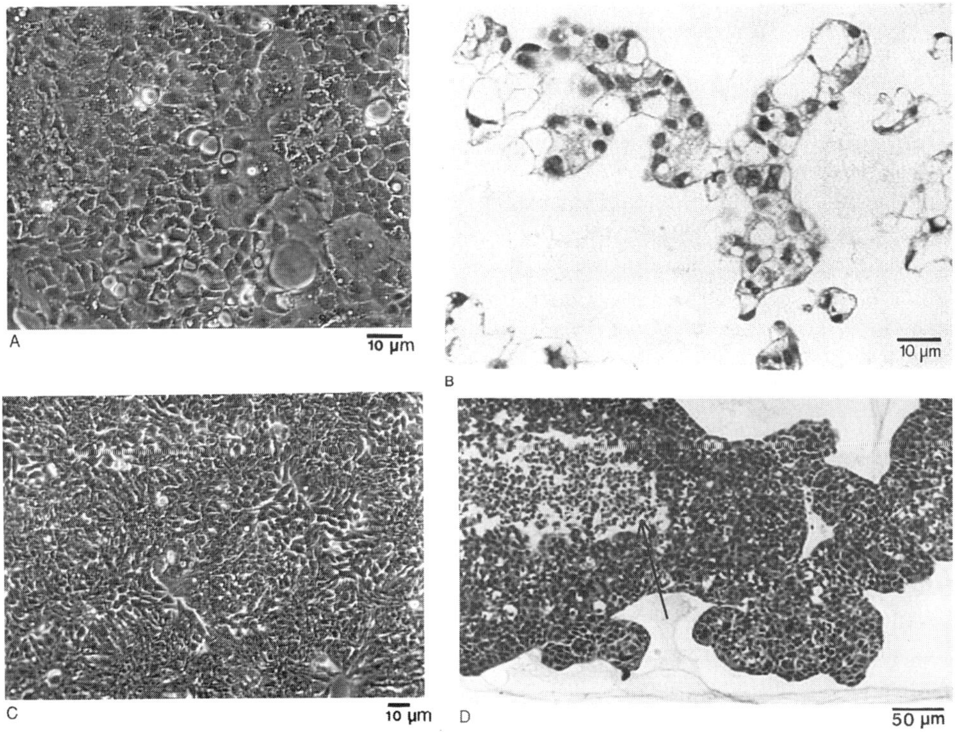

Figure 1. Phase contrast micrographs of intestinal adenocarcinoma cells, Caco-2 (cell monolayer Fig. 1a and cell aggregate Fig. 1b), HT-29 (cell monolayer Fig. 1c and cell aggregate Fig. 1d). Arrow in 1d indicates zone of cell death.

Discussion:

The RWV is a device designed to simulate on earth cell cultivation conditions experienced in microgravity of space. Reduced shear forces and quiescent environment provided by this device allow cells to grow in 3-D form and differentiate to mimic the *in vivo* tissues. Three important requirements are to be secured for normal tissue to develop. (i) Quiescent co-location of dissimilar cell types for cell to cell interactions leading to the exchange of growth factors and bringing about cell to cell linkages. (ii) Mass transfer rates that accommodate molecular scaffolding, supporting mechanical stability and histological changes that will occur as a result. (iii) A microenvironment categorising different types of growth factors needed for development and maturation (5).
The acceleration of the fluid inside the rotating vessel due to viscous coupling (6) induced a rotational fluid flow field, trapping the cells inside the vessel such that the effects of gravity was randomised (7). Inside the RWV, cells continue to proliferate and form larger aggregates. Wolf et al. (9) compared the fluid dynamic of the vessel to that of a microcarrier or large cell aggregate particle having a free fall in a large column of fluid, never reaching the bottom. In the RWV, in orbiting mode, cells are co-located promoting 3-D orientation which enables spontaneous and progressive cell aggregation (8)
The cell lines Caco-2 and HT-29 originally established by Fogh et al.(4) have shown their capabilities of enterocytic differentiation under conventional *in vitro* conditions (i.e. 2-D culture). The Caco-2 intestinal cells in simulated microgravity conditions (i.e. 3-D culture) did not however express the typical brush border morphology, which is characteristic of morphologic differentiation and tight junctions that are associated with polarisation of the cells (2). Histological analysis of HT-29 cell aggregates formed in the RWV with glucose medium has showed no sign of differentiation as found in 2-D culture. When Caco-2 cells were grown inside the RWV, they formed aggregates with large vacuoles inside them which may indicate some kind of cell stress. They also showed evidence of glycolysis: glucose consumption and lactate production. This was observed in both plastic T-flasks and RWV although metabolic activity was slightly reduced in the RWV. However, in the case of HT-29 cells, whilst this expected pattern was seen for cultures in T-flasks, in the RWV, glucose consumption was minimal whilst lactate production showed the 2-D profile. Stable levels of glucose and lactate would indicate low cell proliferation and reduced metabolic activity within. However, the increase in lactate level for HT-29 cells in RWV may be a consequence of altered metabolism in RWV conditions, whereby glutamine is converted to glucose via the TCA cycle intermediates and therefore gluconeogenesis seemed to be the biochemical pathway followed in this case. This data is yet to be confirmed by repeat experiments.
Formation of large structures of HT-29 cells with small central zone of dead cells indicates improved survival of deep tissue structures in the RWV. This highlights the potential of this technology in representing structures more representative of *in vivo* tissue. The RWV provides an environment for cell culture enabling novel cell aggregate

configurations which may be usefully applied in the field of pathogenesis of infection and tissue engineering.

Acknowledgements:
The Authors would like to thank representatives of Cellon Sarl, Luxemboug for excellent back up support for use of the RWV system.

References:

1- Arturson P (1991) Cell cultures as models for drug absorption across the intestinal mucosa. Critical Reviews in Therapeutic Drug Carrier Systems 8 (4):305-330.
2- Pinto M, Appay MD, & Simon-Assmann P (1982) Enterocytic differentiation of cultured human colon cancer cells by replacement of glucose by galactose in the medium. Biol. Cell 44:193-196.
3- Hidalgo JI, Raub TJ, Borschardt RT (1989) Characterisation of human colon carcinoma cell line (CaCo2) as a model system for intestinal epithelial permeability. Gastro. 96:736-749.
4- Fogh J, Wright WC, & Loveless J D (1977) Absence of Hela cell contamination on 169 cell lines derived from human tumours. J. Natl. Cancer Inst. 58:209-214.
5- Spaulding GF, Jessup JM, and Goodwin TJ (1993) Advances in cellular construction. J. Cell. Biochem. 51:249-251.
6- Pinto M, Robine-Leon S, & Appay MD (1982) Enterocyte-like differentiation and polarisation of the human colon carcinoma cell line CaCo2 in culture. Biol. cell 47:323-330.
7- Freed LE, Vunjak-Novakovic G (1995) Cultivation of cell-polymer tissue constructs in simulated microgravity. Biotec. and Bioeng. 46:306-313.
8- Schwarz R P, Goodwin TJ, & Wolf DA (1992) Cell Culture for Three-Dimensional Modelling in Rotating-Wall Vessels. In: An application for simulated microgravity: J. Tiss. Cult. Meth. 14:51-58.
9- Wolf D, Schwarz RP, Trinh T October (1987) Controlled turbulance bioreactor. NASA Technical Briefs :74.

Removal of animal cells and cell debris by continuous two-stage centrifugation

H. Tebbe, F. Gudermann, D. Lütkemeyer and J. Lehmann

Institute of Cell Culture Technology, University of Bielefeld, P.O. Box 10 01 31, 33501 Bielefeld, Germany

Abstract

The development of a continuous two-stage centrifugation process for the removal of hybridoma cells and cell debris from culture broth is reported. The aim was to generate supernatant, which can directly be used for ultrafiltration or chromatography without any filtration steps.

The first centrifugation step for the removal of hybridoma cells from repeated batch fermentation in 100 L scale was performed by an improved disc stack separator. The centrifuge was equipped with a hydrohermetic feed system, which allowed a damage free separation of the cells by moderate centrifugal force.

The clarification for mammalian cells in this step was in the range of 99 to 99.9 %. The amount of subcellular particles in the clarified liquid phase was reduced up to 50 % of the particle content in the feed stream.

In a second centrifugation step the removal of the remaining amount of cell debris and other small particles was carried out. It was performed with centrifugal forces up to 18000 · g in a high speed centrifuge equipped with a continuous flow system.

The monoclonal antibodies in the resulting supernatant were directly concentrated by ultrafiltration and chromatographic methods. In this process no fouling of the UF-membranes or chromatographic membranes was observed.

Costs for membrane replacement can be drastically reduced by employing the two-stage centrifugation process.

Material and Methods

Figure 1 shows the general scheme of the production process. After the fermentation, the supernatant was clarified in two continuous centrifugation steps: lysis-free cell separation by a disc stack separator and removal of cell debris in a high speed

M.J.T. Carrondo et al. (eds.), Animal Cell Technology, 391–397.

centrifuge. The clarified supernatant was directly applied to ultrafiltration or membrane ion exchanger.

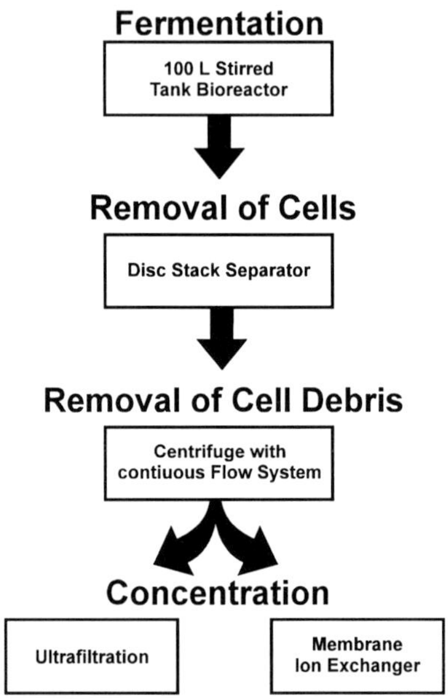

Fig. 1: Schematic diagram of the production process.

Cell line:	Mouse-mouse hybridoma; producing a monoclonal IgG antibody
Medium:	DMEM/F12, 1:1; 10 mg/l Insulin; 10 mg/l Transferrin; 0.5 g/l Albumin [1; 2]
Bioreactor:	100 l stirred tank reactor (B. Braun Biotech)[3]
Cell Separation:	Disc stack Separator (CSA1, Westfalia Separator) [4;5]
Removal of Debris:	Centrifuge equipped with continuous flow system (Contifuge 17RS, Heraeus)
Ultrafiltration:	Crossflow filtration system (SP20, Amicon with module S10Y10, cut-off 10kDa; area: 0.9 m^2)
Ion exchanger:	Membrane ion exchanger (Sartobind, Sartorius) [7]

Particle size distributions were determined using the CASY cell counter and analyser sytem equipped with a 60 μm cappilary (Schärfe System) [6].

Results

The bioreactor was inoculated with 2.8 * 10^5 cells/ml in 35 liters medium (Fig.2). After one day the reactor was filled to the final working volume of 100 liters. The bioreactor was harvested three times at cell densities in the range of 2-3*10^6 cells/ml. Throughout the fermentation more than 90 % of the cells were viable. Consequently there was only a small amount of LDH activity in the supernatant (500-800 nkat/l; maximum LDH activity after destruction of all cells 14000-18000 nkat/l)

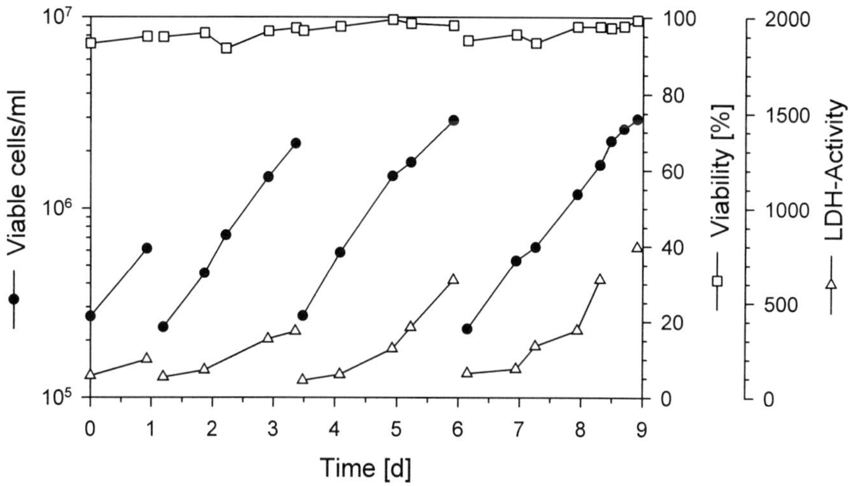

Fig. 2: Time course of the repeated batch fermentation of mouse-mouse hybridoma cells in 100 liter scale.

The particle size distributions presented in figure 3 are revealing the removal of particles from the bioreactor harvest. The total amount of solids in the harvest suspention was in the range of 4.5 to 7.5 µl/ml. After the first centrifugation the cells were completly removed and the amount of subcellular particles was reduced to 50 % of that detected in the harvest suspension. In the second centrifugation at high centrifugal force (18000 · g) there was a further significant reduction of cell debris in the fluid.
For the concentration of the monoclonal antibodies the by two-stage centrifugation clarified cell culture supernatant was directly applied to ultrafiltration or membrane ion exchanger.

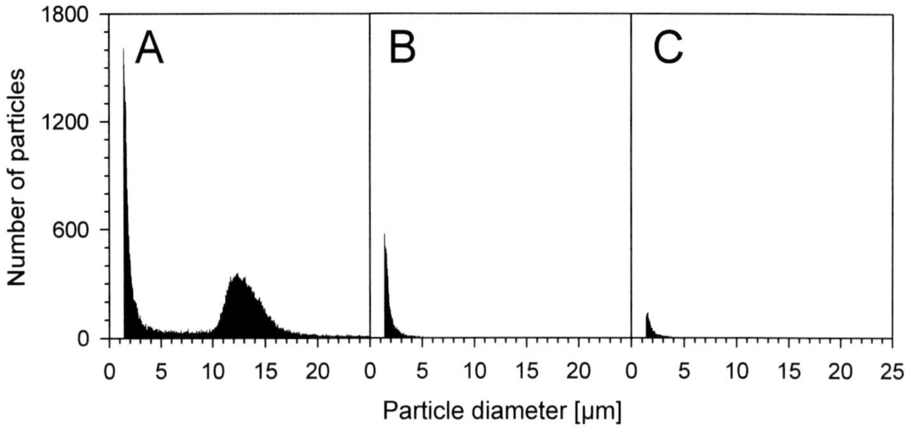

Fig. 3: Particle size distributions **A**: in the bioreactor before harvest; **B**: in the clarified liquid phase after cell separation; **C**: after the second centrifugation at high speed. (Class range: 0.04 µm; Sample volume: 1.6 ml).

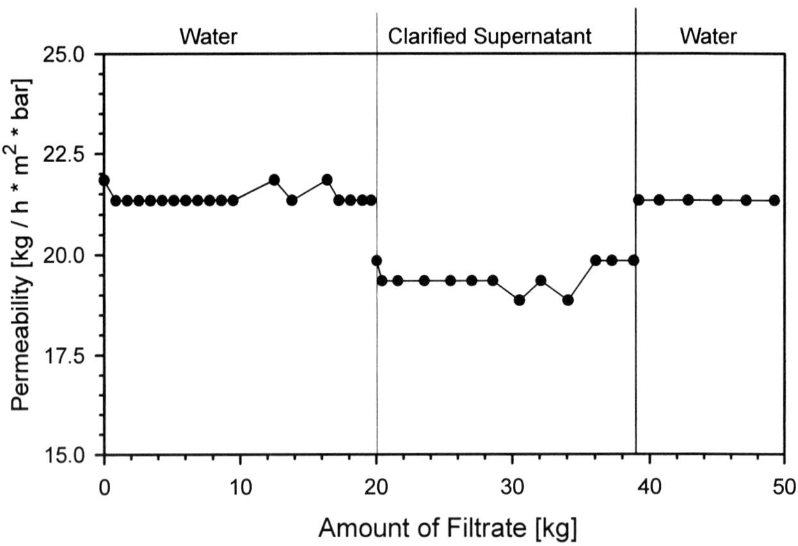

Fig 4: Ultrafiltration of the clarified supernatant
First, the permeability of the membrane was determined for water. Then 21 liters of supernatant, clarified by two-stage centrifugation were concentrated ten-fold. After this procedure the permeability for water was at the same level as before.

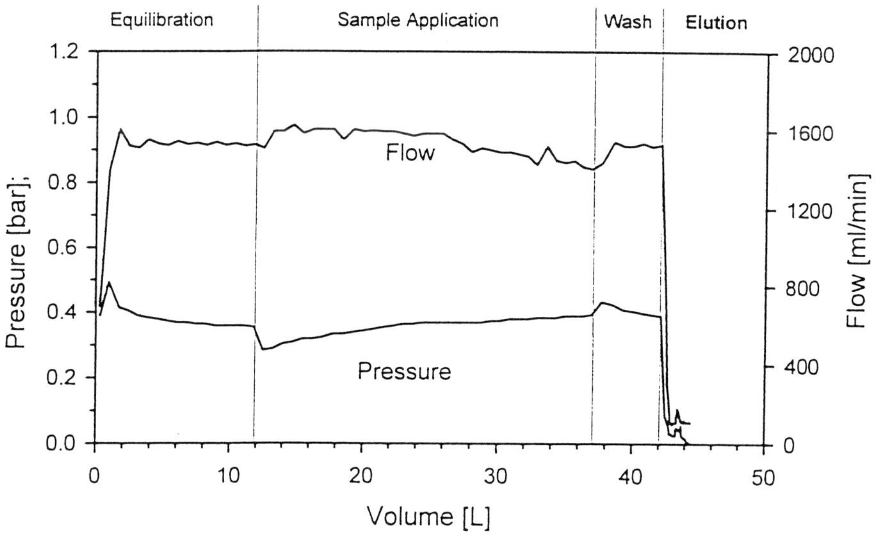

Fig 5: Concentration of MAbs by membrane ion exchange chromatography
During the application of 26 liters of clarified supernatant at constant flow of 1.6 l/min, there was no considerable increase of pressure.

Conclusions

- -- Cells can be efficiently removed from cell culture broth by disc stack separators.
- -- Once the cells are removed, the remaining amount of debris can be separated by centrifuges in continuous mode at high centrifugal forces.
- -- The resulting supernatant can be directly applied to chromatographic or ultrafiltration systems.
- -- Two-stage centrifugation enables a clarification of cell culture broth without the necessity of an additional microfiltration.

Acknowledgement

We would like to thank Westfalia Separator AG and Sartorius AG for supplying the disc stack separator and the membrane ion exchanger system.

References

1. Büntemeyer H, Lütkemeyer D & Lehmann J (1991)
Optimization of serum-free fermentation processes for antibody production. Cytotechnology 5: 57-67

2. Jäger V, Lehmann J & Friedel P (1988)
Serum free growth medium for the cultivation of a wide spectrum of mammalian cells in stirred bioreactors. Cytotechnology 1: 319-329

3. Kuhlmann W (1987) Optimization of a membrane oxygenation system for cell culture in stirred tank reactors. Dev. Biol. Standard. 66: 263-268

4. Kempken R, Preissmann A & Berthold W (1995)
Clarification of animal cell cultures on a large scale by continuous centrifugation. J. Indust. Microbiol. 14: 52-57

5. Tebbe H, Gudermann F, Lütkemeyer D, Heidemann R & Lehmann J (1996)
Lysis-free separation of hybridoma cells by continuous disc stack centrifugation
submitted to Cytotechnology

6. Glauner, B (1991)
Vitalitätskontrolle in der Zellkulturtechnik. BioTec 3(5): 58-61

7. Lütkemeyer D, Ameskamp N, Tebbe H, Bracht K & Lehmann J
Direct capture of monoclonal antibodies using high capacity membrane ion exchangers in pilot scale. Poster presented at the ESACT Meeting 1996 Vilamoura, Portugal

Large Scale Bubble Bed Aeration for Animal Cell Cultures

M. Trocha and Ph. Rudolf von Rohr
Institute for Process Design and Cryogenics, ETH Zurich, CH8092 Zurich

Keywords: *bubble bed reactor, bubble aeration, foaming, cell damage, mass transfer*

1. Introduction

In chemical process technology large gas/liquid-transfer rates have been achieved in large reactors of several hundred m^3 by high fluid exchange devices such as bubble columns. However, the transfer of these *bubble aeration* methods to cell cultures has suffer from a severe loss in cell viability. Detrimental bubble/cell interactions have been revealed as the major cause for these difficulties. Although the lethal interaction of bubbles can be reduced by surface active agents such as fetal calf serum or Pluronic F68 and thus oxygen supply to reactors of several 1000L has been acquainted, the addition of theses supplements is often not desired (contamination, down-stream processing problems). For large scale cell cultures without those protective agents, sufficient aeration still remains a problem. Bubblefree aeration such as surface or membrane aeration is gentle but bears the disadvantage of limited scale-up, on the other hand conventional bubble aeration reactors such as air-lift reactors cause large cell death.

Recently a new approach for gentle oxygenation has been suggested [1]. The adsorption of cells onto the bubble surface depends on the amount of surface active agents on the surface. A freshly introduced bubble caused cell bursting presumably because of surface tension effects, while with increasing surfactant saturation the adsorption of cells declined from irreversible bonding to no adsorption at all. Jordan et alt. [2,3] suggest two different mechanism to be responsible for lethal bubble/cell interactions:

1. Cell bursting at surfactantfree-surfaces due to surface tension effects.
2. Cell damage in the foam region due to floating of attached cells.

This is consistent with results of other researchers who found that the specific death rate in serumfree-media is proportional to the sparging rate [4,5]. Therefore a reactor where the bubbles rise to the surface is largely prevented has been developed by Sucker et alt. [6,7]. Approximately 90% of the sparged oxygen is actually solved by this method. The novel reactor design, the *bubble bed reactor* (BBR), requires only a tenth of the sparging rate of conventional air-lift reactors to sufficiently oxygenate cell cultures. Furthermore, *foaming* problems can be reduced significantly.

M.J.T. Carrondo et al. (eds.), Animal Cell Technology, 399–404.

2. The Principle of the Bubble Bed Reactor

The bubble bed reactor is a modified loop reactor with some interesting new properties (*Fig.1*). The cross-section of the draft-tube increases from a maximum at the outlet to a minimum at the inlet. Forced by the impeller, the fluid flows downwards in the draft-tube (downcomer) and upwards in the outer loop (riser).

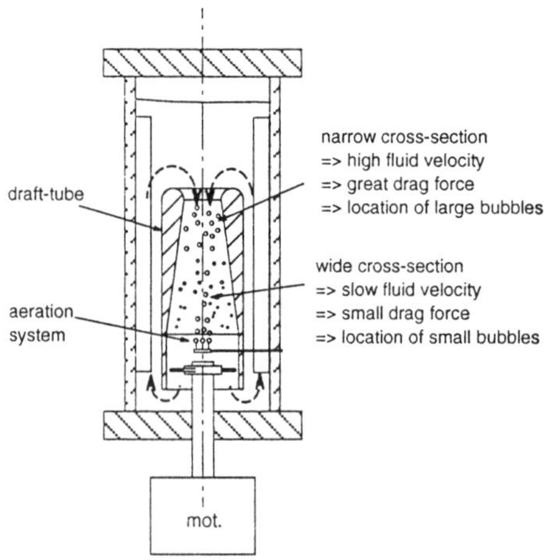

Fig.1 *The principle of the bubble bed reactor(left) and the bubble accumulation in the 2.5L BBR (right).*

The bubbles are sparged into the draft-tube above the impeller by a capillary. Depending on their size, the rising bubbles are balanced by the countercurrent flow at different cross-sections in the draft-tube. By choosing a suitable impeller speed and sparging bubbles, the buoyancy force of initially large bubbles can precisely be counterbalanced by drag forces of the fluid in the narrow region of the loop. *Mass transfer* from the bubbles to the fluid reduces the size and the buoyancy of the bubbles. Thus, the drag force overwhelms the buoyancy force, and forces the bubbles into regions of lower fluid velocities. The specially designed conical draft-tube allows a wide range of bubble sizes to be balanced in the reactor.

The culture is sparged by oxygen Air sparging leads to poor yields since the nitrogen saturation prevents the bubbles from diminishing their size.

3. Results and Discussion

3.1 DESIGN AND SCALE-UP CONSIDERATIONS

A variety of different draft-tube designs and two reactor scales have been studied. Sucker suggested three different modifications of the loop reactor to establish bubble bed aeration [1,6,7]. The conical nozzle design has been chosen for the scale-up since it is easy to realize in existing industrial plants. By mounting the specially designed draft-tube, adapting the impeller and the rate of revolution, and using a suitable aeration, most reactors can readily be modified to a BBR.

Since the bubble bed is limited by the draft-tube volume, the narrow cross-section (*see Fig.1*) needs to exceed to a critical value to achieve a desired mass transfer ($k_L \cdot a$). The angle of the conical nozzle is limited, due to boundary layer instabilities. Even considering this criterion, those instabilities occurs to a lesser extent due to the two-phase flow in the draft-tube, i.e. the rising bubbles at the wall.

Uniform bubble generation leads to the highest yields. If the BBR has been well designed, the maximum oxygen yield can readily be estimated using the following equation (yield = fraction of sparged O_2 which can actually be solved):

$$\text{yield} = 1 - \left[\frac{\text{narrow cross section area}}{\text{wide cross section area}}\right]^3 \qquad (Eq.1)$$

The size of the bubbles determines the required countercurrent fluid flow in the narrow cross-section and thus the power consumption of bubble bed operation. The required impeller power increases for turbulent flows with the third power of the circulating rate:

$$\Delta p = \frac{\rho}{2} \cdot v^2 \cdot \varsigma \quad \wedge \quad P = \Delta p \cdot \dot{V} \quad \Rightarrow \quad P \propto \dot{V}^3 \qquad (Eq.2)$$

Beside of meeting the integral oxygen demand of the culture, the reactor has been analyzed in regard to mixing times. The results suggest that at least up to 250L no critical conditions, especially no oxygen-toxicosis, occur in a well designed BBR.

3.2 ANIMAL CELL FERMENTATION AND CO_2 REMOVAL

Jordan et alt. [3] have cultivated Hybridoma, CHO, and Namalwa cells. They compared vital, death cells, and the antibody production of a 2.5L BBR with a 0.25L (bubblefree-aerated) spinner culture. They found no significant difference between vital cell density and antibody production and only a small increase in death cell density for the BBR in serum- and proteinfree cultivation (*Fig.2*). These results suggest that the bubble related cell death rate can be neglected in the BBR.

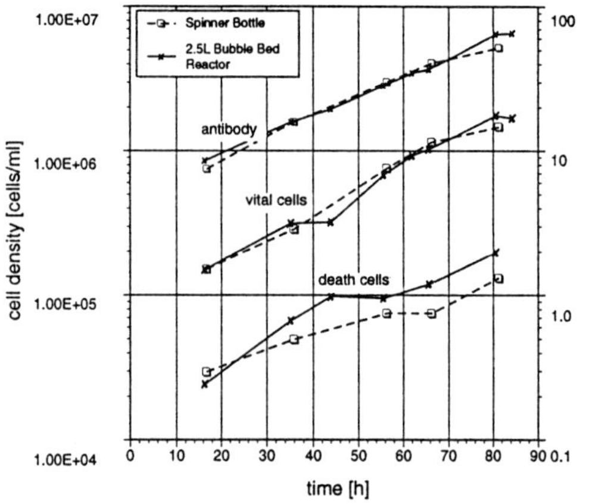

Fig.2 *Surfactantfree cultivation of hybridoma cells in the 2.5L BBR (full symbols) and in a (bubblefree aerated) spinner bottle (Jordan et alt. [3]).*

Due to high yield of the gas bubbles, the dissolved CO_2 cannot be controlled by the aeration only. However, by addition of small amounts of NaOH the partial pressure of CO_2 and the pH can be kept constant. To prevent an unphysiological high osmolarity in the medium, the $NaHCO_3$-buffer has to be reduced in the feeding. The pH and CO_2 is thereby successfully controlled in the 2.5L BBR [2].

Results of the 50L and 250L BBR animal cell cultivation are in progress and results will be published soon.

3.3 MASS TRANSFER AND OXYGEN YIELD

Different modifications and scales (2.5L and 250L) have been intensely studied at our laboratory. *Fig.3* shows results of the oxygen yield for the 250L BBR for different sparging rates. Additionally, the correponding specific energy input of the impeller has been calculated with the Newton correlation, and is depicted in the figure. The bubble bed aeration dissolves up to 95% of the sparged oxygen in the aqueous medium. Too high rates of revolution cause bubble loss through the wide cross-section of the draft-tube, too small rates of revolution lead to bubble escape at the small cross-section. Increasing sparging rates and thus increasing gas hold-up require high energy inputs of the impeller. This can be explained by the increasing fluid energy consumption of the floating bubble bed. However, the calculated specific energy input was in most cases below the critical value of $0.01 W/kg^3$.

While the oxygen yield of the 250L BBR for small sparging rates is even higher than those of the 2.5L BBR, the reactor decreased in efficiency for higher gas loads. The sparger design has been detected as one reason for this reduced efficiency. A modified 250L BBR is believed to transfer even more efficiently the gas into the liquid phase.

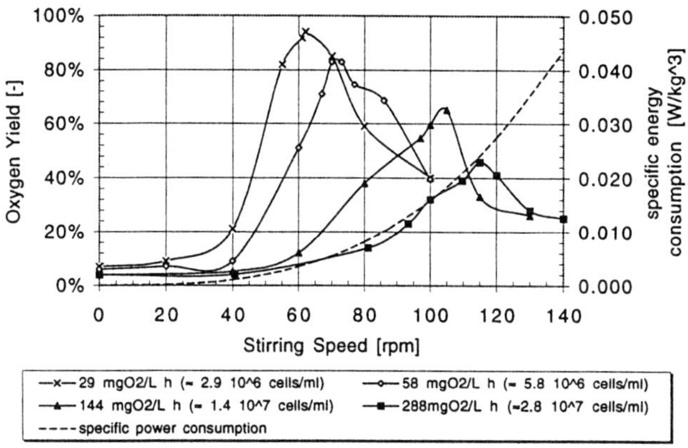

Fig.3 *The oxygen yield for different sparging rates in the 250L bubble bed reactor as a function of the impeller energy input.*

The specific mass transfer of the BBR has the potential to supply high cell densiies up to 10^7 cells/ml sufficiently with oxygen. *Fig.4* shows a comparison between the specific mass transfer of different types and scales of BBRs and air-lift reactors. It is evident, that in comparison to the air-lift reactors, the specific sparging rate of the BBR can be reduced by 1 to 2 orders of magnitude Thus the bubble related death rate should be reduced tremendeously.

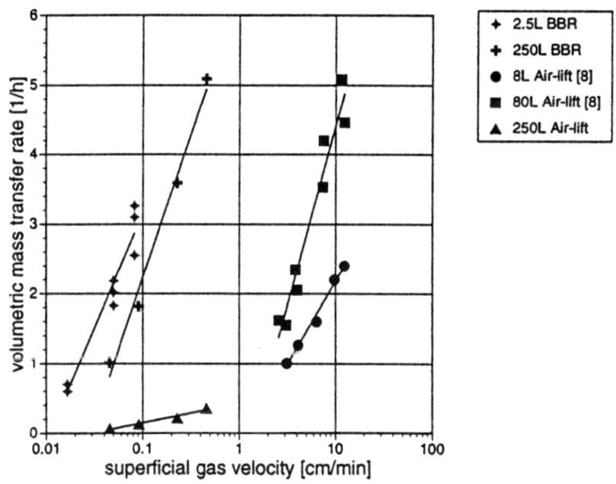

Fig.4 *Specific mass transfer rate of bubble bed reactors and air-lift reactors as a function of the superficial gas velocity (= gas flow / reactor cross-section area). The results of the 8L and 80L air-lift reactors have been taken from published data (Katinger et alt. [8]).*

4. Summary

To reduce lethal cell-bubble-interaction, a novel reactor design has been developed. This patent modified loop reactor balances oxygen bubbles in the downcomer, and has thereby a 10 to 100 fold higher oxygen uptake than standard air-lift reactors. Intense mass transfer and fermentation studies have shown efficient aeration rates and viable cell cultures in the BBR. This reactor type is especially advantageous if the addition of protective agents into the culture is not desirable.

A scale-up of 100 in volume to 250L has been realized, and confirmed the results achieved in the laboratory scale reactor. Oxygen uptake (up to 95%) and the specific mass transfer ($k_L a \approx$ up to $5h^{-1}$) are comparable to the results of the 2.5L BBR. Currently, fermentations in 50L and 250L BBR are conducted with our industry partner

5. Symbols

Δp	$[N/m^2]$	pressure loss	P	[W]	power
v	[m/s]	velocity	$\dot{V}$	$[m^3/s]$	flow
ρ	[kg/m]	density	ζ	[-]	resistance factor

5. References

[1] Sucker H.G. et alt. (1992) Verfahren zum Begasen einer Zellsuspension, Swiss Patent No. CH 681808 A5.

[2] Jordan, M. (1993) Die Rolle von Serum bei der hydrodynamischen Belastung von tierischen Zellen im Bioreaktor; Möglichkeiten der Serum Reduktion, Ph.D. thesis ETH Zurich.

[3] Jordan M., Sucker H.G., Widmer F., and Eppenberger H.M. (1994) Interactions between animal cells and gas bubbles: The influence of serum and Pluronic F68 on the physical properties of the bubble surface, *J. of Biotech. and Bioeng.* **43**, p.446-454.

[4] van't Riet K. and Tramper J (1991) Basic Bioreactor Design. Dekker, New York.

[5] Aunins J.G. and Henzler H.-J. (1993) Aeration in Cell Culture Bioreactors. In Bioprocessing, edit. Rehm H.-J. and Reed G., VCH, D-Weinheim

[6] Sucker H.G., Jordan M., Eppenberger H.M., and Widmer F. (1994) Bubble bed reactor: A reactor design to minimize the damage of bubble aeration on animal cells, *J. of Biotechn. and Bioeng.* **44**, p.1246-1254.

[7] Sucker H.G. (1992) Entwicklung des Blasenbettreaktors zum Vermeiden der Zellschädigung bei der Begasung serumfrei kultivierter tierischer Zellen, Ph.D. thesis ETH Zurich.

[8] Katinger W.D, Scheirer W., and Krömer E. (1978) Blasensäulenfermenter für die Massensuspensionskultur tierischer Zellen. Chem.-Ing.-Tech. **50**, p.472-473.

6. Acknowledgements

The authors are grateful to Dr. H.G. Sucker and Dr. M. Jordan who did most of the work in the 2.5L scale, and contributed to this project in many discussions. This project is supported by the Swiss National Science Foundation.

Contribution to the Understanding of Retention of Animal Cells in Spinfilters

M. Trocha[1], Ph. Rudolf von Rohr[1], and Z. Sümeghy[2]

[1] Institute for Process Design and Cryogenics, ETH Zurich, CH-8092 Zurich

[2] TRD-BT, Sandoz Pharma Ltd., CH-4200 Basel

Keywords: *spinfilter, cell retention, fouling, filter cake, centrifugal forces*

1. Introduction

In the last years progress in biotechnology has implanted *continuous* animal cell cultures into industrial productions. In order to generate sufficient cell densities in suspended perfusion cultures, cell retention is required. Filtration is one of the most frequently used technique to separate cells from the dilution stream. However, the operation time is very much negatively affected by fouling of the filter surface.

Increasing the shear rate at the filter surface can delay this effect. The spinfilter principle is a means to generate flow independent shear rates in filtration devices. The filter is concentrically mounted in the cylindrical housing (*see Fig.1*). By rotation of the filter, the ambient suspension is forced into rotation in the annulus. Thus *a shear gradient is generated at the outer filter surface*. Two different configuration are used in industry: Either the spinfilter is mounted in the reactor (in-situ) or the spinfilter is mounted into an external bypass (ex-situ).

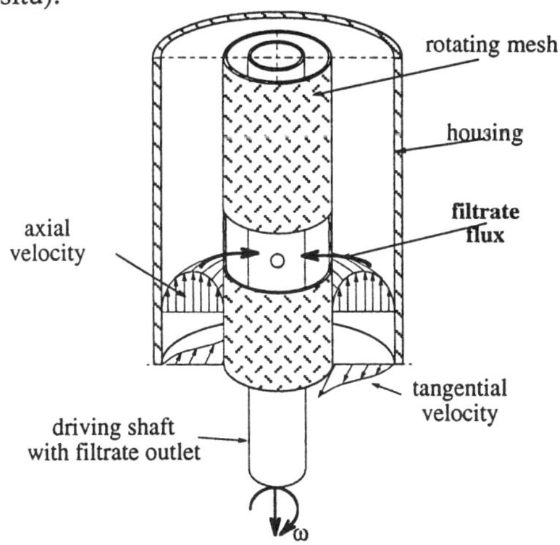

Fig.1 *The design of an ex-situ spinfilter.*

M.J.T. Carrondo et al. (eds.), Animal Cell Technology, 405–409.

Although spinfiltration is widely known, and applied in biotechnology plants, the rejection mechanism is poorly understood, and controversially discussed [1-3]. The scale-up of spinfilters is rather empirically than scientifically achieved. Furthermore complete pore plugging and thus system failure endangers long time perfusion operation. *Fig.2* depicts the cell growth of a suspended hybridoma cell cultivation in a perfusion system with an external spinfilter system. The cell density declines reveal the time when the spinfilter failed due to pore-clogging. To allow further perfusion, each time it was necessary to switch to a redundant spinfilter. This lack of a reliable process and of sufficient design parameters has been the motivation of this investigation.

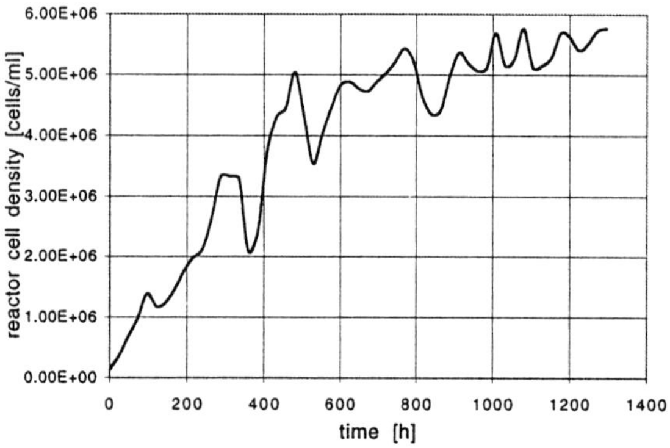

Fig.2 *Cell density decline due to spinfilter failure during hybridoma cell cultivation in a perfusion system with external spinfilter systems*

2. Materials and Methods

In order to investigate more thoroughly into the fluid dynamic aspects of spinfilter retention, a series of abiotic experiments has been carried out in a 100L pilot plant with external spinfilters of different sizes. The model suspension was an aqueous solution of 10^5parts/ml 15µm polystyrene beads (SOURCE 15RPC, Pharmacia, CH-Dübendorf) with supplements of Pluronic F68 (2g/L) , NaHCO3 (2g/L), and NaCl (7g/L). Particles were counted in a Coulter® Counter or by microscopy.

Laser-Particle-Tracking has been carried out by video recording of streamlines of 23µm flurescent latex beads (Fluoresbrite plain YG25, Polyscience Inc., D-Eppelheim) with a high speed CCD-camera (shutter: 1/50s).

The suspended hybridoma cells have been cultivated in 50L and 250L perfusion systems.

3. Results and Discussion

3.1 ABIOTIC FILTRATION EXPERIMENTS USING LATEX BEADS

In *Fig.3,* the solute retention at 600rpm is significantly higher, indicating that there is an influence of the rotation rate of the filter, presumably due to centrifugal effects. On the other hand, the time dependency of the retention can only be explained by filter fouling (assuming there is no significant difference in the residence time).

A radial force balance between the centrifugal force F_Z and the drag force F_D (due to the filtration flux) at a spherical solute shows that the centrifugal forces is dominant for rates of revolution higher than 400rpm:

$$\frac{F_Z}{F_D} = \frac{(\rho_S - \rho_L)\cdot \varpi^2 \cdot R \cdot d_S^2}{18 \cdot \eta \cdot v_F} \qquad (Eq.1)$$

where the density difference between solute and fluid $(\rho_S - \rho_L) \approx 50\text{kg/m}^3$, the spinfilter radius $R = 25\text{mm}$, the solute size $d_S = 15\mu\text{m}$, the viscosity of water $\eta \approx 0.001$ kg/m s, the filtration flux $v_F \approx 3\cdot 10^{-5}$m/s.

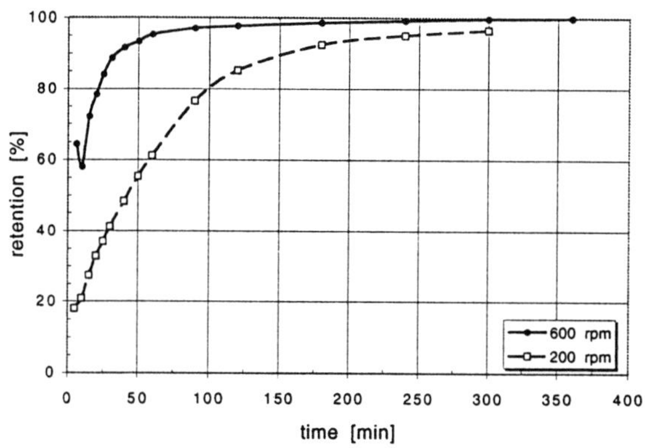

Fig.3 *Filtration of 15µm beads by a 20µm stainless steel mesh in a Ø50mm spinfilter.*

To study the influence of the centrifugal force in the annulus of spinfilters, *laser-particle-tracking* experiments have been carried out. *Fig.4* shows the radial velocities in the annulus arisen by particle tracking. There is no evidence for a preferential direction of the solutes (although these particles were even larger than hybridoma cells and had approximately the same density). The chaotic distribution suggests rather ideal mixing in the annulus than a separation effect due to centrifugal forces.

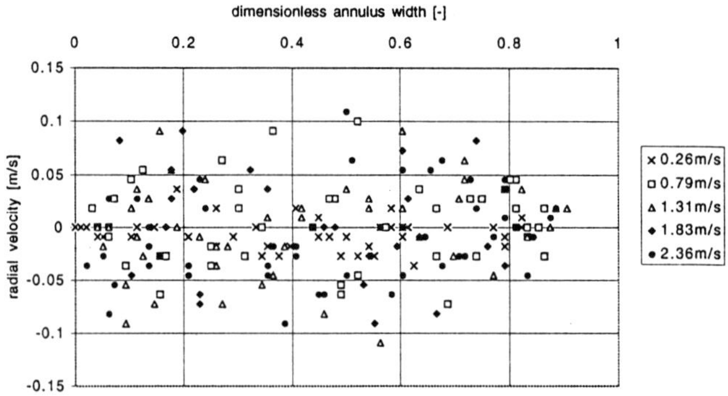

Fig.4 *Radial Velocities in the spinfilter annulus (width=15mm) for different rates of revolution (100-900rpm, corresponding to filter surface velocities between 0.26m/s and 2.36m/s) arisen by laser-particle -tracking.*

The laser-particle-tracking method has shown that centrifugal effects can not contribute to cell separation *in the annulus*. However, the *undisturbed centrifugal field in the lumen* of the spinfilter is sheltered against fluid turbulence. Moreover, there is no shear rate at the inner surface. The centrifugal field increases solute concentration at the inner surface of the filter, and thus helps to form a filter cake.

3.2 FILTRATION OF SUSPENDED HYBRIDOMA CELLS

Although the abiotic system is helpful to indicate fluid dynamic reasons for cell retention, both hybridoma cell retention and maximum operation time need to be studied in the biological system. Especially the wide particle size distribution of cell culture broth causes severe additional problems.

Further investigations in both the abiotic and the biotic system indeed showed accumulations of particles in the spinfilter lumen (as has been suggested above). In the case of hybridoma cell cultivation, the inner surface fouling can generate a very strong flow resistance. This filter cake is responsible for the high retention of spinfilters with large pores at high rotation rates.

Relatively small shear rates (≈0.5-1.0m/s) were sufficient to prevent the formation of filter cakes at the mesh/annulus interface, however for higher rates of revolution (≈2m/s), the centrifugal field in the lumen causes a filter cake onto the mesh/lumen interface.

Therefore the choice of the filter mesh becomes crucial for the duration of the filtration process. In *Fig.5*, mesh 2, a relatively densely woven mesh, allowed a filtration of only 1 hour until the filter was completely plugged. Mesh 1, a more open structure with larger pores, achieved comparable retention but operated successfully for almost 100 hours.

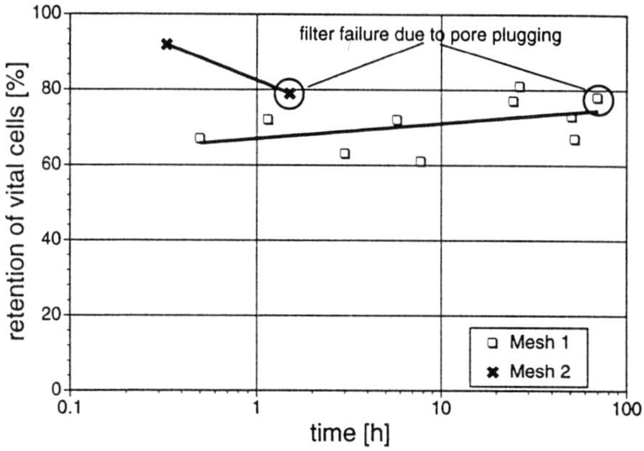

Fig.5 *Pore plugging and subsequent filter failure depends on the mesh (mesh 1: 15µm open structured mesh, mesh 2: 12-14µm woven mesh)*

4. Summary

Experiments using both polystyrene beads and hybridoma cell cultures have revealed a previously unknown form of filter fouling in spinfilters. While smaller rates of revolution (400rpm, 1.05m/s) were sufficient to prevent the formation of a filter cake at the outer surface (in the annulus), a filter cake in the spinfilter (in the lumen) was generated for high revolution rates (900rpm, 2.36m/s). This additional separation layer and fouling in the pores are believed to be the most important reason for high cell retention. However, the same phenomena are responsible for system failure after a certain filtration time due to complete plugging of the filter.

Laser-particle-tracking results give evidence that the centrifugal field in the annulus is overcome by turbulence. Intense mixing make centrifugal cell separation in the annulus rather unlikely while the centrifugal field in the lumen is relatively little disturbed, and can therefore be responsible for solute accumulation in this compartment.

5. References

[1] Kroner K.H. and Nissinen V. (1988) Dynamic Filtration of Microbial Suspensions Using an Axially Rotating Filter. *J. Membrane Sci.* **36,** p.85-100.

[2] Holeschovsky U.B. and Cooney C.L. (1991) Quantitative Description on a Rotating Filtration Device. *AIChE Journal* **37/8**, p.1219-1226.

[3] Favre E. (1993) Constant Flowrate Filtration of Hybridoma Cells Suspensions. *J. Chem. Tech. Biotechnol.* **58**, p.107-112.

6. Acknowledgements

The authors would like to thank V. Bösch and T. Wolewinski for their contributions to this work (polystyrene beads filtration and laser-particle-tracking, respectively). This project is supported by the Swiss National Science Foundation.

Rotating Disk Microfiltration of Cell Culture Broth - Effect of Shear Force and Cell Stability on Filtration Performance

J. H. Vogel and K.-H. Kroner

Bioseparation Techniques Department
Gesellschaft für Biotechnologische Forschung mbH; Mascheroder Weg 1
38124 Braunschweig, Germany

Abstract

Dynamic microfiltration techniques are capable of yielding particle-free supernatants out of cell culture broth, which is often a pre-requisite for subsequent purification procedures. For controlling performance it is essential to generate sufficient shear forces at the membrane surface. This, on the other hand, is limited by the shear sensitivity of animal cells. Furthermore, tangential flow techniques are limited in performance because of their inevitable pressure drops coupled with shear force generation. A novel approach is the application of dynamic shear filters allowing an independent control of shear force and operating pressure. In this study a rotating disc filter was used, analysing the effect of shear on cell viability and performance filtering brothes of rBHK cells. Experiments show that a loss of viability at high shear speeds up membrane clogging and that the adjustment of an optimal shear rate at each part of the membrane is essential. Introducing a conical rotor a uniform shear field can be established. With this novel approach stable filtration was obtained with minor viability loss up to shear rates reported for the cell-physiology determined limits. Due to the high and stable filtration rates this filter concept should be well suited for continuous perfusion/cell recycling.

Materials and Methods

Measurement of viability

The viability was measured by using the trypan blue exclusion method.

M.J.T. Carrondo et al. (eds.), Animal Cell Technology, 411–416.

Experimental Set-up

Recombinant BHK cells (BHKpSVIL2) were cultivated in a 1.6l continous perfusion reactor with double membrane stirrer [1]. When a cell density of 10^7 cells/ml was reached, the desired volume of cell suspension was transferred to the pre-sanitized microfiltration system (fig. 1) for performing off-line filtration experiments. Both retentate and filtrate were recycled in order to keep concentrations constant during filtration. Piezoresistive pressure meters (Jumo, Germany) were used for exact pressure measurements. All filtrations were performed at constant flux, which was adjusted via the filtrate pump. The microfiltration cell was a BIOPEM system (Braun, Germany) for the preliminary experiments and a self-made rotating disk filter for all other experiments.

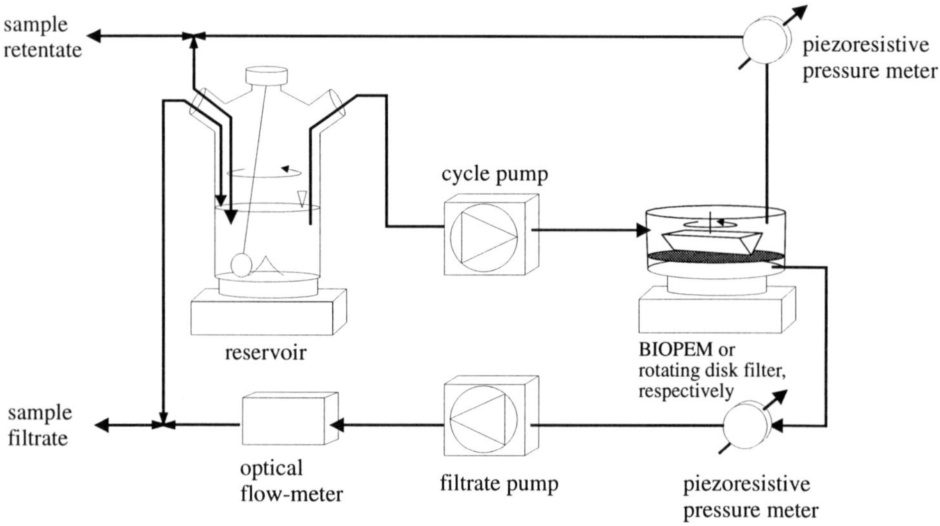

Fig. 1: Experimental set-up

Results and Discussion

Preliminary experiments using a BIOPEM stirring cell

In filtration experiments with a BIOPEM stirring cell (Braun, Germany), certain values of wall-shear stress were applied while filtering the suspension at a constant flux of 30 $l*h^{-1}*m^{-2}$ (see fig. 2). The viability decreased quickly in all cases because of the low shear stability of the cells. The highest shear level applied caused the highest viability-loss. Obviously, this loss of viability directly correlates with a dramatic increase in transmembrane pressure, which is the result of membrane clogging by cell debris, possibly aggregates of proteins, DNA and small membrane fragments. This result clearly demonstrates that higher wall shear stress can only improve filtration performance if the cells are not damaged. Thus optimization of microfiltration requires optimized hydrodynamics.

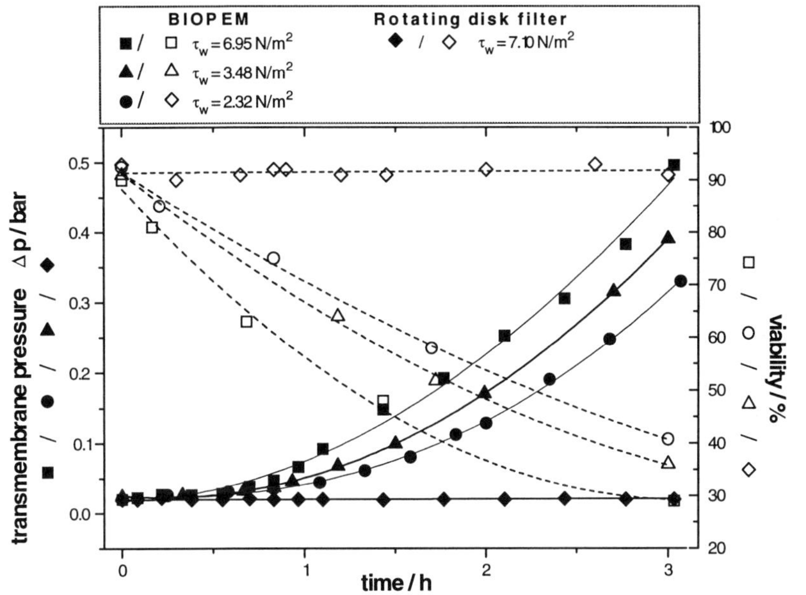

Fig. 2: Correlation between viability decline and membrane clogging in the BIOPEM stirring cell and comparison to the rotating disk filter

Rotating disk microfiltration

In rotating disk filters shear is created in the gap between the membrane and a rotating disk. The hydrodynamics in a rotating disk filter can be described with two Reynoldsnumbers:

$$Re_s = \frac{\omega\, s^2}{\nu}$$ (Re_s describes the influence of the gap-width s)

$$Re_r = \frac{\omega\ r^2}{\nu}$$ (Re$_r$ describes the influence of the radius r)

Operating at Re$_s$ > 4 and Re$_r$ < 210 000 leads to a laminar flow pattern with two boundary layers (see fig. 3) [2]. The bulk fluid rotates with constant velocity while the shear is applied mainly in the boundary layers at the rotor and the membrane surface, leading to high wall shear rates which improve filtration performance [3]. Due to the hydrodynamic lift effect [4] the cells are driven away from the high shear zone at the membrane surface. To compensate for the dependence of the average shear rate on the radius we introduced a conical rotor, which allows the establishment of a uniform shear field. Using this system, high wall shear rates can be applied without damaging the cells, thus stable filtration can be obtained (see fig. 2).

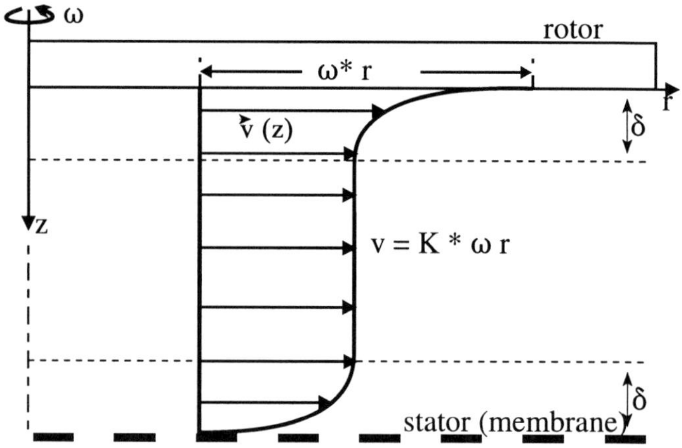

Fig. 3: laminar flow pattern with two boundary layers in the rotating disk filter

Influence of applied wall shear stress on filtration performance

To study the influence of the applied shear rate on filtration performance we measured the steady state transmembrane pressure (TMP) for certain flux values (fig. 4). Three different values of wall shear stress were applied as indicated in the figure. As can be seen, a linear relationship between flux and TMP exists for lower flux values. From a certain flux value on, which depends on the applied wall shear stress, the TMP increases much more rapidly because the shear enhanced back-transport of cells and macromolecular compounds can no longer compensate the convective deposition on the membrane. In the linear range of the pressure/flux-curve the membrane can supposed to be free of particulate matter, therefore stable filtration is possible. Because this linear range is exceeded when higher shear rates are applied, it is obvious that a shear rate just below the maximum shear stability of the particular cell-line should be chosen.

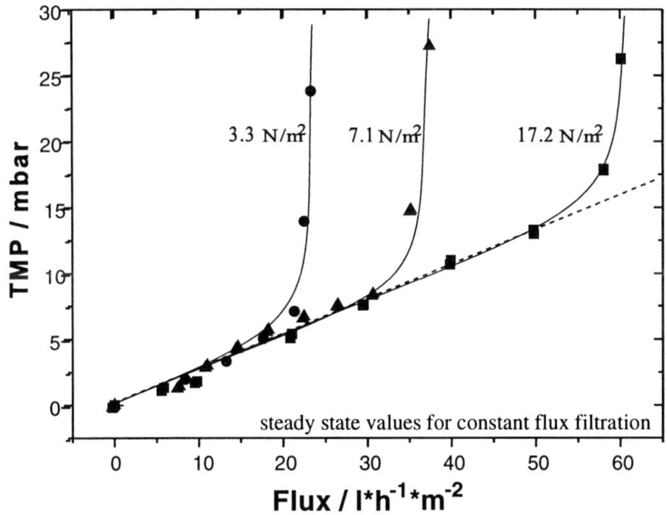

Fig. 4: steady state values of transmembrane pressure for filtration experiments at constant flux

Concluding remarks

The maximum shear rate tolerable by the rBHK cells used in this study is about 5000 s^{-1}. In the rotating disk filter, an average shear rate of 5000 s^{-1} would correspond to a wall shear stress of over 54 N/m^2. The adjustment of such a high wall shear stress, although not realizable in this study because of the limited power of the lab-scale magnetic drive, should further enhance the filtration performance to a stable flux of over 100 $l{*}h^{-1}{*}m^{-2}$.

References

1. Ryll, T., Lucki-Lange, M., Jäger, V., Wagner, R. 1990. Production of recombinant human interleukin-2 with BHK cells in a hollow fibre and a stirred tank reactor with protein-free medium. J. Biotechnol. 14:377-392

2. Ketola, H. N., McGraw, J.M. 1968. Pressure, frictional resistance and flow characteristics of the partially wetted rotating disc. Trans. ASME, J. of Lubr. Techn. 90, 4: 395.404

3. Kroner, K.-H. 1994. Crossflow Filtration of Biological Suspensions. in: Crespo, J. G. and Böddecker, K. W.: Membranes in Separation and Purification. NATO ASI Series E. 272. 59-83.

4. Zydney, A.L., Colton. C. K. 1986. A Concentration Polarization Model for the Filtrate Flux in Cross-Flow Microfiltration of Particulate Suspensions. Chem. Eng. Commun. 47: 1-21.

PROCESS CONTROL OF ANTITHROMBIN III PRODUCTION BY NEAR-INFRARED SPECTROSCOPY

Sven Harthun, Kathrin Matischak and Peter Friedl
Institut für Biochemie, Technische Hochschule Darmstadt,
Petersenstr. 22, D-64287 Darmstadt, Germany

Abstract

A rapid and easy method to simultaneously determine the concentrations of cell culture process metabolites is the analysis with Near-Infrared Spectroscopy. We used a multivariate Partial Least Squares (PLS) algorithm to calculate a model for the prediction of glucose, lactate, glutamine, ammonia, glutamate and human antithrombin III in a production process with recombinant CHO cells.

Once calibrated for a given system (medium composition, concentration ranges) with a sufficient number of samples, the pre-treatment, measurement and prediction of these substances from one spectrum takes only 10 minutes.

Introduction

For controlling biotechnological processes, it is important to know about the concentrations of substrates, products and by-products. In animal cell culture the most common metabolites monitored are glucose, lactate, glutamine, ammonia, glutamate and the respective expression products. Analysis of these substances should be rapid, easy and suitable for automation. Hitherto, assays for these metabolites were based on electrodes, enzymatic assay or antibody recognition. Most of these methods are time consuming in sample pre-treatment and measurement, and many metabolites require specific technical devices.

A promising method for the simultaneous determination of several constituents of liquid, viscous or solid matter from only one sample is near-infrared spectroscopy (NIR). NIR is a non-invasive, rapid and long-term stable method and is widely used for quality control in agriculture and food technology (1). In recent times, it is gaining increasing importance also in pharmaceutical and chemical process control. Covering the wavelength range from 780 to 2700 nm, the spectra can be gained by using either transmission or reflection. Modern fibre-optic probes and fibre bundles allow distances of

M.J.T. Carrondo et al. (eds.), Animal Cell Technology, 417–421.

up to 200 m between spectrometer and sample, and various chemometrical algorithms based on multiple regression analysis support a quantitative interpretation of the spectra (2).

However, NIR measurements in an aqueous matrix are more difficult to realize, since the investigated specific combination and overtone maxima of the dissolved substances overlap with the absorption bands of the water (1379 nm, 1875 nm, 2663 nm) (3,4,5). Therefore, a short optical pathlength and the calculation of an average spectrum from several co-added scans are necessary to improve the signal-to-noise ratio.

In the present study we show that the concentration of six metabolites can be predicted simultaneously from near-infrared spectra of culture broth that are drawn from both a batch and a continuous process.

Materials and Methods

The human antithrombin III producing recombinant cell line CHO SS3 A2 (kind gift of D. Lütkemeyer, Bielefeld) was grown in serum-free CHO-T1 medium containing 10 μM methotrexate. For continuous cultivation we used a 150 ml fluidized bed reactor (6) with 30 ml CultiSpher macroporous gelatine beads (Percell, Lund, Sweden). Data for the batch process were gained from a 50 ml spinner flask suspension culture. At regular intervals we determined the cell concentration and stored a sample of culture broth at -20°C.

All spectra were collected using a Bruins Omega 20 double-beam near-infrared spectrometer (Bruins Instruments, Puchheim, Germany) in transmission mode. We used a 0.2 mm path length Suprasil flow-through quartz cuvette (Hellma, Müllheim, Germany), which was placed in a brass-jacketed sample holder to keep the solution temperature at 25±0.1°C.

The samples were thawed and centrifuged to remove cell remnants. After addition of 0.2 M HEPES pH 8.0 buffer to balance differences in pH, the solution was injected into the cell and allowed to adjust temperature. The final spectrum was created by co-adding 25 sample scans from 1000 to 2500 nm and subtracting a water spectrum. Although every component has only a few characteristic absorption bands, we found that analysing the entire wavelength range yielded the best prediction results.

To calibrate the system, we determined the concentrations of the metabolites by conventional methods (enzymatic assays and ELISA). These values were correlated with the spectral data using a multivariate partial least square (PLS) algorithm including baseline correction and outlier detection (Grams/386™ software from Galactic Industries, Salem, NH, USA).

After establishing the correlation model between spectra and concentrations, we were able to predict from one spectrum the concentrations of glucose, lactate, glutamine, ammonia, glutamate and antithrombin III.

Results

We plotted the conventionally measured concentrations of the respective substances versus the NIR prediction results, and obtained correlations between $r^2=0.947$ and $r^2=0.988$. The calibration curve for antithrombin III is shown in Figure 1. The standard error of prediction (SEP) was below 5 % of the maximum concentrations for all metabolites.

	Concentration Range	r^2	Standard Error of Prediction
D-Glucose [mmol/l]	0 - 30	0.980	1.487
L-Lactate [mmol/l]	0 - 45	0.984	1.648
L-Glutamine [mmol/l]	0 - 6.5	0.988	0.178
L-Glutamate [mmol/l]	0 - 3	0.947	0.118
Ammonia [mmol/l]	0 - 3	0.951	0.123
hAT III [µg/ml]	0 - 18	0.985	0.747

Table 1: Concentration ranges of metabolites as determined by conventional methods, correlation coefficients for calibration curves and SEP of cross-validations.

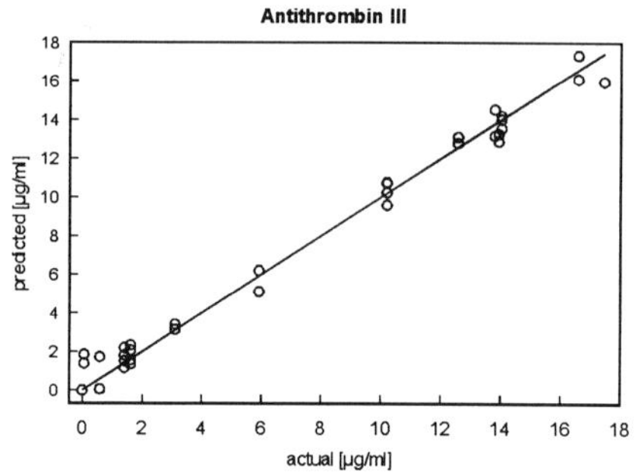

Fig.1:Calibration curve for NIR prediction of hATIII. (*Actual* values were determined by ELISA.)

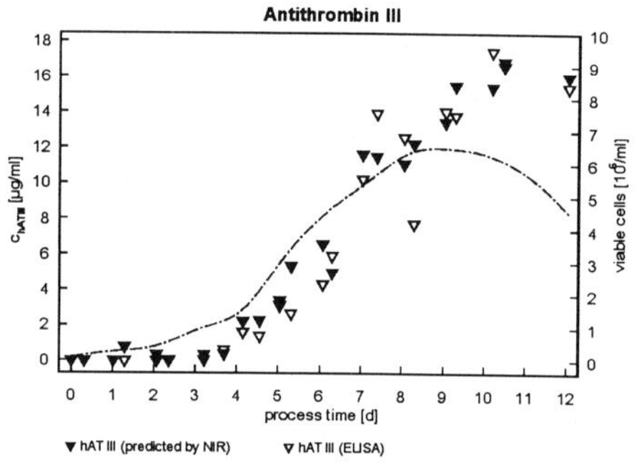

Fig.2: Production of hATIII in a batch culture. Comparison of NIR and ELISA. (Dash-dotted line: number of viable cells).

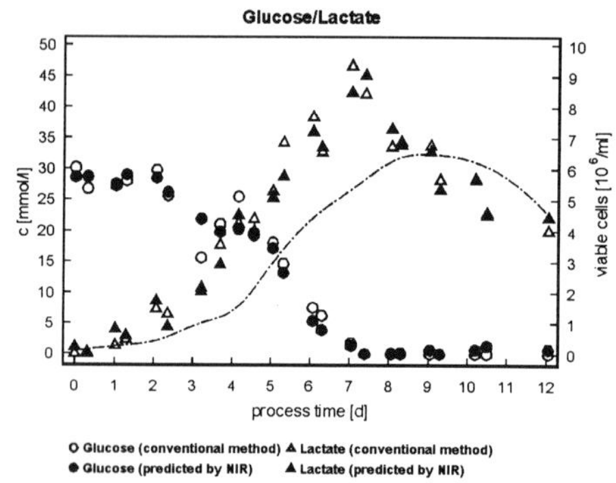

Fig.3: Determination of glucose and lactate in a batch culture. Comparison of NIR and enzymatic assay. (Dash-dotted line: number of viable cells).

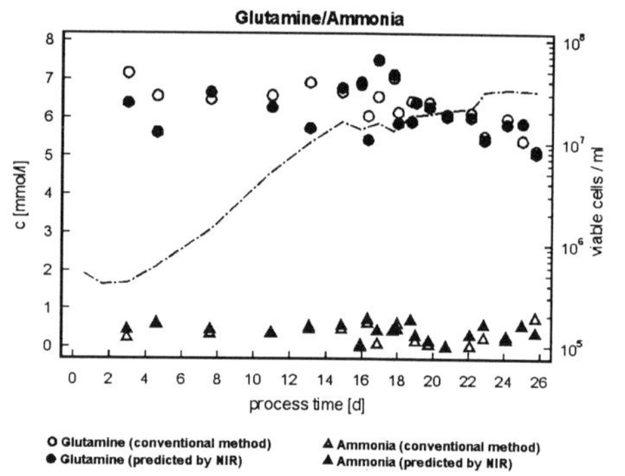

Fig.4: Determination of glutamine and ammonia in a perfusion culture. Comparison of NIR and enzymatic assay. (Dash-dotted line: number of viable cells).

NIR prediction coincided well with the conventionally measured results as shown in Figure 2 to 4. We therefore conclude that near-infrared spectroscopy is a tool for the quick and easy determination of metabolites. (Although, its versatility is limited by the time consuming calibration procedure). In combination with fibre-optic probes it might be suitable for the automation of cell culture processes.

References

1. Molt K (1992), Grundlagen und Anwendungen der modernen NIR-Spektroskopie. *GIT Fachz. Lab.*, **36**, 107-113 and 353-362.

2. Thomas EV and Haaland DM (1990), Comparison of Multivariate Calibration Methods for Quantitative Spectral Analysis. *Anal. Chem.*, **62**, 1091-1099.

3. Chung H, Arnold MA, Rhiel M and Murhammer DW (1995), Simultaneous Measurement of Glucose and Glutamine in Aqueous Solutions by Near Infrared Spectroscopy. *Appl. Biochem. Biotechnol.*, Vol. **50**, 109-125.

4. Chung H, Arnold MA, Rhiel M and Murhammer DW (1996), Simultaneous Measurements of Glucose, Glutamine, Ammonia, Lactate, and Glutamate in Aqueous Solutions by Near-Infrared Spectroscopy. *Appl. Spectrosc.*, Vol. 50, No. **2**, 270-276.

5. Yano T and Harata M (1994), Prediction of the Concentration of Several Constituents in a Mouse-Mouse Hybridoma Culture by Near Infrared Spectroscopy. *J. Ferment. Bioeng.* , Vol. 77, No. **6**, 659-662.

6. Lüllau E, Dreisbach C, Grogg A, Biselli M and Wandrey C (1992), Immobilization of animal cells on chemically modified carrier. In: Spier RE, Griffiths JB and MacDonald C (eds.), Animal Cell Technology: Developments, Processes and Products, Butterworth-Heinemann, Oxford, 469-475.

Acknowledgements

The authors would like to thank the European Commission (DG XII) for granting a bursary to present these results at the 14th ESACT Meeting in Vilamoura, Portugal.

This work was funded by the German Bundesministerium für Bildung, Wissenschaft, Forschung und Technologie, and is part of the Dr.-Ing. dissertation of Sven Harthun, Technische Hochschule Darmstadt (D17).

SELECTION OF AN IMMOBILIZATION METHOD FOR A PERFUSION BIOREACTOR WITH HYBRIDOMA CELLS

Albert Tintó, Pau Vila, Carles Casas, Francesc Gòdia. Departament d'Enginyeria Química. Universitat Autònoma de Barcelona. 08193 Bellaterra (Barcelona). Spain.

KEYWORDS

Immobilization, microcapsules, hybridoma, perfusion, fluidized-bed bioreactor.

ABSTRACT

The selection of a suitable method for the immobilization of the hybridoma cell line KB26.5 is described. Among the various techniques tested, the formation of liquid core microcapsules of calcium alginate presents the best results.

Preliminar runs with these type of particles are promising, and a three-phase fluidized-bed bioreactor has been designed and built to carry out continuous perfusion experiments.

INTRODUCTION

Perfusion systems offer atractive potential advantages in mammalian cell culture, such as higher cell densities, continous production, and cell protection from shear stress.

Immobilization in gel beads or microcapsules is widely known. However, the application to the immobilization of animal cells can have a number of restrictions: toxicity of the reagents used, mechanical stability of the support materials, regular shape and size, and diffusional limitations, that make the bead preparation a not straight forward methodology.

This work is focused on the adaptation of different methods of immobilization to the hybridoma cell line KB26.5 producing a monoclonal antibody IgG_3 against A_1 antigen of the red blood cells. Three immobilization methods have been tested: entrappement in

M.J.T. Carrondo et al. (eds.), Animal Cell Technology, 423–428.

calcium alginate gel beads, and two different kinds of microcapsules with an internal liquid core: Calcium alginate microcapsules and Cellulose-sulfate (CS)/Poly-diallyl-dimethyl- amoniumchloride (PDADMAC) microcapsules. The second part of the work consist in the set-up of a three-phase fluidized-bed bioreactor to be used in perfusion cultures with these particles.

MATERIAL AND METHODS

Cell line: A mouse-mouse hybridoma cell line KB26.5 was used in the work. This cell line produced a monoclonal antibody IgG_3 against A_1 antigen of the red blood cells.

Medium: Culture medium was Dulbecco's Modified Eagle Medium (DMEM, Gibco Laboratories) supplemented with 2% fetal calf serum (FCS), 6 mM glutamine, 44 mM $NaHCO_3$, 5.2 mM $CaCl_2$, 0.05 mM ß-mercaptoethanol, Insuline (0.002 UI/ml), Streptomycine (0.167 g/l), and penicilline (0.073 g/l). In bioreactor cultures 40 mg/l of Antifoam A (Sigma) were added.

Immobilization methodology: Droplets of a cellular suspension were formed by extrusion through a 22-gauge stainless steel needle by an external air jet. Droplets were collected in a stirred solution with a gellification reagent. Microcapsules or beads were formed by ionotropic polymerization when cell suspension reacted whit a gellification solution.

Calcium alginate gel beads: One part of 3% (w/v) alginate (Keltone LV, Kelko) in 0.9% (w/v) NaCl solution was mixed with one part of cell suspension in culture medium; the resulting solution had 8.0×10^6 cells/ml on 1.5% (w/v) alginate solution. The alginate/cell suspension was pumped into a 1.5% (w/v) $CaCl_2$ solution (pH=7.2); Ca^{2+} ions were the gellification agent. After 10 min. $CaCl_2$ solution was removed and beads were resuspended with culture medium. So cells remain entrapped in a calcium alginate gel matrix.

Calcium alginate microcapsules: Two parts of a 75%(w/v) dextran (40.800 MW, Sigma) in 1.6% (w/v) $CaCl_2$ solution (pH=7.2) were mixed with one part of cell suspension in 1.6% (w/v) $CaCl_2$ solution (pH=7.2), the resulting solution had 3.0×10^5 cells/ml on a 50% (w/v) dextran and 1.6% (w/v) $CaCl_2$ solution. Droplets of a cell suspension solution were collected in a 1% (w/v) alginate (Keltone LV, Kelko) in isoosmotic solution: 0.85% (w/v) NaCl, 0.05% (w/v) KCl and 0.1% (w/v) glucose, (pH=7.2). The interfacial alginate gellation around the cell suspension droplet created a microcapsule with a thin calcium alginate wall and a liquid core, in this way cells were immobilized in a liquid enviroment. Dextran is added as a thickening agent for improve de spherical shape of microcapsules, and after the immobilization diffuse quickly. After 10 min. of reaction the microcapsules were removed from the alginate solution and whased twice with isoosmotic solution to reduce the possibility of capsule agregation. Finally microcapsules were tansferred into a hardening solution: 1.6% (w/v) $CaCl_2$, 0.1% (w/v) glucose and 0.088% glutamine (pH=7.2), to improve their strenght. After 15 min. hardening solution was removed and microcapsules were resuspended with culture medium.

CS/PDADMAC microcapsules: One part of cell suspension in culture medium was mixed with nine parts of 3.5% (w/v) NaCS solution (Bioverfahrenstechnik GbR) (pH=7.2), the resulting solution had $5.0x10^5$ cells/ml on a 3.15% (w/v) NaCS solution. This polyanion was dripped in a 2% (w/v) PDADMAC (Bioverfahrenstechnik GbR) in isoosmotic solution (pH=7.2). Microcapsules were formed when CS molecules reacted with a PDADMAC polycation. After 45 min. microcapsules were removed to the PDADMAC solution and washed twice with isoosmotic solution and finally were transfered to culture medium. This kind of microcapsules had a CS-PDADMAC wall and cells were immobilized in an intracapsular liquid enviroment with remaining NaCS solution.

Preliminar cultures: Three different immobilization methods were tested in spiner flask cultures. 20 ml of particles with immobilized cells were cultured in a 100 ml of working volume; the culture medium was replaced daily. Cell growth and final particle characteristics were compared for the three methods.

Perfusion bioreactor: A 600 ml three-phase fluidized-bed bioreactor was used in perfusion cultures. A mixing of Air/N_2/CO_2 was sparged through a sinterred glass at the bottom of the reactor, at a flow rate of 0.07 l/min. to allow microcapsule fluidization. pO_2 was maintened at a level of 40% controlling air/N_2 ratio. pH was stabilized at a level of 7.2 by $NaHCO_3$ buffer and CO_2 sparging or 0.1 M NaOH addition. Temperature was maintained at 37°C by a water jacket. Operational parametres were controlled with a Digital Control Unit (Braun Biotech). Experimental assembly for the perfusion bioreactor is shown in Fig.1.

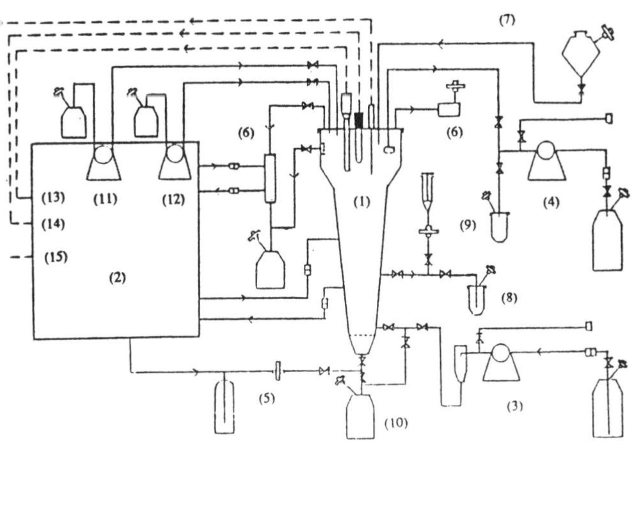

Fig. 1: Experimental assembly for the perfusion cultures: 1) Fluidized-bed bioreactor. 2) Digital control unity and gas mixing station. 3) Fresh medium inlet. 4) Culture medium outlet. 5) Gas inlet. 6) Gas outlets (principal and auxiliary). 7) Inoculum inlet. 8) Solids sample outlet. 9) Medium sample outlet. 10) Drain outlet. 11) Antifoam solution inlet. 12) NaOH solution inlet. 13) pO_2 probe connection. 14) pH probe connection. 15) Temperature probe connection.

Analytical methods: Cell growth was monitored by counting viable cells with a hemocytometer. Viable cells were distiguished from dead cells by the trypan blue dye exclusion method. Glucose and lactate were measured using a glucose/lactate analyzer YSI (Yellow Springs Instruments, model 2000).

RESULTS AND DISCUSSION

Physical properties of immobilization particles: Experimental conditions used in the immobilization procedure: viscosity of the polymer solutions, cell suspension solution extrusion rate, air flow rate, distance between the needle and the gellification solution, stirring of collecting solution, and reaction time of the solutions, determined the physical properties of resulting particles: size, shape, wall thickness, strenght and diffusional properties. The results obtained with the differents immobilization methods tested are shown in Table I.

Physical properties	Calcium alginate beads	Calcium alginate microcapsules	CS/PDADMAC microcapsules
Size	1.3 - 1.5 mm	2.5 - 3.1 mm	1.5 - 1.7 mm
Wall thickness	-	0.14 - 0.21 mm	0.01 - 0.02 mm
Shape	Spheric	Spheric	Spheric
Strenght	+	+	++
Immob. procedure	Simple	Difficult	Simple

Table I : Physical properties of three different particles obtained in the immobilization experiments.

Preliminar cultures: In order to select the immobilization method that provided the best conditions for the growth of immobilized hybridoma cells, the three methods were tested in spiner flask cultures. Table II.

	Ca-Alginate beads	Ca- Alginate microcapsules	CS/PDADMAC microcapsules
Inoculum	$8.0x10^6$ (*)	$3.0x10^5$ (**)	$5.0x10^5$ (**)
Final cell concentration	$3.5x10^7$ (*)	$1.3x10^7$ (**)	$0.7x10^5$ (**)
Viability after immobilization	83.3%	81.8%	81.1%
Final viability	36.2%	65.3%	10.6%
Culture duration	336 h.	270 h.	48 h.

Table II : Results obtained in preliminar cultures. (Conc. viable cells/ml bead, ** Conc. viable cells/ml intracapsular volume).*

Cultures performed whit calcium alginate gel beads showed that it is necessary to have a high inoculum concentration ($5.0x10^6$ - $1.0x10^7$ cells/ml bead), to obtain initial growth of the immobilized cell population; a important drop of viability was always observed after three days of culture (45.9% at 72 h.). It is possible that the contact with the liquid alginate solution in the immobilization procedure affects the cellular viability in the initial stage of culture. At the end of culture bead strenght was affected for calcium chellating anions in the medium.

Immobilized cells cultured in calcium alginate microcapsules did not appeared to be negatively affected by the immobilization procedure; the innoculum required was $2.5x10^5$ - $1.0x10^6$ cells/ml intracapsular volume. Only a gentle drop of viability was detected after 202 h. of culture, and the final viability was 65.3% (270 h.). The microcapsule strenght was also affected by calcium chellating anions at the end of culture.

The results obtained with CS/PDADMAC showed a toxicity problem in this method; probably the PDADMAC solution used or the NaCS remainig inside the microcapsules were toxic for the hybridoma cells. The microcapsule strenght and ellasticity did not seem to be affected for the culture conditions.

In view of these results, the immobilization method selected for perfusion cultures was calcium alginate microcapsules.

Perfusion cultures: A perfusion culture with hybridoma cells immobilized in calcium alginate microcapsules was carried out in the fluidized-bed bioreactor. Inoculum used was $2.0x10^5$ cells/ml intracapsular volume and initial viability 82.5% . The residence time of the culture was 24 h. Results obtained are shown in Figure 2.

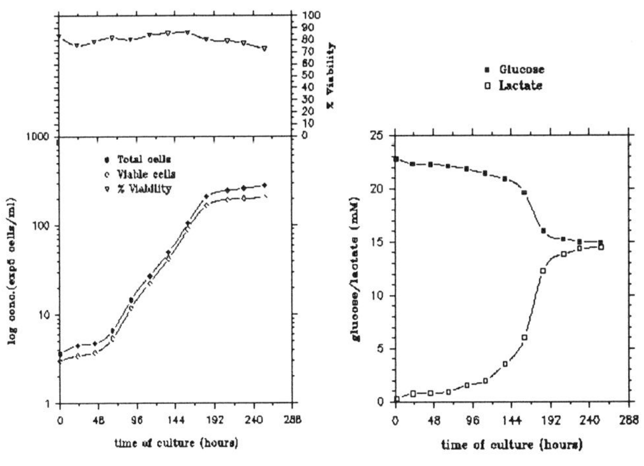

Fig.2 : Evolution of viable and total cell concentration, viability percentage, glucose and lactate concentration in a perfusion culture.

The growth of cells was satisfactory; the cell concentration only remained stable when reached the stationary state. The final cell concentration was $2.0x10^7$ cells/ml intracapsular volume and the viability 71.8% (265 h.); that sugest than the immobilization procedure did not affect cell viability. Glucose and lactate concentration never became a limitant factor for cell growth.

After 265 hours of culture an important percentage of microcapsules were broken by the continous action of chellating anions of the medium and shear stress produced by gas sparging. That sugest than its necessary to improbe the microcapsule strenght or reduce the effect of calcium chellating compounds in medium.

CONCLUSIONS

From the immobilization systems tested, calcium alginate microcapsules with a liquid core appears to have the best suitability regarding the properties required to develope a perfusion system.

The three-phase fluidized-bed bioreactor has been successfully used for a preliminar experiment of short duration.

Future work will be focused on the enhancement of particle formulation and bioreactor operation, and the characterization and modelization of perfusion culture system at different conditions.

ACKNOWLEDGMENTS

Finantial support from CICYT (project BIO94-0288) is gratefully acknowledged. This research was carried out in the frame work of the Centre de Referència de Biotecnologia (Generalitat de Catalunya).

REFERENCES

1. de la Boise, D., Noiseux, M., Massie, B. and Lemieux, R. Hybridoma perfusion systems: A comparison study. *Biotechnology and Bioengineering* 40: 25-32, 1992.

2. Matthew, H.W., Salley, S.O., Peterson W.D. and Klein M.D. Complex coacervate microcapsules for mammalian cell culture and artificial organ development. *Biotechnology Progress* 9: 510-519, 1993.

3. Merten, O.W., Dautzenberg, H. and Palfi, G.E. A new method for the encapsulation of mammalian cells. *Cytotechnology* 7: 121-130, 1991.

4. Nigam, S.C., Tsao, I., Sakoda, A. and Wang, H.Y. Techniques for preparing hydrogel membrane capsules. *Biotechnology techniques* 2: 271-276, 1988.

5. Spiekermann, P., Vorlop, K.D. and Klein, J. Animall cell encapsulated within Ca-Alginate hollow-spheres. *Press 1th European Congress on Biotechnology* 3: 590-593, 1987.

ON-LINE MONITORING OF GLUTAMINE AND AMMONIUM IN MAMMALIAN-CELL CULTURES

Eva Balslev Jorgensen, Antoni Casablancas, Glòria González, Francesc Gòdia.

Departament d'Enginyeria Química. Facultat de Ciències. Universitat Autònoma de Barcelona. 08193 Bellaterra. Spain

Introduction

The on-line monitoring of key nutrients and products is a very useful tool to optimise processes based on animal cell culture. Compounds of great influence on animal cell cultures are L-glutamine and ammonium. L-Glutamine is required as nutrient in a wide number of cell lines, and cells can be very sensitive to low levels of glutamine. Ammonium ions are often inhibitors of cell growth. They are produced by the direct decomposition of glutamine, or as the end product of the amino acid metabolism of the cells. Therefore, developing a reliable on-line system for ammonium and L-glutamine detection will enable the optimisation of cell culture experiments with efficient use of L-glutamine and hence minimisation of inhibition due to ammonium accumulation.

In order to develop a control system for mammalian cell cultures the on-line monitoring of key nutrients is required. Flow Injection Analysis (FIA) has been used in this work, as this method of analysis requires very few initial steps of sample preparation, in comparison to e.g. HPLC. This work shows the different steps taken in order to set-up and test a FIA system for the sequential analysis of ammonium and L-glutamine.

Principles of Flow Injection Analysis

A schematic diagram of the FIA system for sequential analysis of ammonium and L-glutamine is shown figure 1. To measure ammonium the sample is injected into a stream of NaOH (INJV#1). The reaction of the sample with NaOH generates ammonia gas, which diffuses through a selective membrane (GVHP 09050, Millipore) into a carrier stream containing pH indicator (bromo cresol purple). The reaction between ammonia and the pH indicator leads to a change in colour that can be detected spectrophotometrically at $\lambda = 605$ nm. Figure 2 gives the fiagram of an ammonium calibration using triple determinations on 5 standards. Characteristic for the FIA system is a high degree of sample reproducibility.

M.J.T. Carrondo et al. (eds.), Animal Cell Technology, 429–434.

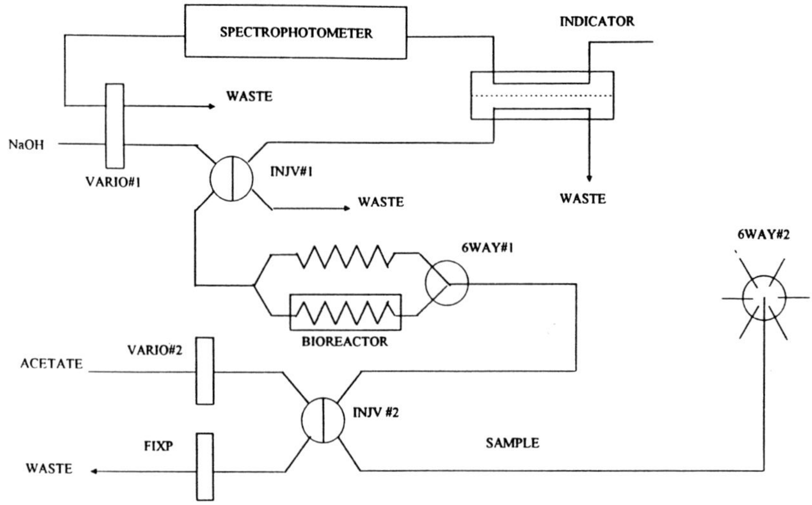

Figure 1.- Diagram of glutamine and ammonium FIA system.

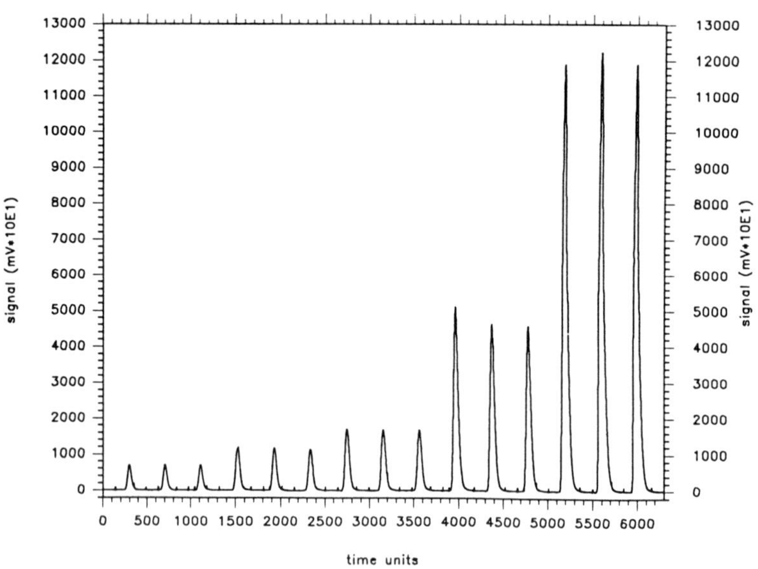

Figure 2.- Fiagram of ammonium calibration standards.

This principle is reused in the detection of L-glutamine, as the introduction of a initial step to the system, being the deamination of L-glutamine by L-glutaminase, makes sequential analysis possible. The sample is led through an enzyme cartridge containing immobilised L-glutaminase, where the deamination takes place.

$$\text{L-glutamine} + H_2O \rightarrow \text{L-glutamate} + NH_4$$

Cell free samples, containing ammonium and L-glutamine, are initially mixed with an acetate buffer carrier stream (INJVV#2) in order to adjust pH to be in the optimal area for the enzymatic reaction (fig 1). Ammonium ions in the culture samples are analysed directly by leading the sample around the enzyme cartridge (6WAY#1), as described above, using ammonium calibrations. Flowing the sample through the cartridge makes the detection of L-glutamine possible. The obtained signal corresponds to the sum of the original ammonium concentration and the concentration of ammonium generated by the enzymatic deamination of L-glutamine. The difference in signals thus is correlated to the amount of L-glutamine, and the exact concentration is calculated from the corresponding calibration. Examples of calibration curves for ammonium and L-glutamine are given figure 3 and 4.

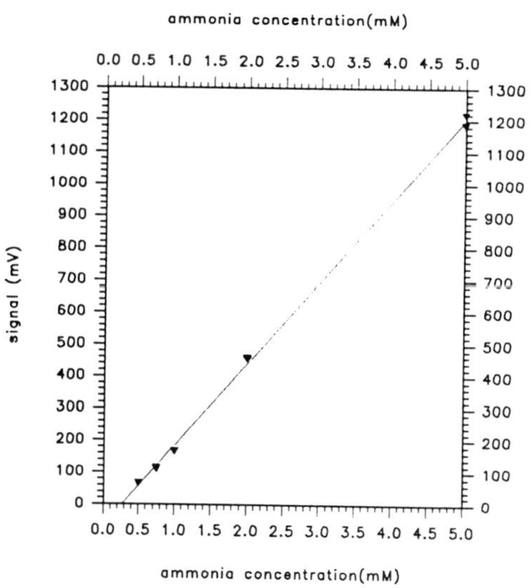

Figure 3.- Ammonium calibration curve.

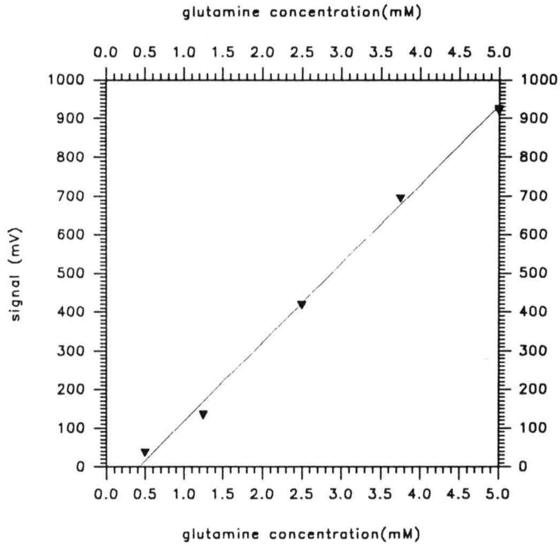

Figure 4.- Glutamine calibration curve.

Immobilised L-glutaminase

The immobilisation procedure is derived from the urease immobilisation, described by Mascini et al. (1983). L-glutaminase (EC 3.5.1.2, SIGMA) is immobilised by covalent attachment in 1 m long manifold nylon tubing. The immobilisation procedure includes 3 reaction steps. First step is nylon alkylation using dimethylsulphate with the formation of an imido ester. Next step is to react the imidoester with an amino group of the spacer, which in this case is hexamethylene diamine. In the final step, the free amino group of hexamethylene diamine is functionalised with glutaraldehyde, and then coupled to L-glutaminase.

Physical characterisation of the enzyme cartridge

To evaluate the robustness and performance of the enzyme cartridge, studies of thermal deactivation and the influence of pH on enzymatic activity were carried out. The thermal stability of the free enzyme was shown to be very poor at 37°C (50% loss of activity in 2 hours), and improves radically at 4°C (30% loss in 10 days). Long time performance test for immobilised L-glutaminase were done at 22°C. Figure 5 shows the variation of the percentage of glutamine conversion with time for various standard solutions. The results are satisfactory from the point of view of maintenance of the activity for a reasonable period of time, and the life-time can be prolonged by storing the enzyme cartridge at 4°C overnight. The daily calibrations kept the linearity, even at the end of the experiment (day 10), which shows the robustness of the system. However, the fluctuations in the signal obtained make evident the need for periodic calibrations.

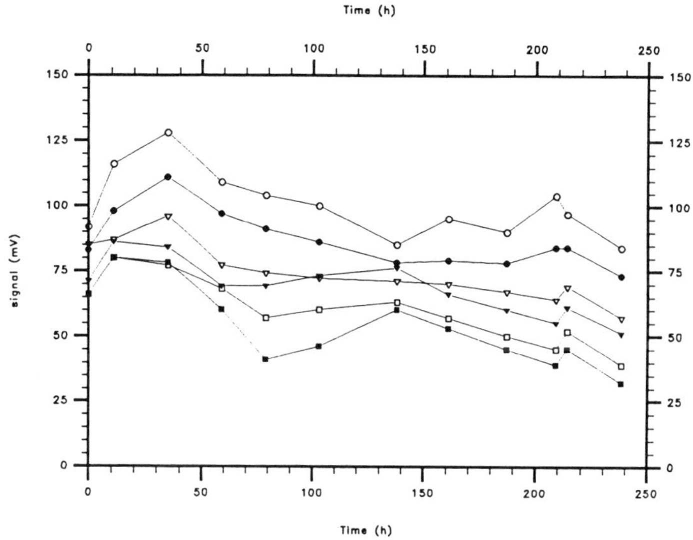

Figure 5.- Thermal stability at 22°C.

Effect of pH on the activity of immobilised L-glutaminase was studied, and as can be seen in figure 6, the effect of pH changes is not very noticeable between 4-5, which will be the range of operation used.

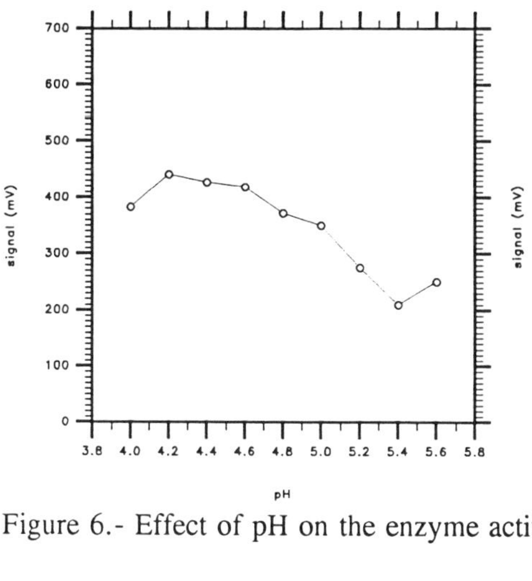

Figure 6.- Effect of pH on the enzyme activity.

Conclusions

A sequential FIA analysis for ammonium and glutamine has been developed and tested with samples from a hybridoma cell line culture. Future work will be focused on the physical implementation of this equipment in a control loop, in order to test different fed-batch culture strategies.

Acknowledgements

Eva Balslev received a HCM fellowship from EBEN (European Biochemical Engineering Network) to perform this work. Finantial funding to the project was provided by CICYT (BIO940288). This research was developed in the framework of the Centre de Referència en Biotecnologia (Generalitat de Catalunya).

References:

Mascini M, Ianello M, and Palleschi G. (1983)
Enzyme electrodes with improved mechanical and analytical characteristics obtained by binding enzymes to nylon nets.
Analytica Chimica Acta 146: pp 135-148

STUDY FOR THE PRODUCTION OF A RECOMBINANT MONOCLONAL ANTIBODY IN STIRRED TANK.

Boggiano T., Fernandez A., Hermida C., Villán J., Ojito E., Hidalgo Y., Arias M., Gomez J.
Center of Molecular Immunology, P.O. Box 16040, Havana City 11600, Cuba.

Keywords: Stirred tank, NS0 myeloma, shear stress, scale-up, kla.

ABSTRACT

Stirred tank is one of the main techniques for scaling-up cultures for production of recombinant proteins for therapeutical use. A study of culture conditions for the production of a humanized monoclonal antibody against EFG receptor was performed.
A basal medium selection was done. The influence of stirring in cell growth was tested. The result showed that there have not influence until 150 rpm in 250 ml spinners flask. An extrapolation to higher volume vessel let velocities of 400 rpm without cell damage was possible.
Paralel determination of oxygen transfer for different levels of gas flow and stirring was done.
A series of suspension batch cultures in spinners flasks and in 4 l fermenter were made with the prestablished conditions, the cell growth and productivity were improved .

INTRODUCTION

The production of humanized monoclonal antibody is getting increasing importance because it enables to apply those antibodies to pacients without the obtention of HAMA response (human antibodies against mouse antibodies). The culture of mammalian cells in stirred tank fermentors with direct sparging requires the optimization of different parameters in order to improve cell growth and productivity because those lines have low level of secretion and tendency to be damaged by shear forces. The selection of culture medium, the kinetic study of the behavior of the cell line in spinners flask and the choice of the best conditions of stirring and sparging using the existing production equipment at CIM, were the initial goals to develop an in vitro culture process with enhanced product yield.

M.J.T. Carrondo et al. (eds.), Animal Cell Technology, 435–440.

MATERIALS AND METHODS

Cell Line : TSTD2, NSO myeloma transfected with the gene of a humanized monoclonal antibody against EGF receptor.
Culture medium : The basal mediums were supplemented with 8% of foetal calf serum, GIBCO, 4 mM of glutamine, 0.2 % of Pluronic F-68, (cultures in the fermenter), 0,11 g/l Sodium Piruvate from SIGMA.
System specifications: The initial cultures were done in spinners flask, TECNOMARA, and the fermenter ones were performed in a stirred tank of 4 litters of working volume with a marine impeller and a stainless steel sinter sparger (100 µm porus size).
Measurement of kla values were done using the static gas-out method.

RESULTS AND DISCUSSION

Selection of medium
Different basal mediums from SIGMA (Table 1) were tested on six well plates.

Medium	Reference	Medium	Reference
DME-a	D-5523	F12	N-6760
DME-b	D-5648	IMD	I-7633
DME/F12	D-8900	IMD/F12	(1:1)

Table I. Basal mediums used and its references.

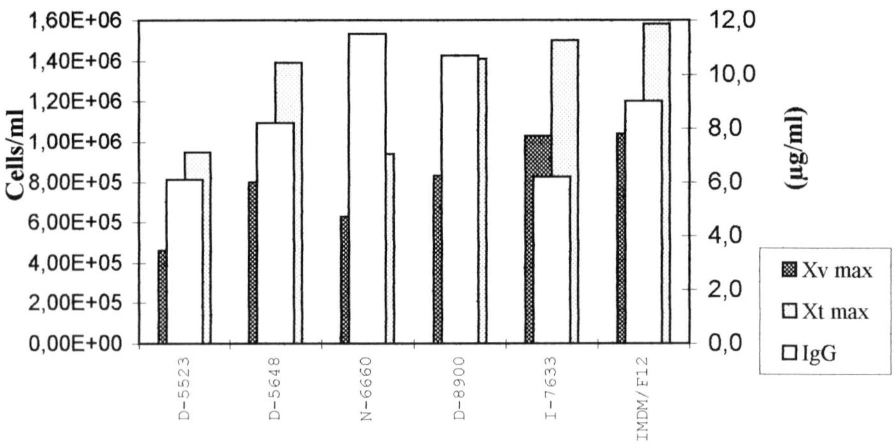

Figure 1. Cell growth and production in mediums tested.

The highest production were obtained in F12 and DME/F12 mediums. Despite, not so high cell growth were achieved. (Fig. 1) The negative growth association of the antibody production has been refered by other authors (1). Further studies were done using DME/F12 as basal medium.

Effect of stirring on cell density

In order to study the effect of shear stress in cell growth several batch cultures in 250 ml spinners flask were performed at different stirring rates, (30, 60, 100 and 150 rpm).
As can be seen in Fig. 4 there is not significant differences in the produced biomasse despite of stirred speed. This information can be used to define the stirring condition that can be used in a production fermenter.

$$Fg = \frac{6.28 * Ni\ Di}{D - Di}$$

Fg : shear factor
Ni : stirrer velocity
Di : impeller diameter
D : vessel diameter

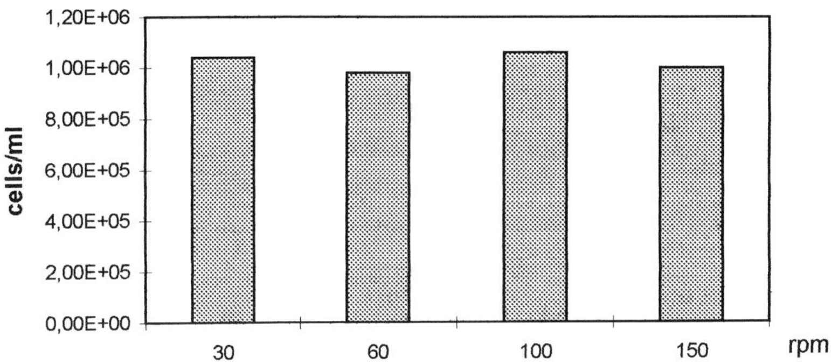

Figure 4. Maximun cell variation with respect to stirring speed.

For same shear factor obtained for 150 rpm in spinners, in the fermenter is posible to increase the speed up to 400 rpm. It was proved lately.

Oxygen transfer in the fermenter

Using the static gas out method (2) the kla values for different stirring velocities (100, 200, 400 and 500 rpm) and aeration rates (0.005, 0.01, 0.02, 0.05 vvm) were obtained and represented in Fig 5.

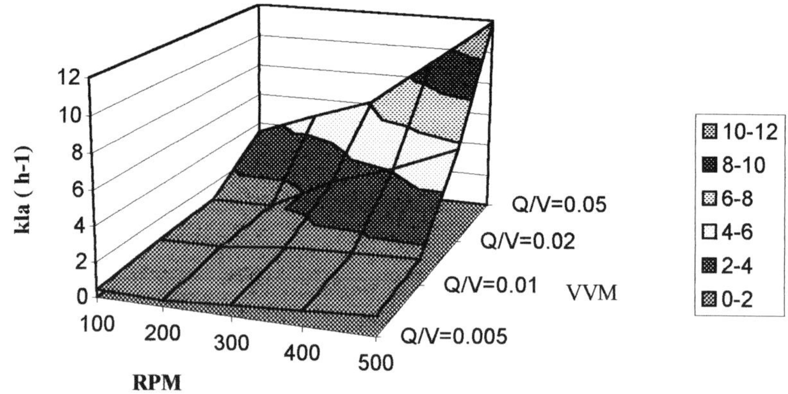

Figure 5. Kla values with respect to stirred speed and aeretion rates.

The final conditions selected for further work were 300 rpm or lower and 0.01 vvm . This represent a compromise between reasonable stirring speed and a kla value (around 0.7 - 1.3 h^{-1}), avoiding an excesive bubbling of air that could affect cell integrity.

Growth and production kinetic in suspension cultures

Kinetic experiments had an initial density of $2 \cdot 10^5$ cells/ml. Shown in fig. 2 and fig. 3 the mean values of cell density, inmmunoglobulin secretion, glucose consumption and ammonia, lactate production during the three cultures. The specific growth and production will be compared with those obtained in fermenter.

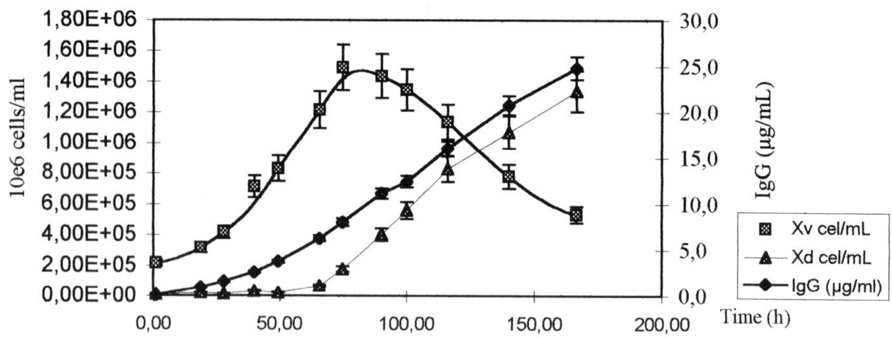

Figure 2. Profiles of cell density antibodies in suspension cultures.

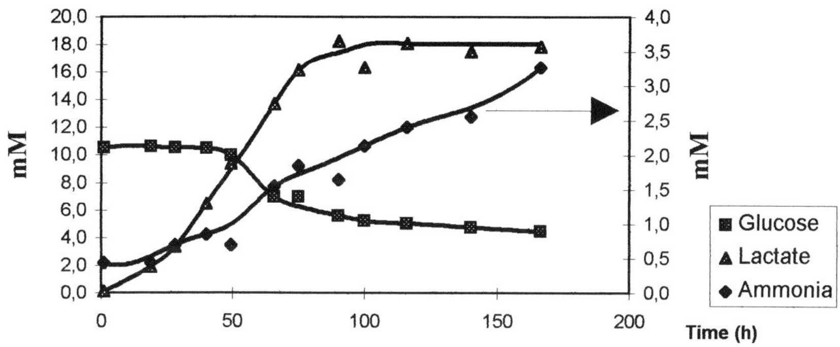

Figure 3. Profiles of glucose, ammonium and lactate concentration.

Cultures in fermenter.

Batch cultures in the fermenter were carried out using the same cell seeding density as spinner cultures.
Three stirring conditions were employed (150, 200 and 300 rpm) and the values of specific growth velocity and productivity are compared with those in spinners flask, Fig. 6. As can be observe there is no influence of tip speed , but there are difference, between the fermentor and the spinner cultures.

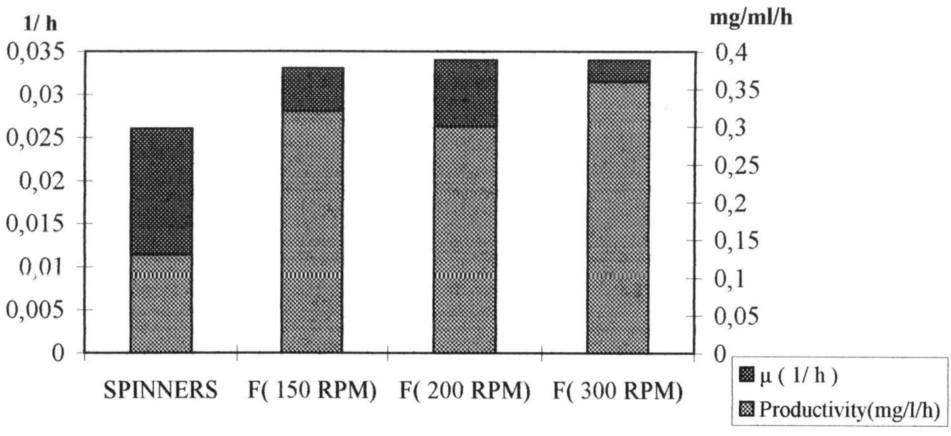

Figure 6. Specific growth rate and productivity in spinners and fermentor.

The better performance of fermenter let obtain higher cell density and antibody titers. This suggest that probably there was an oxygen limitation in spinners cultures. Optimization of medium composition (3) as well as the development of a fed-batch strategy will permit improve the cultures result.

CONCLUSIONS

The conditions for the improvement the production of IOR-R3h monoclonal antibody using TSTD cell line, have been determined. With an stirring rate of 300 rpm and 0.01 vvm of aeration rate is possible obtain satisfactory cell culture output in terms of cell and product yield, without deleterious effects on the cells.

REFERENCES

1. Franek F, Dolníková J (1991) Hybridoma Growth and Monoclonal Antibody Production in Iron-rich Protein Free Medium: Effect of Nutrient Concentration. Cytotechnology 7:33-38.
2. Atkinson B, Mavituna F (1983) Biochemical Engeneering and Biotechnology Handbook, MacMillan Publishers. Ltd.
3. Keen M.J. (1995) The Culture of Rat Myeloma and Rat Hybridoma Cells in a Protein Free Medium. Cytotechnology 17: 193-202.

Strategy for Scale Up of Mammalian Cell Culture in the New GMP Manufacturing Facility of the Cuban Biotechnology Industry

Villán J., Hermida C., Boggiano T., Sierra P., Lage A., Vitón P., Arias M. A., Valdés A. .
Center of Molecular Immunology
216th and 15th Street Atabey, P. O. Box 16040, Havana City, 11600, Cuba.
Tel/Fax: (53-7) 333509, 335049

Abstract

The main characteristics of the Strategy for the Scale Up of Mammalian Cell Culture followed by the Cuban Biotechnology Industry is shown and analyzed. In this Strategy are included as key elements: the expertise of the technical personnel and the access to the know how, the necessity to create an adequate environment, that means the construction of a new GMP facility and the state of the art of the technological aspects of this discipline. The main experimental results obtained by this Center, are shown.
Key words: Strategy, Monoclonal Antibodies, Biotechnology, GMP Facility, Hollow Fiber Bioreactors, Stirred Tank Fermenters.

Introduction.

In spite of the efforts and the results achieved in the field of monoclonal antibodies, its medical application and commercialization is still a promise and the expectatives for the nineties have not been fulfilled. Among the possible causes of this gap are the difficulties of scaling up amounts of product for clinical trials and the year by year increasing cost of the process (1).

The Center of Molecular Immunology was created in 1994 potentiating the development of the scientific-production investment of the country in the field of Biotechnology. The main characteristics of this Center are the technological capacity for the scaling up of mammalian cell culture and the projection in cancer and immune system research. In the design and construction of the Center the current Good Manufacturing Practices guidelines adopted by Cuba and those recommended by WHO and by the European Union have been followed (2).

Strategy for the Industrial Scale Up of Mammalian Cell Culture.

The Industrial Strategy for the Scale Up of Mammalian Cell Culture is much more than the specific discussion about the technological aspects and the state of art of Mammalian Cell Culture, it is also the integral analysis about the products and marketing, the national and international regulations, the accepted industrial standards, the expertise of the technical personnel, and the property or possibility to access to the

M.J.T. Carrondo et al. (eds.), Animal Cell Technology, 441–446.

know how. There is more than one way to establish a successful Industrial Strategy for the Scale Up of Mammalian Cell Culture. The best strategy should be specific and different for each group.

Here we will explain the Cuban approach to the Strategy for Industrial Scale Up of Mammalian Cell Culture:

1. Technical Personnel and the access to the Know How.
2. Building of a GMP Manufacturing Facility in Cuba for the Scale Up of Mammalian Cell Culture.
3. Technological Aspects of Mammalian Cell Culture

We will present a general overview about the main aspects of this Strategy:

1. Technical Personnel and the access to the Know How.

The senior staff of CIM started to work on monoclonal antibodies (MAbs) in 1981. One of the first MAbs obtained was an anti T lymphocytes monoclonal antibody (anti CD6). This MAb was used in 1984, in the first clinical trial in Cuba with this type of molecules, for the treatment of cutaneous T cell lymphomas. In this decade an anti CD3 MAb was also obtained and is registered in Cuba since 1989 for prophylaxis and treatment of organ transplant rejection. MAbs for the first clinical trials and marketing were produced in an "in vivo" production facility with a capacity of 100 g.

A network for the generation of monoclonal antibodies has been created and leaded by the CIM, in the universities all along the country. This network accomplishes a double objective, one is the experimental basis for scientific development of the researchers, and second it is another way to obtain new hybridomas. A permanent training program is performed for the specialists of CIM in Germany, France, Switzerland, etc.

2. Building of a GMP Manufacturing Facility in Cuba for the Scale Up of Mammalian Cell Culture.

CIM is basically a 15, 000 m^2 two floor facility at the west side of Havana City. The GMP production area has been designed for maximum product protection. In Table 1 are shown the main CIM's areas and their classification.

Laboratories	Area	Classification
GMP Production:	1100 m^2	Cleanroom
Hollow Fiber Bioreactors (HFB)	77 m^2	100 000 + (LFC)*
Small Stirred Tank Fermenters (STF)	47 m^2	10 000 + (LFC)
Large Stirred Tank Fermenters (STF)	56 m^2	10 000 + (LFC)
Cell Culture Laboratory for HFB	75 m^2	100 000 + (LFC)
Expansion Laboratory for STF	31 m^2	10 000 + (LFC)
Media Quarantine Tanks	35 m^2	10 000
Media Preparation & Filtration Units	52 m^2	Controlled Area & 10 000
Separation/ Concentration	49 m^2	10 000

Table 1

(* LFC= Laminar Flow Cabinet installed)

This GMP production area is divided in two sections. The alive section where the activities of cell expansion and the technologies of fermentation in hollow fiber and stirred tank bioreactors are performed. The dead section contains the chromatographic purification technology, lyophilization and the vial filling and labeling.

Physical barriers guarantee the independence of each area, activities and technologies avoiding the risks of cross contamination. Additionally, the organizational procedures for a unidirectional stream of personnel and materials have been studied and applied, permitting the establishment of a multiproduct cleanroom.

Equipment	Qty	Total Capacity
Demineralizing Units	2 Units	10 000 L/h
Multistep Stills Units	2 Units	1 500 L/h
Clean Steam Generator	1 Unit	5 80 Kg /h
Clean in Place Units	3 Units	22 m^3/h
Double Door Autoclaves	3 Units	2 800 L
Class 100 Oven	1 Unit	500 L
Biowaste Sterilizat.. Units	2 Units	9 m^3
Oil Free Air Compress.	2 Units	12.6 m^3/min
WFI loops	8	—

Table 2

The CIM has powerful ancillary technical systems (Table 2). Stainless steel AISI 316L pipes with orbital welding were used for the circulation of water for injection, pure steam, cell culture medium transferring and for Cleaning In Place.

3. Technological Aspects of Mammalian Cell Culture

The possibility of commercial production of monoclonal antibodies produced by hybridoma cells has also stimulated further development and optimization of bioreactors for cell culture.

The large scale production of MAbs is a subject of many reviews. Currently, several technologies exist for large-scale mammalian cell culture, but not one of these technologies fulfill successfully, for the different hybridomas, with the criteria that some authors establish as optimal for the large scale mammalian cell production.

Equipment	Qty	Capacity
HollowFiber Bioreactors	10 units	Experimental, Medium & Industrial
Stirred Tank Ferm.	12 units	7, 14, 41, 75, 150,450L
Sep./ Conc.	3 Units	Small, Medium, Big
Medium Preparation	3 Units	100, 600 L
Media Quarantine	4Tanks	2 500 L

Table 3

We decided to install at CIM the two leading scale up technologies , the Stirred Tank Fermenters and the Hollow Fiber Bioreactors (Table 3).

The ultimate choice between the two systems therefore largely depends on various factors such as the characteristics of the antibody, amount of antibody that is requested, the regulations and demands of the market, the existing capacity of the facility infrastructure and the background training and flexibility of the available human resources.

Once the technology is decided, the success of the Scaling Up of the Mammalian Cell Culture in CIM will depend on the ability to handle successfully the problems related with the growth media, the strains, the efficient environmental control, the process of the supply of oxygen and other nutrients to the cells and removing toxic metabolites, mixing, and the use of perfusion systems to achieve high cell densities.

Specific strategies were established in the following aspects:

1. Glucose, lactic acid, glutamine and ammonia control.
2. Oxygen transfer increase.
3. Shear force effect decrease
4. Use of spin filters.
5. Low/Free serum medium, low protein medium.
6. Obtaining industrial strains.
7. Particular and specific exploitation of both technologies.

Experimental results

During these 2 years of GMP facility working, different results have been obtained in the technology assimilation and development (3).

Main experimental results obtained:

A) Kinetic metabolism study and obtainment of industrial strains.

A hybridoma sub-clone (r3-AL) capable to proliferate in culture medium supplemented with 8 mM sodium lactate was isolated through conventional in vitro manipulation from a murine hybridoma (ior-r3), secreting IgG2a monoclonal antibody against epidermal growth factor receptor.

Kinetic and metabolic coefficients didn't show major differences for both hybridoma strains, except for specific ammonia production rate and specific immunoglobulin secretion rate which were higher for sub clone r3-AL (Table 4).

	ior-r3	r3-AL
μ (h^{-1})	0.033	0.032
qGlu (µmoles/g/min.)	0.95	1.09
qLac (µmoles/g/min.)	1.38	1.32
qGln (µmoles/g/min.)	0.33	0.44
qAm (µmoles/g/min.)	0.37	1.12
qAb (pg/g cells/min.)	1.89	2.72

Table 4

Biomass yields from glucose or glutamine as also immunoglobulin yield from cells were also enhanced for subclone r3-AL in relation to original cell population.(Table 5)

	ior-r3	r3-AL
$Y_{X/Glu}$	6.33×10^{-6}	8.15×10^{-6}
$Y_{X/Gln}$	2.0×10^{-6}	3.50×10^{-6}
$Y_{Ab/X}$	1.37×10^{-5}	3.31×10^{-5}

Table 5

B) Selection and adaptation in serum free and protein free media.

Serum free medium adaptation of hybridoma secreting ior-t3 monoclonal antibody was performed. Different metabolic variables were monitored in relation to a control . Metabolic behavior (specific growth rate) of this hybridoma in serum free medium was similar to control. (Figure 1) (4, 5).

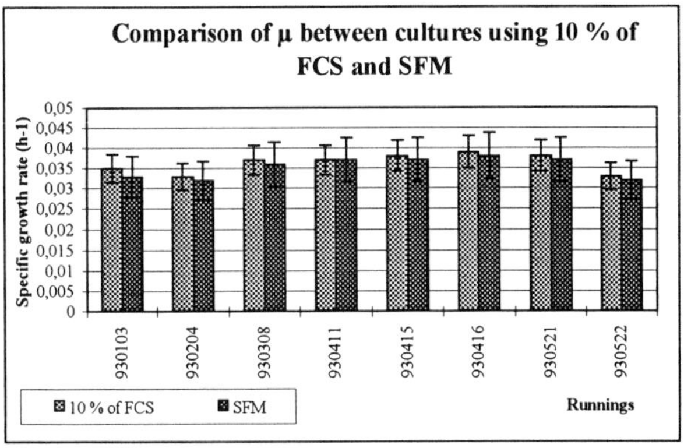

F igure 1

C) Comparison between different culture strategies.

The MAb concentrations obtained in batch cultures in 1.5 L bioreactor remained at the same level as in static and spinner cultures. However, these concentrations in perfusion culture were higher, and the highest MAb concentration was collected from the last exponential growth during one of perfusion culture. (446 ± 25 μg/mL).

The perfusion system had the highest reactor productivity, produced the most concentrated MAb (Table 6) and used the lowest amount of media per gram of MAb.

	Spinner (500 mL)	Batch cult. (1400 mL)	Perfusion cult. (1400 mL)
MAb product. (mg/10^6 cells/d)	13,14 ± 067	10,83 ± 0,60	426 ± 84
System productivity (mg/L/d)	8.76 ± 0,67	7,06 ± 0,60	22,62 ± 5.21
Harvested Mab conc. (μg/ml)	61,39 + 3,07	50,60 ± 2,81	195 ± 22
MAb final (mg)	92,08 ± 4,56	75,90 ± 2.22	6071 ± 115

Table 6

D) In vitro production of MAbs and establishment of a strategy using hollow fiber bioreactors

The influence of the secretory rate, the inoculum cellular density of and the runnings length were tested during the growth phase in runnings for monoclonal antibody production from the hybridoma cell line ior-t3 using hollow fiber bioreactors.

In Table 7 results of five experimental runnings in hollow fiber bioreactors, using different subclones of ior-t3 are shown. An increase in productivity was found for high density inoculum (6).

Running	1	2	3	4	5
Conc. (μg/mL)	4.27	8.1	14.6	42	8.1
Inoculum (10^6 cells)	224	2512	500	430	500
Time (days)	20	20	20	20	20
Final Product (mg)	145	548	152	288	180

Table 7

Conclusions:

The results of the strategy followed for the mammalian cell culture in the Center has permitted the Institution to produce hundreds of grams of MAbs of injectable quality, which are currently used in more than 20 clinical trials in cancer and immune diseases in Cuba and also for the national and international market.

References

1. Lage A., Sierra P., Vázquez A, M. (1993). Anticuerpos monoclonales a escala industrial: Riesgos y Oportunidades. Inmunología (España), Vol. 12, Número 2, Abril-Junio., 64-66.
2. Villán J. (1995). Inmunología Molecular. Escalado de Células Superiores. Avances Médicos de Cuba. Año II. Número 2, 4-7.
3. Villán J. (1995). La importancia del escalado en el desarrollo de anticuerpos monoclonales para uso terapéutico. Lecture at Third International Scale-Up Workshop about Natural Products, Biotech and Chemistry Tech. Cuba.
4. Hermida C., Fernández A., Bouzó L., Alvarez I., Toledo A. L., Peña O., Brito C., Gómez J. A. (1995). Estudio cinético y metabólico del IOR-R3 a diferentes concentraciones de suero. XII Scientific Seminar (CNIC). Cuba.
5. Bouzó L., Vitón P., Jäguer V., González F., Salvo N., Alvarez I. (1994). Evaluación del medio libre de suero en la producción del anticuerpo monoclonal ior- t3. Biotecnología Aplicada. Vol. 11, Número 3, 225-230.
6. Salvo N., Castillo A., Fernández A., Bouzó L., Torres K., González F. (1994). Optimización de la producción del anticuerpo monoclonal ior-t3 en bioreactores de fibra hueca. Biotecnología Aplicada. Vol. 11, Número 2, 160-164.

PROCESS DEVELOPMENT FOR THE LARGE SCALE PRODUCTION OF CLINICAL GRADE HUMAN MONOCLONAL ANTIBODY USING A CYTOPILOT FLUIDIZED BED REACTOR

G. Klima, G. Kreismayr, D. Müller, C. Schmatz, S. Wiederkum, W. Steinfellner, A. Assadian, G. Lhota, G. Blüml, O. Doblhoff-Dier, H. Katinger

Institute of Applied Microbiology (IAM)
University of Agriculture Vienna
Nussdorfer Lände 11, A-1190 Vienna, Austria

Introduction

Monoclonal antibodies for passive immunotherapy demand for cheap large scale manufacturing technologies. Previous studies showed that fluidized bed reactors (FBR) are a promising production system considering scale up potential and volumetric productivity.
r-CHO cells were cultivated on 200-400 ml settled volume of macroporous microcarriers (Cytoline 1, Pharmacia Biotech) in a Cytopilot-Mini FBR to optimize relevant parameters for large scale production of clinical grade material in a 120 l FBR (20-30 l of packed bed). Main focus was put on i) balancing medium composition to enable reasonable perfusion rates, ii) optimized carrier preparation, iii) high efficiency of cell attachment during inoculation iv) fast change from serum supplemented propagation medium to protein free production medium.

Materials & Methods

r-CHO cell line
Human anti-HIV-1 monoclonal antibody IAM CHO2F5
Cytopilot fluidized bed reactor (Pharmacia Biotech)
Cytoline 1 macroporous microcarriers (Pharmacia Biotech)
DMEM :HAMS-F12 1:1
4-l stirred tank reactor (STR) for inoculum preparation (Chemap AG)
HPLC analysis of glucose and amino acid concentration

Carrier Preparation

The microcarriers were prepared and sterilised outside the FBR (e. g. in a modified 50 l vessel) and subsequently transferred into the bioreactor. This simplified the handling of large quantities of microcarriers (20-30 l) as well as the sterilisation of the pilot scale fermentation system (e. g. homogenous temperature distribution in the fermenter, separate sterility testing of fermenter and microcarriers).

M.J.T. Carrondo et al. (eds.), Animal Cell Technology, 447–450.

Balancing Medium Composition

Preoptimization was carried out in T-flask batch cultures.

Due to different amino acid consumption rates in different culture systems the final formulation (Tab. 1) was determined during continuous culture in a stirred tank reactor. No amino acid limitations occurred in the FBR down to residual glucose concentrations of ~500 mg/l.

amino acid	DMEM:HAMS F12	supplement
Glutamine	363 mg/l	580 mg/l
Glutamic acid	8 mg/l	90 mg/l
Aspartic acid	7 mg/l	90 mg/l
Asparagine	7 mg/l	90 mg/l
Serin	27 mg/l	20 mg/l
Tryptophane	9 mg/l	5 mg/l

Tab. 1: Standard and supplemented amino acid concentrations

Inoculation

To achieve the large number of cells needed for the inoculation of a FBR (~$2*10^{10}$ cells for a 20 l carrier bed) a stirred tank reactor - operated as fed batch - was used for inocula preparation. The cells were growing as aggregates and were transferred into the FBR through the bottom valve.

To achieve a shorter lag phase the STR was fed batched again and the FBR was inoculated a second time. After reaching a minimum cell density of $>1*10^7$ cells/ml packed bed protein free perfusion was started (Fig. 1, Tab. 2).

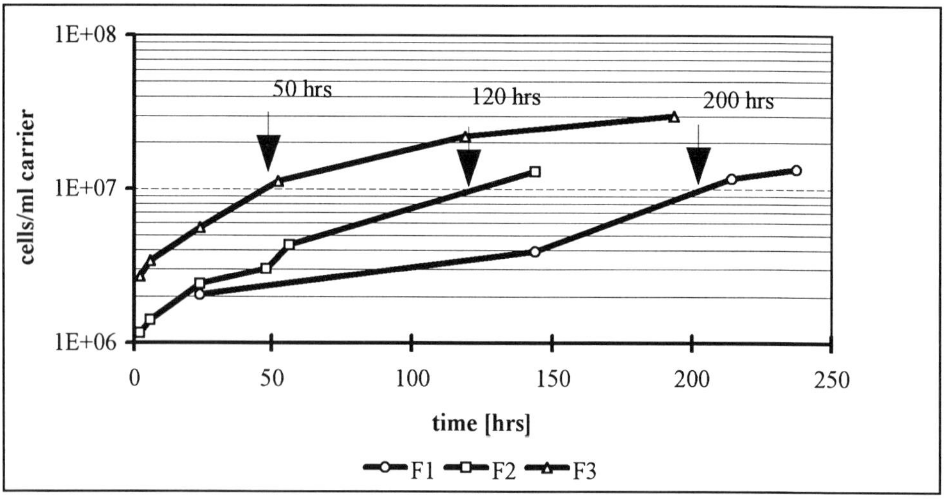

Fig. 1 : Comparison of starter culture development during the first 250 hrs of different fluidized bed fermentations.

--- minimum cell density for protein free perfusion

↓ minimum time for protein free perfusion

fermentation	inoculum	24 hours	attachment rate	2nd inoculum
F1	3,5 E6	2,0 E6	60 %	no
F2	2,4 E6	2,4 E6	95 %	yes
F3	5,6 E6	5,6 E6	80 %	yes

Tab. 2: Cell number/ml carrier at inoculation and after 24 hrs and calculated cell attachment rate after 24 hrs.

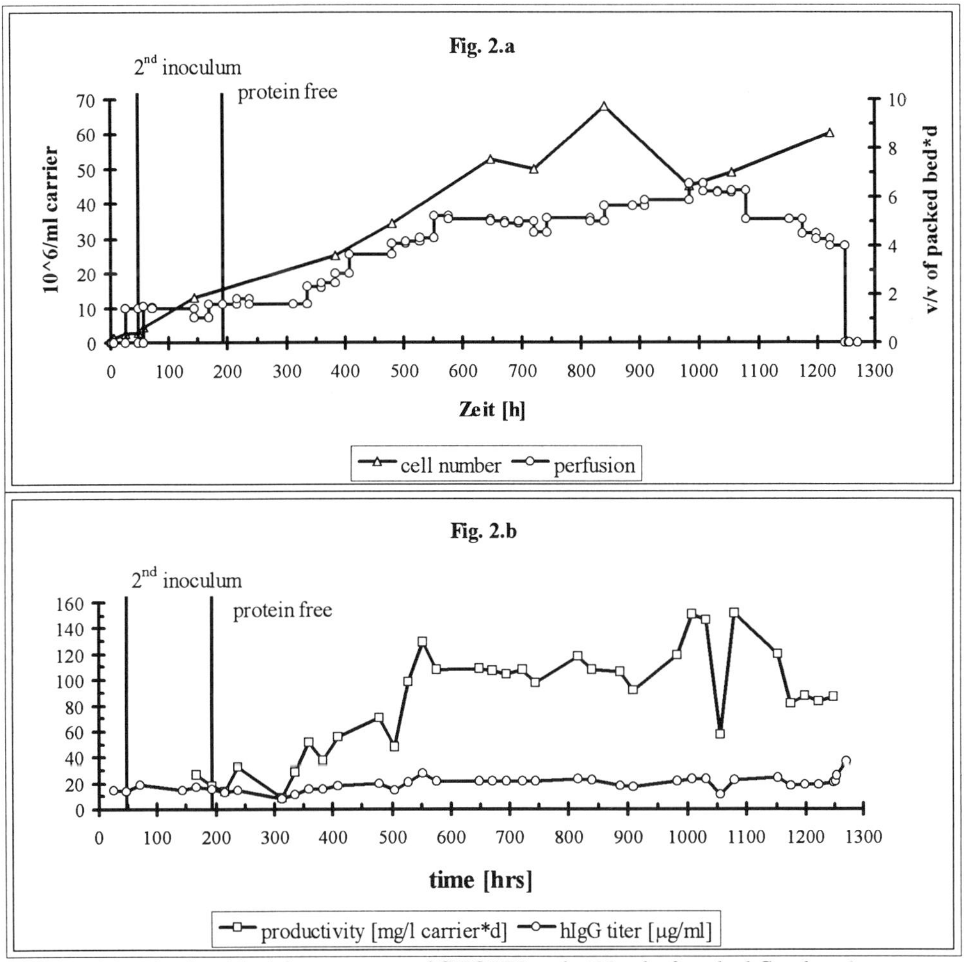

Fig. 2.a & b: Fluidized bed fermentation of CHO2F5 with 400 ml of packed Cytoline 1
Fig. 2.a: Cell number and perfusion rate versus time
Fig. 2.b: Productivity and hIgG titer versus time

Summary

Several fermentations lasting between 700-1200 hours were performed in Cytopilot fluidized bed reactors, achieving cell densities around $4\text{-}7*10^7$ cells/ml of settled carrier volume and a productivity of approximately 100-120 mg/l carrier*d.

Perfusion of serum supplemented medium can be reduced to 50-100 hours depending on the cell density of the starter culture.

Acknowledgments

Part of this work was supported by Pharmacia Biotech

SCALE-UP OPTIMIZATION BASED ON THE DEVELOPMENT OF A PERFUSION RATE CONTROL FOR CHO CELLS USING A CYTOPILOT FLUIDIZED BED REACTOR

D. MÜLLER, G. KREISMAYR, G. KLIMA, G. LHOTA, A. ASSADIAN, C. SCHMATZ, S. WIEDERKUM, W. STEINFELLNER, G. BLÜML, O. DOBLHOFF-DIER, H. KATINGER
Institute of Applied Microbiology, University of Agriculture, Nußdorfer Lände 11, A-1190 Vienna, Austria

Abstract

To optimize cell density and productivity several lab-scale fermentations in a fluidized bed reactor (Cytopilot-Mini™, Pharmacia Biotech) have been carried out with recombinant CHO cells immobilized on macroporous microcarriers (Cytoline™ 1, Pharmacia Biotech) producing a clinical grade human monoclonal antibody directed against HIV 1. After the cells were propagated in a 2% FCS medium to a cell density of at least 10^7 cells/ml carrier, they were able to grow and produce in a protein free medium reaching cell densities of about $5x10^7$ cells/ml carrier. Media compositions were optimized to reduce glucose- and amino acid-limitations during high-cell density perfusion. Oxygen uptake rates were monitored to characterize different states of the fermentation process. The pO2-controller output (aeration rate) was successfully used as an additional estimation for the oxygen uptake rate. The current development of a control strategy for the perfusion rate is based on the oxygen uptake rate and on the glucose consumption rate. This control strategy is designed to avoid nutrient limitation and simplify the fermentation handling in large-scale production.

Introduction

The perfusion rate of previous fluidized bed fermentations was dependent on glucose and amino acid concentrations measured by means of HPLC. The perfusion rate then was either increased or decreased manually due to the measured values. This kind of control strategy was often not adequate to avoid nutrient limitations, especially during exponential growth conditions because of the long period of time between sampling and analysis. The current, new strategy for perfusion rate control is based upon the oxygen uptake rate (OUR) and the glucose consumption correlated with the pO2-controller output (aeration rate). The frequency of pO2-controller output signals in this case is a criterion for oxygen uptake. For example, decreasing intervals between output signals describe an increase in oxygen uptake. A reproducible correlation may eliminate the necessity of determining

M.J.T. Carrondo et al. (eds.), Animal Cell Technology, 451–454.

quantities of consumed oxygen and glucose for shifting the perfusion rate. Several lab-scale-fermentations were tested out to ascertain a reproducible relationship between these parameters. The results are presented on this poster beginning with a general survey of the different states of development.

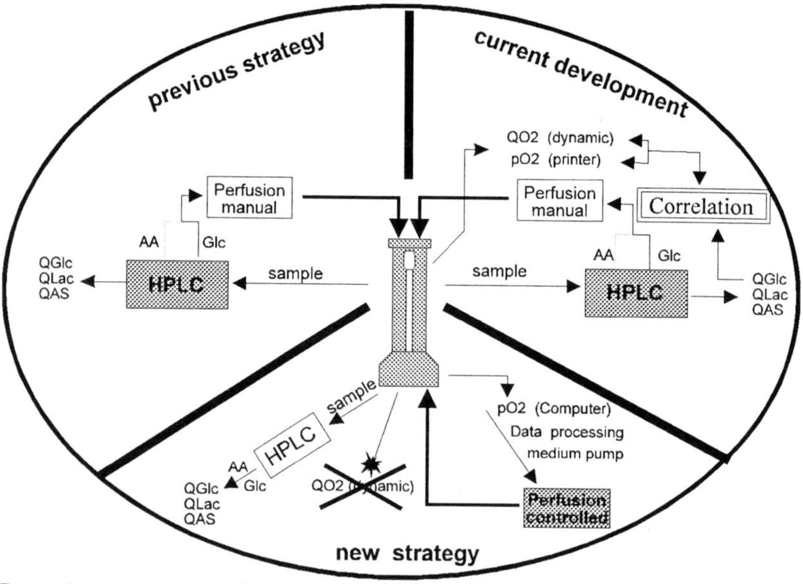

Fig. 1: General survey of the different states of development; AA: aminoacids, Glc: Glucose, Lac: Lactate, Q...: volumetric rate of consumption or production

Results

The measurement of volumetric OUR's compared with the corresponding cell density displays two states of constant specific oxygen uptake. The first state represents exponential growth. The second state representing maintenance metabolism is accompanied by an OUR decrease downward to values of about 60% as shown in Fig.2. The specific glucose consumption rate was found to be constant along the whole fermentation (Fig.2). These results show that there are no significant changes in glucose metabolism and that the change in oxygen metabolism is clearly definable. As a consequence the increase of the volumetric rates during exponential growth only depends on cell proliferation. The decrease of specific oxygen uptake also results in a two-state-relationship regarding glucose consumption. During exponential growth the volumetric glucose consumption rate rises depending on increasing OUR. With the beginning of maintenance metabolism the glucose consumption is constant and uncoupled from the oxygen uptake.

These relationship between cell density, oxygen uptake and glucose consumption puts up the active principle on which the on-line control for the perfusion rate is developed. The on-line determination of process conditions is based on the measurement of the pO2 in the fermentation fluid which is monitored by a computer the same way as the pO2-controller output. As a consequence of discontinuous aeration by sparging pure oxygen into the fluid the periodical controller signal is a criterion for the consumption of oxygen, of glucose and a criterion for the cell growth rate. Considering the profile of the controller signal within a quasi stationary time period (e.g. one hour) it is obvious, that the intervals between two controller signals are constant, as shown in Fig.3. Therefore these data can be processed.

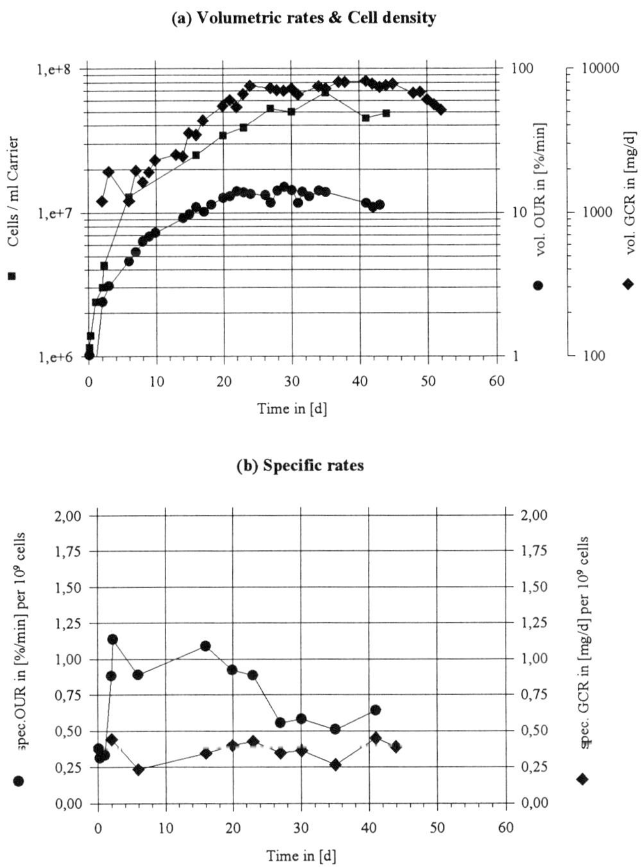

Fig.2 : Volumetric and specific rates; OUR: oxygen uptake rate, GCR: glucose consumption rate, vol.: volumetric, spec.: specific, C.: Carrier

Discussion

An accurate interpretation of these results realized by a data processing unit which is able to convert the incoming on-line data into a corresponding medium flow will be the concept for upcoming lab-scale fluidized bed fermentations. This control strategy avoids nutrient

limitations, reduces the expenditure of work and does not influence the running process because dynamic measurements of OUR's are not necessary any longer. The current strategy shall be applied to large-scale fluidized bed reactors and so simplifies the fermentation handling as well in large-scale production.
It should be stated that the correlations are specific for the cell-line, for the fluidized bed reactor and for the chosen controller parameters.

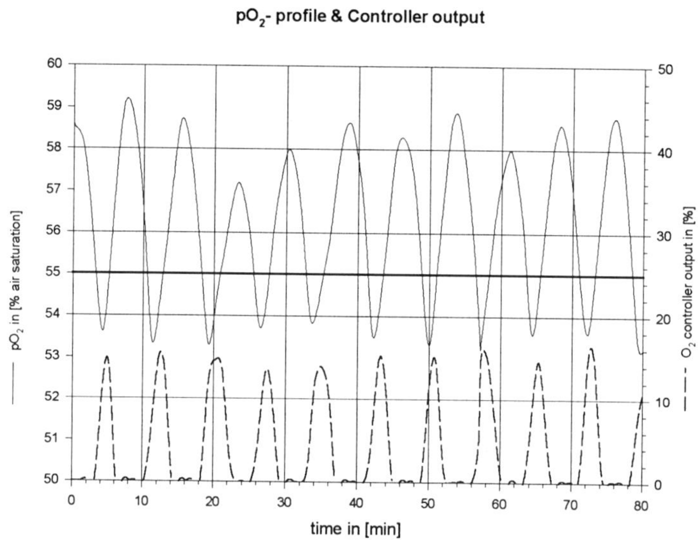

Fig.3: pO2-profile and controller output within a quasi stationary time period; setpoint: 55% air saturation

Development of CHO cell serum-free culture process for expression of recombinant viral glycoprotein.

M.M. Gonze, L. Pierard, R. Fehrenbach, Y. Ghislain and L. Fabry
SmithKline Beecham Biologicals s.a., 1330-Rixensart, Belgium

ABSTRACT

Serum-free CHO cell culture provides several advantages for an improved control of recombinant viral glycoproteins production and for the process scale-up. Culture medium was designed to replace the major serum components by synthetic or recombinant compounds so that the growth/production medium is free of elements derived from animal origin. The medium was supplemented with amino acids, vitamins, Fe carrier, lipid precursor, anti-shear and anti-oxidant agents to adapt CHO cell lines to suspension culture and to develop a batch production process for the recombinant glycoproteins. In these conditions, the biochemical characteristics of the secreted glycoproteins were assessed using isoelectric focusing technique followed by Western blotting on supernatant samples. The glycoprotein profile given by this technique confirmed the protein integrity data obtained by N-terminal sequence analysis performed on the purified protein. The method was used for the optimization of the serum-free culture process for large scale development.

INTRODUCTION

The objectives of the study were to develop a cell culture process in a medium devoid of animal derived components for viral recombinant vaccines production and to better control the cell culture conditions for the production of a constant glycoprotein quality. The first step consisted of the development of the serum free culture medium and the adaptation of the cell line of interest to suspension culture in the new medium. During the second step, the glycoprotein profile was analysed "in process" by isoelectric-focusing/Western blot technique (IEF/WB), with particular attention paid to the N-terminal processing of the protein.

MATERIAL AND METHODS

Cell lines and culture media: The recombinant CHO cell lines expressing the viral glycoprotein of interest were gradually adapted to grow in suspension in SB proprietary serum free medium.

Analytical assays: The cells were counted using a hemocytometer and the viable cell concentration was determined using the trypan blue exclusion method. Glucose/lactate concentrations were determined with an YSI-2700 analyser. Ammonia and the amino acid concentrations were determined by HPLC (fig 1).

Fig. 1 <u>Amino acids consumption as followed after 24H/48H/72H/96H of culture by HPLC analysis</u>:

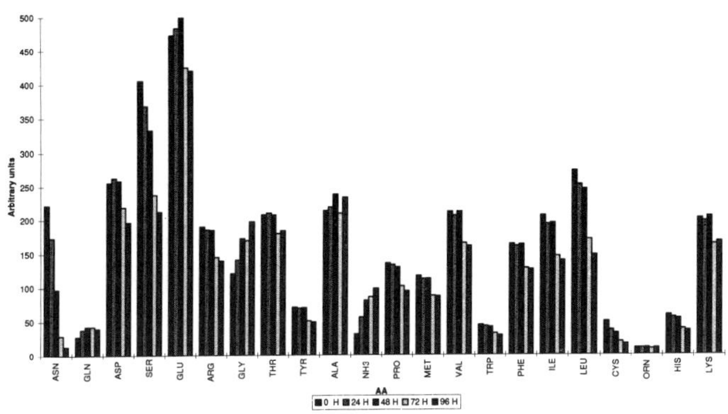

M.J.T. Carrondo et al. (eds.), Animal Cell Technology, 455–459.

Antigen concentration was measured by ELISA. For the isoelectric focusing and W.B. detection, the glycoprotein samples were concentrated 15 fold by ultrafiltration. Gel electrophoresis was performed under denaturating conditions as reported by O'Farrel (1975) and Lognonné (1994) in a vertical slab gel unit; a pH gradient between 3 and 6 was used; migration was stopped after 30 000 Vxhours. After successive washes in a fixing solution, gels were blotted to PVDF membranes in the presence of SDS using a semidry transblot system. After blocking, either a N-terminal monoclonal antibody or a rabbit polyclonal antibody.was used for detection.

RESULTS AND DISCUSSION

Cell line adaptation to suspension/serum free culture and glycoprotein production:

Fig 2A shows the adaptation of an adherent recombinant CHO cell line to suspension culture in serum free medium in parallel and with control culture in medium containing 7.5% FBS (fig 2B). Compared to culture with FBS, the SF adapted cell line displays the same high viability (>95%) and the same or better growth profile (4-5 10^6 cells/ml). Yields of secreted viral glycoprotein are at least equal to those obtained in serum containing medium. The same cell line characteristics are maintained for over 100 generations with and without selective agent.

Fig. 2A <u>Adaptation to suspension/serum free culture: growth/viability/production</u>

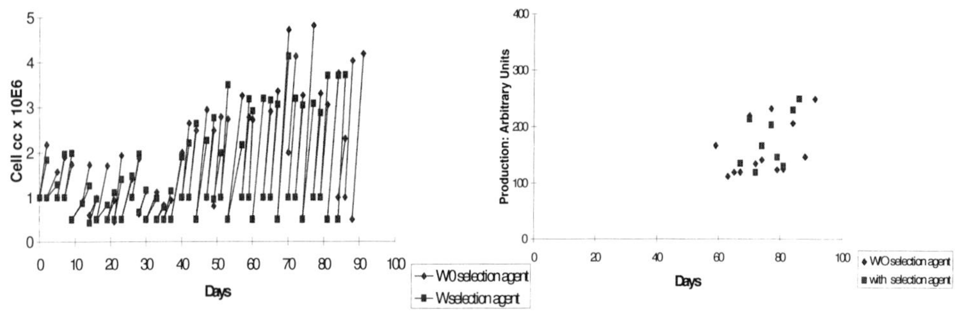

Fig. 2B <u>Adaptation to suspension culture (control with 7.5% serum): growth/viability/production</u>

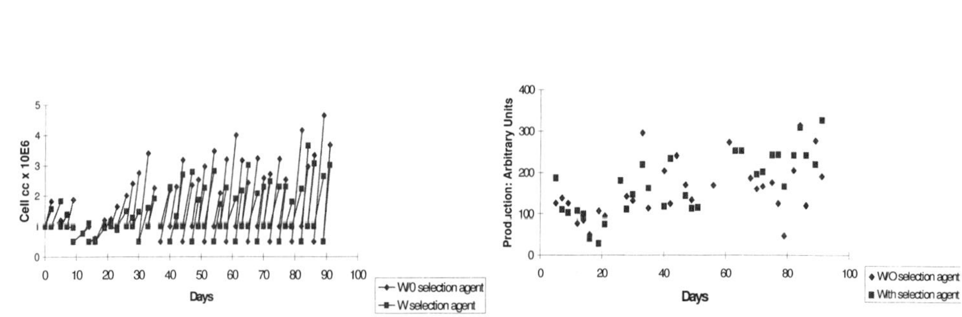

The cultures produced in 15 L batch mode (4 days) give similar results (Fig. 4) as in spinner flask 1L (fig.3).

Fig. 3 Batch production with recombinant CHO cells adapted to the culture with SF medium SB : SP1000

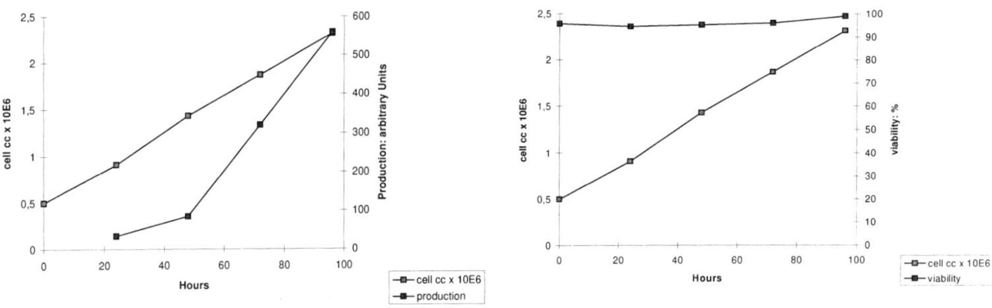

Fig. 4 Batch production with recombinant CHO cells adapted to culture in SF medium : Bioreactor 15L:

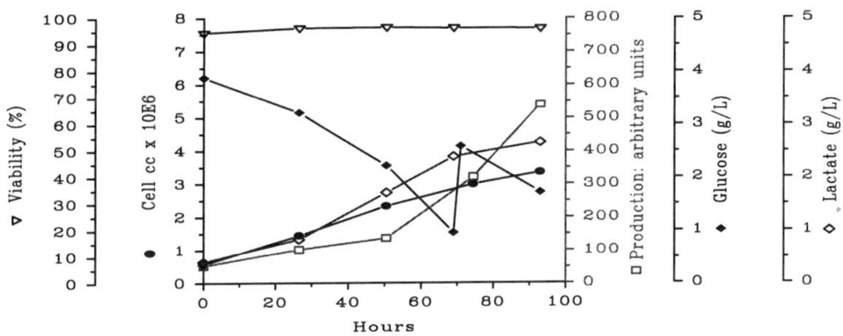

Analysis of purified recombinant glycoprotein samples by IEF/WB

Glycoprotein integrity was evaluated using the IEF/WB technique to discriminate between samples displaying different N-terminal. processing profiles (figure 5). The IEF/WB of a purified recombinant glycoprotein sample with a low percentage of N-terminal. modifications revealed six major bands at a pI between 4.91 and 5.28 (fig 5 lane C: bands 2', 4, 6, 7, 8, 9). For the sample with a high percentage of N-terminalinal modifications (figure 5 lanes a and b), one additional basic band was detected by the polyclonal antibody (bands 11) as well as 5 additional acidic bands (bands 1, 2, 2', 3, 5). The additional bands are not detected by the N-terminal monoclonal antibody. Sequence analysis of the additional bands showed indeed that they correspond to secondary sequences resulting from N-terminal. processing.

The IEF/WB technique was used to further improve the cell culture/production process. Several culture conditions have been tested including lower temperature and change in culture conditions, eg. perfusion vs batch. In both cases the intensity of the additional acidic bands is reduced (data not shown).

Fig. 5 <u>IEF/WB profile after detection by N-terminal monoclonal antibody and polyclonal antibody</u>

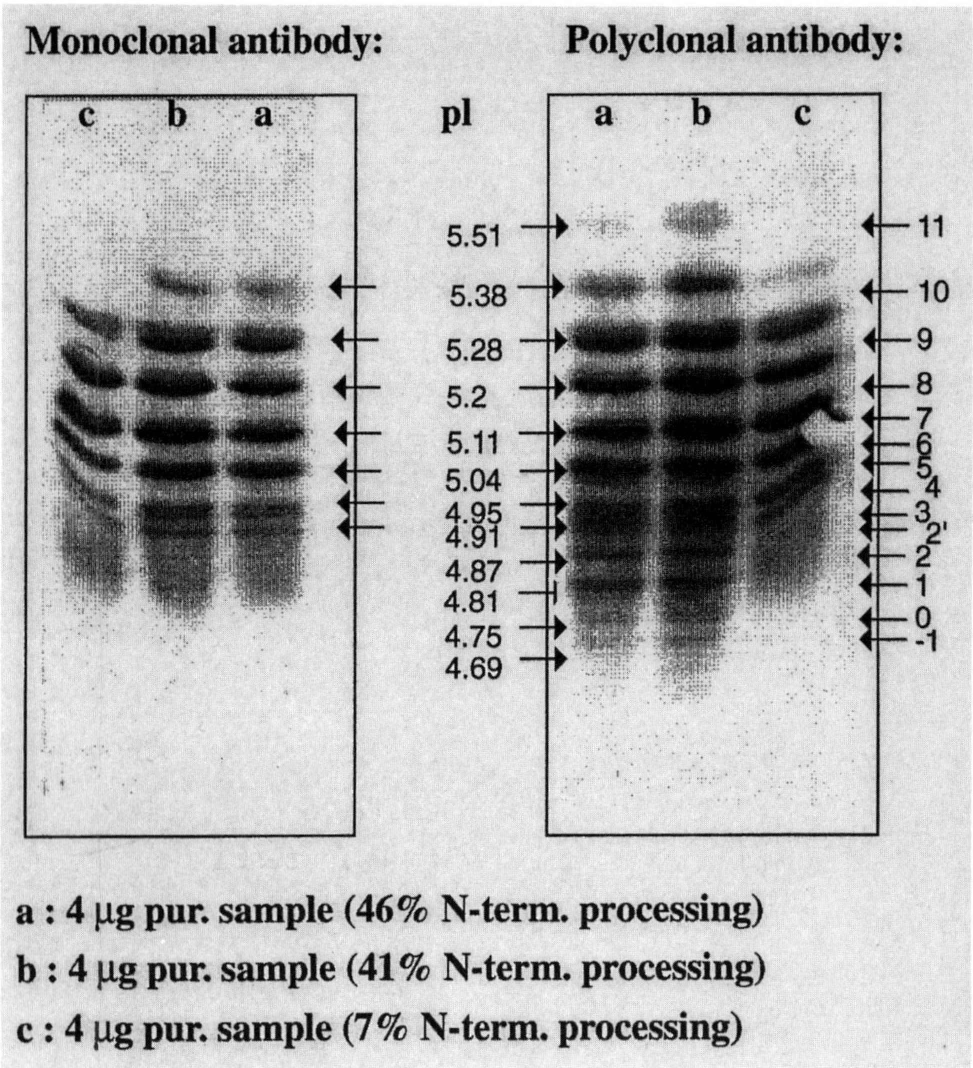

CONCLUSIONS

Serum free culture media devoid of components of animal origin were developed to adapt recombinant CHO cell lines to suspension culture and for the production of viral glycoprotein. Reproducible cell growth profiles and high cell viability, as well as a constant glycoprotein expression level can be maintained during long term propagation in serum free suspension conditions with and without selective agent.

Isoelectric focusing technique with western blotting immunodectection can be used to evaluate N-terminal. processing of the glycoprotein secreted in the culture supernatant. The technique is found to be very useful for the monitoring of glycoprotein quality, for the control of consistency in different culture conditions and for process optimization.

ACKNOWLEDGEMENTS

We thank C. Capiau, D. Plainchamps and M. Loiseau for valuable discussions and for sample sequencing.

REFERENCES

Cumming, D.A. (1991) : Glycosylation of recombinant protein therapeutics : control and functional implications; Glycobiology, 1, 2, 115-135.

Lognonné, J.L. (1994) : 2D-PAGE analysis : a practical guide to principle critical parameters. Cell. Mol. Biol.40(1),41-55.

O'Farrel, P.H. (1975) : High resolution two-dimensional gel electrophoresis of proteins. J.Biol.Chem. 250,4007-4021

CELL PRODUCTIVITY IS OVERESTIMATED BY THE CRYSTAL VIOLET NUCLEI COUNTING TECHNIQUE

J.M. BERRY[1], E. HUEBNER[2] AND M. BUTLER[1]
Departments of Microbiology[1] and Zoology[2], University of Manitoba, Winnipeg, Manitoba CANADA R3T 2N2.

Abstract

Nuclear counts determined by crystal violet staining from samples of stationary or microcarrier cultures of hybridomas, CHO or Vero cells were consistently and significantly higher than cell concentrations determined by the trypan blue or Coulter counter methods. This difference was attributed to the presence of a significant proportion of binucleated cells, which are estimated at 35% of the cell population in the stationary phase of Vero cultures. The proportion of such cells during exponential growth was variable. However, continuous sub-culture of these cells induced a degree of synchrony during growth which resulted in a cyclic variation of the difference between the cell and nuclei counts. This data indicates that cell productivity may be overestimated if cultures are monitored solely by crystal violet counting.

Introduction

A number of cell counting techniques have become standard protocols for monitoring mammalian cell cultures. Haemocytometer counting using the trypan blue dye exclusion method (1) is ideal for suspension cultures. This allows the total cell concentration and % viability to be determined.. However, adherent cells lend themselves less well to this method as they first have to be physically or enzymatically removed from the growth surface. Hence, the most common method used with adherent cells has been the crystal violet nuclei stain procedure (2) widely applied to microcarrier cultures. This involves haemocytometer counting of stained nuclei isolated from lysed cells.

In this paper we compare the nuclei counting method of monitoring cultures to the trypan blue and Coulter counting methods. Three different cell lines were used - Vero, Chinese hamster ovary (CHO) and a murine hybridoma (CC9C10). These were grown by various culturing strategies including stationary cultures and microcarrier cultures which were used to grow the adherent cells (CHO and Vero).

M.J.T. Carrondo et al. (eds.), Animal Cell Technology, 461–466.

Materials and Methods

Cell Lines A murine hybridoma (CC9C10) which secretes a monoclonal antibody against insulin (HB123), Vero (CCL 81) and Chinese hamster ovary (CHO) K-1 strain (CCL 61) were obtained from the ATCC. Cells were grown in DMEM supplemented with 5% v/v iron-enriched calf serum (Gibco BRL).
Cell Counting: (a) Crystal violet method. Cell or microcarrier suspensions (0.5 ml) were treated with 0.5 ml crystal violet reagent (0.2M citric acid, 0.2% crystal violet + 2% Triton X-100). This mixture was incubated at 37°C for 1.5 h prior to counting of stained nuclei using a Neubauer haemocytometer. **(b) Trypan Blue method.** Cell suspensions (0.5 ml) were treated with 0.5 ml trypan blue reagent (0.5 % w/v trypan blue in D-PBS) and immediately counted with a haemocytometer. Cells were trypsinized from microcarriers by incubation with 1 ml trypsin/ EDTA for 5 min. Complete medium (0.6 ml) was added and a cell suspension was collected from the supernatant following sedimentation of microcarriers. Microscopic examination indicated that all cells were released from the microcarrier surface. **(c) Coulter counting.** Cell suspensions (0.1 ml) were treated with 20 ml Isoton and were counted with a Coulter counter (Model D Industrial).

Results

1. **Stationary cultures.** The growth of hybridoma (CC9C10), Vero and CHO cells was monitored in T-flasks by three independent counting methods (Fig. 1). The % viability of each culture was determined by trypan blue exclusion. Data at each time point is based on the extraction of cells from an individual T-flask. For all three cultures the counts determined by the crystal violet method were consistently and significantly higher than either the counts from the trypan blue or the Coulter methods. The greatest difference was shown in the Vero cultures where count differences of 30-50% were observed. For the hybridoma culture there was rapid loss of cell viability (apoptosis) following the cessation of growth, a point at which there appeared to be no significant difference between counts. However, the Vero and CHO cells showed a prolonged stationary phase in which a high cell viability was maintained. Differences between the cell and nuclear counts were greatest for these cultures during the stationary phase. No differences were observed between the trypan blue and Coulter counts for cultures during the growth phase. However, the slightly lower Coulter counts obtained in the stationary phase of Vero and CHO cultures can be accounted for by the presence of cell clumps (5-20 cells).

2. **Microcarrier cultures** The growth of Vero and CHO cells were also monitored in microcarrier cultures (Fig. 2) by the crystal violet and trypan blue methods. Coulter counting was not used because of interference caused by the microcarrier beads. These cultures were re-fed daily in order to promote growth to high cell densities. Without re-feeding the maximum cell densities of these cultures were 40-50% lower but the

Fig. 1: Stationary cell cultures. Cells were grown in 25 cm^2 T-flasks containing serum-supplemented medium (10 ml). Counts were by the trypan blue method (●), the crystal violet method (■) or by Coulter counter (▲). The % viability (♦) is also shown. Data points (±SEM) are means of 2 samples counted two to four times by each method.

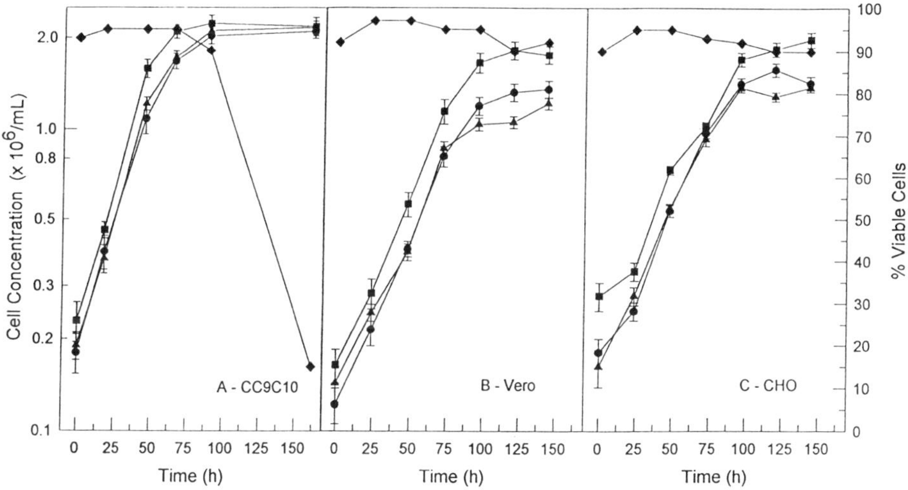

Fig. 2: Microcarrier cultures. Vero cells were inoculated into cultures (100 ml) containing (A) 1 g/l Cytodex-1 or (B) 1.7 g/l Cultispher-G. (C) CHO cells were inoculated into 1 g/l Cytodex-3. The media (0.5 vol) was changed daily from day 3. Counts were performed by the trypan blue (●) or the crystal violet (■) method . Data points (±SEM) are the means of 2 cultures and each sample was counted twice.

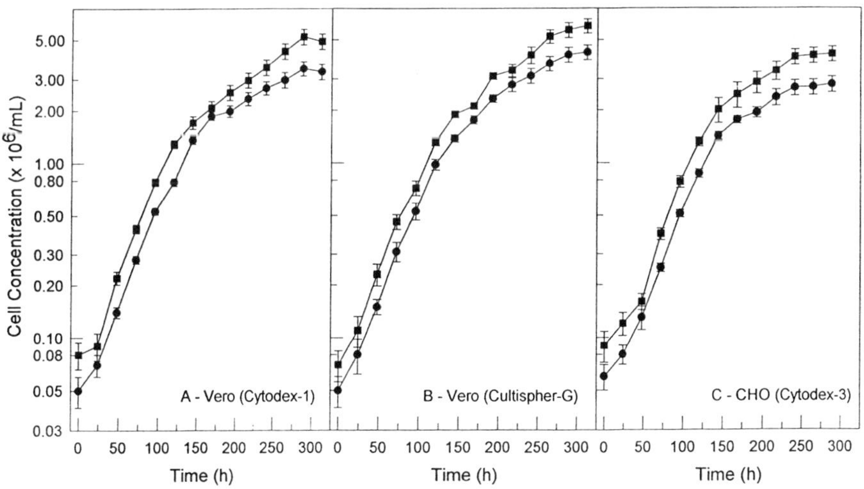

relative differences in counts obtained by the two methods were unaffected. The growth curves indicated that the counts obtained by the crystal violet method were consistently and significantly higher (10-50%) than those determined by the trypan blue method.

3. Time-dependence of count difference

The percentage difference in counts obtained by the crystal violet and trypan blue methods appeared to vary between samples. This prompted an experiment to monitor this difference over short time intervals during the progress of a culture. Vero cells were inoculated into a series of 25 cm^2 T-flasks which were incubated for varying lengths of time up to 160 h. At time intervals of 6-10 h, counts from an individual T-flask were determined independently by the crystal violet and trypan blue methods.

Fig. 3 shows data from one of a series of such experiments in which, consistent with previous cultures, the crystal violet counts were significantly higher than those determined by trypan blue. However, the actual difference in counts appeared to vary in cycles of approximately 24 h before reaching the maximum cell density at the 100 h time point. During these cycles, which occurred in the growth phase, the difference in count values varied from 0-100%. In the stationary phase the difference in count values was constant at approximately 35%. It appeared that the cyclic effect of count differences was more pronounced for cells at a passage number >140, that had been cultured continuously for at least 15 passages from a frozen working cell bank. This suggests that there may be a degree of cell cycle synchrony in these cultures.

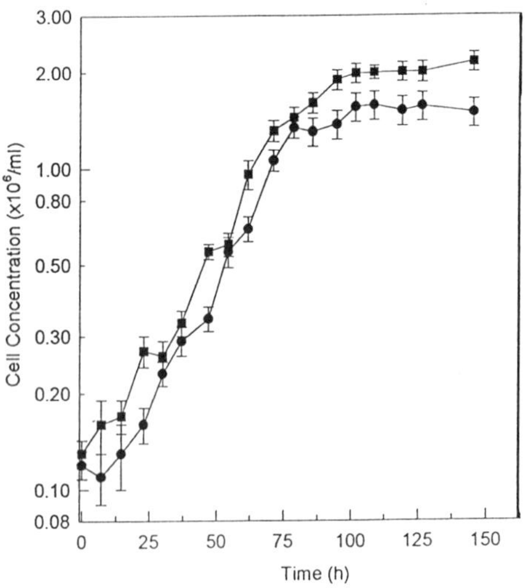

Fig. 3: Vero culture (at passage #142) in 25 cm^2 T-flasks containing serum-supplemented medium (10 ml). Counts were performed every 6-10 h using either the trypan blue (●) or the crystal violet (■) method. Data points (±SEM) represent the means of two independent cultures and each sample was counted twice.

4. Microscopic analysis

Consistent differences between nuclei and cell counts suggested the possibility that there may be a proportion of binucleated cells in the cultures analyzed. We attempted to obtain evidence for the presence of such cells by microscopy. Examination of the Vero cell cultures with both bright field and differential interference contrast (DIC) microscopy showed that a number of cells contain two nuclei in a common cytoplasm.

In some instances cells appeared superficially to be binucleated but careful examination at high magnification and comparison between bright field and DIC images showed they were overlapping flat cells. There are, however, other cells which are clearly binucleated with the cell boundaries around a cytoplasm that harbours two nuclei. Examples of binucleated cells are shown in Fig. 4.

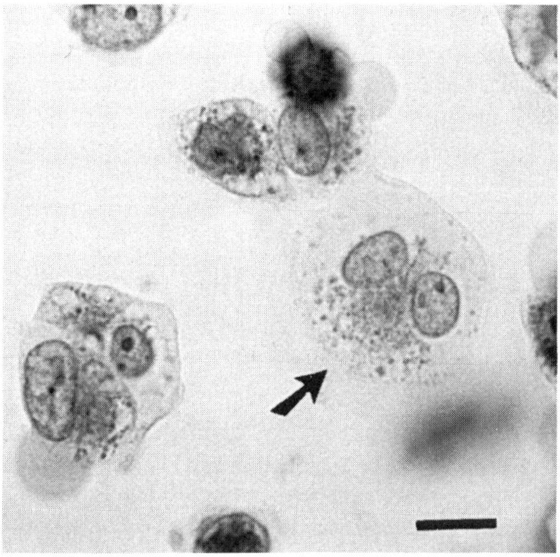

Fig. 4: Photomicrograph of a binucleated Vero cell. Cells were attached to a plastic surface by incubation for 1h and stained with 0.2%w/v crystal violet in D-PBS for 20min. The photomicrograph was obtained using Normaski DIC optics. Note the binucleated cell indicated by the arrow. Bar = 5 μm.

Discussion

We believe that the difference observed between the cell and nuclear counts from the three cell lines used in our study was due to a variable proportion of binucleated cells in these cultures. Frequent sampling from Vero cell cultures revealed that the difference followed a cyclical pattern that correlated with the generation time of cell growth (24 h). The cyclical nature of the count difference in the growth phase was particularly noticeable for cells grown continuously in culture for several passages. Possibly the cell population attained a degree of synchrony that caused the proportion of binucleated cells to vary as the cells progressed through the cell cycle. At the stationary phase of CHO and Vero cultures, the difference between the cell and nuclei counts stabilised at 30-35%. This suggests that a significant proportion of these cells were arrested prior to cytokinesis. Cell division was completed when the cells were re-plated in the presence of serum. However, continued growth required the addition of further nutrients.

Binucleated cells are found *in vivo* under normal physiological conditions. The liver of normal adult rats has an estimated population of ~30% binucleated hepatocytes (3,4). The binucleated cells appear to be formed after normal mitosis but with subsequent inhibition of cytokinesis. Engelhardt and Jen-Hao (5) showed that Vero cell division was halted by a serum limitation causing the cells to accumulate in the late S or G_2 phase of the cell cycle. The DNA content of these cells increased without cell division. Fibronectin was likely the critical serum factor as Orly & Sato (6) showed the presence of a high proportion (50-65%) of binucleated epithelioid cells in its absence. The addition of fibronectin eliminated the binucleated cells and restored normal cell growth.

In conclusion, our data shows that consideration of the interplay between nuclear division relative to cytokinesis is important in the analysis of mammalian cell culture growth. Monitoring cell growth solely by nuclear counting using the crystal violet method requires some caution because it can lead to the overestimation of cell concentrations in culture.

References

1. Philips HJ (1973) Dye exclusion tests for cell viability. In: Kruse PF and Patterson MK (eds), Tissue Culture: Methods and Applications. (pp. 406-408), Academic Press, NY.
2. Sanford KK, Earle WR, Evans VJ, Waltz HK & Shannon JE (1951) The measurement of proliferation in tissue cultures by enumeration of cell nuclei. J. Natl. Cancer Inst.11: 773-795.
3. Wheatley DN (1972) Binucleation in mammalian liver: Studies on the control of cytokinesis in vivo. Exp. Cell Res. 74: 455-465.
4. Mossin L, Blankson H, Huitfeldt H & Seglen P (1994) Ploidy-dependent growth and binucleation in cultured rat hepatocytes. Exp. Cell Res 214: 551-560.
5. Engelhardt DL & Jen-Hao M (1977) A serum factor requirement for the passage of Vero cells through G2. J. Cell. Physiol. 90: 307-320.
6. Orly J & Sato GW (1979) Fibronectin mediates cytokinesis and growth of rat follicular cells in serum-free medium. Cell 17: 295-301.

Acknowledgements

We acknowledge financial support from NSERC, Apotex and Rh Pharmaceuticals. We would like to thank Dr E. Rector for valuable discussion of this work.

Residual DNA from Hybridoma Cultures: Decorrence of Apoptosis?

Maria Teresa A. Rodrigues, Isaias Raw & Ana Maria Moro
Centro de Biotecnologia, Instituto Butantan, 05503-900, São Paulo, SP, Brasil. E-mail: anammoro@usp.br

Abstract

One of the main concerns about therapeutic products derived from mammalian cell cultures is the residual DNA content. The maximum DNA content per dose is established as 100 pg for European and American authorities. For monoclonal antibody production we have been cultivating hybridomas in a packed-bed bioreactor with internal aeration with continuous medium perfusion for some months. The DNA residual content of the crude supernatant is very high and decreases with the downstream processing, but is usually above the allowed by the authorities. However, our findings indicate that the size of the residual DNA is very small, around 180 base pairs, suggesting that the released DNA is consequence of apoptosis. These results allow: i.to speculate over the importance of the limits established by the regulatory agencies, as the smaller oncogene has 1kb size; ii.to reduce the residual DNA content by trying to control apoptosis.

Introduction

New therapeutic reagents have been obtained in the last few years, thanks to hybridoma and recombinant DNA technology. Many of these products require mass culture of mammalian cells, when bacteria and yeast cannot synthesize proteins displaying the same configuration as the natural protein. Monoclonal antibodies, growth hormones, erithropoetin, blood coagulation factors and plasminogen activators are some examples of proteins that need a mammalian cytoplasm to be properly assembled in order to exert the expected activity. One of the major risks associated to the utilization of products derived from continuous mammalian cell lines is the possibility that the released DNA from dead cells, if present in enough amounts and in biologically active form, could transmit the genetic information for pathogenic characteristics to the patients receiving the products. The Food and Drug Administration (1) and the World Health Organization (2) regulations establish that mammalian cells derived products should not contain more than 100pg residual DNA

M.J.T. Carrondo et al. (eds.), Animal Cell Technology, 467–472.

per dose and that the downstream processing must be effective in reducing this contaminant burden. We examined four preparations of anti-CD3 monoclonal antibody production - intended for reversion of graft's rejection - using different culture media and purification procedures, evaluating the level of residual DNA in the final form of the obtained products, as well as defining which purification steps effectively contributed for its reduction.

Material and Methods

A murine hybridoma producing anti-CD3 monoclonal antibody was cultivated in a 3L bioreactor with airlift design (Bellco Glass, Inc. Vineland, NJ, USA), filled with hollow glass cylinders for cells settling, configuring a packed-bed version (3). The culture media used were: DME Hybri-Max® (Sigma Cell Culture™, USA) supplemented with fetal calf serum (Cultilab, Brazil) for preparations I and II; Hybridoma-SFM® (Gibco® Lab., USA) for preparation III and D-MEM (Gibco® Lab., USA) supplemented with young horse serum (I. Butantan, Brazil) for preparation IV. The downstream processing consisted of either 30,000 molecular weight cut-off tangential ultrafiltration (Amicon or Millipore, USA) or cation-exchange chromatography (S-Sepharose®-FF) followed by affinity chromatography through Protein-A Sepharose® 4-FF. The resins were from Pharmacia, Sweden. Few pyrogenic eluates were further processed through Polymixin B insolubilized on cross-linked 4% beaded agarose (Sigma® Chem., USA). The DNA quantification in the final products and downstream steps was carried out through hybridization with radiolabelled hybridoma genomic DNA linked to nylon membranes. The samples were treated with proteinase K and extracted with phenol-chloroform. The aqueous phase was concentrated by 2-butanol and the DNA was precipitated with ethanol (4). A hybridoma genomic DNA, handled as the samples, was used to obtain the standard curve. The culture supernatants, treated as described, were submitted to 0.8% agarose electrophoresis for DNA size evaluation.

Results and Discussion

The Table I present the DNA contents, measured as pg/dose, of the supernatant cultures, intermediate downstream processing steps and final products. The main reductional factor was obtained by the protein-A affinity chromatography, 189, 156 and 208 fold for preparations I, III and IV respectively. The ionic-exchange chromatography resulted in 5 fold reduction in one step (process IV) and when done twice was 10 times more effective (process III). The polymixin chromatography promoted a reduction near 10 fold reduction on the DNA content (II and III). The number of chromatography steps is important for the residual DNA reduction, as the final products from preparations II and III presented the lower DNA content, being 4,381 and 369 pg/dose respectively.

Table I - Downstream processing steps of the supernatants and DNA reduction factors.

Process	Purification steps	DNA (pg/dose*)	reduction factor
I	concentrated supernatant	100,698,457	
	protein A-sepharose eluates pool	533,514	189
	final product	237,964	2.2
II	concentrated supernatant	n.t.	
	protein A-sepharose eluates pool	n.t.	
	polymixin B-agarose eluate	36,784	
	polymixin B-agarose eluates pool	4,278	8.6
	final product	4,381	
III	supernatant	27,008,801	
	S-sepharose eluates pool	n.t.	
	S-sepharose eluate	511,615	53
	polymixin B-agarose eluate	57,615	9
	protein A-sepharose eluate	n.t.	
	final product	369	156
IV	supernatant	116,994,471	
	S-sepharose eluates pool	25,051,152	5
	protein A-sepharose eluates pool	n.t.	
	final product	120,121	208

n.t.) not tested

* one dose is the injected volume of monoclonal antibody solution/day necessary to attain the therapeutic effect

The media for the hybridoma cultivation were: DME-fetal calf serum for preparations I e II, HSFM for preparation III and DME-horse serum for preparation IV.

The *in-vitro* cell cultures generate variable levels of stress, causing death by necrosis or apoptosis (5). The monoclonal antibodies obtained in the processes described herein were produced through continuous medium perfusion of long-term cultures. This condition allowed to achieve a rather homogeneous physiological environment, in which the culture reached a steady-state with a better balance between cellular proliferation and death. In spite of the DNA quantifications are higher than the recommended for therapeutic monoclonal antibodies, it was verified that the DNA size was smaller than 500 base pairs, presenting a decrescent ladder to approximately 180 base pairs (Figure 1). This result leads to the supposition that the cellular death during long-term hybridoma culture in bioreactors is decorrence of apoptosis during the months of culture.

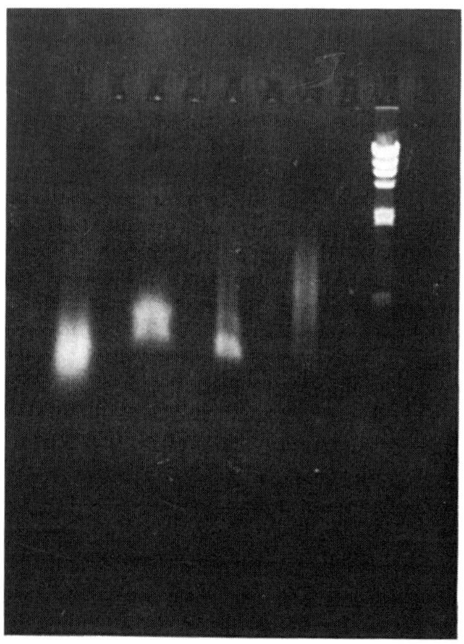

Figure 1: Agarose gel electrophoresis stained with ethidium bromide. Samples of supernatants treated as described in Material and Methods for quantification of residual DNA. Lanes 1, 2 and 3: preparations I, III and IV, respectively; lane 4: genomic hybridoma DNA purified and sonicated; lane 5: nucleic acids size marker (phage λ DNA digested by Hind III) in bp (base pairs).

Notwithstanding the fragmentation of DNA into 180 base pairs is considered to be the hallmark of apoptosis from the biochemical point of view (6), for the serum-free preparation (III) fragments of 300 base pairs were generated (Figure 1). Studies by Oberhammer et al.(7) indicate the same degree of fragmentation when the cells were deprived of serum. The serum-free preparation (III) presented also the lowest DNA content in the crude supernatant (Table I), probably consequence of a lesser degree of cellular death, although by apoptosis.

The limit of residual DNA content in hybridoma or recombinant DNA technology derived products is, according to international regulations (1,2), 100pg/dose/day. A more real perspective from the European quality control authorities suggests that the DNA content must be reduced to acceptable levels in 5 preparations of the product, guaranteeing the downstream processing consistency, instead of simply fix a limit (8). The WHO orientation is to take into account the risk/benefice to the patient receiving a preparation with more than 100pg residual DNA (2) while the FDA establishes the same limit, using the expression *when possible* (1). Since the products of modern biotechnology were introduced, Berthold and Walter (9) consider that the limits were chosen arbitrarily and are below a potential biological risk. A corroborating fact is that no incident case is registered linking DNA content of a biotechnological product and adverse effects. The results presented here, in relation to DNA size, strongly suggest that the risk to the patient is lower than previously supposed, as the smaller cellular oncogenes is around 1kb size (10). Further studies should be done in order to try to control the apoptotic process. The lower DNA content obtained in the serum-free preparation could constitute a starting point to it.

References

1. United States. Food and Drug Administration (1994) Center for Biologics Evaluation and Research. Points to Consider in the Manufacture and Testing of Monoclonal Antibody Products for Human Use. p.1-21, Rockville, USA.
2. World Health Organization (1992) Expert Committee on Biological Standardization. Guidelines for Assuring the Quality of Monoclonal Antibodies for Use in Humans. p.47-67. [Technical Report Series n. 822], Geneva, Switzerland.
3. Moro AM, Rodrigues MTA, Gouvea MN, Silvestri MLZ, Kalil JE & Raw I (1994) Multiparametric Analyses of Hybridoma Growth on Glass Cylinders in a Packed-Bed Bioreactor System with Internal Aeration. Serum-Supplemented and Serum-Free Media Comparison for MAb Production. J.Immunol. Methods,176, 67-77.
4. Sambrook J, Fritsch EF & Maniatis T (1989) Molecular Cloning: A Laboratory Manual. 2nd ed. Cold Spring Harbor: Cold Spring Harbor Laboratory Press,USA.
5. Cotter TG & Al-Rubeai M (1995) Cell Death (Apoptosis) in Cell Culture Systems. TIBTECH, 13,150-155.

6. Wyllie AH (1980) Glucocorticoid-Induced Thymocyte Apoptosis is Associated with Endogenous Endonuclease Activation. Nature, 284, 555-556.
7. Oberhammer F, Wilson JW, Dive C, Morris ID, Hickman JA, Wakeling AE, Walker PR & Sikorska M (1993) Apoptotic Death in Epithelial Cells: Cleavage of DNA to 300 and/or 50 kb Fragments prior to or in the Absence of Internucleosomal Fragmentation. EMBO J.,12,3679-3684.
8. Petricciani JC & Horaud FN (1995) DNA, Dragons and Sanity. Biologicals, 23, 233-238.
9. Berthold W & Walter J (1994) Protein Purification: Aspects of Processes for Pharmaceutical Products. Biologicals, 22, 135-150.
10. Petricciani JC (1985) Regulatory Considerations for Products Derived from the New Biotechnology, p.79-86. In Feder J & Tolbert WR (Eds.) Large-Scale Mammalian Cell Culture, Academic Press, Orlando, USA.

Acknowledgments

The authors are very grateful to Luzia M. Ioshimoto and Cosue Miyaki for helpful technical assistance and to Dr. Oswaldo Sant'Anna for critical reading of the manuscript. This work was in part sponsored by PADCT-FINEP, Brazil (3030757/SBIO). A.M.Moro is recipient of a research fellowship from CNPq, Brazil (300576/90-0) and received grants from FAPESP, Brazil (96/1337-1).

SESSION V

Recombinant Proteins: Biosynthesis and Processing

Expression of human α2,6-Sialyltransferase in BHK-21A cells increases the sialylation of coexpressed human erythropoietin: NeuAc-transfer onto GalNAc(β1-4)GlcNAc-R motives

Peter Schlenke, Eckart Grabenhorst, Roland Wagner*, Manfred Nimtz** and Harald S. Conradt, Department of Gene Regulation and Differentiation, *Cell Culture Techniques, ** Molecular Structure Research, GBF Mascheroder Weg 1, D-38124 Braunschweig, Germany

1. Abstract

We have characterized two BHK cell lines (BHK-21B and BHK-21A) with different glycosylation properties. N-glycans synthesized by BHK-21B cells contain the typical N-acetyllactosamine motif (Gal(β1-4)GlcNAc-R) whereas BHK-21A cells bear to a high amount nonsialylated terminal GalNAc(β1-4)GlcNAc-R moieties [1]. Due to the incapability of the endogenous α2,3-sialyltransferase to transfer NeuAc to the GalNAc(β1-4) GlcNAc-R structure recombinant glycoproteins produced by BHK-21A cells are undersialylated.

Therefore we have transfected BHK-21A cells harbouring a plasmid encoding human EPO with the human Golgi enzyme CMP-NeuAc:Gal(β1-4)GlcNAc-R α2,6-sialyltransferase (ST6N). Detailed structural analysis of oligosaccharides from the affinity purified recombinant EPO (HPAEC-PAD-mapping and MALDI/TOF-MS) revealed a significant increased NeuAc content when compared to the parent BHK-21A cells without ST6N activity. Methylation analysis corroborated these results. The newly introduced α2,6-sialyltransferase recognizes the terminal GlcNAc-R motif as a substrate. The cell line obtained thus exhibits a 'human kidney-type' glycosylation characteristic [2].

Keywords: recombinant human erythropoietin, BHK-21 cells, glycosylation engineering, human α2,6-sialyltransferase, tissue-type N-glycosylation, carbohydrate mapping, recombinant glycosyltransferase expression

2. Introduction

Erythropoietin (EPO) is a glycoprotein hormone that stimulates the proliferation and differentiation of erythroid progenitor cells into mature erythrocytes. It is synthesized mainly in the adult kidney. The protein has a molecular mass of 34-39 kDa. About 40% of its molecular mass is provided by the four different carbohydrate structures (three N-

M.J.T. Carrondo et al. (eds.), Animal Cell Technology, 475–480.

glycosylation sites at Asn24, Asn38, and Asn83 and a single O-glycosylation site at Ser126). The decrease of the EPO production by the kidney due to chronic renal failure causes anemia. Therefore the availability of recombinant human EPO as a therapeutic reagent has provided a major advance in the long-term treatment of patients with renal deficiencies.

EPO has been produced by recombinant gene technologies in various heterologous mammalian host cell lines including BHK and CHO cells. Detailed carbohydrate structural analysis of N-linked glycans of EPO produced from these cells have revealed that NeuAc is exclusively α2,3-linked since these host cells lack a sialyltranferase that transfers NeuAc in an α2,6-linkage [3,4] which is the major linkage of human serum glycoproteins. Therefore, recombinant human proteins when produced from e.g. BHK or CHO cells may differ from their native human counterparts with respect to terminal sialylation.

Differences in the glycosylation patterns of recombinant proteins destined for therapeutically use might result in altered biological activity, e.g. clearance, antigenicity, targeting to specific receptors. For EPO it has been shown that removal or modification of the carbohydrate chains is accompanied by altered in vivo and in vitro activity. The in vivo bioactivity is completely abolished when NeuAc is absent [5,6,7].

3. Material and Methods

BHK-21A cells transfected with the plasmid pABEpo harbouring a cDNA encoding human EPO [8] were transfected with the plasmid pABSial encoding the human Golgi enzyme CMP-NeuAc:Gal(β1-4)GlcNAc-R α2,6-sialyltransferase together with the vector pSV2pac [9] containing the puromycin acetyltransferase gene by using the calcium phosphate precipitation method. Puromycin-resistent cells were selected using several medium exchanges with DME containing 5μg/ml puromycin dihydrochloride during a time period of 2-3 weeks. Cell culture supernatants of confluently growing monolayers were then analyzed for EPO and α2,6-bound NeuAc by Western blotting using a polyclonal anti-EPO antiserum and Sambucus nigra agglutinin, respectively. Clone mixtures of transfected BHK-21A cells exhibiting the most abundant expression of EPO and α2,6-linked NeuAc were used for the production which was performed in 175 cm^2 T-flasks in serum-free DME medium. EPO was purified by affinity chromatography using a monoclonal antibody [8] coupled to Sepharose 4B. The purity of the preparation was verified by SDS/PAGE and Coomassie staining. N-glycans were liberated from dialyzed glycoproteins by PNGase digestion as described previously [3]. Quantitative liberation was checked by SDS/PAGE. The glycans were desalted by gel filtration and were characterized before and after enzymatical removal of NeuAc by analytical HPAEC-PAD-mapping [3]. Native oligosaccharides were structurally characterized by MALDI/TOF-MS and by methylation analysis. (The BHK-21A cells transfected with EPO as well as the plasmid pABSial and the monoclonal antibody used for the purification were kindly provided by G. Zettlmeissl, Behringwerke/Marburg).

4. Results and Discussion

We have previously reported that two BHK cell lines can be distinguished based on their different N-glycosylation of recombinant proteins [1]. N-glycans of glycoproteins synthesized by BHK-21B cells show the typical α2,3-sialylated N-acetyllactosamine-motif (NeuAc(α2-3) Gal(β1-4)GlcNAc-R) whereas BHK-21A cells produce proteins with considerable amounts of nonsialylated GalNAc(β1-4) GlcNAc termini. Due to the obvious incapability of the endogenous α2,3-sialyltransferase to transfer NeuAc to GalNAc residues, oligosaccharides containing terminal GalNAc are found to be undersialylated.

Therefore we have transfected BHK-21A cells expressing human EPO with the plasmid pABSial encoding the human Golgi enzyme α2,6-sialyltransferase which transfers NeuAc in an α2,6-linkage to oligosaccharides. Stable cell clones were isolated and secreted EPO was purified by affinity chromatography. HPAEC-PAD-mapping of the enzymatically liberated oligosaccharides revealed a significantly higher sialylation state when com-

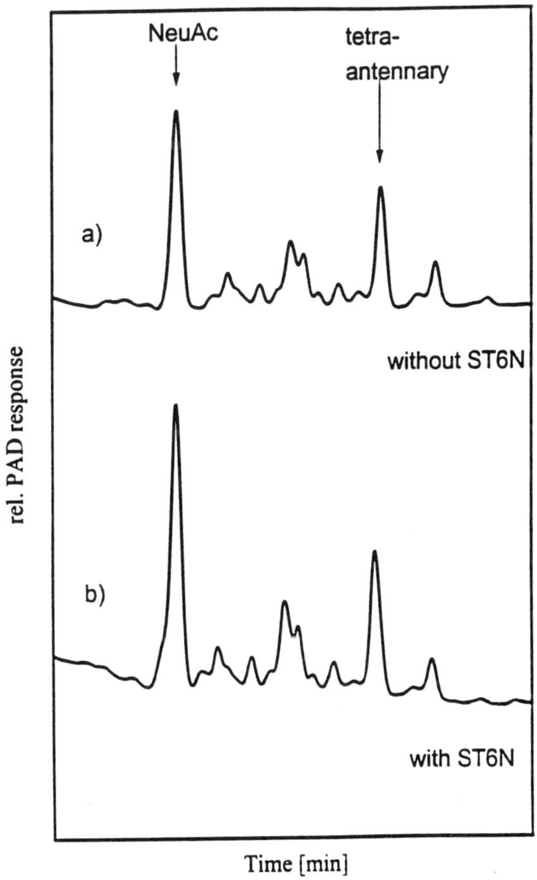

Fig.1. HPAEC-PAD profiles of EPO N-glycans obtained from cells before (a) and after (b) transfection with human α2,6 sialyltransferase

pared to the parent cells transfected only with EPO (see peak areas for NeuAc and the tetraantennary structures in Fig.1). Except for the increased NeuAc content the elution profile of the neutral oligosaccharides with or without ST6N activity was very similar. These results were further corroborated by MALDI/TOF-MS analysis of the native N-glycans. Molecular ion signals corresponding to fully sialylated di-, tri- and tetraantennary N-glycans with GalNAc(β1-4)GlcNAc motives were observed. By contrast, fully sialylated N-glycans bearing terminal GalNAc residues were absent in the product from cells without ST6N activity.
In methylation analysis of the EPO N-glycans obtained after transfection with ST6N, 2,4,6-Tri-O-methyl-galacitol as well as 2,3,4-Tri-O-methy-galacitol were detected indicating the NeuAc being linked α2,3 as well as α2,6 to Gal. The detection of 3,4-Di-O-methyl-2-N-methylacetamido-2-deoxygalacitol unequivocally revealed the presence of GalNAc residues substituted at position 6, indicating an α2,6-linkage of NeuAc (Table 1). The presence of these derivates was verified by the corresponding fragmentation patterns detected by GC-MS (not shown).

Table 1. Methylation analysis of EPO N-glycans. Only the relative amounts of the Gal and GalNAc derivates are shown.

	BHK-21A wild-type	**BHK-21A with ST6N**	**substituted at position**
Galacitol			
2,3,4,6-Tetra-O-methyl-	++	+	terminal
2,4,6-Tri-O-methyl-	++++	++	3
2,3,4-Tri-O-methyl-	-	++	6
2-N-Methylacetamido-2-deoxygalacitol			
3,4,6-Tri-O-methyl-	++	+	terminal
4,6-Di-O-methyl-	-	-	3
3,4-Di-O-methyl-	-	++	6

Therefore the observed increase in NeuAc content of the EPO N-glycans obtained after transfection of the BHK-21A cells with the human ST6N clearly results from the substrate specificity of the enzyme acting on Gal as well as on GalNAc residues.
These findings are in agreement with the previously reported in vitro substrate specificity of the bovine colostrum α2,6-sialyltransferase which is capable of recognizing GalNAc(β1-4)GlcNAc in addition to Gal(β1-4)GlcNAc units [10]. Therefore it is likely that in vivo the human α2,6-sialyltransferase is also involved in the synthesis of NeuAc(α2-6)GalNAc-R structures found in human glycoproteins [2,11].

In conclusion, by transfection of the BHK-21A cell line with the human CMP-NeuAc:Gal(β1-4)GlcNAc-R sialyltransferase we have genetically engineered a novel host cell line with 'human kidney-type' glycosylation properties characterized by the capability to transfer NeuAc in an α2,6-linkage to terminal Gal and GalNAc residues of N-glycans from recombinant glycoproteins in addition to the common α2,3 transfer of NeuAc to Gal termini.
The transfection of heterologous mammalian host cell lines with plasmids encoding glycosyltransferases of defined specificity provides a valuable tool for manipulating the glycosylation characteristics for the production of 'human-type' glycosylated proteins. The construction of host cell lines which exhibit 'human-type' glycosylation properties will aid to improve the safety and quality of biotechnologically produced recombinant proteins destined for clinical use. Eventually glycoprotein products obtained from such a cell line may lead to novel pharmacological properties of glycoproteins.

This work was supported by the EU grant "Animal Cell Biotechnology" (BIO2-CT94-3069) to H.S.Conradt.
The excellent technical assistance of Susanne Pohl is greatfully acknowledged.

References:

[1] Grabenhorst, E., Hoffmann, A., Nimtz, M., Zettlmeissl, G., Conradt, H.S. (1995) Eur. J. Biochem. 232, 718-725
[2] Bergwerff, A.A., Van Oostrum, J., Kamerling, J.P., Vliegenthart, J.F.G. (1995) Eur. J. Biochem. 228, 113-126
[3] Nimtz, M., Martin, W., Wray, V., Klöppel, K.-D., Augustin, J., Conradt, H.S. (1993) Eur. J. Biochem. 213: 39-56
[4] Takeuchi, M., Takasaki, S., Miyazaki, H., Kato, T., Hoshi, S., Kochibe, N., Kobata, A. (1988) J. Biol. Chem. 263: 3657-3663
[5] Goldwasser, E., Kung, K.H., Eliasson, J. (1974) J. Biol. Chem. 249, 4202-4206
[6] Fukuda, M.N., Sasaki, H., Lopez, L., Fukuda, M. (1989) Blood 73, 84-89
[7] Spivak, J.L., Hogans, B.B. (1989) Blood 73, 90-99
[8] Fibi, M.R., Hermentin, P., Pauly, J.U., Lauffer, L., Zettlmeissl, G. (1995) Blood 85 (5), 1229-1236
[9] Vara, J. A., Portela, A., Ortín, J., Jiménez, A. (1986) Nucleic Acids Res. 14, 4617-4624
[10] Nemansky, M., Van den Eijnden, D.H. (1992) Biochem. J. 287, 311-318
[11] Weisshaar, G., Hiyama, J., Renwick, A.G.C., Nimtz, M. (1991) Eur. J. Biochem 95, 257-268

Discussion

Miller: You said that there are glycosylation sites present on the glycosylating enzymes. Are these enzymes themselves glycosylated or not?

Schlenke: Yes, they are.

Miller: Have you characterised these enzymes? If your sialyl transferases are themselves glycosylated and you have, I presume, isolated the enzyme responsible for that, you have a chicken and egg situation. You have an enzyme that puts sialyl residues on your protein, and the enzyme itself has glycosylation sites. Have you analysed the glycosylation of this enzyme?

Schlenke: No, we have not. I do not think that the glycosylation properties of the enzyme affect the function.

Thomas: How much natural variability between BHK clones do you see in total sialic acid content, or moles of sialic acid to moles of protein?

Schlenke: In the BHK 21A cell line we use I am not aware of variability.

Sinacore: I was wondering if you are able to measure the relative expression level of the recombinant sialyl transferase relative to the endogenous? Also, do you believe you can ultimately override the synthesis of the α2-3 linkages merely by amplifying the expression of the human sialyl transferase?

Schlenke: First question. We did not measure the expression level of the enzyme, only the sialyation state of the product. Second question, yes it should be possible.

Construction of novel BHK-21 cell lines coexpressing Golgi resident or soluble forms of human α2,6-sialyltransferase and α1,3/4-fucosyltransferases together with secretory glycoproteins

Eckart Grabenhorst, *Júlia Costa and Harald S. Conradt
Department of Gene Regulation and Differentiation, GBF, Mascheroder Weg 1, D-38124 Braunschweig, Germany, *Instituto de Tecnologia Química e Biológica, Universidade Nova de Lisboa, Rua da Quinta Grande, 6, Apartado 127, 2780 Oeiras, Portugal

1. Abstract

The most frequently used mammalian host cell lines (BHK-21, CHO) for the production of recombinant therapeutic glycoproteins are incapable of producing α2,6-sialylated or sialyl Lewis X containing oligosaccharide structures. Therefore, we have constructed novel stable BHK-21 cell lines by cotransfection of cells with plasmids encoding human secretory glycoproteins (β-trace protein, erythropoietin, antithrombin III) and the *membrane-bound* or *soluble forms* of human glycosyltransferases α2,6-sialyltransferase (ST6Gal), α1,3-fucosyltransferase III (FT3) or VI (FT6). The secreted recombinant glycoproteins were purified from culture supernatants and were analyzed in detail using HPAE-PAD, mass spectrometry and NMR-analysis of their oligosaccharide chains. The soluble forms of the recombinant glycosyltransferases showed *in vitro* activity with oligosaccharides, glycolipids as well as glycoproteins. However, no modification of coexpressed secretory glycoproteins was detected. Thus a proper targeting/localization of the glycosyltransferases into the appropriate Golgi compartments is required for their *in vivo* activity on secreted recombinant glycoproteins. Such cell lines can be successfully used for the production of recombinant proteins with novel/tailored glycosylation characteristics that might alter and improve their biological *in vivo* properties (e.g. stability, tissue addressing).

keywords: recombinant human glycosyltransferases; genetic engineering of mammalian host cells, novel glycosylation properties, therapeutic glycoproteins, intracellular targeting

2. Introduction

The expression of recombinant human glycoproteins destined for clinical applications in mammalian host cells is now widely used for their production in large quantities in biotechnological processes. Despite the efforts required to obtain stable recombinant mammalian cells, the expression of recombinant glycoproteins in these hosts is still favourable over other expression systems: (i) prokaryotes are unable to produce N- or O-glycosylated proteins as they are usually found in eukaryotes [1], (ii) yeasts and insect cells are unable to synthesize complex-type glycan structures on proteins [2,3], (iii) in baculovirus-infected insect cells, the overexpressed protein is often denatured and deposited

M.J.T. Carrondo et al. (eds.), Animal Cell Technology, 481–487.

in inclusion body-like particles [3], (iv) only mammalian cells provide a proper machinery for protein glycosylation and secretion very similar to that in human cells.

On the other hand, it has been shown by several groups that mammalian cells can be successfully transfected with foreign glycosyltransferase genes [4,5]. Therefore, it is plausible to assume that a recombinant glycoprotein, when coexpressed together with a recombinant glycosyltransferase in the same cell, should be modified by the foreign enzyme in a predictable way. Since the recombinant enzyme may compete with endogenous activities for the same donor and acceptor substrate(s), it is necessary to correctly target the newly introduced glycosyltransferase into the proper location within the Golgi apparatus of the host cell.

In order to address this issue, we have constructed several secretory and membrane-bound forms of human glycosyltransferases (α2,6-sialyltransferase, α1,3-fucosyltransferases III and VI). These genes were coexpressed together with plasmids encoding secretory human glycoproteins (antithrombin III, β-trace protein, erythropoietin) in stable BHK-21 cells. Here we report on our results obtained on the glycosylation characteristics of the newly constructed cell lines and on the substrate specificity of the genetically engineered glycosyltransferase forms.

3. Materials and Methods

Cells and plasmids: For coexpression studies, the extensively characterized BHK-21B cell line (see P.Schlenke et al., this volume, for further details) has been used throughout. The human α1,3-fucosyltransferase VI (FT6) cDNA [6] was cloned from genomic DNA of the human A172 glioma cell line (Grabenhorst et al., unpublished work). The cDNAs for human α2,6-sialyltransferase (ST6Gal) and human erythropoietin (EPO) were obtained from Dr. G. Zettlmeissl, Marburg; the human α1,3-fucosyltransferase III (FT3) cDNA [7] was kindly provided by Dr. S. Gonski, Hoechst AG, Frankfurt; the cDNA for human β-trace protein (β-TP) was obtained as described [8]. BHK-21B cells coexpressing human antithrombin III (AT III) and ST6Gal were generated as described previously [8]. All variants of the glycosyltransferase genes were generated by PCR-based site-directed mutagenesis and TA-cloning into the eukaryotic expression vector pCR3 (Invitrogen). Mutant forms of the Golgi enzymes were obtained by replacing the cytoplasmic, transmembrane and part of the stem region with an IL-2 signal peptide. The following variants were generated: s2ST6 (Δ[1-64]) of ST6Gal, s2FT3 (Δ[1-46]) of FT3 and s2FT6 (Δ[1-51]) of FT6.

Coexpression in BHK-cells: Stable BHK-21B cell lines already expressing the human secretory reporter proteins β-trace protein or erythropoietin following the selection with puromycin were cotransfected with different glycosyltransferase forms by using the calcium precipitation method. Stable clone mixtures were obtained following selection with G418. Cells with the highest transferase activities in the supernatant or in the microsomal fractions, respectively, were used for production in 175 cm^2 tissue-culture flasks using DME containing 2% fetal bovine serum. Recombinant glycoprotein products were purified from cell culture supernatant by immunoaffinity chromatography as described [8,9]. The purity of the recombinant glycoproteins was checked by SDS-PAGE and Western blotting. The

secreted fucosyltransferases s2FT3 and s2FT6 were purified from cell supernatants by affinity chromatography using a GDP-Fractogel column as will be detailed elsewhere.
Glycosyltransferase assays: Glycosyltransferase activity was tested in reaction mixtures containing 1 mM DTE, 3 mM sodium azide, 10 mM $MnCl_2$, 20 mM MES, pH 6.8 and 50 mM NaCl, with or without 0.2% Triton X-100. Sialyltransferase or fucosyltransferase activity was assayed using 10 g/l asialofetuin (Sigma) and 0.05 mM CMP-NeuAc + CMP-[^{14}C]NeuAc or 0.05 mM GDP-Fuc + GDP-[^{14}C]Fuc, respectively, in the incubation mixture and adding 1/10 of the total volume of enzyme solution. The incorporation of radioactive monosaccharides was monitored by precipitation of aliquots of the reaction mixture onto glass fiber filters followed by liquid scintillation counting as described earlier [3]. Fucosylation of substrates for subsequent carbohydrate structural analysis was achieved by long-term incubation using purified fucosyltransferase preparations and unlabeled GDP-Fuc (3-5-fold molar excess over potential acceptor sites).
Isolation and characterization of N-glycans: The N-glycans of the purified reporter glycoproteins were obtained by PNGase F-digestion of the native polypeptides or from tryptic glycopeptides as described previously [10,11]. Quantitative release of N-glycans was verified by SDS-PAGE and by carbohydrate compositional analysis of de-N-glycosylated proteins and glycopeptides, respectively. The oligosaccharides were desalted and analyzed by HPAEC-PAD (both in native or desialylated form), MALDI/TOF-MS and methylation analysis as described [8].

4. Results and Discussion

In order to obtain mammalian cell lines with human-type or human tissue-type-specific N-glycosylation properties, we have generated several stable BHK-21 cell lines following cotransfection of cells with plasmids encoding the human secretory glycoproteins β-trace protein, erythropoietin or antithrombin III and the intact or secreted forms of the human glycosyltransferases CMP-NeuAc:Gal(β1-4)GlcNAc-R α2,6-sialyltransferase and the α1,3-fucosyltransferase III and VI.

Recombinant β-trace protein coexpressed with ST6Gal was purified from cell culture supernatants and was subjected to carbohydrate structural analysis. The enzymatically liberated oligosaccharides were found to consist of 90% of biantennary chains. About 90% of total oligosaccharides were recovered in the monosialo and disialo fractions in a ratio of 1:7. The monosialylated oligosaccharides contained NeuAc in α2,6- or α2,3-linkage in the same ratio. 60% of the disialo oligosaccharides were found to contain NeuAc in both α2,3- and α2,6-linkage to Gal(β1-4)GlcNAc-R. From ^{1}H-NMR analysis, the α2,6-linked NeuAc was shown to be attached exclusively to the Gal(β1-4)GlcNAc(β1-2)Man(α1-3)-branch of the biantennary structure. Recombinant β-TP expressed from wild-type BHK-21B cells contained biantennary oligosaccharides with α2,3-linked NeuAc only. This observed *in vivo* specificity of human ST6Gal supports the previously reported *in vitro* branch specificity of the bovine colostrum ST6Gal activity [12], demonstrating the suitability of genetically engineered mammalian host cell lines with novel glycosylation properties for the production of "human-type glycosylated" secretory recombinant polypeptides. This result was confirmed in coexpression experiments using human AT III as a model glycoprotein [8] and for human EPO (see contribution by Peter Schlenke et al., this volume).

Stable cell lines expressing the Golgi forms of human α1,3-fucosyltransferase III or VI (FT3, FT6) were obtained upon transfection with corresponding plasmids.

Soluble forms of human α2,6-sialyltransferase (s2ST6) and α1,3/4-fucosyltransferases III and VI (s2FT3, s2FT6, compare Table 1), containing the human IL-2 signal peptide sequence, were shown to be secreted from stably transfected BHK-21 cells into the culture medium. Contrary to the results with the membrane-bound form, in the case of the soluble form of ST6Gal that was coexpressed with β-TP, no α2,6-linked NeuAc could be detected in the secreted glycoprotein product although its activity was demonstrated in *in vitro* assays. The glycosyltransferase variant and the reporter glycoprotein should use a common secretory pathway and the enzyme should have access to its donor substrate CMP-NeuAc en route. Presumably, a correct orientation of the glycosyltransferase for exerting its catalytic activity within the cell is required and is naturally provided by its insertion into the Golgi membrane. Similarly, the soluble forms of FT6 did not modify the coexpressed β-TP or EPO.

Table 1: Human glycosyltransferase variants expressed in BHK-21 cells: *in vivo* and *in vitro* enzymatic activities. The asterisks denote that membrane fractions were enzymatically active *in vitro*; n.d. = not determined

glycosyltransferase constructs		enzymatic activity in cell culture supernatant	in vivo modification: of coexpressed secretory proteins		in vivo modification: of cell surface membrane proteins
			BTP	EPO	
ST6Gal	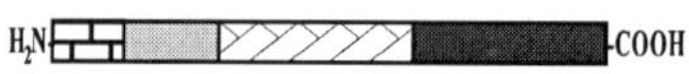	-	+	+	+
s2ST6		+	-	n.d.	-
FT3, FT6	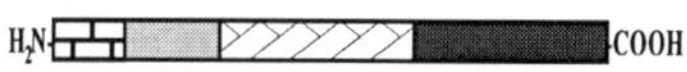	-	n.d.*	n.d.*	n.d.
s2FT3, s2FT6	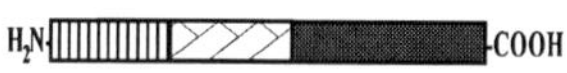	+	-	-	n.d.

key: cytoplasmic domain; transmembrane domain; stem region; catalytic domain; signal peptide (human IL-2)

The two secreted fucosyltransferase variants were purified and were analysed for their activities towards a number of substrates as summarized in Table 2. With asialofetuin as a substrate, both enzyme preparations showed a similar incorporation of radiolabelled fucose in a standard incubation assay as described under Material and Methods. Fetuin contains mostly di- or triantennary lactosamine-type N-linked oligosaccharides and about 10-15% of

Gal(β1-3)GlcNAc-R (type-1 structures) motifs.

When the recombinant glycoproteins AT III, EPO or β-TP were used, almost no incorporation into these proteins was observed with s2FT3 whereas in the case of s2FT6 in all cases greater than 10% of the potential acceptor sites (Gal(β1-4)GlcNAc, type-2 structures) were modified in long-term incubations with the native proteins. The recombinant glycoproteins are modified by α2,3-linked NeuAc and each protein bears its own set of N-glycans with characteristic antennarity (see ref. 8 and literature cited therein).

Table 2: Comparison of the substrate specificity of soluble forms of recombinant human α1,3/4-fucosyltransferase III (s2FT3) and VI (s2FT6) expressed in BHK-21B cells. The oligosaccharides of asialofetuin, antithrombin III, β-trace protein and erythropoietin were structurally characterized as described [8]. The di-, tri- and tetraantennary N-glycans were obtained from the same proteins. n.d. = not determined

Acceptor substrate	**Relative enzyme activity (%)**	
	s2FT3	**s2FT6**
Native glycoproteins		
asialo fetuin (bovine)	++++	++++
antithrombin III (CHO)	-	++++
β-trace protein (BHK)	+	++++
erythropoietin (BHK)	-	++++
Oligosaccharides		
Gal(β1-3)GlcNAc(β1-3)Gal(β1-4)Glc	38	n.d.
Gal(β1-4)GlcNAc(β1-3)Gal(β1-4)Glc	40	n.d.
NeuAc(α2-3)Gal(β1-3)GlcNAc(β1-3)Gal(β1-4)Glc	42	n.d.
disialo biantennary proximally fucosylated	n.d.	50
tetrasialo tetraantennary proximally fucosylated	n.d.	45
tetrasialo tetraantennary proximally fucosylated + 2 lactosamine repeats	n.d.	50

EPO contains almost exclusively tetrantennary chains with large amounts of lactosamine repeats, β-TP contains more than 90% of biantennary oligosaccharides and AT III bears a mixture of di-, tri and tetraantennary glycans. FT3 has been described to act on type-1 as well as on type-2 structures [13]. When we evaluated the specificity of our enzyme variant with small acceptor substrates similar results were obtained (Table 2). Both type-1 and type-2 structures were fucosylated to about 40% and similar fucosylation was detected for the α2,3-sialylated type-1 chain. S2FT6 activity was also tested towards free oligosaccharides and in this case it was observed that these substrates were fucosylated efficiently.

From these *in vitro* studies it could be inferred that FT3 would not act on N-glycoproteins *in vivo* also and probably is implicated in the biosynthesis of O-linked carbohydrates and/or glycolipids. Definite proof should be obtained by coexpression of the membrane-bound form of the enzyme with model glycoproteins.

It will be interesting to construct mutant forms of this enzyme (and others) with variations in membrane and stem regions and to compare their substrate specificity in coexpression experiments. The interchange of membrane and stem region between different fucosyltransferases might provide a means to control their *in vivo* activity and specificity due to the importance of these domains for the intracellular targeting of the glycosyltransferases.

This work was in part supported by an EU-grant (BIO2-CT94-3069) to H.S.C.

5. References

[1] Lechner J, Wieland F, 1989. Structure and biosynthesis of prokaryotic glycoproteins. *Ann. Rev. Biochem.* **58**: 173-194

[2] Kukuruzinska MA, Bergh MLE, Jackson BJ, 1987. Protein glycosylation in yeast. *Ann. Rev. Biochem.* **56**: 915-944

[3] Grabenhorst E, Hofer B, Nimtz M, Jäger V, Conradt HS, 1993. Biosynthesis and secretion of human interleukin-2 glycoprotein variants from baculovirus-infected Sf21 cells. Characterization of polypeptides and posttranslational modifications. *Eur. J. Biochem.* **215**: 189-197

[4] Lee EU, Roth J, Paulson JC, 1989. Alteration of terminal glycosylation sequences on N-linked oligosaccharides of Chinese hamster ovary cells by expression of β-galactoside α2,6 sialyltransferase. *J. Biol. Chem.* **264**: 13848-13855

[5] Meier W, Leone DR, Miatkowski K, Lobb R, Goelz SE, 1993. The ELAM ligand fucosyltransferase, ELFT, directs E-selectin binding to a secreted scaffold protein: a method to produce and purify large quantities of specific carbohydrate structures. *Biochem. J.* **294**: 25-30

[6] Weston BW, Nair RP, Larsen RD, Lowe JB, 1992. Isolation of a novel human α(1,3)fucosyltransferase gene and molecular comparison to the human lewis blood group α(1,3/1,4)fucosyltransferase gene. Syntenic, homologous, nonallelic genes encoding enzymes with distinct acceptor substrate specificities. *J. Biol. Chem.* **267**: 4152-4160.

[7] Kukowska-Latallo JF, Larsen RD, Nair RP, Lowe JB, 1990. A cloned human cDNA determines expression of a mouse stage-specific embryonic antigen and the Lewis blood group α(1,3/1,4) fucosyltransferase. *Genes Dev.* **4**: 1288-1303

[8] Grabenhorst E, Hoffmann A, Nimtz M, Zettlmeißl G, Conradt HS, 1995. Construction of stable BHK-21 cells coexpressing human secretory glycoproteins human Gal(β1-4)GlcNAc-R α2,6-sialyltransferase. α2,6-Linked NeuAc is preferentially attached to the Gal(β1-4)GlcNAc(β1-2)Man(α1-3)-branch of diantennary oligosaccharides from secreted recombinant β-trace protein. *Eur. J. Biochem.* **232**: 718-725

[9] Fibi MR, Hermentin P, Pauly JU, Lauffer L, Zettlmeissl G, 1995. N- and O-glycosylation muteins of recombinant human erythropoietin secreted from BHK-21 cells. *Blood* **85**: 1229-1236

[10] Nimtz M, Martin W, Wray V, Klöppel K-D, Augustin J, Conradt HS, 1993. Structures of sialylated oligosaccharides of human erythropoietin expressed in recombinant BHK-21 cells. *Eur. J. Biochem.* **213**: 39-56

[11] Hoffmann A, Nimtz M, Wurster U, Conradt HS, 1994. Carbohydrate structures of β-trace protein from human cerebrospinal fluid: evidence for "brain-type" N-glycosylation. *J. Neurochem.* **63**: 2185-2196

[12] Joziasse DH, Schiphorst WECM, Van den Eijnden DH, Van Kuik JA, Van Halbeek H, Vliegenthart JFG, 1987. Branch specificity of bovine colostrum CMP-sialic acid:Galβ1-4GlcNAc-R α2-6 sialyltranferase. *J. Biol. Chem.* **262**: 2025-2033

[13] de Vries T, Srnka CA, Palcic MM, Swiedler SJ, van den Eijnden DH, Macher BA, 1995. Acceptor specificity of different length constructs of human recombinant α1,3/4-fucosyltransferases. *J. Biol. Chem.* **270**: 8712-8722

Discussion

Hendricks: Do you have any evidence that your α2-6 sialyl transferase is targeted to late Golgi rather than early Golgi compartments?

Grabenhorst: The only evidence we have is that it is acting efficiently on such glycoproteins. We did not look at the cellular level.

Hendricks: Do you expect, or see, any physiological effects on the BHK of the expression of the α2-6 and modification of some of the BHK proteins?

Grabenhorst: No. We do not expect this since the productivity was the same as before.

James: What is the overall increase in sialyation over the basal 2-3?

Grabenhorst: It did not change.

James: What exactly are the potential advantages of the 2-6 linked?

Grabenhorst: The advantage is that the human-type glycosylated structure does not conflict with antigenicity.

James: So there is nothing specific?

Grabenhorst: It is specific for human serum glycoproteins.

James: As there is no increase in sialyation you might expect it to be cleared at the same rate.

Grabenhorst: I do not know if there is a difference in clearance rates. It has not been analysed.

METABOLIC ENGINEERING OF N-LINKED GLYCOFORM SYNTHESIS SYSTEMS IN CHINESE HAMSTER OVARY (CHO) CELLS

James E. Bailey*, Pablo Umaña, Sherrill Minch#, Michael Harrington+, Martin Page°, and Adriana Sburlati-Guerini

Institute of Biotechnology, ETH Zürich, CH-8093 Zürich, Switzerland
#Department of Chemical Engineering, California Institute of Technology, Pasadena, California 91125, USA,
+Beckman Institute, California Institute of Technology, Pasadena, California 91125, USA,
°Wellcome Research Laboratories, Beckenham, Kent, BR3 3BS, United Kingdom
* Corresponding author

Abstract

Successful metabolic engineering of glycoform distribution to obtain a preferred glycoprotein product can be aided by a systems approach which considers the multiple input (enzyme and cosubstrate levels and localization)-multiple output (glycoform distributions on multiple glycoproteins) nature of oligosaccharide biosynthesis. Elements of such a systems approach now under development in our laboratories include a new mathematical framework for computer simulation of N-linked glycosylation which can be used to suggest effective glycosylation engineering strategies. Also, two-dimensional protein electrophoresis protocols for resolution of hundreds of Chinese hamster ovary (CHO) cell glycoproteins have been developed and applied to study global changes in oligosaccharide modification in wild-type CHO and in an β-N-acetyl glucosaminyl transferase III (GnTIII)-expressing mutant CHO cell cultures. The action of GnTIII creates a bisected oligosaccharide and reduces the *in vitro* activity of at least five other enzymes involved in the same glycosylation pathway, giving this enzyme an important role in controlling the biosynthesis of complex and hybrid oligosaccharides. These studies show that many CHO glycoproteins are modified by the action of GnTIII.

Levels of several glycosyltransferases must be genetically modified in order to achieve certain glycoform distributions, such as altered patterns of oligosaccharide branching. As an initial step towards cell lines providing these capabilities, we have cloned GnTIII into a ß-interferon (IFN-ß) overproducing CHO cell line. Four individual clones produce a novel glycoform of recombinant human IFN-ß containing oligosaccharides which bind the lectin E-PHA and thus likely contain bisected structures.

Introduction

Metabolic engineering of the oligosaccharide biosynthesis pathway to change oligosaccharides on glycoproteins potentially can reduce the antigenicity and improve the stability, the pharmacokinetics, the targetting, and the specific activity of a glycoprotein product [1-3]. Novel glycoproteins, inaccessible by other techniques, can be obtained by this type of metabolic engineering. By creating new biochemical entities within an industrially important class of molecules, metabolic engineering can contribute directly to discovery and development of new pharmaceuticals and diagnostic reagents.

Glycosylation is a complex system. There are multiple inputs in the form of many enzyme activities which participate in oligosaccharide biosynthesis. There are also multiple outputs in terms of the population of different glycoproteins produced by cultured cells. These inputs and outputs are coupled through a complex network of transport and catalytic steps. Effective engineering of such a complex system to achieve a desired objective will

M.J.T. Carrondo et al. (eds.), Animal Cell Technology, 489–494.

be maximized by careful consideration, and characterization, of the multiple responses of the system to changes in one or more inputs.

We have developed two different general tools useful for systems engineering of oligosaccharide structure: a mathematical model for N-linked oligosaccharide biosynthesis, and a two-dimensional electrophoresis procedure for resolving many individual glycoproteins. Here these tools are summarized, and their applications to a particular theme in the metabolic engineering of oligosaccharide structures on glycoproteins are presented.

A Mathematical Model for Biosynthesis of N-Linked Oligosaccharides

The purpose of this model is to provide qualitative guidance for metabolic engineering to achieve a desired glycoform distribution. Thus, the model is intended to indicate qualitative trends resulting from changes in the amounts of different enzymes involved in N-linked oligosaccharide biosynthesis.

N-linked oligosaccharides are built in a series of enzyme-catalyzed reactions which occur in the Golgi apparatus. Enzymes which catalyze oligosaccharide synthesis are typically localized in particular compartments of the Golgi. This model represents the Golgi apparatus as four successive well-mixed compartments through which glycoproteins, with their oligosaccharides, move at a rate determined by the overall rate of transit through the secretory apparatus. This representation of Golgi trafficking has been used by previous investigators interested in describing other properties of the secretory system (4).

Figure 1 shows the reactions involved in synthesizing N-linked oligosaccharides in mammalian cells. The pathway is complicated, with many branch points at which a single substrate can serve as a substrate for two different enzymes, leading to two different structures which may then initiate elaborated pathways of further reactions (5).

Bisected biantennary glycoforms can be desirable in some applications. Producing these requires action of β-N-acetyl glucosaminyl transferase III (GnTIII). However, the action of GnTIII has potentially major effects on the glycoform distribution which is obtained, since GnTIII can convert many different oligosaccharides to the corresponding bisected form. Furthermore, the products of GnTIII action are not substrates for further elaboration of branching and subsequent modifications of the oligosaccharide structure. In particular, GnTIII competes, in the reactions indicated in the upper section here, with reactions catalyzed by mannosidase II (ManII) and β-N-acetyl glucosaminyl transferase II (GnTII), taking hybrid oligosaccharides "off pathway" and committing them to remain hybrid. Hybrid oligosaccharides are cleared rapidly from the circulation.

CHO cells do not normally express GnTIII, so it is interesting to examine the consequences of GnTIII expression in CHO cells, from several different systems engineering perspectives. This will be the particular system considered here.

Some of the parameters in this model are available from previous research, and many others may be reasonable estimated based on studies of related systems. The remainder have been chosen based upon the closest analogous situation. This procedure allows 69 of the model's 86 parameters to be estimated with some confidence. One initial question about the model, then, is the sensitivity of its simulations to modifications in the relatively uncertain parameters. With the base parameters estimated for the model, it simulates well the distribution of oligosaccharides observed experimentally for interferon-ß expressed in CHO cells.

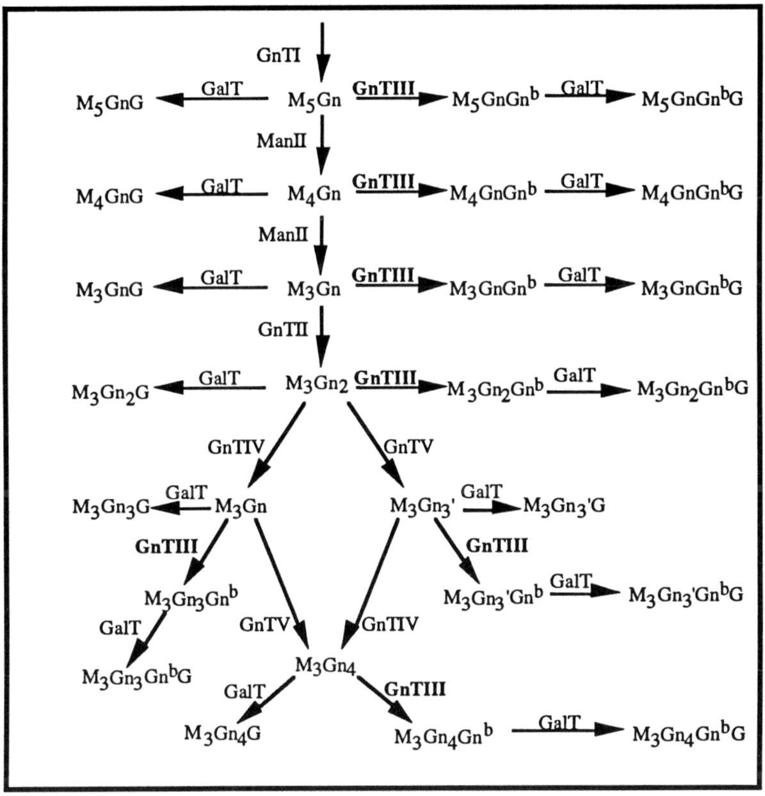

Figure 1. N-linked branching pathway including GnTIII

The model has been used to simulate the influence of expressing different quantities of GnTIII in CHO cells, which in their wild-type state have negligible GnTIII activity (Figure 2). Here the fraction of bisected structures which are biantennary (taken as the desired form here) and hybrid (undesired due to rapid clearance) have been calculated as a function of the GnTIII level in the cells. For reference, the GnTIII activity in cell extracts is less than 5 pmol/h/10^6 cells for wild-type CHO cells, about 200 in a CHO mutant called LEC10 which overexpresses GnTIII (6), and around 4000 (7) in mouse melanoma cells stably transfected with an expression vector for GnTIII.

GnTII forms a biantennary structure, and this reaction competes with a bisecting reaction catalyzed by GnTIII which commits the oligosaccharide to remaining hybrid. Thus, higher GnTII activity should increase the bisected, biantennary population. There is no way to estimate this without doing the experiment, or simulating the experiment using a mathematical model. The model which we have developed indicates an increase in the biantennary fraction of the bisected oligosaccharides from about 70% to 80% when the GnTII activity is tripled relative to the base case. Also, with higher GnTII activity, the sensitivity of the biantennary fraction to GnTIII activity variations is significantly reduced in the range of high (>1000) GnTIII activities.

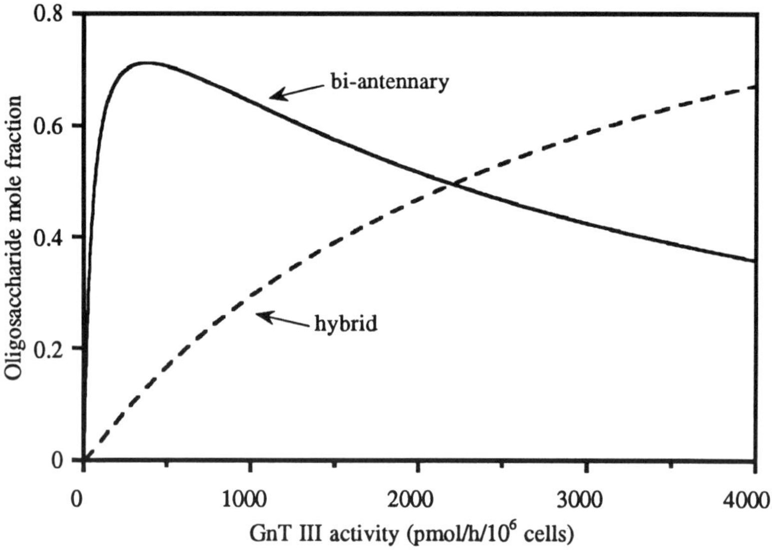

Figure 2. Effect of GnT expression on the bisected glycoform distribution

K_ms for the model were estimated from the available literature data which, for GnTIII, is based on use of free oligosaccharide substrates. Oligosaccharides on glycoproteins will, in general, have lower accessibility to GnTIII resulting in a larger K_m. At a K_m for all GnT III-catalyzed reactions 10-fold greater than the base case, there is no longer a GnTIII activity which maximizes the bisected biantennary fraction in the range of GnTIII activity studied here. For the greater K_m, this fraction increases to a plateau as GnT III activity is increased.

Still, a significant quantity of bisected forms are produced, according to the model, over a broad range of GnTIII activities, independent of the K_m values. This result suggests that the response of the glycoform distribution is not highly sensitive to the protein context of the oligosaccharide which serves as a substrate for GnTIII, at least in certain ranges of GnTIII activities. This suggests that many glycoproteins will be modified if GnTIII is overexpressed in a cell which has little or no GnTIII activity originally.

Two-Dimensional Electrophoresis Analysis of Glycoprotein Patterns in Cell Extracts and Supernatants

If standard protocols are used for CHO cell protein extraction and 2D electrophoresis, no glycoproteins are reproducibly resolved (see Figure 3A). A special protocol was developed based upon systematic variation of many experimental parameters which provides excellent separation of both extracted (e.g., Figure 3B) and supernatant glycoproteins form CHO cell cultures. The details of this protocol will be presented elsewhere; the most important elements of the protocol are use of low concentrations of organic solvents in the extraction and isoelectric focusing (IEF) steps, and use immobilized pH gradient material for the IEF separation.

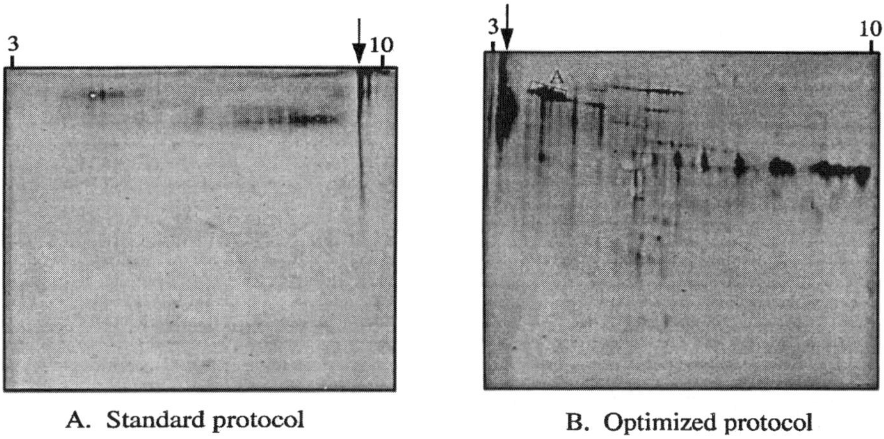

Figure 3. Significance of an optimized extraction and electrophoresis protocol. Lec10 cell extracts analyzed by 2D electrophoresis and detected with E-PHA-alkaline phosphatase.

This method has been used to determine protein patterns which bind the erythroagglutinating lectin of *Phaseolus vulgaris* (E-PHA), which primarily interacts with bisected oligosaccharides, in cell extracts and in culture supernatants from cultures of CHO LEC10 (a mutant which exhibits significant GnTIII activity) and its parental cell line CHO Pro5 (ATCC CRL-1781; this cell line lacks GnTIII activity detectable by E-PHA binding to the cell surface). Comparison of E-PHA binding patterns in Pro5 and LEC10 CHO cell extracts reveals many more labelled spots in LEC10 cells (Minch et al., in preparation). This demonstrates that modification of the set of oligosaccharide biosynthesis enzymes affects the resultant glycoforms of many host cell glycoproteins. Similarly. E-PHA binding patterns in Pro5 and LEC10 CHO cell culture supernatants show many more labelled glycoproteins from the LEC10 culture. Many labelled spots belong to strings of apparently related spots which show more basic pI and decreasing molecular weight; these are presumably multiple glycoforms, with decreasing oligosaccharide complexity and sialic acid content.

Many prior studies of glycosylation in cell culture have considered the influence of culture conditions on glycosylation. We have done some first experiments in this connection, applying this 2DE method. In one such experiment, E-PHA binding patterns were compared in extracts from cells harvested after one, two, and three days in culture. Most of the spots observed do not change intensity nor shift position over time. There are some significant intensity changes for a few particular proteins, but this is more likely due to a change in amount of these proteins, rather than to a change in GnTIII activity, since global changes in labelling intensity are absent (Minch, in preparation).

We see that overexpression of GnTIII modifies many secreted CHO glycoproteins. This suggests that engineering a recombinant CHO cell line which secretes a heterologous glycoprotein to also overexpress GnTIII may well produce novel, GnTIII-modified forms of that heterologous protein. An experiment to test this hypothesis is summarized next.

Overexpression of Cloned GnTIII in a Recombinant CHO Cell Line Engineered for Efficient Expression of Human Beta-Interferon

The cell line CHO M-159 was constructed to express an amplified gene encoding human interferon-β (8). This cell line was transfected with an expression vector for GnTIII which was constructed by cloning the cDNA for rat GnTIII behind the CMV promoter in the vector pcDNA3 (Invitrogen). Stable clones from this transfection, selected on the basis of neomycin resistance, were grown in FMX-8 medium supplemented with 10% FCS. The cells were transferred to FMX-8 protein-free medium and the supernatant was harvested 60 hours later for analysis by electrophoresis. Figure 4 shows blots of an SDS-PAGE gel of concentrated supernatant proteins. Part A shows a blot by E-PHA-alkaline phosphatase conjugate; part B shows the response to anti-interferon-β antibody visualized by a peroxidase-secondary antibody conjugate.

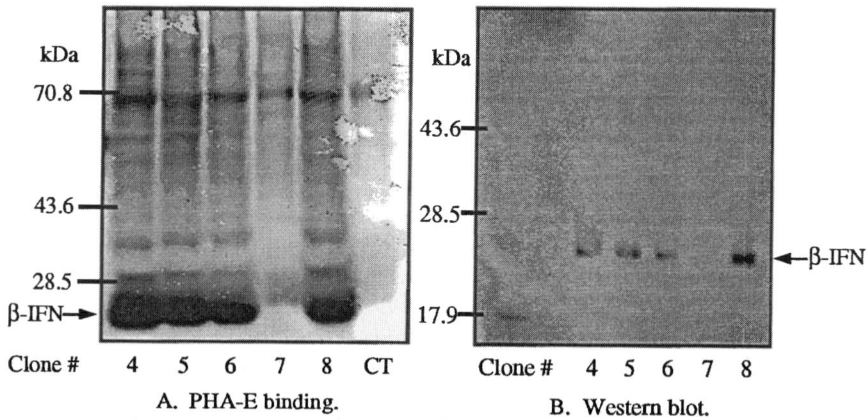

Figure 4. Lectin and antibody blots of supernantants from different clones of CHO M-159:pGnTIII.

Conclusions

1. A mathematical model is useful for discovering qualitative trends resulting from changing levels of biosynthetic enzymes.
2. 2DE shows that alteration of GnTIII activity in CHO cells changes glycosylation of many cell and secreted CHO proteins.
3. An E-PHA-binding form of interferon-β can be produced by further engineering of an interferon-β producing CHO cell line with a GnT III expression vector.

References

1. Bailey JE (1991) Science 252, 1668-1675.
2. Stanley P (1992) Glycobiology 2, 99-107.
3. Minch SI, Kallio PT & Bailey JE (1995) Biotechnology Progess 11,348-351.
4. Bibila TA & Flickinger MC (1991) Biotechnology and Bioengineering 38, 767-780.
5. Schachter H (1985) Biochem. Cell. Biol. 64, 163-181.
6. Campbell C & Stanley P (1984). J. Biol. Chem.261, 13370-13378.
7. Yoshimura M, Nishikawa A, Ihara Y, Taniguchi S & Taniguchi N (1995) J. Biol. Chem. 92, 8754-8758.
8. Page MJ (1985) Gene 37, 139-144.

EXTRACELLULAR PROPEPTIDE PROCESSING OF RECOMBINANT HUMAN FACTOR IX BY A SECRETED FORM OF THE ENDOPROTEASE, PACE.

Maureen A. Hamilton and Timothy S. Charlebois, Mammalian and Microbial Cell Sciences, Genetics Institute, Inc., One Burtt Rd, Andover, MA, 01810, USA

Abstract

Biological activity of human coagulation factor IX is dependent upon proteolytic removal of an 18-amino acid N-terminal propeptide domain. At expression levels of recombinant factor IX (rFIX) suitable for commercial production, Chinese Hamster Ovary (CHO) cells were limited in their ability to cleave proFIX. Subsequent introduction of an expression construct encoding PACE-SOL, a subtilisin-like endoprotease, dramatically reduced the proportion of proFIX relative to mature rFIX secreted from these cells. PACE-SOL cleaves proFIX intracellularly and can cleave proFIX which escapes intracellular cleavage in the conditioned medium. Only very low PACE-SOL expression levels are required to ensure full propeptide cleavage due to the activity of PACE-SOL in the conditioned medium during cell culture and during downstream processing steps of cell removal and medium concentration. Our results demonstrate that coexpression of PACE-SOL and rFIX in CHO cells is an important component of an efficient system for producing high-quality rFIX: augmenting the intracellular propeptide cleavage capacity of the cell and adding an extracellular cleavage component unavailable to cells without PACE-SOL, provide for a flexible and robust system for proFIX cleavage, which extends beyond the culture harvest point into downstream processing steps.

Keywords: Factor IX, propeptide processing, PACE-SOL, endoprotease, CHO cells, recombinant, hemophilia B, extracellular, blood coagulation

Introduction

Factor IX is an important component of the blood coagulation cascade and is deficient in patients with Hemophilia B. One challenge associated with making recombinant human factor IX (rFIX) in Chinese Hamster Ovary (CHO) cells for the treatment of Hemophilia B is achieving efficient propeptide removal from the molecule. Factor IX is synthesized as a precursor molecule containing an eighteen amino-acid propeptide which directs the vitamin K-dependent gamma-carboxylation of twelve glutamic acid residues in the amino-terminal portion of the mature molecule. Subsequent to gamma-carboxylation, the propeptide must be removed in order for FIX to achieve a biologically active conformation (1). At amplified rFIX

M.J.T. Carrondo et al. (eds.), Animal Cell Technology, 495–501.

expression levels, CHO cells are limited in their capacity to fully cleave all the proFIX synthesized, and therefore, a significant amount of proFIX can be detected in the conditioned medium. To augment the capacity of the CHO cells to process proFIX, a secreted soluble form of the endoprotease PACE, PACE-SOL (2), was coexpressed with rFIX. Because PACE-SOL is secreted, and is able to cleave proFIX both inside the cell and in conditioned medium, only very low expression levels of PACE-SOL are required to provide for efficient proFIX cleavage during rFIX production, even if some proFIX escapes intracellular cleavage.

Materials and Methods

Construction of Cell Lines :
rFIX-expressing CHO cell lines (A, B, C, D, E) were obtained by transfection of the pMT2-IX expression construct (2) into a DUKX-B11 host cell line (3). After initial selection in nucleoside-free medium, cell lines expressing higher levels of rFIX were obtained by amplification of the dhfr/rFIX genes with methotrexate (MTX). rFIX/ PACE-SOL coexpressing cell lines were obtained by transfection of an expression construct containing the PACE-SOL gene into each of the amplified rFIX cell lines. Resultant cell lines expressed amplified levels of rFIX and unamplified levels of PACE-SOL.

Cell culture and secretion:
Monolayer cultures were grown in a fortified DMEM/F12 (50:50, v:v) medium supplemented with 10% dialyzed heat-inactivated fetal bovine serum (FBS), and 500 ng/ml vitamin K_1. Suspension cultures were grown in spinner flasks in a fortified DMEM/F12 (50:50, v:v) medium suitable for suspension growth of CHO cells supplemented with 100 ng/ml Vitamin K_1. After 24 hours of growth, medium was replaced with fresh secretion medium. At indicated time points, conditioned medium was sampled and frozen at -80°C.

Determination of rFIX Concentration in Conditioned Medium:
Total rFIX protein concentration was determined in a sandwich-type enzyme-linked immunosorbent assay (ELISA) using conformation-independent rabbit-anti-FIX polyclonal antibodies for both capture and detection of rFIX. The detection antibody was biotinylated. Avidin-horseradish peroxidase colorimetric detection was used to quantitate amounts of rFIX per well from optical density readings.

Total rFIX Immunoprecipitation:
A conformation-independent monoclonal antibody to FIX, coupled to Sepharose CL4B was used to immunoprecipitate total rFIX protein from conditioned medium samples. Total rFIX was eluted from the CL4B/antibody complex in 2X Sample Buffer (2% SDS, 20% glycerol, 0.02 g/l bromophenol blue, 125 mM Tris-HCl), by heating at 95°C for 5 minutes.

SDS-PAGE and Western Blot Analyses:
Proteins in conditioned medium plus Sample Buffer, or in 2X Sample Buffer alone after elution following immunoprecipitation, were resolved by SDS-PAGE under non-reducing conditions using a 10-20% acrylamide gradient. Proteins were transferred to nitrocellulose. Western blot analysis was performed to detect total rFIX protein, proFIX, or PACE-SOL. Primary antibodies used for Western blot analyses are as follows:

anti-proFIX: rabbit polyclonal raised against the eighteen amino acid propeptide plus the first two amino acids of mature human FIX fused with thioredoxin protein.

anti-total FIX: rabbit polyclonal conformation-independent antibody raised against total FIX

anti-PACE: rabbit polyclonal raised against a 13 amino acid peptide in the N-terminal portion of the mature human PACE molecule.

Antibody-antigen complexes on all blots were visualized using enhanced chemiluminescence (ECL) reagents (Amersham Life Science, Arlington Heights, IL).

Results

rFIX expressed at economically useful levels in CHO cells is poorly processed with respect to propeptide removal. Figure 1 shows a western blot analysis of proFIX in conditioned medium from four cell lines before and after rFIX gene amplification with MTX. In each case, the percentage of proFIX detected in the conditioned medium is greatly increased at the higher rFIX expression level. Even at a non-amplified level, the "A" cell line already displays a processing limitation, as some proFIX can be detected in the 0 MTX lane. At higher rFIX expression levels, each of the four lines is limited in its ability to fully process proFIX, and so secretes a significant amount into the conditioned medium (20 and 50 MTX lanes).

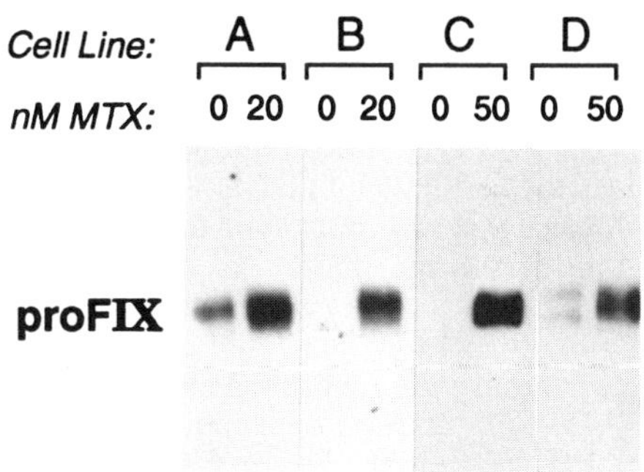

Figure 1. Effect of rFIX gene amplification on proFIX levels in conditioned medium. Conditioned medium was collected from 24 hour secretions of cell lines before and after rFIX gene amplification. SDS/PAGE - western blot analysis was performed to detect proFIX as described in Materials and Methods.

PACE-SOL, when coexpressed in the amplified rFIX cell lines, augments the cellular proFIX cleavage capacity. In each case, cell lines expressing amplified levels of rFIX secreted approximately 30% proFIX as detected by the conditioned medium western blot in Figure 2. After subsequent introduction of PACE-SOL, the same cell lines were able to fully process the proFIX synthesized, and none is detected in the conditioned medium at 24 hours.

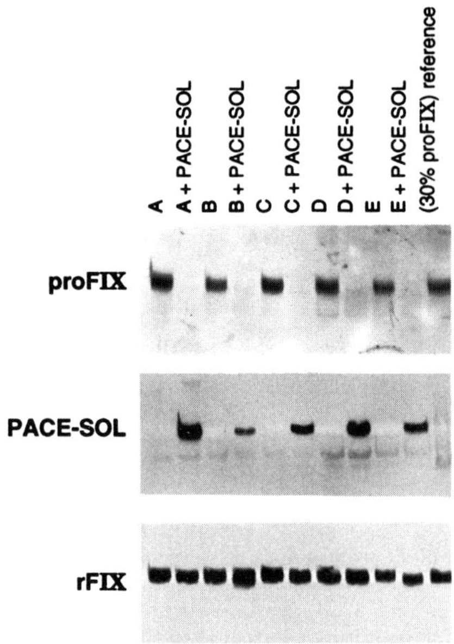

Figure 2. Effect of coexpression of PACE-SOL in rFIX cell lines on proFIX in conditioned medium. Conditioned medium was collected from amplified rFIX expressing cell lines before (lanes 1,3,5,7, and 9) and after PACE-SOL coexpression (lanes 2,4,6,8, and 10). rFIX concentration for each sample was determined by ELISA, and SDS/PAGE-western blot analysis was performed as described in Materials and Methods. N-terminal sequence analysis had been previously performed on conditioned medium from an rFIX-expressing cell line, (lane 11). Approximately 30% of the total rFIX secreted by this cell line contains the propeptide. Conditioned medium from this cell line is included on this western blot for reference (lane 11).

Depending on the expression level of PACE-SOL, some proFIX could theoretically escape intracellular cleavage in rFIX/ PACE-SOL coexpressors. In such cases, extracellular proFIX processing by secreted PACE-SOL might contribute to the low level of proFIX observed in the rFIX/PACE-SOL coexpressors. PACE-SOL can cleave proFIX in the conditioned medium (2, 4), and this extracellular activity can be inhibited with EDTA, due to the Ca^{++} dependence of the PACE protease (4, Hamilton and Charlebois, submitted for publication). Two cell lines (shown in Figure 2 as D+ PACE-SOL, and E + PACE-SOL) were chosen to further investigate whether extracellular PACE-SOL activity is required for complete proFIX processing by the rFIX/ PACE-SOL coexpressors. The E.100P (shown in Figure 2 as E + PACE-SOL) cell line secretes approximately 5-fold less PACE-SOL than does the D.50P cell line (shown in Figure 2 as D + PACE-SOL). When western blot analysis was performed to detect secreted proFIX after a 3 hour secretion period in the presence and absence of 1 mM EDTA, the results shown in Figure 3 are different for these two cell lines. In the higher PACE-SOL expressor, D.50P (panel a), no proFIX is detected in the conditioned medium after 3 hours (-EDTA lane), similar to the 24 hour result in Figure 2 (lane 8). EDTA inhibition of extracellular PACE-SOL is inconsequential to the D.50P cell line because it cleaves all the proFIX synthesized before the rFIX is secreted. For the D.50P cell line, extracellular PACE-SOL is redundant. In contrast, there is some proFIX in the conditioned medium of the low PACE-SOL expressor, E.100P (panel b, -EDTA lane) at this early time point. This result is different from the 24 hour secretion with this cell line where no proFIX was detected in the conditioned medium (Figure 2, lane 10). Apparently, some proFIX is continuously secreted from this cell line but is only detected at relatively early points during cell culture before secreted PACE-SOL has a chance to cleave it. This result is corroborated by the dramatically higher amount of proFIX detected in the conditioned medium of the E.100P cell line when extracellular PACE-SOL is inhibited with EDTA (panel b, + EDTA lane). Thus, for the low PACE-SOL expressor, the low percentage of proFIX typically seen at the end of a batch is the net result of both intracellular cleavage and extracellular cleavage of proFIX over time.

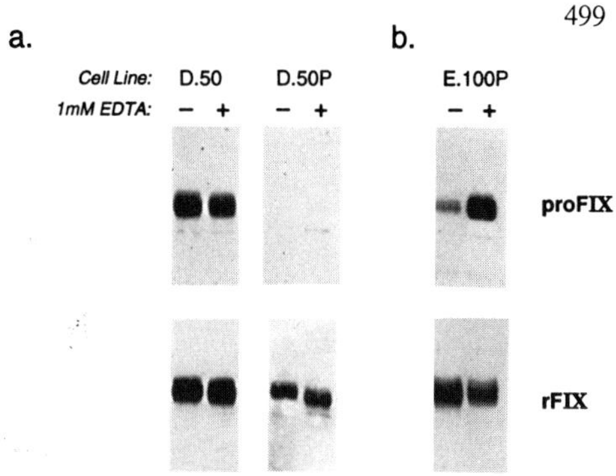

Figure 3. Effect of EDTA on proFIX levels in conditioned medium of rFIX/PACE-SOL coexpressors. Cell lines D.50P, (panel a) and E.100P (panel b), were grown in the presence or absence of 1mM EDTA. Conditioned medium was collected after 3 hours. Total rFIX concentration was determined by ELISA. rFIX was immunoprecipitated, subjected to SDS/PAGE, and western blot analysis was performed as described in Materials and Methods. The D.50 cell line (expresses only rFIX, not PACE-SOL), is included as a control in panel "a".

To understand better the critical parameters of the rFIX/ PACE-SOL coexpression system, and to assess the potential robustness of the rFIX/ PACE-SOL coexpression system, we further investigated the extracellular PACE-SOL activity in the E.100P low PACE-SOL expressor. The typically low level of proFIX seen at the end of a 24 or 72 hour secretion from this cell line is seen in Figure 4, lane 1. When these cells are allowed to secrete for two hours into fresh medium, a significant amount of proFIX is detected in the conditioned medium which over time is progressively cleaved by extracellular PACE-SOL (lanes 2 through 5). If the high-proFIX conditioned medium is incubated at 37°C for 24 hours after cell removal, the proFIX is still cleaved (lane 8).

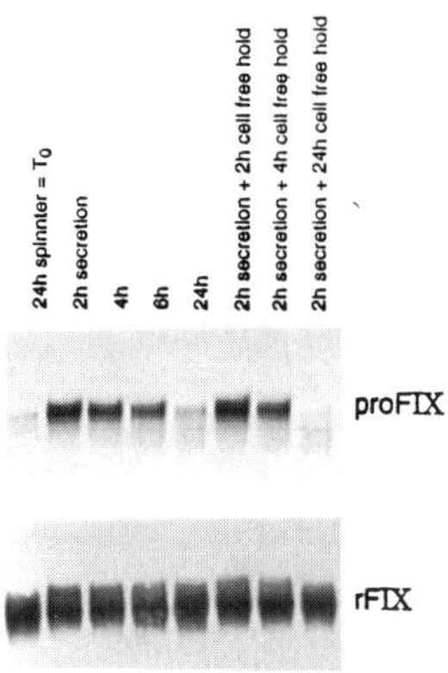

Figure 4. ProFIX levels in conditioned medium during cell culture and during cell-free incubation of rFIX/PACE-SOL-containing medium. E.100P cells were grown in suspension culture. Conditioned medium was collected at 2, 4, 6, and 24 hours, and immediately frozen at -20°C. A sample of the 2 hour cell-free conditioned medium was incubated at 37°C for an additional 2, 4, and 24 hours. Total rFIX concentration was determined by ELISA and rFIX protein was immunoprecipitated. 1 μg total FIX per lane was subjected to SDS-PAGE and western blot analysis as described in Materials and Methods.

The above data indicate that the use of a secreted form of PACE affords the possibility of extracellular cleavage both throughout the course of cell culture and in the harvest medium after cell removal. Following cell removal, the rFIX manufacturing process involves an ultrafiltration (medium concentration) step. An experiment was performed to assess whether proFIX cleavage might continue during ultrafiltration before the PACE-SOL is separated from the rFIX in purification (Figure 5). High proFIX-containing material was generated from the E.100P cell line in a 6 hour secretion. This conditioned medium was then subjected to a 20-fold concentration. A control sample-concentration was done in the presence of EDTA to inhibit PACE-SOL activity in the conditioned medium. The initial unconcentrated material contained approximately 30% proFIX (Figure 5, compare to reference material). During the 1 hour concentration, the level of proFIX was dramatically reduced. When PACE-SOL activity in the medium was inhibited by EDTA, no significant reduction in proFIX was observed (+ 1mM EDTA lane).

Figure 5. ProFIX cleavage by PACE-SOL during a 20-fold concentration of cell-free conditioned medium. E.100P cells were grown in suspension culture. Conditioned medium was collected after 6 hours and either frozen immediately or concentrated 20-fold in the presence or absence of 1mM EDTA using a Centricon 10 ultrafiltration device (Amicon, Danvers, MA). Total rFIX concentration was determined by ELISA. rFIX protein was immunoprecipitated and 500 ng total rFIX per lane was subjected to SDS-PAGE, and western blot analysis as described in Materials and Methods. A dilution series of reference conditioned medium from an rFIX cell line previously determined by N-terminal sequence analysis to secrete approximately 30% proFIX is included on this blot.

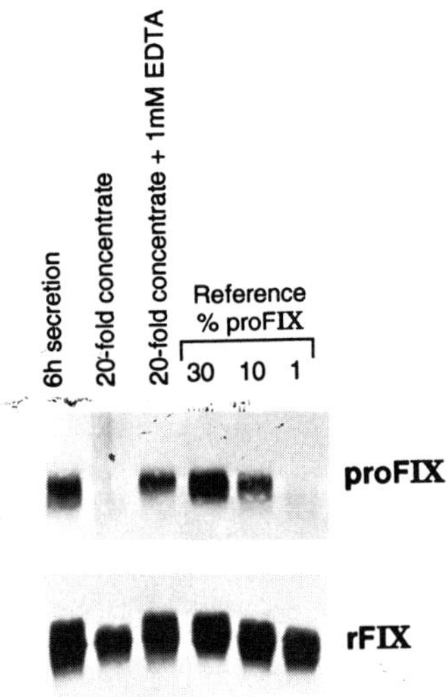

Discussion

The rFIX/PACE-SOL coexpression system we have employed is efficient and robust with respect to propeptide removal from the rFIX molecule. The robustness of this system stems in large part from the expression of a secreted form of PACE which retains activity in the conditioned medium and can therefore cleave proFIX which may escape intracellular cleavage due to low PACE-SOL expression levels. Additionally, PACE-SOL has opportunity to act on proFIX during downstream processing steps of cell removal and ultrafiltration. Relatively low PACE-SOL levels have been shown to efficiently cleave proFIX during downstream processing even when challenged with higher levels of proFIX than would be expected at the conditioned medium harvest point. The rFIX/ PACE-SOL coexpression system is therefore a very effective and reliable one for the production of high-quality rFIX at economically useful rFIX expression levels.

References

1. Furie, B., and Furie, B.C. 1990. Molecular basis of vitamin K-dependent γ-carboxylation. *Blood* 75:1753-1762.

2. Wasley, L.C., Rehemtulla, A., Bristol, J.A., Kaufman, R.J. 1993. PACE/furin can process the vitamin K-dependent pro-factor IX precursor within the secretory pathway. *J. Biol. Chem.* 268:8458-8465.

3. Urlaub, G. and Chasin, L.A. 1980. Isolation of Chinese hamster cell mutants deficient in dihydrofolate reductase activity. *Proc. Natl. Acad. Sci. USA* 77:4210-4220.

4. Hatsuzawa, K., Murakami, K., and Nakayama, K. 1992. Molecular and enzymatic properties of furin, a Kex2-like endoprotease involved in precursor cleavage at Arg-X-Lys/Arg-Arg sites. *J. Biol. Chem. (Tokyo)* 111:296-301.

Discussion

Lodish	Since PACE will process many proteins, have you tried it on any others?
Hamilton:	We are hoping to try it in others which have the same problem of phenopropeptide cleavage.
Cunningham:	Did you get an improvement in specific productivity with improved processing?
Hamilton:	Yes, I was unable to go into this but in cases where in cell lines all of the other post-translational modifications are happening efficiently and the only problem is propeptide cleavage. When you express PACE cells you do see a boost in the percentage of active Factor IX, so you do get an increase in what you are trying to make.
Cunningham:	Is it significant?
Hamilton:	It can be significant if they are only secreting 30-50% pro-IX then you can get a 30-50% boost in the good stuff you are making.
Thomas:	Was there different cell performance between the high and the low expressions or was it a burden to express higher levels of the PACE?
Hamilton:	We did not observe that because we transfected the PACE under selection. The levels of PACE I showed are relative and even the high expressors have a relatively low unamplified level. The cells do not seem to mind it at all so there was no real difference in the cells.

Glycosylation patterns of recombinant therapeutic proteins produced in two mammalian cell lines

Kristina Kopp, Michael Schlüter, Wolfgang Noé and Rolf G. Werner
Dr. Karl Thomae GmbH, 88397 Biberach / Riss, Germany

Abstract

To elucidate the quality of glycosylation patterns when changing the host cell, the desired protein or the extracellular environment, we expressed human therapeutic glycoproteins in Chinese Hamster Ovary (CHO) and mouse myeloma (NS0) cells under different fermentation conditions. sICAM consistencies produced in both cell lines were shown to differ significantly. Their dissimilar oligosaccharide structures became apparent with all applied electrophoretic and chromatographic methods. In the case of CHO-expressed proteins, the altered sICAM, t-PA and IFN-ω glycosylation patterns resulting from varied fermentation conditions were rather small and limited on quantitative changes of preexisting oligosaccharide structures. These changes in product consistencies were therefore difficult to evaluate with conventional carbohydrate analytical methods such as the HPAEC / PAD mapping technique. However, differences became evident after capillary electrophoretic separation and mapping of aminobenzamide (AB)-labeled oligosaccharides.

Introduction

N-glycosylation is a very common post-translational modification found in eukaryotic secretory proteins. For glycoprotein therapeutics, the carbohydrates are important for properties such as stability, solubility, in vivo activity and immunogenicity [1]. A detailed knowledge of synthesized carbohydrate structures is therefore inevitable and requires sophisticated analytical methods. Since there is no template that determines glycosylation patterns, several factors contribute to carbohydrate synthesis. First of all, the host cell with its glycosylation enzymes determines the range of possible structures. In addition, the amino acid sequence defines glycosylation sites and influences synthesized structures. Third, different kinds of process parameters can affect glycosylation.
The aim of this work was twofold: First, we were interested in the influence of CHO versus NS0 host cell lines on sICAM oligosaccharide structures. Second, we investigated the effect of several process parameters on the glycosylation of CHO-expressed sICAM, IFN-ω and t-PA.

M.J.T. Carrondo et al. (eds.), Animal Cell Technology, 503–509.

Materials and Methods

Suspension-adapted cells expressing human sICAM, IFN-ω and t-PA were cultivated in 2 L fermentors as previously described [2]. In the course of this, process relevant fermentation conditions were varied. Glycoprotein variants were purified by standardized purification procedures and analyzed by electrophoretic and chromatographic methods.

Results and Discussion

The influence of the host cell on the consistency of sICAM is shown to be huge. We found that NS0 cells provide a completely different glycosylation pattern when compared with the CHO product. Isoelectric focusing as well as SDS-PAGE were able to distinguish between the proteins derived from the two cell lines (Fig.1).

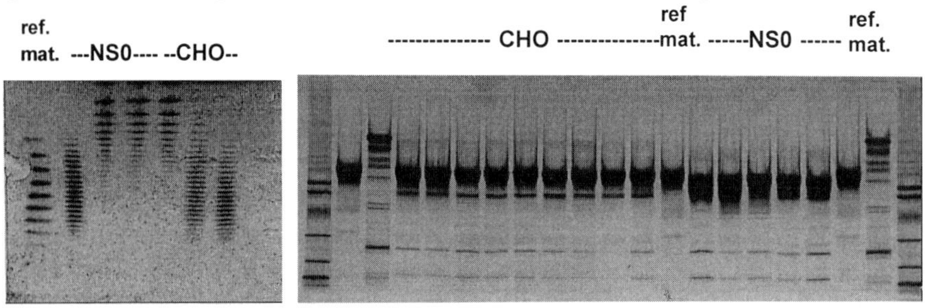

Fig. 1: Isoelectric focusing and SDS-PAGE analyses of sICAM expressed in two cell lines (CHO and NS0)

Chromatographic methods verified these results. HPAEC of desialylated oligosaccharides showed a different and more heterogeneous composition of oligosaccharide structures for sICAM derived from NS0 cells, compared with the product from CHO cells (Fig.2).

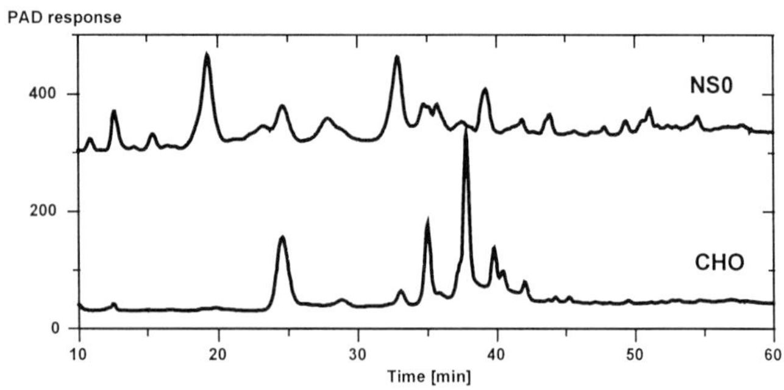

Fig. 2: HPAEC oligosaccharide mapping of sICAM expressed in NS0 and CHO cells

Concerning the types of sialic acids, we found that N-acetylneuraminic acid - important for the clearance rate of circulating glycoproteins - is nearly absent in NS0 cells. The latter synthesize N-glycolylneuraminic acid instead, a monosaccharide that is not expressed in humans and therefore is considered to be immunogenic (Fig.3).

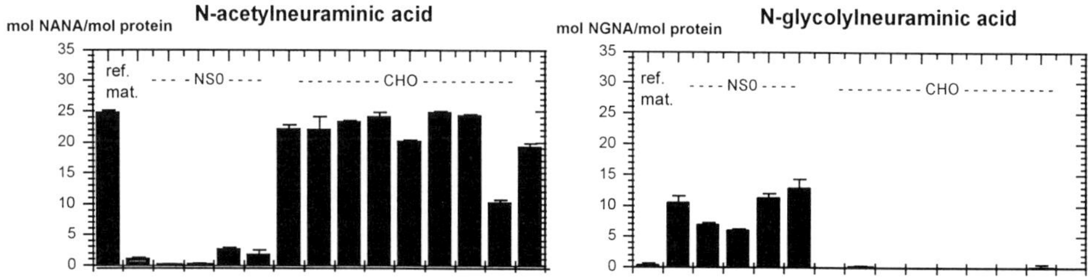

Fig. 3: Sialic acid content of sICAM expressed in CHO and NS0 cells

To elucidate process parameters that are able to affect glycosylation, we expressed sICAM, t-PA and IFN-ω in CHO cells varying in

- cell parameters (clone, cell age)
- media parameters (pH, butyrate, ammonia)
- fermentation parameters (scale, process time, adherent / suspended cultivation)

Our results from these investigations were the following:

IEF, SDS-PAGE as well as peptide mapping revealed no differences; HPAEC-PAD showed only insignificant deviations from the fermentation variants investigated.

A new HPLC mapping method of aminobenzamide (AB)-labeled desialylated oligosaccharides and capillary electrophoresis however were able to detect clear differences in some of the glycosylation patterns investigated. Some examples for HPAEC-mapped oligosaccharides are shown in Figure 4. The sICAM variants result from fermentations over a shortened process time, without butyrate, under increased ammonia concentrations, and from adherent cells.

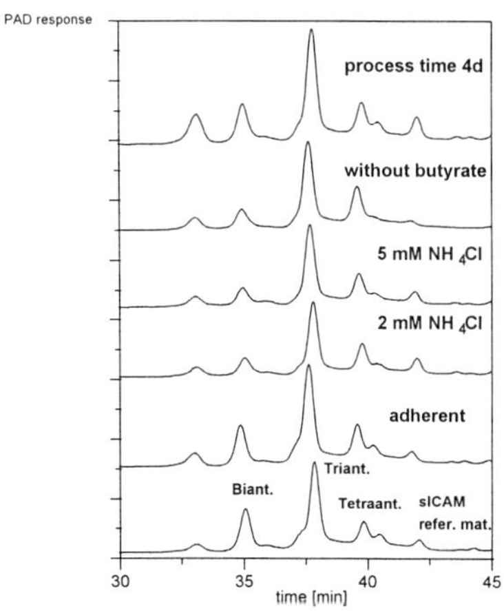

In the case of all variants, profiles were similar. Some showed small shifts from biantennary towards higher-branched structures. It however became clear that HPAEC is not optimal for typical CHO structures, since different oligosaccharides coelute in this system. In this context, we developed an HPLC method for AB-labeled, desialylated oligosaccharides. Figure 5 shows a direct comparison of the two mapping methods.

Fig. 4: HPAEC of sICAM variants expressed in CHO cells

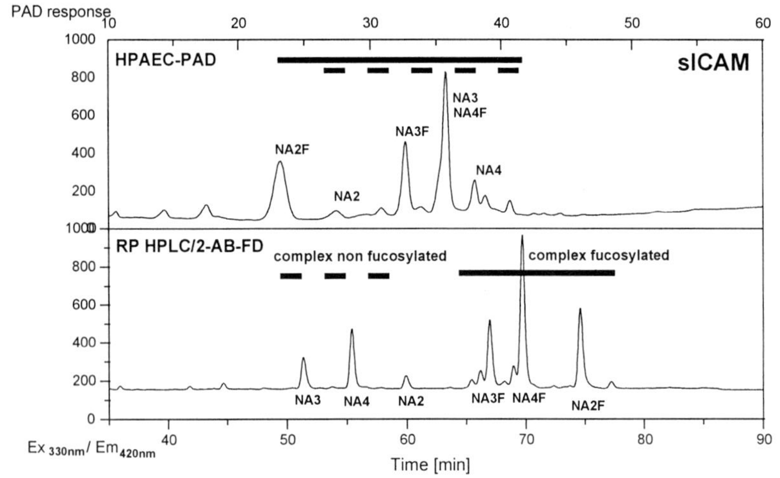

Fig. 5: Desialylated oligosaccharides derived from CHO-expressed sICAM. Two alternative mapping methods are compared.

The HPAEC profile consists of several relatively broad peaks resulting from the fact that complex-type fucosylated structures coelute with their non-fucosylated counterparts. On the other hand the HPLC mapping method is characterized by a high resolution and sharp peaks above a very low baseline noise.

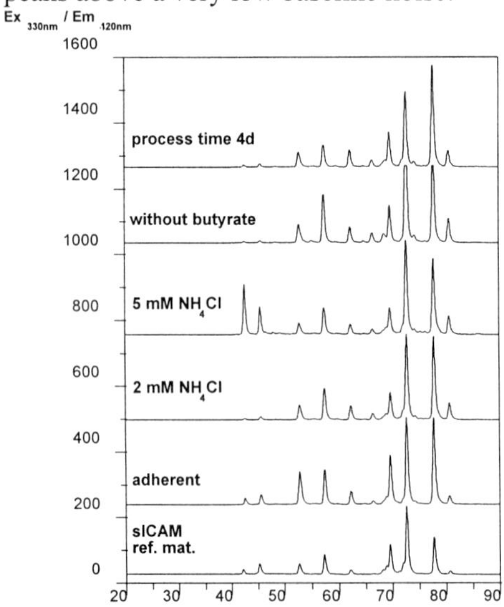

Fig.6: Map of AB-labeled oligosaccharides derived from sICAM fermentation variants

It also provides baseline-separated peaks for the fermentation variants and therefore allows identification and an easier quantification of occuring structures (Fig. 6). The second analytical method that was able to detect changes in some of the investigated glycosylation patterns is capillary electrophoresis.

Figure 7 shows electropherograms of IFN-ω glycoforms. The product profiles derived from a 4 day- and a 8 day- fermentation differ from the more than 12 day-fermentation that provides the reference material - they suggest a distinct influence of the process time. Likewise, the mode of cultivation seems important for the characteristic glycosylation pattern of IFN-ω.

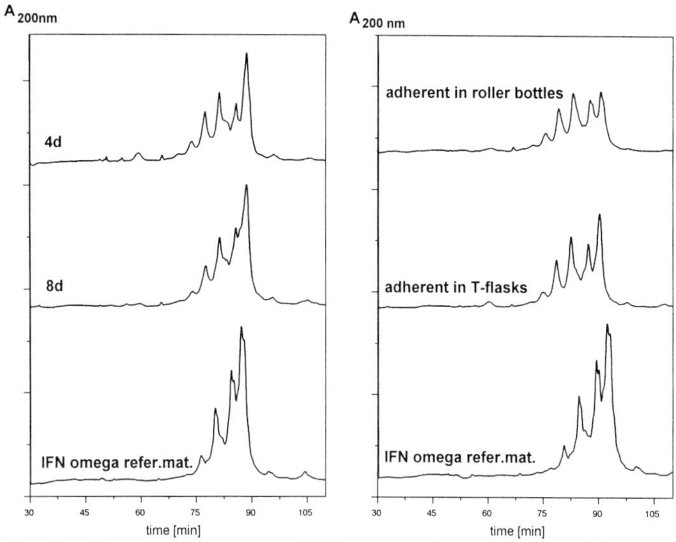

Fig. 7: Electropherograms of IFN-ω glycoforms influenced by process time and cultivation mode.

Capillary electrophoresis also detected alterations induced by increased ammonia concentrations. This parameter is also known in literature to affect glycosylation. In the case of both IFN-ω and t-PA, changed product profiles were revealed (Fig. 8).

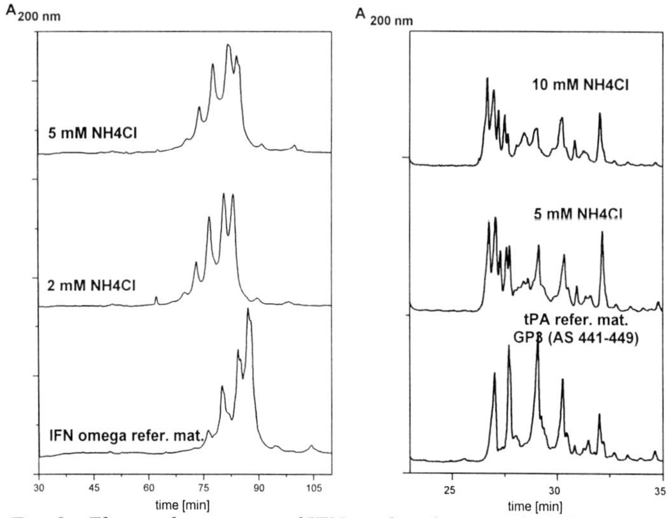

Fig. 8: Electropherograms of IFN-ω glycoforms and t-PA tryptic glycopeptides influenced by increased ammonia.

On the basis of these fermentation variants it became clear that the kind and extent of changes that occurred were protein-specific. While t-PA was shown to have a very stable glycosylation pattern, considerably more glycosylation alterations were detected in the case of sICAM and IFN-ω.

In the course of our studies, most of the parameters investigated influenced glycosylation, but to a different extent (Fig. 9).

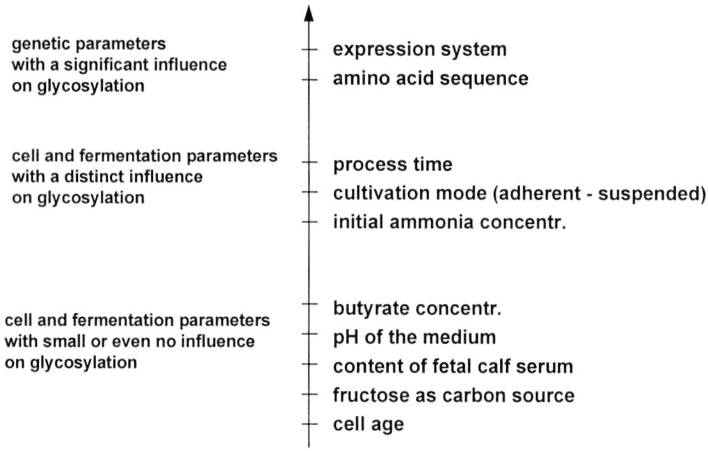

Fig. 9: Investigated parameters arranged by order of influence on glycosylation.

We conclude from our results:

1. Several factors can influence glycosylation profoundly. Therefore mammalian cell fermentations in biotechnical production of pharmaceuticals must be performed under defined and constant conditions.
2. Alterations of glycosylation patterns can not be generalized. They have to be investigated on a case-by-case-basis.

References

1. Elbein, AD (1991) The role of N-linked oligosaccharides in glycoprotein function. Tibtech **9** 346-352
2. Kopp, K et al. (1995) The role of physiological cell parameters and culture conditions on product consistency and glycosylation pattern of recombinant CHO expressed glycoproteins. In: Beuvery E.C.et al. (eds.) Animal Cell Techology: Developments towards the 21st century, Kluwer, 403-407

Acknowledgement

The authors are grateful to Iris Eisenmann for expert technical assistance.

Discussion

Sinacore: Were the changes in glycosylation when you varied the operational parameters of the cell culture systems correlated with any changes in the growth phenotype of the cells you were testing?

Kopp: No, there is no general correlation.

Sinacore: You suggested there was protein specific variability in the stability of glycosylation. It seemed the molecules you were looking at were also different in the extent of glycosylation they received. I wondered if the degree of stability was also related to the degree to which proteins were modified?

Kopp: There is no correlation. For example, IFN omega is not heavily glycosylated at all, only 1 glycosylation site, and it changed quite a lot.

James: I was interested in your results with sodium butyrate addition to the culture. In a number of systems sodium butyrate has been shown to decrease sialyation but increase the branching of N linked oligosaccharides associated with recombinant proteins. Did you observe this?

Kopp: We have developed this new HPLC method I showed to you and now we have to re-evaluate all our results because so far we only used HPE mapping. This is not optional.

Steiner: You showed that the CHO cell line had high expression of NANA where the NSO had a low one. Is this a 'yes' or 'no' situation, or could you, by varying media and culture conditions, bring one cell line up and lower the other one so both could express similar amounts of NANA?

Kopp: I do not think we can do that. NSO synthesises only glycomuraminic acid and in principle, as far as I know, all mouse cell lines synthesise a lot of N glycomuraminic acid so it may not be suitable for production of human proteins.

CONSTRUCTION OF RECOMBINANT CELL LINES WITH DEFINED PROPERTIES USING FLP RECOMBINASE DRIVEN GENE REPLACEMENT

Simone Karreman, Christiaan Karreman[1] and Hansjörg Hauser

GBF, Department of Gene Regulation and Differentiation, Mascheroder Weg 1, D 38124 Braunschweig. [1])present address: GSF, Department of Clinical Molecular Biology, Marchioniniweg 25, D, 81377 München.

SUMMARY

A pilot experiment for the construction of a high producer mammalian cell line using site specific recombination is described. In this experiment, chromosomal loci with intrinsic high expression characteristics were identified by infection with a recombinant retrovirus containing double FRT sites and subsequent screening for overproduction of an encoded reporter gene. In a second step, these sites were targeted with an expression cassette that recombined with the aid of FLP recombinase from *Saccharomyces cerevisiae* yielding cells that had the second construct at the exact position of the first. By using retroviral constructs with double and single FRT sites repectively, stable cell clones were created that no longer can be excised with FLP.

INTRODUCTION

The chromosomal locus of an artificially introduced expression element is one of the most important parameters for the level and stability of expression of the introduced genes. Currently, the creation of stable and well expressing cell lines is therefore marked by long periods of screening. As an alternative we have designed a two-step method, wherein only one step of extensive screening and selection for a whole series of high producer cell clones is neccessary. The primary cell line could then be used repeatedly for the construction of all kinds of derivates by replacing the originally inserted reporter construct with an expression cassette of choice. The construction of a high producer cell line is reduced to this replacement step without further screening, since the chromosomal surrounding is a constant instead of a variable.

This article contains a short review on the concept and the results received so far. Details and an extensive description of the plasmids and methods are described by Karreman et al., 1996 (1).

RESULTS

Creation and selection of reporter gene expressing cells

The use of retroviruses at low infectious doses is an effective way of introducing single

M.J.T. Carrondo et al. (eds.), Animal Cell Technology, 511–517.

copy genes. A prerequisite for retroviral manipulation of cells is that they can be infected efficiently. We have shown earlier that a substrain of BHK-21 cells (BHK-A) is infectable with ecotropic murine retroviruses (2). Since BHK-21 cells are frequently chosen for overexpression of transferred genes, own a highly specific glycosylation pattern (3) and can be grown in protein free media in fermentors (4), we regard them as a highly versatile object for scientific and technological purposes which now include retroviral gene transfer.

The use of the FRT/FLP recombination system from the yeast *Saccharomyces cerevisiae* (5) allows an exchange of cassettes. In order to eliminate the retroviral sequences as much as possible, we had to introduce the FRT-sequences into the LTRs. The replacement of the primary introduced screening cassette demands the use of two FRTs with different sequences (6). These FRTs will recombine with another FRT with exactly the same sequence, but they cannot recombine with each other. The reproductive cycle of retroviruses in which both LTRs of the provirus are derived from the 3' LTR would not allow for the use of two different FRTs in the LTR's. Such a construct would be reduced to a provirus with only one kind of FRT and with its intrinsic instability towards FLP. For this reason a new approach was chosen with double FRTs, one carrying a wild type FTR sequence and the second a mutant, tandemly integrated into the LTR. This leads to a minimum of left residues after replacement (Fig. 1).

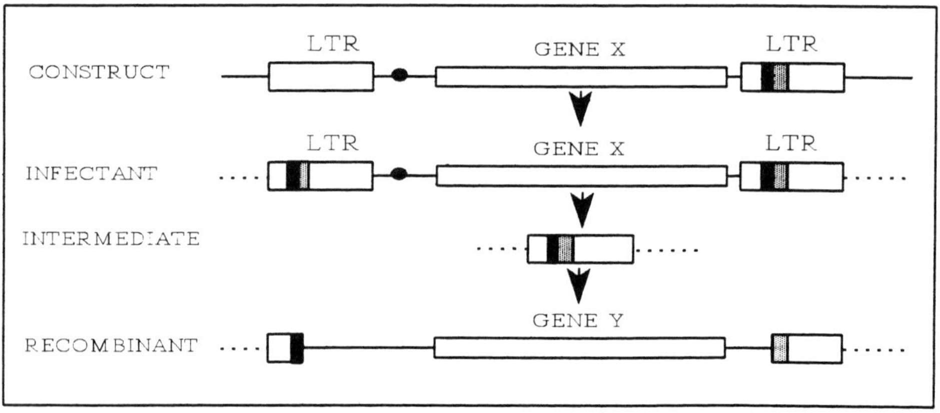

Fig.1.: ***Schematic representation of the fate of the reporter retroviral construct containing the double FRT sequence***

A double FRT sequence consisting of the wild type (black bar) and a mutant (M5)(dotted bar) is integrated into the 3'LTR of the retroviral construct. After replication of the retrovirus the provirus (infectant) will contain the duplicated FRT site in both LTRs. Excision as a result of FLP activity will end up with a single LTR sequence tagged with a double FRT sequence. FRT sequence recombination of a tagged expression cassette with wt and mutant FRT interrupted by the gene of interest should be stable with respect to FLP excision.

We constructed an MPSV based retroviral bicistronic vector (pAPTKNEF; Fig. 2) and infected 10^7 BHK-A cells with an moi of 2×10^{-4}. Two thousands of the resulting neomycin resistant clones were assayed using the SEAP screening method (7). Of those clones that expressed SEAP at visible level, clones were picked with low SEAP levels, with intermediate levels and with high levels. Furthermore, random clones were picked as controls.

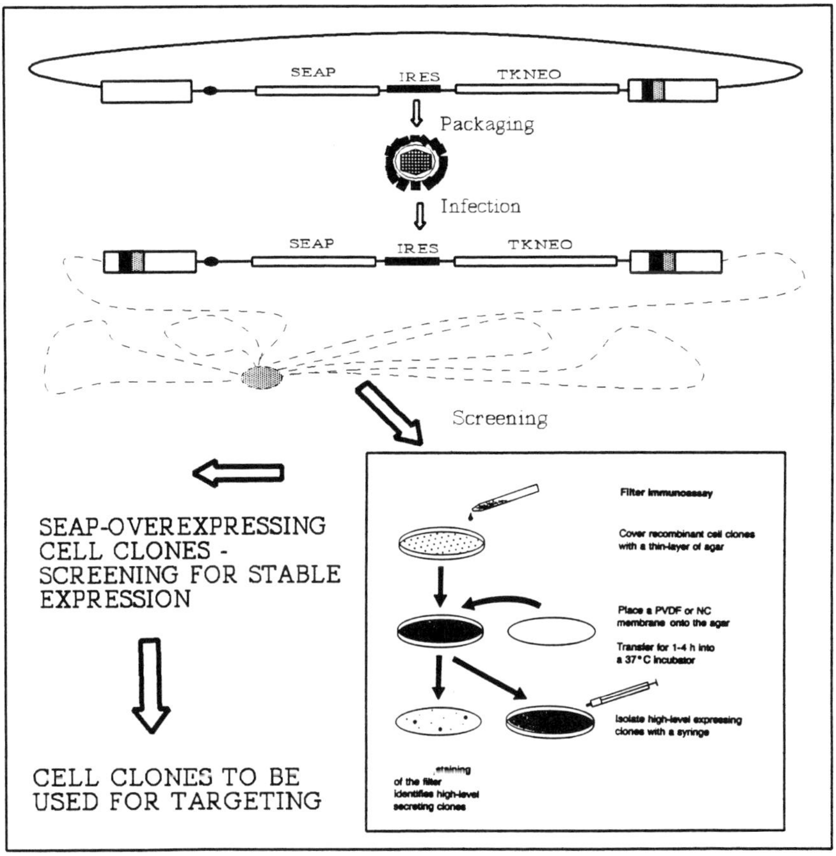

Fig. 2: ***Screening for cell clones overexpressing the proviral reporter construct***
The retroviral vector pAPTKNEF is depicted in the upper part. The vector was transfected into ecotropic helper cells and the produced virus was used to infect BHK-A cells. The provirally tagged genome of a recipient cell (infectant) is shown in the middle part. When the provirus is sufficiently expressed these infectants are resistant to G418. Grown in cell culture plates, the filter overlay assay (lower right) is used to identify the overproducers of secreted alkaline phosphatase (SEAP) by making use of the enzyme activity on the filter membranes. Cell clones of interest are picked and expanded for screening of expression stability.

The chromosomal DNA of these clones were subjected to Southern blot analysis. Only cell clones containing one fully integer provirus were further used.

Functionality of the proviral FRTs

To confirm the functionality of the FRT sites after the retroviral replication circle an excision experiment was performed, for this the remaining cell clones were transfected with an FLP expressing plasmid, and selected with GANC for the loss of the TkNeo moiety. As controls transfections with empty cloning vector and mock transfections were performed. The results show that the efficiency of excision differs more than 50-fold between various cell clones. Excision was also probed by Southern blotting using inverse PCR derived neighbouring chromosomal DNA fragments. As expected, the results show that after FLP expression the part of the provirus between the FRTs was removed (Fig. 1). All assayed clones gave the same, predicted band.

Targeting the proviral locus

The targeting construct (pGETHEAD; Fig 3) was introduced by Ca-phosphate cotransfection of FLP recombinase expressing plasmid. Selectional criteria for correct integration of the second construct was firstly puromycin resistance for the insertion of the contruct. Secondly, on the selection with GANC to select against the co-integration of the HygTK moiety. The GANC selection also serves as selection for the removal of the inserted provirus.

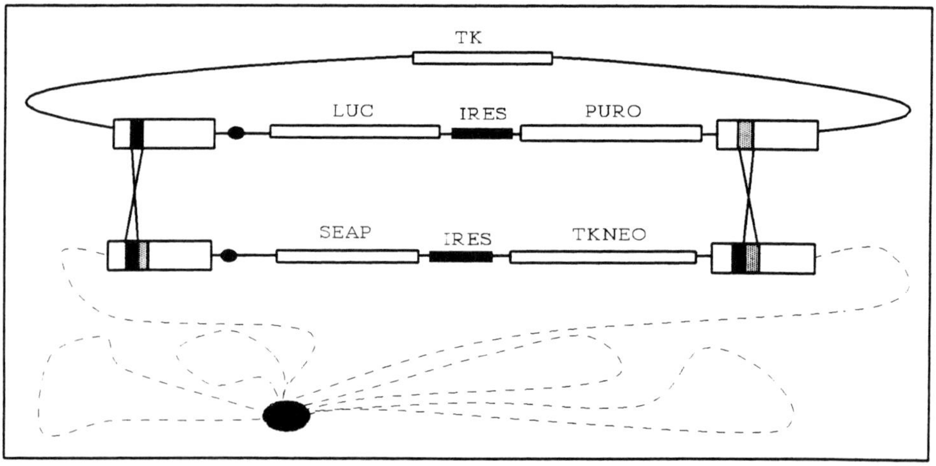

Fig 3: ***Expression cassette targeting with the aid of FLP recombinase***
The targeting plasmid pGETHEAD (upper) contains the expression cassette (here the luciferase gene and the puromycin resistance gene in a bicistronic arrangement) between the FRT sequences which are integrated into the LTR sequences. This plasmid is cotransfected together with the FLP expression plasmid into the provirally tagged reporter cell clone (lower). In several steps (compare Fig. 1) the expression cassette replaces the reporter cassette.

Two days after the targeting the cells were split and put on medium containing puromycin. Two days later GANC was added. Only from two of four primary infectants puromycin/GANC resistant clones were found. Eleven of these cell clones produced luciferase. These were tested for all markers, and found to be Neo$^-$ .SEAP$^-$, hygromycin-sensitive, GANC-resistant, and puromycin-resistant, indicating that they all showed the expected phenotype.

The probe for the bordering host sequences (see above) was used to test the derivates of one of the infectants by Southern blotting. The derivates of the other infectant were analysed by a restriction panel followed by Southern blotting with a neomycin and a puromycin specific probe respectively. In 10 of 11 cases only excision of the retroviral construct had taken place, only in one case an integration was detected. However, using the sequence of the puromycin gene as a probe a strong signal was found in all derivates, indicating that many copies of the targeting vector are present in these cells, even in the cell line where recombination has taken place.

DISCUSSION

The concept of using double FRT sites in retroviruses and its intrinsic advantage with respect to stability in the presence of FLP (Fig. 1) is proven to be feasible by our work. As demonstrated by excision and integration, the double FRT is fully functional. The filter overlay assay has enabled us to screen thousands of individual clones for high expression of the proviral DNA in a very short time. The experiments demonstrate that the number of high producers is rather small. Only ~50 % of the obtained cell clones showed any reaction in the filter overlay test and the SEAP production levels of the transfectant pool is ~25-fold below that of best expressing cell clones (Fig. 2). The influence of the proviral integration site on the intraviral excision frequency was thought to be linked to the expression level of both the provirus and its immediate surroundings (8). Interestingly, we found no direct correlation between the SEAP-expression from the provirus and the frequency of recombination. The cell clones with the highest expression have very different excision frequencies.

The main problem we encountered was the appearance of multi-copy integrates of the targeting construct. The puromycin resistance gene as selection marker may have been the main reason for this problem: a small subpopulation of BHK-A cells shows a high natural tolerance toward puromycin. To reduce the background high concentrations of puromycin have to be used. The majority of BHK-A cells will only survive if they show a strong PAC expression which is accomplished in most of the cases by cells with multi-copy integrates.

With this work we have shown that a procedure for targeting single copy proviral sites is feasible. We have further defined the main problems for selection of these infrequent events. The problem might be overcome either by using another cell line or with a different selection method. It might also be possible to eliminate puromycin-tolerant BHK-A cells by subcloning. We still believe that this cell line, with its many advantages, would be an excellent tool for this approach.

The fundamental question which we have attempted to answer with the site-specific recombination system remains: are certain chromosomal sites (e.g. those which have been tagged with the screening procedure) indeed determining the strength and stability of expression of different genes when inserted? To investigate this, a range of genes, probably even controlled by various promoters, have to be targeted into various loci. The results will also answer questions concerning the influence of chromosomal loci versus transferred promoters and enhancers. The establishment of a whole series of cell clones to be targeted, ranging in their expression from high to almost nothing will make these experiments statistically meaningful. Apart from these scientific questions we hope to develop the process as outlined in this work until it becomes a reliable and fast routine method for the generation of biotechnologically interesting cell lines.

ACKNOWLEDGEMENTS

The work was supported by an EC grant, PL 133069.

REFERENCES

1. Karreman S, Hauser H & Karreman C (1996) On the Use of Double FLP Recognition Targets (FRTs) in the LTR of Retroviruses for the Construction of High Producer Cell Lines. Nucleic Acids Res. 24, 1616 - 1624.
2. Karreman C, Karreman S & Hauser, H (1996) Retroviral Infection of Syrian Hamster BHK Cells Depends on Age and Susceptibility towards Sialidase. Virology, 220, 46 - 50.
3. Grabenhorst E, Hoffmann A, Nimtz M, Zettlmaißl G & Conradt H S (1995) Construction of Stable BHK-21 Cells Coexpressing Human Secretory Glycoproteins and Human Galβ1-4GlcNAc-R α2,6-Sialyltransferase: α2,6-linked NeuAc is Preferably Attached to the Galβ1-4GlcNAcβ1-2Manα1-3-Branch of Biantennary Oligosaccharides from Secreted Recombinant β-trace Protein. Eur.J.Biochem., 232, 718-725.
4. Racher A J, Moreira J L, Alves P M, Wirth M, Weidle U H, Carrondo M J T & Griffiths J B (1994) Expression of Recombinant Antibody and Secreted Alkaline Phosphatase in Mammalian Cells. Influence of Cell Line and Culture System upon Production Kinetics. Apl. Microbiol. Biotechnol. 40, 851 - 856.
5. O'Gorman S., Fox D T & Wahl G M (1991) Recombinase-Mediated Gene Activation and Site-Specific Integration in Mammalian Cells. Science 251, 1351 - 1355.
6. Schlake T & Bode J (1994) Use of Mutated FLP Recognition Target (FRT) Sites for the Exchange of Expression Cassettes at Defined Chromosomal Loci. Biochemistry 33, 12746-12751.
7. Kirchhoff S, Köster M, Wirth M, Schaper F, Gossen M, Bujard H & Hauser H (1995) Identification of Mammalian Cell Clones Exhibiting Highly Regulated Expression from Inducible Promoters. Trends Genet. 11, 219-220.
8. Thyagarajan B, Johnson B L & Campbell C (1995) The Effect of Target Site Transcription on Gene Targeting in Human Cells in vitro. Nucleic Acids Res. 23, 2784-2790.

Discussion

Wurm: Do you know the number of retroviral copies you get into the cell with your method? I think it would have some impact on the probabilities of your events.

Hauser: The way we did the infection was with a low MOI to get a low copy per cell, in most cases just one copy. The structure was also as expected in most cases.

Wurm: On infection of the cells how many viral particles really go in and deliver your DNA, or RNA in this case, to the cell?

Hauser: This depends upon the virus preparation. You must determine the number of infectious particles, never just the number. In retroviruses it depends very much on the construct and the packaging system. Not all particles are infectious.

HIGH LEVEL EXPRESSION OF FOREIGN GENE IN CHINESE HAMSTER OVARY CELLS USING MODIFIED TETRACYCLINE-REGULATED SYSTEM

Yingpei Zhang, Yoshinori Katakura and Sanetaka Shirahata
Graduate School of Genetic Resources Technology, Kyushu University,
6-10-1 Hakozaki, Higashi-ku, Fukuoka 812-81, Japan

ABSTRACT

High level inducible production of recombinant proteins in CHO cells was achieved by modifying the tetracycline (tet)-regulated system. In this system, both the recombinant gene and the gene encoding transactivator (tTA) composed of both the tet repressor and the activating domain of viral protein VP16 of herpes simplex virus were controlled by the enhancer-less promoter PhCMV*-1, a minimal promoter from human cytomegalovirus (hCMV) fused to tet operator sequences. In the presence of tetracycline, the production of recombinant proteins was not detectable. After removing the tetracycline, the production was induced over one thousand times. This system is relatively independent of cellular regulation signals, and may be used in various cell types.

INTRODUCTION

In most cases, the productivity of recombinant protein producing mammalian cells is significantly related to the transcriptional efficiency of the target gene. Activity and specificity of a promoter are determined by the interactions of transcriptional factors between themselves and with various *cis*-regulation sequences contained in the promoter in response to cellular signals (Tjian and Maniatis, 1994). The signaling network is subtly regulated, suggesting that the activity of a promoter may vary in different cell types and that even a strong promoter may not be fully activated. To overcome these problems, we have an attempt to construct a novel recombinant protein expression system using transactivators which can specifically activate the transcription of the target gene relatively independent of cellular regulation signals.

Gossen and Bujard (1992) have developed an inducible expression system using the transactivating domain of herpes simplex virus VP16 protein and regulation elements from *tet*-resistance operon from *Escherichia coli* transposon Tn10. In their system, the target gene was controlled by tetracycline responsible promoter PhCMV*-1 and the transactivator tTA was controlled by the strong hCMV promoter. Since the expression of the target gene

M.J.T. Carrondo et al. (eds.), Animal Cell Technology, 519–523.

depends on the amount of tTA, and the activity of the hCMV promoter controlling the *tTA* gene is regulated by many cellular signals, the system is dependent on the cell types. If the *tTA* gene is also controlled by the tetracycline responsive promoter PhCMV*-1, the system will be relatively independent of cellular signals because the PhCMV*-1 promoter contains little *cis*-regulation sequences. Another advantage of this modification is that the toxicity caused by constitutive over-expression of *tTA* can be partially avoided, therefore more transfectants can be obtained, from which high producing clones can be screened out. The principle of the system is described in Figure 1A.

MATERIALS AND METHODS

The *hIL-6* expression vector pUIL6-8 (Figure 1B) was constructed by inserting the human *IL-6* cDNA fragment with *Eco*RI-*Bam*HI ends (0.65 kb) at the downstream of PhCMV*-1 promoter in pUHD10-3 (Gossen and Bujard, 1992). To construct the inducible *tTA* expression vector pUTA-1(Figure 1C), we replaced the constitutive hCMV promoter in pUHD15-1 (Gossen and Bujard, 1992) with a 0.46 kb *Xho*I- *Eco*RI fragment containing the PhCMV*-1 promoter derived from pUHD10-3.

CHO cells were co-transfected with pUTA-1, pUIL6-8 and pSV2/bsr using the calcium phosphate co-precipitation method. Transfected CHO cells were selected with 5 μg/ml blasticidin S in the presence of 2 μg/ml tetracycline for 2 weeks, then were cloned with either cloning rings or the limiting dilution method.

RESULTS AND DISCUSSION

Thirty-one out of 61 clones obtained were found to produce hIL-6, although the productivity varied. The highest producing clone, UU63, was expanded and further characterized. In the presence of 2 μg/ml tetracycline, hIL-6 could not be detected in the culture supernatant of UU63 cells by sandwich enzyme-linked immunosorbent assay (sandwich-ELISA). After removing tetracycline from the medium, hIL-6 production was detected. As shown in Figure 2A, more than 3 μg/ml of hIL-6 was accumulated in the supernatant of 2 ml dishes during a 120 hour induction period. It is difficult to estimate the exact induction fold because the hIL-6 production before induction was below the limit of the detection method. Since the 1.62 ng/ml standard sample is still distinguishable from the blank control, the hIL-6 production was induced over one thousand times. Figure 2B shows that the hIL-6 production was almost completely inhibited upon incubation with a very low level (0.25 μg/ml) of tetracycline, indicating that the production of hIL-6 is strictly controlled by tetracycline in the culture medium. Figure 2C shows the productivity curve of recombinant CHO cells after induction. The highest productivity was observed 72 hours after induction, and reached a peak value of 2.3 μg/10^6 cells/day. Figure 2D shows the comparison of the productivity of UU63 cells with another hIL-6 high producing CHO cell lines (A7 cells). A7 cell line was established by co-transfecting CHO cells with the *hIL-6* expression plasmid under the control of the hCMV promoter and pSV2/bsr. From the 42 single clones isolated, clone A7 demonstrated the highest productivity. The productivity of hIL-6 in UU63 cells was similar to that of A7 cells. It is surprising that an expression system without strong authentic enhancers could produce recombinant proteins comparable to the strong hCMV promoter-based system, demonstrating that novel efficient mammalian expression system can be established by direct and specific recruitment of transactivators to a promoter.

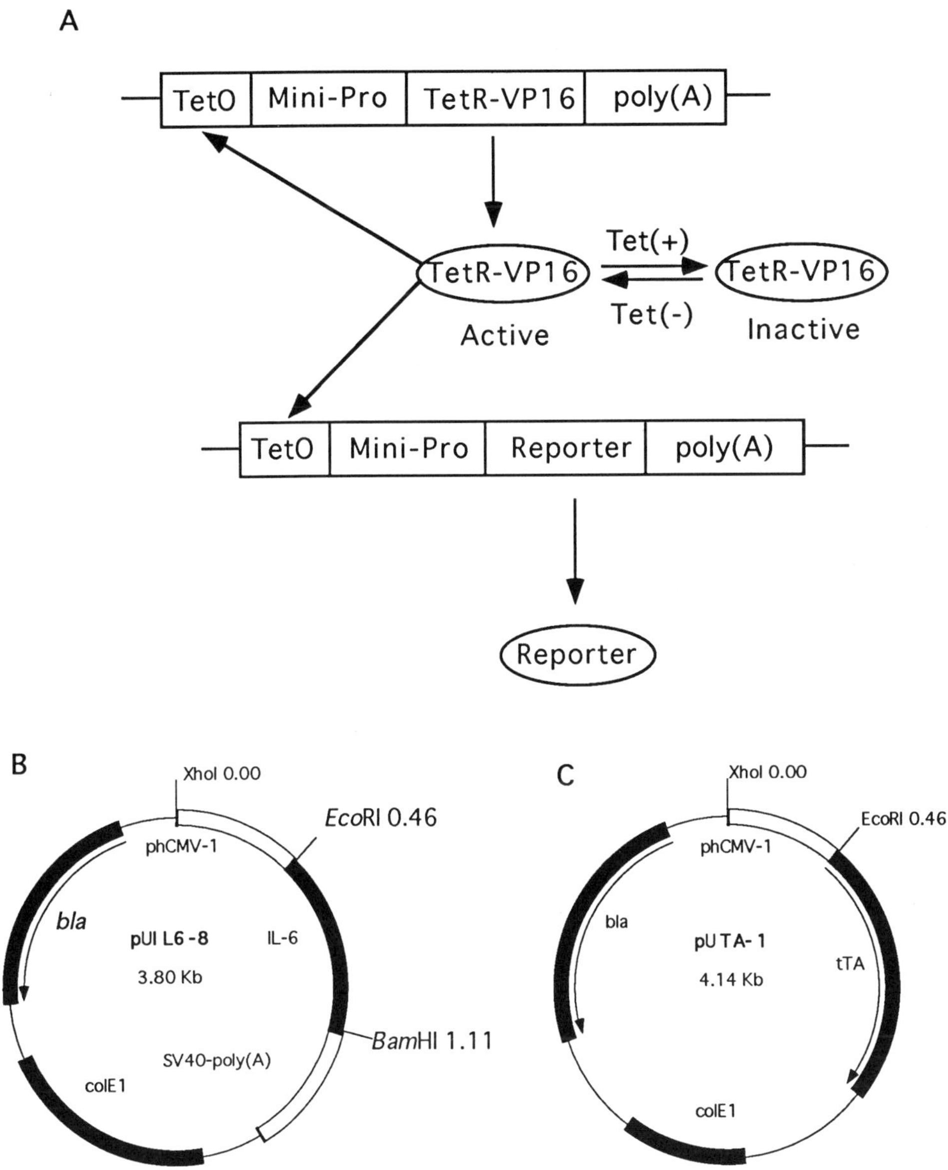

Figure 1. Principle of tetracycline-regulated expression system and the maps of the vectors. A, the principle of the the system. B, reporter plasmid. C, transactivator expression plasmid. TetO, tetracycline operator. Mini-Pro, minimal promoter. TetR-VP16, chimeric transactivator composed of tetracycline repressor and the activating domain of VP16.

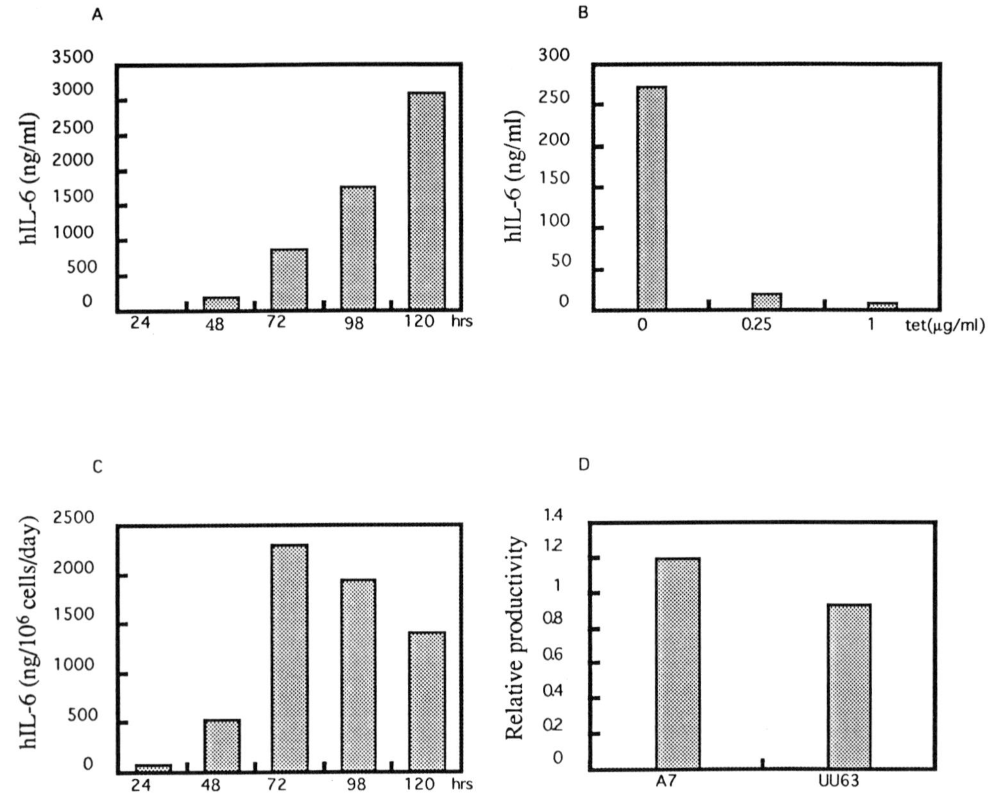

Figure 2. Production of hIL-6 in UU63 cells. A. 10^6 cells were inoculated in 5 ml dishes in Dulbecco's modified eagle medium (DMEM) supplemented with 10% fetal bovine serum (FBS) in the presence of 2 μg/ml tetracycline. Two days later, the cells were washed t times with PBS and maintained in tetracycline-free medium. The medium was then changed every 24 hours and the hIL-6 production was examined with ELISA. B. The cells which hav been induced for 48 hours were plated into 24 well plate in DMEM medium supplemented with 10% FBS and various concentrations of tetracycline. After overnight incubation, the amount of hIL-6 secreted into the supernatants was determined. C. 10^5 cells were plated in 2 ml dishes in the presence of 2μg/ml tetracycline. After overnight incubation, the cells were washed three times with PBS, and maintained in tetracycline-free medium. Every 24 hours, the cells were counted and hIL-6 production was examined. D. A7 and UU63 cells which have been induced for three days were plated in 5 ml dishes at about $2x10^5$ cells /ml and incubated in tetracycline-free DMEM medium supplemented with 10% FBS. After overnig the hIL-6 production were determined.

Several advantages of this strategy are envisaged. The low level of basal expression will make it possible to establish permanent mammalian cell lines expressing toxic products. The relative independence from the host cellular regulation signals will allow this system to be used in various cell types. Combination with a gene amplification marker will probably result in much higher productivity. Since it has been demonstrated that transactivating ability of VP16 could be further potentiated by some coactivator such as PC4 (Ge and Roeder, 1994), combination of these factors might improve the expression level of recombinant proteins.

REFERENCES

1. Ge H and Roeder RG (1994) Purification, cloning, and characterization of a human coactivator, PC4, that mediates transcriptional activation of class II genes. Cell 78, 513-523.
2. Gossen M and Bujard H (1992) Tight control of gene expression in mammalian cells by tetracycline-responsive promoters. Proc. Natl. Acad. Sci. USA 89, 5547-5551.
3. Tjian R and Maniatis T (1994) Transcriptional activation: a complex puzzle with few easy pieces. Cell 77, 5-8.

Expression of Triabin, a Novel Highly Potent Inhibitor of Thrombin, in Sf9 and High Five Insect Cells using a Recombinant Baculovirus

Thomas Petri, Birgit Neukamm, Marina Isernhagen, Brigitte Ockert, Christiane Noeske-Jungblut,
Research Laboratories of Schering AG, D-13342 Berlin, Germany

Abstract

Triabin is a novel thrombin inhibitor originally isolated from the saliva of the hematophageous bug *Triatoma pallidipennis*. The mechanism of thrombin inhibition by triabin is different from other known inhibitors like hirudin or heparin/anti-thrombin III, but the potency of triabin is similar to that of hirudin. Using the cloned cDNA of triabin, a recombinant baculovirus expressing this secreted protein was constructed. However, only low expression rates were obtained in Sf9 insect cells infected with this virus (1-3 mg per l culture). Therefore, the expression of triabin in High five insect cells was also investigated and it was found that from these cells 3-8 times higher yields could be obtained. Recombinant triabin purified from the insect culture supernatants had a biological activity similar to that of natural salivary triabin. The availability of an efficient production process and the unique biological properties of triabin suggest that this protein has a potential as a novel therapeutic.

Introduction

Thrombin plays a central role in the primary as well as in the secondary hemostasis. Therefore, thrombin inhibitors such as hirudin or heparin/antithrombin III are of great importance in the prevention of venous thrombosis, lung embolism,myocardial infarction and stroke. Recently the isolation of a novel thrombin inhibitor, called triabin, from the saliva of the hematophageous bug *Triatoma pallidipennis* has been described, as well as the cloning of the triabin cDNA and its heterologous expression in *E. coli* (1). However, the yields of active heterologous protein obtained from *E. coli* expression were extremely low. Therefore a recombinant baculovirus for the triabin expression in cultured insect cells was constructed. Here the optimization of the recombinant protein production as well as some biochemical data of the insect cell derived recombinant protein are described.

Materials and Methods

The cDNA encoding the complete 160 amino acid long triabin molecule including a potential 18 amino acids long signal sequence was inserted into the baculovirus transfer vector pVL 1392. Starting from this plasmid a recombinant baculovirus was constructed and isolated using standard techniques (2).

M.J.T. Carrondo et al. (eds.), Animal Cell Technology, 525–528.

Sf 9 cells were grown serumfree in Sf 900 II medium and High five cells in supplemented Grace medium containing 10 % fetal calf serum (all media: Life Technologies). Because at the time of these experiments the High five cells had not been adapted to growth in suspension all experiments were performed in adherent cultures for a better comparison.

Triabin was determined in the cell culture supernatants by measuring the inhibition of the thrombin induced fibrin production from fibrinogen against a triabin standard. Pure cell culture media and supernatants from uninfected or wildtype baculovirus infected Sf9 respectively High five cells were also analyzed as controls. None of these contained a significant thrombin inhibiting effect.

Results

Time course of expression of triabin in Sf 9 and High five cells. 50 % confluent monolayers of Sf 9 cells (2,1 x 10E5/well) or High five cells 2,7 x 10E5/well) in 6 well plates were infected with a moi of 4 and the triabin content in the supernatant was analysed at the indicated times. Triabin could be detected in the supernatant of the infected Sf9 cells from 40 h p.i. onwards and the triabin content increased for up to 120 h p.i. to a value of about 3 mg/l. In the supernatant of the correspondingly infected High five triabin was found already at 24 h post infection and reached an maximun of about 20 mg/l at around 64 h p.i. These results are summarized in fig. 1.

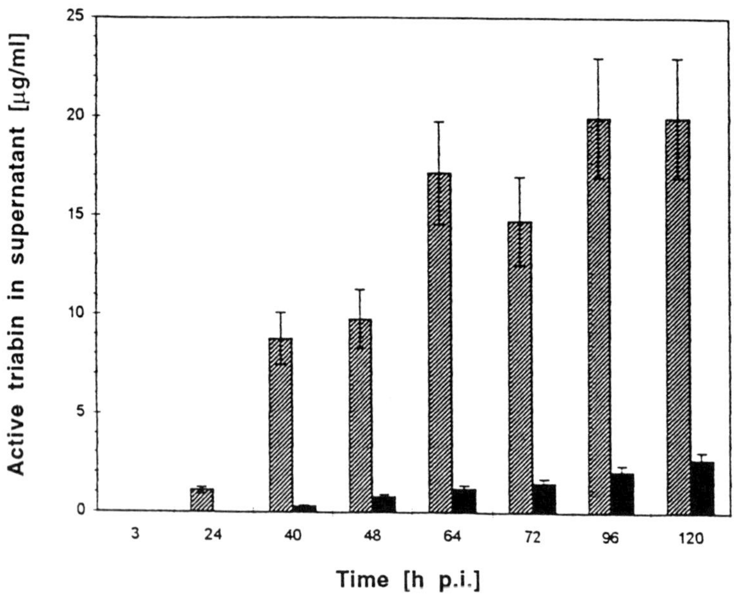

Fig. 1 Time course of triabin expression in Sf9 (black columns) or High five inscet cells (hatched columns) infected with a recombinant baculovirus. 50 % confluent monolayers of Sf 9 cells (2,1 x 10E5/well) or High five cells 2,7 x 10E5/well) in 6 well plates wereinfected with a moi of 4 and the triabin content in the supernatant was analysed at the indicated times.

Inhibition of the thrombin by triabin and hirudin. To characterize the insect cell derived triabin its effect on the thrombin-induced platelet aggregation and amidolytic activity of thrombin towards a low molecular weight synthetic substrate was investigated. As shown in fig. 2A, triabin inhibits thrombin-induced platelet agggregation with a potency similar to that of hirudin. The same inhibitory effect was noted on the the thrombin-catalyzed fibrin formation (data not shown). However, in another assay using the synthetic low molecular weight thrombin substrate S2288 triabin only shows a weak thrombin inhibiting activity even at tenfold higher molar concentration than the concentration of hirudin which gives a 90% inhibition (fig. 2B).

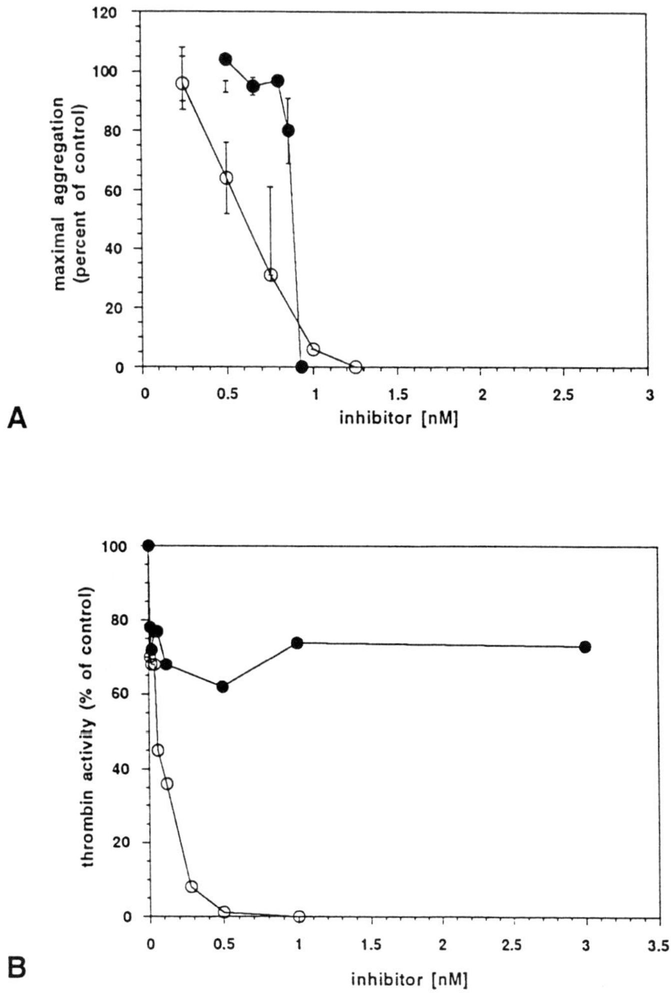

Fig. 2. Effects of triabin (closed circles) and hirudin (open circles) on thrombin measured by the inhibition of the thrombin induced platelet aggregation (A) or by the effect on the amidolytic activity of thrombin towards a synthetic substrate (S2238) (B). Thrombin was preincubated with the indicated concentrations of either hirudin or triabin. Subsequently washed human blood platelets were added and maximum aggregation determined (A) or the synthetic substrate S2238 was added and the adsorbance at 405 nm measured (B).

Discussion

These data show that triabin can be very efficiently produced (approx. 20 mg/l) in adherent cultures of High five insect cells using the baculovirus expression system. The yields in Sf9 cells under similar conditions were for unknown reasons considerably lower. The biological activity of the insect cell-derived recombinant protein was identical to the natural protein under the conditions tested. In the meantime the High five cells have been adapted to growth in serum free medium as well as to growth in suspension culture and it can be expected that the yield of triabin will further increase corresponding to the higher cell densities which are obtained in suspension culture.

The biological data of triabin presented here support the hypothesis that triabin does not block the catalytically active center of thrombin but exerts its inhibitory effect by binding to a different domain which may be the anion binding exosite.

References

1. Noeske-Jungblut, Ch., Haendler, B., Donner, P., Alagon, A., Possani, L., Schleuning,, W.-D. (1995) Triabin, a highly potent inhibitor of thrombin. J. Biol. Chem. 270 (48), 28629-28634

2. Whiting, J.P., Wafford, K.A., Pribilla, I., Petri, T.: Channel cloning, mutagenesis and expression - ion channel expression in insect cells infected with recombinant baculoviruses. In: "Ion channels, a practical approach." R. Ashley (ed.) Oxford University Press-IRL Press, Oxford 1995

Acknowledgement:

The authors thank Dr. Pohl (Benjamin-Franklin-Klinikum, FU Berlin) for providing the High five insect cell line.

Expression Augmenting Sequence Element (EASE) Isolated From Chinese Hamster Ovary Cells

Arvia E. Morris, Chi-Chang Lee[1], Karmen Hodges, Teri L. Aldrich, Carol Krantz, Pauline S. Smidt, and James N. Thomas

Immunex Corporation, 51 University Street Seattle, WA 98101
[1]Regeneron Pharmaceuticals Inc. 777 Old Sawmill River Rd., Tarrytown, NY 10591-6707

Abstract

We have isolated a 5.7kb Chinese hamster ovary (CHO) genomic sequence that can improve expression of reporter proteins in stable cell pools when inserted upstream of the CMV promoter/enhancer. This expression augmenting sequence element (EASE) was identified by cloning the integration site of a unique expression cassette encoding recombinant dimeric Tumor Necrosis Factor Receptor (TNFrFc) from a cell line expressing high levels of this protein. The EASE has been incorporated into expression vectors and used to express high levels of recombinant protein in selected cell pools 5 to 9 weeks post transfection using an efficient two-step selection protocol. *Key words: expression vectors, CHO cells, gene expression*

Introduction

One widely used method for generating stable cell lines expressing high levels of recombinant protein is co-amplification of the dihydrofolate reductase gene (*dhfr*) and the gene of interest. Though this method is able to reliably generate stable cell lines with high expression it often involves time consuming amplification and cloning steps. In the present study we have cloned an expression vector integration site from a cell line making high levels of recombinant protein from a single integration cassette. The clone contains 14.5kb of CHO genomic DNA upstream of the expression cassette. We demonstrate that this integration site when transfected into CHO cells will confer an increased frequency of high expression with low selective pressure. Mapping studies indicate that a 5.7kb sub-region of the integration site has similar activity compared with the entire 14.5kb sequence. This fragment, Expression Augmenting Sequence Element (EASE), has been incorporated into an expression vector and used to generate CHO cell pools expressing high levels of recombinant protein using a rapid and efficient methotrexate (MTX) selection protocol. We show that RNA encoding recombinant protein is not limiting expression in these cells and discuss the application of this protocol to express a variety of recombinant proteins.

M.J.T. Carrondo et al. (eds.), Animal Cell Technology, 529–534.

Procedure

Materials and Methods

Tissue culture. Dihydrofolate reductase (DHFR) deficient DXB11 CHO cells (1) were maintained in Dulbecco's minimal essential medium F12 (DMEM F12) supplemented with 7.5% fetal bovine serum (FBS) (Hyclone, Logan, Utah or Sigma, St. Louis, MO), 2mm L-glutamine, 90μM thymidine (T), 90μM hypoxanthine (H) and 120μM glycine (G). For DHFR selection studies and methotrexate selections, cells were cultured in DMEM F12 lacking GHT and supplemented with 7.5% dialyzed FBS, 6mM L-glutamine and 1mM asparagine. For methotrexate (MTX) (Lederle, Pearl River, NJ) selections, MTX was added to the selection media at appropriate concentrations. When neomycin selection was employed, 400μg/ml of G418 (Gibco, Grand Island, NY) was added to the media. The cells were transfected using calcium phosphate transfection (2) or Lipofectamine™ transfection as recommended by the supplier (GIBCO BRL, Gaithersburg, MD).

Genomic library screening and subcloning. Genomic DNA was isolated from 2a5-3 CHO cells, partially digested with Mbo1 and cloned into a lambda FIX II cloning vector (Stratagene custom genomic library, La Jolla, CA) to form a library. The p80 TNF receptor coding sequence (3) along with 14.5kb of cellular flanking sequences, was cloned from the library essentially as recommended by Stratagene. The filters were probed with random primed Not1 PvuII DNA fragment encoding the cell surface portion of p80 TNF receptor extracellular. This clone, which was designated 2a5-3λ, was used in all subsequent analyses. Various inserts derived from 2a5-3λ were subcloned into the vector Pgem1 (Promega, Madison, WI) to create plasmids PG5.7, PG5.7ΔS and PG2.2. using standard methods. To generate the 2a5I vector, a Not1 Bsab1 fragment containing the encephalomyocarditis virus internal ribosomal entry site (nt260-nt824) and part of the murine *dhfr* coding sequence was isolated from pCDE (4) and cloned into a derivative of PG5.7 containing an expanded multiple cloning site.

RNA blot analysis. Total cellular RNA was isolated from cells using RNAeasy total RNA kits and a QIAshredder (Qiagen, Chatsworth, CA), analyzed using formaldehyde agarose gel electrophoresis (5) and transfered to Zetaprobe filters (Bio-Rad Laboratories, Richmond, CA). Filters were hybridized in conditions recommended by the supplier. Probes were prepared using random priming kits (Boehringer Mannheim, Indianapolis, IN). TNFrFc encoding sequences were detected using the same probe as described for the library screen.

ELISA assays for TNFrFc production. Production of recombinant proteins can be monitored by antibody sandwich enzyme-linked immunosorbent assay (ELISA) (as described in reference number 6). For detection of TNFrFc two monoclonal antibodies (mAb) against different epitopes of TNFrFc were used. The first mAb was adsorbed into 96 well plates overnight, and the peroxidase (HRP) conjugated second antibody was added after a wash step. In several experiments quantities of between 0.78 and 50ng/ml TNFrFc were detected.

Results and Discussion

Cloning of the expression augmenting sequence element (EASE)

We have identified a 5.7kb CHO genomic DHA segment that confers high frequencey of high expression when tested in transfection experiments. This fragment called EASE is a subdomian of a 14.5kb CHO genomic fragment isolated from CHO cell genomic library. A transformed CHO cell line (2a5-3) expressing high levels of TNFrFc was selected for preparation of a genomic library since Southern blot analysis indicated that the high expression of TNFrFc expression observed for this cell line is driven by a single integration of an expression cassette encoding TNFrFc (A.E. Morris, unpublished). A clone designated 2a5-3λ was isolated from the library and found to contain the bicistronic TNFrFc expression cassette comprising the CMV promoter/enhancer, TNFrFc coding sequence and *dhfr* cDNA. In addition, 5' to the expression cassette the clone contained 14.5kb of CHO genomic DNA (Fig. 1).

Fig.1 CHO genomic clone 2a5-3λ

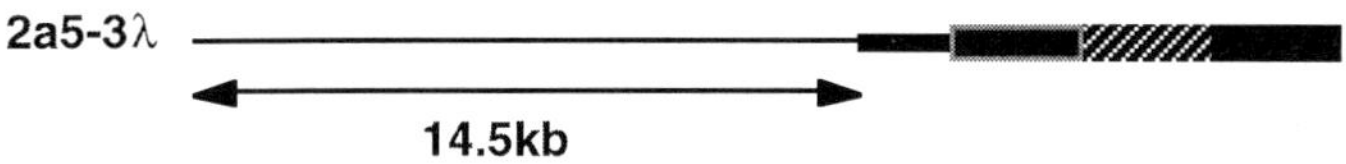

Thin line: CHO genomic flanking sequence; small box: CMV promoter/enhancer; grey box: adenovirus tripartite leader; diagonal stripes: TNFrFc encoding sequence; large solid box: *dhfr* cDNA.

To determine if CHO sequences 5' to the TNFrFc integration site in the 2a5-3λ phage clone could confer high expression when randomly integrated into DXB11 cells transfection experiments were performed. In addition, the activity of 2a5-3λ DNA was compared with derivatives containing the TNFrFc expression cassette and various lengths of the 5' region cloned into Pgem1 (Fig 2A). DXB11CHO cells were co-transfected with the 2a5-3λ DNA or the Pgem derivatives and pSV3NEO DNA at a ratio of one to one using calcium phosphate transformation. The amount of test DNA was normalized so that 0.16μg of DHFR coding sequence was used for each case. Pgem 1 DNA was added to each DNA mix as necessary to ensure that the same amount of DNA was used for each transfection.

After a 48 hr recovery period post transfection, cells were selected sequentially for G418 resistance and DHFR expression. Pools of 1-3 doubly selected colonies were picked and plated in 24-well vessels and the cultures were sampled at confluence. The frequency of high expression (> 0.5 units/mL) was scored for pools derived from transfection with each test construct (Fig. 2B). Transfection with the PG2.2 construct resulted in pools with a low frequency of high expression compared with the other constructs tested indicating that the longer lengths of CHO genomic sequence in 2a5-3λ and PG5.7 and PG5.7ΔS improved expression of recombinant protein.

Fig. 2 Relative Production of Recombinant Protein by Cells Transformed with 2a5-3λ DNA and subclones

A.

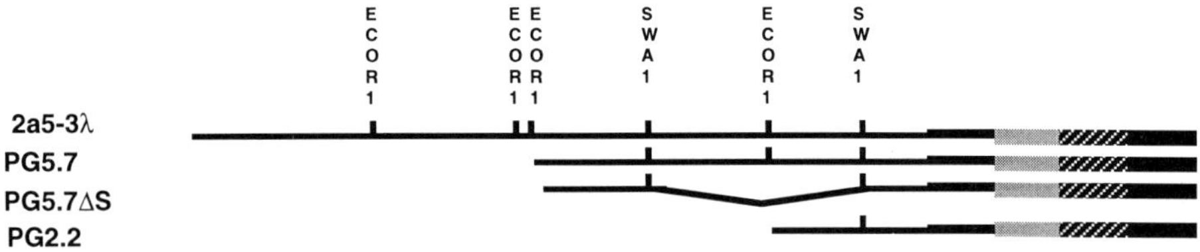

2a5-3λ TNFrFc expression cassette and subclones used for transfection experiments. The subclones, PG5.7, PG5.7ΔS, and PG2.2 are in Pgem1. PG5.7 includes 5.7kb of CHO genomic flanking sequence; PG5.7ΔS contains the same 5.7kb of flanking sequence with a deletion of the internal Swa1 fragment; PG2.2 contains 2.2kb of flanking sequence. Thin line: CHO genomic flanking sequence; small box: CMV promoter/enhancer; grey box: adenovirus leader; diagonal stripes: TNFrFc encoding sequence; large solid box: *dhfr* cDNA

B.

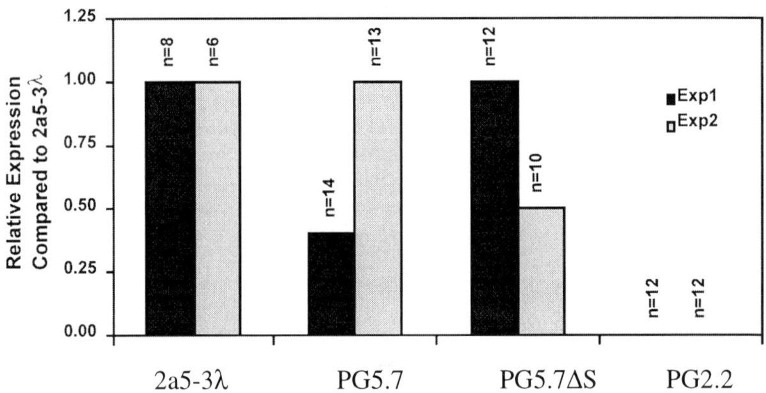

Relative expression of TNFrFc from two transfection experiments. Data is normalized to results from the 2a5-3λ DNA. n= the number of confluent cell pools analyzed for each case.

Expression of recombinant protein using an expression vector containing the EASE

Since the PG5.7 plasmid conferred a greater frequency of high expression relative to PG2.2, this plasmid was modified to make 2a5I (Fig. 3), an expression plasmid which includes an internal ribosomal binding site (IRES) up stream of *dhfr.* The IRES sequence has been shown to stabilize expression of recombinant proteins expressed in CHO cells from bicistronic expression cassettes under high selective pressure (7).

Fig. 3 Expression vector with EASE

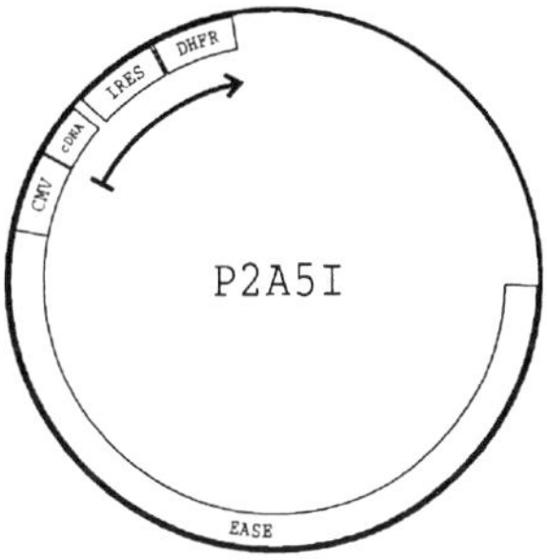

EASE: expression augmenting sequence element; CMV: Cytomeglovirus Promoter/Enhancer; cDNA: coding sequence for any gene of interest to be expressed; IRES: internal ribosomal entry site; DHFR: dihydrofolate reductase encoding sequence. Arrow indicates the direction of transcription of the bicistronic mRNA encoding the gene of interest and DHFR.

To test if the 2a5I vector could be used rapidly to develop pools of cells expressing high levels of recombinant protein, the TNFrFc coding sequence was inserted into the 2a5I vector to make 2a5I TNFrFc. This plasmid was used to transfect DXB11 CHO cells that were subsequently selected in -GHT. The -GHT selected cells were then pooled and plated in selective mediun containing various concentrations of MTX (25nM, 50nM, 150 nM 250nM, and 500nM) in parallel. After the cultures were confluent the specific productivity (μg of TNFrFc/10^6 cells/day) was determined from duplicate cultures. In addition RNA was isolated from each culture and analyzed by Northern blot analysis. Relative TNFrFc protein expression and mRNA levels were compared (Fig. 4). Protein expression in cultures selected at 150nM, 250nM and 500nM MTX were similar to each other but greater than expression in cultures selected at 25nM and 50nM MTX. Interestingly, the level of RNA continued to increase as the selective pressure increased; thus, the RNA was not limiting expression in these populations. Similar results for RNA and protein were found for another recombinant protein analyzed in this way (T.L. Aldrich, unpublished).

Fig. 4 mRNA is not limiting in pools expressing TNFrFc

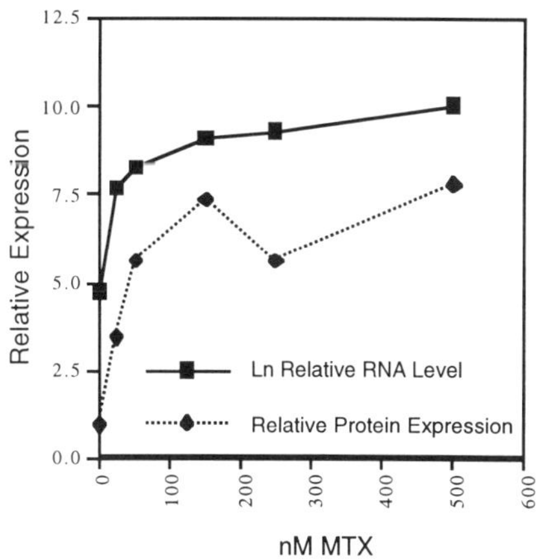

RNA encoding recombinant TNFrFc was quantitated by RNA blot analysis and subsequent densitometry. TNFrFc protein values were determined by ELISA. Relative TNFrFc RNA levels are expressed on a logarithmic scale to facilitate graphing with relative protein values. Diamonds: relative RNA values; squares: relative protein values.

Similar protocols have been used to express a variety of proteins using the 2a5I vector and in each case high expressing pools were generated in two steps without screening clones (C. Krantz, unpublished). The pools have been used for small scale production (1-10L scale) and are stable for at least four months. In general, pools can be generated and initial production runs performed in a 5- to 9-week time frame. Use of the 2a5I EASE containing vector with the two-step selection protocol has facilitated small scale production of CHO derived recombinant protein by eliminating labor intensive screening and time consuming step amplifications. Cell lines suitable for clinical manufacturing can be easily cloned from high expressing cell pools.

Acknowledgments
We wish to thank Jacques Peschon for discussions during the course of the work and Alison House for help in preparation of the manuscript.

References

1. Urlaub G & Chasin LA (1980) Isolation of Chinese hamster cell mutants deficient in dihydrofolate reductase activity. Proc. Natl. Acad. Sci. USA 77, 4216-4220.
2. Wigler M, Pellicer A, Silverstein S, Axel R, Urlaub G & Chasin L (1979) DNA-mediated transfer of the adenine phosphoribosyltransferase locus into mammalian cells. Proc. Natl. Acad. Sci. USA 76, 1373-1376.
3. Mohler KM, Torrance DS, Smith CA, Goodwin RG, Stremler KE, Fung VP, Madani H & Widmer MB (1993) Soluble tumor necrosis factor (TNF) receptors are effective therapeutic agents in lethal endotoxemia and function simultaneously as both TNF carriers and TNF antagonists. J. Immunol. 151, 1548-1561.
4. Brasel K, McKenna HJ, Morrissey PJ, Charrier K, Morris AE, Lee CC, Williams DE & Lyman SD (1996) Hematological effects of FLT3 ligand *in vivo* in mice. Blood in press.
5. Sambrook J, Fritsch EF & Maniatis T (1989) Molecular Cloning: A Laboratory Manual, Cold Spring Harbor Laboratory Press, Plainview, New York.
6. Engvall E & Perlman P (1971) Quantitative assay of immunoglobulin G. immunochemistry (ENGLAND) 8, 871-874.
7. Kaufman RJ, Davies MV, Wasley LC & Michnick D (1991) Improved vectors for stable expression of foreign genes in mammalian cells by use of the untranslated leader sequence from EMC virus. Nucleic Acids Res 19, 4485-4490.

CONTINUOUS PRODUCTION OF THE EXTRACELLULAR DOMAIN OF RECOMBINANT HUMAN CA^{++} RECEPTOR FROM HEK 293 CELLS USING NOVEL SERUM FREE MEDIUM.

J. Shiloach[1], J. Kaufman[1], L. Trinh[1] and C. Kemp[2]

[1] NIH,NIDDK Biotechnology Unit, Bethesda MD 20892, USA. [2] Kemp Biotechnologies, Frederick MD 21701

INTRODUCTION

Large amounts of the extracellular domain of human parathyroid Ca^{++} receptor were needed for structural studies and clinical evaluation[1]. The domain was cloned into human embryonic kidney cells (HEK 293) which produced the protein constitutively and secreted it to the media. To obtain the quantity of protein required for evaluation an optimal production procedure was developed. Since the cells cannot grow in suspension, solid support was required. Several solid support arrangements are currently available and these arrangements may be divided into two major groups. The first group consists of supports that allow growth in a stirred tank reactor (i.e microcarriers)[2]. The second group of supports allow growth in arrangements other than a stirred tank reactor, for example hollow fibers[3] and flat sheets. When cells grow on solid supports, medium replacement (perfusion)[4] is the method of choice to sustain growth and protein production. Medium replacement operation maintains a relatively constant supply of growth factors such as carbon source, amino acids and vitamins, while lowering the concentration of inhibitory metabolites such as lactic acid and ammonia[5]. The perfusion may be performed in several ways: if small variations in metabolite concentration affects cellular metabolism and protein biosynthesis, continuous perfusion is the optimal method. However, if cellular metabolism and protein biosynthesis are stable during changes in nutrient concentration and metabolite formation, batch or semi batch replacement may be performed. In each case, the important parameter is the amount of medium needed per cell, per unit time.

The following parameters should be determined during the course of the process development work:

a. Type of reactor.
b. Type of solid support and growth system.
c. Strategy of medium replacement.
d. Medium composition and frequency of alternating between serum containing medium and serum free medium (if required).

In the authors' opinion, the stirred tank reactor is the preferred system when continuous production from anchorage dependent cells is required. Compared to other systems, the initial cost may be higher; however, the system provides optimal control of growth parameters, easy operation, convenient medium replacement and the ability to obtain large amounts of biomass in a relatively small volume. Therefore, for this application the stirred tank reactor arrangement was considered first.

M.J.T. Carrondo et al. (eds.), Animal Cell Technology, 535–540.

When dealing with production and purification of extracellular proteins from mammalian cells, serum containing medium may significantly affect the purification process and consequently serum free medium or protein free medium may be required. Serum free medium formulations may be limited in their ability to support cellular functions. Therefore, continuous monitoring of cellular metabolism followed by a sequential alternation between media containing serum and serum free media is performed. Hence, this production scheme involves two phases, production phase, where serum free medium is used; and maintenance phase where serum containing medium is being used. The production method and the serum free media used will be discussed.

MATERIAL AND METHODS.

a. HEK 293 cells: Transformed cells producing the extracellular domain of the human parathyroid Ca^{++} receptor were generated by NPS Pharmaceuticals, Salt Lake City, Utah.

b. Bioreactor operation: Cells grown on tissue culture plates were transferred to a 5.0 liter bioreactor equipped with an internal retention device (basket) and vertical mixing system, (Celligen plus™, New Brunswick Scientific, NJ). The cells were trapped onto Fibra-Cel™ disks inside the retention assembly (50-75gr). Dissolved oxygen was kept at 30% saturation and pH at 7.0 using a four gas mixing system (air, oxygen, nitrogen, CO_2), and 1N NaOH. The agitation rate was 100 rpm and the growth temp 37°C. Media was replaced when glucose was depleted. The collected media was concentrated 25 times using 10000 MWCO hollow fiber (AG Technology Needham MA) centrifuged and stored at -20°C for further purification.

c. Media: Growth media was DMEM containing 10% fetal bovine serum, 20 mM glutamine, 200 units/ml Hygromycin B and 0.1 % pluronic F-68 (Life Technologies Ind, Grand Island N.Y). Production was performed in DMEM containing 10 ml/liter of supplement solution (Table 1).

Component	Concentration (per liter of 100x solution)
Bovine Insulin	0.50 g
Human Holo Transferin	0.50 g
2-Mercaptoethanol	0.350 ml
Ethanolamine	0.350 g
Triiodo-L-Thyronine	0.065 mg
Glutamine	20.0 g
Ex-Cyte VLE®	10.0 ml
NaCl	8.0 g
Na_2HPO_4	1.15 g
KH_2PO_4	0.2 g

Table 1. Composition of the serum-free medium supplement

d. Analysis of the Ca^{++} binding domain. Western blotting was used to detect the amount of protein produced during the production process. Serial dilutions of concentrated media samples were electrophoresed by 12% SDS acrylamide gel under reducing conditions. The proteins were transferred to nitrocellulose filters at 30 volts for one hour. The resultant blot was blocked with 10% normal goat serum in PBS for 1 hour, the first antibody (rabbit) was applied for 1 hour and the blot was washed with PBS containing 0.1% tween 20. The second antibody, goat-anti rabbit IgG alkaline phosphatase conjugate (Kirkegaard and Perry, Gaithersurg, MD) was applied for 1 hour, the blot was washed and developed using BCIP/NBT phosphatase substrate solution.

e. Metabolites detection. Glucose, acetate and ammonia were analyzed using an IBI Biolyser (Eastman Kodak Rochester, NY). Glutamine was analyzed using the YSI 2700 select biochemistry analyzer (YSI, Yellow Spring, OH).

RESULTS

1. Support selection

Since a stirred-tank reactor was chosen as the preferred production method, suspended supports such as microcarriers and trapped support such as disks were tested. Microcarriers, which may be used in a simple bioreactor were investigated first. Several types of solid microcarriers were examined: a. Cross linked dextran (Cytodex® 1 Pharmacia, Sweden), b. Cross linked dextran coated with collagen (Cytodex® 3 Pharmacia, Sweden), c. Polystyrene (Biosilon®, Nunc, Denmark). The cells did not adhere to any of these microcarriers. Therefore, fiber disks were tested. Unlike microcarriers, the use of disks require a more complicated bioreactor with an internal retention device. The cells adhere to the polyester fiber disks (Fibra-Cel, NBS, NJ), and these disks were chosen as the support for the production process.

2. Serum free media development.

Several proteins in the serum containing medium were similar to the Ca^{++} binding domain (Fig 1) and interfered with the purification process, therefore a serum free medium formulation was required for the production phase. The serum free medium was developed with DMEM as the basal formula. Cell growth and production were maintained by

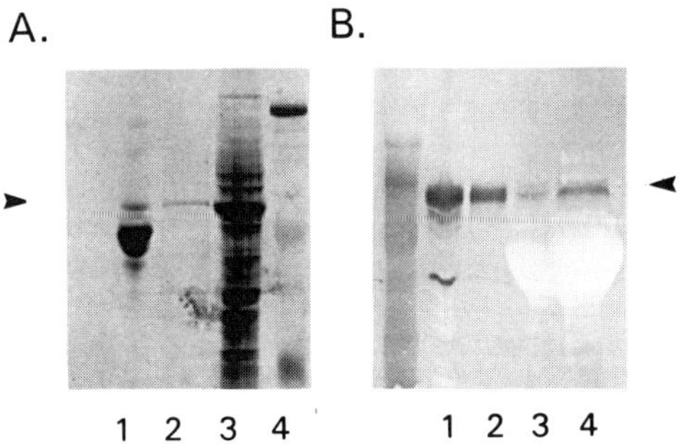

Fig 1. Production of the Ca^{++} binding domain in serum containing medium and in serum free medium:

a. SDS PAGE of various fractions.
1. 1 μl Serum containing medium.
2. 30 μl Serum free medium.
3. 10 μl 25x Serum free medium.
4. Marker.

b. Immunoblot of various fractions.
1. 10 μl 25x Serum free medium.
2. 10 μl Serum free medium.
3. 10 μl Serum containing medium.
4. 30 μl Serum containing medium.

Arrows indicate the position of the Ca^{++} binding protein.

substituting serum with a supplement solution, summarized in Table 1. The production of Ca^{++} binding domain by HEK 293 adhering to the Fibra-Cel disks in serum free and serum containing medium is shown in Fig 2. The serum free medium formula supports the production of the Ca^{++} binding domain better than the serum containing medium. Glucose consumption and lactate production rates were similar in serum containing and serum free medium, but higher consumption of glutamine was observed in the serum free medium.

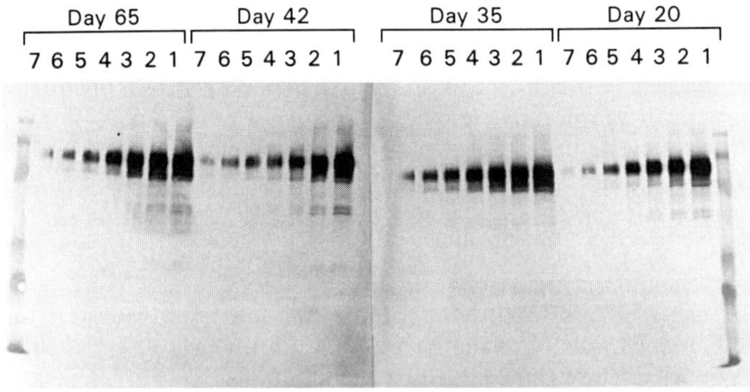

Fig 2. Extracellular Ca^{++} biding domain production. Lane 1-7 represent immunoblot of concentrated sup. (2x serial dilution of 5 μl) at day 20, day 35, day 42 and day 65.

3. Production process.

Cells from 40 confluent tissue culture flasks (175 cm^2) were transferred to a 5.0 liter bioreactor containing 50 g Fiber-Cel disks in the retention device (basket). Since the cells were insensitive to decreasing glucose concentration or increasing lactate or ammonia concentration, a batch replacement strategy based on glucose concentration was adopted. Media (serum containing or serum free) was replaced when glucose was depleted. After inoculation, glucose consumption was increasing probably as a result of the increase in the cell biomass. When the glucose consumption rate in the serum containing medium became constant, the medium was replaced with the serum free formula and the production process started. Serum free medium was exchanged up to 6 times before a decrease in glucose consumption rate was observed. At this point, the cells were transferred to media containing fetal calf serum for 24 hours followed by a return to serum free medium conditions. Daily glucose consumption rate values (mg/l/hr) are shown in Fig 2, and the protein production is shown in Fig 3. Before medium replacement, lactate concentration reached 3.5 g/l and ammonia concentration reached 9 mmol/l. Both are noninhibitory concentrations. Using the above strategy, glucose consumption and protein production values were stable for a period of 65 days. The process was terminated when the glucose and glutamine consumption rate decreased and could not be returned to normal by refeeding. The irreversible decrease in consumption rate was associated with a decrease in ammonia production rate and the loss of adherence by a large portion of the cell population.

DISCUSSION AND CONCLUSIONS

Recombinant protein production utilizing mammalian cells is more complicated and less efficient than the same process performed in bacteria. However, in some cases

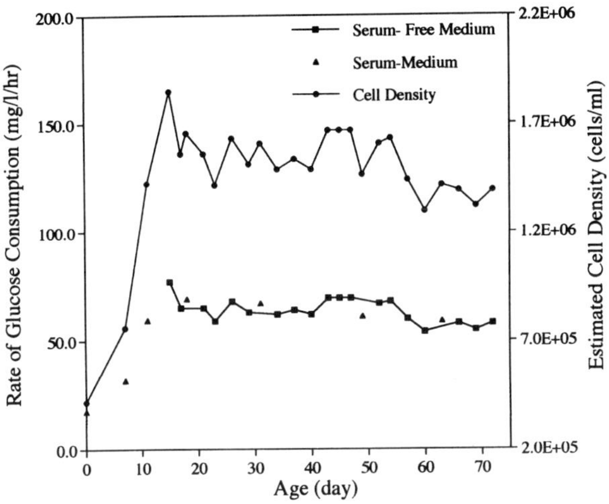

Fig 3. Glucose Consumption Rate and Cell Density during the Production Process

production from mammalian cells is the only way to obtain the proper protein. There are several reasons for the inefficiency of mammalian cells' production of recombinant proteins. The growth rate and the density levels of the culture are low, the growth and production medium are sensitive and expensive, and some cell lines cannot grow in suspension without added support systems. Two main problems were associated with the production process described in this work. First, the producing cells HEK 293, are anchorage dependent so that proper support and a special fermentation system were needed. Second, it was impossible to purify the protein, (the extracellular domain of human parathyroid Ca^{++} receptor) from the serum containing medium, thus serum free medium was required. However, the use of polyester disks (Fibra-Cel) as support for cell growth and a novel serum free supplement solved these two problems.

As was indicated above, medium replacement (perfusion) is the method of choice when adherent cells are involved. Continuous medium replacement is often implemented to minimize the inhibitory effect of metabolites such as lactate and ammonia on the cell activity, and to maintain constant concentration of the various nutrients. In the process described here, glucose consumption rate was not affected by the increase in lactate and ammonia concentrations. Therefore, the perfusion was done using batch mode based on the glucose concentration. Batch replacement offers several advantages compared to continuous mode; the most important is the elimination of a constant pumping system adding fresh medium and the constant processing of conditioned media. Based on the glucose consumption rate (1.0 ng/cell/day) the total number of cells in the 5 liter reactor was calculated to be around 10 x 10^{10} (2 x 10^6/ml). The medium replacement rate was 5.0 liters in 36-40 hours. By increasing the amount of the polyester carriers, the production rate of the basket system can be as much as five times higher. In practical terms this requires the removal of 5.0 liters medium every 6 or 7 hours. Such a rate would exceed the processing capacity of the pilot plant, and would required around-the-clock operation. Therefore the production rate was maintained at the rate of one reactor volume per 36 hours. The production process was stable for a period of 65 days and was terminated as a result of changes in the cell physiology. The serum free medium supports the protein biosynthesis better than the serum containing medium, and maintained constant production levels during six medium replacements.

The way to increase the efficiency of the process is by decreasing the amount of medium used per cell. The current serum free medium contained 4 g/l glucose and the culture was unable to use glucose above this concentration. However, higher concentration of glucose together with various amino acids and vitamins decreased the amount of medium used per unit cell. Nevertheless, the specific production was lower. Further experiments are needed to increase the efficiency of the process.

References

1. Brown E M et al. (1933) Cloning and Characterization of an Extracellular Ca^{++} Sensing Receptor from Bovine Parathyroid. Nature 366, 575-580.

2. Reuveny S (1990) Microcarrier Culture System. In: Lubiniecki AS (ed). Large Scale Mammalian Cell Culture Technology pp271-341. Marcel Dekker. New York.

3. Omasa T et al. (1995) Enhancement of Antibody Production by Growth Factor Addition in Perfusion and Hollow Fiber Culture System. Biotech & Bioeng 48, 673-680.

4 Banik GG & Heath C (1995) Hybridoma Growth and Antibody Production as a Function of Cell Density and Specific Growth Rate in Perfusion Culture. Biotech & Bioeng 39, 289-300.

5. Ozturk SS, Riley MR & Palsson B (1992) Effect of Ammonia and Lactate on Hybridoma Growth, Metabolism and Antibody Production. Biotech & Bioeng 39, 418-431.

SESSION VI

Tissue Engineering and Biomedical Devices

Bioreactor for a Hybrid Liver Support

Jörg C. Gerlach
Chirurgische Klinik, Virchow - Klinikum, Humboldt - Universität Berlin, Germany.

Introduction

Acute liver failure remains a disease with very high mortality. Massive necrosis of hepatocytes destroys synthetic, metabolic, and elimination pathways of the liver. Loss of liver cell function strikes patients dramatically and often leads to death.
Orothotopic liver transplantation changed the prognosis of patients with this clinically and pathophysiologically complex disease and is the only curative treatment in fulminant hepatic failure at present. However, in the case of acute liver failure, donor liver transplantation means replacement a potentially self-regenerating organ with the lifetime danger of immunosupression and complications. Therefore, an extracorporeal therapy, supporting the patient during acute liver failure until regeneration of their own organ is of interest.
Possible therapeutic indications for hybrid liver support have been discussed previously ([1]). Since these developments involve high technology and high cost devices, temporary application in intensive care units rather than therapy of chronic cirrhosis would seem to be realistic.

Consideration of acute liver failure promotes interest in the concept of temporary liver support systems. Systems based on blood purification by membranes or sorbents have been unsuccessful and there is a strong opinion that it is relevant to develop extracorporeal liver support systems utilising cultured hepatocytes. The term hybrid system has been applied to therapeutic devices utilising living biological tissue in an artificial environment ([2]).
Many different hepatocyte culture models and bioreactor constructions are being discussed in the literature. This indicates increasing activity in this field of medical research. Reviews ([3,4]), have focused on the development of hepatocyte culture

M.J.T. Carrondo et al. (eds.), Animal Cell Technology, 543–555.

models for bioreactors, subsequent in vitro studies and animal experiments with hybrid liver support systems.

The overall aim of our work was to further improve the culture and utilisation of hepatocytes within a bioreactor construction relevant for clinical application of hybrid liver support systems (LSS).

Development of a culture model for primary hepatocytes

It is generally agreed that the in vitro immobilisation of hepatocytes requires a material suitable for cell attachment. This material has a strong influence on the construction of a bioreactor, e.g. the use of microcarriers for hepatocyte attachment introduce a large volume of material which contributes nothing to the efficiency of the bioreactor. Different methods of immobilisation have been used, e.g. microcarriers ([5]) and membranes ([6]). The aim of our own work was to establish means of cell immobilisation with a reduced requirement for artificial or heterologous, and potentially immunogenic, biological matrix substrata ([7]) .

There are three general forms of hepatocyte culture for hybrid liver support systems. Each has its disadvantages. The first, suspension culture, is already obsolete. In this form hepatocytes lose function within hours ([8]). However, by stirring the cells this technique gives an optimal distribution of media around the hepatocytes. Low gradients of nutrients, metabolites, toxins and oxygen are therefore possible giving high mass transfer efficiency.

The second, and currently the most common technique, is attachment culture. Here, prolonged survival and metabolic function can be achieved. This technique has been applied to hollow fibre ultrafiltration cartridges containing capillaries which support attachment of primary hepatocytes ([9]). However, in such devices, mass transfer occurs mainly by transmembrane diffusion and is consequently poor.

The third form is hepatocyte spheroid culture ([10,11,12]) in which the advantages of the two previously mentioned forms can be harnessed. Depending on the size of the spheroids, prolonged activity and survival are enabled without the drawbacks of large metabolite gradients and difficulties with oxygen supply and mass transfer. However, encapsulation ([13]) will increase these problems.

Capillary membranes enable different technical functions in bioreactors. They can provide gas exchange, substrate supply and metabolite removal. And, since some capillary membrane materials can support hepatocyte attachment, one interesting prospect would be to integrate them as an artificial framework within a bioreactor to allow partial immobilisation and 3 dimensional reorganisation of hepatocyte aggregates between the membranes. In comparison to the conventional attached cell reactors or to cell encapsulation, the quantity of artificial material can be reduced, but the technical advantages of capillary membranes can still be utilised.

Polyamide ([14]) and polyethersulphone hollow fibres, potted as two discrete capillary membrane systems in a bioreactor ([15]) can be used for homogeneous and decentralised cell perfusion. Two such distinct networks provide a cell compartment, with the hepatocyte aggregates immobilised on the surface of the capillaries, which is perfused between a medium inflow compartment and a medium outflow compartment. With this model, the problematic nutrient and metabolite gradients found in conventional bioreactors might be reduced. Addition of a third capillary network with oxygenation membranes could provide decentralised oxygen supply and CO_2 removal ([16]) within the reactor.

Taken together, the potential advantages of capillary membranes and the aggregate culture technique ([17]), when combined with a modified multicompartment capillary membrane configuration ([18]), results in a novel culture model ([19]): An artificial framework of capillaries, woven in a three-dimensional array, gives a structure in which hepatocytes can be immobilised and reform high density tissue-like structuresby hepatocyte self aggregation ([20]).

Currently used culture models represent a remarkable change in the local conditions of cells in comparison to their in vivo situation. Electron microscopy investigations have demonstrated that major differences in the ultrastructure of hepatocytes, when cultured as a monolayer, include endocytosis of the junctional complexes with secondary effects on the delocalisation of the cytoskeleton and loss of orientation of the cell organelles ([21]). A culture model which enabled high cell densities ([22]) and 3 dimensional cell rearrangement by hepatocyte self aggregation ([23]), would aid the reconstruction of junctional complexes, the relocalisation of the cytoskeleton and the reorientation of the cell organelles. The culture model described above, in which the cells can reorganise themselves into three-dimensional structures may result in a hepatocyte microenvironment more akin to the physiological situation.

Most bioreactor culture models rely on transmembrane diffusion for mass transfer, while in vivo, the hepatic acinus operates under perfusion conditions with transport to the hepatocytes involving convection and active transport. The in vivo gradients of oxygen, carbon dioxide and other metabolites are not usually given sufficient consideration when designing bioreactor culture models. As a result, the mass transfer vectors in such models are often greatly distorted when compared to the in vivo situation. The culture model described above allows the cell aggregates to operate under more physiological conditions. The local flow conditions and smaller nutrient and metabolite gradients may result in a cell macroenvironment more closely related to the physiological situation.

Bioreactor construction

Bioreactors require to be developed if culture models are to be used on a large scale. The first capillary membrane liver cell bioreactor was described in 1972 ([24]). Most of the culture models described above can be reproduced within hollow fibre bioreactors and some techniques, such as microencapsulation ([25]) or the incorporation of special biological matrix layers ([26]), find prospective application in membrane bioreactors. Three general categories of hepatocyte bioreactors have emerged:
(I) bioreactors for suspension culture without cell immobilisation (27): the advantage of hepatocyte suspension bioreactors is their ability to incorporate a large number of cells in constructions which are simple to build and to operate.
(II) bioreactors based on cell immobilisation on non-specific materials ([28,29]).
(III) bioreactors with membranes, which provide both mass transfer and cell immobilisation functions ([30,31,32,33,34,35, 36]):

Our design can be described as follows: The main structure of the reactor consists of interwoven, independent capillary membrane systems which serve different functions. The hepatocytes are seeded on the outer surface of, and between, the capillaries. One specific feature of the bioreactor is that the woven capillaries enter and leave the reactor in sorted, discrete bundles so that they can provide separate functions. Each of many identical capillary units supplies only a few hepatocytes, resulting in perfusion with small gradients. Inclusion of a further capillary system allows intraluminal co- culture with sinusoidal endothelial cells to prevent the negative side effects of the hepatocytes over-growing in the inter-capillary compartment. The parallel arrangement of these units also makes scale-up relatively simple.
Four capillary systems make provision for four functions in the reactor space:
I. plasma inflow (polyamide, Akzo Nobel, Wuppertal, Germany),
II. oxygen supply and carbon dioxide removal (Silastic, Dow Corning, USA),

III. plasma outflow (polyethersulphone, Akzo Nobel)
IV. sinusoidal endothelial co-culture (hydrophilic polypropylene, Plasmaphan, Akzo Nobel).

The capillaries are woven in such a way, that the majority of hepatocytes find all types of membranes in their surroundings. With this capillary array (see fig. 1) decentralised metabolite and gas exchange with small gradients are possible. Due to the independent plasma inflow and plasma outflow compartments, decentralised perfusion of the cells between these capillaries is achieved.
The reactor design allows further independent capillary systems to be added. Therefore, additional functions could be integrated into the module. The incorporation of dialysis membranes or the use of adsorbents is feasible.
A consideration in the development of a bioreactor for hepatocytes is the selection of a material for the reactor housing. Since polyurethane is widely used for the potting of capillary membranes in devices for extracorporeal blood treatments, polyurethane materials were investigated ([37]). The outer housing of the reactor was subsequently fabricated in polyurethane PUR 725 (Morton, Bremen, Germany). The housing (see fig. 2) encloses the capillaries, using capillary dialyser potting techniques. Flow heads distribute the media to, and collect media from the capillaries. Additional openings in the housing enable introduction of cells, inclusion of a heat exchanger and temperature and pressure control.

Evaluation of the bioreactor concept

The results of initial studies, in which 2.5 x 10^9 pig hepatocytes were cultured in the reactors, demonstrate that those metabolic parameters investigated were relatively stable for at least 3 weeks of culture ([38]). A long term study of hepatocyte function in the reactors, in which synthetic function (albumin production), enzyme leakage (LDH, GOT, GPT, γGT), metabolic function (galactose and sorbitol) and cytochrome P450 activity (midazolam metabolism) were quantified, demonstrated that long term use of hepatocytes, for a month or more, is possible ([39]).
The in-vitro amino acid (AA) and a-keto acid (KA) metabolism of the bioreactors was of special interest since the amino acid metabolism of the liver is strongly related to the formation of liver coma. Considerable changes in the culture medium concentrations of AA and KA were measured during the first days in culture ([40]). This is indicative of dynamic cellular metabolism in the initial phase. The concentration of pyruvate

decreased starting from the first day of culture, at the same time, α-ketoglutarate, α-ketoisocaproate, α-ketoisovalerate and α-keto-ß-methyl-n-valerate were synthesised. The long term use of hepatocyte cultures in bioreactors and consequently, a desirable clinical hybrid liver support therapy, appear to be possible since the hepatocytes subsequently switched to steady state conditions with a stable amino acid turnover featuring general AA uptake accompanied by KA release. The release of branched chain KA, in particular that of α-ketoisocaproate, reflects an effective transamination activity in the bioreactor system.

Primary pig hepatocytes cultivated in the bioreactor can sustain amino acid metabolism for at least 30 days. However, during the investigation of long term hepatocyte culture, a general phenomenon of primary hepatocyte in vitro culture became apparent. An initial metabolic disorder following cell isolation was recorded. This may limit immediate utilisation of such systems during this 'pre-recovery' phase. Further investigation of the immediate use of primary hepatocytes after isolation (or possibly directly after freezing and thawing) would seem necessary, because one of the theoretical advantages of hybrid bridging systems, namely immediate availability, may be of less value. On the other hand, the demonstrated maintenance of primary hepatocyte AA and KA metabolism over one month should allow long-term stand-by systems to be maintained in intensive care units for on-demand application. This is described later as part of our first clinical application.

Results of a study, utilising transmission and scanning electron microscopy demonstrate also that primary hepatocyte cultures can be maintained over a period of several weeks and that, at high densities, they reorganise themselves to give tissue-like structures within the artificial system ([41]). The cells spontaneously form aggregates on, and between, the capillaries and reorganise into a three-dimensional array. Rebuilt cell junctions and the reconstruction of bile canaliculi-like structures were demonstrated in all reactors under investigation. Even after seven weeks of perfusion, the cells had a relatively intact ultrastructure. Cell organelles which depend on an intact cytoskeleton, such as the endoplasmic reticulum and the Golgi apparatus, were well maintained.

It appears as a paradox, that the cell morphology of the surviving cells was well preserved while some non surviving cells were also observed. For technical reasons, the locality of these non-surviving cells within the reactors could not be identified. This phenomenon could be due to sub-optimal bioreactor design: The diameter of the capillaries used is relatively large in comparison to the size of the hepatocytes and the liver sinusoids. Therefore, the distances between the packed capillaries are (depending on the localisation of the cells) in some cases in the millimetre range. These technical drawbacks will have to be improved in the future by using specially adapted capillaries.

Other groups, Matsumura et al. (42), Margulis et al. (43), Sussman et al. ([5]) and Rozga et al.([44]) have already reported clinical applications of hybrid liver support systems. Our own development was evaluated experimentally in order to exclude possible side effects in 60- 80 kgBW healthy pigs which were connected via jugular vein catheters to the system ([45]). In this model, technical function and safety was proven.

The efficacy of the system was then investigated in a surgical model of acute liver failure. 50- 70 kgBW pigs were hepatectomised and a cavo-cava-portal shunt was introduced under general anaesthesia. Animals were then allowed to wake up and to move freely in intensive care cages. Within the first four hours following connection to the system with up to 4 x 10^{10} hepatocytes in the reactor, animals awoke and stood freely, ate and drank freely and subsequently developed grade IV encephalopathy after a maximum of 58 hours. In a control group of hepatectomised animals, which were connected to systems without cells in the reactor, grade IV coma developed within a maximum of 24 hours.

After proving safety and efficacy, a first clinical application was prepared under approval of the university ethical committee. Approval from the local health authorities was gained and an independent evaluation was carried out in order to meet the German MedGV and obtain patient insurance.

A 45 year old patient with drug induced grade III hepatic coma received treatment for 40 hours. After announcing at Eurotransplant, Leiden, Netherlands and consulting the relatives, she was treated until an organ for liver transplantation was available. Sonography proved that liver size reduced significantly during development of grade III coma, Duplex sonography and computer tomography demonstrated partial non-perfusion of the organ. This was subsequently confirmed by histology. Following organ explantation, biopsies revealed mainly necrotic hepatocytes in the sinusoids. The patient, haemodynamically unstable before treatment, became stable and cardio-pulmonary and immunological side effects were absent. After follow up in our hospital within the normal time frame she is now at home and well.

It is still not clear whether a single therapy with a hybrid liver support system can provide all the desired (and potentially some presently unknown) liver functions for the treatment of acute liver failure. Since with intensive care alone the expected spontaneous survival rate of fulminant hepatic failure lies between 1-30%, hybrid liver support must produce a significant increase in this survival rate. A recently published paper ([46]) suggested that only a controlled clinical study, comparing other modern strategies with that of hybrid artificial liver assist devices will provide sufficient information to judge the benefit of such developments.

Literature

[1] Gerlach J, Ziemer R, Neuhaus P: Fulminant liver failure: relevance of extracorporeal hybrid liver support systems. Int. J. Art. Org. 1996;19: 7-13

[2] Chang TMS. The role of biotechnology in bioartificial or hybrid artificial cells and organs. Biomater Artif Cells Artif Organs 1988; 16 : v-vi

[3] Gerlach J. Hepatocyte culture and bioreactor design for hybrid liver support systems. In: Williams R, Lee WM. Acute Liver Failure, Cambridge 1996

[4] Nyberg SL, Shatford RA, Hu WS, Payne WD, Cerra FB: Hepatocyte culture systems for artifical liver support: implications for critical care medicine. Crit. Care med. 1992; 20,8: 1157-1168

[5] Demetriou AA, Arnaout WS, Backfisch G, Moscioni AD. Immobilized isolated liver cells on a biomatrix. Brunner G; Mito M (eds), Artificial liver support, Springer Verlag, Berlin, Heidelberg, New York, 1993: 283-295

6 Wolf CFW, Munkelt BE. Bilirubin conjugation by an artificial liver composed of cultured cells and synthetic capillaries. Trans Am Soc Artif Int Organs 1976; 21: 16-27

7 Dunn JCY, Yarmush ML, Koebe HG, Tompkins RG. Hepatocyte function and extracellular matrix geometry: long-term culture in a sandwich configuration. FASEB 1989; 3 : 174-177

[8] J.Gerlach, H.H.Schauwecker, K. Klöppel, Tauber R, Müller Ch, Bücherl E.: Use of hepatocytes in adhesion and suspension cultures for liver support bioreactors. Int. J. Artif. Org. 12/1989: 788-793

[9] Gerlach, J., Stoll, P., Schnoy, N. and Bücherl, E.S.B. (1990). Membranes as substrates for hepatocyte adhesion in liver support bioreactors. Int. J. Artif. Org. 13/7 , 788-793

10 Landry J, Bernier D, Ouellet C. Speroidal aggregate culture of rat liver cells: histiotypic reorganization, biomatrix deposition, and maintainance of functional activities. J Cell Biol 1985; 101: 914-923

[11] Tong JZ, Bernard O, Alvarez F. Long term culture of rat liver cell spheroids in hormonally defined media. Exp Cell Res 1990; 189: 87-92

[12] Koide N, Sakagucci K, Koide Y: Formation of multicellular spheroids composed of adult rat hepatocytes in dishes with positive charged surfaces and under other nonadherent environments. Exp. Cell Res 1990; 186: 227-235

[13] Takabatake H, Koide N, Tsuji T: Encapsulated multicellular speroids of rat hepatocytes produce albumin and urea in a spouted bed circulating culture system. Artif. Org. 1991; 15: 474-480

[14] Gerlach J, Stoll P, Schnoy N, Bücherl E.: Membranes as substrate for hepatocyte adhesion in liver support bioreactors. Int J Artif Org 1990; 13/7: 436-441

[15] Gerlach J, Stoll P, Schnoy N, Neuhaus P. Capillary membranes for liver support bioreactors. Int. J. Artif. Org.1996; : submitted

[16] Gerlach J, Klöppel K, Stoll P, Vienken J, Müller Ch, Schauwecker HH.Gas supply across membranes in liver support bioreactors. Artif. Org. 1990; 14: 328-333

[17] Gerlach J, Klöppel K, Müller C, Schnoy N, Smith M, Neuhaus P. Hepatocyte aggregate culture technique for bioreactors in hybrid liver support systems. Int J Art Org 1994; 16: 785-788

[18] Gerlach J, Smith M, Neuhaus P. Hepatocyte culture between woven capillary systems - a microscopy study. Artif Org 1994; 18: 226-230

[19] Gerlach J, Neuhaus P. Culture model for primary hepatocytes. In Vitro Cell. Dev. Biol. 1994; 30A : 640-642

[20] Yuasa C, Tomita Y, Shono K, Ishimura K, Ichihara A. Importance of cell aggregation for expression of liver functions and regeneration demonstrated with primary cultured hepatocytes. J Cellular Phys 1993;156: 522-530

[21] Berry MN, Edwards AM, Barrit GJ: Isolated hepatocytes preparation, properties and applications. In: Burdon RH, van Knippenberg PH eds. Laboratory techniques in biochemistry and molecular biology. Amsterdam: Elsevier, 1991

[22] Clayton DF, Harrelson AL, Darnell JE. Dependence of liver- specific transscription on tissue organisation. Mol Cell Biol 1985;5: 2623-2632

[23] M Gleisner

24 Knazek RA, Gullino PM, Kohler PO, Dedrick RL. Cell culture on artificial capillaries: an approach to tissue growth in vitro. Science 1972; 178: 65-67

25 Matthew HW, Salley SO, Peterson WD, Deshmukh DR, Mukhopadhyay A, Klein MD. Microencapsulated hepatocytes. Prospects for extracorporeal liver support. ASAIO Trans 1991; 37(3): 328-330

26 Akaike T, Tobe S, Kobayashi A, Goto M, Kobayashi K. Design of hepatocyte specific extracellular matrices for hybrid artificial liver. Gastroenterol Jpn 1993; 28 (Suppl 4): 45-56

27 Olumide F, Eliashiv A, Kralios N, Norton L, Eiseman B. Hepatic support with hepatocyte suspension in a permeable membrane dialyzer. Surgery 1977; 82: 599-606

28 Uchino J, Matsue H, Takahashi M, Nakajiama Y, Matsushita M, Hamada T, Hashimura E. A hybrid artificial liver. Function of cultured monolayer pig hepatocytes in plasma from hepatic failure patients. ASAIO Trans 1991; 37: 337-338

29 Yanagi K, Ookawa K, Mizuno S, Ohshima N. Performance of a new hybrid artificial liver support system using hepatocytes entrapped within a hydrogel. ASAIO Trans 1989; 35: 570-572

30 Jauregui HO, Naik S, Driscoll JL, Solomon BA, Galletti PM. Adult rat hepatocyte cultures as the cellular component of an artificial hybrid liver. In: Biomaterials in Artificial Organs, Proceedings of a seminar on biomaterials held at the University of Strathclyde, Glasgow (Edit. by Paul JP, Gaylor JDS, Courtney JM, Gilchrist T) 1983: 130-140

31 Rozga J, Holzmann MD, Ro MS, Griffin DW, Neuzil DF, Giorgio T, Moscioni A, Demetriou AA. Development of a hybrid bioartificial liver. Ann Surg 1993; 217(5): 502-511

32 Dixit V. Development of a bioartificial liver using isolated hepatocytes. Art. org. 18; 1994: 371-384)

33 Hu WS, Nyberg SL, Shatford RA. Cultivation of hepatocytes in a new entrapment reactor: a potential bioartificial liver. In: Animal cell culture and production of biologicals. Sasaki R, Ikura K (Eds). Dordecht, The Netherlands, Kluwer Academic Publishers, 1991: 75-80

34 Nyberg SL, Russel A, Shatford, Madhusudan V, Peshwa, White JG, Cerra FB, Hu WS. Evaluation of a hepatocyte-entrapment hollow fiber bioreactor: a potential bioartificial liver. Biotech Bioengineering 1993; 41: 194-203

35 Shatford RA, Nyberg SL, Meier SJ, White JG, Payne WD, Hu WS, Cerra FB. Hepatocyte function in a hollow fiber bioreactor: a potential bioartificial liver: J Surg Res 1992; 53: 549-557

36 Scholz M, Hu WS. A two compartment cell entrapment bioreactor with three different holding times for cells, high and low molecular weight compounds. Cytotechnology 1991; 4: 127-137

37 Gerlach J, Planck H, Schauwecker HH.Cytocompatibility of Polyurethanes. Planck H ed., Polyurethane in Biomedical Engineering 3, Elsevier 1989

38 Gerlach J, Encke J, Hole O, Müller C, Ryan CJ, Neuhaus P. Bioreactor for larger scale hepatocyte in - vitro perfusion. Transplantation 1994; 58: 984-988

39 Gerlach J, Encke J, Hole O, Müller C, Courtney J, Neuhaus P. Hepatocyte culture between three dimensional arranged biomatrix- coated independent artificial capillary systems and sinusoidal endothelial cell co-culture compartments. Int. J. Artif. Org. 1994; 17: 301-306

40 Gerlach J, Fuchs M, Smith M, Bornemann R, Neuhaus P, Riedel E. Is a clinical application of hybrid liver support systems limited by an initial disorder in cellular amino acid and a-keto acid metabolism, rather than by later gradual loss of primary hepatocyte function? Transplantation 1996; 62: in press

41 Gerlach J, Schnoy N, Encke J, Müller C, Smith M, Neuhaus P. Electron microscopic studies on a hepatocyte culture model with woven multicompartment capillary systems. Hepatology 1995; 22: 546-552

42 Matsumura KN, Guevara GR, Huston H, Hamilton WL, Rikimaru M, Yamasaki G, Matsumura MS. Hybrid bioartificial liver in hepatic failure: Preliminary clinical report. Surgery 1987; 101(1): 99-103

43 Margulis MS, Erukhimov EA, Andreiman LA, Kuznetsov KA, Viksna LM, Kuznetsov AI, Devyatov VV. Hemoperfusion through suspension of cryopreserved hepatocytes in a treatment of patients with acute liver failure. Res Surg 1990; 2: 99-102

44 Rozga J, Podesta L, Le Page E, Hoffman A, Morsiani E, Sher L, Woolf GM, Makowka L, Demetriou AA. Control of cerebral oedema by total hepatectomy and extracorporeal liver support in fulminant hepatic failure. Lancet 1993; 342: 898-899

45 Janke J, Böhmer C, Kardassis D, Rossaint R, Neuhaus P, Gerlach J. Exclusion of cardipulmonary side effects of a hybrid liver support system. Int. J. Artif. Org.1996; submitted

46 Cattral M, Levy G. Artificial liver support - pipe dream or reality ? New England Journal of Medicine 1994; 268-269

Discussion

Kitagawa: The polarisation of liver cells where the apical cells face the bile duct, I am sure is something that you are able to take advantage of because the apical side only connects to the bile duct.

Gerlach: The secretion of bile products from the cells might not be that necessary since patients with a complete obstruction of the bile duct due to a tumour can survive for several weeks. On the other hand for long term maintenance of the cells, it is interesting to study this specific function. I mentioned this false compartment (in my talk) which is also arranged de-centralised in the bioreactor system. A system in the States works on the specific binding of the *gal-pol* of hepatocytes to such a capillary system. In theory it is possible in the future to have a decentralised removal of bile but no group is working on a technical solution for that. Actually, I believe it is not that important because we think in terms of short term applications only in the care units.

Noe: A question concerning the technological status of your bioreactor. Can you comment on the sterility status of your bioreactor? What would happen if you lost sterility when connected to a patient?

Gerlach: Of course, in such an *in vitro* culture of hepatocytes there are no immunologically competent cells which can arrest bacterial contamination. So sterility is very important. The bioreactor is constructed like a dialysis module with an outer housing and capillaries that are sterilised by formaldehyde or ethylene oxide. Special pathogen-free pigs for cell isolation were used.

Ijima: Can you explain why the biluribin concentration decreased?

Gerlach: Biluribin is conjugated to be eliminated later on. This is not the pathway that is involved in bile secretion.

Ijima: Is your model a total or partial hepactectomy?

Gerlach: No, we used a total hepactectomy model. By surgical means the organ was removed and connected to the vascular system and used extracorporeally with the bioreactor device.

Degouys: Which kind of medium is perfused in the bioreactor?

Gerlach: A modification of William's E medium.

Degouys: No protein in it?

Gerlach: This is an important question. Normally one uses around 5-10% foetal calf serum with hepatacytes. Since we are using heterologous pig cells we did not want to include serum from another species. Therefore we used serum from the animal from which the cells were obtained at 5%.

Degouys: It is possible to preclude biofouling of your system?

Gerlach: Biofouling of the membranes is a problem for long term running of bioreactors. We do not have this problem for the length of our cultures. For hepatocyte culture there is no suitable protein free medium. For clinical application we rinse the systems with sodium chloride to remove the culture medium, including serum, 5 minutes before clinical application.

Brown: A comment. I believe the bioreactor was set up fundamentally in the same way that Tolbert's Static Maintenance Reactor was set up many years ago. A case for history repeating itself. Different materials, different context, different application but the same technology!

Gerlach: I do not know about this reactor.

Lehmann: There is a knowledge gap between the medical doctors and the original biotechnologists and we need to bridge this gap.

Bonarius: Have you tried a standard artificial kidney apparatus where you would externally aerate the serum, instead of this complex apparatus you are using now?

Gerlach: Two other groups use such commercially available systems. The problem is that hepatocytes need a very special cultural surrounding which cannot be supplied by a simpler system. A dialyser will allow a 6-8 hr treatment. We believe it is necessary to support the cell functions more than for this 6-8 hrs. Firstly because of the clinical relevance a longer running system is needed. Secondly, such systems do not enable a standby to operate for patients.

Bonarius: The hepatocytes will not only break down certain components but they will generate a lot of products as well. How can you be sure you do not introduce into the patient certain compounds that might result in a immune response later on?

Gerlach: We use semi-permeable membranes and the cell compartment is closed in the direction of the patient. The membrane cut-off is 200,000 ie cells cannot reach the patient but cell products can.

Factors Affecting Megakaryocytic Expansion in Hematopoietic Cultures for Transplant Therapies

J.A. LaIuppa, W.M. Miller, E.T. Papoutsakis Northwestern University, Evanston, IL 60208 USA

Abstract
Ex vivo expansion of platelet-producing megakaryocytes (Mk) is clinically important in reducing the period of low platelet counts which occurs after chemotherapy. Several cytokines have been shown to be important in expanding cells of the Mk lineage, especially Thrombopoietin (TPO). Other culture conditions such as oxygen tension also play a role in hematopoietic cell proliferation and differentiation. We examined the effect of oxygen tension and cytokines on Mk progenitors (CFU-Mk), total number of Mks, and size of Mks. Both the number and size of Mks are important factors for determining platelet production. We also evaluated the number of progenitors of the granulocyte/monocyte lineage (CFU-GM) and erythroid lineage (BFU-E) in these cultures. CFU-Mk was enhanced under 5% oxygen tension, while the total number and size of Mks was increased at 20% O_2 tension compared to 5% O_2 tension. Reduced oxygen tension enhanced CFU-GM expansion. We examined various combinations of the following cytokines: Interleukin-3 (IL-3), Stem Cell Factor (SCF), IL-6, Granulocyte-Macrophage Colony Stimulating Factor (GM-CSF), G-CSF, Erythropoietin (EPO), and TPO. Under 20% oxygen tension, the combination of IL-3, SCF, IL-6, and TPO produced the greatest number of CFU-Mk, while the combination of IL-3, SCF, EPO, and TPO produced the greatest number of total Mks.

Introduction
The study of megakaryocytopoiesis has significant direct applications in the clinical setting. Knowledge of the proper conditions for expansion of cells of the Mk lineage would be of immediate consequence, since platelet recovery times after chemotherapy are longer and much more variable than neutrophil recovery times. However, before the recent isolation and purification of Thrombopoietin (TPO) in 1994, the study of megakaryocytopoiesis *in vitro* was difficult. Results from different studies were inconsistent because Mks are rare (0.1% of bone marrow cells) and almost all studies used either serum or plasma (usually aplastic plasma) in their cultures. In order to achieve more consistent and reproducible results, well-defined culture systems are needed. We aimed to increase the number of Mks produced in liquid cultures using serum-free medium and multiple cytokines.

While cytokine combinations play the major role in determining the lineage and maturation stage of the cells produced in culture, other culture parameters can make a significant contribution. Oxygen tension greatly influences hematopoietic cell cultures. Long-term liquid cultures of cord blood mononuclear cells (MNC) on irradiated stromal

M.J.T. Carrondo et al. (eds.), Animal Cell Technology, 557–563.

layers, cord blood MNC, and bone marrow MNC grown under low oxygen tension all have an improved and prolonged maintenance of hematopoietic progenitor cells [1, 2]. In murine bone marrow liquid cultures, different maturation stages of hematopoietic progenitors display varying degrees of sensitivity to hypoxia. The most primitive progenitors show the least sensitivity to severe hypoxia [3]. Although the effects of oxygen tension on CFU-GM and BFU-E have been studied extensively, no studies have examined the effect of oxygen tension on Mk maturation and only one study has addressed the effect of oxygen tension on CFU-Mk formation [4].

In order to determine the effect of oxygen tension and cytokines on the differentiation pattern of the Mk lineage, we evaluated our serum-free cultures of mobilized peripheral blood (PB) CD34+ cells using Mk-specific assays (CFU-Mk, CD41a+ cells, and Mk size).

Materials and methods

Cells and cell separation procedures

Patient samples were provided by Response Technologies (Memphis, TN, USA). Samples were collected after informed consent under protocols approved by the respective Institutional Review Boards. Apheresis products were collected from cancer patients following stem cell mobilization with G-CSF, GM-CSF, and/or cyclophosphamide. Positive selection of CD34+ cells was performed using one of two methods. The MiniMacs column was used according to directions of the manufacturer (Miltenyi Biotec, Auburn, CA, USA). The other previously described method [5] used antibody-coated magnetic beads which bind anti-CD34 antibody. The average purity of the CD34+ populations was approximately 90%.

Culture and harvest of cells

All cultures were performed with X-VIVO 20 (Biowhittaker, Walkersville, MD, USA) supplemented with 1.5 ng/ml IL-3 (R&D Systems, Minneapolis, MN, USA) and 50 ng/ml SCF (R&D Systems). Cytokines were added to the base combination as indicated: 83 ng/ml EPO (Amgen, Thousand Oaks, CA, USA), 1.5 ng/ml G-CSF (Amgen), 10 ng/ml IL-6 (R&D Systems), 4 ng/ml GM-CSF (Immunex, Seattle, WA, USA), and 25 ng/ml TPO (Genentech, San Francisco, CA, USA). CD34+ cells were plated at $2x10^4$ cells/ml in the center 8 wells of 24-well plates in 1 ml volumes. Cells were cultured for 10 days at 37°C under fully humidified conditions in an atmosphere of 5% CO_2 and 5% O_2 in N_2 or 5% CO_2 in air. Cultures were fed at day 6 by doubling the culture volume with fresh growth factor-containing medium. Medium was equilibrated in the incubator at the appropriate oxygen concentration for several hours before feeding in an attempt to keep the low oxygen cultures at 5% oxygen. Cultures were harvested at day 10 by removing the cell suspension from the culture plate, rinsing the plate with cell dissociation solution (Sigma, C-5789) to remove any adherent cells, and then washing the plate with PBS. All reagents were obtained from Sigma Chemical (St. Louis, MO, USA) unless otherwise stated.

Colony assays

CFU-Mk assays were performed as described by Hunt et al. [6]. Cells were plated at $2x10^4$ cells/ml in X-VIVO 20 supplemented with 1.5 ng/ml IL-3, 50 ng/ml SCF, and 25 ng/ml TPO. Aliquots of 300 μl were plated in triplicate into 96-well plates for 12 days at 37°C under fully humidified conditions in an atmosphere of 5% CO_2 in air. The outer wells of the plates were filled with water to minimize evaporation from the inner wells. After 12 days, colonies in the assays were fixed with gluteraldehyde and the Mks were stained according to the cytochemical staining procedure using anti-CD41a antibody as described below. Colonies of greater than 3 cells were scored as CFU-Mk.

Cytospins

Approximately $2x10^4$ cells from each sample were centrifuged for 5 min at 850 rpm onto a slide using a Cytospin 2 (Shandon, Pittsburgh, PA, USA) and then fixed immediately with acetone.

Cytochemical staining

Cytospins of cells were stained for the CD41a antigen using a Histomark Strepavidin-horseradish peroxidase (HRP) kit. The staining procedure was performed at 25°C. Slides were soaked in a 15:1 solution of PBS and 30% H_2O_2 (Sigma) for 10 min to block endogenous peroxidase and were then washed with PBS. Slides were incubated with normal goat serum for 15 min in a humidified environment to block non-specific staining. Excess serum was wicked off the slides before incubation with anti-CD41a antibody (1:500 dilution in PBS) (AMAC, Westbrook, ME, USA) for 1 hour. Slides were washed with PBS before incubation with biotinylated secondary antibody for 30 min. Slides were rinsed with PBS before incubation with strepavidin-peroxidase for 30 min. Slides were washed with PBS and a chromogen solution (Peroxidase chromogen kit, Biomeda, Foster City, CA, USA) was added for 2-10 min for color development. Slides were rinsed with millipore water before background staining with hematoxylin (Biomeda) for 1 min. Slides were rinsed for a final time with tap water and air dried. Slides were coated with Crystal Mount (Biomeda) and dried overnight before applying a coverslip using Accumount (Scientific Products). Approximately 400 cells per slide were counted to determine the percentage of $CD41a^+$ cells in the cultures.

Image analysis

Slides of cultured cells stained with anti-CD41a antibody were imaged at 400X magnification using a Kodak EktaPro video system (Eastman Kodak Company, San Diego, CA, USA) connected to an Olympus microscope (Model BHTU B061). The video out signal was directly connected to a Power Macintosh 7100/80 AV and images were captured as PICT files using the Video Monitor program (Apple Computer), achieving a final digital resolution of 240 x 192 pixels. Images were analyzed using the program Image (NIH) to obtain pixel area measurements of the Mks. An average of 100 $CD41a^+$ cells were measured for each culture condition for each experiment. A slide micrometer was used to convert pixel area measurements to μm^2 area measurements.

Statistical analysis

Data were analyzed using the paired Student's t-test. Points marked on the graphs were statistically significant, as indicated in the figure legends.

Results and Discussion

In control cultures with IL-3 and SCF, there was no significant difference in the number of Mks at the different oxygen tensions (Figure 1). All of the cytokine combinations examined stimulated an increase in the number of Mks over the level in control cultures under both 5% and 20% oxygen (Figure 1). TPO was the single best factor for increasing Mk numbers. Oxygen tension also had a significant impact on the number of Mks produced in these cultures. Cultures with TPO and/or IL-6 under 20% oxygen produced 1.2-1.4-fold more Mks than those with the same cytokines under 5% oxygen. The addition of both TPO and IL-6 to cells increased the number of Mks over that found in cultures with TPO alone, but this effect was only apparent under 20% oxygen.

TPO combined with GM-CSF did not result in an increase in Mk numbers over that for TPO alone at either oxygen tension.

Since Mks undergo endomitosis and increase in size as they mature, Mk cell areas were measured to evaluate the state of Mk maturation. There was no difference in the size of Mks at the different oxygen tensions in control cultures. However, a significant increase in Mk size was observed in cultures supplemented with TPO, and larger Mks were produced in cultures with TPO under 20% oxygen compared to 5% oxygen (Figure 2).

Oxygen tension also plays an important role in CFU-Mk expansion. An increase in the number of CFU-Mk over that in control cultures occurred in all cultures containing TPO, IL-6, or GM-CSF. The number of CFU-Mk produced in these cultures containing TPO was 1.2-fold greater under 5% than under 20% oxygen. When TPO was combined with IL-6 or GM-CSF under low oxygen, the number of CFU-Mk produced was higher than that seen in cultures with TPO alone. The greatest increase in CFU-Mk numbers occurred when TPO and IL-6 were combined.

We also found that CFU-GM were enhanced under 5% oxygen for all cytokine combinations examined and BFU-E expansion was enhanced under 5% oxygen in most cases. The inverse relationship between the number of CFU-Mks and total Mks at the different oxygen tensions is consistent with that found in the *in vivo* bone marrow (BM) microenvironment. In murine BM, the more primitive progenitors were found concentrated in the low oxygen areas of the BM along the bone, while the more mature cells were located primarily in the higher oxygen areas where the blood enters the BM circulation [7]. Our results are also consistent with a study which investigated the effect of oxygen tension on liquid cultures of murine BM cells [3]. This study showed that the total number of cells produced in the culture was directly proportional to the oxygen concentration of the culture. The greatest number of cells was produced under ambient oxygen conditions, while the smallest number was produced under 1% oxygen.

Results of our study indicate that TPO in combination with IL-6 results in increased stimulation of the Mk lineage over that found with TPO alone. Our results are consistent with previous studies which showed that TPO in combination with IL-6 [8] results in stimulation of Mk progenitors. Broudy et al. did not observe an increase in CFU-Mk over that found with TPO alone upon addition of both TPO and GM-CSF to murine BM cultures [8], which is also consistent with our results at 20% oxygen.

Traditionally, growth factors were thought to have exclusive control over proliferation and differentiation in the hematopoietic system. While these factors are necessary, results from our experiments and others show that other culture parameters such as oxygen tension can greatly influence cell lineage and maturation stage. These culture parameters need to be carefully selected and controlled in *ex vivo* expansion systems designed for clinical use in order to consistently produce the desired cell type at a specific cell stage.

Acknowledgments

Supported by National Institutes of Health (USA) grant HL48276. We would like to also thank Genentech for their generous donation of TPO. We are grateful to Response Technologies for providing us with PB samples. We thank Dr. James Bender and Steven Smith at Baxter Healthcare for supplying us with reagents for $CD34^+$ cell selection.

References

1. Koller M, Bender J, Miller W, and Papoutsakis E (1992) Reduced oxygen tension increases hematopoiesis in long-term culture of human stem and progenitor cells from cord blood and bone marrow. Exp Hematol 20: p. 264-270.

2. Koller M, Bender J, Papoutsakis E, and Miller W (1992) Effects of synergistic cytokine combination, low oxygen, and irradiated stroma on the expansion of human cord blood progenitors. Blood 80: p. 403-411.

3. Cipolleschi M, Sbarba P, and Olivotto M (1993) The role of hypoxia in the maintenance of hematopoietic stem cells. Blood 82(7): p. 2031-2037.

4. Katahira J and Mizoguchi H (1987) Improvement of culture conditions for human megakaryocytic and pluripotent progenitor cells by low oxygen tension. Int J Cell Cloning 5(5): p. 412-420.

5. Smith S, Bender J, Maples P, Unverzagt K, Schilling M, Lum L, Williams S, and VanEpps D (1993) Expansion of neutrophil precursors and progenitors in suspension cultures of CD34+ cells enriched from human bone marrow. Exp Hematol 21: p. 870-877.

6. Hunt P, Li Y, Nichol J, Hokom M, Bogenberger J, Swift S, Skrine J, Hornkohl A, Lu H, Clogston C, Merewether L, Johnson M, Parker V, Knudten A, Farese A, Hsu R, Garcia A, Stead R, Bosselman R, and Bartley T (1995) Purification and biologic characterization of plasma-derived megakaryocyte growth and development factor. Blood 86(2): p. 540-547.

7. Lord B, (1990) The architecture of bone marrow cell populations. Int J of Cell Cloning 8: p. 317-331.

8. Broudy V, Lin N, and Kaushansky K (1995) Thrombopoietin (c-mpl ligand) acts synergistically with erythropoietin, stem cell factor, and interleukin-11 to enhance murine megakaryocytic colony growth and increases megakaryocyte ploidy in vitro. Blood 85(7): p. 1719-1726.

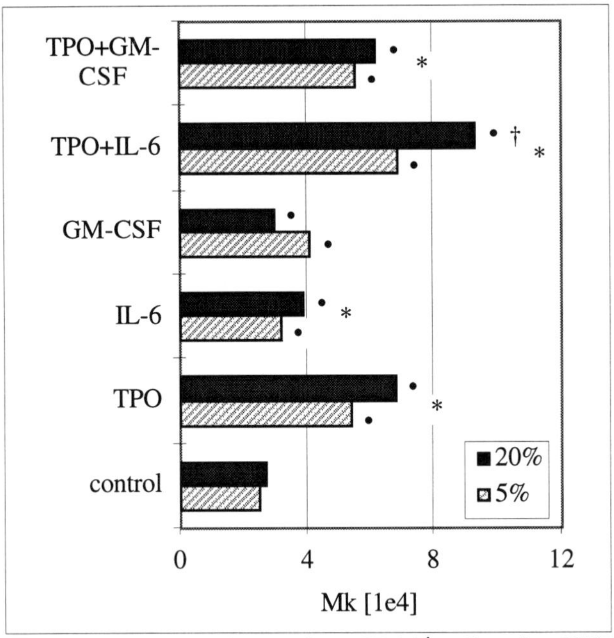

Figure 1: Total Mks from day 10 cultures of PB $CD34^+$ cells using XVIVO-20 supplemented with IL-3, SCF and the indicated cytokines. Control cultures contained IL-3 and SCF. The means of 8 experiments under both 5% and 20% oxygen tension are graphed.
* represents significant difference ($p<0.05$) between 5% O_2 and 20% O_2.
• represents significant increase ($p<0.05$) over control cultures.
† represents significant difference ($p<0.05$) between the cytokine combination and each factor in the combination alone.

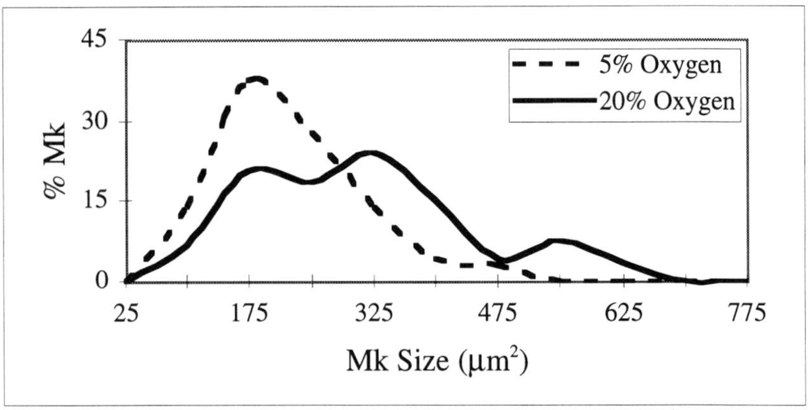

Figure 2: Size distribution of Mks from day 10 cultures of PB $CD34^+$ cells using XVIVO-20 supplemented with IL-3, SCF, and TPO under both 5% O_2 and 20% O_2. One representative experiment of five is shown.

Discussion

Spier: Can you fill us in on TPO. What is it and what are its characteristics?

Papoutsakis: It is a protein with a good deal of homology with EPO. It was discovered simultaneously by 3 different companies so the patent position is not clear. It is an important differentiation factor. It is in late clinical stage trials, third stage and is a cytokine that should be used shortly for human therapy. It is called thrombopoieten..

Al-Rubeai: Is there any relationship between oxygen uptake rate and tension or any relationship with the cell size?

Papoutsakis: We have some measurements on how the stage of differentiation affects the oxygen utilisation but we have no data on whether this is related to size. One would guess it probably is.

The possible role of the tumor suppressor protein p53 in neonatal and adult cardiomyocytes

Peter G. David[1,2], James E. Bailey[2], Hans M. Eppenberger[1]
Institutes of Cell Biology[1] and Biotechnology[2], ETH, CH-8093 Zurich, Switzerland

Abstract

During early development cardiomyocytes proliferate while in the adult cardiomyocytes cell division does not occur. DNA synthesis in rat heart stops shortly after birth and increase in organ mass only occurs by growth of cell size (hypertrophy). Examination of cell cycle control in cardiomyocytes in culture would be of great importance and an identification of components of the control mechanisms would be a first step towards understanding how such factors control either proliferation or initiation of terminal differentiation. Overcoming the strong inhibition of the cell cycle will be crucial for inducing new cell production or for producing a cardiac cell line. Here we show first indications that the tumor suppressor protein p53 might play an important role in keeping the cardiomyocytes out of the cell cycle. Presence of p53 associated with cytoskeletal structures could be demonstrated as well as possible induction of cyclin-cdk inhibition during cardiomyocyte differentiation.

Introduction

During the second half of gestation, rat cardiac myocytes progressively withdraw from the cell cycle and DNA synthesis is no longer detectable in these cells by neonatal day 17. It is generally accepted that the adult mammalian heart does not have any regenerative capacity (1) Therefore cardiomyocyte loss as a result of disease or injury is irreversible. The mechanisms that exclude these terminally differentiated cells from the cell cycle are poorly understood. Recent findings support the notion that proteins exhibiting intrinsic growth suppressive properties in both myogenic and nonmyogenic cells might play a central role (2, 3). Among these proteins, generally known as tumor

M.J.T. Carrondo et al. (eds.), Animal Cell Technology, 565–570.

suppressors, are members of the retinoblastoma (Rb) family of pocket proteins (4), which also include p107 and p130.

Here we report the appearance of another protein which is generally regarded to play a crucial role in cell cycle arrest: the tumor suppressor p53. p53 is a transformation-related protein that is synthesized at high levels in transformed cells and at very low levels in non-transformed cells (5-7). Exerting its role as a tumor suppressor gene in a wide variety of tissue types, p53 performs several different critical tasks in regulating cell growth, replication and death. It can positively regulate transcription by binding to specific DNA consensus sequences. These sequences are associated with several known genes including the human ribosomal cluster, muscle creatine kinase gene, WAF-1/CIP-1/p21, cyclin G and mdm-2 (reviewed in 8).This sequence-specific transcriptional activation is associated with the induction of growth suppression (9).

Materials and Methods

Isolation and Culture of Cells. Ventricular cardiac muscle cells of 2 month old Sprague-Dawley Javonas rats were isolated by retrograde perfusion of hearts with collagenase (Worthington) as described earlier in detail (10). Cells were first plated for 2h in medium 199 (Animed, Basel), containing 1% fetal calf serum (Readysystem, Zurzach, Switzerland) and then on dishes coated with 0.1% gelatine (Fluka) in medium 199/20% fetal calf serum/1% penicillin/streptomycin stock (Animed, Basel)/20 mM creatine (Merck). To inhibit fibroblast and smooth muscle cell growth, 10μM 1-β-D-arabinofuranosyl-cytosine was added throughout the culture period. The medium was changed after 2 and 7 days.

Antibodies and Fluorescence Reagents. F-actin in the I-bands of myofibrils and in actin filament bundles was stained with rhodamine-conjugated phalloidin (Molecular Probes). The monoclonal antibody specific against wild-type p53 (PAb246, Santa Cruz) has been described by (11).

As second antibody fluorescine-conjugated goat anti-mouse IgG (Cappel) was used.

Immunochemistry and Microscopy. Cells were treated for immunostaining as adapted from (12) and (13). Cultured cells were washed twice with solution A (Sol A, 100 mM KCl, 1 mM EGTA, 5 mM EDTA, pH 7.0) and permeabilized for 8 min using TGA (2% Triton X-100, 50% glycerol in Sol A). Fixation was done for 15 min in 3% paraformaldehyde in Sol A. Free aldehyde groups were saturated with 100 mM glycine in Sol A for 10 min and unspecific protein binding was blocked with 2% BSA in Sol A for 20 min. After washing cells were incubated with the anti-p53 antibody (1:50 in Sol A with 2% BSA) and then with the second antibody (1:100 in Sol A with 2% BSA) in a humidified chamber for at least 1 hr.
The imaging system consisted of a Zeiss Axiophot fluorescence microscope, a Biorad MRC-600 confocal scanner and a Silicon Graphics Indigo 2 workstation. The images were produced using a Zeiss Neofluar x40/1.3 objective lens and processing was done using Imaris software (Bitplane, Switzerland).

Results and discussion

The subcellular localization of the p53 molecule was studied by immunofluorescent staining of fixed cells using an anti-p53 monoclonal antibody and subsequent analysis by confocal light microscopy. Representative immunofluorescent staining of 12 day cultured adult rat cardiomyocytes is illustrated in Figure 1. By using PAb246 as first antibody followed by Texas Red conjugated anti-mouse immunoglobulin, we observed that p53 is present mainly in the cytoplasm, closely associated with the cytoskeletal actin filaments.
The notion, that wild-type p53 bears the ability to interact with the cytoskeleton has been under discussion several times during the past. Earlier publications suggested that p53 in non-transformed fibroblasts is represented mainly in a cytoskeleton-associated form, since it can not be removed by treating the cells with Triton X-100 and that it can be found closely associated with cytosekeletal actin filaments (14, 15).
Here we demonstrate that the tumor suppressor protein, which is normally expressed only at minute levels in non-transformed cells, can be distinctly detected in primary adult rat cardiomyocytes. While commonly found in numerous types of transformed cells, only a few types of non-transformed cells, where p53 is detectable, were known (15). We

suggest that the relatively high levels of p53 expression in cardiomyocytes reflects a possible mechanism by which these cells retain homeostasis for a lifetime, i. e. by controlling cell cycle regulation, repair mechanisms and cell elimination (apoptosis). Work is in progress in our laboratory to identify the factors responsible for reactivation of the cell cycle in postmitotic cardiomyocytes.

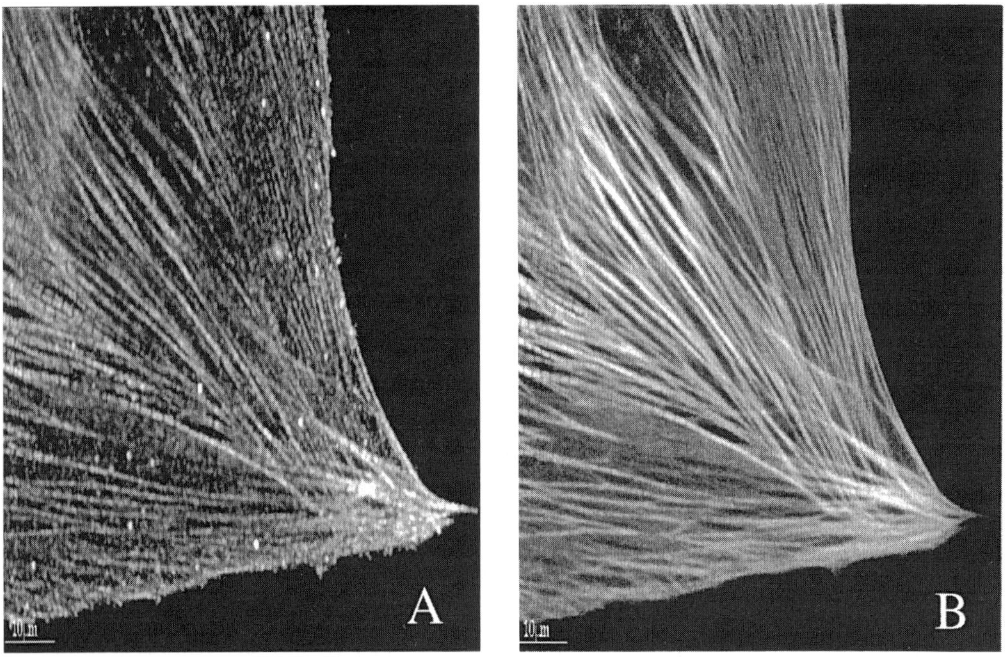

Figure 1. (A) Indirect immunofluorescent staining of p53 protein in adult rat cardiomyocytes grown 12 d in culture. Cells were permeabilized with Triton X-100, fixed with paraformaldehyde and processed for immunofluorescence with PAb246 anti-p53 antibody and fluorescine conjugated goat anti-mouse antibody as detailed in Materials and Methods. (B) A perfect correlation of the immunostaining and the labelling of F-actin with rhodaminated phalloidin could be observed.

Literature

1. Rumyantsev PP (1991) In: Carlson BM, Ed., Growth and Hyperplasia of Cardiac Muscle Cells, pp 3-68, Harwood, New York.
2. Sorrentino V, Pepperkok R, Davis RL, Ansorge W & Philipson L (1990) Cell Proliferation Inhibited by MyoD1 Independently of Myogenic Differentiation. Nature (Lond.) 345, 813-815.
3. Crescenzi M, Fleming TP, Lassar AB, Weintraub H & Aaronson SA (1990) MyoD Induces Growth Arrest Independent of Differentiation in Normal and Transformed Cells. Proc. Natl. Acad. Sci. USA 87, 8442-8446.
4. Gu W, Schneider JW, Condorelli G, Haushal S, Mahdavi V & Nadal-Ginard B (1993) Interaction of Myogenic Factors and the Retinoblastoma Protein Mediates Muscle Cell Commitment and Differentiation. Cell 72, 309-324.
5. DeLeo AB, Jay G, Apella E, Dubois GC, Law LW & Old LJ (1979) Detection of a Transformatio-related Antigen in Chemically Induced Sarcomas and other Transformed Cells of the Mouse. Proc. Natl. Acad. Sci. USA, 76, 2420-2424.
6. Jay G, DeLeo AB, Appella E, Dubois GC, Law LW, Khoury G & Old LJ (1979) A Common Transformation-related Protein in Murine Sarcomas and Leukemias. Cold Spring Harbor Symp. Quant. Biol., 44, 659-664.
7. Rotter V, Witte ON, Coffman R & Baltimore D (1980) Abelson Murine Leukemia Virus-induced Tumors Elicit Antibodies Against a Host Cell Protein, P50. J. Virol., 36, 547-555.
8. Elledge RM & Lee WH (1995) Life and Death by p53. BioEssays 17, 11, 923-930.
9. Pietenpol JA, Tokino T, Thiagalingam S, El-Deiry WS, Kinzler KW & Vogelstein B (1994) Sequence-specific Transcriptional Activation is Essential for Growth Suppression by p53. Proc. Natl. Acad. Sci. USA 91, 1998-2002.
10. Eppenberger-Eberhardt M, Flamme I, Kurer V & Eppenberger HM (1990) Reexpression of α-Smooth Muscle Actin Isoform in Cultured Adult RAt Cardiomyocytes. Dev. Biol. 139, 269-278.
11. Yewdell JW, Gannon JV & Lane DP (1986) Monoclonal Antibody Analysis of p53 Expression in Normal and Transformed Cells. J. Virol. 59, 444-452.

12. Eppenberger ME, Hauser I, Baechi T, Schaub MC, Brunner UT, Dechesne CA & Eppenberger HM (1988) Immunocytochemical Analysis of the Regeneration of Myofibrils in Long-Term Cultures of Adult Cardiomyocytes of the Rat. Dev. Biol. 130, 1-15.
13. Koch-Schneidemann S, Gehr P, Rutishauser B & Eppenberger HM (1994) Attachment of Adult Rat Cardiomyocytes (ARC) on Laminin and Two Laminin Fragments. J. Struct. Biol. 113, 107-116.
14. Rotter V, Abutbul H & Ben-Ze'ev A (1983) P53 Transformation-Related Protein Accumulates in the Nucleus of Transformed Fibroblasts in Association with the Chromatin and is found in the Cytoplasm of Non-Transformed Fibroblasts. EMBO 2, 1041-1047.
15. Katsumoto T, Higaki K, Ohno K & Onodera K (1995) Cell-Cycle Dependent Biosynthesis and Localization of p53 Protein in Untransformed Human Cells. Biol. Cell. 84, 167-173.

Characterization of Extracellular Matrix in Three-Dimensional Cultures of DU 145 Human Prostate Carcinoma Cells

K. O'Connor[1], R. Enmon[1], A. Primavera[1], R. Dotson[2] and S. Clejan[2]
Tulane University, Chemical Engineering Department, New Orleans, LA 70118, USA[1] and Tulane University School of Medicine, Pathology Department, New Orleans, LA 70112, USA[2]

Abstract

Three-dimensional cultures of DU 145 human prostate carcinoma cells grown in the High Aspect Rotating-Wall Vessel (HARV), spinner flask and Transwell insert had dramatically different content of extracellular matrix proteins within cell aggregates. The HARV culture stained more intensely than the Transwell culture by at most a factor of 6 for collagen IV and 3 for laminin. The spinner-flask culture yielded results between these two extremes. Future research will determine if the accumulation of extracellular matrix in the HARV culture inhibits cancer-cell migration and aggressiveness.

Introduction

DU 145 human prostate carcinoma cells are a moderately metastatic and androgen-independent cell line [1]. In patients being treated for metastatic prostate cancer by androgen deprivation, there is frequently a transition in the cancer from androgen-dependent to androgen-independent growth [2]. After this transition occurs, survival is usually less than a year as so little is known about the growth regulatory pathways in androgen-independent prostate cancer [3].

We are in the process of developing DU 145 cell models of varying aggressiveness to provide insight into these pathways. It is our goal to influence aggressiveness mostly through the extent of three-dimensional growth. At present, we have three cell models: one grown in the High Aspect Rotating-Wall Vessel (HARV), another in a spinner flask and the third in a Transwell insert. The first two are suspension cultures of aggregated microcarrier beads; Transwell cultures are static and form a bi- to trilayer of cells.

In our characterization of these three cell models to date, we have found that the HARV culture behaved like a well differentiated solid tumor; Transwell culture, an aggressive invading tumor margin; and spinner-flask culture, an intermediate between these two extremes [4]. The HARV culture had the most three-dimensional growth, greatest fraction of G_1-phase cells, and strongest staining for select cytoskeletal proteins. Our current investigation extends this characterization to extracellular matrix proteins.

M.J.T. Carrondo et al. (eds.), Animal Cell Technology, 571–575.

Materials and Methods

DU 145 cells were propagated at 37 °C, 95% relative humidity and 5% CO_2 in GTSF-2 complete medium at pH 7.4. Experiments were performed in a HARV (Synthecon, Houston, TX) rotated at 15 rpm, Bellco spinner flask (Vineland, NJ) mixed at 60 rpm, and Transwell-Clear 3450 insert (Cambridge, MA) as previously described [4]. In the HARV and spinner flask, DU 145 cells were grown attached to Cytodex-3 beads (Sigma, St. Louis, MO) at an initial seeding of 5 viable cells/bead. A hemocytometer was used for counting single-cell suspensions of Transwell cultures treated with trypsin and HARV and spinner-flask cultures treated with DNase-free RNase (Boehringer Mannheim, Indianapolis, IN) and NONIDET P-40 (Sigma) [4]. For immunocytochemistry, DU 145 cells were fixed, paraffin blocked, sectioned and stained by standard procedures. Cell aggregates were assayed by incubation with mouse monoclonal IgG antibodies for the presence of human collagen IV and laminin (DAKO, Carpenteria, CA).

Results and Discussion

In our previous investigation, scanning electron microscopy revealed that the development of intertwined filopodia and microvilli-like structures on the surface of cell aggregates was more pronounced in the HARV than in the spinner flask [4]. In our current investigation, we examine this structural phenomenon on a more molecular level, characterizing the relative content of select extracellular matrix proteins in HARV, spinner-flask and Transwell cultures.

Experiments were conducted over 17 days to promote three-dimensional growth and matrix formation. By day 17, the Transwell culture formed a bi- to trilayer of cells on average (Figure 1A), while the HARV and spinner flask exhibited more three-dimensional growth by a factor of 2 to 3. While growth profiles for the two suspension cultures were identical, the extent of three-dimensional growth was greater in the HARV. As noted in our earlier work, the HARV contained larger bead aggregates than the spinner flask and, consequently, had a greater fraction of cells residing in the interior of aggregates [4].

Collagen IV and laminin accumulated in all three culture chambers over time (Figures 1B and C). Starting on day 11 for collagen IV and day 14 for laminin, more matrix protein had accumulated in the HARV culture. On day 17, the staining intensity for collagen IV was greatest in the HARV(by a factor of 3 to 6), intermediate for the spinner flask (by a factor of 1.5 to 3), and least for the Transwell. For laminin, the staining intensities for the spinner flask and Transwell more closely approximated each other. Nevertheless, the intensity score for the HARV culture was again the highest by the end of the experiment.

The trends in the growth and immunostaining data are consistent. The HARV culture had the greatest accumulation of matrix proteins and the most extensive three-dimensional growth. This accumulation would favor increased cell-matrix interactions. It is well known that three-dimensional growth mediates its effects on culture performance in part through such interactions.

The extracellular matrix synthesized by DU 145 is typical of other epithelial cells in its composition. It contains collagen IV and laminin as major components and forms the basal lamina which separates the epithelium and stroma in normal prostatic tissue [5]. During benign-to-malignant transformation, the basal lamina begins to degrade [6], becoming less organized and focally discontinuous [5]. Also, the staining intensity for collagen IV is reduced in highly aggressive prostatic tumors relative to well

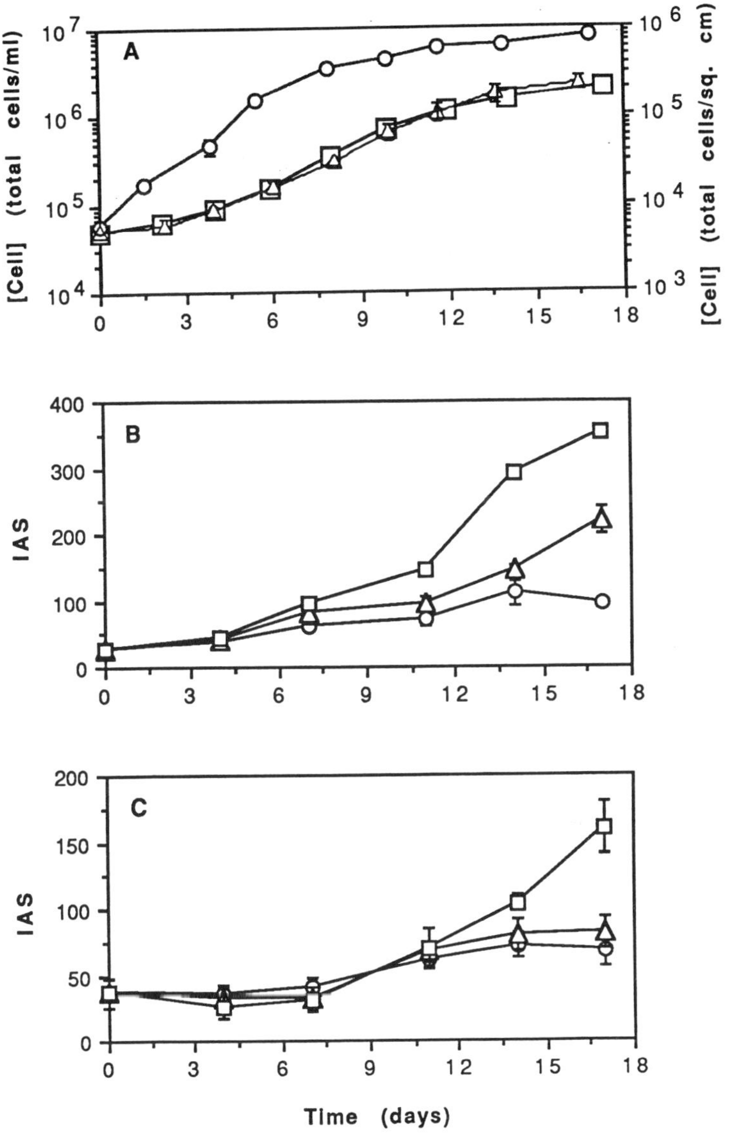

Figure 1. Growth profiles (A), collagen IV staining (B) and laminin staining (C) for DU 145 cells cultivated in the HARV (□), spinner flask (Δ) and Transwell (O). Cell concentrations are given on a per milliliter basis for suspension cultures and on a per square centimeter basis for the static Transwell culture. Staining intensity is reported on a 0 to 400 scale as an intensity-times-area score (IAS).

differentiated ones [7]. This deterioration may favor metastasis by enabling neoplastic cells to escape the epithelium and invade the stroma.

In our previous investigation, the HARV culture of DU 145 cells behaved as the least aggressive of the three cultures examined. The strong immunostaining for the HARV culture is consistent with this description.

Acknowledgments

This research was supported by grants from NASA (NAG9-826, KO), the Tulane Cancer Center (KO) and the Department of Defense (DoD 536198, SC). K. O'Connor, R. Enmon and S. Clejan are members of the Tulane Interdisciplinary Program in Molecular and Cellular Biology. Also, Professors O'Connor and Clejan are members of the Tulane Cancer Center.

References

1. Stone, KR, Mickey, DD, Wunderli, H, Mickey, GH & Paulson, DF (1978) Isolation of a Human Prostate Carcinoma Cell Line (DU 145). Int. J. Cancer 21, 274-281.
2. Isaacs, JT & Coffey, DS (1981) Adaption Versus Selection as the Mechanism Responsible for the Relapse of Prostatic Cancer to Androgen Ablation Therapy Studied in the Dunning R-3327H Adenocarcinoma. Cancer Res. 41, 5070-5075.
3. Gibbons, RP (1987) Prostate Cancer. Cancer 60, 586-588.
4. Clejan, S, O'Connor, KC, Cowger, NL, Cheles, MK, Haque, S & Primavera, AC (1996) Effects of Simulated Microgravity on DU 145 Human Prostate Carcinoma Cells. Biotechnol. Bioeng. 50, 587-597.
5. Nagle, RB, Hao, J, Knox, JD, Dalkin, BL, Clark, V & Cress AE (1995) Expression of Hemidesmosomal and Extracellular Matrix Proteins by Normal and Malignant Human Prostate Tissue. Am. J. Pathol. 146, 1498-1507.
6. Fuchs, ME, Brawer, MK, Rennels, MA & Nagle, RB (1989) The Relationship of Basement Membrane to Histologic Grade of Human Prostatic Carcinoma. Mod. Pathol. 2, 105-111.
7. Sinha, AA, Gleason, DF, DeLeon, OF, Wilson, MJ, Limas, C, Reddy, PK & Furcht, LT (1991) Localization of Type IV Collagen in the Basement Membranes of Human Prostate and Lymph Nodes by Immunoperoxidas and Immunoalkaline Phophatase. Prostate 18, 93-104.

Discussion

Aunins: You mentioned pH was the same. Has the HARV been characterised for oxygen levels?

O'Connor: Yes, it was on the slide containing growth conditions. The DO is measured with a microprobe as the medium comes out of the syringe. Feed was based on glucose and pH. This was as much as we could do with the design of these vessels.

Noe: What relevance have your findings for modified therapy?

O'Connor: Too early for a comment. It looks as if it has the potential to be a good cell model to study these growth regulatory pathways in a metastatic cell line because they are not well studied at all. We need to do further characterisation of the model to characterise metastatic potential.

Grammatikos: I thought your differential activity assessment a bit arbitrary perhaps. Are there definite differentiation markers?

O'Connor: The classical clinical marker is cytokeratin 8 and 18 and these were examined but the data is not presented here.

Grammatikos: In terms of the aggressiveness of the cells in the HARV system have you tried to inject them into mice to test this?

O'Connor: As I indicated, we are in the process of looking at metastatic potential. That particular assay you cannot do as the cells will adapt back.

SPHEROID FORMATION OF PRIMARY DOG HEPATOCYTES USING POLYURETHANE FOAM AND ITS APPLICATION TO HYBRID ARTIFICIAL LIVER

H. IJIMA[1], T. MATSUSHITA[1], N. NAKAZAWA[1], S. KOYAMA[1], T. GION[2], K. SHIRABE[2], M. SHIMADA[2], K. TAKENAKA[2], K. SUGIMACHI[2] and K. FUNATSU[1]

Department of Chemical Engineering [1] and Surgery [2], Kyushu University, 6-10-1 Hakozaki, Higashi-ku, Fukuoka 812-81, Japan

Abstract

Primary dog hepatocytes formed spheroids in the pores of polyurethane foam (PUF) which was used as a cell culture substratum. The dog/hepatocytes spheroid maintained liver specific activities better than in monolayer culture.

We designed a novel large artificial liver module, multi-capillary PUF packed-bed, for the dog experiment. Furthermore, we developed a hybrid type artificial liver support system utilizing this module. This system is effective for ammonia detoxification and maintenance of blood glucose concentration and ketone body ratio in a warm ischemic liver failure dog.

Introduction

There is a huge number of hepatic failure patients in the world. But, it is difficult to support liver specific functions because the liver has many kinds of metabolism in the body. Recently, many researchers have tried to develop an effective artificial liver module[1),2),3),4)], and some researchers applied their modules for human medical care as a bridge use of liver transplantation. We also developed a hybrid artificial liver module using rat hepatocytes.

We already indicated that primary rat hepatocytes formed multicellular aggregate (spheroid) in the pores of polyurethane foam (PUF) which was used as a cell culture substratum[5),6)]. Liver specific functions such as albumin secretion, urea synthesis and so on can be maintained for a long time in PUF/spheroid culture *in vitro*[6),7)]. So, we developed a multi-capillary (MC) PUF packed-bed type artificial liver module[8)] utilizing the PUF/spheroid culture system. From *in vitro* experimental results, the volume of the MC-PUF module required for saving the life of hepatic failure patients was calculated to be

M.J.T. Carrondo et al. (eds.), Animal Cell Technology, 577–583.

only 1500 cm^3.

In this paper, we indicate the results of *ex vivo* experiments using extracorporeal circulation of a hepatic failure animal.

Materials and methods

Experimental animal We used a D-galactosamine induced hepatic failure male wistar rat (7 weeks, about 200 g), and a warm ischemic liver failure dog (about 20 kg) which was made by porto-caval shunt.

Hepatocyte Primary rat and dog hepatocytes were prepared by collagenase perfusion method[9]. Cell viabilities were about 90 % evaluated by trypanblue dye exclusion method.

Module MC-PUF/spheroid packed-beds as an artificial liver module for the rat and dog experiments are indicated in Fig.1. PUF volumes of the rat and dog experiments were 4.5 cm^3 and 294 cm^3, and initial immobilized cell densities were 5 × 10^7 cells/module and 0.6~1.0 × 10^9 cells/module, respectively. Serum free medium[6),10)] based on Williams'E medium was used *in vitro* experiment.

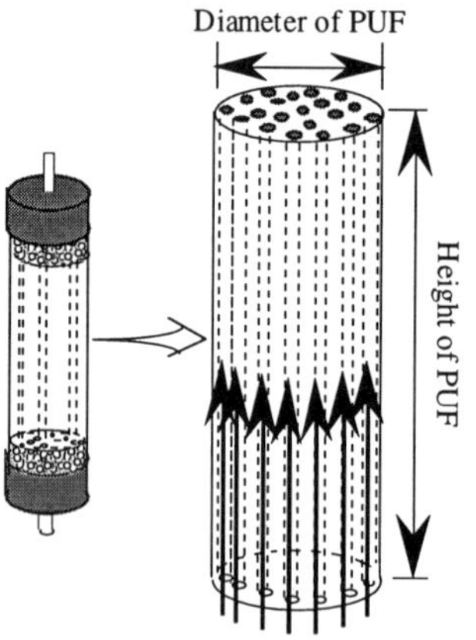

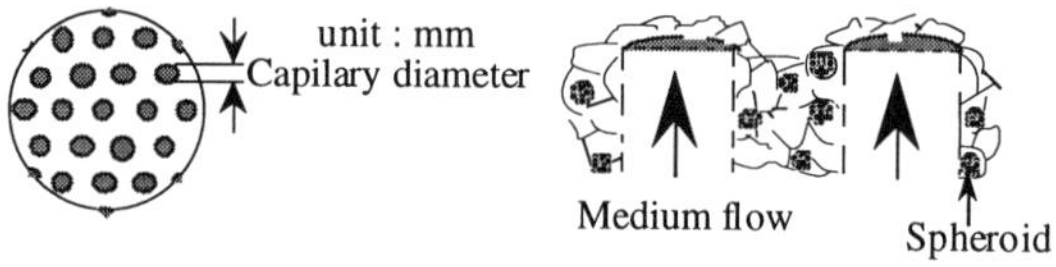

	Rat experiment	Dog experiment
Height of PUF	: 50 mm	: 170 mm
Diameter of PUF	: 12 mm	: 54 mm
Capillary diameter	: 1.5 mm	: 2.0 mm
Pitch of capillaries	: 3.0 mm	: 4.0 mm
PUF packing ratio	: 79 %	: 77 %
Capillary number	: 14	: 168
Volume of PUF	: 4.5 cm^3	: 294 cm^3

Fig.1 Details of Multi-Capillary PUF (MC-PUF) packed-bed for rat and dog experiments

Liver support Liver functions of hepatic failure animals were supported using extracorporeal circulation[11)] with artificial liver module. Blood circulation flow rates in the rat and dog experiments were 0.5 ml/min and 120 ml/min, respectively. And blood volume in these circulation lines in rat and dog experiments were 1.4 ml and 150 ml, respectively.

Results and discussions

Spheroid formation Primary dog hepatocytes formed spheroids the same as rat hepatocytes in the pores of PUF for 5~10 days culture as shown in Photo.1. The spheroid diameters were in the range of 100~150 μm. Dog hepatocytes/spheroid in the stationary culture maintained lidocaine metabolic activity (de-ethylation) for more than 1 week and the activity (max ; 0.6mg/10^6viable-cells/day) was higher than that of hepatocytes/monolayer.

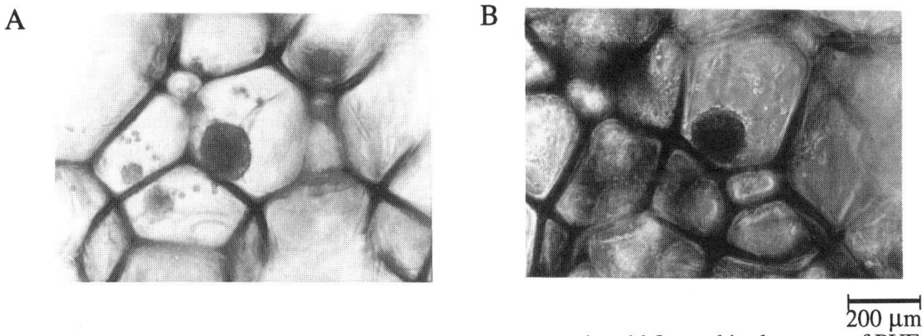

Photo. 1 Appearance of primary rat and dog hepatocytes spheroid formed in the pores of PUF
A : Rat hepatocytes , B : Dog hepatocytes

Rat extracorporeal circulation[12)] Fig.2 indicates the changes in blood ammonia concentration in the hepatic failure rat during and after extracorporeal circulation. Ammonia concentration in the blood was dramatically increased during three hours circulation in the control experiment and all rats died. On the contrary, The artificial liver support system could metabolize the ammonia and their concentrations in the blood (n=5) were maintained at a much lower level than in the control experiments, and these recovered to normal levels after about one day.

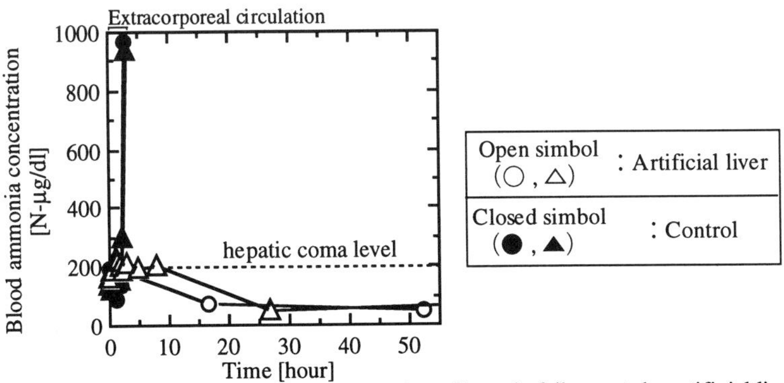

Fig. 2 Changes in blood ammonia concentration of hepatic failure rats by artificial liver

All results are summarized in Table 1. Four rats recovered from hepatic failure by application of our hybrid artificial liver support system in five experiments, but all rats died in three control experiments. So, it was indicated that our system was effective for recovery from hepatic failure in the rat experiment.

Table 1 Results of the extracorporeal circulation using D-galactosamine induced hepatic failure rats to estimate the performance of hybrid artificial liver (MC-PUF)

	Prothrombin time [sec] (normal)	Prothrombin time [sec] (preoperation)	Result
Control	20.5	25.2	Death
	18.8	31.2	Death
	19.1	36.7	Death
Artificial Liver	22.0	32.3	Recovery
	20.1	28.9	Death
	20.2	26.8	Recovery
	19.4	33.9	Recovery
	19.3	43.4	Recovery

MC-PUF module for dog Albumin secretion rate of dog hepatocytes in a larger MC-PUF module was 26 μg albumin/10^6 cells/day, which was almost the same value as in the PUF/spheroid stationary culture. Furthermore, dog hepatocyte/spheroids were also formed in the MC-PUF module. From these results, we could consider that scaling-up of the MC-PUF module was performed successfully.

Dog extracorporeal circulation Results of dog experiments are summarized in Table 2. In the case of control experiments, the blood ammonia concentration increased dramatically and the value reached 602 N-μg/dl after 6 hours of circulation. On the contrary, the blood ammonia was metabolized well by using our artificial liver support system and its concentration was 149 N-μg/dl after 6 hours of circulation.

Table 2 Blood ammonia and glucose concentrations of ischemic hepatic failure dogs treated by MC-PUF type artificial liver support system

Circulation time [hour]		0	6
Ammonia [N-μg/dl]	Control (n=3)	173±46	602±318
	Artificial Liver	74	149
Glucose [mg/dl]	Control (n=3)	221±39	19±23
	Artificial Liver	128	68

The blood glucose concentration in the control experiment decreased dramatically and the value was 19 mg/dl after 6 hours of circulation. On the contrary, it could be maintained by using our artificial liver support system and the concentration was 6 mg/dl after 6 hours of circulation. As to ketone body ratio, the same phenomenon was observed (data is not shown).

These results indicate that our hybrid type artificial liver support system is effective for treatment of hepatic failure in dogs as well as rats.

Conclusion

Primary dog hepatocytes could also form spheroids in the pores of PUF as well as rat hepatocytes, when we used PUF as a cell culture substratum.

We developed a hybrid type artificial liver support system utilizing PUF/spheroid culture of primary dog hepatocytes. This system was effective for ammonia detoxification and maintenance of blood glucose concentration and ketone body ratio in a warm ischemic liver failure dog.

References

1. Uchino J., Tsuburaya T., Kumagai F., Hase T., Hamada J., Komai T., Funatsu A., Hashimura E., Nakamura K., and Kon T. (199?) A hybrid bioartificial liver composed of multiplated hepatocyte monolayers. *Trans Am. Soc. Artif. Intern. Organs* **34**, 972-977.
2. Rozga J., Williams F., Ro M.S., Neuzil D.F., Giorgio T.D., Backfisch G., Moscioni A.D., Hakim R and Demetriou A.A. (1993) Development of a bioartificial liver: Properties and function of a hollow-fiber module inoculated with liver cells. *Hepatology* **17**, 258-265.
3. Sussman N.L., Gislason G.T., Conlin C.A., and Kelly H.H. (1994) The hepatix extracorporeal liver assist device ; Initial clinical experience. *Artificial Organs* **18**, 390-396.
4. Gerlach J.C., Encke J., Hole O., Muller C., Ryan C.J. and Neuhaus P. (1994) Bioreactor for a larger scale hepatocyte in vitro perfusion. *Transplantation* **58**, 984-988.
5. Matsushita T., Ketayama M., Kamihata K. and Funatsu K. (1990) Anchorage-dependent mammalian cell culture using polyurethane foam as a new substratum for cell attachment. *Appl. Microbiol. Biotechnol.* **33**, 287-290.
6. Matsushita T., Ijima H., Koide N. and Funatsu K. (1991) High albumin production by multicellular spheroid of adult rat hepatocytes formed in the pores of polyurethane foam. *Appl. Microbiol. Biotechnol.* **36**, 324-326.
7. Matsushita T. Ijima H. and Funatsu K. (1992) Development of a hybrid artificial liver utilizing three dimensional culture of adult hepatocytes. *Jpn. J. Artif. Organs* **21**,

1050-1054.
8. Ijima H., Matsushita T. and Funatsu K. (1994) Development of a hybrid artificial liver using multi capillary PUF/spheroid packed bed. *Jpn. J. Artif. Organs* **23**, 463-468.
9. Seglen P.O. (1976) Preparation of isolated rat liver cells. In: Prescott M.M. (ed.), *Methods Cell Biol.* **13**, 29-83, Academic Press, New York.
10. Lazar A., Mann H.J., Remmel R.P., Shatford R.A., Cerra F.B. and Hu W.S. (1995) Extended liver-specific functions of porcine hepatocyte spheroids entrapped in collagen gel. *In Vitro Cell Dev. Biol. Anim.* **31**, 340-346
11. Ijima H., Matsushita T., Taga M., Wada S., Hamazaki K., Koide N. and Funatsu K. (1994) Establishment of a rat extracorporeal circulation system to estimate the performance of a hybrid artificial liver. In: Kobayashi T., Kitagawa Y. and Okumura K. (eds.), *Animal Cell Technology: Basic & Applied Aspects*, **6**, 395-399.
12. Matsushita T., Ijima H., Wada S. and Funatsu K. (1995) Estimation of the performance of a PUF/spheroid packed-bed type artificial liver by using an extracorporeal circulation with hepatic failure rats. *Jpn. J. Artif. Organs* **24**, 815-820.

Discussion

Gerlach: I suppose the size of the cell aggregates can be regulated by the diameter of the pore of the capillaries? Do you know how many cells can be supported in an aggregate?

Ijima: In our case the spheroid diameter was regulated by the inoculum cell density, and was 100 microns. If the spheroids are bigger, we have to have a special oxygenation procedure.

The bilaminar membrane (BLM) bioreactor for hepatic tissue engineering with primary porcine hepatocytes

[1]Bader A., [1]Rocker D, [2]Wagner R, [1]Acikgöz A, [1]Sewing K.-Fr.
[1]Institut für Allgemeine Pharmakologie, Medizinische Hochschule Hannover, [2]GBF Braunschweig, Germany

Abstract

Techniques for whole pig liver cell isolation and consecutive sandwich culture have not been available. Autoaggregation of cells, as it frequently occurs with traumatized hepatocytes, must be prevented to obtain a monolayer with bilateral attachment. The development of a gentle large scale isolation technique is described. For the culture of primary porcine hepatocytes a bioreactor reconstructing normal in vivo matrix geometry and sinusoidal analogues is presented. A 4 step isolation protocol was designed, which resulted in complete digestion of a pig liver. Viability was 88-94 %. These cells were maintained as cell plates within two layers of hydrated collagen in the bioreactor. The bioreactor is based on the use of bilaminar membranes (BLM), which are separated by a spacer. These structures contain oxygen supply channels. Upscaling is achieved by stacking of individual modules.

Introduction

The aim of the development of the bilaminar membrane (BLM) bioreactor has been to create a device, which allows the mass culture of hepatocytes under minimal spatial conditions and in a state which comes as close as possible to an "in vivo" state of the liver. In vivo hepatocytes (the so called parenchymal cells) are arranged as monolayer cell plates. Hepatocytes have bilateral attachment with two opposing sinusoidal cell surfaces to extracellular matrix, the so called space of Disse, which is rich in collagen. Non parenchymal cells such as endothelial cells of the liver are positioned on the outside of the space of Disse facing the vascular rooms (sinusoids). This means that each hepatocyte attaches to two opposing layers of collagen. On both sides of such hepatocellular plates vascular rooms are present. A possible reason for the failure of other devices to maintain hepatocytes in an in vivo like state may be deficiencies in the microenvironment provided for these cells in the respective devices (1). For primary hepatocytes standard approaches are apparently not sufficient to maintain the cells in an in vivo like state (2). This is proven by functional differences which vary by a factor of 10-1000 if different approaches are compared (1). Also matrix geometry has been shown to be important to normalize hepatocellular morphology in vitro (3; 5). In addition techniques for whole pig liver cell isolation and consecutive sandwich culture in a BLM bioreactor have not been available. Autoaggregation of cells, as it frequently occurs with traumatized hepatocytes, must be prevented to obtain a monolayer with bilateral attachment. The study describes the development of a gentle

M.J.T. Carrondo et al. (eds.), Animal Cell Technology, 585–587.

large scale isolation technique and the consecutive culture of these cells in a BLM bioreactor.

Large scale porcine hepatocyte isolation

A 4 step gravity driven isolation protocol was designed, which resulted in complete digestion of the pig liver. Viability was 88-94 %. These cells were maintained as cell plates within two layers of hydrated collagen in the BLM bioreactor.

Cell organization and matrix geometry in the bilaminar membrane bioreactor

The BLM - bioreactor is based on a highly specific organization of the cell culture. It describes neither a conventional monolayer culture configuration with one sided cellular attachment to extracellular matrix of the cells nor does it correspond to a gel entrapment technique. For realization of the above cited morphological criteria the following design was chosen: A sinusoidal analogue is reconstructed on both sides of a plate of primary porcine hepatocytes being attached to extracellular matrix via both sinusoidal surfaces. Non parenchymal can be located on top of the matrix layers enclosing the hepatocytes (4). It is important to realize that these hepatocytes are located as coherent two dimensional layers within two distinct and opposing layers of collagen. The supportive bilaminar membrane plate has two opposite surfaces with membranes and contains oxygen supply structures. Plates are arranged in a modular mode for upscaling.

Conclusion: Adequate isolation techniques will advance the use and availability of high quality primary porcine hepatocytes. The bilaminar membrane (BLM) bioreactor reconstructs miniature sinusoidal analogues and respects matrix geometry of the in vivo liver. Up-scaling is facilitated by the modular design of the device.

References:

1. Bader A, Böker K, Knop E, Frühauf N, Oldhafer K, Ringe B, Pichlmayr R & Sewing K-F (1995) Reconstruction of liver tissue *in vitro* - Geometry of characteristic flat bed -, hollow fiber- , and spouted bed bioreactors with reference to the in vivo liver. Artificial Organs 19, 941-950.

2. Bader A, Knop E, Böker K, Frühauf N, Schüttler W, Oldhafer K, Pichlmayr R & Sewing K-F (1995) A novel bioreactor design for in vitro reconstruction of in vivo liver characteristics. Artif Organs 19, 368-374.

3. Bader A, Knop E, Böker KHW, Crome O, Frühauf N, Gonschior A, Christians U, Esselman H, Winkler M, Pichlmayr R & Sewing KF (1996) Tacrolims (FK 506) metabolism in primary rat hepatocytes depends on extracellular matrix geometry. NS Arch Pharm 353, 1-13.

4. Bader A, Knop E, Kern A, Böker K, Frühauf N, Crome O, Kempka G, Esselmann H & Sewing K-F 3-D coculture of hepatic sinusoidal cells with matrix overlaid parenchymal liver cells -design of an organotypical model. Exp Cell Research in press.

5. Knop E, Bader A, Böker K, Pichlmayr R & Sewing KF (1995) Normal ultrastructure of primary hepatocytes depends on bipolar attachment to extracellular matrix. Anatomical records 242, 337-349.

DIFFERENTIATION OF BONE MARROW CELLS IMMOBILIZED ON MICROCARRIERS IN A FLUIDIZED BED REACTOR

B. Schröder[1], C. Herfurth[1], P. Meißner[1], M. Biselli[1], H. Link[2], C. Wandrey[1]
1) Institut für Biotechnologie, Forschungszentrum Jülich GmbH, 52425 Jülich
2) Medizinische Hochschule Hannover, Abt. Hämatologie/Onkologie, 30623 Hannover

ABSTRACT
The cultivation of hematopoietic cells was demonstrated in a fluidized bed reactor. The hematopoietic cells were immobilized on microcarrier, cocultivated with bone marrow stromal cells. The progenitor cells can be cultivated for more than 2 weeks and can be removed from the carrier without destroying their ability to form colonies.

KEYWORDS: BONE MARROW, CD34, BIOREACTOR, MICROCARRIER

INTRODUCTION
The *ex vivo* expansion of hematopoietic cells is of great interest for a variety of applications[1,2]. To mimic the *in vivo* hematopoietic system maintenance and differentiation of hematopoietic progenitor cells must be provided by an *ex vivo* system.
A functional layer of adherent stromal cells permits production of extracellular matrix, cytokines and cell-to-cell contacts between stromal and progenitor cells[3]. Culture systems such as perfusion chambers, hollow-fibre reactors, and airlift reactors have been investigated[4,5,6].

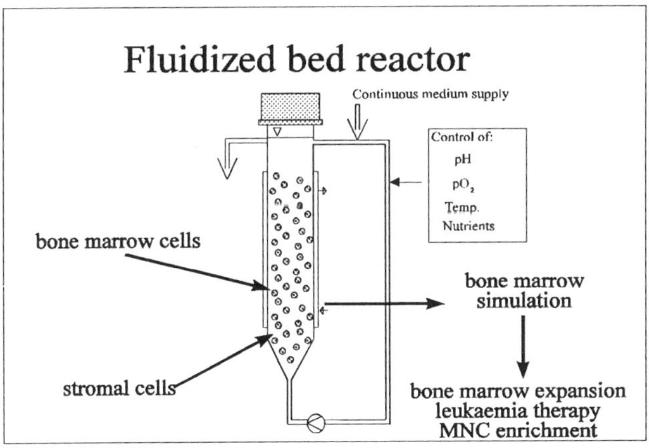

Fig. 1: Design of carrier mediated cocultivation

In our approach fluidized bed technology is used for the cocultivation of bone marrow cells or enriched CD34 cells with stromal cells immobilized in open porous microcarriers.

M.J.T. Carrondo et al. (eds.), Animal Cell Technology, 589–593.

MATERIALS AND METHODS

Bone marrow cells were obtained from hematologically normal patients. Mononuclear cells were selected after ficoll gradient.

Stromal fibroblast cells have been established by cultivation of bone marrow for 2-3 weeks in Long-Term-Cultivation-Medium until stromal layers reached confluence.

Cells were collected by trypsin from the carrier for methylcellulose assays (15 min, 37°C).

Cell culture medium: McCoy-Medium, 10% FCS, 10% HS, 1% $NaHCO_3$, 1% sodium pyruvat, 0,4% MEM nonessential amino acid solution, 0,8% MEM essential amino acid solution, 1% vitamin solution, 10^{-4} M α-thioglycerol, 10^{-6} M hydrocortisone

For **immunoflourescence staining** for FACS analysis cells were labelled for 30 min at 4°C with FITC- or PE-conjugated antibodies and then washed twice with PBS.

To remove the adherent cells from the carriers, they were washed twice with PBS and were incubated with dispase / collagenase solution for 5 min (37°C). The proteases were stopped by EDTA.

Carrier: Gelatine modified porous glass carriers (size 400-700 μm) were used[7]. Such carriers have successfully been used for cultivation of permanent stromal cell lines, Hybridoma, CHO, BHK..., where cell densities above 10^8 cells /ml carrier were observed.

Fluidized bed reactor (Fig. 2): A reactor with total volume of 50-180 ml containing 15-50 ml carrier (settled volume) was used. The reactor consisted of a thermostated reactor tube with a conical bottom inlet, leading to a homogeneous fluidization of the carriers. The reactor was inserted into a recycle loop containing the recirculation pump, DO probes and internal or external oxygenator for bubble free aeration by silicon tubes.

Fig. 2: Fluidized bed reactor

RESULTS

Maintenance and differentiation of hematopoietic progenitor cells in long term bone marrow culture critically depends on the presence of a functional layer of stromal cells. For the cocultivation a stromal layer was established on the carrier before inoculation of bone marrow mononuclear cells. Samples of carrier were taken daily. Fig. 3 shows the cell density per ml carrier plotted versus time.

50% of the inoculated stromal cells could be immobilized on the carrier reaching a cell density of $6*10^4$ cells/ml carrier. One day later $1.5*10^8$ mononuclear bone marrow cells were added. 19% of these cells could be immobilized on the stromal feeder layer.

Fig. 4 and 5 show SEM-pictures of one carrier.

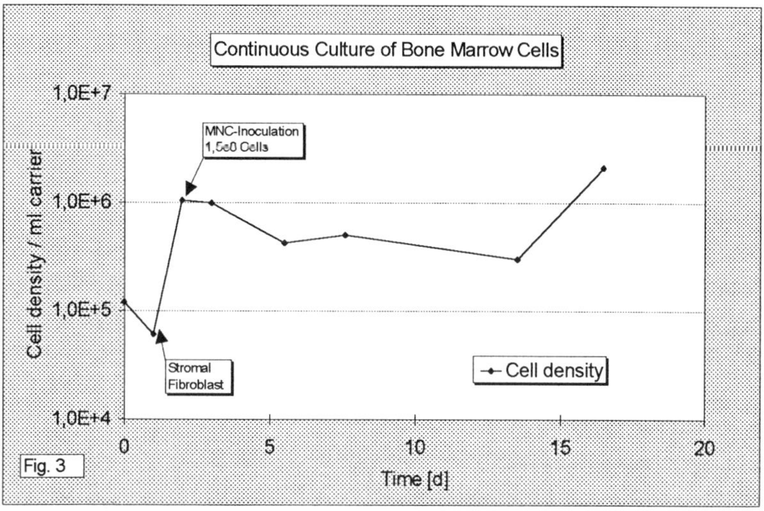

Fig. 3

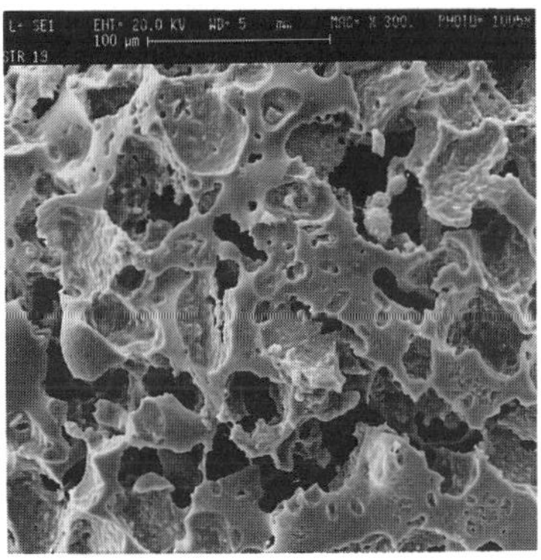

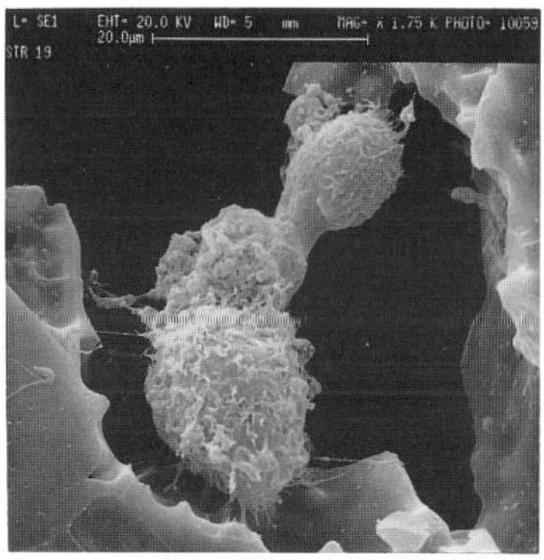

Fig. 4 Stromal feeder-layer with mononuclear cells

Fig. 5 Detail of Fig. 4:

Every day cells were collected by trypsinization for methylcellulose assays and by dispase/ collagenase treatment for FACS-analysis. Figure 6 and 7 point out the results of the different cell typs (measured by FACS), which could be found on the microcarrier.

There is a high affinity of erythroid cells (Gly A) and a good affinity of leucocytes (CD45). The percentage of CD14 (monocytes), CD15 (granulocytes), CD34 (progenitor cells), and CD45 positiv cells increased in the first days. The culture conditions did not support the growth of lymphocytes.

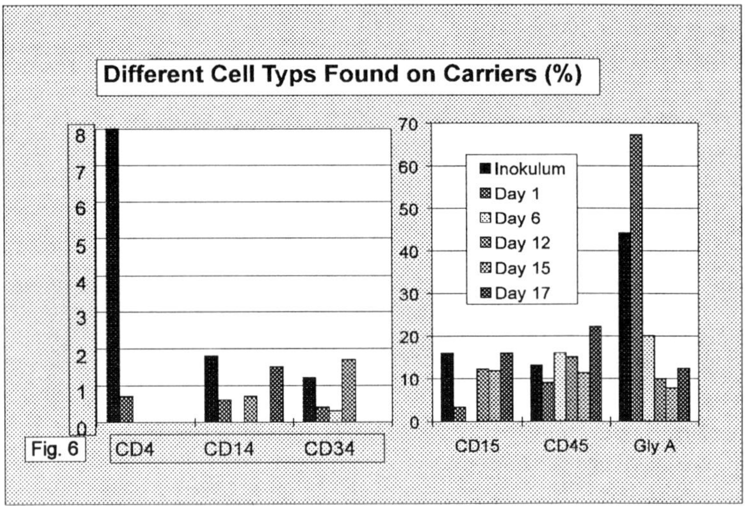

Fig. 6

Additionally Fig. 7 shows the density of the different cell typs on the carriers.
19% of the cells could be immobilized on the carriers. As a result different cell typs adhere with different percentages:

CD4	CD14	CD15	CD34	CD38	CD45	GlyA
1,8%	6,4%	4%	6,4%	15%	13%	29%

The total number of CD34 cells was stable during the cultivation time. The cell density of CD15 and CD45 cells increased due to the maturing of the progenitor cells.

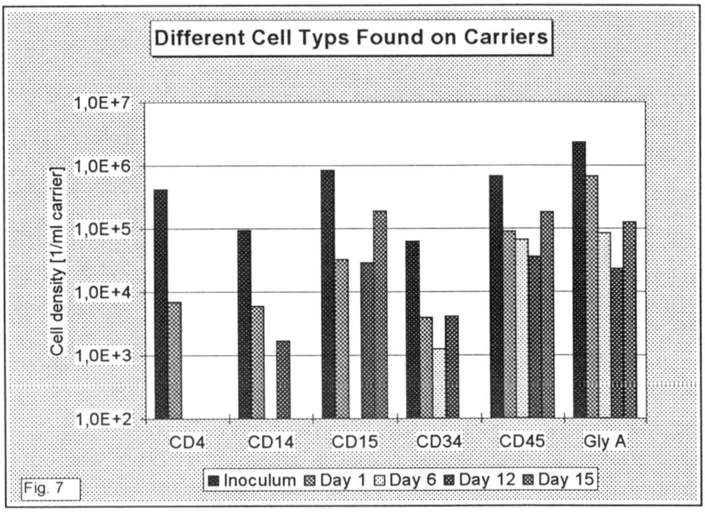

Fig. 7

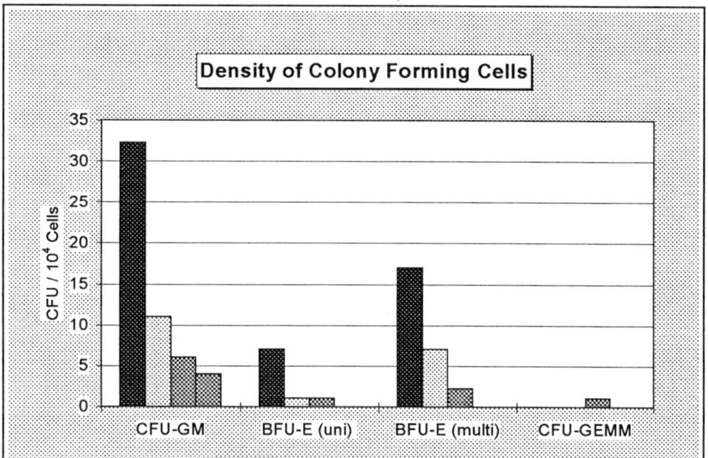

Fig. 8

The number of colony forming cells which were found on the carrier are presented in Fig. 8. For example from 10^4 cells released from the carrier 32 cells were able to form CFU-GM at day 1. Thus we could show that the removal of cells from the carrier does not destroy their ability to form colonies. The different cell lineages were cultivated without preference. Colony forming cells could be identified even after 12 days.

We conclude that the cocultivation of bone marrow cells could be used to immobilize mononuclear cells without loss of vitality. During the cultivation the mature cells lost the adherence and were washed out into the medium. Therefore only immature cells were bound to the carrier. The expansion of stem cells on the carrier has to be examined in future experiments.

The combination of glass carriers and the fluidized bed reactor has proved to be suitable for *ex vivo* cultivation of bone marrow mononuclear cells.

SUMMARY

- Cocultivation of bone marrow mononuclear cells and stromal cells is possible.
- Progenitor cells (CD34) could be detected by FACS on the carrier over a long time.
- Colony forming cells could be identified even after 12 days.
- Cells could be removed from carriers without destroying their ability to form colonies.

References

1. Edgington SM: New horizons for stem cell bioreactors. Biotechnology 10: 1099, 1992
2. Koller MR, Palson BO: Tissue engineering: Reconstitution of human hematopoiesis ex vivo. Biotech/Bioeng. 42: 909-930, 1993
3. Thalmeier K et al. Establichment of two permanent human bone marrow stromal cell lines with long-term post irradiation feeder capacity. Blood 83: 1799-1807, 1994
4. Koller MR et al. Bioreactor expansion of human bone marow: Comparison of unprocessed, density-separated and CD34-Enriched cells. J. Hematol 4: 159-169, 1995
5. Sardonini CA, Wu Ying-Jye. Expansion and differentiation of human hematopoietic cells from static cultures through small scale bioreaktors. Biotech. Prog. 9: 131-137, 1993
6. Wu Tzuu-Yi et al. A novel three-dimensional long term bone marrow bioreactor culture system. Animal cell technology 6, 115-120, 1994
7. Lüllau E et al. Immobilzation of animal cells on chemically modified siran carrier, Ed.: Spier RE, Griffiths JB, MacDonald C, Butterworth-Heinemann, Oxford pp469-479, 1992

THE BIOCHIP- A NEW CELL FACTORY FOR THE CELL CULTIVATION IN DEFINED THREE-DIMENSIONAL LAYERS

T. Seewöster, S. Wilmsmann, A. Werner, J. Lehmann

Institute of Cell Culture Technology, Faculty of Technical Sciences,
University of Bielefeld, POB 10 01 31, D-33501 Bielefeld, Germany

Introduction

Mass cell cultivation of mammalian cells still is under formative influence of industrial applications for the production of economic relevant proteins. Due to newly developed genetically engineered cell lines and upscaling of continuous processes during the last decade, the bioreactor volumes could be reduced. Nevertheless, today mass cell cultivation of mammalian cells enters new fields of innovations as in the shadow of gene therapy direct medical applications of animal cells become reality. Especially for the replacement of complex organ functions by tissue engineering techniques [1-10], the need for large amounts of defined and active cells is tremendous. For example, to ensure a sufficient metabolic capacity by a treatment with primary hepatocytes at acute liver failures, approximately 30 - 40% of native liver cell mass had to be used. These are 4.5 - 6.0 x 10^{10} active cells in total. A standard stirred tank suspension bioreactor with an optimal cell density of 3.0 x 10^6 cells/ml would have a dimension of 20 L. As only small blood or plasma volume is available for a treatment with a bioartificial liver, a large active surface had to be provided by the bioartificial devices to ensure efficient exchange between cells and plasma, respectively blood. These two parameters, a high amount of cells combined with a small liquid volume, indicate the development of high cell density systems, which in addition allows optimal nutrient supply by perfusion techniques.

A suitable model that reflects to all these conditions is the three-dimensional structure of glands that show high metabolic activity. Such tissues have a laminar structure infiltrated with a brought network of vascular capillaries in which each tissue unit has a cell layer thickness of 30 - 90 µm to ensure an optimal flux of metabolites through the tissue.

Here we report the current state of investigations towards the development of a new culture system based on the native three-dimensional microarchitecture of gland tissues. The main aims are a reproducible free adjustable three-dimensional microarchitecture in the range of 30 - 90 µm layer thickness, which allows free nutrient flow through the layer. As it is

M.J.T. Carrondo et al. (eds.), Animal Cell Technology, 595–600.

expected that in a few years immortalised cells or recombinant cell lines will replace primary cells, a recombinant model cell line should be used, as they show no inhibition at cell-cell contacts. Besides the development of a high density culture system itself, the problems regarding a reproducible standardised long-term provision of a viable cell culture should be solved.

Material and Methods

Cell Line

A recombinant CHO cell line expressing the humanised anti-lysozyme antibody D1.3 was used. The cell line was already adapted to serum-free conditions, and the optimised culture medium based on a 1:1 mix of DMEM/F12 allows densities in standard batch cultures up to 3.5×10^6 viable cells/ml. Detailed description of medium composition, cell counting and analysis of metabolites are given elsewhere [11].

Cultivation

To ensure a high viable cell culture in the high density system a two-step cultivation process was developed. The first step consisted of a continuous suspension culture in chemostat mode for continuous cell growth with a stable cell metabolism. This simple, reliable and reproducible cultivation process allows a formation of steady-state conditions controlled only by the medium flow rate through the system (= dilution rate). A 1.5 L bioreactor system was equipped with a tumbling stirrer carrying a hydrophobic polypropylene membrane for bubble-free aeration. Air saturation was adjusted at 40%, pH at 7.2, temperature at 37 °C, and stirrer at 40 rpm. Medium flow was controlled by scales. In a second step the cells were packed into the high cell density system to achieve tissue cell densities. This system should prevent the cells from further growth by physical limitations of space. For cell maintenance the cell unit was enclosed in a loop system consisting of a 1 L conditioning tank equipped with pH, pO_2 and temperature sensors and perfused continuously in cross-flow mode with flow rates up to 900 ml/min.

Biochip System

To copy the laminar microarchitecture of organs, the new cell factory should enclose a supporting structure to obtain a defined layer thickness over a wide area. To prevent further cell growth by space limitations and to ensure a sufficient nutrient flow around the cells, microfiltration membranes should be used, acting only as a kind of fence for the cells. Both components had to be fixed together, but also consider the inoculation of the system. Besides the mentioned demands all components of the system should be commercially available, cheap, sterilizable, and biocompatible.

Results and Discussions

Figure 1 shows a cross-section of the high cell density unit that consists of three different components. The cell grid, a platinum coated metal foil, similar to shaving blades, defines the cell layer thickness. It can be manufactured exactly by galvanic techniques in 0.01 μm steps. Because of the platinum layer, the metal is biocompatible in a wide range of different oxygen-containing or salt-containing liquids. To form separated cell chambers or metabolic units within the cell grid, two microfiltration membranes were placed on both sides of the metal.

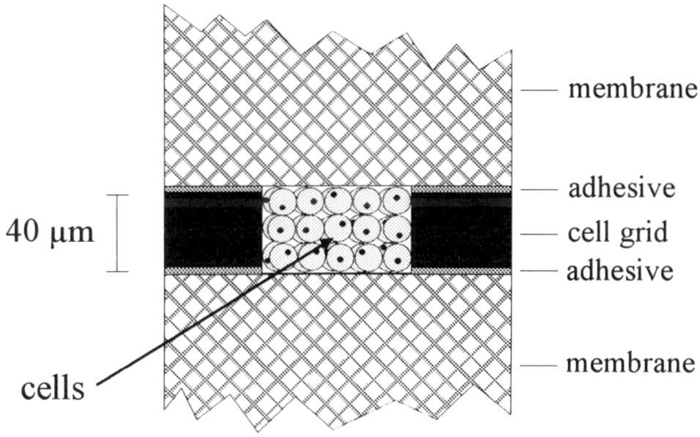

Figure 1: Schematic cross-section of biochip chamber. Original prototype consists of 7000 chambers, each 1 mm in diameter x 40 μm.

Although preliminary screening of different commercially available membranes (3 celluloseacetate, 2 cellulose, and 2 polyethersulfone membranes in a pore size range of 0.2 - 5.0 μm) showed that all membranes are biocompatible and provide a sufficient trans-membrane flux, the celluloseacetate membrane (0.45 μm pore size) was favoured, because of no cell adsorption and a homogeneous surface. Membrane and metal were stuck together with a biocompatible pressure-sensitive adhesive, commonly used at plasters. Because of the pressure-sensitive characteristics, the adhesive can be used several times. All three components were put together, fixed in a modified membrane holder, normally used for medium filtration, and were then steam-sterilised. For transferring the cells from chemostat to biochip system the membrane holder was opened, the top membrane was removed and a defined amount of cells was given onto the cell grid. Medium was then slowly removed through the bottom membrane so that the cells totally were sucked into the cell grid. The cell grid was then closed again with the top membrane and the filter holder containing the cells was continuously perfused with medium on both sides in cross-flow mode.

For a continuous provision of a high viable cell culture, the chemostat parameters had to be adjusted on the CHO cell line. During several long-term cultivations the dilution rate was varied to obtain a high viability within the steady-state. A sufficient dilution rate was found at 0.55 d^{-1} (Fig. 2) keeping a viability of 98%. Higher dilution rates near the maximum growth rate of 0.8 d^{-1} led to instability of the culture.

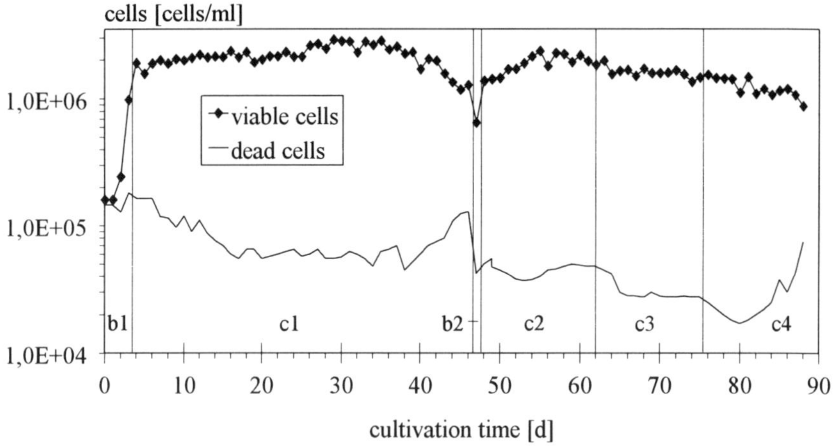

Figure 2: Continuous cell provision with chemostat system. b1, b2: batches; c1: chemostat D = 0.37 d^{-1}; c2: D = 0.37 d^{-1}; c3: D = 0.55 d^{-1}; c4: D = 0.71 d^{-1}.

The used biochip had a diameter of 140 mm with an area of 5700 mm^2, respectively 228 μl available for cell inoculation. Preliminary experiments in centrifuges showed that the bulk volume for this cell line is 2.4 μl x 10^{-6} cells, whereby all cells are then still alive. The calculated maximum cell amount for one layer of the biochip will be 9.5 x 10^7 cells. Figure 3 shows the cultivation over four days in the biochip system. For monitoring the metabolism, glucose, lactate, ammonia and amino acids were measured once a day. The cell concentration only was measured before inoculation of the biochip and afterwards by resuspending all cells of the biochip. The increasing lactate concentration and the decreasing glucose concentration show that the cells are still viable and active. Amino acid concentrations changed also during the cultivation (data not shown), but viability of the culture decreased from 98% to 56%. Nevertheless, the cells survived these first runs of the new high cell density system. A few problems occurred during the transfer into the biochip and during the cultivation, but these are not insoluble. First, the inoculation of the biochip system had to be more reproducible and reliable, especially for upscaling of the system. The adhesive had to be stronger to keep the membranes fixed on the cell grid over a long period of peristaltic pressure, and last the membrane holder had to be optimised to have smallest dead volume and to ensure an optimal flow.

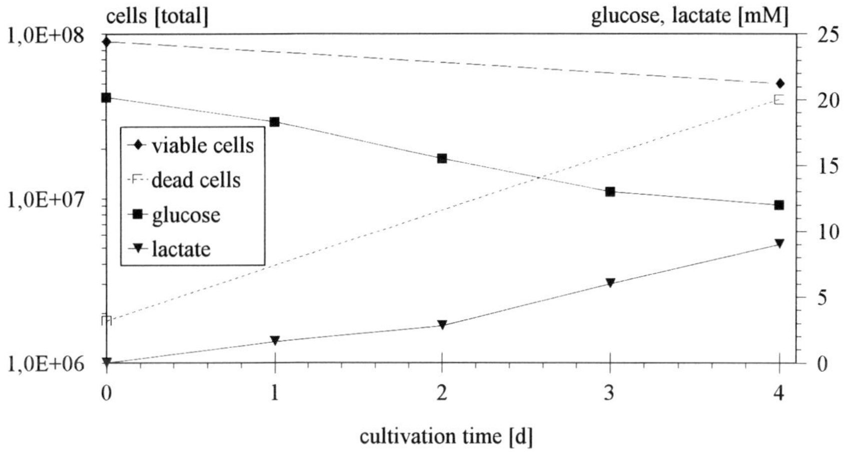

Figure 3: Cultivation in biochip system. Cells were counted before and after the run.

Conclusions

The new high cell density system, called biochip, has gone successfully through the first tests. It is possible to move cells from a continuous suspension culture system where the cells stay at low densities and show continuous growth into a system where they stay at tissue-like cell densities and are restricted from further growth by space limitations. Due to a sufficient nutrient supply through that bioartificial structure, a long-term maintenance of cell metabolism seems to be possible.

Acknowledgement

This work is supported by the European Union (BIO2-CT94-3069) and is done in cooperation with the Sartorius AG, Germany. We would like to thank Mr. H. Brinkmann for excellent manufacturing of the biochip prototypes and the additional equipment.

References

1. Bader A, Knop E, Fruehauf N, Crome O, Boeker K, Christians U, Oldhafer K, Ringe B, Pichlmayr R & Sewing KF (1995)
Reconstruction of liver tissue in vitro: Geometry of characteristic flat bed, hollow fiber, and spouted bed bioreactor with reference to the in vivo liver. Artificial Organs 19, 941-950.
2. Colton CK (1995)
Implantable biohybrid artificial organs. Cell Transplantation 4, 415-436.
3. Cauwels JM (1986)
The body shop: Bionic revolutions in medicine, Mosby, St. Louis.
4. Galletti PM (1991)
Organ replacement: A deram come true. In: Johnsson-Hegyeli R & Marmont du Haut Champ AM, Discovering new worlds in medicine, pp 262-277, Farmitalia Carlo Erba, Milan
5. Galletti PM (1992)
Bioartificial Organs. Artificial Organs 16, 55-60.
6. Gerlach J, Klöppel K, Schauwecker HH, Trauber R, Müller C & Bücherl ES (1989)
Use of hepatocytes in adhesion and suspension cultures of liver support bioreactors. Int. J. Artificial Organs 12, 788-792.
7. Li AP, Barker G, Beck D, Colburn S, Monsell R & Pellegrin C (1993)
Culturing of primary hepatocytes as entrapped aggregates in a packed bed bioreactor: A potential bioartificial liver. In Vitro Cell. Dev. Biol. 29A, 249-254.
8. Nyberg SL, Shatford RA, Peshwa MV, White JG, Cerra FB & Hu WS (1993)
Evaluation of a hepatocyte-entrapment hollow fiber bioreactor: A potential bioartificial liver Biotechnol. Bioeng. 41, 194-203.
9. Palsson BO, Paek SH, Schwartz RM, Palsson M, Lee GM, Silver S & Emerson SG (1993)
Expansion of human bone marrow progenitor cells in a high cell density continuous perfusion system. Bio/Technology 11, 368-372.
10. Rozga J, Holzman MD, Ro MS, Griffin DW, Neuzil DF, Giorgio T, Moscioni AD & Demetriou AA (1993)
Development of a hybrid bioartificial liver. Ann. Surg. 217, 502-511.
11. Seewöster T & Lehmann J (1995)
Influence of targeted asparagine starvation on extra- and intracellular amino acid pools of cultivated Chinese hamster ovary cells. Appl. Microbiol. Biotechnol. 44, 344-350.

SESSION VII

Cell and Physiology Engineering

ANIMAL CELLS AND SERUM AS NATURAL SOURCES OF CHEMOKINES

J. VAN DAMME, Laboratory of Molecular Immunology, Rega Instituut for Medical Research, University of Leuven, Belgium

Chemokines comprise a novel family of chemotactic cytokines that selectively attract and activate leukocytes. In general, α-chemokines (e.g. interleukin-8) predominantly activate neutrophil granulocytes, whereas β-chemokines (e.g. MCP-1) are rather chemotactic for monocytes and lymphocytes. All chemokines act through structurally related receptors belonging to the family of seven transmembrane G protein-linked receptors. In conjunction with adhesion molecules (selectins, integrins) chemokines are the key mediators for leukocyte migration to sites of tissue damage. Chemokines are also reported to function as stimulators or inhibitors of angiogenesis and haematopoiesis. Recently, some chemokines have been identified as HIV-suppressive factors. Over the past years we have isolated and identified a number of novel animal chemokines (human, mouse, bovine) from ex vivo leukocytes, from in vitro cultured normal and tumor cells or from serum. These include the α-chemokines IL-8, NAP-2 and GCP-2 and the β-chemokines MCP-1, MCP-2 and MCP-3. The cellular source is different for each of these chemokines in that some (NAP-2) are solely derived from platelets, whereas others (IL-8) can be induced in many cell types by a variety of proinflammatory stimuli. For the characterization of chemokines, in vivo migration tests and in vitro chemotaxis and activation (e.g. enzyme release) assays have been used. It can be concluded that each chemokine has a separate set of producer and responder cells.

M.J.T. Carrondo et al. (eds.), Animal Cell Technology, 603–605.

Discussion

Spier: You said some of these cytokines are involved in the suppression of HIV.

Van Damme: Yes, in binding of HIV.

Spier: No, in the suppression of HIV.

Van Damme: We know that people have isolated molecules that induce HIV suppression. These are called HIV suppressing factors and have been isolated from T-cells. The structures are identical to some chemokines. Recently the secondary receptor for HIV 1 has been cloned and, big surprise, the receptor had 37% homology with IL-8. The second receptor in addition to CD4 for HIV is maybe a chemokine receptor and this explains why they have a so-called HIV suppression effect. However, this is speculation at the moment.

Bidawid: In which chromosomes do you think you will discover the next 50 cytokines? Also, it was not clear to me what the average rate of productivity per litre is.

Van Damme: The 2 sub-groups of chemokines Alpha and Beta, are located on different chromosomes. The alpha group (IL-8) is on human chromosome 4. The Beta group are clustered on chromosome 17. Assuming 100% recovery productivity is in the high microgram range eg 10-100 mg.

Bernard: Why did you not consider going to recombinant sources? We get tens of milligrams per litre from recombinant cells. Is there anything wrong with the activity of the recombinant chemokines?

Van Damme: My paper was a shift from biotechnology to basic research. The only reason for our work is to discover chemokines. We have, in fact, stimulated *in vitro* cells to discover them, rather than the company approach of molecular cloning. Weissmann was the first to sequence IL-8 by cloning but they cannot tell anything about the biology. It was just an induced sequence. Our studies were to identify IL-8 purely from the biological activity.

Schmid: Could you comment on the nature of the processing enzymes involved in the N-terminal truncation and the modulation or activation of activity?

Van Damme: This is a phenomenon which is not fully understood. One thing for sure is that it is not artificial as a result of purification. Different cell types, same purification, different truncated forms. The generation of protease's is important and there is a whole bunch of them. Trypsin, chemotrypsin, catapsin G, plasmin all play a role in cleaving the N.terminus So it is a mixture of protease's. Part of the molecule is loosely attached so it is easily accessed by protease's.

De novo formation in mice of adipose tissue and muscle fibers at the site of injecting Matrigel and growth factors

Y. Kitagawa†, N. Kawaguchi†, K. Toriyama‡, S.Torii‡

†*Nagoya University BioScience Center, Chikusa, Nagoya 464-01,*
‡*Department of Plastic and Reconstructive Surgery, School of Medicine, Nagoya University, Showa, Nagoya 466, Japan*

Augmentation of lost soft-tissue is a major goal of the plastic surgery but autografting of fat pad has been only practical operation. This operation in human was first tried by Neuber (1) in 1893, in which multiple small grafts were used to fill a soft-tissue depression. Lexer reported in 1909 the autogenous fat graft as large as 12x12 cm from the abdomen to treat a depression of the malar infraorbital area and to build out a receding chin (2). Many other uses of free autogenous fat grafts in the treatment of depressed regions or scars of the zygomatic arch, temporalis muscle, malar area, forehead, parotid area, orbital regions, and hemifacial and facial atrophies had been successful. To maximize donor-fat survival, many improvement of the technique have been tried such as local treatment of the implantation bed with insulin, mincing the grafts into small pieces, and atraumatic antiseptic technique. However, many researchers became dissatisfied with progressive absorption of the graft. Neuhof in 1923 (3) suggested that autogenous adipocytes of the graft would die and be totally replaced by either newly formed host adipocytes or host fibrous tissue. When free fat autograft used to treat a brain and skull defect was removed and microscopically examined at 10 days, necrotic adipose tissue was observed in some areas but regenerative proliferation of original fat cells were observed in other areas (4).

Since the behavior of free fat transplantation is deeply linked to adipocyte precursors and their differentiation into mature adipose cells, investigation of the preadipocyte gave a basic understanding of fat tissue as a transplantable material. In 1948, Wertheimer and Shapiro reported that fat develops from very primitive adipose cells much like that of fibroblasts of connective tissue (5). The study of such preadipocyte was accelerated by the development of adipose tissue culture technique. Smith in 1971 first described fibroblast-like cells grown in tissue cultures of human adipose tissue (6). Van et al. reported cytological and enzymatic characterizations of adult human adipocyte precursors (7). Green and Kehinde in 1979 established an preadipocyte cell line that was susceptible to adipose conversion (8). When they injected subcutaneously in the lower chest and upper abdomen of nude mouse, a soft fat pad of mature adipose cells was formed after 6 weeks. This is the first example demonstrating that a cell line of adipocyte precursors differentiated *in vivo* and they opened the door of transplanting cultured cells for augmentation of soft-tissue.

In this project, we first tried to reproduce the result of Green and Kehind. Figure 1 shows the fat pad formed in a nude mouse 5 weeks after injecting 10^7 3T3-F442A cells. Because of the lack of significant endogenous adipose tissue, we injected the cells subcutaneously over the sternum. When the skin was reflected from abdomen,

M.J.T. Carrondo et al. (eds.), Animal Cell Technology, 607–612.

well marked fat pad with abundant capillaries was observed at the site of injection. Probably due to movement of the needle during the injection, two centers of fat pad were formed. This fat pad began to be visible from outside the body 1 week after the injection and increased in size as the days passed. Figure 2 shows the histological section of the pad. A typical adipose tissue composed of well differentiated adipocytes developed and the size of exogenous adipocytes was larger than endogenous adipocytes observed under the epidermis. 3T3-F442A cells can differentiate *in vitro* into adipocyte-like cells with intracellular lipid separated into small droplets. In contrast, the cells injected in the mice formed typical adipocytes with single lipid droplet filling whole space of a cell. In addition, newly former adipose tissue was well delivered by abundant capillaries. When the number of F442A cells to be injected was reduced to 1/5, 1/25 or 1/125, implantation site showed no sign of adipose tissue formation. Other adipogenic cell line, 3T3-L1 or C3H 10T1/2 cells were also injected. These cell lines can convert into adipocyte *in vitro* with much lower frequency than 3T3-F442A. All mice injected with 10^7 3T3-L1 cells or 10T1/2 cells gave rise to no visible fat pads.

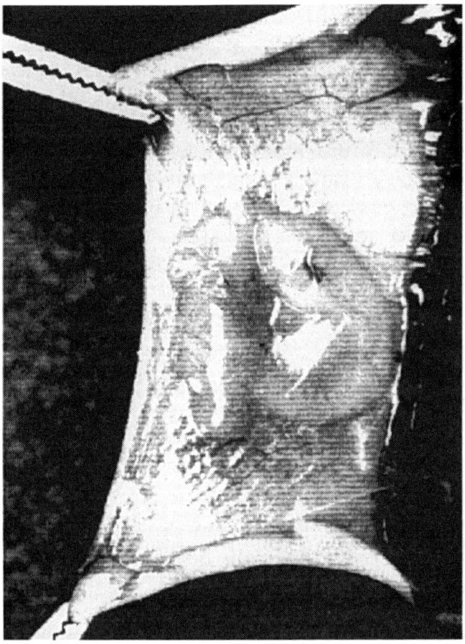

FIGURE 1 Fat pad formed in nude mouse 5 weeks after subcutaneous injection of 3T3-F442A preadipocytes

We have shown that the synthesis of basement membrane components is enhanced during adipose conversion of 3T3-L1 cells *in vitro* (12). Electron microscopic observation have revealed that adipocyte is surrounded by basement membranes in fat tissue (15-17). Since these observations suggest that the components of basement membrane are deeply related to the adipose tissue formation, we tried to co-inject F442A cells with reconstituted basement membrane, Matrigel. For this, Matrigel was prepared

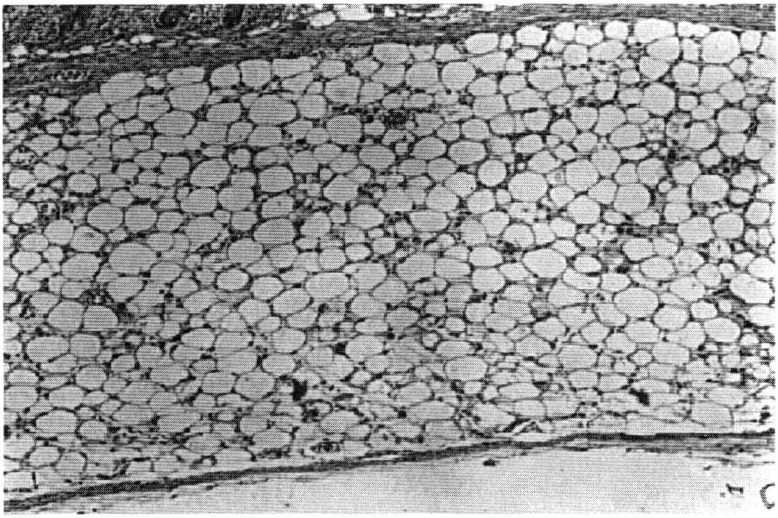

FIGURE 2 Histological section of adipose tissue fromed by injection of 3T3-F442A preadipocytes

from the transformed mouse EHS tumor as described by Kleinman (18). The tumor was excised from the mice, washed with high salt to remove blood components and then extracted with 2 M urea. Even if urea was dialyzed out of this Matrigel extract, it remained as liquid as far as the temperature kept below 4°C. However, the components started to

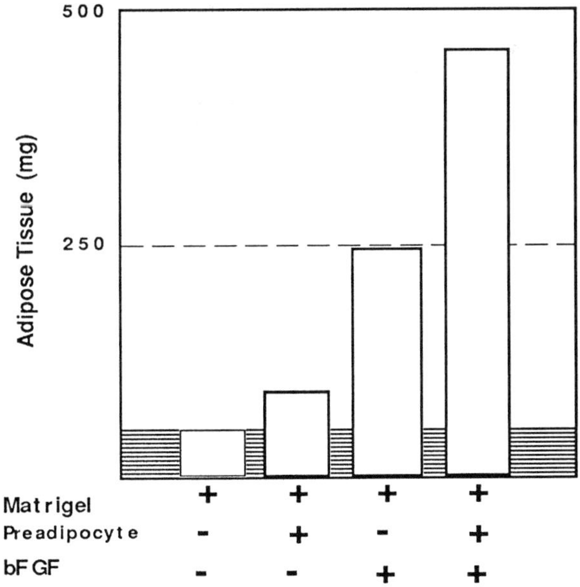

FIGURE 3 Weight of fat pad fromed 5 weeks after injecting Matrigel in combination with 3T3-F442A cells and bFGF

polymerize and reconstituted the network of basement membrane within 30 min after raising the temperature to 37°C by injecting the liquid into mice. When F442A cells were injected as suspension in liquid Matrigel, the weigh of newly formed pad increased two fold of the injection of Matrigel alone (Fig.3). Addition of bFGF to this cell suspension at final concentration of 1μg/ml markedly increased the weight of newly formed fat pads (Fig.3). This suggested that bFGF (directly or indirectly) stimulated proliferation and differentiation of F442A. To our surprise, injection of Matrigel in combination with bFGF alone (without F442A cells) could also give rise to adipose tissue formation. Figure 4 shows a histological section of adipose tissue formed by injection of Matrigel and bFGF. Although the size of cells was smaller than endogenous adipocyte, the differentiation of well developed adipocytes was observed within reconstituted basement membrane network formed by injected Matrigel. This result demonstrates that Matrigel in combination with bFGF created new local condition in connective tissues and the endogenous precursor of adipocytes can migrate, proliferate and differentiation to form adipose tissue *de novo*. In experiment not shown, we confirmed that injection of bFGF alone did not show this effect, indicating that the reconstituted basement membrane can retain and stabilize bFGF and stimulate the precursor cells.

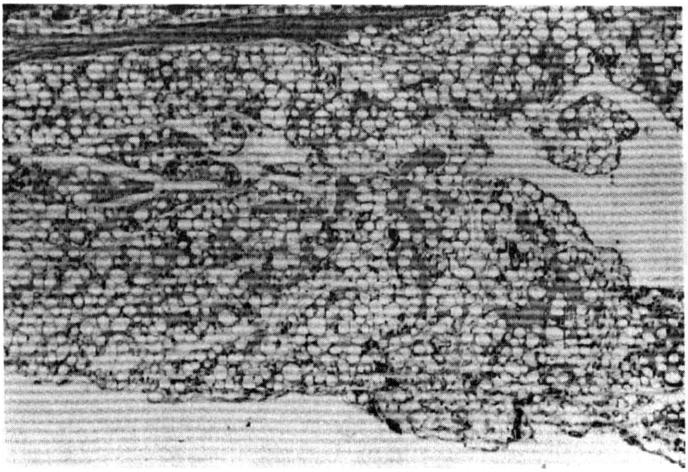

FIGURE 4 Histological section of adipose tissue formed by injecting Matrigel and bFGF

We tested the injection of matrigel in combination with other growth factors such as PDGF, growth hormone, insulin and IGF-1. Among these growth factors, bFGF formed the largest pad. To our surprise, however, growth hormone, insulin and IGF-1 induced muscle fiber formation in addition to adipose tissue.

References

1. Neuber G.*Verhandl. d. deutsch. Gesellsch. f. Chir.(Berl.)* **22**, 159, (1910)
2. Lexer, E. *Dtsch. Med. Wochenschr.* **36**, 640 (1910)
3. Neuhof, H.*The transplantation of Tissue*, p.73, New York: Appleton (1923)

4. Marchand,F. *Beitr. z. Path. Anal. u.z. Allg. Path*, **66**, 32, (1920)
5. Wertheimer, E. and Shapiro, B. *Physiol. Rev.*, **28**, 451, (1948)
6. Smith,V. *Anat. Rec.* **169,** 97 (1971)
7. Van, R. L. R., Bayliss, C. E. and Roncari, D. A. K. *J. Clin. Invest.* **58,** 699 (1976)
8. Green. H. and Kehinde, O. *J. Cell Physiol.*, **101,** 169 (1979)
9. Morikawa, M., Nixon,T. and Green, H. *Cell*, **29**, 783 (1982)
10. Goldberg, B. *Cell*, **11**, 169 (1977)
11. Aratani, Y. and Kitagawa, Y. *FEBS Lett.*, **235**, 129 (1988)
12. Aratani, Y.,and Kitagawa, Y. *J. Biol. Chem.*, **263,** 16163 (1988)
13. Ono, M., Aratani, Y., Kitagawa, I.and Kitagawa, Y. *Exp. Cell Res.*, **187**, 309 (1990)
14. Aratani, Y., Sugimoto, E. and Kitagawa, Y. *FEBS Lett.* **218**, 47 (1987)
15. Chase, W. H. *J. Ultrastruct. Res.*, **2**, 283 (1959)
16. Napolitano, L. *J. Cell Biol.*, **18**, 663 (1963)
17. Cushman, S.W. *J. Cell Biol.*, **46**, 326 (1970)
18. Kleinman, H.K., McGarvay, M.L., Hassell, J.R. et al. *Biochemistry,* **25**, 312 (1986)

Discussion

Spier: If you change the composition of the matrigel do you get a different kind of differentiation?

Kitagawa: We tried to identify the effective component in the matrigel. It contains laminin, collagen etc so we tried to separate the constituents and inject each fraction separately. Unfortunately separation destroyed this function. If we change the components it is difficult to get this result.

Overexpression of the apoptosis suppressor gene, *bcl-2*, in hybridoma cells prolongs viability and increases productivity in batch culture

Nicholas H. Simpson and Mohamed Al-Rubeai
Centre for Biochemical Engineering, University of Birmingham, Edgbaston, Birmingham, UK. B15 2TT

Keywords: Apoptosis, metabolic engineering, BCL-2

Abstract

Apoptosis is a cell autonomous death process that is the most prevalent form of cell deletion in hybridoma cultures. Apoptosis can be induced by a variety of stimuli found in bioreactors that include nutrient depravation, hydrodynamic stress and oxygen excess or depletion. Bcl-2, in common with other proto-oncogenes, can antagonise the apoptotic pathway. We have cloned a *bcl-2* gene into a hybridoma cell line in order to examine the effect on cell survival in culture conditions. We have demonstrated that overexpression of bcl-2 significantly prolongs cell viability by inhibiting, rather than eliminating, apoptosis and also increases monoclonal antibody productivity of hybridoma cells in: batch culture, conditions of hypoxia and hyperoxia and nutrient deprivation. Bcl-2 transfected cells also displayed near instantaneous adaptation to agitated culture conditions and enhanced viability in serum-free medium. Under conditions of growth inhibition the bcl-2 transfectants retained high viability and expressed enhanced productivity.

Introduction

The classic morphology of apoptotic cell death, i.e. the condensation of nuclear chromatin, in hybridoma cells grown in bioreactors was first demonstrated by Al-Rubeai et al 1990. Further studies in this laboratory compared the incidence of apoptosis in hybridoma cultures with other important cell lines, along with the environmental factors which induce apoptosis (Al-Rubeai et al 1994, 1995). In addition to nutrient limitation, hydrodynamic forces and cytotoxic products, one of the main factors that influence culture yield is oxygen. Culture intensification strategies, such as cell immobilisation and *in vitro* culture perfusion, lead to oxygen gradients and zones of anoxia. Oxygen is not only essential for cell growth but also for the maintenance of cell viability. On the other hand, excess levels of oxygen may enhance cell damage through the emergence of high levels of reactive oxygen species (ROS) and induce cell death. In a spin-filter perfusion system, supra-optimal levels of dissolved oxygen

M.J.T. Carrondo et al. (eds.), Animal Cell Technology, 613–620.

has been found to inhibit hybridoma cell growth, but cells could recover from this stress after reduction of the dissolved oxygen level to 50% (Emery et al 1995).

1. Identified at the t(14;18) break-point found in human follicular lymphoma
2. The protein is localised at the mitochondrial membrane, endoplasmic reticulum and nuclear envelope
3. Bcl-2 onco-protein suppresses apoptosis induced by a variety of cell types in response to a variety of signals
4. Its action is not universal
5. Member of an extended family of proteins

Figure 1. Summary of the *bcl-2* gene (reviewed by Cory 1995).

The most comprehensively studied anti-apoptotic gene is *bcl-2* (Fig 1). Recent models propose (Fig. 2) that the apoptotic programme begins with the activation of a CED-3/ICE family of proteases which cleaves poly(ADP-ribose) polymerase (PARP), an enzyme involved in the coordination of genome structure and integrity in stressed cells. Loss of the PARP function may render the Ca^{2+}/Mg^{2+} dependent endonucleases highly activated. Anti-apoptotic genes such as *bcl-2* may affect the rate at which any one cell begins apoptosis through their involvement in an upstream regulatory mechanisms.

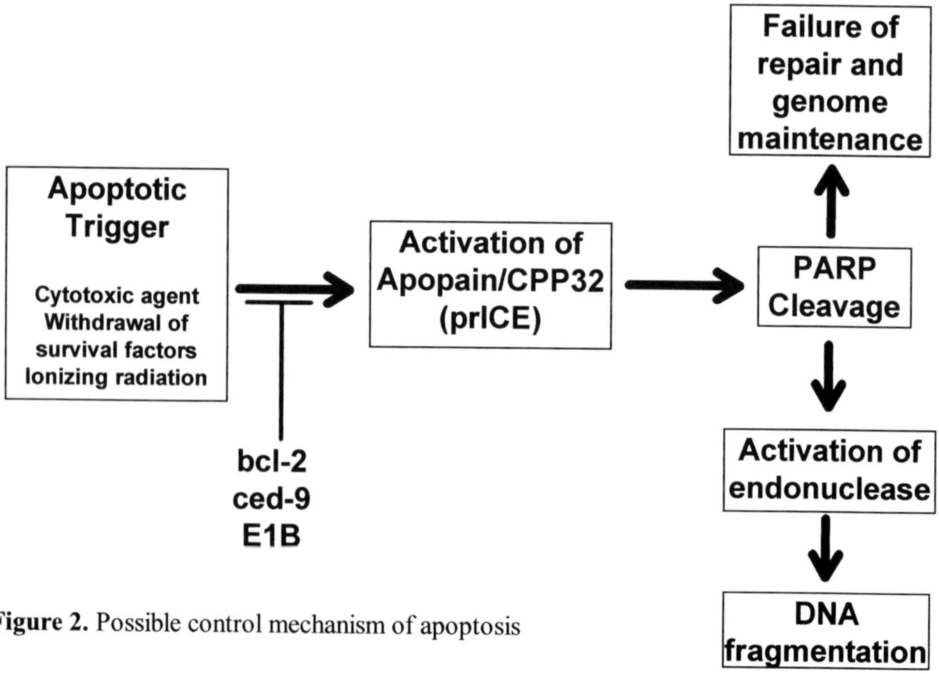

Figure 2. Possible control mechanism of apoptosis

Materials and Methods

Cell Line and Culture Conditions. The mammalian cell line used throughout this work was TB/C3, an NS1-derived murine hybridoma, which produces immunoglobulin G (IgG) monoclonal antibody (MAb) specific to the hapten Cγ2 domain in the Fc region of human IgG. Cells were cultured at 37^0C in RPMI 1640 basal medium (Gibco, Paisley, UK) supplemented with 5% (v/v) foetal calf serum (APP, Birmingham, UK). Cultures were grown in 25 cm^2 T-flasks (Nunc, Denmark) and 50 mL or 250 mL Duran-bottles (Schott, Germany) stirred by magnetic followers. Cell density and viability were determined by haemacytometer count and trypan blue exclusion. Cultures were also cultured in hypoxic conditions (90% nitrogen/10% carbon dioxide) and hyperoxic conditions (95% oxygen/ 5% carbon dioxide).

Transfection with Human bcl-2 Gene. The bcl-2 (pEF bcl2-MClneopA) and control (pEF-MClneopA) plasmids were kindly donated by Christopher Gregory, Department of Immunology, Birmingham University, UK, with the permission of Suzanne Cory, Royal Melbourne Hospital, Australia. Transfections were performed by standard electroporation methods. Selection was carried out in the presence of G418 cytotoxin.

Propidium Iodide/Acridine Orange Staining and Fluorescence Microscopy. Cell suspensions were mixed with an equal volume of staining solution containing 10 μg mL^{-1} acridine orange and 10 μg mL^{-1} propidium iodide. Cells were classified based on morphology outlined in figure 3.

Viable	**Excludes PI. Diffuse chromatin. Membrane intact.**
Apoptotic	**Excludes PI. Condensed chromatin. Membrane intact.**
Secondary necrotic	**Includes PI and retains apoptotic morphology. Condensed chromatin. Membrane damaged.**
Necrotic	**Includes PI. Non- condensed chromatin. Membrane damaged.**

Figure 3. Summary of cell morphology after staining with propidium iodide and acridine orange.

Results and Discussion

Two cell lines were isolated after the transfection protocol:

TB/C3 pEF control cell line with blank vector
TB/C3 bcl2 cell line which overexpresses the Bcl-2 protein

Interestingly, the *bcl-2* gene was found to be auto-selective in semi-continuous cultures at very low dilution rates (results not shown). In these conditions, cells without the *bcl-2* gene were deleted through apoptosis due to the depletion of nutrients and accumulation of toxic metabolites in the culture medium.

	Viable cell index (Integral of viable cell density with respect to time) (cell hours mL^{-1})	Monoclonal antibody titre (ug mL^{-1})
TB/C3 pEF	5.01 x 10^7	52.78
TB/C3 bcl2	12.91 x 10^7	73.03

Table 1. Viable cell index and monoclonal antibody titre for TB/C3 pEF and TB/C3 bcl2 cells in batch culture.

	Viable (%)	Apoptotic (%)	Secondary necrotic (%)	Necrotic (%)
TB/C3 pEF	9	9	72	10
TB/C3 bcl2	22	1	20	57

Table 2. The proportions of cells based on morphology after fluorescent staining at the end of batch cultures.

In batch culture the viability of the TB/C3 bcl2 cells was significantly prolonged compared with that of TB/C3 pEF cells (Table 1). As a consequence of the prolonged viability the production period was also increased and the mono-clonal antibody titre was increased by 38%. The predominant mechanism of death for the TB/C3 pEF cells was apoptosis (Table 2) compared with necrosis for the TB/C3 bcl2 cells. The apoptotic pathway had been significantly inhibited by the *bcl-2* gene which resulted in the slower decline of the TB/C3 bcl2 cultures due to attritional death, i.e. necrosis, rather than a cell autonomous death.

	Hypoxic conditions	Hyperoxic conditions
TB/C3 pEF	32 hours	20 hours
TB/C3 bcl2	48 hours	32 hours

Table 3. Length of viable cell period for TB/C3 pEF and TB/C3 bcl2 cells in conditions of hypoxia and hyperoxia.

	Viable (%)	Apoptotic (%)	Secondary necrotic (%)	Necrotic (%)
TB/C3 pEF	10	21	57	12
TB/C3 bcl2	74	3	1	22

Table 4. The proportions of cells based on morphology after fluorescent staining at the end of batch cultures under hypoxic conditions.

	Viable (%)	Apoptotic (%)	Secondary necrotic (%)	Necrotic (%)
TB/C3 pEF	9	12	65	14
TB/C3 bcl2	4	7	9	80

Table 5. The proportions of cells based on morphology after fluorescent staining at the end of batch cultures under hyperoxic conditions.

In conditions of either oxygen limitation or excess the culture duration was significantly increased for the TB/C3 bcl2 cells compared with the TB/C3 pEF cells (Table 3). In both cases the bcl-2 gene suppressed apoptosis (Tables 4 and 5). Excess oxygen (at a dissolved level of approximately 500% compared with the partial of oxygen in air) was far more toxic to both cell lines compared with oxygen deprivation. However, in intensive culture strategies, where oxygen demand would exceed supply, it should be possible to increase the oxygen levels delivered to cells transfected with anti-apoptotic genes without eliciting an adverse apoptotic response.

Conclusions

The conclusions are outlined as follows:

1. Bcl-2 suppresses apoptosis and prolongs hybridoma batch culture duration
2. Bcl-2 by enhancing survival reduces the need for frequent subculturing
3. The *bcl-2* gene is autoselective
4. Bcl-2 increases MAb productivity
5. Bcl-2 enhances cell survival during growth inhibition and increases specific productivity
6. Bcl-2 significantly prolongs viability in hypoxic and hyperoxic conditions

References

1. Al-Rubeai, M., Mills, D., Emery, A. N. 1990. Electron microscopy of hybridoma cells with special regard to mono-clonal antibody production. Cytotechnology **4**: 13-28.
2. Al-Rubeai, M., Singh, R. P., Goldman, M. H., Emery, A. N. 1995. Death mechanisms of animal cells in conditions of intensive agitation. Biotech. Bioeng. **45**: 463-472.
3. Emery, A. N., Jan, D. C. H., Al-Rubeai, M. 1995. Oxygenation of intensive cell culture systems. App. Microbiol. Biotech. **43** 1028-1033.
4. Singh, R. P., Al-Rubeai, M., Gregory, C. D., Emery, A. N. 1994. Cell death in bioreactors: a role for apoptosis. Biotechnol. Bioeng. **44**: 720-726.
5. Cory, S. 1995. Regulation of lymphocyte survival by the bcl-2 gene family. Annu. Rev. Immunol. **13**: 513-543.

Acknowledgements

We thank Anne Milner, Department of Immunology, University of Birmingham, for help and advice in transfecting the cell lines.

Discussion

Spier: You may be comparing a heterogeneous population of high and low producers in what is basically a cloned population which happens to have a high production.

Simpson: What I did not mention was that there was no end-point dilution or selection of clones as such. The cells were just electroporated in the presence of the vector and selected on drug and we just grew them up.

Spier: The point I am making is that the natural variability within a population may encompass the kind of specific effect you have by adding the BCL-2 gene. You have only a 30% increase, not a 10-fold or something significant like that.

Simpson: Transfections have been done 3 times with a consistent result.

Spier: That does not answer the question.

Simpson: I cannot answer further.

Spier: You need to subclone your parent population to find out what the variability is, then you can say you have a difference or not.

Sinacore: I was wondering if you looked to see in your BCL-2 cells whether the cells that expressed high levels of BCL-2 were capable of achieving higher cell densities and thus increasing volumetric productivity?

Simpson: We are trying that now but we have not got an increased cell density yet.

Shirahata: Does BCL-2 work by suppressing or eliminating apoptosis?

Simpson: Suppresses, we still see some apoptosis.

Shirahata: I wondered if the cells produce the same quality of antibody?

Simpson: Product heterogeneity is being worked on. It is possible that the quality as measured by ELISA has changed and that this is responsible for the increase.

Noe: Have you any evidence or data for the effects of BCL-2 on fed batch cultures of hybridomas?

Simpson: No. I guess that semi-continuous is not that far removed from fed batch so it will probably work quite well.

Bergemann: Can this phenomenon be generalised to other cell lines, hybridomas and other productive lines like CHO cells?

Simpson: CHO probably not as one does not see much apoptosis on that cell line. Generally, NSO derived cell lines will, insect cells will not. As a general rule look for apoptotic cell death before using BC-2.

Merten: If you transfect MDCK cells with BCL-2 you can produce more virus but the glycosylation is different and the virus is unable to infect other cells. In your case you may have a modification of the glycosylation. It is very important that you look at this.

Simpson: Thank you, we are.

DEREGULATED EXPRESSION OF CLONED TRANSCRIPTION FACTOR E2F-1 IN CHINESE HAMSTER OVARY CELLS SHIFTS PROTEIN PATTERNS AND ACTIVATES GROWTH IN PROTEIN-FREE MEDIUM

Kelvin H. Lee, Adriana Guerini-Sburlati, Wolfgang A. Renner and James E. Bailey, Institute of Biotechnology ETH-Zürich, CH-8093, Zürich, Switzerland

Engineering of the cell cycle can be an effective means for bypassing growth factor requirements of animal cells. Cloned human E2F-1 from Nalm 6 cells was subcloned into pRc/CMV and transfected into Chinese hamster ovary (CHO) cells. Ten stable transfectant clones isolated from cells cultured under neomycin resistance selection pressure all expressed significantly higher amounts of E2F-1 than control cells as determined by Western analysis. Confocal immunofluorescent microscopy and Southern analysis of several clones also provided evidence for the expression of cloned E2F-1 in these cells. CHO K1:E2F-1 cells are able to proliferate on well-defined serum- and protein-free basal medium and exhibit an S-phase extended by 65% compared to CHO K1 cells mitogenically stimulated by basic fibroblast growth factor (bFGF). Two-dimensional electrophoresis of the intracellular proteins of E2F-1 clones shows an increase in 236 gene products compared to CHO K1 control cells, further verifying a functional regulatory role of cloned E2F-1 in CHO cells. Among these upregulated species is the cell-cycle regulatory protein cyclin A, which has already been shown to be regulated by E2F-1 in human fibroblasts. Overexpression of cloned E2F-1 in CHO cells is a potentially useful new strategy for bypassing serum requirements in mammalian cell culture. Furthermore, such cell-cycle control stimulus-protein pattern response data can contribute to a clearer understanding of complex multigene networks involved in mammalian cell cycle regulation.

M.J.T. Carrondo et al. (eds.), Animal Cell Technology, 621–623.

Discussion

Miller: Your E2F1 is an initiation factor in translation, I presume?

Bailey: It is a transcription factor. It activates transcription.

Miller: I am surprised to see the different morphology of cells stimulated with cyclin E and by the transcription factor. Are these cells stimulated with cyclin E and do they become malignant as they are so rounded compared to normal CHO K1 cells? Do they by-pass contact inhibition for instance? You have a clear difference between these cells treated differently. Do cyclin E treated cells show malignant characteristics compared to E2F cells?

Bailey: The characteristics are different. Cyclin E cells move almost immediately into suspension. If you add serum they will adhere to the surface and grow in the normal way for surface attached cells. This is a convenient way to manipulate them. In protein-free medium the phenotypes are completely different. In view of the difference in levels of 300 proteins we cannot assign these phenotypic differences only to these 2 different proteins.

Morris: Since a lot of the DNA proof reading and quality control occurs at the borders of S phase, do you think altering them as you have could potentially create cells that are more likely to have a higher mutagenesis rate? Is this a problem?

Bailey: It is hard to say in view of the information we have now about how many protein levels are changed. In externally stimulated proliferating cell cultures we have elevated levels of these factors and we make the transition from G1 to S, so there is nothing fundamentally unnatural about making this transition. Whether forcing it this way with the concomitant side effects could hare an influence at this level is not known.

Ryll: A question in the 2-D gel technique. How do you normalise, and did you validate by ELISA based assays or sequencing of spots to ensure it reflects the reality?

Bailey: Spot identification was done by antibody labelling. This is routinely applied now in 2-D electrophoresis. No microsequencing is necessary I think. In every case, spot intensity was normalised by the actin intensity on every gel so it is the scale of intensities that we compare in making these analyses.

The influence of long-chain bases (sphingolipids) on early stages of T lymphocyte activation.

Elena A. Martinova
Institute of Experimental Cardiology, Cardiology Research Center, 121552 Moscow, Russian Federation, fax (095) 414-6699, e-mail csc@adonis.ias.msk.su

indexing key words: • sphingosine • ceramide • fumonisin B1 • T lymphocyte • CD antigens

I. Introduction: Known Immunobiological Effects of Some Sphingolipids.

The alteration of immune cell function by long-chain bases (sphingolipids) may be determined by **1**)the inhibition of protein kinase C (PKC) by sphingosine (Spho) [13], and activation of phosphatases by ceramide (Cer) [22,29], **2**)the inhibition of cell growth by an arrest in the G_0/G_1 phase of the cell cycle [8,12,13], **3**)the function of Cer and Spho as the endogenous modulators mediating the apoptotic signal [12,22,28], **4**) the activation Raf/MAP kinase kinase pathway by a G-protein-dependent mechanism by Spho, Sphingosine-1-phosphate (Spho-1-P) and Cer [31,34] and **5**)sphingolipid modulation of GPI-anchored cell surface antigens [2,10,20].

Spho down-regulates c-myc expression, inhibits IL-8-induced human lymphocyte migration and both the MHC class II - induced adhesion and adhesion via LFA-1-dependent pathway. Spho, Cer, and sphingomyeline (SM) suppress both the IL-2-dependent proliferation of cell lines HT-2 and CTLL, and the cell proliferation in MLC [3,5]. Spho-1-P inhibits integrin-dependent cell motility, but not integrin-dependent adhesion. Cer induces IL-6 gene expression, modulates endocytosis, inhibits PKC-a translocation that may be involved in regulating PLD activity [4,16,19]. C2-Cer (N-acetyl-D-erythro-sphingosine) is a cell-permeable analog mimicking a lot of function of natural Cer, including apoptosis, and suppresses the expression of cytochrome P450, enhances superoxide generation by neutrophil in low doses [33].

Mitogenic action of Spho, Spho-1-P, and Cer has been shown only on Swiss 3T3 fibroblast line [14, 26, 34]. Increasing cAMP concentration in these cells causes both the MAP kinase activity and cell differentiation, and the change from G1 to S-phase of cell cycle arisen from cAMP augmentation. In normal T cells and T helper lines, selective activation of cAMP-dependent protein kinase type I (cAKI), but not type II, is sufficient to mediate an inhibition of T cell replication induced through the TCR/CD3 complex [32]. Spho and fumonisin B1 (FB1) (known inhibitor ceramide synthase) have been shown to suppress DNA synthesis in normal mouse spleen cells [23].

Abbreviations: Ab - antibody; Ag - antigen; CD - clasters designations; Cer - ceramide; CTL - cytotoxic lymphocyte; FB1 - fumonisin B1; GlsCer - glucosylceramide, IL - interleukin, MAPK - mitogen-activated protein kinase; mAbs - monoclonal antibodies; MLS - mixed lymphocyte culture, MHC - major histocompatibility complex, PBS - phosphate buffered salution, PHA - phytohemagglutinin, PLA2 - phospholipase A2, PLC - phospholipase C; PLD - phospholipase D, PTK - phosphotyrosine kinase; PTP -phosphotyrosine phosphatase; SM - sphingomyeline; SMase - sphingomyelinase; Spha - sphinganine; Spho - sphingosine;

M.J.T. Carrondo et al. (eds.), Animal Cell Technology, 625–632.

II. Sphingolipid turnover: the implication for lymphocyte activation.

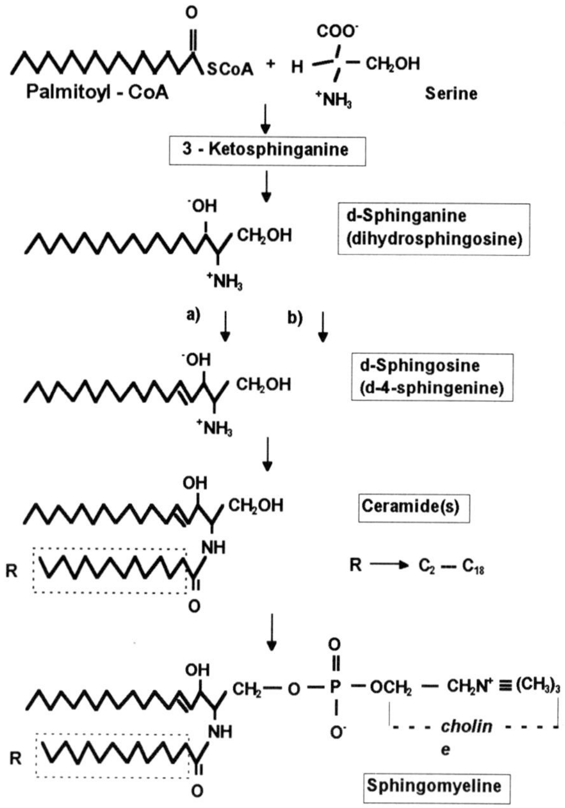

Fig.1. Pathway of sphingolipid biosynthesis *de novo* [Merrill et al, 1993]

Sphingolipids *de novo* synthesized in the Golgi complex and transported to the plasma membrane, may be involved into signal transduction pathway generated via TCR/CD3 complex. Hydrolysis of SM by neutral sphingomyelinase (nSMase) has been shown to be in plasma membrane, whereas acid SMase activity is inherent in the lysosomes. nSMase activity is dependent on the cell type, cell activation, cell cycle phase, and can be stimulated by multiple extracellular stimuli (IL-1, TNF-a, and others) which initiate the signal pathway termed the sphingomyelin cycle generating the lipid second messengers, in particular Cer, mediating of extracellular signals [1,3,6,9,20].

TCR/CD3 complex binding by specific antigen (Ag) or other ligands can activate nSMase in the plasma membrane to generate Cer. We have previously demonstrated *in vivo* the nSMase activation in splenocytes and thymocytes by fumonisin B1 (FB1) administered i.p. that caused the modulation of common T lymphocyte antigen (CD3) expression [25]. *In vitro*, bacterial SMase (EC 33.1.4.12) has been shown to modulate CD3, CD4, and CD8 antigen expression in the spleen and thymus cells [23]. In our last experiments, the addition of anti-CD3 Abs to the culture medium caused the twofold increasing nSMase activity in thymocytes, and 1.5-fold elevation in splenocytes. Anti-CD8 Abs caused the 4.5-fold and 1.5-fold increasing nSMase activity in thymocytes and splenocytes, respectively [Martinova, in press].

Immune effects of sphingolipids have raised a number of issues regarding T lymphocyte activation, as well as the early events contributing to the TCR/CD3 complex binding by the foreign antigen processing peptide in the complex with MHC molecules. This recognition causes the tyrosine phosphorylation initiated by $p56^{lck}$, src, and syk/ZAP-70 phosphotyrosine kinases (PTK), intracellular calcium $[Ca^{+2}]_i$ mobilization, protein kinase C (PKC) and phospholipase D (PLD) activation, initiation of Ras cascade, nuclear factors induction, c-fos, c-mic, c-jun oncogenes activation, IL-2 production, and the T cell proliferation accompanied by the other T cell surface antigen expression and the cytokine releasing [7,17,29,35]. Distant events of T lymphocyte activation are dependent upon the establishment of a positive feedback loop involving PKC and PLD which would result in the prolonged activation of PKC reqiured for certain lymphokine production. These events are modulated by Cer and Spho [4,16].

III. Influence of Sphingosine on Early Stage of T Lymphocyte Activation *in vivo* and *in vitro*.

CD4 and CD8 are glycoproteins expressed on T lymphocytes, serve as the co-receptors for TCR/CD3 complex, and bind to monomorphic MHC class II and class I determinants, respectively [7,27]. Ag-induced signal transduction via TCR/CD3 and CD4 (or CD8) receptors leads to the IL-2 and IL-2R synthesis, and some modifications to the

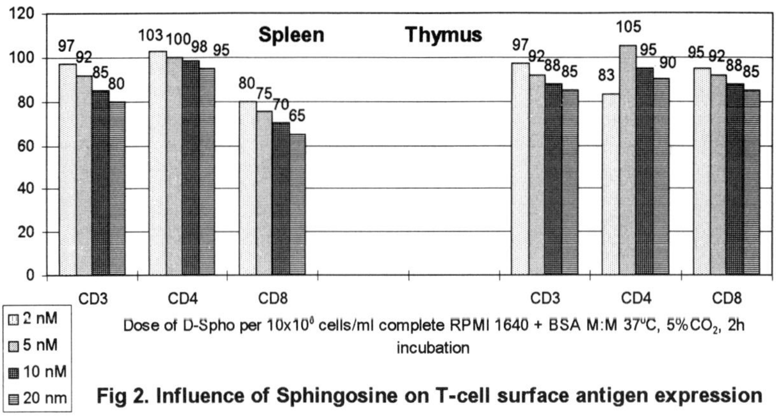

Fig 2. Influence of Sphingosine on T-cell surface antigen expression

The spleen and thymus cells from BALB/c mice (20g weight, 8 weeks old) were cleared from erythrocytes by hypotonic lysis, washed 3 times, diluted in complete RPMI 1640 supplemented by 2mM glutamine, and $10x10^6$cells/ml were placed into each culture tube. D-Sphingosine ("Sigma") was added to this volume in the dose 2nM (5nM or 10nM) mixed with BSA ("Sigma") mole/mole. Cells were incubated for 2h, at 37°C and 5% CO_2, and then washed 3 times in the cold PBS. After that, fluorochrome-conjugated mAbs reactive with CD3 (Thyl,2), CD4 (L3T4), and CD8 (Let2) ("Becton Dickinson") were added for flow cytometer analysis in common use. The cell population was gated by size 5.000 events were measured. The fluorescence was analysed with excitation wavelength of 488 nm and an emission wavelenth of 530 nm for FITC and 575 nm for PE. Data is expressed as percentage positive cells per gated event, and present as percentage of same CD-antigen expression on the cells without treatment. The results are given as mean ± S.E

T cell surface Ag expression [7,8,29,35]. Within the first 4 hours of activation, the linear range of the kinetic response, and the density of CD Ag expression are proportional to the time of activation. CD antigens expression can be measured in terms of either fluorescence intensity (antigen density) or positive cells which may be independently regulated during lymphocyte activation.

Spho added to the culture medium for 2h modulates the CD3, CD4, and CD8 Ags expression on spleen and thymus cells in dose-dependent manner **(Fig.2)**. Mature spleen T cells expressed CD8 Ag have been found to be more sensitive to Spho exposure *in vitro* compared with immature thymocytes. Because we did not separate double-positive (DP) thymocytes ($CD4^+CD8^+$) from single-positive cells ($CD4^+$ or $CD8^+$), we can only note the affinity of Spho for CD8 Ag *in vitro*, although it is possible to assume the influence of Spho on the DP thymocytes differentiation into $CD4^+$ cells.

Spho i.p. administration causes the significant modification to the CD3, CD4, and CD8 Ag expression on the spleen and thymus cells. *In vivo*, thymus has been found to be more sensitive compared with spleen **(Fig.3)**. An activation of the thymocytes causes either a positive selection or apoptosis which can be induced by Cer (and Spho in some cases) during SM cycle activation [12,22,28]. In the spleen T cells, an activation leads to the proliferation or the anergy dependent upon co-stimulation [7,18,21,29].

Previously we have shown the CD45 PTP activation by FB1 or Spho administration *in vivo* [Martinova EA & Merrill AH, *in press*]. We found also that IL-2R expression may be modulated by Spho *in vivo* and *in vitro* (data not shown). Taking together, this data indicates that Spho disrupts of signal provided via TCR/CD3 and CD4 receptors, and down-regulates the T lymphocyte activation.

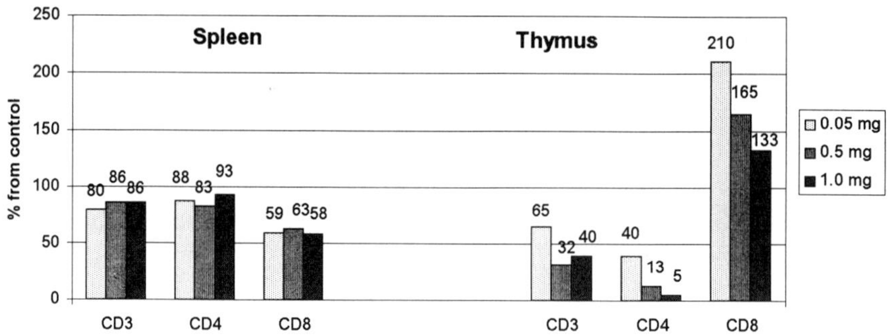

Fig 3. Dose-dependent effect of Sphingosine on CD-antigen expression *in vivo*

Sphingosine (0.05; 0.5; 1.0μg/ g body weight in 0.5ml PBS) was injected i.p. into male BALB/c mice (20g weight, 8 weeks old). After 2h, the spleen and thymus were removed, the cells cleared from erythrocytes by hypotonic lysis were washed 3 times and incubated with FITC-conjugated mAbs reactive with CD3, CD4, and CD8 cell surface antigens. The cell populations gated by size 5.000 events were measured by flow cytometer analysis. Data is expressed as percentage positive cells per gated event, and present as a percentage of same CD-antigen expression on the cells obtained from control mice. The results are given as the mean ±S.E.

Influence of *endogeous* formed Spho on the signal transduction is dependent on source of Spho and its location in the special compartments connected with signal pathways. *Exogenous* Spho can be incorporated into plasma membrane and generate its own signal pathway including Spho-1-P, or be metabolized to Cer (and then to GlsCer within several minutes) and regulate the CD Ag expression by this means too [2,10, 34].

IV. Ceramide influence on T lymphocyte activation.

Exogenous added Spho may be converted into Cer which acts as a second messenger in the intracellular signaling pathway and across the plasma membrane. Cer can be *de novo* synthesized via activation of the ceramide synthase,or generated by SM hydrolysis in the lysosome (by a-SMase) and in the plasma membrane (by n-SMase). A distinction needs to be drawn between two pools of Cer specify toward signal pathways provided through TCR/CD3 complex.

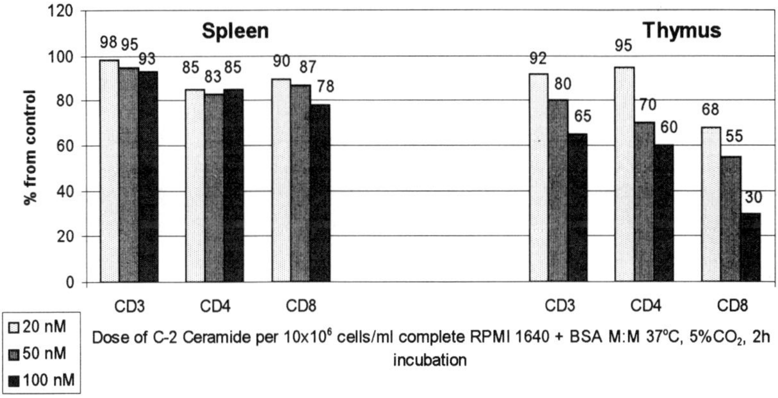

Fig 4. Influence of C-2 Ceramide on T-cell surface antigen expression

The spleen and thymus cells from BALB/c mice were incubated with C2-Ceramide ("Sigma") in the dose 20nM (5nM or 100nM) as described for **Fig.2**. CD antigen expression was measured as described for **Fig.2**. Data is present as percentage positive cells from control level. The results are given as the mean ± S.E.

When a data on Cer influence on T cell surface Ags *in vitro* (**Fig.4**) has been compared with that of Spho effect, C2-Cer has been found to be less effective in same doses. However, C2-Cer impaired CD8 Ag expression on thymocytes.

According other authors, C2-Cer alone has been shown not to change cytosolic free Ca^{+2} levels, but inhibited Ca^{+2} and Mn^{+2} influx in fmlp-activated neutrophils. In contrast, Spho enhanced Ca^{+2} entry [33]. Spho assumed to be more effective than C2-Cer for T lymphocyte activation.

One of possible explanation is that plasma membrane-bound alkaline ceramidase produces Spho from Cer releasing during SM cycle initiation. This one can be inhibited by Spho, SM, and phosphatidylcholine [6]. Flowing equilibrium in the sphingolipid metabolism may regulate the accumulation of Spho and Cer in special cell compartments and, by this means, regulate intracellular signal transduction pathways.

FB1 blocks ceramide synthase activity to accumulate Spha and Spho [26], and can serve as a model system to study immunobiology of sphingolipids. We have shown previously that i.p. administration of FB1 for 2h decreased the contents of Cer in the immune organs in the dose-dependent manner (about 1.4-fold for spleen, and 2.5-fold for thymus), in spite of nSMase activation and SM cycle initiation by FB1 [25].

In vitro, FB1 added to culture medium, causes the different effects on spleen and thymus cells. CD4-Ag expression was more than 3-fold decreased on splenocytes, and only a little changed on thymocytes (**Fig.5**). CD4 and CD8 antigens have been found to appear the diverse affinity to FB1 supplementation. Compared with Spho and Cer effects on CD antigens expression *in vitro*, FB1 has been shown to more affect the mature spleen cells, whereas Spho and Cer displayed more activity in the thymocytes.

FB1 molecular weight (721) in not enough to exhibit classic antigen capacity and to generate immune response, nonetheless, we postulate some antigen effects of FB1 [24] that can cause the antigen recognition of FB1 by TCR/CD3 complex that led to the different response of CD4 and CD8 Ags to FB1 supplementation *in vitro* compared with Spho and Cer.

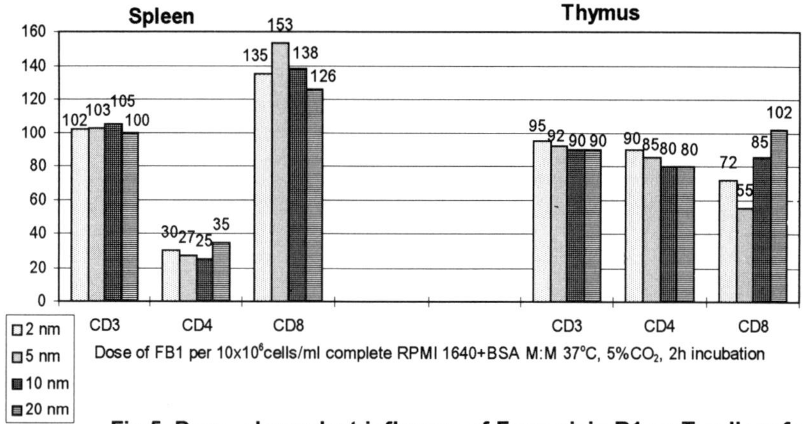

Fig 5. Dose - dependent influence of Fumonisin B1 on T-cell surface antigen expression

The spleen and thymus cells from BALB/c mice were incubated in the whole culture medium RPMI 1640 supplemented by FB1 in the dose 2nM; 5; 10; or 20nM as described for **Fig.2**. CD antigen expression was measured as described for **Fig.2**. Data is present as percentage positive cells from control level. The results are given as the mean ± S.E.

V. Possible Mechanism of Sphingolipid Modification to T Lymphocyte Activation.

Exogenous sphingolipids can modificate to immune response through incorporation into plasma membrane that leads to the PKC inhibition by Spho and down-regulation of signals provided via TCR/CD3 complex. From the other side, the long acyl chain of sphingolipid is not interdigitated, but rather must terminate at the bilayer center, that may force the carbohydrate headgroup of the glycosphingolipid farther above the bilayer surface, allowing it to be recognized better by various carbohydrate binding ligands and proteins [10,15,20]

In the same time, TCR/CD3 complex can be modulated by extramembrane Spho to initiate the CD45 PTP activation, and PTKs cascades [21]. Changing plasma membrane lipid composition can modulate the activity of some phospholipases. Sphingolipid modification to PLA2, PLD may be important for lipid-mediated signaling pathways [4,30].

Spho and Cer incorporation into plasma membrane may regulate nSMase activity and SM cycle product releasing.. Because nSMase is activated by many extracellular signals, as well as anti-CD3 and anti-CD8 antibodies, we can assume an linkage between the signal transduction pathways initiated through TCR/CD3 complex from one side, and nSMase and SM cycle product from other side. Signal pathway provided through SM cycle can be linkage also with cAMP signal transduction pathway.

Because cAMP blocks signal from TCR/CD3 [32], but activates Raf/MAPK pathway in the cell lines exhibited reverse response to cAMP, the Cer and Spho signal transduction pathways are presumed to intensify (or, at least, interact) cAMP signal pathway to enhance the down-regulative effects of cAMP to lymphocyte activation.

Conclusions: Sphingolipids, in particular Sphingosine and Ceramide, down-regulate early stage of T lymphocyte activation, and may act as a one of signal pathways served into cells to stop the proliferative signal.

Sphingolipid modulation of the major signal transduction pathway generated by triggering of TCR/CD3 complex

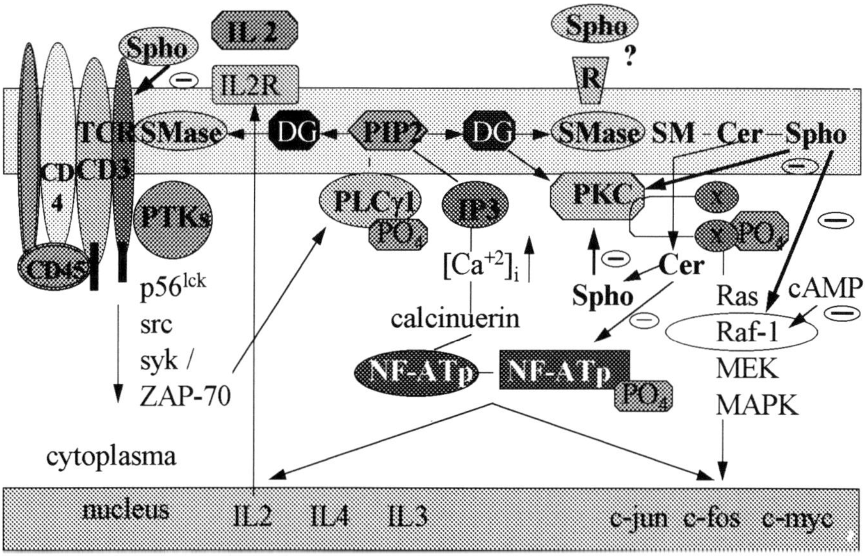

References:

1. Anel A, Richieri GV & Kleinfeld AM (1993) Membrane partition on fatty acids and inhibition of T cell function, *Biochemistry*. 32:530-536.

2 .Bell LM, Solomon KR, Gold JP & Tan K-N (1994) Cytoplasmic tail deletion of T cell receptor (TCR) b-chain results in its surface expression as glycosylphosphatidylinositol - anchored polypeptide on mature T cells in the absence of TCR-a, *J.Biol.Chem*. 269:22758-22763.

3. Borchardt RA, Lee WT, Kalen A, Buckley RH, Peters C, Schiff S & Bell RM (1994) Growth-dependent regulation of cellular ceramides in human T-cells, *Biochim. Biophys. Acta*. 1212:327-336.

4. Cao YZ, Mastro AM, Eskew ML, Hildenbrandt G, Reddy PV & Reddy CC (1995) The mechanism of sphingosine enhancement of phorbol ester-mediated phospholipase D activation in lymphocytes, *Biochem Biophys Res Commun*. 217(3): 908-915.

5. Chu JW & Sharom (1991) Effect of micellar and bilayer gangliosides on proliferation of interleukin-2-dependent lymphocytes, *Cell Immunol*. 132(2):319-338.

6. Coroneos E, Martinez M, McKenna S & Kester M (1995) Differential regulation of sphingomyelinase and ceramidase activities by growth factors and cytokines: implications for cellular proliferation and differentiation, *J.Biol.Chem*. 270: 23305-23309

7. Dasgupta JD, Granja CB, Yunis EJ & Relias V (1994) MHC class I antigens redulate CD3-induced tyrosine phosphorylation of proteins in T cells, *Int.Immunol.* 6(3):481-489.
8. Dbaibo GS, Wolff RA, Obeid LM & Hannun YA (1995) Activation of a retinoblastoma-protein-dependent pathway by sphingosine, *Biochem.J.* 310(Pt.2):453-459.
9. Fitzgerald V, Blank ML & Snyder F (1995) Molecular species of sphingomyelin in sphingomyelinase-sensitive and sphingomyelinase-resistant pools of HL-60 cells, *Lipids* 30: 805-809.
10. Futerman AH (1995) Inhibition of sphingolipid synthesis: effects on glycosphingolipid-GPI-anchored protein microdomains, *Trends Cell Biol.* 5:377-379.
11. Gouy H, Debre P & Bismuth G (1995) Triggering of a sustained calcium response through a p56lck-dependent pathway by exogenous ganglioside GM1 in human T lymphocytes, *J. Immunol.* 155(11): 5160-5166.
12. Gulbins E, Bissonnette R, Mahboubi A, et al (1995) Fas-induced apoptosis is mediated via a ceramide-initiated RAS signaling pathway, *Immunity* 2(4):341-351.
13. Hannum YA & Linardic CM (1993) Sphingolipid breakdown products: anti-proliferative and tumor-suppressor lipids, *Biochem. Biophys. Acta* 1154(3-4):223-236.
14. Hauser JML, Buehrer BM & Bell RM (1994) Role of ceramide in mitogenesis induced by exogenous sphingoid bases, *J.Biol.Chem.* 269: 6803-6809.
15. Inooka S & Toyokuni, T (1996) Sphingosine transfer in cell-to-cell interaction, *Biochem Biophys Res Commun.* 218(3): 872-876.
16. Jones MJ & Murray AW (1995) Evidence that ceramide selectively inhibits protein kinase C - alpha translocation and modulates bradykinin activation of phospholipase D, *J.Biol.Chem.* 270(10):5007-5013.
17. Kelly K & Siebenlist U (1995) Immediate-early genes induced by antigen receptor stimulation, *Current Opin.Immunol.* 7:327-332.
18. Kemeny D, Noble A, Holmes B & Diaz-Sanchez D (1994) Immune regulation: a new role for the CD8+ T cells, *Immunol. Today* 15: 107-110
19. Laulederkind SJ, Bielawska A, Raghow R, Hannun YA & Ballou LR (1995) Ceramide induces IL-6 gene expression in human fibroblasts, *J.Exp.Med.* 182(2):599-604.
20. Laurenz JC, Gunn JM, Golly CA & Chapkin RS (1996) Alteration of glycerolipid and sphingolipid-derived second messenger kinetics in ras transformed 3T3 cells, *Biochim. Biophys. Acta* 1299: 146-154.
21. Leitenberg D, Constant S, Lu DD, Smith BR & Bottomly K (1995) CD4 and CD45 regulate qualitatively distinct patterns of Calcium mobilization in individual CD4+ cells, *Eur. J. Immunol.* 25: 2445-2451.
22. Lucas M & Sanchez-Margalet V (1995) Protein kinase C involvement in apoptosis, *Gen.Pharmacol.* 26(5):881-887.
23. Martinova EA (1996) Fumonisin B1 - immunological effects. The influence of FB1 on the early stage of immune response, in: Singh RB & Tu AT, *Natural Toxins II*, pp.331-342, Plenum Press, NY
24. Martinova EA & Merrill AH, Jr (1995) Fumonisin B1 alters sphingolipid metabolism and immune function in BALB/c mice: Immunological responses to fumonisin B1, *Mycopathologia* 130: 163-170.
25. Martinova EA, Solovjev AS, Khrenov AV, Zabotina TN & Alesenko AV (1995) Fumonisin B1 modulates sphingomyelin cycle product levels and the expression of CD3 receptors in immunocompetent organs, *Biochemistry (Moscow)* 60 (4): 461-465.
26. Merrill AH, Wang E, Gilchrist DG & Riley RT (1993) Fumonisins and other inhibitors of *de Novo* sphingolipid biosynthesis, *Adv.Lipid Res.* 26: 215-234.
27. Nakayama Ki, Nakayama K, Nagishi I, Kuida K, Louie MC, Kanagawa O, Nakauchi H & Loh DY (1994) Requirement for CD8-β chain in positive selection of CD8-lineage T cells, *Science* 263:1131-1133.
28. Ohta H, Sweeney EA, Masamune A, Yatomi Y, Hakomori S & Igarashi Y (1995) Induction of apoptosis by sphingosine in human leukemic HL-60 cells: a possible endogenous modulator of apoptotic DNA fragmentation occuring during phorbol ester-induced differentiation, *Cancer Res.* 55(3):691-697.
29. Okumura M & Thomas ML (1995) Regulation of immune function by protein tyrosine phosphatases, *Current Opin. Immunol.* 7:312-319.
30. Perillo MA, Guidotti A, Costa E, Yu RK & Maggio B. (1994) Modulation of phospholipase A2 and C activities against dilauroylphosphorylcholine in mixed monolayers with semisynthetic derivatives of ganglioside and sphingosine, *Mol. Membr. Biol.* 11(2): 119-126.
31. Sasaki T, Hazeki K, Hazeki O, Ui M & Katada T (1995) Permissive effect of ceramide on growth factor-induced cell proliferation, *Biochem J.* 311(Pt 3):829-834.
32. Skalhegg BS, Tasken K, Hansson V, Huitfeldt HS, Jahnsen T & Lea T (1994) Location of cAMP-dependent protein kinase type I with the TCR/CD3 complex, *Science* 263:84-87.
33. Wong K, Li XB & Hunchuk N (1995) N-acetylsphingosine (C2-ceramide) inhibited neutrophil superoxide formation and calcium influx, *J.Biol.Chem.* 270(7):3056-3062.
34. Wu J, Spiegel S & Sturgill TW (1995) Sphingosine-1-phosphate rapidly activates the mitogen-activated protein kinase pathway by a G protein-dependent mechanism, J.Biol.Chem. 270(19):11484-11488.
35. Yakura H (1994) The role of protein tyrosine phosphatase in lymphocyte activation and differentiation, *Critical Reviews Immunol.* 14:311-336.

METABOLIC FLUX ANALYSIS OF HYBRIDOMA CELLS USING MASS BALANCES, $^{13}CO_2$ MASS SPECTROMETRY AND NMR SPECTROMETRY

Towards metabolic flux analysis without using isotopic tracer methods

Hendrik P.J. Bonarius[1,4*], Ahmet Özemre[2], Bram Timmerarends[1], Peter Skrabal[3], Elmar Heinzle[2], Cornelis D. de Gooijer[1], Johannes Tramper[2] and Georg Schmid[4]

[1]*Department of Food Science, Food and Bioengineering Group, Wageningen Agricultural University, The Netherlands.*
[2]*Department of Chemical Engineering, Federal Institute of Technology Zürich, Switzerland.*
[3]*Department of Polymer Chemistry, Federal Institute of Technology Zürich, Switzerland.*
[4]*Department Pharmaceutical Research Preclinical - Biotechnology, F. Hoffmann-La Roche Ltd., Basel, Switzerland.*
[*]*Corresponding author. Present address: Novo Nordisk A/S, Health Care Group, Gentofte, Denmark. E-mail:* Bona@novo.dk

The analysis of the intracellular metabolic fluxes in mammalian cells can be used for optimizing culture media, for tracing metabolic bottle-necks and for toxicological and metabolic studies. It is desirable to determine metabolic fluxes without using isotopic tracers. This could potentially be achieved by using the material balances of the relevant metabolites. For each metabolite a mass balance is set up using the known biochemical stoichiometry and its measured uptake or production rate. Together, these mass balances form a linear set of equations, containing the unknown fluxes. However, such a set of linear equations is *underdetermined* due to the presence of *cyclic* pathways in mammalian cell metabolism. Fluxes in such pathways have to be quantitated using isotopic tracers.

Instead of determining fluxes in cyclic pathways experimentally using isotopic tracers, additional theoretical assumptions can be applied to obtain an estimate for these fluxes. In this work, both methods have been applied. Fluxes that were estimated using different theoretical constraints were compared to experimentally obtained flux values using isotopic tracers. Continuously cultured hybridoma cells were fed with 1-^{13}C-glucose supplemented medium. By analyzing the $^{13}CO_2/CO_2$ ratio (by using mass spectrometry) and the ^{13}C/C ratio of 2-C- and 3-C-lactate and alanine in the extracellular medium (by using ^{13}C-NMR spectrometry), fluxes of the pentose shunt, the malate shunt and the TCA cycle were measured. Additionally, the cellular composition and the uptake and production rates of all relevant metabolites were measured. The mass balances that were derived from these measurements were either supplemented with different theoretical constraints or with data generated from the isotopic tracer experiment. The theoretically estimated fluxes were finally compared with the experimentally measured fluxes. The effect of the various theoretical assumptions that are required in order to generate an (over)determined system on the quality of the estimation is discussed.

The results give experimental evidence for a fundamental solution for underdetermined metabolic networks. The developed approach can serve as a tool for a wide range applications, also beyond cell culture engineering.

INTRODUCTION

The interest in metabolic flux analysis has been growing rapidly in the last decade, because of its potential use as a tool for optimizing production yields[9]. In the past, intracellular metabolic fluxes have been quantified by using radioactive[4] or stable[7] isotopes. Isotopic tracer methods have however some major disadvantages. They have practical limitations, are extremely costly and are not applicable for each type of cultivation. Alternatively, metabolic fluxes can be estimated by applying the material balances of the relevant metabolites[2,3,8,9].

M.J.T. Carrondo et al. (eds.), Animal Cell Technology, 633–638.

The major disadvantage of this alternative method is that the mass balances provide insufficient information to determine certain fluxes. In particular, fluxes in *cyclic* pathways can not be quantitated by only measuring the metabolic rates of the relevant metabolites. The linear set of equations that is defined by the mass balances is therefore *underdetermined*. This fundamental problem in metabolic flux balancing is addressed here. Various theoretical solutions are given and discussed. Additionally, the intracellular fluxes in hybridoma cells were measured experimentally using an isotopic tracer. Flux values based on different theoretical approaches to estimate fluxes in underdetermined networks are finally compared with flux data from the tracer experiment.

THEORY

The fundamental problem of underdetermined metabolic networks is visualized in Figure 1A (See also reference 8). In order to quantify metabolic flux x_1 and x_2 it is not sufficient to determine the extracellular metabolic rates of metabolite A and B (r_A and r_B). The mass balance of A and the mass balance of B yield the same information. When co-metabolites are converted in one of the reactions, it is sometimes possible to quantify fluxes in cyclic pathways. In contrast to reaction x_1 in Figure 1A, reaction x_1 in Figure 1B co-produces metabolite C. Now, by measuring the production rate of C, both fluxes can be determined. However, most co-metabolites are

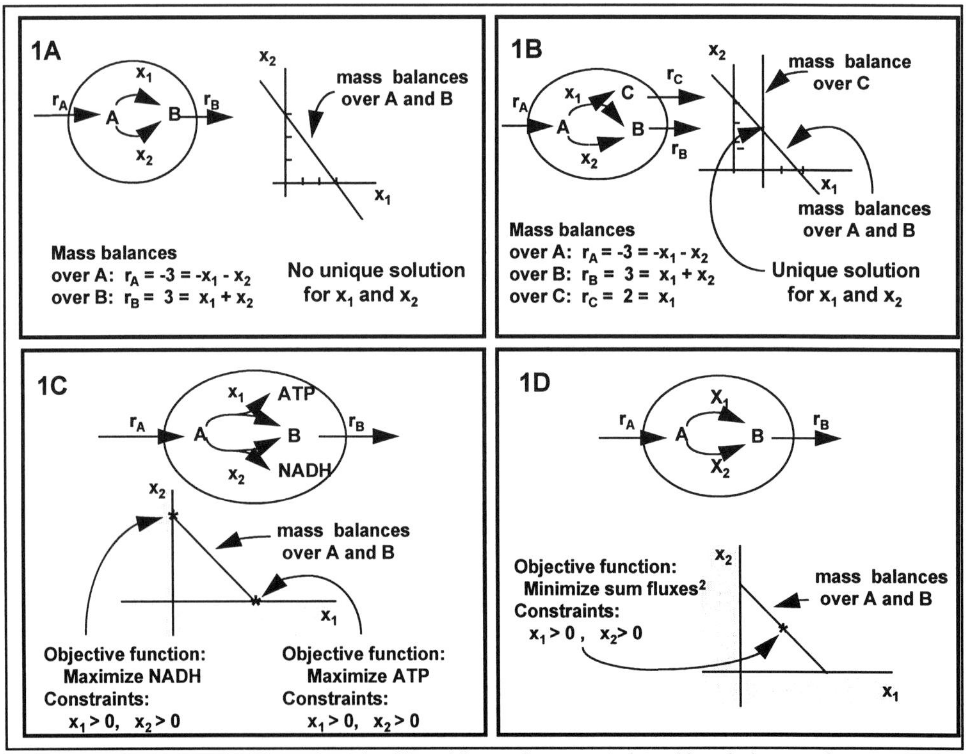

Underdetermined networks (A) can be solved by adding various constraints: Mass balances of co-metabolites (B), objective functions (C), or the minimum norm constraint (D).

converted in more than one cyclic pathway (Figure 2), as a consequence of which the linear system remains underdetermined[2]. In order to determine such unbalanceable fluxes without using isotopic tracers, various theoretical methods are available. For example, Savinell and Palsson used linear optimization techniques. They showed that various objective functions, which satisfy certain metabolic strategies, such as "maximize the ATP production" (Figure 1C), can be used to estimate a metabolic flux distribution[6]. In Figure C two different objective functions are used to solve an underdetermined metabolic network. In this simple example an inherent disadvantage of linear optimization is visualized: only *extreme values* at the boundaries of the solution space that are determined by the constraints (here, x_1 and $x_2 > 0$), are allowed solutions. In sharp contrast, this is not the case when the minimum-norm constraint is applied[2,3] (Figure 1D). This constraint, that gives the solution with the minimum Euclidean norm, which here is the minimum length of the square of the flux vector, yields the most "average" solution. Extreme flux values are avoided when the minimum-norm constraint is applied (See also Hatzimanikatis[3]).

In Figure 2 a distinction is made between two types of metabolic fluxes: (1) fluxes that *can* be determined by using the material balances (dotted lines) and (2) fluxes in cylcic pathways that *can not* be determined using mass balances (fluxes 1 to 21). Their quantition needs more information than the material balances equations only. In order to estimate these fluxes either additional theoretical assumptions, such as given above, or experimental data are required. In this work, both methods are applied: Fluxes that were estimated using various theoretical constraints are compared with the experimentally obtained flux values using 1-^{13}C-glucose.

2

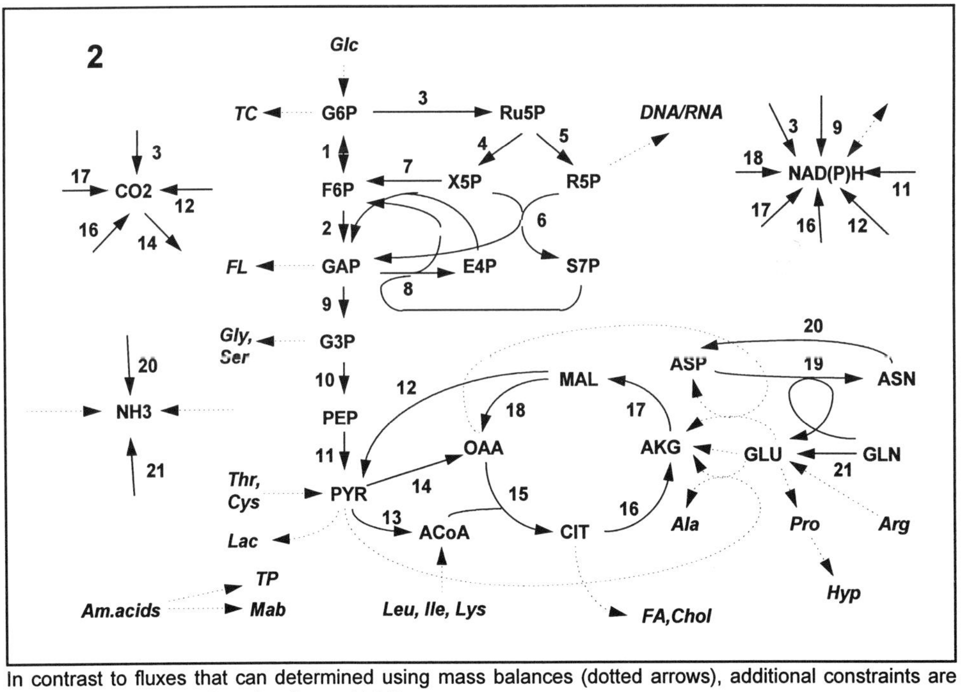

In contrast to fluxes that can determined using mass balances (dotted arrows), additional constraints are required to determine the other fluxes (1-21).

MATERIALS AND METHODS

A detailed description of most of the metabolite analyses was published before[2]. Briefly, glucose and lactate were determined with automated enzymatic assays, ammonia using an ion-selective electrode, and amino acids by HPLC. The cellular composition was measured as follows: the total lipid fraction was determined by weight after chloroform/methanol extraction, total carbohydrates were analyzed by the phenol reaction method, and total cellular protein was estimated using the Biuret assay.

For experiments with labeled substrates it is desirable to work on a small scale, because of the high costs of ^{13}C-labeled substrates. Therefore, a very small bioreactor (working volume of 100 ml) was specially designed in order to cultivate cells in a continuous mode (D= 1.0 d^{-1}) and yet using small amounts of medium. The O_2 uptake rate, and the CO_2 and $^{13}CO_2$ production rate were determined by analyzing the in- and outlet gas composition using mass spectrometry. In order to account for $^{(13)}CO_2$ accumulation in the liquid phase and for CO_2 coming from the bicarbonate in the culture medium the following equation was derived[1],

$$CER = x^{o}_{CO_2}(t)\,[F^{i}_{g}+F_{l}(\frac{P}{H^{CO_2}}+\frac{F^{i}_{g}}{0.89k_{l}^{O_2}a\;V_{l}})\,(1+\frac{K_1}{10^{-pH}})]\;-\;\beta F_{l}C^{i}_{A}(0)$$

where the volumetric CO_2 production rate (CER) is a function of the measured fraction of CO_2 in the off-gas ($x_{CO2}{}^{o}$). The term $ßF_lC_A{}^{i}(0)$ refers to the correction for the bicarbonate in the ingoing medium and is required to determine the CO_2 production rate. In contrast, for the $^{13}CO_2$ production rate this correction is *not* necessary, because the bicarbonate in the ingoing medium is not ^{13}C-labeled. See also Bonarius[1].

Fractional labeling ($^{13}C/(^{12}C+^{13}C)$) in 2-C- and 3-C- lactate and alanine was determined by quantifying the ^{13}C-satellites in the ^{1}H-spectrum of the extracellular medium. Fluxes were calculated using a method based on the principles of a method that was developed by Zupke[6]. In this method, the information from isotopic tracer experiments is combined with mass balances in order to generate an (over)-determined metabolic network.

Here, the use of mass spec-trometry, additional to NMR spectrometry, makes it possible to determine fluxes in *three* different cyclic pathways in only *one single independent experiment.* By analyzing the fractional labeling of lactate and/or alanine it is possible to determine the malate shunt and TCA cycle activity (Figure 3). During the same experiment the pentose shunt activity can be determined by measuring the fractional labeling of carbon dioxide using mass spectrometry (Figure 3). Therefore, reliable CER data[1] are not only critical for metabolic flux analysis using mass balances[2], they are also important for flux analyis using isotopic tracers[4].

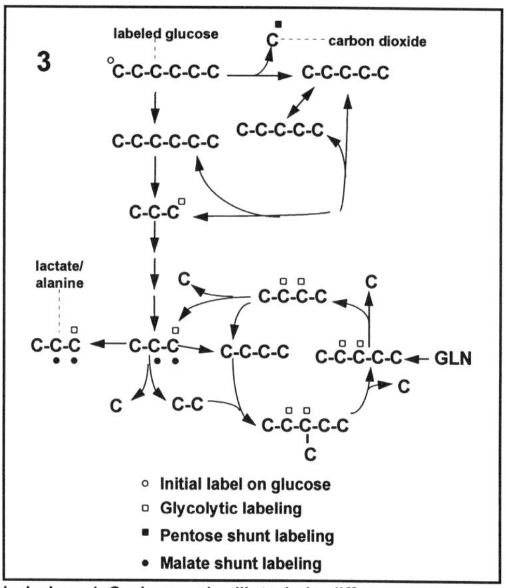

Label on 1-C-glucose is diluted via different pathways. See also Mancuso[5] for a more detailed isotope map.

RESULTS AND DISCUSSION

It is not possible to estimate the intracellular fluxes by merely using the material balance equations, because such a linear system is underdetermined. Therefore, additional theoretical assumptions are required. Here, estimations of fluxes using three different types of theoretical constraints are compared with experimentally obtained flux values. The intracellular fluxes of 1-^{13}C-glucose-fed hybridomas are determined by measuring the fractional labeling of lactate, alanine and carbon dioxide. The three different types of theoretical assumptions that have been evaluated are:

I. the assumption that cells optimize their metabolism by minimizing the sum of the fluxes (as calculated using the "minimum-norm constraint" (Figure 1D)),

II. the assumption that cells optimize certain specific metabolic parameters, such as the ATP production (as calculated using linear optimization techniques (Figure 1C)).

III. the assumption that certain fluxes are negligible.

In Figure 4 the deviations of the estimated flux distributions as compared with the experimentally determined flux distribution are shown. A small deviation indicates a close estimate compared to the isotopic tracer experiment. Of all the used theoretical assumptions, the minimum-norm constraint gives the best possible estimation for the investigated culture conditions. However, the values given here are mean deviations. For certain single fluxes the deviation is still significant. More flux data for various culture conditions using 1-^{13}C-glucose have been collected. In the near future, these will be compared to flux values estimated using the material balances and different theoretical constraints. This strategy will hopefully allow us to develop a method to determine fluxes without the need for isotopic tracers that is generally valid for a variety of culture conditions and a variety of cell lines.

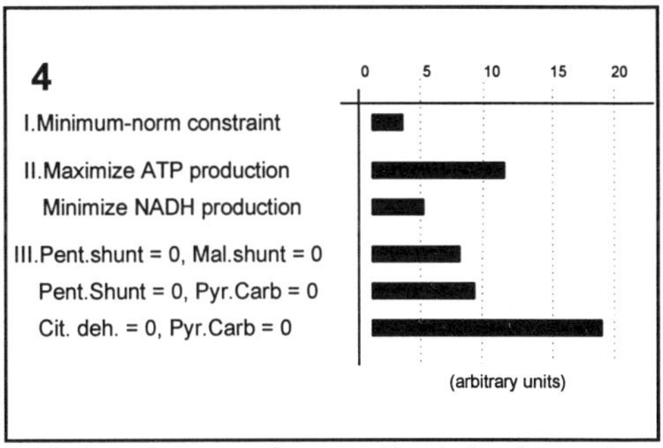

Deviations between fluxes estimated using various theoretical constraints, and experimentally determined fluxes.

LITERATURE

1. Bonarius, H.P.J., De Gooijer, C.D., Tramper, J., and Schmid, G. (1995) Determination of the respiration quotient in mammalian cell culture in bicarbonate-buffered media. *Biotechnol.Bioeng.* 45: 524-535.
2. Bonarius, H.P.J., Hatzimanikatis, V., Meesters, K.P.H., Schmid, G., De Gooijer, C.D. and Tramper, J. (1996) Metabolic flux analysis in mammalian cells in different media using mass balances. *Biotechnol.Bioeng.* 50: 299-318.
3. Hatzimanikatis, V. (1996) PhD thesis, California Institute of Technology, Pasadena, CA.
4. Katz, J. and Wood, H.G. (1963) The use of $^{14}CO_2$ yields from glucose-1- and -6-^{14}C for the evaluation of the pathways of glucose metabolism. *J.Biol.Chem.* 238:517-523.
5. Mancuso, A., Sharfstein, S.T., Tucker, S.N., Clark, D.S., and Blanch, H.W. (1994) Examination of primary metabolic pathways in a murine hybridoma with carbon-13 nuclear magnetic resonance spectroscopy. *Biotechnol.Bioeng.* 44: 563-585.
6. Savinell, J.M., and Palsson, B.O. 1992. Optimal selection of metabolic fluxes for in vivo measurement. II. Application to *Escherichia coli* and hybridoma cell metabolism. *J.Theor.Biol.* 155: 215-242.
7. Shulman, R.G., Brown, T.R., Ugurbil, K., Ogawa, S., Cohen, S.M., Den Hollander, J.A. (1979) Cellular applications of ^{31}P and ^{13}C nuclear magnetic resonance. *Science.* 205: 160-166.
8. Vallino, J.J., and G. Stephanopoulos. (1989) Flux determinations in cellular bioreaction networks. In: *Frontiers in Bioprocessing.* (Sikdar, S.K., Bier, M. and Todd, P. eds.) pp.205-219. Boca Rotan, FL, CRC Press.
9. Varma, A., and Palsson, B.O. (1994) Metabolic flux balancing: Basic concepts, scientific and practical use. *Bio/Technology.* 12: 994-998.
10. Zupke, C., and Stephanopoulos, G. (1994) Modelling of isotope distributions and intracellular fluxes in metabolic networks using atom mapping matrices. *Biotechnol. Progr.* 10: 489-498.

Discussion

Zeng: One of your conclusions was that the system is undetermined. I can only agree with you partly. The question is how much detail you want to consider eg metabolic pathways, details in the literature from MIT (Stephanapoulas). If you simplified the metabolic pathways you can still get a good estimation of the intracellular metabolic fluxes.

Bonarius: I think I actually showed this even if you don't do the experiments with labelled glucose, you can get very good approximations.

Ozturk: There is a lot of information on oxygen and lactate. Can your system predict the published observations?

Bonarius: No, the aim of this model is to find which constraints are additionally necessary to determine the metabolic fluxes. No assumptions or kinetics have been modelled to get the effects you mentioned.

Ozturk: Will you extend it to accommodate other people's data?

Bonarius: This would be possible but then you would have to include kinetics.

MODEL-BASED CONTROL OF ANIMAL CELL CULTURES

R. Biener and E.D. Gilles
Institut für Systemdynamik und Regelungstechnik,
Universität Stuttgart, Pfaffenwaldring 9,
70550 Stuttgart, Germany

J. Haas, W. Noé, P. Schorn and D. Schuller
Dr. Karl Thomae GmbH, Biotechnische Produktion
88397 Biberach, Germany

Abstract

A metabolically structured mathematical model was developed. This model describes the dynamics of hybridoma growth and monoclonal antibody production. The model is applied for model-based monitoring and control strategies. It will be demonstrated that for a monoclonal antibody production process the process efficiency can be improved remarkably.

1 Introduction

The growth and production behaviour of animal cells in culture is influenced by the environmental condition of the cells. Cell growth is limited by the depletion of nutrients (e.g. glucose, amino acids and vitamins) or by the accumulation of waste products (e.g. ammonia and lactate). Therefore, for the efficient production of proteins with animal cell cultures the cell environment has to be controlled. Mathematical models can help to understand the complex behaviour of cells in culture. Model-based methods of monitoring and control are powerful tools for the efficient design of animal cell culture processes. A metabolically structured mathematical model was developed. It will be demonstrated how this model can be applied for the model-based control of an antibody production process.

2 Metabolically structured mathematical model

The mathematical model described here is based on the metabolic processes in a hybridoma cell. A schematic diagram of the hybridoma metabolic pathways is shown in Figure 1 [1]. A medium for the cultivation of animal cell cultures contains beside the substrates glucose and glutamine a large variety of other substrates. All of them are necessary for the cells to grow in vitro. Therefore, in addition to glucose and glutamine the model also includes amino acids as nutrients. ATP is produced via glycolysis and oxidative phosphorylation. The metabolites ammonia, lactate and alanine are also included in the model. Figure 1 describes the stoichiometric part of the model. In the kinetic part only the rate limiting steps of the network are included. For glutaminolysis and glycolysis these are assumed to be the first steps

M.J.T. Carrondo et al. (eds.), Animal Cell Technology, 639–645.

of the respective pathways [2]. For the rest of the reactions a pseudo steady state assumption was made. Inhibitory effects of metabolites to cell growth are part of the kinetic expressions [3].
Figure 2 shows experimental and simulation results of a fed-batch fermentation. The simulation is compared to the experimental data. The simulation agrees very well with the experimental data.

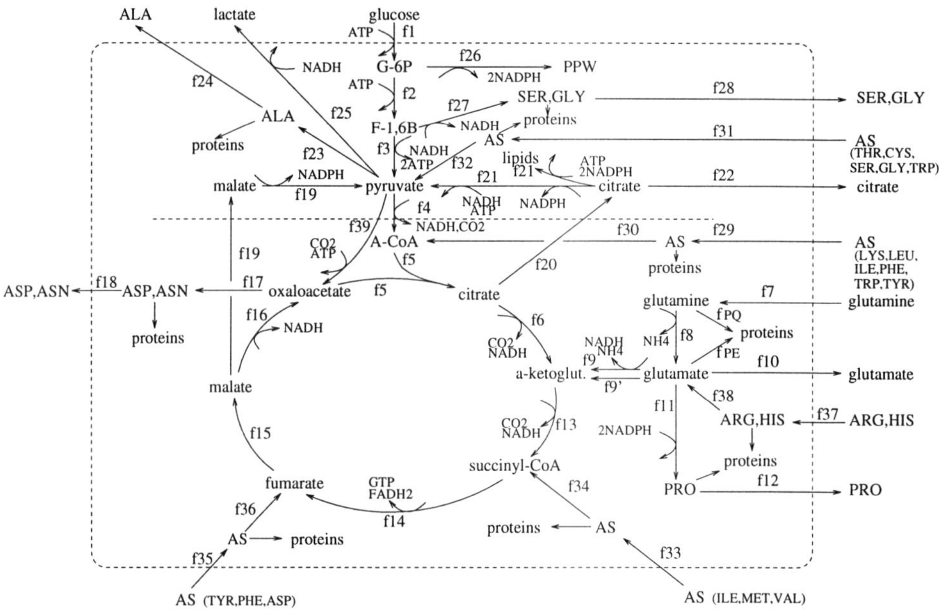

Figure 1: Hybridoma pathways for glucose and amino acids [1]

3 Model-based control

Usually control of a process is done by maintaining measurable signals, e.g. pH or temperature at constant values or trajectories. The control of substrate concentrations in general can not be achieved by conventional controllers due to the fact that these concentrations are usually not measured online. To overcome this problem model-based measuring methods can be used in conjunction with controllers [3]. Model-based measurement systems such as Kalman filters allow an estimation of unmeasured state variables based on existing measurement information [4]. Based on the mathematical models, methods of process control are developed and applied to the antibody production process. Figure 3 gives an overview of the developed advanced model-based process control system. As measurement information for the model-based process control system the oxygen uptake rate is available online and glucose and lactate are measured once or twice per day.

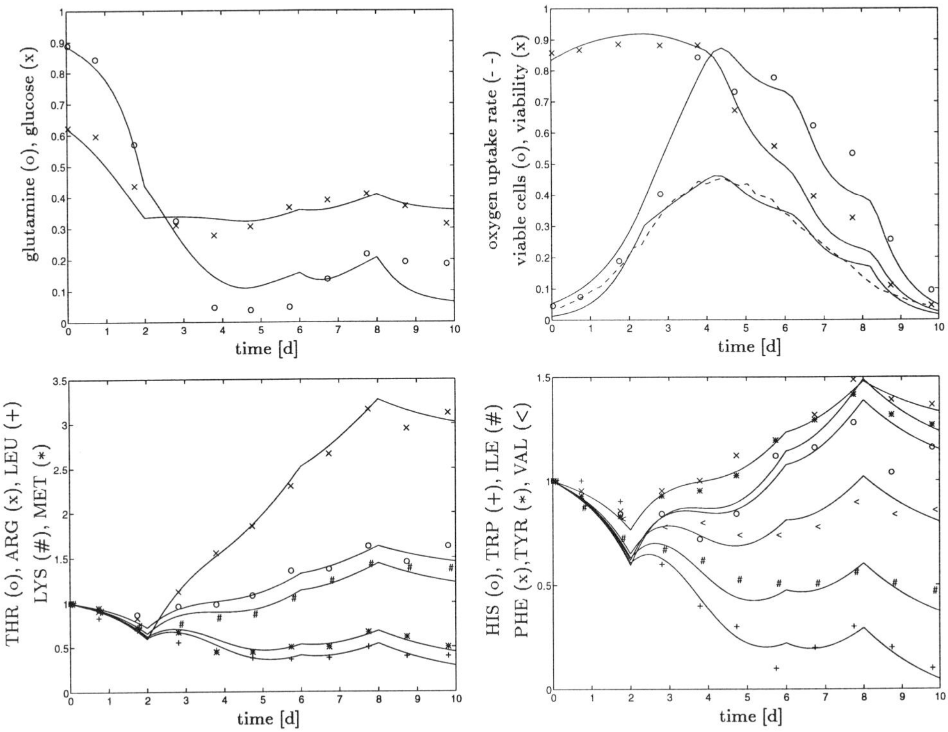

Figure 2: Glucose, glutamine concentration, oxygen uptake rate, viable cell density, viability and amino acid concentrations during a fed-batch run (relative units). The model simulations (solid lines) are plotted along with the experimental results (o,x,- -).

The model was applied to design the feed medium composition with respect to amino acids. This is demonstrated in Figure 4. The left plot shows the concentration profiles of the essential amino acids for a fed-batch fermentation before the medium design. Some amino acids accumulate and some are close to exhaustion. This imbalance may lead to growth inhibition. With the model it is possible to calculate the feed medium composition for the essential amino acids such that the amino acid concentration profiles are balanced. This is shown on the right plot for a fed-batch fermentation after the medium design.
A second important application of the model is the calculation of optimal time profiles for substrate feeding of fed-batch fermentations [5]. Based on model simulations a feeding strategy for a concentrated medium formulation was developed. The goal of this feeding strategy is to optimize a criterion consisting of the product yield and the production costs (substrate costs, fermentation time). Furthermore, plant and model constraints have to be considered. These include ranges for model validity, measurement ranges, range of the reactor volume, and ranges for feeding rates. This leads to a highly complex nonlinear optimization problem. For the solution of this optimization problem a direct shooting method using piecewise linear control profiles was applied [5].

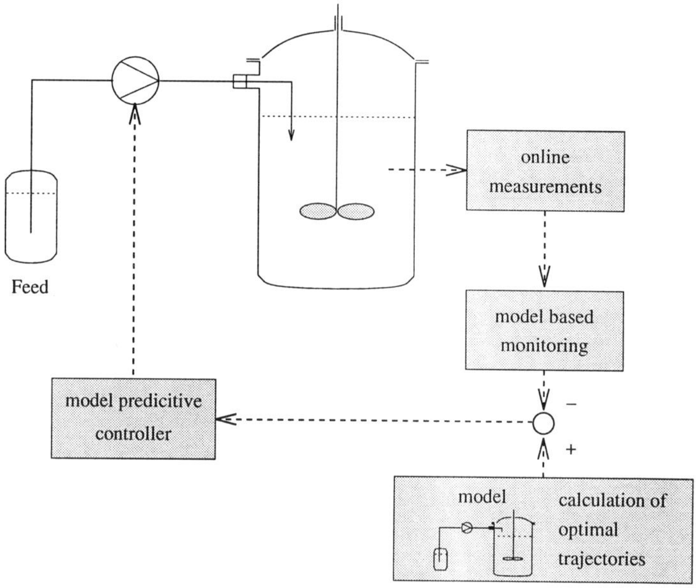

Figure 3: Model-based monitoring and control of a fed-batch bioprocess

In the practical application the process will not exactly follow the predicted trajectories. This is due to disturbances of the process, measurement disturbances, and parameter and model uncertainties. Therefore, a controller is needed to keep the real process on the desired trajectory. Furthermore, the controller has to take into account constraints on state variables (e.g. substrate concentration) and output variables (e.g. maximal feeding rates). A model predictive controller and a nonlinear adaptive controller [6] were used for this control problem.
Figure 5 shows the optimal feed trajectory to maximize the total amount of monoclonal antibody (MAb) produced. The performance of the model-based optimized fed-batch process is shown in Figure 6. This process is compared to an empirically optimized fed-batch process [7]. Compared to this fed-batch process the final MAb-concentration can be improved by 40 %. In addition to that the substrate costs could be reduced by 40 %.

4 Acknowledgement

This work was supported by the State Government of Baden-Württemberg (Germany) and the Dr. Karl Thomae GmbH. The authors thank the Dr. Karl Thomae GmbH and the employees of the Thomae department "Biotechnische Produktion".

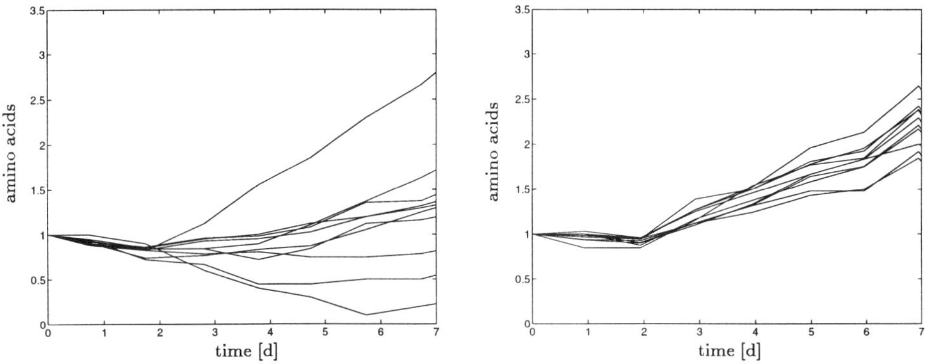

Figure 4: Concentration profiles of essential amino acids (not named in the graph) during a fed-batch run. The concentrations are normalized to the values at time 0. The left plot shows the concentration profiles before the feed medium design for the essential amino acids. The right plot shows the concentration profiles after the feed medium design for the essential amino acids.

References

[1] *Sharfstein, S. T., Tucker, S. N., Mancuso, A., Blanch, H. W., und Clark, D. S.:* Qantitative In Vivo Nuclear Magnetic Resonance Studies of Hybridoma Metabolism. Biotechnology and Bioengineering 43 (1994), S. 1059–1074.

[2] *Fitzpatrick, L., H.A.Jenkins, und M.Butler:* Glucose and glutamine metabolism of a murine b-lymphocyte hybridoma grown in batch culture. Applied biochemistry and biotechnology 43 (1993), S. 93–116.

[3] *Biener, R. K., Waldraff, W., Noé, W., Haas, J., Howaldt, M., und Gilles, E. D.:* Model-based monitoring and control of a monoclonal antibody production process. Proc. Recombinant DNA Biotechnology III: The integration of Biological and Engineering Sciences (Hrsg.: Asenjo, J. A. und Andrews, B. A.), Volume 782 of *Annals of the New York Academy of Sciences*, Engineering Foundation, New York Academy of Sciences, New York 1996, S. 272–285.

[4] *Krebs, V.:* Nichtlineare Filterung. R. Oldenbourg Verlag, München Wien, 1980.

[5] *Waldraff, W.:* Modellgestützte Überwachung und Führung von Fed-Batch-Bioreaktoren. Dissertation, Universität Stuttgart, 1996.

[6] *Estler, M.:* Neue Ansätze zur adaptiven nichtlinearen Regelung von Fed-Batch-Bioprozessen. Dissertation, Universität Stuttgart, 1995.

[7] *Noé, W., Schorn, P., Bux, R., und Berthold, W.:* Fed-batch strategies for mammalian cell cultures. Proc. Animal Cell Technology: Products of today, prospects of tommorrow (Hrsg.: Spier, R., Griffiths, J., und Berthold, W.), European Society for Animal Cell Technology, 12th Meeting in Würzburg/Germany May 1993, Butterworth-Heinemann, Oxford 1993, S. 413–418.

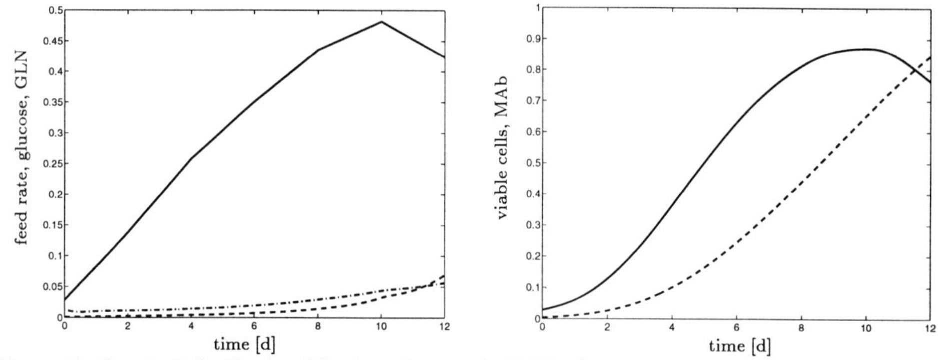

Figure 5: On the left: Optimal feed trajectory (solid line) to maximize the total amount of MAb produced and the corresponding concentration profiles for glucose (- -) and glutamine (- .-) in relative units. On the right: The corresponding profiles of viable cell density (solid line) and monoclonal antibody concentration (- -) in relative units.

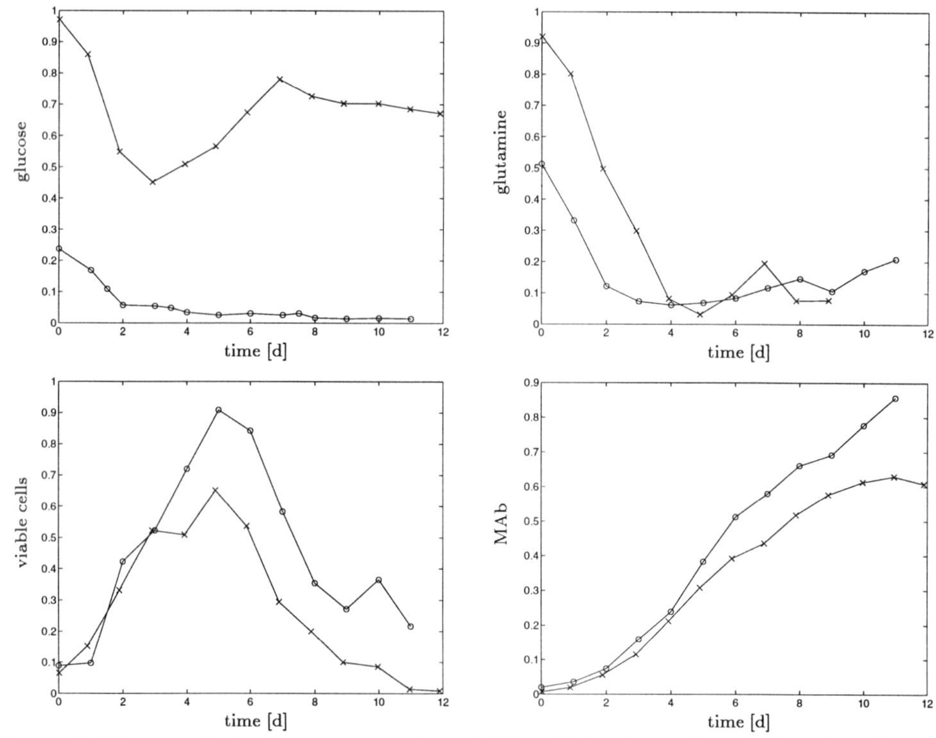

Figure 6: Glucose, glutamine concentration, viable cell density and monoclonal antibody concentration (relative units) for the empirically optimized fed-batch fermentation (x) and for the model-based fed-batch fermentation (o).

Discussion

Anon: You presented this scheme of a metabolic map and had over 100 parameters if you include kinetic information. You stated you determined the rate limiting steps. I think that is not only dependant on the kinetic parameters but on the concentrations that are going into each specific step. How did you deal with that problem? Also, how did you determine the parameters of the remaining equations and how many of those did you actually use?

Biener: The kinetics of the rate limiting steps eg first step in glycolysis is modelled as Michaelis Menten kinetics. This includes glucose concentration in the media. We have a lot of parameters to identify (around 20) and for this we use at least square algorithms. To get information on identifiability we have to use optimal experimental design methods to get these parameters identified. This is basically our approach, and it is very time consuming. We carry out 5 experiments simultaneously to get the results.

Ozturk: Did the specific oxygen consumption rate change with the optimal feeding strategy and the regular culture?

Biener: The difference is that glucose and glutamine are consumed very highly and glucose is almost totally converted to lactate. The specific oxygen rate is at first high and then decreases slowly. In the optimised fed batch culture it is more on a lower level.

Ozturk: How about the antibody production?

Biener: The specific antibody production was almost constant.

Aunins: The empirical culture you are referring to, did that have any feed-back?

Biener: There was no feed-back, just fed-batch.

ALTERATION IN GLYCOSYLATION: CONTROLLING THE DIVERSITY OF RECOMBINANT GLYCOPROTEINS

Thomas Ryll, George Dutina, Riley Speckart, Mary Sliwkowski and Tina Etcheverry
Cell Culture and Fermentation R&D, Genentech, Inc., 460 Point San Bruno Blvd., South San Francisco, CA 94080

N-linked oligosaccharides on a recombinant protein play a critical role in the metabolic behavior of the molecule. The extent of glycosylation on the protein can determine clearance rates and the ultimate pharmacokinetic profile. The amount of terminal sialylation is particularly important in this respect. Changes in the glycosylation pattern due to changes in culture conditions will be shown for a model glycoprotein produced in CHO cells. The parameter determining the sialic acid content will be discussed together with a method to control for this type of heterogeneity.

M.J.T. Carrondo et al. (eds.), Animal Cell Technology, 647–649.

Discussion

Ozturk: How did you change the productivity? Through the growth rate perhaps?

Ryll: I showed you a long list of parameters you can use. We use some of them but I am not able to say which ones.

Gray: Could you tell us something about glycoprotein X. Is it a multi-chain glycoprotein or single chain, the degree of glycosylation?

Ryll: It contains 4 N linked sides but I cannot give you more specifics at the moment. It is a bichain molecule linked by disulphide bonds.

Gray: The number of chains and degree of assembly in the ER would very much influence the different results you got compared to some of our studies.

Ryll: Yes. In general people believe the rate-limiting step is in the transport from the ER to the Golgi for secreted proteins. If you have that type of limitation, usually if the protein enters the Golgi, it is secreted very fast. So the time it spends in the ER, or in transport is important. If you overwhelm the system to secrete a lot more polypeptides then you might end up seeing limitations occurring in the Golgi probably.

Brown: Was the material produced under the higher specific activity a more desirable product from its performance standpoint?

Ryll: As I pointed out the pharmokinetics indicate you should go for a highly sialyated molecule if possible so it is less desirable. There are maybe applications where it is desirable eg if you have a molecule which should clear fast from the patient.

Aunins: You had multiple points for your productivity in batches A through to H etc. Were the time points from the same batch?

Ryll: No, repeated experiments. They are done at 1•5L working volumes, some up to pilot scale so we have repeated a lot of these experiments. If you run an experiment 10 times in a not highly controlled manufacturing environment you get some fluctuations. All the points were put in.

Al-Rubeai: How can you separate the specific productivity from the culture environment? When you talk about serum-free media as soon as the cells start to produce then protein will be in the medium. This changing environment is going to affect the specific productivity.

Ryll: Yes, you can separate between the culture environment and specific productivity. All experiments were in serum free environments and so there is no basic change going from serum to serum-free. When you design a process you define the environment. There are some physicochemical parameters such as salt concentration in which there are a lot of ways of changing osmolality. The productivity should be the same for a particular osmolality even if you use different salts to achieve it, as long as they are not toxic of course.

INHIBITORS OF CELL GROWTH: ACCUMULATION AND CONCENTRATION

Heino Büntemeyer, Sonja Siwiora and Jürgen Lehmann
Institute of Cell Culture Technology, University of Bielefeld, P.O. Box 100131, 33501 Bielefeld, Germany

Abstract

Animal cells cultivated in conditioned medium stop growing, even if a sufficient nutrient supply is guaranteed. Autocrine cell growth inhibitors secreted into the culture medium are responsible for this effect.
The objective of this investigation was the determination of components of conditioned medium which cause low growth rates and cell death. We found that other secreted inhibitors must exist than the metabolites ammonia and lactate.
In order to provide considerable amounts of inhibitory cell culture supernatant for characterisation and purification of the growth inhibitors, a special fermenter system was used. The supernatant produced in this system was concentrated and fractionated. The fractions were tested in a bioassay for inhibition. A dose-dependant effect was observed. Very low concentrations and high efficacy were determined. Several methods were used to obtain a high concentration of these substances. The experiments showed a molecular weight of the inhibitor(s) of less than 5000 Da.

Materials and Methods

Cells:	CHO SS3A2, producing rec. hu. AT III, (Behringwerke, DE) [1]
Medium:	DMEM/F12, serum-free, low protein (Insulin, Transferrin)
Bioreactor:	Biostat BF2, 2 litre, bubble free aeration, internal microfiltration membrane for perfusion [2] (B. Braun Biotech International, DE)
Separation:	Stationary phase: 120 ml of Superdex 30 prep grade (Pharmacia,SE) in XK 16/70 column, 60 cm, (range 0-10000 Da), mobile phase: Basal medium DMEM/F12, flow rate 1 ml/min, sample volume: 7 ml
Concentration:	cross flow ultrafiltration, cut off 1000 Da (Pall Filtron, DE)

MTT-Assay

The MTT - assay is a bioassay which determines the activity of dehydrogenases by a colorimetric reaction (Mossmann 1983, Hansen et al. 1989). Under controlled conditions the test system can be used for indirect determination of cell density. The assay was

M.J.T. Carrondo et al. (eds.), Animal Cell Technology, 651–655.

performed in 96 well microtiter plates as follows: The test fractions from the Superdex separations were supplemented with a Insulin/Transferrin stock solution and inoculated with cells ($1.5 \bullet 10^5$ cells/ml). After a 2-3 days proliferation period the MTT reagent was added. After incubation for 3 hours the reaction product, an insoluble formazan, has to be resolved overnight in 20% aq. SDS - DMF (1:1). The 96 well plate was analysed in a multiwell reader at 570 nm. For each sample 6 wells were assay in parallel (all data show mean values). Controls were made by assaying growth medium on each plate.

RESULTS

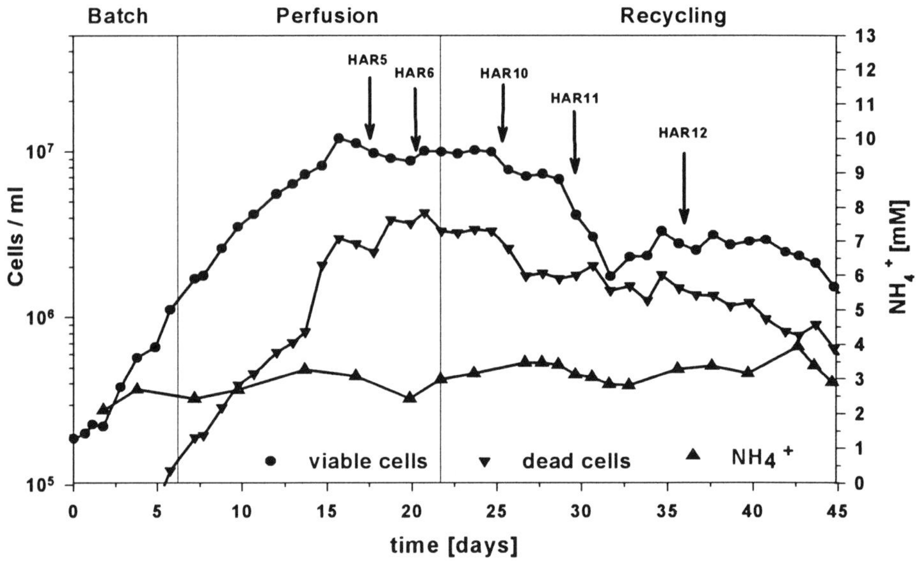

Figure 1: Time course of continuous fermentation for inhibitor production
A continuous perfusion fermentation was performed in a 2 L BIOSTAT BF2 bioreactor. After inoculation and a 6 day batch phase the perfusion process was started with a dilution rate of 0.89/d. A viable cell density of approx. $1 \bullet 10^7$ cells/ml could be obtained. To control viable and dead cell density a cell bleed of 0.16/d was started at day 13 resulting in a stable viable cell density for 7 days. Two cell free harvest pools (HAR5, HAR 6) were collected from the perfusion stream. At day 24 the continuous recycling procedure was started. The perfusion stream (about 65%) was joined with a nutrient concentrate stream and led back to the bioreactor. (All substrates were monitored throughout the complete fermentation. No limitation occured.) For more than 20 days this procedure was continued, resulting in slowly decreasing viable cell density. During different stages cell free harvest pools were taken and analysed (HAR10-12).

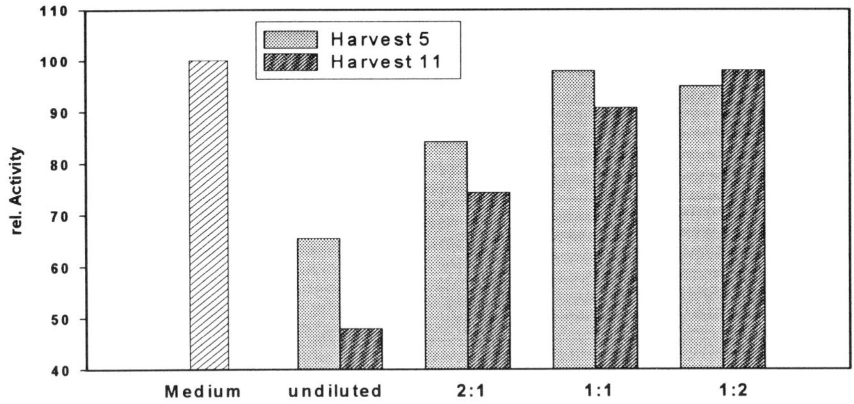

Figure 2: **Activity of HAR 5 and 11 in MTT - assay**
Cell free harvest pools HAR 5 and 11 were investigated directly in the MTT - assay. Different dilutions of HAR 5 and 11 with fresh growth medium were analysed in comparison to growth medium. Inhibition of up to 60% (Harvest 11) was found which could be overcome by 1:1 dilution with medium. The effect of the inhibitor is dose-dependant. A minimal inhibitor level seems to be neccessary to be effective.

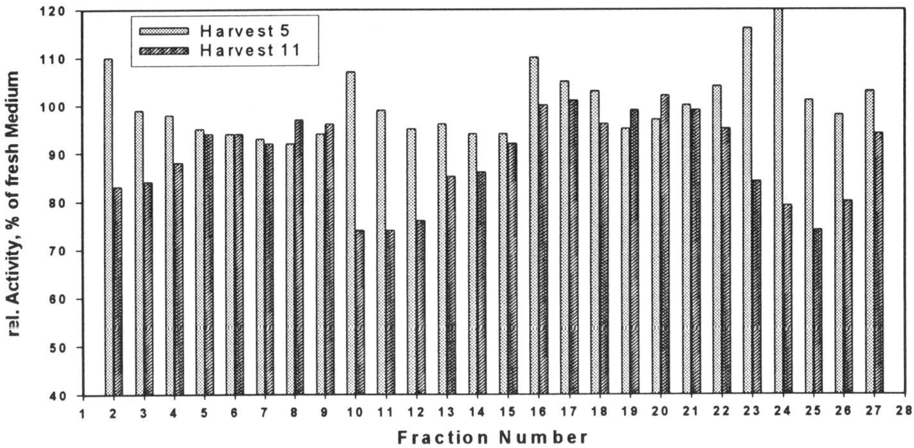

Figure 3: **Activities of Superdex 30 fractions of HAR 5 and 11**
HAR 5 and 11 were separated on a Superdex 30 gelfiltration column. Each fraction was analysed in the MTT - assay. No fraction of HAR 5 separation showed an inhibition although the total HAR 5 was inhibiting (see Fig.2). In contrast to HAR 5, two ranges of inhibiting fractions were found when

separating HAR 11, in the low molecular range (fractions 23-26) and in a higher molecular range (fractions 10-12).

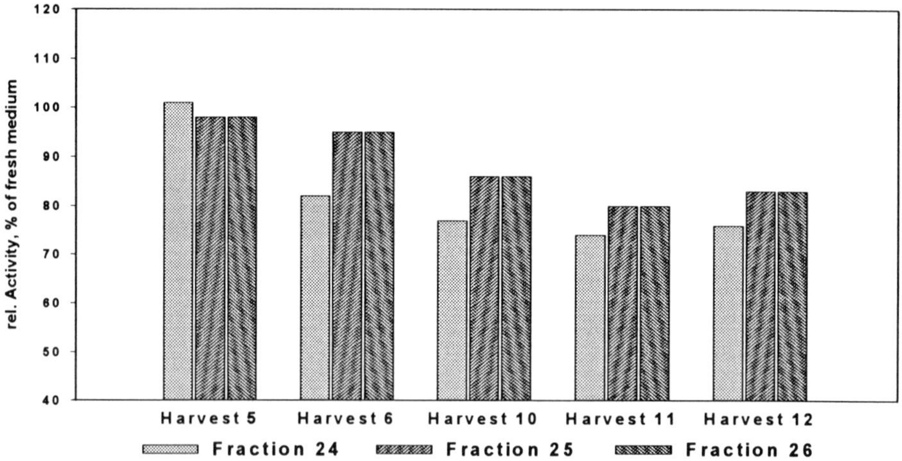

Figure 4: **Inhibitory activities of certain fractions**
HAR 5 to 12 were separated on a Superdex 30 gelfiltration column. Fractions 24-26 of all separations are shown in comparison. Fractions from harvests coming from perfusion mode (HAR 5,6) show no or very low inhibition effect whereas fractions from harvest coming from recycling show clearly inhibitory activity.

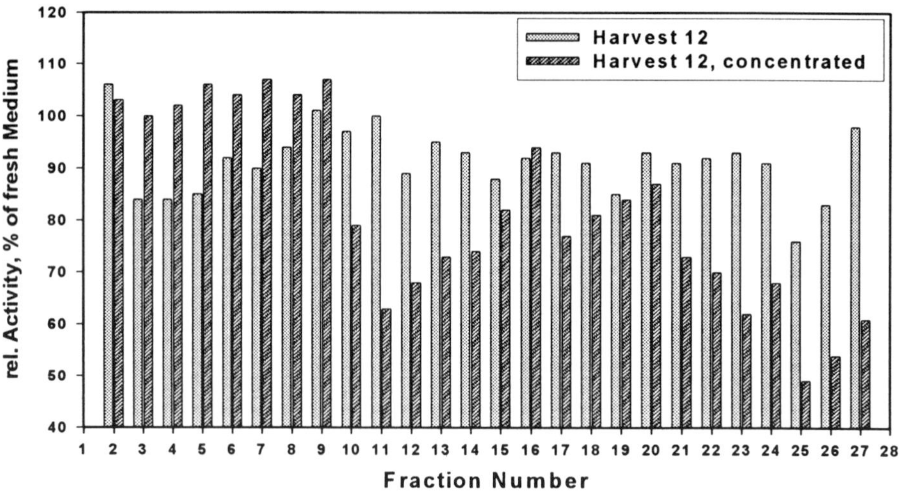

Figure 5: **Effect of harvest concentration**

An aliquot of harvest 12 was concentrated with a 1000 Da ultrafiltration membrane (Filtron) in a cross flow separation device. The original harvest 12 and the concentrated harvest 12 were gelfiltrated on Superdex 30. Each fraction from both separations was tested in the MTT-assay for activity. With this method it is possible to concentrate inhibitor substances.

Conclusions

The supernatant from recycling/perfusion processes can contain cell growth inhibitors. The inhibitor effect is dose-dependant. The inhibitor is of a molecular weight less than 5000 Daltons and can be concentrated by ultrafiltration. The inhibitor found in this study is not ammonia nor lactate!

References

1. Zettmeissl G, Ragg H, Karges HE (1987) *Expression of biologically active human Antithrombin III in chinese hamster ovary cells.* Biotechnology **5**, 720-725
2. Büntemeyer H, Wallerius C, Lehmann J (1992) *Optimal medium use for continuous high density perfusion processes.* Cytotechnology **9**, 59-67
3. Hansen MB, Nielsen SE, Berg K (1989) *Re-examination and further development of a precise and rapid dye method for measuring cell growth/cell kill.* J. Immunol. Meth. **119**, 203-210
4. Mossmann T. (1983) *Rapid colorimetric assay for cellular growth and survival: Application to proliferation and cytotoxicity assays.* J. Immunol. Meth. **65**, 53-63

Acknowledgements

This study was supported by the Bundesministerium für Bildung, Wissenschaft, Forschung und Technologie, BMBF (Nr. 0310754), Germany and Boehringer Mannheim, Penzberg, Germany.

OPTIMISATION OF THE FATTY ACID COMPOSITION OF A HYBRIDOMA ENHANCES MAB PRODUCTION AND PERTURBS ENERGY METABOLISM

M. BUTLER AND N. HUZEL
Department of Microbiology, University of Manitoba,
Winnipeg, Manitoba, CANADA R3T 2N2

Abstract

Unsaturated fatty acids stimulate growth and antibody productivity of serum-free hybridoma cultures. However continuous culture passage with fatty acids results in a gradual decline in Mab productivity. The inability of cells to regulate uptake results in the accumulation of exogenous fatty acids in all cellular lipid fractions. The saturated/ unsaturated fatty ratio of the polar lipid fraction changed from 1.39 to 0.13 following one culture passage with linoleic acid.

The increased linoleic composition of polar lipids is associated with a decreased permeability of membranes for glutamine. Consequently the cells rely increasingly on glucose for energy metabolism. This results in a significant increase in the specific lactate production rate and the molar yield coefficient, $Y_{lac/glc.}$ The metabolic changes are reversed by growing the cells in fatty acid-free media. This also restores the cells to high Mab productivity. The optimal intracellular lipid content is finely balanced between a reduced and over-loaded state. The fatty acid composition of the cells has been analysed at the point of maximum Mab productivity.

Introduction

Fatty acids are often included in serum-free cultures of antibody-secreting hybridomas to replace the lipid component of serum. However, the effect of fatty acids on the growth of cultured mammalian cells is variable. There are cases in which fatty acids have been observed to stimulate or inhibit cell growth (1,2).

In a previous report (3) we showed that linoleic or oleic acid enhances significantly the cell yield and monoclonal antibody productivity of a B-lymphocyte hybridoma (CC9C10). However, continued culture passage with the unsaturated fatty acids leads to a lipid-loaded state in which cells maintain a high capacity for growth but a decreased capacity for antibody production. The mechanism of growth-promotion of these fatty acids is unclear but may be related to their importance in the synthesis of cellular membranes (4,5). Changes in the fatty acid composition of phospholipids can have a significant effect on membrane fluidity (6) which may affect permeability to energy

M.J.T. Carrondo et al. (eds.), Animal Cell Technology, 657–662.

substrates. We now present data to show that there are major differences in the energy metabolism of the B-lymphocyte hybridoma following prolonged growth in the presence of linoleic acid. These differences are dependent upon the fatty acid composition of the intracellular lipid fractions and are partially reversible.

Materials and Methods

Cell line: The murine B-lymphocyte hybridoma (CC9C10), which secretes a monoclonal antibody (IgG_1) against insulin, was obtained from the ATCC (No HB123). The cells were adapted to a serum-free medium over several passages and grown for at least 10 passages in this medium prior to the described experiments.
Culture: The basal medium consisted of DMEM: Ham`s F12 (1:1v/v) supplemented with a serum-free mixture of hormones and micronutrients (3). Cultures were also supplemented, where indicated, with fatty acid-free bovine serum albumin (0.1 mg ml^{-1}) complexed to a specific concentration of oleic or linoleic acid. Cultures were monitored by cell counting using a Coulter counter or haemocytometer. Viability was determined by use of trypan blue (0.2%).
Lipid extraction and separation. Lipids were extracted from cells by organic solvents as previously described (7). The cellular lipid extract was fractionated by a bonded phase column consisting of 100 mg aminopropyl glass beads (Sigma) between fibreglass plugs in a Pasteur pipette. Fractions were separated by sequential elution with 3 ml chloroform/ isopropanol ((2:1), 3ml diethyl ether/ acetic acid (98:2) and 4 ml methanol.
Fatty acid analysis. Lipid fractions were saponified by 0.27 M methanolic KOH for 20 min at 80^0C and derivatised with p-bromophenacyl bromide (7). The derivatised fatty acids were separated by HPLC (Pharmacia/ LKB) using a C-8 column (25 cm x 4.6 mm; Beckman). Margaric acid (17:0) was used as an internal standard. Quantification was by a spectral detector set at 254 nm and an associated integrator.

Results and Discussion

1. The effect of individual fatty acids on cell growth

Twelve fatty acids were tested individually for their ability to promote cell growth (Fig. 1). Only oleic and linoleic acid caused significant and substantial growth enhancement (>300%) compared to the fatty acid free control cultures. Oleic acid was optimal at a concentration of 50 μM and linoleic acid at 25 μM. In the presence of these fatty acids the doubling time of the cells was 17 h as opposed to 28 h for control cultures. At a higher concentration of either of these unsaturated fatty acids (>75 μM), the maximum cell density fell below the level of the control cultures.

Four fatty acids (caproic, lauric, margaric and stearic acids) had no effect on cell growth over the concentration range tested. Four fatty acids (arachidonic, linolenic, caprylic and myristic acids) caused a concentration dependent inhibition of cell growth. Two fatty acids (palmitic and capric acids) enhanced cell growth marginally (9%) but significantly at 25 μM.

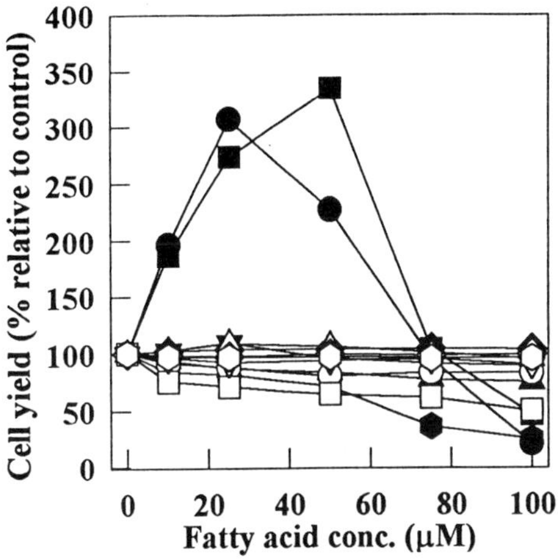

Fig. 1: The effect of fatty acids on the growth of CC9C10 cells. Serum-free cultures (2 ml) were established in 24-well plates in the presence of one of the following fatty acids: linoleic (18:2) (●), oleic (18:1) (■), linolenic (18:3) (⬣), arachidonic (20:4) (□), caproic (6:0) (⬡), caprylic (8:0) (▲), capric (10:0) (▼), lauric (12:0) (♦), myristic (14:0) (○), palmitic (16:0) (△), margaric (17:0) (◇) or stearic (18:0) (▽) acids. Cell yields were determined after 3 days growth. Each point is a mean (±SEM). n=3.

2. Continuous passage in fatty acid-based media.

The hybridomas were grown over several passages in media supplemented with linoleic acid (25 μM), oleic acid (25 μM) or an equimolar mix of oleic/ linoleic (25 μM). Culture over 5 passages in medium containing the unsaturated fatty acids indicated a consistent enhancement of cell yield (Fig. 2). Linoleic acid enhanced growth more than oleic acid but a mixture of oleic and linoleic acid provided the greatest growth stimulation. The growth enhancement effect was shown to be reversible. When fatty acid-grown cells were re-introduced into fatty acid-free medium, the growth advantage over control cultures was lost within 2 to 3 passages.

Fig. 2: Cell and Mab yields over consecutive culture passages. Control cultures (●) were compared with those supplemented with 25 μM linoleic (■), oleic (▲) or oleic/ linoleic mix (▼). Yields at each passage were determined after 4 d growth in 75 cm^2 T-flasks. At the arrow cells were re-introduced into fatty acid-free medium. Points are means (±SEM) for n = 2 or 3.

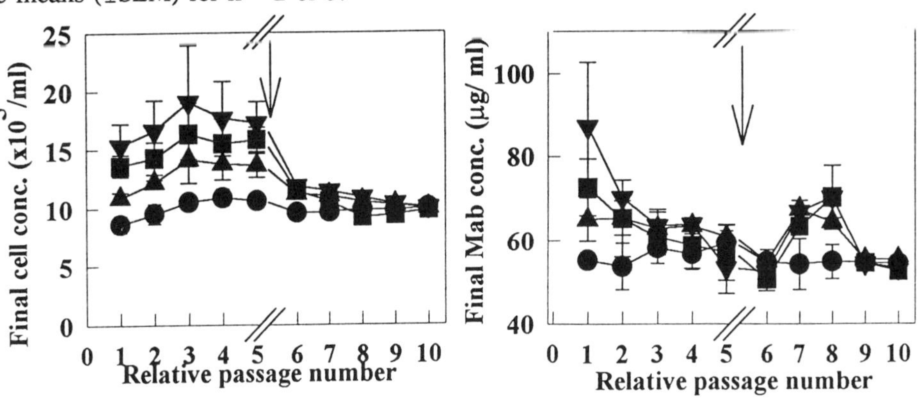

The Mab yield increased significantly at the first passage of growth in the presence of fatty acids. However, after 3 passages there was no difference in Mab yield from the control culture. The effect of removing the fatty acids after 5 passages was to temporarily restore the higher Mab yield of the cells previously grown in fatty acid. However, this advantage was eroded after 4 passages.

3. Optimal fatty acid loading of cells

During continuous passage the cultures assumed a transitory state in which the cell and Mab yields were maximised. This coincided with passage 1 and a partial restoration of this state occurred at passage 7-8. (Fig. 2). It was our objective to analyse the fatty acid content of the cells during this state of maximum productivity. Fatty acids accumulated in 3 intracellular fractions - polar, non-polar lipid and free fatty acids (Table 1). The data shows that in fatty acid-free control cultures, >90% of the cellular content of fatty acids could be accounted by linoleic, palmitic, oleic, stearic and arachidonic acids. Of these major fatty acids, the proportion of linoleic acid varied from 0-7% whereas oleic acid was present at 24-31%.

Growth in the presence of linoleic acid resulted in the content of this fatty acid to increase in all three lipid fractions. The increase was particularly significant in the polar lipid fraction which increased from an undetectable level in the control cells to 84.5% of the total fatty acid content after 1 passage. The balance of fatty acids also changed significantly as evidenced by the tenfold decrease in saturated/ unsaturated fatty acid ratio in the polar lipid fraction of linoleic-grown cells.

TABLE 1 Fatty acid distribution in 3 intracellular fractions (μmole/10^6 cells). Cells were grown for 1 passage in media supplemented with 25 μM linoleic acid. Control cultures contained no added fatty acid. Values are means (n=2)

Fatty acid	Control cells			Linoleic-grown/ passage #1		
	free fatty acids	neutral lipids	polar lipids	free fatty acids	neutral lipids	polar lipids
16:0	1.32	0.99	0.49	1.06	0.47	0.49
18:0	1.72	0.74	0.65	0.57	0.17	0.19
18:1(n-9)	1.29	0.91	1.12	0.49	0.27	0.32
18:2(n-6)	0.38	0.24	0	3.90	4.83	8.10
20:4(n-6)	0.22	0.24	0.05	0.19	0.00	0.02
TOTAL	4.93	3.12	2.31	6.21	5.74	9.12
satd/unsatd ratio	1.79	1.35	1.39	0.47	0.16	0.13

4. The effect of fatty acids on glucose and glutamine metabolism

The specific consumption of glucose and glutamine was determined from analysis of the media harvested during continuous passage of the cells. For each culture the rate of decrease of the concentration of glucose and glutamine was linear over 4 days of culture and neither substrate was completely exhausted from the media. The glucose decreased from an initial value of 17.5 mM to 0.7-3.8 mM whereas the glutamine decreased from an initial 6 mM to 0.5-2.9 mM.

Analysis of the specific glucose consumption rate (q_{glc}) showed that the presence of a fatty acid in the medium (passage 1-5 in Fig. 3) reduced the q_{glc} by a small but significant percentage ($P<0.01$). However re-inoculation of the fatty acid grown cells in fatty acid-free medium caused an increase of the q_{glc} up to the level of the control .

The specific glutamine utilisation rate (q_{gln}) decreased significantly and substantially in fatty acid containing cultures (Fig. 3) by a factor of x1.9. Furthermore, this change was reversible as shown by the increase in q_{gln} when the fatty acid-grown cells were inoculated into a fatty acid-free medium. Short-term radioactive studies showed that this change in glutamine consumption is associated with a x5 decrease in membrane transport of glutamine for cells grown in linoleic acid.

Fig 3: Specific consumption/ production rates of glucose, glutamine, lactate and ammonia (q_{glc} , q_{gln}, q_{lac} and q_{amm}) were determined from concentration measurements made at the end of each of the cultures described in Fig. 2. At the arrow cells were re-introduced into fatty acid-free medium. Error bars were smaller than the symbols used (n-=3).Control cultures (●) linoleic (■), oleic (▲) or oleic/ linoleic mix (▼).

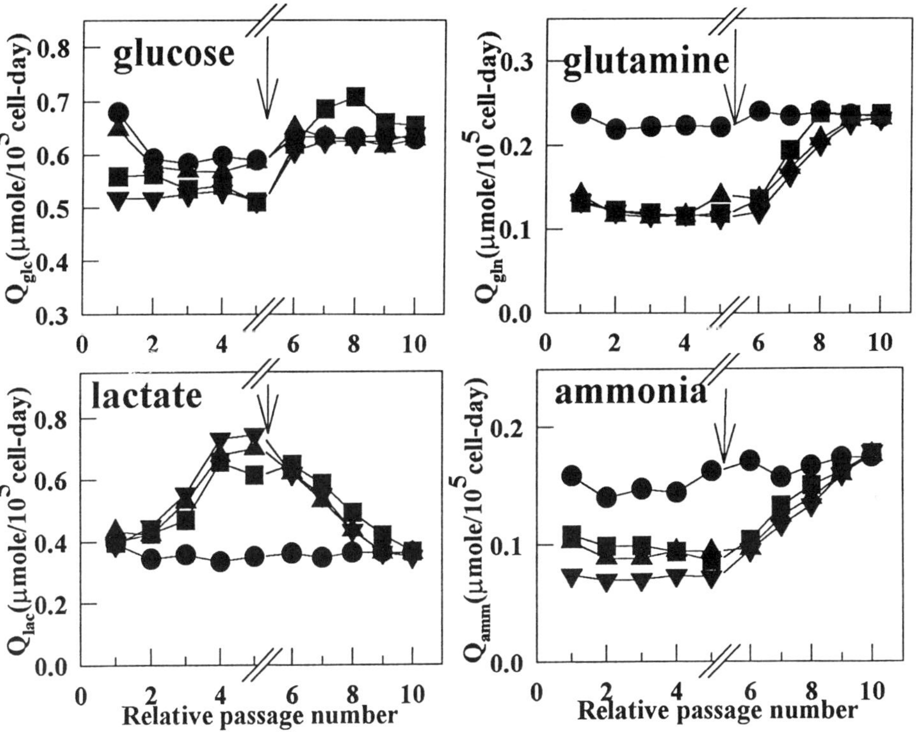

Measurements of the specific rates of production of the metabolic by-products, ammonia and lactate (q_{amm} and q_{lac}) gave an indication of a significantly different metabolism occurring in the fatty acid-grown cells (Fig. 3). The q_{amm} decreased significantly in the presence of fatty acids whereas the q_{lac} increased significantly. The increased lactate production resulted in a lower pH at the end of each culture. These changes were reversible on removal of the fatty acids.

The molar yield coefficients ($Y_{amm/gln}$ and $Y_{lac/glc}$) have often been used as indicators of patterns of energy metabolism. It is noteworthy that the $Y_{amm/gln}$ did not differ significantly between fatty acid-grown and control cells whereas the $Y_{lac/glc}$ was substantially higher in the presence of fatty acids. This suggests that although the glutamine consumption declines significantly in the presence of fatty acid, the pattern of its metabolism is unaffected. On the other hand, although the glucose consumption rate was only altered slightly, the pattern of glucose metabolism appeared to be altered significantly.

Acknowledgements

We acknowledge financial support from NSERC, Apotex and Rh Pharmaceuticals.

References

1. Grammatikos, S.I., Subbaiah, P.V., Victor, T.A. and Miller, W.M. (1994) Diverse effects of essential (n-6 and n-3) fatty acids on cultured cells. Cytotechnology 15, 31-50.
2. Jager, V., Lehmann, J. and Friedl, P. (1988) Serum-free growth medium for the cultivation of a wide spectrum of mammalian cells in stirred bioreactors. Cytotechnology 1, 319-329.
3. Butler,M. and Huzel,N. (1995) The effect of fatty acids on hybridoma cell growth and antibody productivity in serum-free cultures. J. Biotechnol. 39, 165-173.
4. Rockwell, G.A., Sato, G.H. and McClure, D.B. (1980) The growth requirements of SV40 virus transformed Balb/c-3T3 cells in serum-free monolayer culture. J. Cell Physiol. 103, 323-331.
5. Rintoul, D.A, Sklar, L.A. and Simoni, R.D. (1978) Membrane lipid modification of Chinese hamster ovary cells. Thermal properties of membrane phospholipids. J. Biol. Chem. 253, 7447-7452.
6. Calder, P.C., Yaqoob, P., Harvey , D.J, Watts, A. and Newsholme, E.A. (1994) Incorporation of fatty acids by concanavalin A-stimulated lymphocytes and the effect on fatty acid composition and membrane fluidity. Biochem. J. 300, 509-518.
7. Puttmann, M., Krug, H., Vonochsenstein, E., Kattermann, R. (1993) Fast HPLC determination of serum-free fatty acids in the picomole range. Clinical Chemistry. 39, 825-832.

Effect of Power Input in Virus Like Particles Production

Pedro E. Cruz[1], Cristina C. Peixoto[1], José L. Moreira[1], Manuel J. T. Carrondo[1,2]
1 - Inst. Biol. Exp. Tecn. / Inst. Tecn. Quim. Biol. IBET/ITQB, Ap. 12, 2780 Oeiras, Portugal
2 - Lab. Eng. Bioq., FCT/UNL, 2825 Monte da Caparica, Portugal

Abstract

In this work Sf-9 cells previously adapted to SF900II serum free medium were grown in 125 cm^3 stirred tanks at different agitation rates in order to study the effects of both medium and power input in cell growth, infection and production. The maximum cell concentration in serum free medium increased from 3 x 10^6 to 5 x 10^6 cells/cm^3 when the agitation rate increased from 80 to 170 rpm, clearly indicating that cell growth is limited by oxygen. The product concentration at 120 hours post infection (hpi) was higher in SF900II medium at 170 rpm than in serum containing medium. It is expected that the specific productivity in SF900II can still be increased by increasing oxygen transfer; this hypothesis results from the observation that the productivity increased 3.5 fold when agitation rate was increased from 80 to 170 rpm. Thus, in serum free medium both cell growth and cell infection are oxygen limited and can be further optimized.

Key words: Insect cells, baculovirus, infection, virus like particles, power input

Introduction

In the last decade the insect cell / baculovirus expression system has been used successfully for production of many eucaryotic proteins. In recent years, the use of recombinant baculovirus expressing several proteins became increasingly important since with this system the recombinant proteins form complex quaternary structures (core and virus like particles) that can be used as human and veterinary candidate viral vaccines (1,2,3). Although many studies of optimization of insect cell infection with single expression baculovirus have been performed (4), there is reduced published data concerning optimization of multiple expression baculovirus infection. Regarding the production and purification of proteins produced by insect cells, several serum free media were developed (5). Due to the high oxygen demand of insect cells after infection (6) and high sensitivity to shear stress (7), for optimization and scale-up it is essential to know the influence of power input upon cell growth and infection.

M.J.T. Carrondo et al. (eds.), Animal Cell Technology, 663–668.

Materials and Methods

Cell growth and infection
Sf-9 cells (ECACC no. 89070101) previoulsy adapted to SF900II (Gibco, UK) serum free medium were grown in 125 cm^3 Wheaton ball impeller stirred vessels (Wheaton, USA) at 27°C. Cells were maintained in stirred vessels at 170 rpm and passed twice a week with an inoculum of 2-4 x 10^5 cells/cm^3. Pluronic F-68 (Gibco, UK) was added to the medium in the concentration of 0.1 and 0.2% (w/v, final concentration) for serum containing and serum free medium, respectively. The infection was studied using baculovirus that generate virus like particles, provided by Dr. Polly Roy (IVEM, Oxford, UK) with titres determined by plaque assay. The infection was performed in the mid exponential phase of cell growth with a multiplicity of infection (MOI) of 2-5 pfu/cell. Samples were taken at several times post infection in order to evaluate product quantity and quality.
Analytical
Cell concentration was determined by hemacytometer cell counts and cell viability was evaluated by trypan blue exclusion dye. Product was quantified by using specific ELISA tests provided by N.V. Innogenetics S.A. (Ghent, Belgium).

Results and Discussion

Effect of power input on cell growth
The influence of power input on cell growth was studied by varying the agitation rate between 50 and 170 rpm in TC100 with 10% FBS (control) and SF900II. The results are depicted in Figure 1.

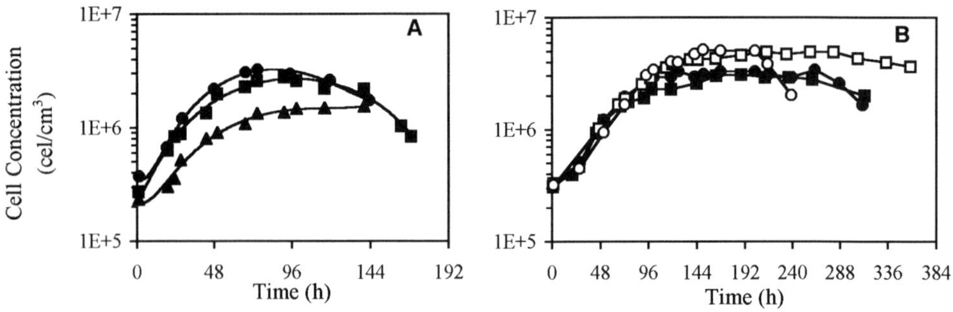

Figure 1 - Variation of the viable cell concentration with time and specific growth rates of Sf-9 cells in (A) TC100 supplemented with 10% serum at 50 rpm (▲), 80 rpm (■) and 170 rpm (●) and (B) SF900II at 80 rpm (□) and 170 rpm (O).

As can be observed from Figure 1 in serum containing medium, cell growth and maximum concentration remain approximately constant for agitation rates larger than

50 rpm. At 50 rpm cell growth is reduced; this can be associated with cell oxygen transfer limitations through the liquid surface since it has been proved that Sf-9 cell growth is strongly inhibited under oxygen limitations (8). In serum free medium the specific growth rates obtained were similar to the ones in serum containing medium at both 80 and 170 rpm (data not shown). From Figure 1 it can also be observed that in serum free medium the maximum cell concentration increased almost two fold with an increase in the agitation rate from 80 to 170 rpm; this can be associated to an increase in oxygen transfer into the liquid through the gas/liquid interface. In serum containing medium this increase in maximum cell concentration with agitation rate was not verified due to medium depletion since it was observed that cell viability starts to decrease when the cell concentration of 3 x 10^6 cells/cm^3 is reached, independently of the agitation rate (data not shown). In fact, up to now in similar experimental conditions a maximum cell concentration of 2.5 x 10^6 cells/cm^3 in TC100 with 10% serum was reported (9). Therefore, it can be concluded that in TC100 with 10% serum there is no oxygen limitation at 80 rpm or an increase in cell concentration would be observed at 170 rpm as was observed in serum free medium; cell growth is, in this medium and for agitation rates higher or equal to 80 rpm, limited by medium components.

In serum free medium maximum cell concentrations of 5 x 10^6 cells/cm^3 were achieved with this cell concentration remaining approximately constant for 6 days. Therefore, it should be possible to obtain even larger cell concentrations if the system is further optimized in terms of oxygen supply. This can be performed either by increasing power input or using sparged aeration, as previously described for mammalian cells (10).

Effect of medium and power input in cell infection

In order to study the influence of the medium and power input in cell infection the cells were infected with a MOI of 2 in SF900II and TC100 + 10% FBS at 80 and 170 rpm (Figure 2).

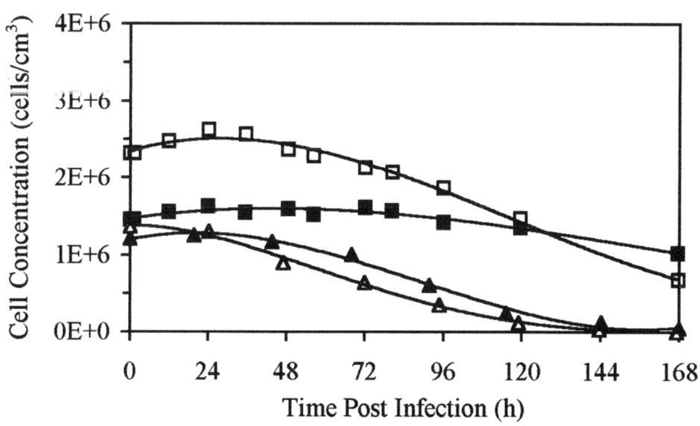

Figure 2 - Variation of viable cell concentration with time post infection at 80 rpm and 170 rpm in SF900II (■,□) and TC100 supplemented with 10% serum (▲,Δ).

As can be observed from Figure 2, the infection process seems to be faster in serum containing medium since the viable cell concentration decreases more rapidly at both 80 and 170 rpm. This can be explained by the higher oxygen transfer in the serum containing medium since it is known that lipids and surfactants like Pluronic F68 (which exist in higher concentration in serum free medium than in serum containing medium) cause a decrease in k_la value (10).
It can also be observed that in serum containing medium the increase in the agitation rate from 80 to 170 had only a small influence in the infection process while in SF900II medium the same increase in power input caused a dramatic increase in the rate of cell death. This indicates that for cell infection in serum containing medium the process is limited by medium components, as observed in cell growth experiments (Figure 1), while in SF900II it is limited by oxygen transfer. In SF900II the increase in oxygen transfer caused by the increase in power input can explain these observations, thus indicating that the infection process can be further optimized. This observation has also important consequences in scale-up since it indicates that oxygen transfer and power input are probably the scale-up criteria to use in larger volume stirred tanks.

Effect of power input in product concentration
The variation of product concentration in the conditioned medium with time post infection was determined for the media tested at 80 and 170 rpm (Figure 3).

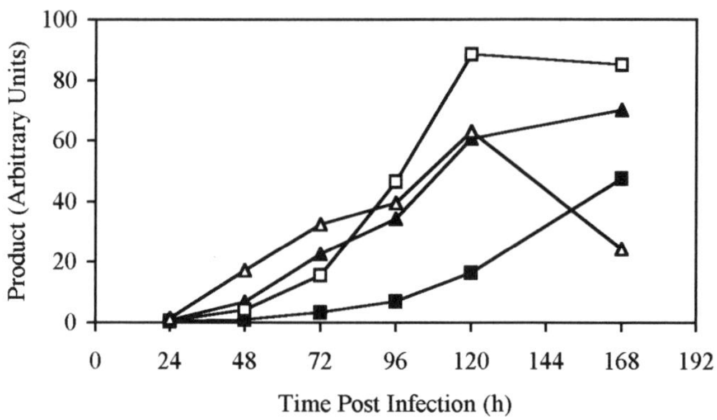

Figure 3 - Variation of product concentration with time post infection at 80 rpm and 170 rpm in SF900II (■,□) and TC100 supplemented with 10% serum (▲,Δ).

The data shown in Figure 3 clearly indicates that for the same agitation rate the specific productivity in serum containing medium is higher than in serum free medium, confirming the previous observations presented in Figure 2. In fact, it has been observed that protein production is largely influenced by oxygen limitations (11). The final product concentration obtained at 170 rpm in SF900II at 120 hpi was much higher than in serum containing medium. This means that for the same volume more product is being produced particularly in the initial times. In addition, the fact that at 120 hpi the overall specific productivity in TC100 medium (at both agitation rates) was larger than in SF900II medium at 170 rpm (since in this case cell concentration at the time of infection was two fold higher) is an indication that there are still oxygen limitations in SF900II medium and therefore infection can be further optimized by improved oxygenation. The effect of the increase in agitation rate in SF900II medium is quite evident as the specific productivity at 80 rpm is 3.5 fold lower than at 170 rpm. This result is in perfect accordance with the previous observations concerning cell infection (Figure 2).

At 170 rpm in both TC100 and SF900II media a decrease in product concentration was observed between 120 and 168 hpi; this can be associated with product degradation caused by proteases resulting from cell lysis.

Therefore, the optimal harvest time should consider not only product titre but also product quality since candidate viral vaccines should be intact particles, easy to separate from contaminants.

Conclusions

In serum free medium higher cell concentrations were achieved than in serum containing medium.

In serum free medium the maximum cell concentration was further increased when higher agitation rates were used, indicating that cell growth is oxygen limited; this was not observed in serum containing medium which indicates that in this case cell growth is limited by medium components.

The final product concentration was higher in SF900II medium at 170 rpm than in serum containing medium.

The productivity increased 3.5 times when the agitation rate was increased from 80 to 170 rpm. This means that in serum free medium both cell growth and cell infection are oxygen limited and should be further improved by increasing oxygen transfer.

A careful optimization of the harvest time based on product quality is required since significant product degradation resulting from cell lysis was observed at later times post infection.

References

1. Roy P, Bishop DHL, LeBlois H and Erasmus BJ (1994) Long Lasting Protection of Sheep Against Bluetongue Challenge after Vaccination with Virus-Like Particles: Evidence for Homologous and Partial Heterologous Protection. Vaccine, 12, 805-811.
2. Zhao Y, Jones IM, Hockley DJ, Nermut MV and Roy P (1994) Complementation of Human Immunodeficiency Virus (HIV-1) Gag Particle Formation. Virology, 199, 403-408.
3. Tsao EI, Mason MR, Cacciuttolo MA, Bowen SH, Folena-Wasserman G (1996) Production of Parvovirus B19 Vaccine in Insect Cells Co-infected with Double Baculovirus. Biotechnology Bioengineering, 49, 130-138.
4. Licari P and Bailey JE (1992) Modeling the Population Dynamics of Baculovirus Infected Insect Cells: Optimizing Infection Strategies for Enhanced Recombinant Protein Yields. Biotechnology Bioengineering, 39, 432-441.

5. Schlaeger E-J, Foggetta M, Vonach JM and Christensen K (1993) SF-1, a Low Cost Culture Medium for the Production of Recombinant Proteins in Baculovirus Infected Insect Cells. Biotechnology Techniques, 7, 183-188.
6. Reuveny S, Kim YJ, Kemp CW and Shiloach J. (1993) Effect of Temperture and Oxygen on Cell Growth and Recombinant Protein Production in Insect Cell Cultures. Applied Microbiology and Biotechnology, 38, 619-623.
7. Murhammer DW and Goochee CF (1988) Scaleup of Insect Cell Cultures: Protective Effects of Pluronic F-68. Bio/Technology, 6, 1411-1418.
8 Wang M-Y, Kwong S and Bentley WE (1993) Effects of Oxygen/Glucose/Glutamine Feeding on Insect Cell Baculovirus Protein Expression: A Study on Epoxide Hydrolase Production. Biotechnology Progress, 9, 355-361.
9. Akhnoukh R, Kretzner G and Schügerl K (1993) Investigation of the Cultivation of Spodoptera frugiperda in Spinner Flasks and in a 10 l Bioreactor with Bubble Free Aeration. Bioprocess Engineering, 8, 229-233.
10. Moreira JL, Cruz PE, Santana PC, Feliciano AS, Lehmann J and Carrondo MJT (1995) Influence of Power Input and Aeration Method on Mass Transfer in a Laboratory Animal Cell Culture Vessel. Journal of Chemical Technology and Biotechnology, 62, 118-131.
11. Lindsay DA and Betenbaugh MJ (1992) Quantification of Cell Culture Factors Affecting Recombinant Protein Yields in Baculovirus Infected Insect Cells. Biotechnology Bioengineering, 39, 614-618.

Acknowledgments

The authors acknowledge and appreciate the financial support received from the European Commission (VALUE CE-CTT-535) and Junta Nacional de Investigação Científica e Tecnológica - Portugal (EUREKA PUEM/C/EU and BD/4545/94) and the partnership with N.V. Innogenetics S.A..

ISOLATION AND CHARACTERIZATION OF A CHO CELL LINE EXPRESSING rhFSH IN LOW PROTEIN AND PROTEIN-FREE MEDIA

Carol Luchette, Frank Deer, Kim Gurnett, Judy Rosenthal, Erik Johnson, Angelo Polito and Mark Cunningham.

Ares Advanced Technology, 280 Pond Street, Randolph, Massachusetts 02368, USA

Key words: recombinant protein; serum-free media; bioproduction; adherent; suspension.

SUMMARY

A recombinant CHO cell line expressing rhFSH, CL1, was isolated from a parental clone using an unconventional approach. CL1 was shown to possess a contrasting phenotype as well as several process advantages when compared to the parental clone. These included production of rhFSH in basal protein-free IMDM/F12, a characteristic never exhibited by the parental clone. The CL1 rhFSH production level in basal medium was 4-fold greater than that observed with the parental clone in serum-containing medium. The recombinant product derived from cultured CL1 displayed identical "*in vitro*" bioactivity to rhFSH produced by the parental clone. The potential of CL1 as a production cell line was demonstrated when both adherent and suspension perfusion cultures produced rhFSH at levels in excess of 14 mg/l reactor/day.

INTRODUCTION

The Chinese hamster ovary cell (CHO) is an established host in expression of recombinant proteins (1). Methods to select clonal lines containing genes amplified to high copy numbers are now routine (2). As a production tool the advantages of CHO cells include authentic protein folding and posttranslational modification, ease of culturing and regulatory acceptance (3). However, one disadvantage of this mammalian cell line is the complex nature of the production conditions required to express recombinant proteins at levels suitable for commercial development (4). The use of a chemically defined (protein-free) production medium, has obvious advantages over the use of serum or serum fractions (5). However, developing such media requires extensive research activities and is often production clone specific (6). Even after such optimization efforts, it is not always possible to adapt an existing high expressing production cell line to a protein-free production environment. We transfected, amplified and selected a stable CHO clone to produce recombinant human follicle stimulating hormone (rhFSH), a heterodimeric protein requiring glycosylation for bioactivity (7). This parental clone produced rhFSH in serum-containing medium but when placed in a protein-free environment, the clone failed to survive and rhFSH expression was lost. This paper describes a strategy to isolate, from the parental clone, a cell line whose phenotype is more desirable i.e. expression of rhFSH in a chemically defined, protein-free, medium. We employed a 'sink or swim' approach, that favored the survival of a phenotype which survived and expressed rhFSH in the protein-free basal medium IMDM/F12. Not only did CL1 produce rhFSH in basal

M.J.T. Carrondo et al. (eds.), Animal Cell Technology, 669–674.

medium, it did so at levels greater than the parental clone cultured in serum containing medium. When bioactivity of the rhFSH expressed by CL1 was compared to that produced by the parental clone it was found to be equally as active. In addition, we show that CL1 has process utility. Evaluation in bioreactors using both adherent and suspension based perfusion processes demonstrated that CL1 had the characteristics of a production cell line.

MATERIALS AND METHODS

Cell Lines

The parental clone was generated using standard techniques. CL1 was isolated from a T-flask culture of the parental clone initially grown to confluence in serum-containing medium. The confluent culture was washed with sterile PBS and fed every three to four days with basal IMDM/F12 for a period of 30 days. Cell death was observed within the flask after 1 week. Dead cells were continuously lost with each feeding. One, of only two colonies, that survived at the end of the 30 day study, was recovered using a cotton swab soaked in trypsin/EDTA solution and transferred to a well of a 24 well plate containing DMEM/F12 with 1% FBS. Serum was added at this point to neutralize the action of trypsinization. From this point CL1 was grown in this low serum-containing medium. After expansion, the culture was cryopreserved, banked and designated CL1.

Media

The parental clone and cell line CL1 were routinely grown in T-flasks using DMEM/F12 (Gibco/BRL, Grand Island, NY) containing 5% and 1% FBS respectively (Gibco/BRL, Grand Island, NY). For production of rhFSH, the parental clone was switched to DMEM/F12 supplemented with 1% FBS whereas CL1 was switched to protein-free IMDM/F12 (JRH Biosciences, Lenexa, KS). For the suspension bioreactor, CHO-S-SFMII (Gibco/BRL, Grand Island, NY) was used. CL1 was adapted to CHO-S-SFMII using sequential passaging in a 100 ml spinner. Thereafter, the suspension adapted CL1 was maintained in spinner cultures.

FSH Immunoassay

rhFSH levels were assessed, according to manufacturers instructions, using a commercial immunoradiometric assay - IRMA (MAIAclone from Biodata S.p.A, Rome, Italy). A specific activity assignment for the kit standard of 10,000 IU hFSH= 1 mg hFSH was used.

FSH Y1 Bioassay

The FSH Y1 bioassay was run as described previously (8).

Microcarrier based bioreactor run

A 5 l bioreactor was set-up to permit continuous perfusion of a microcarrier culture (Asahi @ 2g/L). Four confluent rollers, containing ~4 x 10^9 cells total, were trypsinized and used to seed the reactor. At this point the culture was maintained in batch mode to expedite the attachment of cells to the microcarriers. The residual glucose concentration was monitored to give an indication of metabolic activity. At ~1 g/L residual glucose, a continuous feed (DMEM/F12 with 5% FBS) was initiated to optimize the conditions for cell growth on the microcarriers. The perfusion rate was increased until the cell number in the culture stabilized. At this point the feed rate was 10 L/day (2.5 bioreactor volumes). This prompted initiation of the production phase where the medium was switched to basal IMDM/F12 and the perfusion rate was decreased to 4 L/day (1 bioreactor volume). These conditions were maintained throughout the term of the run.

Suspension based bioreactor run

A 3 l bioreactor was set up to promote suspension growth using a marine impeller and in addition, an agitation shaft mounted spin filter (10 μ pore size). Once sterilized, the vessel was filled with 2.7L of SFMII and left to reach parameter set points. Once equilibrated, the system was inoculated with 300 ml of a spinner culture of suspension adapted CL1 cells grown in SFMII to a cell density of ~1 x 10^6/ml. Following inoculation, the culture was maintained in batch mode until the residual glucose level fell to below 1g/L. At this point, a continuous perfusion feed was initiated. Fresh SFMII was fed to the culture at a constant rate of ~1.5L/day (0.5 reactor volumes). After a few days of perfusion, the temperature of the reactor was lowered to 32° C to slow cell division but maintain the cells in a metabolically active state.

RESULTS

After isolation, the rhFSH specific production rate of CL1 was determined and compared to its parental clone in low serum-containing medium. Table 1 indicates that CL1 had a greater specific production rate (pg/cell/day) than the parental clone.

TABLE 1 - SPECIFIC rhFSH PRODUCTIVITY OF PARENTAL CLONE AND CL1

Cell Line	Production Media	rhFSH specific production rate (pg/cell/day)
Parental clone	DMEM/F12 with 1% FBS	2.46
CL1	DMEM/F12 with 1% FBS	7.06

This initial investigations demonstrated that CL1 expressed rhFSH in serum-containing medium. An experiment was then completed in T-flasks to determine if CL1 retained the ability to express rhFSH in protein-free media. The results demonstrated that CL1, unlike its parent, possesses the ability to express rhFSH in basal IMDM/F12. The data (not shown) also indicated stable production of rhFSH in both media over 30 days. Following this, we investigated the biological activity of rhFSH protein expressed by CL1 in the Y1 bioassay (Fig. 1). We compared parental clone derived rhFSH (cultured in serum-containing medium) to CL1 derived rhFSH (cultured in basal IMDM/F12) and a purified preparation of rhFSH (Serono - Gonal F®).

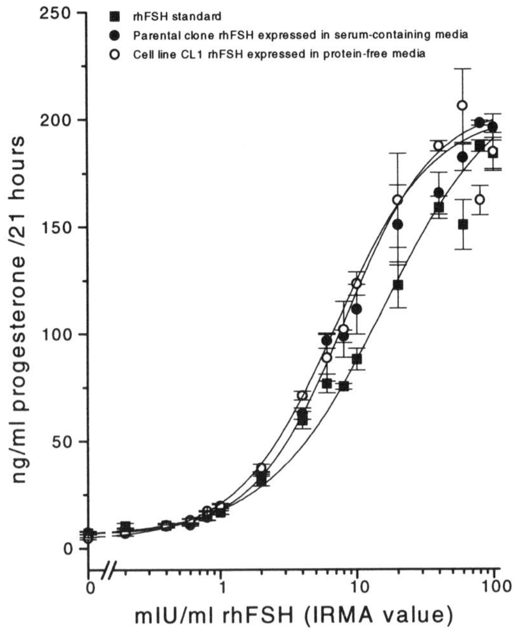

FIG. 1 Y1 bioassay comparing conditioned medium taken from the parental clone in serum-containing medium, CL1 in protein-free medium and a purified preparation of recombinant rhFSH (Serono - Gonal F®).

CL1-derived rhFSH and parental clone derived rhFSH, had identical behavior in the Y1 bioassay indicating no differences in bioactivity between the two proteins produced under the two different conditions. The Gonal-F standard (purified from serum-containing media) also demonstrates similar activity.

Once it was determined that CL1 expressed bioactive rhFSH in a stable manner when cultured in a protein-free environment, the emphasis of the investigations turned towards the use of CL1 in a bioreactor production process. The original production of rhFSH from the parental clone was performed using a microcarrier based perfusion process. To assess CL1 using a similar process the cell line was tested for the potential to produce in a bioreactor. Fig. 2 shows the kinetics of growth and rhFSH production when the CL1 cell line was tested at 10 l scale.

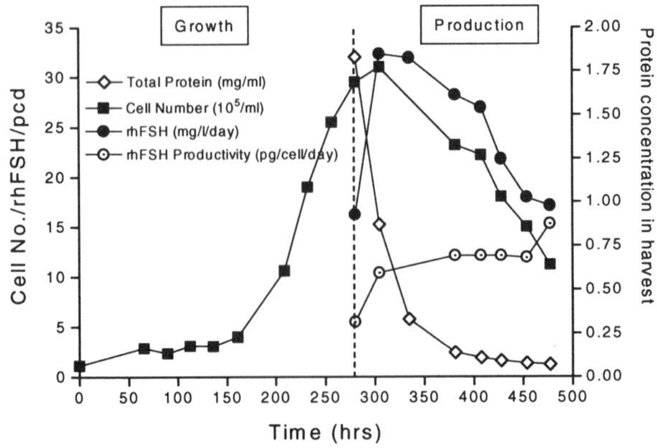

FIG. 2. Microcarrier perfusion culture of CL1.

In the bioreactor CL1 grew well on Asahi microcarriers. However, the introduction of basal IMDM/F12 as the production media resulted in a drop in attached cell number. As a result there was a concomitant drop in volumetric rhFSH production which ranged between ~30 and 15 mg/l/day. Regardless, during this production phase the cells that remained attached to the microcarriers expressed rhFSH at a steady rate of ~11 pg/cell/day. This production level, confirmed in T-flasks (data not shown), is 4-fold higher than that observed for the parental clone. In parallel with the microcarrier studies a suspension adapted CL1 cell line was grown and scaled to a 3 l bioreactor. Again a perfusion based process was used. Fig. 3 illustrates the culture characteristics associated with this process.

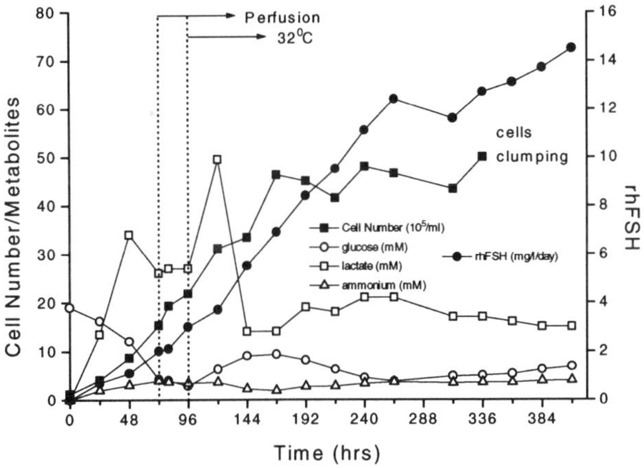

FIG. 3. Suspension based bioreactor run using suspension adapted CL1 and a serum-free medium to produce rhFSH.

The data demonstrate the potential of CL1 to grow and produce rhFSH in a suspension based process. The initial batch phase of the culture indicates metabolism of the carbon source glucose and production of lactic acid and ammonia. The application of perfusion culture at 32° C successfully maintains provision of glucose in addition to preventing the build up of the potentially toxic waste products. In this environment growth continues until a density of ~4.5 x 10^6 cells/ml is reached. Although cell clumping was observed the production of rhFSH was not compromised with levels rising continuously until at the end of the run the production rate was ~14 mg/l/day.

DISCUSSION

There are increasing pressures to pursue a route of recombinant protein expression in which the production system does not contain undefined agents of animal origin (9). Our study has shown that the cell line CL1, isolated from a parental CHO clone, satisfies this production criteria and thus possesses advantages over its parent, which required low levels of serum for expression of recombinant human follicle stimulating hormone (rhFSH). CL1, which produced rhFSH in the protein-free medium basal IMDM/F12, was selected using simple cell culture techniques, from a clonal population that lacked the ability to survive in this basal formulation. The 'forced selection' method used (sink-or-swim) demonstrated that an alternative procedure to the more frequently used adaptation protocols could be successful in generating new cell lines with altered and novel phenotypes (10). Following recovery, CL1 was grown in a low-serum containing medium until it was banked. The ability to secrete stable levels of rhFSH over a period of 30 daily harvests of IMDM/F12, indicated that this newly acquired characteristic was not transitory. In fact our investigations show that CL1 exhibited a higher specific productivity level (pg rhFHS/cell/day) than the parental clone when both were cultured in low serum-containing medium. Despite the differences in production host cell and expression media, the results obtained with the *in vitro* Y1 bioassay indicated that the rhFSH expressed by the parental clone and CL1 had equal bioactivity. These findings prompted us to evaluate CL1 in production environments. However, in an adherent-based bioreactor, cells were merely stripped from the microcarrier to which they were attached and were subsequently washed from the system. This was probably a direct result of using basal IMDM/F12 in the system where total protein levels fell to less than 100 mg/l and no protective surfactant was present (11). The production level of ~11 pg rhFSH/cell/day which resulted in volumetric productivities of ~20 mg rhFSH/l reactor/day. We also adapted CL1 to a serum-free formulation where it continued to express rhFSH. Using this medium, and a perfusion system, CL1 was grown in suspension to high cell densities where it produced rhFSH at levels in excess of 12 mg rhFSH/l reactor/day. These two production systems, adherent and suspension, gave an indication of the process utility of CL1. We hypothesize that since clonal CHO cell populations are not static systems (12), simple selection procedures such as the work described here (or even subcloning) could be used to isolate cell lines with novel and advantageous characteristics.

REFERENCES

1. Cartwright, T.; Crespo, A. In: Spier, R. E.; Griffiths, J. B.; Meignier, B., eds. Production of biologicals from animal cells in culture. Oxford, Butterworth-Heinemann; 1991: 669-475.
2. Kaufman, R. J.; Wasley, L. C.; Spiliotes, A. J., et al. Molecular and Cellular Biology 5: 1750-1759; 1985.
3. Hodgson, J. Biotechnology 11: 887-893; 1993
4. Zang, M.; Trautmann, H.; Gandor, C., et al. Biotechnology 13: 389-392; 1995.
5. Broad, D.; Boraston, R.; Rhodes, M. Cytotechnology 5: 47-55; 1991.
6. Jager, V. In: Spier, R. E.; Griffiths, J. B.; Meignier, B., eds. Production of biologicals from animal cells in culture. Oxford, Butterworth-Heinemann; 1991: 155-166.
7. Ryan, R. J.; Charlesworth, M. C.; McCormick, D. J.; et al. J. FASEB 2 (11): 2661-2669; 1988.
8. Kelton, C. A.; Cheng, S. V.; Nugent, N. P., et al. Molecular and Cellular Endocrinology 89: 141-151; 1992.
9. Jayme, D. W.; Epstein, D. A.; Conrad, D. R. Nature 334: 547-548;1988.
10. Tilkins, M. L.; Judd, D. A.; Weiss, S. A., et al. Focus 14: 95-98; 1994
11. Cherry R. S.; Papoutsakis, E. T. Bioprocess Engineering 1: 29-41; 1986.
12. Federspiel, N. A.; Beverley, S. M.; Schilling, J. W., et al. Journal of Biological Chemistry 259: 9127-9140;1984.

UNBALANCED MEDIA FOR HYBRIDOMA CELL CULTURE - AN ALTERNATIVE REALITY

F.FRANĚK and K. ŠRÁMKOVÁ
Institute of Molecular Genetics, Fundamental Cytotechnology Laboratory, Academy of Sciences of the Czech Republic, CZ-14220 Praha 4, Czech Republic

Abstract. Starvation-induced apoptosis can be prevented by certain amino acids added to the starving culture at unusually high concentrations. Consequently, the media containing these protective amino acids display a large unbalance among individual medium components. The adddition of 2 mM L-alanine to a medium diluted to 40% resulted in higher viable cell density and higher viability, eventhough alanine was still produced by the cells, and production of serine increased almost threefold. In another experiment alanine, serine, asparagine or glutamine were added to a medium diluted to 75%. In the presence of these protective amino acids the viable cell density and viability values exceeded the values found in the standard undiluted medium. In contrast to the beneficial effect of protective amino acids on the viability and longevity of the culture, the monoclonal antibody production in unbalanced media never reached the level found in the standard medium. It is concluded that in media containing an excess of a protective amino acid the cells largely abandon their tendency to undergo starvation-induced apoptosis and suppress the production of the secretory protein, while maintaining cell growth.

Introduction

Apoptotic character of cell death in hybridoma cultures has been revealed several years ago [1,2]. Further studies have elucidated the kinetics of spontaneous apoptosis onset and the correlation with exhaustion of nutrients [3-8]. In order to dissect the effect of nutrient exhaustion from the toxic effect of accumulating metabolites we developed an experimental setup in which the cells were inoculated into media diluted with saline [9,10].

The hybridoma culture on the verge of starvation-induced apoptotic death proved to be sensitive to additions of nutrient mixtures or of individual nutrients. Cells starving in media diluted to 50% or less could be protected against apoptosis by certain components of the basal nutrient medium [10]. Further investigation revealed the following features of the phenomenon of protection of starving cells against apoptosis by additions of amino acids [9,10; Franěk & Šrámková, submitted]:

M.J.T. Carrondo et al. (eds.), Animal Cell Technology, 675–680.

1. Cells starving in media diluted with saline can be protected against apoptosis not only through addition of amino acid mixtures (e.g. MEM essential amino acids, MEM nonessential amino acids), but also through addition of certain individual amino acids.
2. The protective effect has its maximum in the range of millimolar concentrations of added substances. The composition of the resulting medium is thus strongly unbalanced.
3. The effect of protective amino acids is apparent even in media diluted down to 10%.
4. The effect is selective. Significant protective activity is a property of eight coded L-amino acids (glycine, alanine, serine, threonine, proline, asparagine, glutamine, histidine) and of some related substance (e.g. β-alanine).
5. The specificity pattern of the phenomenon corresponds conspicuously to combined specificities of amino acid transport systems A and N.

The present work extended the knowledge on the above outlined phenomenon. The analysis of amino acid consumption and production in a culture conducted in 40% medium revealed continuous production of alanine, glycine and serine . In a moderately diluted medium, i.e. diluted to 75%, the viable cell densities achieved upon addition of protective amino acids were higher than the densities observed in the standard undiluted medium.

Materials and Methods

Mouse hybridoma PVA -187 producing an IgG1 monoclonal antibody to potato virus A was cultured in RPMI 1640 / DMEM / F 12 (2:1:1) or in RPMI 1640 / IMDM (1:1) media supplemented with the iron-rich protein-free growth-promoting mixture containing 200 μM ferric citrate [3,4]. For starvation experiments the media were diluted with saline supplemented with the same iron-rich mixture. The additives dissolved in saline formed 10% (v/v) of the culture volume. The cultures set up in 25 cm^2 T-flasks were kept at 37°C in a humidified atmosphere containing 5% CO_2.

The total and viable cell densities were counted in a haemocytometer. The apoptotic index, i.e. the percentage of apoptotic cells relative to total cells was determined by microscopic counting of cells displaying apoptotic morphology, i.e. shrunked cells with ruffled membrane [9]. The experimental errors of cell density and of apoptotic index were ±15% to ±20%, when taking into account the variability among parallel culture flasks.

Monoclonal antibody (mAb) concentration was determined by immunoturbidimetry [9]. For determination of glucose concentration the diagnostic kit Oxochrome Glucose (Lachema, Brno, Czech Republic) was used. The concentrations of amino acids were determined on a Durrum D-500 analyzer in culture fluid aliquots deproteinized by ultrafiltration.

Results

Excess of L-alanine added to a medium diluted to 40%

The growth in an RPMI 1640 / DMEM / F12 (2:1:1) medium diluted to 40% with saline was characterized by a drop of viable cell density to ~60% of the density achieved in the standard medium. The apoptotic index increased by ~50% (Table 1). The final mAb concentration in the diluted medium represented ~60% of the concentration found in the standard medium. The beneficial effect of L-alanine, added to the diluted medium, manifested itself by an increase of viable cell density to 118% of the control and, mainly, by the decrease of the apoptotic index to the level found in the standard medium. However, the mAb concentration showed an opposite trend than the viable cell density.

Table 1. Batch culture in diluted medium with 2 mM L-alanine

	40 % Medium			40% Medium +2mM alanine			*Standard medium*		
	0 h	72h	Δ	0 h	72h	Δ	0 h	72h	Δ
Viable cells, $ml^{-1}x10^{-3}$	155	862	*+707*	155	988	***+833***	*155*	*1405*	***+1250***
Dead cells, $ml^{-1}x10^{-3}$	16	116	***+100***	16	80	***+64***	*16*	*122*	***+106***
Apoptotic index, %	9	12	*+3*	9	8	***-1***	*9*	*8*	***-1***
mAb, μg/ml	<2	12	***+10***	<2	8	*+6*	*<2*	*20*	***+18***

Our main concern were the changes of amino acid consumption and production upon addition of L-alanine. Alanine and serine were found to be produced in all three experimental situations (Fig. 1). In the diluted medium a significant production of glycine could be also observed. In the diluted medium with alanine excess the production of alanine still overrided its consumption. A rather unexpected phenomenon was the highly enhanced production of serine, and the inversion of the flux of proline from consumption to production.

Excess of protective amino acids added to a medium diluted to 75%

Parameters of cultures conducted in a moderately diluted medium were determined in order to investigate the effect of protective amino acids under conditions of limited extent of apoptotic death. The cultures carried out in RPMI 1640/IMDM (1:1) medium diluted to 75% with saline were examined at the decline phase, i.e. after 192 hours.

The viable cell density found in the diluted medium represented 78% of the density value achieved in the standard medium. The viability values were identical within the limits of error. The residual concentration of glucose in the diluted medium was correspondingly lower (1.14 g/l) than the residual concentration in the standard medium (1.50 g/l). A substantial difference was found between the final mAb concentration values. The mAb concentration in the diluted medium represented only 55% of the value found in the standard medium (Fig.2).

Upon additions of protective amino acids the viable cell densities arose in all cases, being in the interval from 134% to 183% of the control. A lower increase of viability was found, from 113% to 137%. With respect to the error involved in viability determination, only the values >120% could be taken as statistically significant. Glucose concentrations were in the interval between 0.96 and 1.14 g/l. The addition of protective amino acids influenced positively also the mAb concentrations. These were found to be in the range from 116% to 165% of the control.

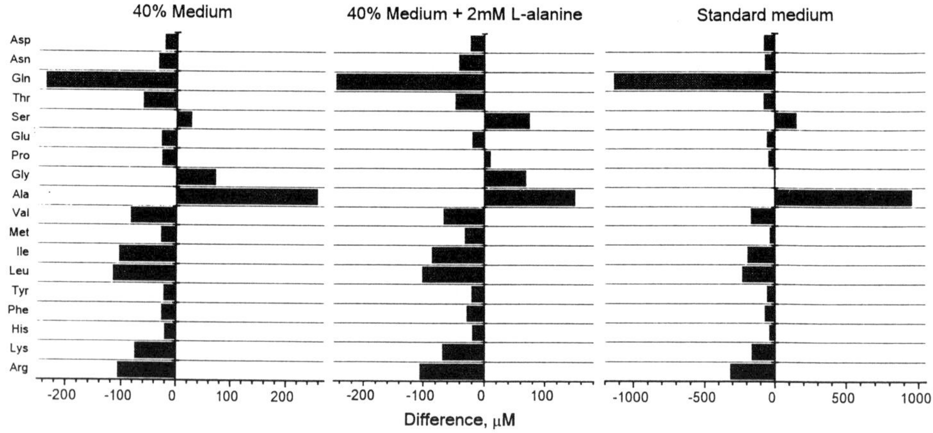

Fig. 1. Consumption and production of amino acids during a 72h culture period.

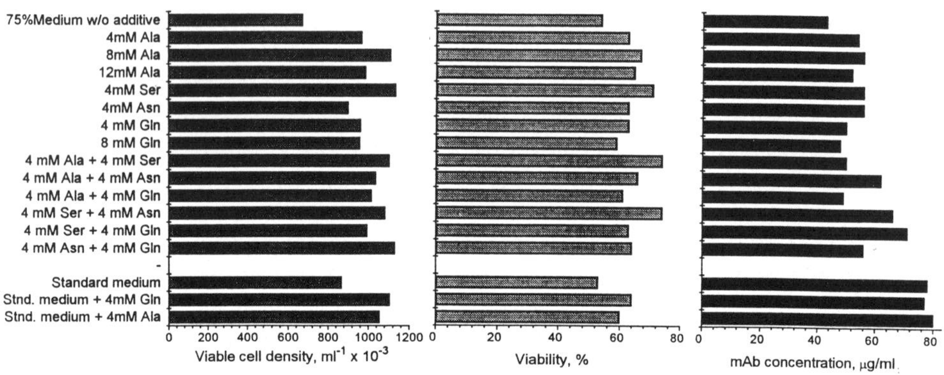

Fig. 2. Culture parameters in diluted and standard medium with protective amino acids.

The most important conclusions could be drawn from the relation of the culture characteristics obtained in the unbalanced media with the respective values of the positive control, i.e. the culture in the standard medium. The viable cell density and the viability values were in all cases higher relative to the characteristics found in the standard medium. Although only around one half of the cases could be considered as statistically significant, the results showed a remarkable potential of unbalanced media containing an excess of a protective amino acid. In contrast to the mentioned growth characteristics the final mAb concentration did not reach the mAb concentration found in the standard medium.

The effect of protective amino acids added to the standard medium was examined only marginally. The effects were in all cases beneficial, but below the level of statistical significance.

Discussion

Media diluted with saline and supplemented with a large excess of a single amino acid represent a source of nutrients deviating substantially from well-proven spectra of substances constituting balanced nutrient mixtures. An anomalous growth in such media can be anticipated. The finding that the anomaly is of beneficial nature is somewhat surprising and enjoyable. It is not very surprising, and not entirely new [7-9], that an excess of glutamine can prevent apoptosis in nutrient-poor diluted media. However, the idea that nonessential amino acids, such as alanine, serine or glycine, could protect cells against apoptosis, is clearly contraintuitive. These amino acids are known to be mostly eliminated from cultured cells as redundant metabolites, rather than being utilized as substrates. To get a deeper insight into the mechanism of protection by alanine, we analyzed the amino acid concentrations in the culture carried out in an unbalanced medium with excess of alanine. The results confirmed the beneficial effect of 2 mM alanine on viable cell density and viability (Table 1). The increase of outward flux of serine and proline can be explained on the basis of countertransport phenomena [11]. When the inward flux of alanine is accelerated, due to external elevation of this amino acid, the exodus of all amino acids using the same transport system , i.e. system A [12] in this case, is accelerated as well. The effect is highest in the case of serine, because a supply of redundant serine is available in the pool of intracellular amino acids. The present data add weight to our previous conclusion that the protective amino acids act as signal molecules, besides their nutritive function, and that the signal has a character of a survival factor.

A question may arise, why the increasing level of alanine, normally taking place in every batch culture, is not sufficient to prevent the spontaneous onset of apoptosis. Our studies on the concentration dependence of the protection effect show that maximum protection is achieved at ~4 mM alanine concentration [10]. However, the concentration of alanine at 72 hours culture duration in 40% medium is ten times lower [Franěk, unpublished data] and , thus, insufficient to protect the culture significantly.

In our previous work the prevention of apoptosis has been studied in media diluted below 50% [6,9,10]. In the present work we have complemented the investigation of the activities of protective amino acids by the study of the effect of amino acid additions to moderately diluted media in which apoptosis is not much higher than in the standard medium. The main outcome of the set of experiments carried out in the 75% medium (Fig. 2) is the finding tha the viable cell density and viability values achieved in unbalanced media can exceed the values found in the standard medium. The analysis has been done at the decline phase of batch cultures. The beneficial effect of protective amino acids on the culture longevity is thus demonstrated.

The above reported beneficial effect of protective amino acids on the viability and longevity of the hybridoma culture is not associated with a parallel intensification of mAb production. The antibody production seems to be more sensitive to starvation than the viability of the cell population. Obviously, determinants different from those regulating apoptosis control the intensity of mAb synthesis and secretion. Investigation of the intracellular levels of mAb is intended to contribute to the understanding of mAb production control.

The examples of hybridoma culturing in unbalanced media showed a potential of non-traditional media formulations. Application of unbalanced media may become an approach of choice when a specific performance of the culture is needed. In the present experiments the hybridoma cells largely abandonned their suicidal tendency , and suppressed the production of the secretory protein, while maintaining cell growth. It is important to keep in mind that this effect has been achieved under conditions of suboptimal levels of basic nutrients. The findings add further weight to the view that the protective amino acids act not only on the level of amino acid metabolism but also on the level of cell signalling.

References

1. Al-Rubeai M, Mills D & Emery AN (1990) Electron microscopy of hybridoma cells with special regard to monoclonal antibody production. Cytotechnology 4, 13-28.
2. Franěk F & Dolníková J (1991) Nucleosomes occurring in protein-free hybridoma cellculture. Evidence for programmed cell death. FEBS Letters 284, 285-287.
3. Franěk F & Dolníková J (1991) Hybridoma growth and monoclonal antibody production in iron-rich protein-free medium: Effect of nutrient concentration. Cytotechnology 7, 33-38.
4. Franěk F, Vomastek T & Dolníková J (1992) Fragmented DNA and apoptotic bodies document the programmed way of cell death in hybridoma cultures. Cytotechnology 9, 117-123.
5. Vomastek T & Franěk F (1993) Kinetics of development of spontaneous apoptosis in B cell hybridoma cultures. Immunol. Letters 35, 19-24.
6. Franěk F (1994) Spontaneous apoptosis contributes negatively to the regulation of hybridoma cell viability. In: Spier RE, Griffiths JB & Berthold W, Animal Cell Technology: Products of Today, Prospects for Tomorrow, pp 192-196, Butterworth-Heinemann, Oxford, UK.
7. Singh RP, Al-Rubeai M, Gregory CD & Emery AN (1994) Cell death in bioreactors: A role for apoptosis. Biotechnol. Bioeng. 44, 720-726.
8. Mercille S & Massie B (1994) Induction of apoptosis in nutrient-deprived cultures of hybridoma and myeloma cells. Biotechnol. Bioeng. 44, 1140-1154.
9. Franěk F (1995) Starvation-induced programmed death of hybridoma cells: Prevention by amino acid mixtures. Biotechnol. Bioeng. 45, 86-90.
10. Franěk F & Chládková-Šrámková K (1995) Apoptosis and nutrition: Involvement of amino acid transport system in repression of hybridoma cell death. Cytotechnology 18, 113-117.
11. Christensen HN (1969) Some special kinetic problems of transport. Adv. Enzymol. 32, 1-20.
12. Kilberg MS (1986) System A-mediated amino acid transport: Metabolic control at the plasma membrane. Trends Biochem. Sci. 11, 183-186.

Acknowledgments. This work was supported by the Grant No. I/68 957 of the Volkswagen Foundation, and by the Support to the 4th Framework Programme of the EU provided by the Government of the Czech Republic through the Contract OK 120(1996).

VIRAL CONTAMINANTS FOUND IN MOUSE CELL LINES USED IN THE PRODUCTION OF BIOLOGICAL PRODUCTS

Stephen J. Froud†, John Birch†. Carol McLean*, Alasdair J. Shepherd* and Kenneth T. Smith*

*†Lonza Biologics (previously Celltech Biologics), 216 Bath Road, Slough, Berkshire, SL1 4EN, U.K. and *Q-One Biotech Ltd., Todd Campus, Maryhill Road, Glasgow, G20 0XA, U.K.*

INTRODUCTION

Murine hybridoma and plasmacytoma cell lines are used in the production of biologics for human use such as therapeutic monoclonal antibodies and recombinant proteins. Most of these cell lines produce retrovirus-like particles and in many cases infectious C type murine leukaemia virus (MLV) which can be either ecotropic (infectious for cell lines of murine origin or a closely related species) or xenotropic (infectious for species other than murine; X-MLV).

Guidelines on tests for retroviruses in cell lines used in the production of biologicals recommend that test samples be assayed by transmission electron microscopy (TEM), reverse transcriptase (RT) assay (both Mn^{2+} and Mg^{2+} preference) and by assays for infectious retrovirus. TEM using negative staining allows enumeration of the total retrovirus load by counting of retroviral particles while the RT assay indicates the presence of retroviral enzyme. Thus, neither of these assays distinguish between infectious and defective retrovirus. Infectious murine retrovirus can be detected by inoculation of susceptible cell lines such as mink S^+L^- (sarcoma positive, leukaemogenic negative) cells or feline PG4 S^+L^- cells for detection of xenotropic or amphotropic MLV, FG10 S^+L^- or SC-1 cells for the detection of ecotropic MLV or *Mus dunni* cells which are susceptible to infection by xenotropic, amphotropic, mink cell focus (MCF) virus and ecotropic retroviruses (with the exception of Moloney MLV).

In this study we report a comparative study on the relative sensitivities of assays used for detection of murine retrovirus and on the dynamics of replication of infectious MLV in fermenter cultures by analysis of sequential samples taken from four separate 200 litre fermenter runs with a recombinant murine plasmacytoma cell line Rec Mab 8. This cell line was created by transfecting the mouse plasmacytoma NSO with a glutamine synthetase (GS) vector expressing a recombinant antibody. The results obtained with this cell line are compared with results obtained from fermenter and cell bank samples from 97 mouse hybridoma or recombinant plasmacytoma cell lines.

M.J.T. Carrondo et al. (eds.), Animal Cell Technology, 681–686.

RESULTS

Testing of MCB, WCB and ECB for Murine Retrovirus

The master cell bank (MCB), working cell bank (WCB) and cells at the limit of *in vitro* cell age (the extended cell bank, ECB) of the cell line Rec Mab 8 were tested for the presence of retrovirus by TEM (of sectioned cell pellets), RT assay and for ecotropic MLV by the XC infectivity assay and X-MLV by mink S^+L^- infectivity assay. All assays were performed by standard methods. The results showed C-type retrovirus being associated with or budding from a high percentage of the cells, RT activity with a preference for the manganese cation, and detection of infectious C-MLV by the mink S^+L^- assay (Table 1). Ecotropic MLV was not detected. Assays on the MCB, WCB and ECB were performed on supernatant harvested 2-3 days post subculture when the cells were mitotically active. No bacteria, fungi, yeasts, mycoplasmas or viruses were detected by the other assays used (sterility, mycosplasma stain and culture, MAP, co-cultivation with MRC-5 and *in vitro* and *in vivo* virus assays).

Table 1 Testing of MCB, WCB and ECB for Murine Retrovirus

Test	MCB	WCB	ECB
S^+L^-*	36	5	207, 221
XC	neg.	neg.	neg.
RT†	47396	127917	49799
TEM‡	61	52	74

neg. no retrovirus detected
* infectious mink S^+L^- ffu/ml
† incorporation of ^{3}H TTP into rA template expressed as dpm (Mn^{2+} dependent)
‡ percentage of cells with budding and/or associated type C retrovirus particles

Retrovirus Dynamics in 200 Litre Air-Lift Fermenters

Four 200 litre air-lift fermentations of cell line Rec Mab 8 were sampled at various times ranging from day 3 to day 13. In all four fermenters total and viable cell concentration increased rapidly until day 7 (approximately 150h), thereafter the viable cell concentration declined steadily. The level of X-MLV detected mink S^+L^- assay followed the viable cell concentration in all four fermenters (Figure 1). Similarly, the level of infectious X-MLV detected by the direct *Mus dunni* assay correlated with viable cell concentration (Figure 1, fermenter 4). A low level of infectious X-MLV was detected in the final bulk harvest in three of the four fermenters.

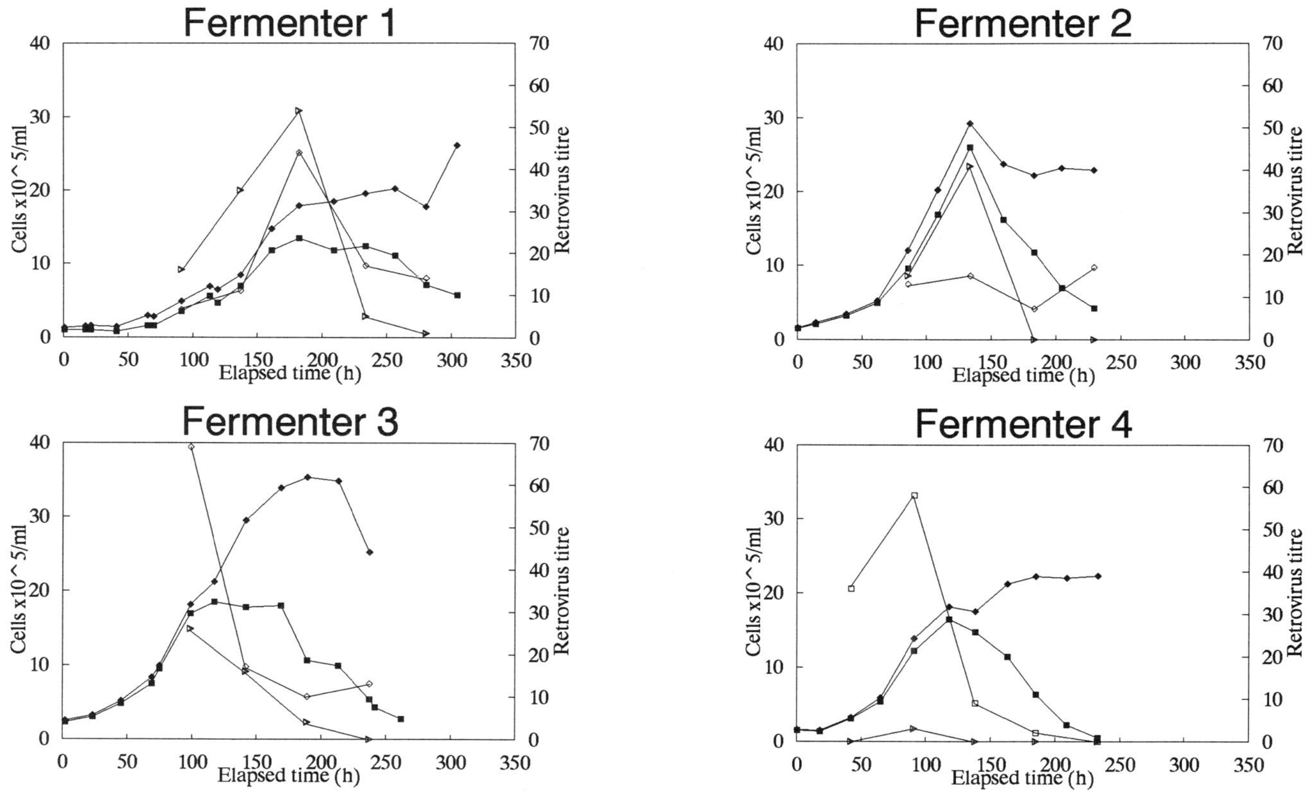

Fig. 1 Cell concentration and infectious xenotropic mouse retrovirus in 200 litre fermenters
■ viable cells ◆ total cells ◇ electron microscopy (millions particles/ml)
▷ mink S+L− assay (ffu/ml) □ Mus dunni assay (ffu/ml)

RT activity with manganese dependence (the cation of preference by the RT enzyme of MLVs) was detected in all four fermenters (data not shown). RT activity was, however, detected only in the early samples (up to day 5) and thereafter decreased to non-significant levels. RT activity was not detected in several samples in which infectious X-MLV was detected by the direct and/or extended infectivity assays.

An estimate of the total retrovirus load was made for fermenters 1, 2 and 3 by enumeration of retrovirus-like particles by negative stain electron microscopy following ultracentrifugation of the fermenter sample in a sucrose density gradient. In each fermenter, the number of retrovirus like particles was between 6.5 x 10^6 and 6.9 x 10^7/ml throughout the process (Figure 1). Fermenters 1 and 3 showed an apparent correlation between infectious X-MLV and retrovirus-like particles, suggesting that approximately one in every 10^6 retrovirus particles were infectious.

Comparative Sensitivity of Retroviral Assays for Fermenter Samples

Comparison of the samples positive in any particular assay indicated that *Mus dunni* cells were the most sensitive of the three cell lines to X-MLV infection (with X-MLV detected in 17 out of the 18 samples), followed by mink S^+L^- cells, with PG4 cells the least sensitive. RT assay activity was positive in only six of the 18 samples and was therefore a poor indicator of presence of X-MLV in these fermenters.

Comparative Results from 97 mouse cell lines used between 1986 and 1996

The results from the cell line Rec Mab 8 are typical for mouse hybridoma and plasmacytoma cell lines that have been used for the production of diagnostic and therapeutic proteins for the last 10 years (Table 2). Although not all of the cell lines were tested in all of the available assays, no bacteria, yeasts, moulds, mycoplasmas or viruses (other than mouse retroviruses) were detected in the 325 fermenter and 113 cell bank samples analysed. Retrovirus particles were detected in every cell line by electron microscopy. A small number of cell lines (5-6%) produced infectious mouse ecotropic retrovirus. Almost all (85%) of the mouse cell lines tested produced a low level of infectious mouse xenotropic retrovirus (X-MLV) when the cell bank was tested. X-MLV was detected less frequently in fermenter samples even though these were taken when the cells were at a high viable cell concentration (see above).

The number of infectious X-MLV particles detected in the production fermenters analysed was 4-6 orders of magnitude lower than the total number of retrovirus-like particles detected by electron microscopy (after sucrose density gradient centrifugation). When detected, the mean number of infectious X-MLV particles in the bulk harvest was 32 particles/ml (SEM 11.1, n=27; range 1 to 290). By contrast the mean number of retroviral-like particles detected was 22 x 10^6 particles/ml (SEM 2.9 x 10^6, n=40) with a relatively narrow range of 1 to 87 x 10^6 particles/ml.

Table 2 Tests for 97 mouse cell lines used between 1986 and 1996

Test	Number of cell lines tested	Percentage of cell lines positive	
		Fermenter samples	Cell bank samples
Mycoplasma	97	0% (n=325)	0% (n=113)
Sterility	65	n/a†	0% (n=74)
MAP	46	0% (n=52)	0% (n=67)
in vitro virus assays*	44	0% (n=125)	0% (n=74)
in vivo virus assays	16	0% (n=1)	0% (n=27)
Electron Microscopy	22	100% (n=44)	100% (n=43)
Reverse transcriptase	45	87% (n=52)	67% (n=169)
XC (direct)	24	5% (n=132)	6% (n=28)
XC (extended)	18	0% (n=25)	6% (n=38)
S^+L^- (direct)	58	28% (n=171)	42% (n=65)
S^+L^- (extended)	21	38% (n=25)	85% (n=41)
Co-cultivation with human cell line	18	not tested	0% (n=25)

n the total number of samples tested in this assay (each cell line was usually tested more than once)
* with MRC-5 and Vero and other cell lines
† occasional fermenter contaminations can be traced to specific caused and are not indicative of contaminants in the cell lines

DISCUSSION

We have determined the relative sensitivities of various assays for murine retrovirus detection in a mouse plasmacytoma cell line derived from the Balb/c cell line NSO. Overall comparison of infectivity assays using fermenter samples indicated that for the retroviruses found in such cell lines the *Mus dunni* is the most sensitive of the three cell lines for detection of infectious murine xenotropic retrovirus (X-MLV), followed by mink S^+L^- and, lastly, feline PG4 cells. Detection of retrovirus by RT assay in fermenter samples proved the least sensitive of the tests employed.

Testing of four separate 200 litre fermenter cultures of this cell line at varying time points showed that infectious X-MLV was present in all four fermenters and followed the viable cell concentration profile, with maximal titres on day 5-7, falling to low levels by the final day, when the bulk harvest was made. This study demonstrates that while it is possible that infectious retrovirus is not detected in the final bulk harvest sample it may have been present in the fermenter in significant amounts up to this point. Detection of infectious X-MLV in fermenters is therefore optimal at the point of high viable cell concentration.

The results from this cell line are typical for mouse hybridoma and plasmacytoma cell lines that have been used for the production of diagnostic and therapeutic proteins for the last 10 years (Table 2). Retrovirus particles were detected in all of the 97 cell lines by electron microscopy. Almost all (85%) of the mouse cell lines tested produced a low level of infectious mouse xenotropic retrovirus (X-MLV) when the cell bank was tested. All of the mouse cell lines commonly used for antibody or recombinant protein production are derived from the MOPC21 tumour of a female Balb/c mouse. The production of mouse xenotropic retrovirus was not restricted to any one of the four cell lines derived from this tumour that account for 98% of the cell lines analysed; X-MLV was detected in samples from cell lines derived from the plasmacytomas P3X63.Ag8.653, NS1, NSO and the hybridoma SP2/0.Ag14.

For mouse cell lines used for protein production for human use the manufacturer's process should be validated for removal and/or inactivation of retroviruses. An indication of the retrovirus load in fermenters is given by enumeration by electron microscopy. The mean total retrovirus-like particle load was significantly higher (by 4-6 $\log_{10}$ units) than infectious MLV titres. Retrovirus particles undetected by the infectivity assays may be defective retrovirus particles that have lost infectivity through inactivation by physical disruption or factors relating to cell death and release of cellular enzymes. It is unlikely that these particles represent other classes of infectious MLV as the *Mus dunni* cell line is susceptible to infection by all MLVs tested except Moloney (which would be detected in the XC assays performed). Comparison of the infectious X-MLV with total retrovirus particle counts by TEM in fermenters 1 and 3 suggests that approximately one in every 10^6 retrovirus-like particles observed by electron microscopy are infective in the mink S^+L^- assay.

Nevertheless evaluations of retroviral removal during purification are usually based upon the ‘worst case’ electron microscopy data. Given however, that murine retroviruses are not pathogenic to man, are relatively liable and can be removed and/or inactivated by standard purification procedures, adequate clearance factors and a margin of safety are obtained routinely for products derived from mouse hybridoma and plasmacytoma cell lines.

ANALYSIS OF ENDOGENOUS RETROVIRUS PRODUCTION IN A MURINE HYBRIDOMA BY FLOW CYTOMETRY

L. Johnson[1], J. Bryant[2], J. P. Barford[3], C. Harbour[1], and A. Fletcher[2]

Departments of [1]Infectious Diseases and [3]Chemical Engineering, The University of Sydney, N.S.W., 2006 Australia, [2]New South Wales Red Cross Blood Transfusion Service, 153 Clarence St., Sydney 2000 Australia.

Key Words: Retroviruses, hybridoma productivity, flow cytometry, cell cycle

Abstract

A variety of different types of endogenous retroviruses are expressed in murine hybridomas. The two commonly produced retrovirus particles are members of the murine leukemia virus family (MuLVs) and the intracisternal A-type particles (IAPs). Previous studies had indicated that such retrovirus expression may present a significant metabolic load to the cell. We therefore aimed to identify relationships between endogenous retrovirus expression and other factors in a well-characterised murine hybridoma (AFP-27). The relationship between IAP production, cell growth in culture, cell cycle and antibody production was examined. Flow cytometric analysis of IAP expression and cell cycle was analysed with a monoclonal antibody to the IAP gag protein. IAP and monoclonal antibody expression were found to change with time in batch culture and vary in different phases of the cell cycle of AFP-27.

Introduction

IAPs are found in virtually all hybridoma cell lines and their myeloma precursors (6). The levels of IAP production are high, approximately 10^4-10^5 particles occurring per cell. The relationship between growth of murine hybridomas and antibody production by these cell lines to IAP expression or production, if any, has not been characterised. There are, however, a number of studies demonstrating the potential importance of IAP elements in modulating the expression of nearby genes (1, 3, 8). Such work has demonstrated the importance of IAP elements in modulating the expression of genes that include those related to cell growth (5).

In these preliminary studies we describe the use of flow cytometry to determine the levels of IAP expression during growth of a murine hybridoma in culture. IAP expression was analysed over time in culture and compared with each phase of the cell cycle using a dual staining technique. IAP expression was also compared with monoclonal antibody production in AFP-27. The techniques developed therefore allowed for an analysis of IAP expression, cell cycle, cell culture growth phase and monoclonal antibody production in a hybridoma cell line.

M.J.T. Carrondo et al. (eds.), Animal Cell Technology, 687–690.

Materials and Methods

The hybridomas AFP-27 and VG10 were grown in DMEM supplemented with 10% bovine serum at 37 ^{0}C, 5% CO_2. Cells were prepared as described elsewhere (9).The antibodies used with the dual staining technique are shown in table I.

Table I: Antibodies Used for Flow Cytometry

Host	Antibody to	Conjugate	Working Dilution
VG10 Rat Monoclonal (IgGIIa)	IAP gag protein	None	Undiluted Supernatant
Rat Monoclonal (IgGIIa)	Mouse LFA-1	None	1:10
Sheep Polyclonal (Whole Molecule)	Rat IgGIIa	FITC	1:50
Sheep Polyclonal F(ab')2 Fragment	Mouse IgG	FITC	1:50
Sheep Polyclonal F(ab')2 Fragment	Rabbit IgG	FITC	1:50
Goat Polyclonal (Whole Molecule)	Mouse IgG	Phycoerythrin	1:50
Goat Polyclonal (Whole Molecule)	Rabbit IgG	Phycoerythrin	1:50

Results

The cell line AFP-27 was subsequently labelled with VG10 and FITC in combination with PI staining to follow IAP gag expression with DNA content. Dual staining with the VG10 monoclonal antibody and PI indicated that the pattern of VG10 labelling varied with DNA content of AFP-27 when compared with the isotype control. Labelling with VG10 was higher ($p < 0.03$) in $G_{0/1}$ phase cells than in S and G_2/M phase cells. Examination of AFP-27 cells at different times in culture revealed a variation of VG10 labelling with cell culture phase. Expression of VG10 was greatest in the earlier (lag-log) phases of culture than at later stages.

The hybridoma AFP-27 was designed to produce a monoclonal antibody to human α-fetoprotein in high yield. It was therefore interesting to examine the synthesis of this monoclonal antibody with IAP expression. AFP-27 IgG production was first examined separately with FITC and PE-conjugated anti IgG antibody to ensure effective PE-staining was possible. IgG production was then examined by dual staining with PI and anti-IgG. Analysis of IgG labelling with the isotype control revealed that IgG expression was significantly higher ($p < 0.01$) in $G_{0/1}$ phase cells. This increase in antibody expression in $G_{0/1}$ cells correlated with the increase in IAP P73 antigen levels in $G_{0/1}$ cells.

Discussion

Studies of IAP expression and production levels have mainly been performed by measuring nucleic acid expression levels (7). A relationship between IAP expression and cell growth has been demonstrated in tumour cell lines, through the insertion of IAP elements into a number of genes that regulate cell growth (2, 4). In this study, we have examined the relationship between cell cycle and IAP production in a monoclonal antibody-producing hybridoma. IAP production was shown to be greatest in cells in the $G_{0/1}$ phase of the cell cycle. Monoclonal antibody production was also greater in the $G_{0/1}$ phase.

We have recently examined the relationship between cellular metabolism and endogenous retrovirus particle production using complex models of animal cell culture and quantitative data on retroviral particle numbers (submitted). The modelling work suggested that high levels of retrovirus particle production might reduce cell growth and antibody yield significantly. It was therefore interesting to experimentally determine the relationship between cell growth and IAP production in the cell line that was simulated. These preliminary results indicate a strong positive relationship between IAP production and cell growth, contrasting with the computer model predictions.

Acknowledgments

This work was supported in part by the Australian Research Council.

References

1. **Aberdam, D., V. Negreanu, L. Sachs, and C. Blatt.** 1991. The oncogenic potential of an activated Hox-2.4 homeobox gene in mouse fibroblasts. Molecular and Cellular Biology. **11:**554-557.

2. **Ben-David, L., D. Aberdam, L. Sachs, and C. Blatt.** 1991. A deletion and a rearrangement distinguish between the intracisternal A-particle of Hox-2.4 and that of interleukin-3 in the same leukemic cells. Virology. **182:**382-387.

3. **Chang Yeh, A., D. E. Mold, M. H. Brilliant, and R. C. C. Huang.** 1993. The mouse intracisternal A particle promoted placental gene retrotransposition is mouse strain specific. Proceedings of the National Academy of Sciences, U.S.A. **90:**292-296.

4. **Kuff, E. L., A. Feenstra, and K. Lueders.** 1983. Homology between an endogenous viral LTR and sequences inserted in an activated cellular oncogene. Nature. **302:**547-548.

5. **Kuff, E. L., and K. K. Lueders.** 1988. The intracisternal A-particle gene family: Structural and functional aspects, p. 183-276. *In* G. Klein and S. Weinhouse (ed.), Advances in Cancer Research, vol. 51. Academic Press, Inc., San Diego.

6. **Lubiniecki, A. S.** 1989. Endogenous retroviruses of continuous cell substrates, p. 85-92. *In* R. E. Spier and J. B. Griffiths and J. Stephenne and P. J. Crooy (ed.), Advances in Animal Cell Biology and Technology for Bioprocesses, vol. Volume 9. Butterworth & Co. Ltd., Oxford.

7. **Lueders, K. K.** 1991. Genomic organization and expression of endogenous retrovirus-like elements in cultured rodent cells. Biologicals. **19:**1-7.

8. **Malstrom, C. E., A. L. Kotsenas, and B. Jones.** 1992. Interaction of H-2Eb with an IAP retrotransposon in the A20/2J B cell lymphoma. Immunogenetics. **36:**238-247.

9. **Schmid, I., C. H. Uittenbogaart, and J. Giorgi.** 1991. A gentle fixation and permeabilization method for combined cell surface and intracellular staining with improved precision in DNA quantification. Cytometry. **12:**279-285.

CONSTRUCTION OF A HUMAN CHONDROMODULIN-I EXPRESSION SYSTEM IN CHO CELLS

Takaharu Kono, Midori Nakajima, Kazuhiro Takahashi, Jun Kondo,
Tatsuro Shibui, Fujio Suzuki* and Yuji Hiraki*
Yokohama Research Center, Mitsubishi Chemical Corporation,
Kanagawa 227, JAPAN
*Department of Biochemistry, Osaka University Faculty of Dentistry,
Osaka 565, JAPAN

Introduction

Growth and maturation of epiphyseal growth-plate cartilage play an important role in endochondral bone formation. Chondromodulin-I (ChM-I) is a 25 kDa glycoprotein and was first purified from fetal bovine epiphyseal cartilage as a cartilage specific functional matrix component. ChM-I stimulates synthesis of proteoglycan and DNA in cultured growth-plate chondrocytes. Its activity on chondrocyte growth is reinforced in the presence of basic fibroblast growth factor (FGF) (1). ChM-I mRNA has only been observed in cartilage.

The human ChM-I (hChM-I) gene exists as a part of a large precursor gene, and mature hChM-I protein consists of 120 amino acids (2). Since the hChM-I protein is believed to be glycosylated, we thought that mammalian cell lines such as CHO or COS cells would be desirable for the production of recombinant hChM-I molecule. To characterize hChM-I molecule, we constructed four types of expression vector in attempting to produce a recombinant hChM-I protein.

Materials and Methods

HChM-I DNA cloning

The nucleotide sequence of bovine ChM-I (bChM-I) gene was determined. Subsequently, a Lambda phage human genomic library was screened to obtain the desired clone. The human ChM-I gene was isolated by PCR, and the amplified fragment was inserted into the cloning vector pCRII (Invitrogen) for sequencing.

Construction of vector 1

PCR primers AM2 (<u>aagctt</u>catgacagagaactccga) and AM3 (atcctgggcatgg<u>tctaga</u>) were synthesized. HindIII and XbaI sites (underlined) were constructed at the 5' and 3' ends of AM2 and AM3, respectively.(Fig.1) Human ChM-I cDNA (0.9 kb) was PCR amplified using pCRII-hChM-I as a template, and inserted into the expression vector, pcDNA1 (Invitrogen). PcDNA1 has HindIII and XbaI sites in multicloning region and functions by CMV promoter. The constructed vector was sequenced by a GENESIS 2000 DNA sequencer (Dupont).

Transfection of CHO cells

M.J.T. Carrondo et al. (eds.), Animal Cell Technology, 691–695.

CHO cells were plated at a density of $1x10^5$ cells/6 well plate one day prior to transfection. After overnight culturing with 5% FCS-eRDF, the medium was harvested. After the cells were washed with PBS, vector and drug resistant marker DNA plasmids were added with the lipopolyamine reagent, Transfectam (IBF) according to the manufacture's protocol. Transfected cells were cultured with 5% FCS-eRDF for 48hr, then selection was started by the addition of selection reagent.

Selection of transfected cells to obtain recombinant hChM-I

Vector transfection and drug resistant selection were performed three times onto the same cells with different reagent, that is, neomysin, blasticidin S and hygromycin, respectively. Each time, hChM-I vector (5 µg) and drug resistant vector (0.5 µg) were cotransfected. Total RNA of drug resistant clones were assessed by dot northern hybridization assay, and several clones were selected. The culture supernatants of these clones were analyzed by western blotting using an antiserum against peptide fragment corresponding to hChM-I(9-33).

Dot northern hybridization

Selection of hChM-I transfectant clones was performed by dot northern hybridization. Transfectant clone cells were expanded to a 6-well plate. Total RNA was extracted from about $1x10^6$ transfected cells and transferred onto the Nylon membrane under vacuum with a dot blot apparatus (1 µg RNA / dot). ^{32}P labeled hChM-I cDNA was then added and allowed to hybridize overnight at 42 °C. The amount of RNA was estimated by BAS2000 imaging analyzer (Fujix).

Results and Discussion

We constructed four types of expression vectors as shown in Fig.2 in order to produce a recombinant hChM-I. The possible problems and the improvement points of these vectors are as follows. Vector 1 was constructed by insertion of PCR amplified fragment into pcDNA1. Vectors 2, 3, 4 were constructed from vector 1. As hChM-I protein may have undesirable effects on the growth of transfectant cells, vector 2 was constructed using heat

(1) AM2

gaattcggctt catgacagagaactccgacaaagttcccattgccctg

MetTheGluAsnSerAspLysValProIleAlaLer

(2) AM3

atcatgccatgtagctggtgggtggcccgcatcctgggcatggtgtaa

IleMetProCysSerTrpTrpValAlaArgIleLeuGlyMetVal***

Fig.1 Nucleotide and amino acid sequences of (1)amino-terminal and (2)carboxy-terminal region of hChM-I. The sequence derived from the vector pCRII is shown in bold. The sequences from PCR primers AM2 and AM3 are underlined.

(1)

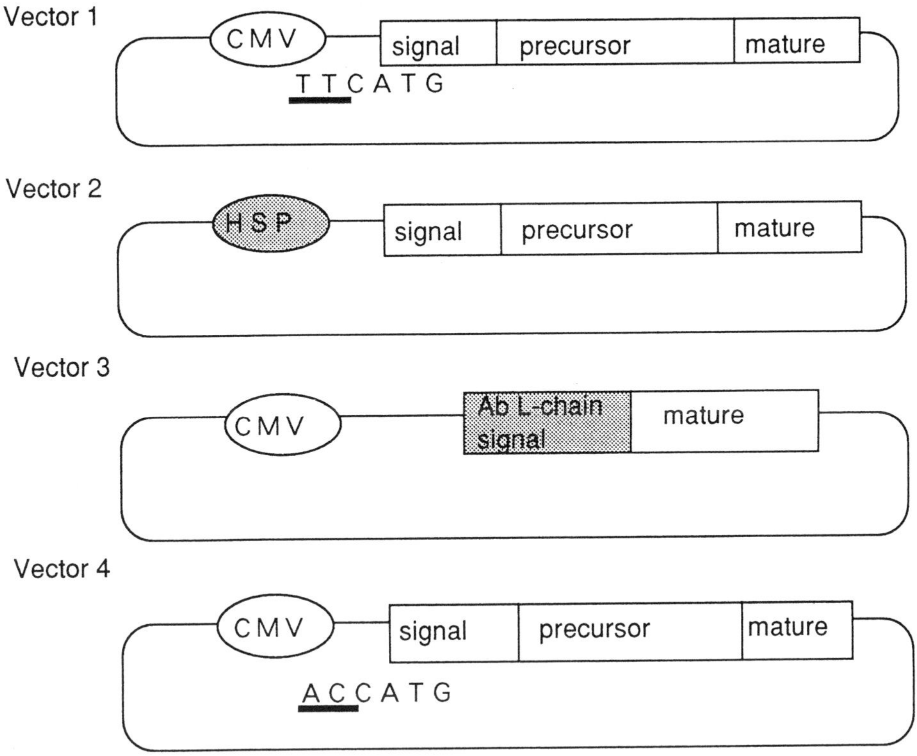

(2)

	mRNA	protein
Vector 1	++	-
Vector 2	++	-
Vector 3	++	+
Vector 4	++	+++

Fig.2 (1) Construction of four types of expression vectors. (2) Summary of results of mRNA expression and protein production using these vectors.

shock promoter, *hsp70*. *Hsp70* is repressed at 37 °C, but is induced at 42 °C. Vector 3 was constructed because the secretion process of product does not go well by the large precursor region of pre-hChM-I. Therefore, signal sequence and a large precursor region of hChM-I were deleted and the mature sequence of hChM-I was directly connected downstream to a signal region of an antibody light chain. Vector 4 was constructed to improve translation from mRNA to protein. Two bases upstream the ATG initiation codon were altered from ggctt**catg** to ggcac**catg** in accordance with Kozak's rule (3). As to vector 1, the cytosine of (-1) position derives from hChM-I cDNA, but the thymine of (-2) position derives from pCRII vector.

CHO cells were transfected with these expression vectors, and the transfectant clones were selected by dot northern hybridization. The amount of hChM-I RNA extracted from twice transfectant was increased greatly compared with once transfectant. By western blot analysis, we could not find significant amount of mature hChM-I protein in culture supernatant of transfected cells using vectors 1, 2 or 3 in spite of the production of considerable amounts of mRNA. With vector 4 transfected cells, recombinant hChM-I protein of diffuse band was detected at 25 kDa by western blotting of 2% FCS-eRDF culture supernatant. The amount of recombinant protein in second transfected cells correlated with the amount of mRNA expression. It was surprising that only a two base substitution upstream the ATG initiator codon improved translation and greatly increased the amount of mature hChM-I. In the case of vector 3, significant amounts of mature hChM-I was not detected in spite of its convenient sequence around the initiator codon of the antibody light chain signal region. It may be that the precursor region of hChM-I is necessary for folding, processing or secretion of mature hChM-I.

(1) Vector 4-rat IgG Fc

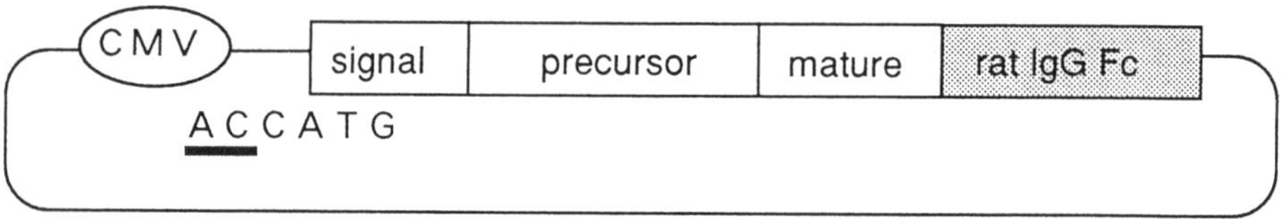

(2) COS (transient)

	0～48 h	48～76 h
Vector 1-rat IgG Fc	0.7	3.1 (ng/ml)
Vector 4-rat IgG Fc	5.5	9.6
pcDNA1	0	0

Fig.3 The amounts of fusion protein produced by transfected cells were quantitated by EIA assay. (1) Construction of vector 4-rat IgGFc. (2) Transient transfection using vector 4-rat IgGFc, about 3-5 times rise of production was detected compared with transfecion using vector 1-rat IgGFc.

Table 1 Summary of the stable transfection using vector 4

	selection reagent	concentration	hChM-I (ug/ml/3day)
1st transfection	neomysin	1000ug/ml	<0.5
2nd transfection	blasticidin S	3	<5
3rd transfection	hygromycin	200	15

The rat immunoglobulin G constant region (IgGFc) was connected to the C-terminus of the mature hChM-I gene in vectors 1 and 4. The amounts of fusion protein produced by transfected cells were quantitated by EIA assay using an antibody against rat IgGFc. Transient expression using COS cells revealed that about 3-5 times more protein was produced using vector 4-rat IgGFc (Fig. 3).

To obtain transfected cells which produce higher amounts of recombinant hChM-I, we transfected the same CHO cells three times using vector 4. First transfection used neomysin resistance, and the clone that produced the highest amount of hChM-I mRNA was selected for the second transfection. The process was repeated for the third transfection, and a cell line which produces apporoximately 15 μg / ml of hChM-I was obtained (Table 1). In order to obtain higher production cells, we are now trying to construct a cosmid vector containing tandem arrays of hChM-I genes.

References

(1) Hiraki Y. et al. (1991) Molecular Cloning of a New Class of Cartilage-Specific Matrix, Chondromodulin-I, which Stimulates Growth of Cultured Chondrocytes. Biochem. Biophys. Res. Commun. 175, 971-977.

(2) Takahashi K. et al. (1994) Molecular Cloning and Functional Expression of Human Chondromodulin-I(hChM-I) and Sequence Analysis of an Upstream Region of hChM-I Gene. International Symposium on Cartilage Metabolism. 84-88.

(3) Kozak M. (1986) Point Mutations Define a Sequence Flanking the AUG Initiator Codon That Modulates Translation by Eukaryotic Ribosomes. Cell 44, 283-292.

Elucidation of metabolic network of mammalian cell line K562 using 2-^{14}C glucose label and stoichiometric modelling

Monika DREWS and Toomas PAALME

Laboratory of Molecular Genetics, Institute of Chemical Physics and Biophysics, Akadeemia tee 23, EE0026 Tallinn, ESTONIA

Abstract

Nutrient consumption and incorporation of the label from 2-^{14}C glucose into metabolic by-products were studied. A compartmentalised metabolic network of K562 cells was assembled and intracellular fluxes using a stoichiometric model were calculated. An equation system of label movement was created. Enrichments of lactate, alanine, glutamate and proline predicted from the model agreed well with those experimentally determined. According to the model 5% of glucose consumed entered the mitochondria. Most of the pyruvate in the mitochondria was derived from glutamine. Alanine was formed in mitochondria.

Keywords: K562 cells, 2-^{14}C glucose, metabolism, stoichiometric modelling

Introduction

Mammalian cell lines are widely used as host systems for the production of recombinant proteins. The intensive glycolysis in cultured mammalian cells as well as in malignant cells has puzzled investigators for several generations. Visualisation of the contribution made by different metabolic pathways in growing mammalian cells will help advance our understanding of metabolic control mechanisms. Intracellular enzyme activities provides information about rate limits (fluxes) of biochemical reactions. Stoichiometric modelling, on the other hand, allows quantitative analysis of complicated metabolic flux patterns. Usually stoichiometric models are analysed on the basis of experimentally determined growth characteristics (1, 2, 3, 4, 5). Stoichiometric modelling together with labelling (6, 7) is a powerful method for determining real intracellular fluxes.

In this study we used macroscopic growth characteristics (specific substrate consumption, growth yields, chemical composition of cell) in combination with ^{14}C labelling method and stoichiometric modelling to construct an intracellular network for the erythroleukemia cell line K562.

Materials and methods

Cell lines, media, cultivation conditions

The human erythroleukemia cell line K562 was used. Cells were grown at 37°C in 25 cm^2 culture flasks and spinner flasks (Wheaton, USA) with a working volume of 100 ml in CO_2 incubator (5% of CO_2). DMEM (Sigma) medium supplemented with 10% of FCS

M.J.T. Carrondo et al. (eds.), Animal Cell Technology, 697–702.

(Gibco) and antibiotics was used for cell maintenance and cultivation in spinner flasks. In labelling experiment JMEM (Sigma) was used.

Analytical methods
Cell numbers were measured by hemocytometer and viability determined by Trypan Blue exclusion. Sugar and organic acid content of the culture medium were analysed by HPLC using Bio-Rad HPX-87H by isocratic elution (0.6 ml min^{-1}) with 0.009 N H_2SO_4. The concentration of amino acids was determined with a "Biotronic" amino acid analyser LC 2000. Protein from the cell biomass was determined by the Folin method (8) using serum albumin as a standard. RNA was determined by the orcinol method (9) and DNA by the Burton diphenylamine method (10). Polysaccharides were determined with anthrone reagent (11).

The measurement of specific radioactivity of excreted compounds
Specific radioactivity was determined by β-liquid scintillation counting. 0.5 ml fractions from HPLC or amino acid analyzer were collected in scintillation vials containing 10 ml toluene and Triton-X100 based scintillant. Radioactivity's of fractions containing individual compounds were totalled. The amounts of the individual amino acids in the combined fractions were determined in the parallel with organic acid and sugars in the same chromatographic run.

Results

The incorporation of nutrients into biomass, and formation of metabolic by-products during the growth of K562 cells on JMEM medium was studied.

Table 1. Amino acid and glucose consumption, incorporation into the different components of cell biomass and by-products. Specific radioactivity of spent media components.

Component	Nr i	Consumption Excretion ($mmol/10^6 cell$) U_{Si}	Incorporation into protein ($mmol/10^6 cell$) p_i	Incorporation into biomass* ($mmol/10^6 cell$) $p_i+s_i+n_i$	Metabolized a_i	Enrichment E_i
Glucose	21	- 16.8	0.26 (Gly)	0.48*		1
Lactate	30	+ 31.92	-	-	-	1/2
Ala	1	+ 0.405	0.27	0.27	-	1/4
NH_3	40	+ 0.72	-	n.d.	-	0
Arg	2	- 0.345	0.09	0.09	0.255	0
Asp	3	0			-	0
Asn	4	0	0.195	0.325	-	0
Glu	5	+ 0.54	0.15	0.15	-	1/8
Gln	6	- 1.95	0.15	0.35	1.6	0
Gly	7	0	0.285	0.335	-	0
His	8	- 0.06	0.06	0.06	-	0
Ile	9	- 0.3	0.09	0.09	0.21	0
Leu	10	- 0.36	0.24	0.24	0.12	0
Lys	11	- 0.255	0.225	0.225	0.03	0
Met	12	- 0.06	0.069	0.069	-	0
Phe	13	- 0.165	0.075 (0.03 Tyr)	0.075	0.06	0
Pro	14	+ 0.3	0.03	0.03	-	1/23
Ser	15	0	0.09	0.09	-	0
Thr	16	- 0.09	0.105	0.105	-	0
Tyr	17	- 0.03	0.06	0.06	-	0
Val	18	- 0.315	0.045	0.045	0.27	0

* - incorporation into polysaccharides (s_i) and nucleic acids (n_i)

Alanine, glutamate and proline accumulated together with ammonia and lactate in the cultivation medium. Incorporation of 2-[^{14}C]-glucose into lactate was 2-fold, alanine 4-fold, glutamate 8-fold and proline 23-fold lower than that of glucose.

A comparison of amino acid consumption (U_{si}) with the amino acid composition of K562 cells (numerically equal to specific incorporation of amino acids into cell protein) showed that the *His*, *Met* and *Thr* were consumed in amounts comparable to their content in cell protein. The consumption of *Gln*, *Arg*, *Ile*, *Leu*, *Lys*, *Phe* and *Val* was higher than required for cellular protein synthesis. 1.9±0.1 moles of lactate were formed from 1 mole of glucose. The two fold lower specific radioactivity of lactate compared with 2-[^{14}C]-glucose suggests that lactate was formed mainly from glucose via anaerobic glycolysis in the cytosol. The lower radioactivity of exogenous alanine compared to lactate suggests that alanine was not formed from cytosolic pyruvate. The high specific radioactivity of glutamate compared to proline in the culture medium can be attributed to the formation of excreted glutamate and proline from different precursors: glutamate arose from labelled mitochondrial α-ketoglutarate while proline arose from mitochondrial glutamate and ornithine.

Combining the experimental data of substrate consumption and label incorporation (Table 1), with analysis of label movement in the citric acid cycle, and on the distribution of metabolic reactions in the cytosol and mitochondria (12) we postulated a two-compartmental model of K562 cells metabolism (Fig.1).

The incorporation of low molecular weight compounds (mmol/10^6 cells) into macromolecular fractions (nucleic acids (n_i), polysaccharides (s_i), proteins (p_i) and lipids (l_i)) was calculated on the basis of macromolecular composition of the dry biomass of K562 cells. The mass-balance equations for every key intermediate assuming constant concentration were written and solved (Table 2).

Table 2. Solutions of mass-balance equations

Cytosol		**Mitochondrion**	
$c_1=U_{21}-s_1-s_2$	$c_1=16.58$	$m_5=a_{13}+a_{17}+a_{31}+a_9+a_{18}+k_7+a_2-k_8$	$m_5=2.165$
$c_2=2c_1$	$c_2=33.16$	$m_1=m_5+k_1-m_2$	
$c_{12}=n_7+p_7+U_7$	$c_{12}=0.335$	$m_2=k_6$	$m_2=0.8$
$c_{11}=c_{12}+p_{15}-U_{15}$	$c_{11}=0.425$		$m_1=2.305$
$c_3=c_2-c_{11}$	$c_3=32.7$	$m_7=m_1+a_9+a_{10}+a_{11}+a_{13}$	$m_7=2.665$
$c_8=c_2-U_{30}+a_{AH}+n_{NA}$	$c_8=2.58$	$m_{12}=m_7-k_4$	$m_{12}=1.065$
$c_{14}=a_{18}+a_9+a_{10}+a_{11}+$		$m_9=m_3+m_{12}+m_6+m_2-k_2$	
$+a_{13}+a_{17}$	$c_{14}=0.69$	$m_6=k_3$	$m_6=1.305$
$c_7=k_4=l_1$	$c_7=1.6$	$m_3=k_3+k_7+a_2-k_6-k_3-k_8$	$m_3=0.775$
	$k_4=1.6$	$m_9=m_3+m_{12}+k_3+k_6-k_2$	$m_9=2.945$
	$l_1=1.6$	$m_{10}=a_9+a_{18}+m_9$	$m_{10}=3.425$
$k_2=c_8-k_4-c_5;$	$k_2=0.98$	$m_{11}=m_{10}+a_{13}+a_{17}+a_{31}$	$m_{11}=3.555$
$c_{13}=c_8-c_7; c_7=k_4;$		$m_8=m_{11}+k_2+k_4-m_5$	$m_8=3.97$
$c_{13}=k_2$	$c_{13}=0.98$	$m_7=m_8-m_6$	$m_7=2.665$
$c_6=c_{13}+c_{14}$	$c_6=1.67$		
$k_3=p_3+p_4+n_3+c_{13}-U_3-U_4$	$k_3=1.3$		
$c_4=c_6+n_6-k_3-U_5-c_{15}-p_5$	$c_4=-0.125$		
$k_6=U_1-c_4+p_1$	$k_6=0.8$		
$k_1=c_{13}-U_{30}-c_4+c_5$	$k_1=0.94$		
$k_7=a_{gln}$	$k_7=1.6$		

Fig. 1. The metabolic network of K562 cells

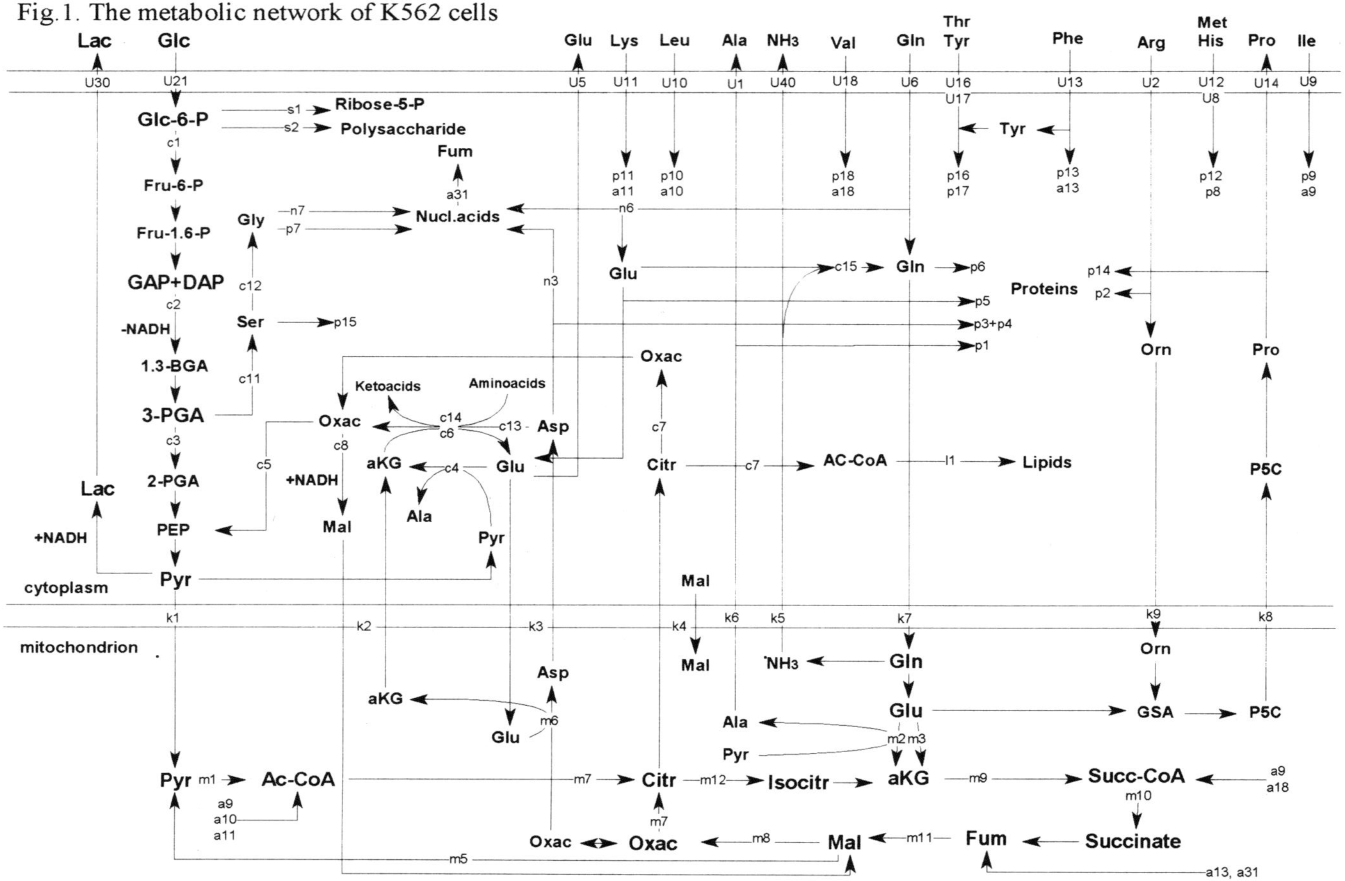

To evaluate the accuracy of the metabolic fluxes (Table 2) the movement of 2-[^{14}C]-glucose label was analysed. The equations describing label movement were developed and solved using values of intracellular fluxes estimated from the stoichiometric model. For example enrichments of the mitochondrial and cytosolic pools of *C-2* of pyruvate can be calculated from the following equation system:

$$[2]Pyr_c = \frac{\frac{1}{2} \cdot [2]Glc \cdot c_3}{c_3 + c_5 - c_4} \tag{1}$$

$$[2]Pyr_m = \frac{k_1 \cdot [2]Pyr_c + m_5 \cdot [2]Mal_m}{k_1 \cdot m_5}; \qquad [2]Mal_m = 0; \tag{2}$$

Theoretical enrichments of alanine, glutamate and proline, calculated from label movement equations were lower than their experimental values (Table 3). The fact can be explained with very high sensitivity of model to lactate formation glucose consumption ratio (U_{30}/U_{21}). At $U_{30}/U_{21}=1.85$ the best match between the experimental points and theoretical enrichments was obtained (Table 3). Taking into account experimental error (±5%) of the U_{30}/U_{21} measurement, the value of $U_{30}/U_{21}=1.85$ falls well into range of the experimental errors.

Table 3. Calculated and experimental enrichment of metabolic intermediates

Compound	Experimental enrichment	Calculated enrichment ($U_{30}/U_{21}=1.9$)	Calculated enrichment ($U_{30}/U_{21}=1.85$)
E_{lac}	0.5±0.05	0.5	0.5
E_{ala}	0.25±0.05	0.177	0.2647
E_{glu}	0.125±0.025	0.054	0.1299
E_{pro}	0.043±0.0086	0.02	0.0553

The sensitivity of the model to errors of determination of biomass lipid composition, glutamine consumption, and the effect of the glutaminolysis was relatively small.

Conclusions

The two-compartmental model of cell metabolism (Fig.1) was in good agreement with experimental results of label incorporation. In growing K562 cells excreted alanine was formed in mitochondria from pyruvate. Mitochondrial pyruvate was formed from exogenous amino acids, mostly from glutamine (~70%), and glucose. 1/4 of mitochondrial pyruvate was converted into alanine, the rest was oxidised in TCA cycle. About 93% of glucose was converted into lactate 0.4% was used in nucleic acid synthesis, 0.3% was used in polysaccharides synthesis in cytosol and 5% entered the mitochondria. In contrast more than 80% of glutamine entered intact the mitochondria, 8% was incorporated into protein and 12% was converted to glutamate during nucleic acid synthesis in cytosol. The glutaminolysis could not be detected. Proline was derived from mitochondrial pool of glutamate and ornithine. Excreted glutamate was derived in cytosol from mitochondrial α-ketoglutarate transported into cytosol by malate-aspartate shuttle.

References

1. Fell DA & Small JR (1986) Fat synthesis in adipose tissue. Biochem J. 238, 781-786.
2. Vallino JJ & Stephanopoulos G (1990). Flux determination in cellular bioreaction networks: applications to lysine fermentations. In: Sikdar SK, Bier M, Todd P, Frontiers in bioprocessing, pp 206-219, CRC press.
3. Palsson BO, Narang A & Joshi A (1989) Computer model of human erythrocyte metabolism. In: The Red Cell: Seventh Ann Arbor Conference, pp133-154, Alan R. Liss, Inc.
4. Galazzo JL & Bailey JE (1990) Fermentation pathway kinetics and metabolic flux contol in suspended and immobilized *Saccharomyces cerevisiae*. Enzyme Microb. Technol., 12, 162-172.
5. Varma A, Boesch BW & Palsson BO (1993). Biochemical production capabilities of *Escherichia coli*. Biotechnol. Bioeng., 42, 57-73.
6. Cohen DM & Bergman RN (1994) Syntax: A rule-based stochastic simulation of the time-varying concentration of positional isotopomers of metabolic intermediates. Computers and Biomedical Research 27, 130-147.
7. Wiechert W, de Graaf AA & Marx A (1995) In vivo stationary flux determination using ^{13}C NMR isotope labelling experiments. In: Munack A & Schüger K, 6th Int. Conference on computer applications in biotechnology, pp 130-135, Garmisch-Partenkirchen, Germany, May 14-17, 1995.
8. Lowry OH, Rosenbrough NJ, Farr Al, Randall RL (1951) Protein measurement with folin reagent. J. Biol. Chem., 193, 265-275.
9. Ceriotty G (1955) Determination of nucleic acids in animal tissues. J. Biol. Chem., 214, 59.
10. Gerhard P et al. (1981) Manual of Methods for General Bacteriology. Washington, DC 2 0006. A American Society For Microbiology.
11. Herbert D, Phipps PJ, Strange RE (1971). Chemical analysis of microbial cells. In: Norris JR & Ribbons DW, Methods in Microbiology, pp 209-344, London, New York. Academic Press.
12. Chuang DT & Shih V (1995) Disorders of branched chain amino acid and keto acid metabolism. In: Scriver CR, Beaudet AL, Sly WS, Valle D, The metabolic and molecular bases of inherited disease, McGraw-Hill, USA.

Expression of a Protein A-Fucosyltransferase III fusion protein in insect cells.

B. Vonach, B. Hess, M. Streiff * and C. Leist

Biopharmaceuticals Development and Production, * Central Research Unit, Ciba-Geigy Ltd, CH-4002 Basel, Switzerland

Abstract

The selectins are a family of cell surface receptors that mediate adhesion of leucocytes to vascular endothelium. Fucosylated oligosaccharides play an important role as ligands for the selectins at the initial stages of leucocyte recruitment in inflammation and almost certainly in the blood borne spread of carcinoma cells. As a tool for chemo-enzymatic synthesis of complex carbohydrates we have expressed a Protein A-Fucosyltransferase III fusion protein using the baculovirus/insect cell system. Sf-9 cells were first adapted to grow and maintained in serum-free medium in T-flasks as well as in suspension cultures. Infection was performed in the mid-log phase ($2x10^6$ cells/ml) with a MOI of 0.5. The recombinant protein (70kD) was secreted into the medium at levels of 12 mg/l by Sf-9 cells and purified from the supernatant by affinity chromatography. Lipids and yeastolate added at the time of infection appeared to be expression promoting substrates. We also used an alternative insect cell line for baculovirus expression, namely the HighFive cells. These cells were grown on glass beads in the presence of serum. After reaching confluency, infection occurred in the absence of serum. These cells provided up to 8-fold higher levels of expression of the fusion protein. Several media were tested for optimal growth and expression of HighFive cells in suspension cultures, also under serum-free conditions.

Key words: *insect cells (Sf-9, HighFive), Protein A-Fucosyltransferase III fusion protein, feeding*

Introduction

Glycosyltransferases form a functional family of membrane bound enzymes that are involved in the coordinate biosynthesis of oligosaccharide chains on glycoproteins and glycolipids. Enzymes responsible for synthesis of terminal structures, and therefore of more immediate biological relevance, like the galactosyltransferases, the fucosyltransferases and the sialyltransferases are more heavily studied than others. We have been particularly interested in such structures as sialyl Lewis X (sia-α3gal-β4(fuc-α3)glcNAc-β), involved in a

M.J.T. Carrondo et al. (eds.), Animal Cell Technology, 703–707.

number of cell adhesion events. Recombinant enzymes can be employed in the enzyme assisted synthesis of oligosaccharides of bio-medical interest and in the *in vitro* remodelling of protein-linked glycans, thus allowing the study of oligosaccharide structure/function relationships. As a tool for enzyme assisted synthesis of complex oligosaccharides, we have expressed a protein A - α 3/4 fucosyltransferase III fusion protein using the baculovirus/insect cell system. We worked with two different cell lines (Sf-9 and HighFive cells) adapted to grow under serum-free conditions and compared the production of the recombinant fusion protein in several different media.

Materials, Results and Discussions:

Expression of the recombinant fusion protein

cDNA encoding the human α1,3/4 fucosyltransferase III (FucTIII) was provided by John Lowe (Univ. Michigan). A soluble and exported proteinA-FucTIII fusion protein was constructed by fusing the coding sequences of the IgG binding domain of protein A to the N-terminally truncated FucTIII (lacking the first 43 amino acids). To ensure secretion, the dog insulin leader sequence was inserted in front of the proteinA-FucTIII coding sequences. The construct was subcloned into the baculovirus transfer vector pVL1392 under the control of the polyhedrin promoter (Invitrogen). Recombinant baculovirus was generated and isolated according to the manufacturer's protocol.

Immunoblotting analysis using an anti-proteinA antibody demonstrated that culture supernatant of infected insect cells contained a highly prominant protein A positive band of approximately 70kD (not shown). The biological activity of the proteinA-FucTIII fusion protein was measured using the Sep-Pack assay described by Palcic *et al.*[1] with radio-labeled GDP-Fucose[C14] as donor and Galβ(1-3) GlcNAcβO-$(CH_2)_8COOCH_3$ as acceptor. The enzymatic activity was shown to be correlated to the protein A concentration measured in an ELISA assay using an anti-proteinA antibody (Fig.1). The recombinant fusion protein was purified by affinity chromatography on Sepharose immobilized GDP.

Production in Sf-9 cells

Sf-9 cells (ATCC CRL-1711) were adapted to grow in the serum-free Sf-900 II medium (Gibco). Cells in spinner flasks (Tecnomara, 70 rpm) or in Erlenmeyer flasks (Corning, 120 rpm) were incubated at 27°C. Highest enzyme yields 3 to 4 days post infection were obtained using cells from the middle exponential growing phase ($2x10^6$ cells/ml) infected with a MOI of 0.5 or 1 (Fig. 2).

Productivity was enhanced by supplementing the culture at the time of viral infection with several substrates (lipids, amino acids) showing that it is the absence of one or more nutrients rather than the accumulation of inhibitory byproducts that limits recombinant protein production in batch cultures with Sf-900 II SFM. The limiting substrate has as yet not been identified. The addition of lipids and yeastolate (Gibco) was almost as effective as a medium exchange which is technically more difficult and time consuming to realise. Optimal concentrations of these substrates are under investigation.

Fig.1 Measurement of proteinA concentration and fucosyltransferaseIII activity.

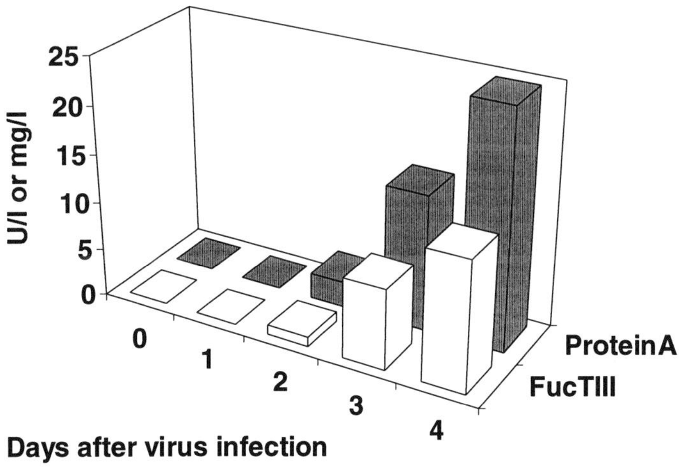

Fig.2 Production of the fusion protein in different media.

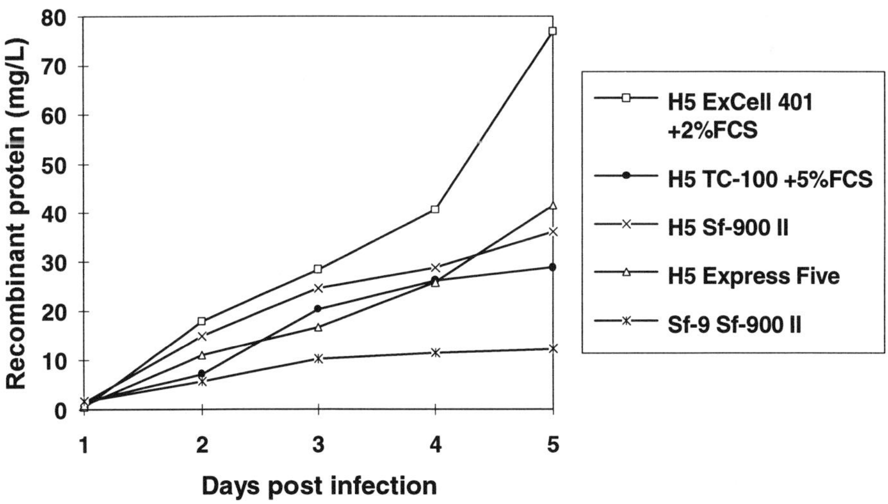

Production in HighFive cells

We also used an alternative insect cell line for baculovirus expression, namely the HighFive cell line, originally growing as adherent cells [2] (inVitrogen B855-02). In a preliminary experiment HighFive cells were grown on glass beads (10 mm diameter) packed in a small column (7 cm high) in TC-100 medium supplemented with 5% FCS (Gibco). After reaching confluency (7 days), cells were infected with the recombinant baculovirus. The column was then continuously perfused with Sf-900 II medium without serum in the presence of 50 μg/ml gentamycin. The fusion protein accumulated in the medium up to a total amount of 1.6 mg in 9 days (Fig.3).

Fig.3 Production of the fusion protein using HighFive cells on glass beads.

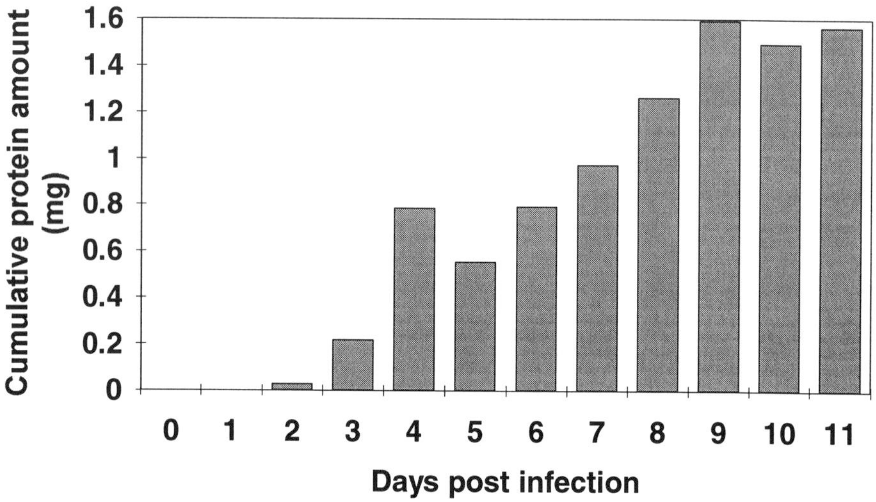

We next tested the production of the cells in several different media. Considerable differences were found (Fig.2). Recombinant protein titers obtained from HighFive cells (36 mg/l) were 3-fold higher than those obtained from Sf-9 cells (12 mg/l) in the same serum-free Sf-900II medium. The best concentration of the recombinant protein (80 mg/l) was obtained in ExCell-400 medium with 2% FCS. HighFive cells were therefore adapted to grow in suspension in ExCell-400 medium in the absence of FCS. These adapted cells were then used for all following experiments. We cultured the cells in a 5L bioreactor equiped with an anchor stirrer (MegaMech, t°: 27°C, 40% pO_2) and measured similar protein concentrations compared to the results obtained in the culture in Erlenmeyer flasks. Here also the production could be increased by addition of glucose, glutamine, lipids and yeastolate (Fig.4). Nutrient addition is clearly an advantage in bioreactor culture eliminating the need for the cell separation step required when replacing medium.

Fig.4 Effect of addition of different substrates using HighFive cells.
A: conditioned medium, B: A + glucose (2 g/l) + glutamine (2mM)
C: B + lipids (0.25%), D: B + yeastolate (2%)
E: B + lipids + yeastolate, F: fresh medium

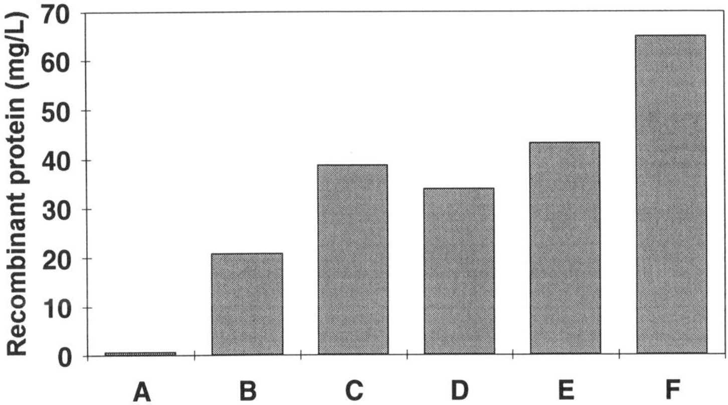

Conclusion

The data presented in this study proved that the structural and functional maturation of the recombinant α 3/4 fucosyltransferase III was retained after fusion with protein A, which simplifies detection and purification and that large amounts of α 3/4 fucosyltransferase III can be generated as a tool for chemo-enzymatic synthesis of complex carbohydrates using the baculovirus/insect cell system. This is of considerable importance for the study of the interaction between selectins and their glycosylated ligands as well as for the design of inhibitor molecules that might become useful anti-inflammatory agents.

Acknowledgements

We thank M. Marangi for assistance in running the bioreactor cultures, F. Kolbinger for the construction of the baculovirus transfer vector, Y. Henriquez for technical assistance and T. Hawthorne for critical reading of the manuscript.

References

1. Palcic M.M., Heerze L.D., Pierce M. and Hindsgaul O. (1988) Glycoconjugate J., 5, 49-63.

2. Wickham T.J., Davis T., Granados R.R., Shuler M.L. and Wood H.A. (1992) Biotechnol. Prog. 8, 391-396.

Adaptation to Serum Free Media of BHK Cells Producing a recIgG

Helder J. Cruz, Elsa M. Dias, José L. Moreira, Manuel J.T. Carrondo*

Inst. Biol. Exp. Tecn./Inst. Tecn. Quím. Biol., IBET/ITQB, Apartado 12, P-2780 Oeiras, Portugal
* also Lab. Eng. Bioq., FCT/UNL, 2825 Monte da Caparica, Portugal

Abstract

In this work, a BHK21 clone producing a fusion protein with potential application in tumor target therapy was adapted to several serum-free media. Three different serum-free media (SFM) were tested (QBSF 56, CHO-S-SFMII, DMEM/Ham's F12 derived medium) as well as a protein-free medium (PFM) (SMIF6). Only SMIF6 medium did not require a gradual adaptation to cell growth in the absence of serum. For all SFM tested, cell productivity was not affected by the decrease in serum concentration during adaptation; however, cell growth was largely affected. Regarding the PFM SMIF6 both cell growth and productivity increased compared to DMEM with 10% FCS; thus, it was chosen for the development of future work.
Using PFM SMIF6 alternatives have been tested to the use of trypsin and FCS inactivation for cell dislodging. These included: non-inactivated trypsin (negative control); trypsin inactivated with FCS (positive control); trypsin inactivated with soybean trypsin inhibitor (STI); cell dissociation solution (CDS) instead of trypsin; dispase instead of trypsin; absence or use of centrifugation was also tested for each alternative. Results indicate that the best method is the use of trypsin inactivated with soybean inhibitor followed by a centrifugation step. For all methods used, to include a centrifugation step always led to improved results. Optimal ratio for total trypsin inactivation is 1 mg of STI to 2 mg of trypsin. Inhibitor cytotoxicity was tested and no toxic effect was observed. Cell growth, viability and productivity using long term subculturing with this new alternative method were not affected.

Key words: BHK cells, serum-free medium adaptation, protein-free medium, cell dislodging methods, trypsin alternatives, trypsin inhibitor.

M.J.T. Carrondo et al. (eds.), Animal Cell Technology, 709–713.

Introduction

Using serum-free media as an alternative to serum-containing media, more controllable and reproducible cultures are obtained, downstream processing is made easier, cost is often lower and there is no potentially harmful complex animal mixture inside the culture system (Butler & Jenkins, 1989). This later factor has extreme importance in the production of human therapeutic products.

Cell productivity of a genetic engineered clone tends to decrease as the operational periods are extended; adaptation to serum-free media is not usually possible in a single step, therefore it should be done in a minimum number of steps. Reduction in serum concentration often leads to a decrease in cell growth and productivity (Barnes & Sato, 1980; Butler & Jenkins, 1989).

Cell subculture using trypsin for cell dislodging requires its subsequent inactivation. With serum-containing media trypsin inactivation is done by the serum components, but under serum-free conditions alternatives must be found. These can consist in finding another inactivation substance (other than serum) or another dislodging agent (other than trypsin) (Barlian *et al*, 1993). The chosen method should avoid the use of serum, even if it could be almost totally eliminated by a centrifugation step. It should also be non toxic and allow high cell growth, productivity, viability and adherent fraction to be maintained after long term subculturing.

Material and Methods

Cell lines and growth systems

BHK 21 cells genetically modified to produce a fusion protein were obtained from Dr. H. Hauser (GBF, Braunschweig, Germany). Cells were at first grown in Dulbecco's modified Eagle's medium (DMEM) (GibCo, Glasgow, UK) supplemented with 10 % fetal calf serum (FCS) (Sigma, St. Louis, USA), 4 mM glutamine and 1 g l^{-1} glucose (Sigma) (all final concentrations). All cultures were performed in 25 cm^2 static flasks (Greiner, Germany) at 37 °C in a 7% CO_2 humidified atmosphere and all media contained puromycin at 5 mg l^{-1} (Sigma). Three different SFM were used: QBSF 56 (Sigma), CHO-S-SFMII (GibCo) and DMEM/Ham's F12-derived medium previously used with BHK cells (Jager *et al*,1988) and a PFM, SMIF6, developed by Drs. K. Scharfenberg and R. Wagner (GBF). Productivity measurements were done by seeding cells in a 25 cm^2 flask with 5 x10^4 cells ml^{-1}. After 72 hours, medium was replaced by fresh one; 24 hours later, suspended and adherent cells (after trypsinization) as well as product concentration were quantified.

Cell dislodging methods

Five cell dislodging methods have been tested, with and without a subsequent centrifugation step (10 minutes at 400 g): non-inactivated trypsin, using a 0.2% (w/v) trypsin/EDTA solution (GibCo); trypsin inactivated with FCS; trypsin inactivated with soybean trypsin inhibitor (STI) at 1 mg/ml (Sigma); cell dissociation solution (CDS)

(Sigma) instead of trypsin; dispase (grade II) at 2.4 U/ml (Boehringer Mannheim) instead of trypsin. All experiments were repeated at least three times and were performed in triplicate. This work was done with cells already growing in PFM SMIF6.

Sampling

Viable cells were determined using the trypan blue exclusion method. Adherent cells were determined after cell release from the static surface by digestion with trypsin. Product quantification was performed using an ELISA protocol obtained from Dr. Erwin Rieke (E. Merck, Darmstadt, Germany).

Results and Discussion

Serum-free medium adaptation

Single step adaptation has only been achieved with PFM SMIF6; the three SFM required a gradual adaptation. For the SFM tested no significant change in cell specific productivity during adaptation was observed; however, cell growth during adaptation decreased with decreasing serum content. Switching from DMEM containing 10% FCS to PFM SMIF6 led to an increase in both cell growth and specific productivity.

Serum-free media comparison

Cell growth in SFM containing low amounts of serum was lower than in the control (DMEM 10% FCS), with specific growth rates ranging from 0.015 to 0.019 h^{-1}, contrasting with 0.024 h^{-1} obtained for the control (table 1). Cell growth for the DMEM/Ham's F12 derived medium was lower than for the QBSF56 medium (data not shown).

The cell specific productivity obtained for PFM SMIF6 was significantly higher than in the control and in the other media (figure 1). The same is also observed for cell growth (0.039 h^{-1}). This medium also has the advantage of not needing any step of adaptation.

Table 1. Summary of the static culture comparison for the different serum free media.

Medium	FCS (%)	Specific Growth Rate (h^{-1})	Cell Specific Productivity (arbitrary units)	Average Viability (%)	Average Adherent Fraction (%)
DMEM	10	0.024	53	90	94
QBSF56	0.5	0.015	53	76	78
CHO-S-SFMII	0.5	0.019	51	79	81
SMIF6	0	0.039	100	87	93

Figure 1. Variation of product concentrations during static cultures for the different protein-free and serum-free media tested.

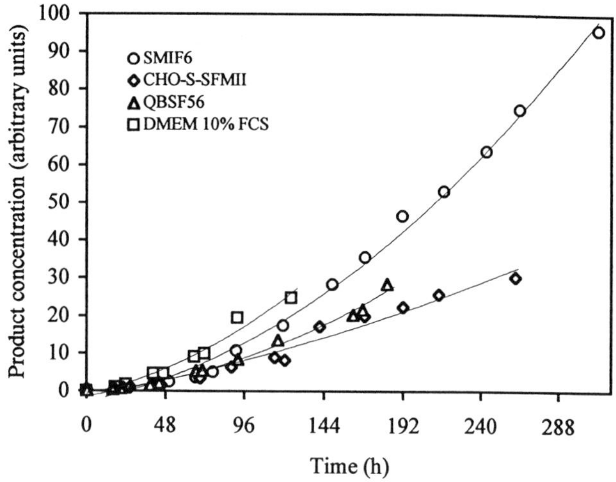

Cell dislodging methods

After cell adaptation to serum-free or protein-free conditions the use of trypsin to cell subculture requires its inactivation after cell dislodging with a substance other than serum. Results of the several inactivation and dislodging methods tested in this work, with cells growing in PFM SMIF6, are presented in figure 2.

Figure 2. Comparison of different cell dislodging methods. These results were obtained after 4 days of static cell culture in PFM SMIF6. Light bars: without centrifugation; dark bars: with centrifugation.

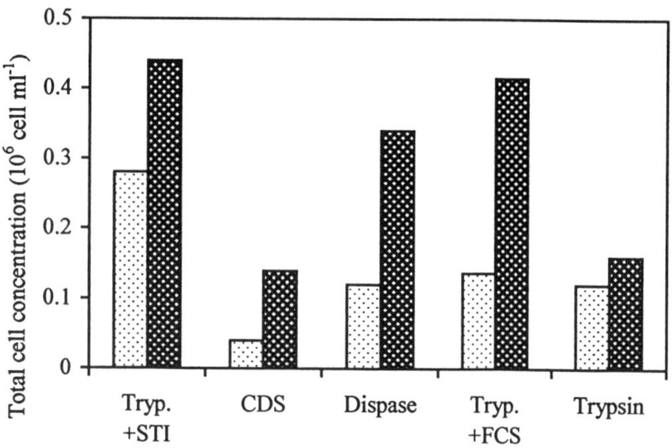

Results indicate that inhibition of trypsin with FCS and STI showed the highest performances; as expected, trypsin without inactivation led to low cell growth; CDS showed even lower performance (figure 2). Centrifugation always led to better results in all cases. Since methods using serum are to be avoided, trypsin plus STI followed by a centrifugation step was chosen. Dispase is also a possible choice, although the cell growth is lower. 1 mg of STI proved to be enough for the complete inactivation of 2 mg of trypsin, which is in accordance with reported values of 1-3 mg of trypsin to 1 mg of STI (Morris, 1995). Long term subculturing using the chosen method did not affect cell growth, viability nor productivity, clearly confirming that this method is suitable to be used in future work.

Conclusions

Single step adaptation has only been achieved with SMIF6, the new PFM used in this work; for the three SFM tested no significant change in cell specific productivity during adaptation was observed; however, cell growth decreased. PFM SMIF6 showed higher cell growth and productivity.

Although the use of dispase can also be considered, trypsin inactivated with STI followed by a centrifugation step showed to be the best dislodging method for subculture under serum-free conditions; 1 mg of STI is enough for complete inactivation of 2 mg trypsin; centrifugation led always to better results, due to removal of cell dislodging agent activity; long term subculturing using the chosen method did not affect cell growth, viability and productivity.

Acknowledgments

The authors acknowledge the financial support received from the European Commission (Biotech PL933069) and Junta Nacional de Investigação Científica e Tecnológica - Portugal (PBICT/BIO/20333/95 and PRAXIS XXI/BD/2764/94).

References

1. Butler M & Jenkins H (1989) J. Biotechnol. 12: 97-110.
2. Barnes D & Sato G (1980) Analytical Biochemistry 102: 255-270.
3. Barlian A, Ganassin RC, Tom D & Bols NC (1993) Cell Biology International, 17, 7: 677-684.
4. Jager V, Lehmann J & Friedl P (1988) Cytotechnology 1: 319-329.
5. Morris CB (1995) In: Doyle A, Griffiths JB & Newell DG (eds.) Cell & Tissue Culture: Laboratory Procedures 1: 4A: 1.4. John Wiley & Sons, England.

Product Quality of a recIgG Under Different Culture Systems of BHK Cells

Helder J. Cruz, Elsa M. Dias, Cristina M. Peixoto, José L. Moreira, Manuel J. T. Carrondo*

Inst. Biol. Exp. Tecn./Inst. Tecn. Quím. Biol., IBET/ITQB, Apartado 12, 2780 Oeiras, Portugal
* also Lab. Eng. Bioq., FCT/UNL, 2825 Monte da Caparica, Portugal

Abstract

In the present work, BHK cells producing a recombinant protein with potential application in tumor target therapy were used in order to study the effect of operating conditions upon growth, production and product quality. Cells were grown in a protein-free medium, SMIF6, as spinner batch cultures with cells immobilized in porous and non-porous supports. Two different support concentrations and two distinct agitation rates were used. Results obtained indicate that increasing support concentration led to slightly higher titers, particularly for non porous supports; increasing agitation rates always led to lower final titers. Product quality decreased with increasing support concentration and agitation rates, this being more evident for non porous support cultures. No significant difference in specific growth rates for the different supports was observed, although higher cell concentrations were obtained with porous supports. Specific productivities were higher for non porous supports, but final titers as well as product quality were superior with porous supports.

Key words: BHK cells, protein-free medium, recombinant proteins, porous and non-porous supports, product quality.

Introduction

The industrial relevance of Baby Hamster Kidney (BHK) cells is due to their relatively high productivity and ability to perform correct post-translational modifications of the desired products. It is known that operating conditions, particularly the culture system, have a large influence upon product quality.

M.J.T. Carrondo et al. (eds.), Animal Cell Technology, 715–720.

The use of supports makes possible to obtain more surface for growth of anchorage dependent cells in relation to the overall culture volume. In such cultures, providing optimized support concentration is necessary to avoid surface limitation of cell growth. Supports settle rapidly under gravity easing cell separation from culture medium. Porous supports provide some protection from mechanical forces in stirred cultures under intense agitation, leading often to higher cell growth than that obtained with non-porous supports (Moreira *et al*,1994).
Culture conditions affect cell productivity as well as product quality (Gawlitzek *et al*, 1994). Thus, different immobilization systems should be compared in order to conclude about the best method to achieve higher product titer and quality. For therapeutical use, for which high value proteins with high quality are needed, product quality is the most important criterion. Better knowledge about the cultivation conditions and their influence on product quality is necessary.
Using serum-free and protein-free media as an alternative to serum-containing media, more controllable and reproducible cultures are obtained, downstream processing is made easier, cost is often lower and there is no potential contaminants; however, these media lack protease inhibitors, usually found in serum, and thus even the smallest amount of proteolytic activity can not be eliminated.

Material and Methods

Cell lines and growth systems
BHK21 cells genetically modified to produce a recombinant protein with potential application in tumor target therapy were obtained from Dr. H. Hauser (GBF, Braunschweig, Germany). Cells were grown in protein-free medium (PFM) SMIF6, developed by Drs. K. Scharfenberg and R. Wagner (GBF). All cultures were performed at 37 °C in a 7% CO_2 humidified atmosphere and all media contained puromycin at 5 mg l^{-1} (Sigma). Trypsinization was performed using a 0.2% (w/v) trypsin/EDTA solution (GibCo) and since methods using serum are to be avoided, trypsin was inactivated with soybean inhibitor followed by a centrifugation step. This was the method of choice for cell dislodging under serum-free conditions, as obtained in previous studies. Centrifugations were performed for 10 minutes at 400 g. Optimal ratio for total trypsin inactivation is 1 mg of STI to 2 mg of trypsin.
Studies were performed in 125 ml spinner flasks (Wheaton, Millville, New Jersey, USA) containing 125 ml of medium and 3 x10^5 cells ml^{-1} as inoculum. Two different immobilization systems were used: a non-porous support, Cytodex 3 (Pharmacia, Uppsala, Sweden) and a macroporous support, CultiSpher G (HyClone, Lund, Sweden). The supports were prepared in accordance with the manufacturer's recommendations (Anon, 1981 and HyClone Product Information, 1991). Studies were done at two different agitation rates (60 and 100 rpm) using two different support concentrations (2 and 4 g l^{-1} for CultiSpher G and 3 and 6 g l^{-1} for Cytodex 3).

Sampling
Viable cells were determined using the trypan blue dye exclusion method. Suspended cell concentration was determined after support sedimentation and adherent cells were determined after cell release from non-porous support or total macroporous support digestion by trypsin.
Product quantification was performed using an ELISA procedure obtained from Dr. E. Rieke (E. Merck, Darmstadt, Germany).

Product quality assays
SDS-PAGE electrophoresis were carried out on 10% polyacrilamide gels with 4% stacking gels (Laemmli, 1970). Protein transfer onto PVDF (Immobilon P, Millipore) was done using a semi-dry system (Bio-Rad). Product immunodetection was performed using an alkaline phosphatase coupled antibody and BCIP/NBT substrate. In all cases 1 μg of product was used as well as the same amount of standard.

Results and Discussion

Non-porous supports
No growth was observed as single cells nor as aggregates in suspension. However, cells did grow in porous as well as on non porous supports; these results were obtained with PFM SMIF6, lacking any adhesion factor, sometimes referred as essential for cell adhesion to a surface.
Cultures show similar growth behavior with non porous support concentration of 3 g l^{-1} showing higher maximum cell concentrations (data not shown); specific growth rates were also similar with lower levels for higher support concentration; a high fraction of adherent cells was observed (Table 1).

Table 1. Summary of the comparison of the different conditions obtained with non-porous support.

Cytodex-3 ($g\ l^{-1}$)	Agitation rate (rpm)	Specific Growth Rate (h^{-1})	Average Viability (%)	Aver. Adher. Fraction (%)
3	60	0.020 ± 0.004	88	88
3	100	0.020 ± 0.003	88	87
6	60	0.017 ± 0.002	91	93
6	100	0.016 ± 0.001	87	87

Production curves show similar behavior; final titers were higher with higher support concentration and lower with increased agitation rate (Figure 1). This may be due to cells being under higher shear stress with higher agitation rates and support concentrations.

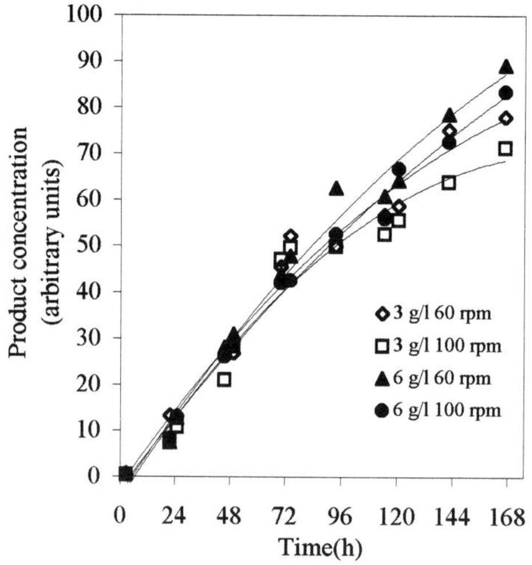

Figure 1. Variation of product concentration during growth on non-porous support (Cytodex 3).

Product quality with non porous supports

As culture time increases, product quality decreases due to an increased product degradation; increasing non-porous support concentration leads to a decrease in product quality (data not shown). This can be due to higher collisions that occur with higher support concentrations or higher agitation rates, leading to cell lysis that will release proteases or incompletely processed product to the supernatant; these aspects are currently under study.

Porous supports

Although similar growth behaviors were observed, maximum cell concentrations were higher for higher porous support concentration and lower agitation rate, the maximum being achieved with 4 g l^{-1} of support at 60 rpm (data not shown); these were also the conditions under which higher titers were achieved (Figure 2).

Table 2. Summary of the comparison of the different conditions obtained with porous support.

CultiSpher G (g l^{-1})	Agitation rate (rpm)	Specific Growth Rate (h^{-1})	Average Viability (%)	Aver. Adher. Fraction (%)
2	60	0.022 ± 0.003	90	83
2	100	0.015 ± 0.004	87	75
4	60	0.017 ± 0.003	92	86
4	100	0.018 ± 0.003	89	81

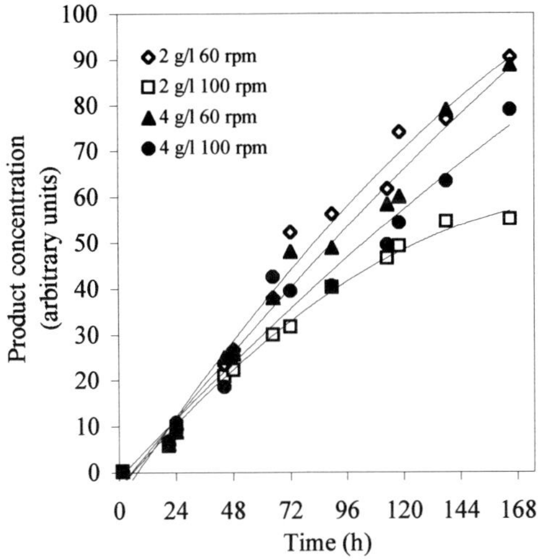

Figure 2. Variation of cell (A) and product (B) concentration during growth on macroporous supports (CultiSpher G).

Product quality with porous supports
As culture time increases, product degradation band also increases, showing a decrease in product quality. Increasing porous support concentration or agitation rate leads to a slight decrease in product quality; however, these effects were not as high as those observed with non porous supports, because porous supports provide the cells with some protection from mechanical forces (Moreira *et al*,1994).

Product quality decreases when comparing CultiSpher-G to Cytodex-3 cultures. This may be due to cells being more exposed in the non porous culture, leading to more cell lysis that will release incompletely processed product as well as proteases that may degrade secreted product. Even the smallest amount of proteolytic activity could not be eliminated by proteases inhibitors which naturally occur in serum (Teige *et al*,1994); 10% of the total amount of serum proteins are protease inhibitors (Travis, 1983). The protein free medium used, SMIF6, does not protect product from possible protease activity.

Conclusions

Increasing support concentration led to slightly higher titers, particularly for non porous supports; increasing agitation rates always led to lower final titers; product quality decreased with increasing support concentration and agitation rates, this being more evident for non porous support cultures because cells are more exposed to mechanical forces, leading to higher cell lysis that will release proteases and incompletely processed product. No significant difference in specific growth rates for the different supports was observed, although higher cell concentrations were obtained with porous supports; specific productivities were higher for non porous supports, but final titers and product quality were superior with porous supports.

Acknowledgments

The authors acknowledge Miss Rosário Clemente for technical assistance and the financial support received from the European Commission (Biotech PL933069) and Junta Nacional de Investigação Científica e Tecnológica - Portugal (PBICT/BIO/20333/95 and PRAXIS XXI/BD/2764/94).

References

1. Moreira, J.L.; Alves, P.M.; Rodrigues, J.M.; Cruz, P.E.; Aunins, J.G.; Carrondo, M.J.T. (1994). Annals N.Y. Acadm. Sci. **745:** 122-133.
2. Gawlitzek, M.; Villers, C.; Verbert, A.; Wagner, R.; Conradt, H.S. (1994). pp. 649-651. In: R.E. Spier, J.B. Griffiths, W. Berthold (eds.), Animal Cell Technology: Products of Today, Prospects for Tomorrow. Butterworth-Heinemann, Oxford.
3. Anon (1981). Microcarrier Cell Culture. Principles and Methods. Pharmacia LKB Biotechnology, Uppsala.
4. HyClone Product Information (1991) CultiSpher G, batch nr. 25021, 05/04/1991.
5. Laemmli (1970) Nature 227: 680-685.
6. Teige M *et al* (1994) Journal of Biotechnology 34, 101-105.
7. Travis J & Salvesen S (1983) Annu. Rev. Biochem. 52, 655-709.

A Potent Growth Factor Supplement For Cell Culture Purified From Whey

G.L. Francis, C. Goddard, D.A. Belford, P.E. Walton, *G.O. Regester, *K. DeSilva, *A. Copeland, *G.W. Smithers, †L. Toneman & F.J. Ballard, CRC for Tissue Growth and Repair and GroPep Pty. Ltd., Adelaide SA 5000; *CSIRO Division of Food Science and Technology, Highett VIC 3190; †Bonlac Bioscience International Pty. Ltd. Melbourne VIC 3001 Australia.

Abstract

We developed a single-step cation-exchange process for enriching growth factors present in cheese whey. Whey-derived growth factor extract (WGFE) contained only 0.5% whey protein, but all the growth factor activity. It stimulated the growth of mesodermal-derived cells and inhibited epithelial cells. Maximum growth of fibroblasts exceeded the response to serum, whereas other lines, such as CHO cells and L6 myoblasts, required the presence of low serum concentrations (1% v/v) to achieve similar growth rates. Specific assays detected IGF-I and -II, FGF-1 and -2, PDGF, and TGF-ß1 and -ß2 at concentrations that partially account for the growth-promoting activity. A combination of these growth factors could not reproduce the maximum effect of WGFE, suggesting the presence of other growth factors. Membrane ultrafiltration of WGFE partitioned and concentrated growth factors, producing fractions enriched in IGF or TGF–ß peptides, with enhancement of potency by up to 30-fold. WGFE as a source of growth factors, has application as a serum supplement or extender, particularly for fibroblasts and other mesodermal-derived commercially important cell lines.

Keywords: growth factors, milk, insulin-like growth factor-I, insulin-like growth factor-II, fibroblast growth factor, transforming growth factor-β, platelet-derived growth factor, serum replacement.

Introduction

Milk and colostrum contain growth factors that can be used in routine cell culture in media containing low serum concentrations (1,2). The growth factors present in milk are retained in cheese whey and we have developed a single-step cation-exchange process to enrich growth factors from whey (3). Whey growth factor extract (WGFE) stimulates the growth of a number of cell types (4). The aim of the present study was to produce fractions with a higher potency in cell culture and to analyse the growth factors responsible for the activity.

M.J.T. Carrondo et al. (eds.), Animal Cell Technology, 721–724.

Materials and Methods

Preparation of WGFE and fractions

WGFE was prepared by cation-exchange chromatography as previously described (3). A proportion of the eluate from the column was acidified to pH 2.5 with HCl (5 mol/l) and incubated overnight to produce an acidified WGFE fraction. This was concentrated, extensively diafiltered against NaCl (150 mmol/l, pH 2.5) using a 100 kDa membrane to produce permeate and retentate fractions. The permeate was concentrated using a 3 kDa membrane and all fractions were freeze dried and tested in cell culture.

Assays

Cell growth was measured using a methylene blue dye-binding assay, as previously described (4, 5). TGF-ß bioactivity was measured in Mv1Lu cells using TGF-ß1 as a standard. TGF-ß1 and -ß2 immunoactivity were measured using a commercially available ELISAs (Promega and R&D systems, respectively). IGF-I and -II were measured by radioimmunoassay (6). Acidic and basic FGF and PDGF were measured by radioreceptor assay in BHK-21 and Balb/3T3 cells, respectively.

Results

Three types of cellular growth responses to WGFE have previously been observed (4). Firstly, fibroblastic cells, in which WGFE produces a greater response than fetal bovine serum (10% v/v), for example, human skin and embryonic lung fibroblasts, and Balb/3T3 fibroblasts. Secondly, cells in which the response to WGFE alone is less than fetal bovine serum, but co-incubation with a low concentration of serum is synergistic, for example, L6 myoblasts, BHK-21, and CHO cells. Finally, in many epithelial cell lines, WGFE inhibits the response to serum, for example, MDCK and Mv1Lu.

To identify the factors responsible for the biological activity of WGFE, we have measured the concentrations of several known growth factors (Table 1). The known growth factors are present in WGFE at concentrations that will elicit biological activities on cultured cells. WGFE also contains IGF binding proteins (classes 2, 3, and 4) and the truncated type II IGF receptor.

Table 1. Concentration of growth factors in WGFE

Growth Factor	Concentration (ng/mg WGFE)
IGF-I	22.7
IGF-II	23.9
FGF-1 (acidic)	0.19
FGF-2 (basic)	0.66
PDGF-BB	4.0
TGF-ß	2 - 5

Acidification of WGFE, followed by membrane ultrafiltration produced three further fractions with different growth characteristics from control WGFE. The first of these fractions, acidified WGFE, is more potent in the inhibition of Mv1Lu cells which is associated with a higher concentration of activated TGF-ß peptides. Most of the TGF-ß activity is acidified WGFE is retained by the 100 kDa membrane, consistent with higher biological activity in the retentate (Table 2).

Table 2. Concentration of TGF-ß Peptides in fractions of WGFE

Fraction	Bioactivity ng/mg	TGF-ß1 ng/mg	TGF-ß2 ng/mg
Control	6.97 ± 0.77	0.29 ± 0.01	7.13 ± 0.16
Acid control	245.41 ± 5.92	15.82 ± 0.72	134.29 ± 0.16
Permeate	8.38 ± 1.41	1.92 ± 0.08	18.97 ± 0.85
Retentate	314.5 ± 35.1	8.14 ± 0.41	146.16 ± 4.61

The 100 kDa permeate of acidified WGFE has >10-fold higher potency in fibroblasts (Fig. 1) and myoblasts compared to control WGFE. This fraction contains much higher concentrations of IGF peptides and probably other smaller molecular weight growth factors (Table 3).

Figure 1. Effect of WGFE, acidified WGFE and membrane fractions of WGFE on the growth of Balb/3T3 fibroblasts

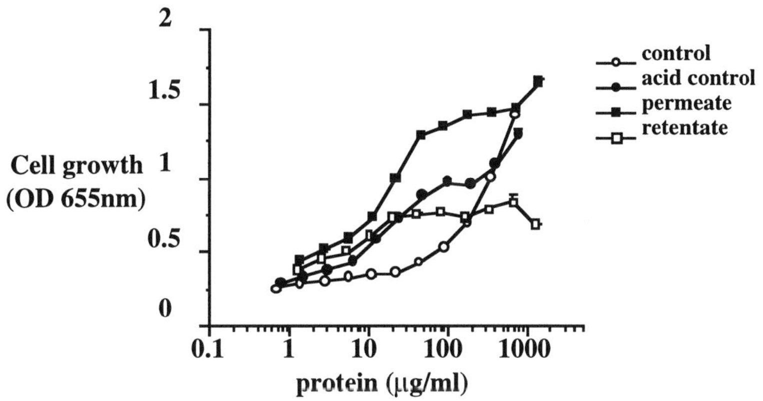

Table 3. IGF Peptides in WGFE Fractions

Fraction	IGF-I ng/mg	IGF-II ng/mg
Control	4.71	8.00
Acid control	4.67	ND
Permeate	130.97	216.72
Retentate	ND	ND

ND = below detectable limit

Discussion

Cation exchange chromatography of cheese whey yields a growth factor enriched preparation that stimulates growth of fibroblasts and other mesodermal cells but inhibits serum-stimulated epithelial cell growth. The concentration of known growth factors suggest that they account for some, but not all, of the cell growth activity. These fractions provide a source of growth factors in low-serum cell culture. Further commercial potential may be elicited by reverse-engineering WGFE to define the mixture of known and unknown factors and then producing these as recombinant products.

References

1. Klagsbrun M (1978) Human milk stimulates DNA synthesis and cellular proliferastion in cultured fibroblasts. Proceedings of the National Academy of Sciences USA 75, 5057-5061.
2. Sereni A, Baserga R (1981) Routine growth of cell lines in medium supplemented with milk instead of serum. Cell Biology International Reports 5, 339-345.
3. Francis GL et al. (1995) Extraction from cheese whey by cation-exchange chromatography of factors that stimulate the growth of mammalian cells. Journal of Dairy Science 78, 1209-1218.
4. Belford DA et al. (1995) Milk-derived growth factors as serum supplements for the growth of fibroblast and epithelial cells. In Vitro Cellular and Developmental Biology 31, 750-758.
5. Oliver MH et al. (1989) A rapid and convenient assay for counting cells cultured in microwell plates: application for assessment of growth factors. Journal of Cell Science 92, 513-518.
6. Owens PC et al. (1991) Developmental changes in growth hormone, insulin-like growth factors (IGF-I and IGF-II) and IGF-binding proteins in plasma of young growing pigs. Journal of Endocrinology 128, 439-447.
7. Growth-promoting agent PCT/AU91/0303
8. Modified milk growth factor PCT/AU95/00237

Development of carbon balances for continuous animal cell cultures using ^{14}-C labelled glucose

Pierre-Alain Ruffieux, Sébastien Mauel, Urs von Stockar and Ian Marison, Institute of Chemical Engineering, Swiss Federal Institute of Technology, CH-1015 Lausanne, Switzerland.

Abstract

Continuous cultures of a hybridoma cell line (*Zac3*) were performed in a stirred tank reactor with protein- free media. At steady state the medium was replaced by one containing U-^{14}C labelled glucose. The fluxes leaving the reactor were analyzed with a scintillation counter to determine the specific activity of each. The gas and solid phases allowed determination of the oxidation and the assimilation ratio of the consumed glucose. The liquid phase was separated chromatographically to determine quantitatively the different species consumed or formed by the cells. Identical experiments were also performed with U-^{14}C glutamine labelled. The major part of the glucose consumed (83.65 %) was found to be converted to lactate. By this method it was possible to complete the carbon balance.

Keywords

Metabolism, ^{14}C labelled substrate, hybridoma, carbon balance

Introduction

Glucose and glutamine are the major energy and carbon sources for hybridoma growth, however little information has been reported concerning the metabolic fluxes and fates of these two substrates. Clearly this information is essential in order to develop control strategies for substrate feeding and toxic product removal in order to obtain optimal growth and antibody production rates.

Some authors (Sharfstein et al., 1994) have used NMR with cells growing in hollow fiber reactors in which an increase of antibody production was observed by reducing the flux of glutamine. However, due to the low sensitivity of NMR methods these measurements were only possible with high cell density cultures such as hollow fibre systems with which it is extremely difficult to obtain homogeneous cell cultures under defined growth conditions. Furthermore batch culture are also difficult to monitor by such methods due to continually changing physico- chemical conditions. In order to overcome these limitations and to enable experiments to be undertaken with low cell density systems, a technique was developed for determining fluxes using ^{14}C- labelled substrates. This technique is extremely sensitive and allows determination of the fates of glucose and glutamine and thus the development of carbon balances to describe cell growth and production.

M.J.T. Carrondo et al. (eds.), Animal Cell Technology, 725–729.

Materials and Methods

Cell line and growth conditions. A murine hybridoma cell line (*Zac3)* made by fusion of the mouse myeloma cell P3X63/AgU.1 with a lymphocyte from Peyer's patches of a BALB/c mouse were kindly provided by Prof. J.-P. Kraehenbuhl (Swiss Institute for Experimental Cancer Research, Lausanne, CH) and used for all experiments. Cells were grown at 36.5°C under a 95% air / 5% CO_2 atmosphere in a completely defined serum- and protein- free medium (FMX-Turbodoma, Cell Culture Technologies, Glattbrugg, CH).

Bioreactors. The culture was performed in a 2 liter bioreactor (Biolafitte, St-Germain-en-Laye, France) operated as a CSTR (see Figure 1). pH was measured and controlled auomatically at 7.3 through the ratio air/CO_2 and the CO_2/H_2CO_3 buffer system used in the media. Bubble- free oxygenation was achieved using tubular Teflon membranes and the pO_2 maintained at 80% air saturation (Schneider et al., 1995). The medium flow rate was controlled by a thermal mass flow meter (Flomega, Brooks, Netherlands) and a constant volume achieved using a constant level device.

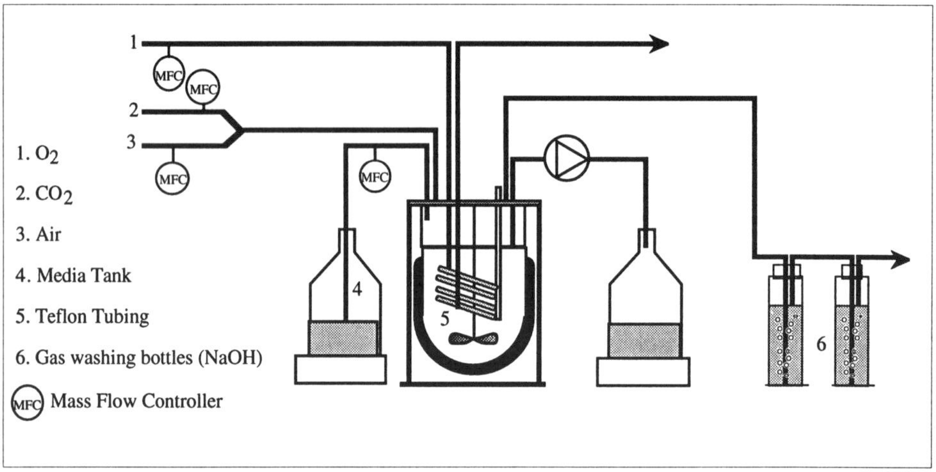

Figure 1: Schematic representation of CSTR system

Analytical methods. Culture samples were removed at intervals and cells counted using a haemocytometer. Viable cell number was determined using the Trypan blue exclusion method. Samples were centrifuged and the supernatant stored at -20°C until required for further analysis. Glucose and lactate was determined using standard enzyme methods (Boehringer Mannheim, Germany) while glutamine and alanine were determined by a HPLC method (Stoll et al., 1994).

Gas phase

The exit gas from the CSTR was bubbled through two reservoirs (250 ml) of NaOH (1M) in order to trap CO_2 as $NaHCO_3$. Culture samples were also acidified in order to release dissolved CO_2 into NaOH solutions. The ^{14}C- labelled $NaHCO_3$ was determined using a scintillation counter and the CO_2 flux calculated.

Solid phase

300 ml culture were harvested, centrifuged, the cell pellet washed with PBS-Tween, lyophilized and the ^{14}C- label accumulating in biomass determined using a scintillation counter and used to calculate the flux of substrate into biomass formation.

Liquid phase

The ^{14}C labeled components of the culture medium filtrate (residual glucose, lactate, etc.) were separated by ion-exchange chromatography determined using a scintillation counter and used to calculate the flux of substrate into products.

Results and Discussion

Biomass activity

Samples of the dried biomass were dissolved in scintillation fluid and divided into aliquots from which the number of counts was determined using a scintillation counter. The results are shown in Figure 2 in which the slope yields the specific activity of the biomass.

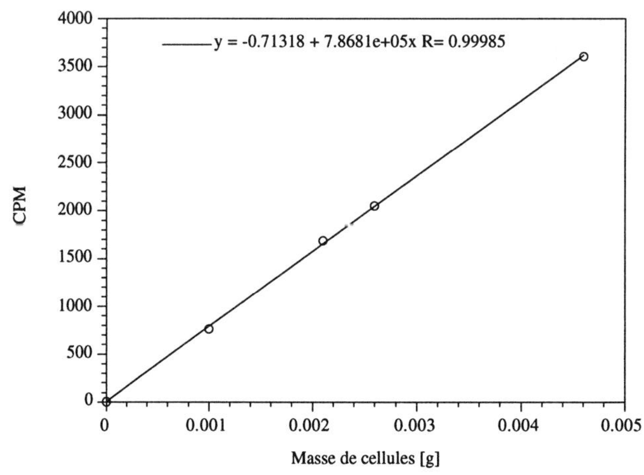

Figure 2: Plot of the counts per minute (CPM) versus biomass.

Analysis of the fractions obtained after ion- exchange chromatography are presented in Figure 3.
The specific activity of the glucose is constant (i.e. the ratio glucose / ^{14}C-glucose) and was determined by analyzing a sample of fresh medium. From the concentration of glucose

in each fraction it was possible to calculate the activity due to glucose. The difference from the total activity yields the activity due to other medium components. The specific activity (count per C-mol) of lactate formed was determined from the lactate concentration and the residual activity. If all the lactate was derived from the glucose, the specific activity must be the same. Any difference is due to the formation of lactate from other compounds such as glutamine.

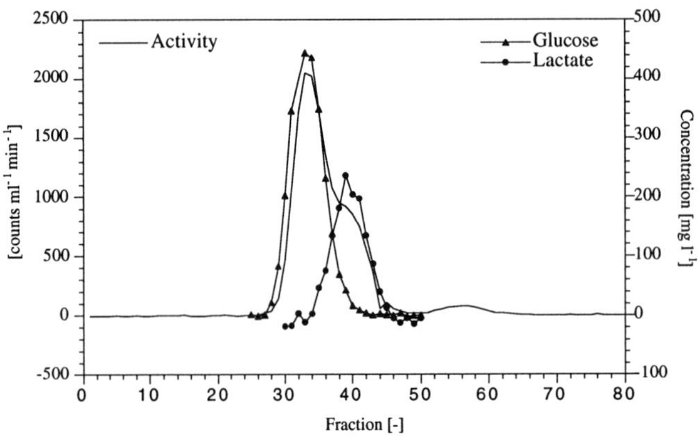

Figure 3 : Analysis of the different fractions after ion- exchange chromatography.

The glucose consumed was found to be transformed mainly into lactate and alanine (Table 1). Only a very low level was oxidized to carbon dioxide with even less being incorporated into biomass.

Table 1: Fate of the glucose in C-mol

Product	**percentage**
Lactate	83.65 %
Alanine	11.13 %
Carbon dioxide	4.11 %
Biomass	0.9 %

The same experiment was performed with ^{14}C-glutamine (results not presented) and 50 % of the consumed glutamine was metabolized into carbon dioxide. With this information concerning the production of CO_2 a carbon balance could be established (the results are presented in Table 2). This method allowed the determination of each flux, even very small ones such as biomass with great accuracy. The next step will be to improve the C- balance by separation of the amino acids in the medium by HPLC techniques and determining the rates of consumption and formation.

Table 2: Consumption or production flux of the major carbon compounds.

Product	Flux [Cmmol l^{-1} h^{-1}]
Alanine	0.09392
Glutamine	-0.24913
Glucose	-1.11569
Lactate	1.01190
Biomass	0.00009
Carbon dioxide	0.18704

The sum of the fluxes out of the reactor is 1.292 C-mmol l^{-1} h^{-1} while that for the fluxes in is 1.364 C-mmol l^{-1} h^{-1}, consequently the C- balance could be closed to within 5.3%.

Conclusions

The use of ^{14}C labelled substrates, such as glucose and glutamine, enables the fluxes of the different substrates through the metabolic pathways to be determined with great accuracy.
Complete carbon balances are normally difficult to develop due to difficulties in the accurate determination of the very low level of CO_2 production by cell cultures, particularly when CO_2/H_2CO_3 buffer systems are used. This has been shown to be overcome by the use of ^{14}C-labelled substrates. Since animal cell cultures usually involve the use of multiple carbon sources, such as glucose and glutamine, it is possible to develop carbon balances by supplying each substrate in a 14C- labelled form and determining the label which is found in the biomass, products and CO_2. In order to close the balance completely it would be necessary to quantify the different amino acids formed.

References

Schneider M., Reymond F. , Marison I W and von Stockar U. (1995) Bubble-free oxygenation by means of hydrophobic porous membranes. Enzyme and Microbial Technology 17, 839-847.

Sharfstein S.T., Tucker S.N., Mancuso A., Blanch H. W. (1994) Quantitative in vivo nuclear magnetic resonance studies of hybridoma metabolism. Biotechnology and Bioengineering 43, 1059-1074.

Stoll T, Pugeaud P, von Stockar U and Marison I W (1994) A simple HPLC technique for accurate monitoring of mammalian cell metabolism. Cytotechnology 14, 123-128.

THE MECHANICAL STRENGTH OF MAMMALIAN CELLS DURING MITOTIC CELL DIVISION

F. Adibah Abdul Majid and Mohamed Al-Rubeai
School of Chemical Engineering, University of Birmingham, Edgbaston, Birmingham, B15 2TT, UK.

Keywords: Hydrodynamic damage, cell cycle, mitosis

Abstract

Recently we have shown that exposure of animal cells to intense hydrodynamic forces can result in preferential destruction of S and G2 cells (Al-Rubeai *et al* 1995b). To further understand the interactions between cell cycle events and the physical environmental factors in large scale bioreaction processes, thereby providing terms for adequate process optimisation, we have evaluated the possible role of mitotic dependent features in shear sensitivity. Synchronised and unsynchronised cultures were exposed to intense hydrodynamic forces exerted in capillary flow and the reduction in interphase and mitotic cells was monitored by flow cytometry and fluorescence microscopy. Destruction of mitotic cells was significantly higher than interphase cells. Cells from the exponential phase of CHO 320 batch cultures, where the percentage of mitotic cells was 5.2 %, was almost entirely destroyed by the laminar flow. At this relatively low stress level, passing of synchronised cells 25 times through the capillary tube resulted in 46% reduction in cell number, respectively. In such conditions, the proportion of G2/mitotic cells (measured by flow cytometry) was reduced from 37.9% to 18.1%, while the mitotic cells (measured by fluorescent microscopy) was reduced from 8.9 % to 1 %.

Introduction

Mitotic division is a complex phenomenon by which duplicated cellular material is separated equally between daughter cells. It represents an essential feature determining the growth rate of cell populations during *in vitro* cultivation of continuous cell lines. Without it, cell arrest ensues leading to growth imbalance and eventually cell death. As such, it presumably is an important process that determines the cell sensitivity to hydrodynamic forces. The understanding of the mechanism by which such forces can regulate cell proliferation and death is therefore essential.

M.J.T. Carrondo et al. (eds.), Animal Cell Technology, 731–736.

Animal cell lines have been shown to exhibit large variations in their morphology, surface area, volume, elastic and viscous resistance to deformation as a single cell progresses through its cell cycle (Needham *et al* 1991). Such cell cycle dependence of cellular properties has implied that the behaviour of proliferating cells in culture can not be characterised by population mean values only. It is now clear that cells in the G2 phase of the cell cycle are more susceptible to mechanical damage than cells in the G1 phase (Al-Rubeai *et al* 1995b) but it is not yet known whether such response is merely due to the larger cell volume of G2 cells or to other cellular changes in the plasma membrane and cytoskeleton structure, as well as in enzymatic activities that are associated with the cell cycle. In this work, we present data to demonstrate that mitotic cells, which in flow cytometric analysis are counted as part of the G2 region, are highly sensitive to shear. These data may explain why the cell number remains low, in spite of a high proliferation rate (represented by high percentage of S phase; Lakhotia *et al* 1992 and by high DNA synthesis; Oh *et al* 1994) seen during the cultivation of animal cells in a relatively high agitation environment.

Materials and methods

Cell line and cell maintenance

The recombinant Chinese Hamster Ovary 320 (CHO 320) cell line was kindly supplied by Glaxo Wellcome (Beckenham, UK). Cells in suspension were maintained in RPMI 1640 medium containing 5 % Foetal Calf Serum (FCS) and 1μM methotrexate.

Viability and viable cell number of random samples were counted using trypan blue dye exclusion in a haemocytometer.

Cell synchronisation with thymidine block

Exponentially grown cells were treated with 3mM thymidine and incubated for 18-24 hours. The cells were then released from thymidine by washing them once with warm thymidine-free medium. Cells were then resuspended in warm fresh complete medium and incubated for another 9 hours before disruption in a capillary flow tube.

Capillary flow experiment

The cell suspensions were circulated through a flow loop of silicone tubing connected to a stainless steel capillary tube of inner diameter, D =1.1 mm; L/D= 46.82 using a peristaltic pump as described by Born *et al* (1992) and Zhang *et al* (1993). The suspension flow rate and Reynolds Number rate were 1.33×10^{-6} M^3/s and 2000, respectively.

Cell cycle and mitotic analysis

To study the cell cycle distribution cells were fixed with ice cooled 70 % ethanol at a concentration of 1×10^6 cells/mL, washed with phosphate buffer saline (PBS) and stained with 50 μg/ mL propidium iodide (PI) solution in PBS for 5 minutes prior to analysis. The flow cytometric analysis was carried out using a Coulter EPICS Elite analyser. PI stained cells were selected by means of dual multiparameter analysis of peak height and integral fluorescence. Cell cycle distributions were deconvoluted using the Multicycle software programme (Phoenix Flow System).

To count the number of cells in mitosis cells were fixed in cold 3:1 absolute ethanol: glacial acetic acid, washed with PBS and stained with PI prior to examination. Cells were examined by fluorescent microscopy and only cells that are in late prophase, metaphase, anaphase and early telophase were counted as mitotic cells. The percentage of these cells in a microscopically examined cell population is called Mitotic Index (MI).

Results and discussion

The results of the experiment carried out with cells obtained from the exponential phase of CHO 320 batch culture are shown in Table 1. After passing the cells 10 times in a capillary tube the cell number was reduced by 13%. However, cell viability as measured by trypan blue exclusion was only reduced by 3%, which indicates the large scale of cell destruction by the shear environment. These results agreed with the findings of Al-Rubeai *et al* (1995a) in conditions of high-energy dissipation rates in agitated bioreactors where the rate of destruction was equal to the rate at which live cells became nonviable. Examining the remaining sub-population it was found that mitotic cells were reduced from 5.2% to 1%, a reduction of nearly 80%. This reduction was higher than that of the flow cytometric G2/M fraction by 2.4%, a discrepancy which may denote the difference between the two techniques in estimation of the fraction of dividing cells.

No. of Passes	Cell No.(x e5) cells/mL	Viability (%)	G1(%)	S(%)	G2/M(%)	MI
0	9.60	96.64	51.8	33.1	15.1	5.2
5	8.53	93.43	51.4	35.4	13.2	1
10	8.33	93.98	52.4	34.3	13.3	1

Table 1: Cell number, viability, cell cycle distribution and mitotic index of mid-exponentially grown CHO cells after passing them through capillary tube.

Next we synchronised the cells by treating them with excess thymidine in order to obtain high mitotic cell population. At 6 and 8 hr after the release of cells from the thymidine they were subjected to 15 passes through a capillary tube. Table 2 shows the effects on cell number, viability, cell cycle distributions and mitotic index. After 15 passes the cell number was reduced by 9% and 23% at 6 and 8 hr from release, respectively. The increasing reduction with time was due to the increased G2 and M cells which were relatively more sensitive to shear than other cells. Such conclusion was supported directly by the results of the G2 and M fractions examined in the same experiment. The reduction in G2 and M cells at both times after release from the thymidine block is clearly indicating that these cells are particularly sensitive to shear stress and such sensitivity is not only related to cell size but also to the changes in cellular structure during mitotic division.

Cell population	**Capillary Flow Test (15 passes)**	**6 hours**	**8 hours**
Cell No.(x e5) cells/mL	Before	2.57	3.48
	After	2.36	2.74
Remaining cells (%)		***91.23***	***76.74***
G2/M (%)	Before	34.9	37.3
	After	34.4	33.8
Remaining G2/M (%)		***98.57***	***90.62***
Mitosis (%)	Before	5.2	13.6
	After	0.6	1.4
Remaining mitotic cells (%)		***11.54***	***10.29***

Table 2. The reduction in G2 and mitotic cells of synchronised CHO cells after passing them through capillary tube.

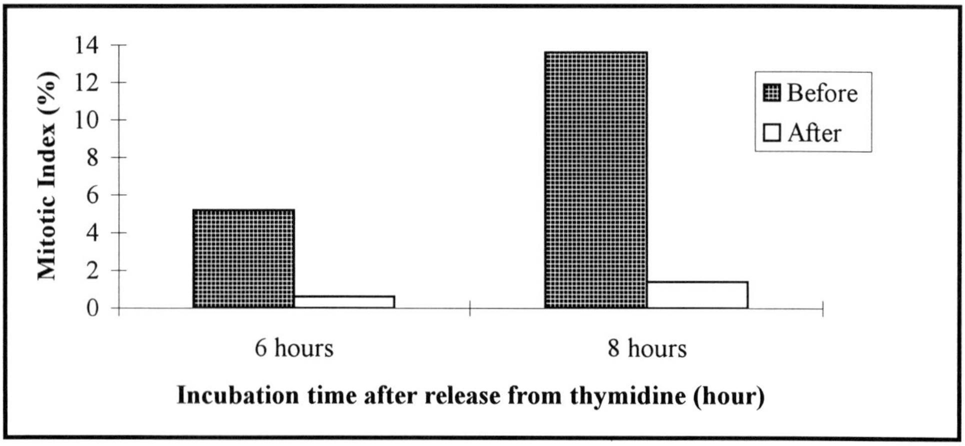

Figure 1: Mitotic index of remaining cells after capillary flow test at various incubation time.

To examine this pattern of destruction by laminar flow, unsynchronised cells were subjected to a longer exposure to shear stress. After passing the cells 25 times through capillary tube the results observed were essentially similar to those found in the previous experiment (Table 3). Most importantly the G2/M fraction as measured by FC was significantly reduced. The reduction from about 38% to 18% in G2 cells was consistent with the similarly higher reduction in mitotic index (7.9%), a pattern which demonstrated the relatively higher sensitivity of mitotic cells in comparison to G2 cells. The latter required longer exposure to laminar flow to be significantly damaged.

No. of Passes	Cell No.(x e5) cells/mL	Viability (%)	G1(%)	S(%)	G2/M(%)	MI
0	6.44	96.89	35.9	26.2	37.9	8.9
25	3.47	77.76	41.7	40.2	18.1	1

Table 3: Cell number, viability, cell cycle distribution and mitotic index of CHO cells subjected to laminar flow by passing them 25 times through capillary tube.

It is possible that the decrease in mitotic cells is due; in addition to their large size, a characteristic that they share with G2 cells; to the cellular changes in the plasma membrane, chromosome morphology and cytoskeleton structure, as well as in enzymatic activities that are associated with mitosis. In summary, the results show that the mitotic phase of the cell cycle is hypersensitive to hydrodynamic forces. This may explain why, during high agitation or sparging rate, when DNA synthesis was reported to increase (Lakhotia *et al* 1992; Oh *et al* 1994) while cell number remains low. Cells undergoing DNA synthesis are progressing from G1 to G2 phase where their cell size increases making them more sensitive to hydrodynamic forces. Surviving cells will eventually enter mitosis where they become even more sensitive to destruction before they destined to divide causing lower growth rate compared to that normally expected with high DNA synthetic rate. This mechanism of cell damage is illustrated in Fig. 2.

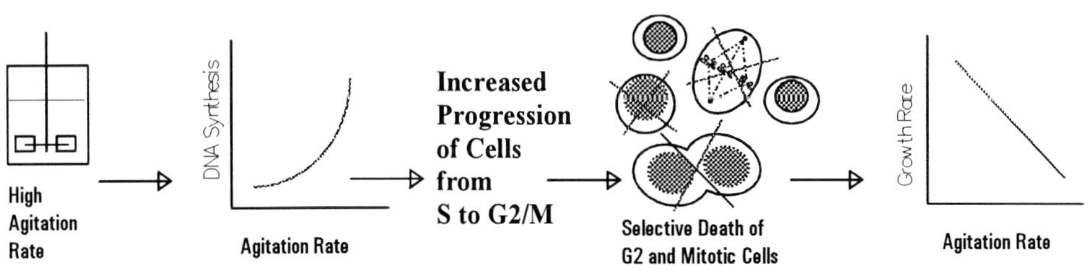

Figure 2: Proposed mechanism for reduced growth rate at high agitation rate

References

1. Al-Rubeai M, Singh RP, Goldman MH & Emery AN (1995a) Death mechanisms of animal cells in conditions of intensive agitation. Biotechnology Bioengineering. **45**: 463-472.
2. Al-Rubeai M, Singh RP, Emery AN & Zhang Z (1995b) Cell cycle and cell size dependence of susceptibility to hydrodynamic shear. Biotechnology Bioengineering **46,** 88-92.
3. Born C, Zhang Z, Al-Rubeai M & Thomas CR (1992) Estimation of disruption of animal cells by laminar shear stress. Biotechnology Bioengineering **40**: 1004-1010.
4. Lakhotia S, Bauer KD & Papoutsakis ET (1992) Damaging agitation intensities increase DNA synthesis rate and alter cell-cycle phase distributions of CHO cells. Biotechnology Bioengineering **40**: 978-990.
5. Needham D, Ting-Beall HP & Tran -Son-Tay R (1991) A physical characterisation of GAP A3 hybridoma cells: morphology, geometry and mechanical properties. Biotechnology Bioengineering **38**: 838-852.
6. Oh S, Al-Rubeai M, Emery AN & Nienow AW (1994) Bioresponses and scanning electron microscopic studies of murine hybridomas subjected to hydrodynamic stresses. In: Animal Cell Technology Products of Today, Prospects for Tomorrow. Eds.: Spier RE, Griffiths JB and Berthold W. Pp 405-407, Butterworth Heinemann, Oxford.
7. Zhang Z, Al-Rubeai M & Thomas CR (1993) Estimation of Disruption of Animal Cells by Turbulent Capillary Flow. Biotechnology Bioengineering **42,** 987-993.

Acknowledgement

Thanks to the Malaysian Government for scholarship support to F. A. Abdul Majid.

Development of Unique Growth Factors for Serum-Free Cell Culture using Fusion Partner Technology

G. L. Francis, C. Goddard, P. E. Walton, Z. Upton and F. J. Ballard,
Cooperative Research Centre for Tissue Growth and Repair and GroPep Pty. Ltd., P.O. Box 10065 Gouger Street, Adelaide SA 5000, Australia.

Abstract

Using fusion partner technology in an *E. coli* expression system designed to give selective attributes, we have produced novel recombinant growth factor analogs specifically for serum-free culture. Long EGF, a human EGF with a 53 amino acid N-terminal fusion partner, and LR^{3}IGF-I, a human IGF-I with an arginine in place of glutamate at position 3 and a 13-residue N-terminal fusion partner, have been developed and tested in serum-free cell culture. Advantages of this approach are a substantially higher yield and improved economy of recombinant peptide production compared to making the native growth factors, and significantly higher potency in cell culture, particularly in the case of LR^{3}IGF-I. Combinations of these growth factor analogs are potently synergistic and provide an important tool for formulating serum-free media.

Keywords: growth factors, insulin-like growth factor, epidermal growth factor, serum-free media, insulin

Introduction

The use of defined serum-free media for commercial cell culture will increase during the next decade for a number of reasons, but in particular to meet regulatory requirements for elimination of animal source proteins in pharmaceutical manufacture. Serum provides a number of essential ingredients for optimum cell growth and protein production rates which must be replaced in synthetic media. These include growth factors, transferrin, fatty acids and attachment proteins. The aim of this work was to develop recombinant growth factor analogs for use in serum-free media. We have initially developed, manufactured and characterised analogs of epidermal growth factor (EGF) and insulin-like growth factor-I (IGF-I).

Materials and Methods

Genetic Engineering
To manufacture Long EGF, a synthetic human EGF gene was modified by site-directed mutagenesis to include cleavage sites, ligated into expression vector p[Met^1]-pGH(1-46)-

M.J.T. Carrondo et al. (eds.), Animal Cell Technology, 737–741.

Val-Asn (1), and transformed into *E. coli* (JM101). After fermentation of the bacteria, inclusion bodies were isolated and solubilized to yield Long EGF. The protein was reduced, desalted, refolded and then purified using Sepharose Fast Flow S and C_{18} reverse-phase HPLC. Long EGF was quantified by HPLC and analysed using N-terminal protein sequencing and mass spectroscopy.

Long R^3 insulin-like growth factor-I (LR^3IGF-I) was produced in a vector [p[Met^1]-pGH(1-11)-Val-Asn], derived from p[Met^1]-pGH(1-46)-Val-Asn, by site-directed mutagenesis in which Arg^3 was substituted for Glu^3 in IGF-I to produce a growth factor with low affinity for IGF binding proteins (2). The expression vector was transformed into *E. coli* (JM101) and, after fermentation, inclusion bodies were isolated and solubilized to yield LR^3IGF-I. The protein was reduced, desalted, refolded and then purified using Sepharose Fast Flow S, C_4 and C_{18} reverse-phase HPLC. It was quantified by HPLC and analysed using N-terminal protein sequencing and mass spectroscopy.

Biological activity
The activity of Long EGF was compared to native EGF in a radioimmunoassay, radioreceptor assay and in the growth of Balb/3T3 fibroblasts and IEC-6 intestinal epithelial cells (3) using a dye-binding assay (4). The activity of LR^3IGF-I was assayed using a radioreceptor assay, by measuring effects on protein synthesis and degradation in L6 myoblasts, and by the binding to IGF binding proteins. The growth of L6 myoblasts in response to LR^3IGF-I and insulin was also compared (2).

The potential for use of Long EGF and LR^3IGF-I in combination in serum-free medium was determined by incubating Balb/3T3 cells with both growth factors (10 and 50 ng/ml, respectively) in the absence or presence of fetal bovine serum (0.5% v/v) and comparing the growth response to that obtained with 5% (v/v) fetal bovine serum alone, using the dye-binding assay (4).

Results

Yield of recombinant fusion proteins
A small-scale fermentation (12.5 litre) of *E. coli* containing the plasmid p[Met^1]-pGH(1-46)-Val-Asn-Gly-Phe-Ala-His-Tyr-EGF yielded 118g wet weight of inclusion bodies. A total of 4.92g of Long EGF was isolated and purified from 25g of these inclusion bodies. Growth factor identity was confirmed by amino-acid sequencing. The purity was greater than 90%, with the residual containing a minor amount of modified Long EGF.

A 100 litre fermentation of *E. coli* containing the plasmid p[Met^1]-pGH(1-11)-Val-Asn-Arg^3IGF-I yielded 647g wet weight of inclusion bodies. A total of 5.2g of LR^3IGF-I was isolated and purified from 170g of inclusion bodies. Growth factor identity was confirmed by amino-acid sequencing. The purity was greater than 90%, with the residual containing a minor amount of modified LR^3IGF-I.

Biological activity
Long EGF was slightly less potent than native EGF in stimulating the growth of Balb/3T3 cells. Half maximum concentrations were 44.2 ± 8.1 and 106.2 ± 12.1 pmol/l, respectively (Fig. 1). Similar potencies were found in IEC-6 cells. Long EGF was between two- and six-fold less potent than EGF in radioimmunoassay and radioreceptor assay.

LR^3IGF-I was more potent than native IGF-I in all the biological assays except in its ability to bind to IGF binding proteins, where it was less potent. It was approximately 1000-fold more potent than insulin in an L6 myoblast cell growth assay and was able to potentiate the effects of serum in the same assay (Fig. 2).

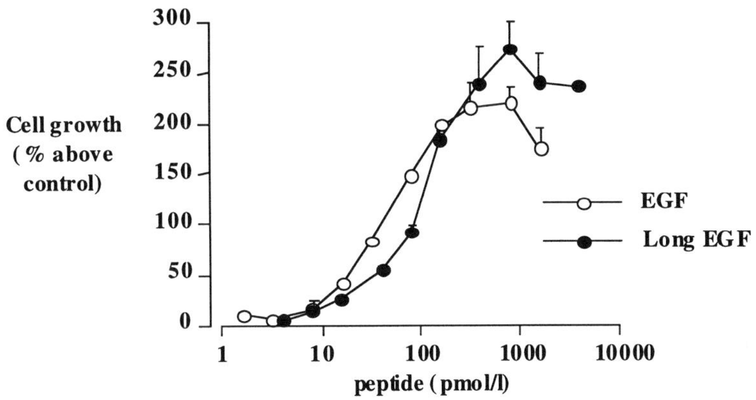

Figure 1. Comparison of the growth of Balb/3T3 fibroblasts in response to increasing concentrations of EGF or Long EGF

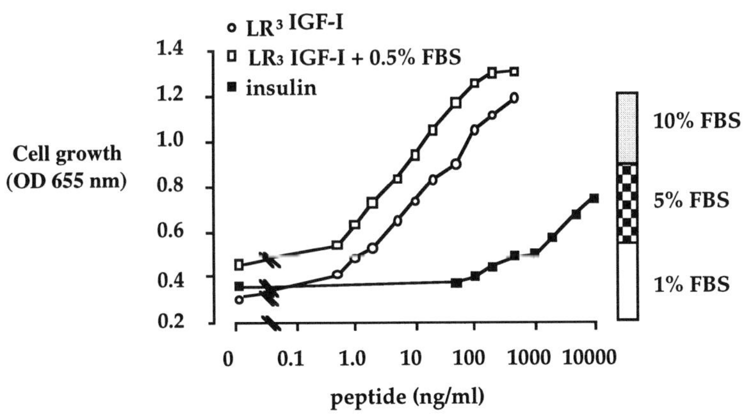

Figure 2. Comparison of the growth of L6 myoblasts in response to increasing concentrations of LR^3IGF-I, LR^3IGF-I containing serum (0.5% v/v) or insulin

A combination of Long EGF and LR^3IGF-I stimulated the growth of Balb/3T3 fibroblasts in serum-free culture and potentiated the effect of a low concentration of serum (Fig. 3).

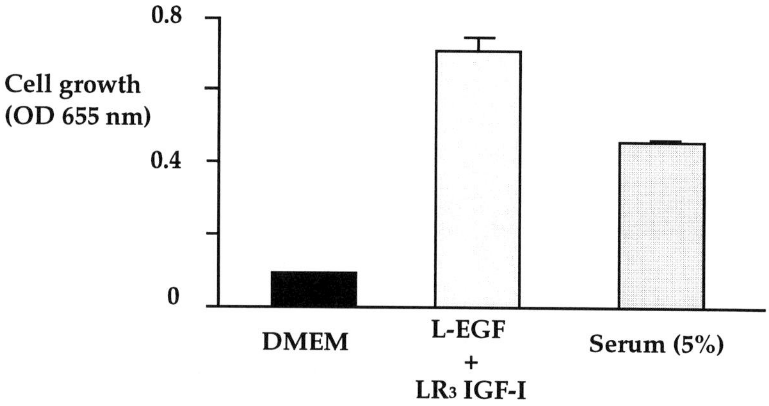

Figure 3. Comparison of the growth of Balb/3T3 fibroblasts in response to medium alone or medium containing LR^{3}IGF-I (50 ng/ml) and Long EGF (10 ng/ml) or fetal bovine serum (5% v/v)

Discussion

This study describes the production of recombinant growth factor analogs which have been developed for use in serum-free media for commercial scale cell culture. These growth factors have a number of special features and selective advantages over native growth factors and insulin.

The Long EGF analog can be produced economically in very high yields because the EGF domain is refolded to a biologically active conformation with the fusion partner without the need for enzymatic or chemical cleavage.

LR^{3}IGF-I is the most potent IGF in cell culture, being 10-fold more potent than IGF-I, and around 1000-fold more potent than insulin. The enhanced activity of LR^{3}IGF-I is the result of insignificant binding to inhibitory IGF binding proteins, compared with IGF-I. Previous studies in commercial CHO cell lines have demonstrated improved production rates of recombinant product and increased cell viability in serum-free media containing LR^{3}IGF-I compared to insulin at up to 1000 times the concentration (6). This growth factor analog, like Long EGF, can also be produced economically in high yields.

Combinations of Long EGF and LR^{3}IGF-I act synergistically, especially in fibroblasts, being more effective than 5% (v/v) fetal bovine serum even though no attempt was made to optimise cell growth and the experiment was performed in the absence of transferrin and albumin. These growth factor analogs provide important tools for the development of specific serum-free formulations.

References

1. King R et al. (1992) Production and characterisation of recombinant insulin-like growth factor-I (IGF-I) and potent analogues of IGF-I, with Gly or Arg substituted for Glu^3, following their expression in *Esherichia coli* as fusion proteins. Journal of Molecular Endocrinology 8, 29-41.
2. Francis GL et al. (1992) Novel recombinant fusion protein analogues of insulin-like growth factor (IGF)-I indicate the relative importance of IGF-binding protein and receptor binding for enhanced biological potency. Journal of Molecular Endocrinology 8, 213-223.
3. Xian CJ et al. (1996) Production of a human epidermal growth factor fusion protein and its degradation in rat intestinal flushings. Journal of Molecular Endocrinology 16, 89-97.
4. Oliver MH et al. (1989) A rapid and convenient assay for counting cells cultured in microwell plates: application for assessment of growth factors. Journal of Cell Science 92, 513-518.
5. Growth hormone fusion proteins, methods of production, and methods of treatment. PCT/AU90/00210; US Patent 5,330,971.
6. Thomas JN and Fung V (1994) Comparison of long R^3 IGF-1 with insulin in the support of cell growth and recombinant protein expression in CHO cells. In Animal Cell Technology, Eds., Spier RE, Griffiths JB and Berthold W, Butterworth-Heinemann.

A Serum-Free Medium Developed to Support the Growth of Several Fibroblastic Cell Lines

T. W. Irish, S. E. Lenk, L. A. Bugner, K. C. Torres and K. J. Etchberger
JRH Biosciences, Inc 1304 W. 107th St. Lenexa KS 66215 USA

ABSTRACT

A serum-free medium, EX-CELL™501, has been developed specifically to support the long term culture of several fibroblast cell lines that are capable of expressing viral and protein products and which also may be used in oncogenic transformation studies. During the development of this medium, human derived proteins, and human recombinant proteins and growth factors were used whenever possible. We have successfully grown NIH 3T3, BALB/c 3T3, Vero, BHK-21 and MDBK cell lines for an extended period of time (> 100 days). Final culture densities and growth rates were comparable to those achieved using a control medium containing 10% FBS. With the addition of 1% FBS to this medium we have also been able to grow immortalized and primary chicken embryo fibroblasts and hepatocytes.

KEY WORDS

serum-free medium, fibroblast, EX-CELL™501, recombinant growth factor.

INTRODUCTION

JRH Biosciences has developed a serum-free medium, EX-CELL™501, that is capable of supporting long term, serum-free propagation of fibroblast cells derived from a number of species including, but not limited to: NIH 3T3, BALB/c 3T3, BHK-21, MDCK, MDBK, A549, 293, VERO, SL-29, and HeLa. All of these cell lines have commercial applications and are currently being used by human and veterinary biopharmaceutical companies to produce vaccines against foot-and-mouth disease, rabies, rubella and influenza. The emerging field of gene therapy utilizes many of these same cell lines for production of retrovirus, adenovirus and adeno associated virus for use in research and clinical studies.

M.J.T. Carrondo et al. (eds.), Animal Cell Technology, 743–748.

Although there have been published serum-free formulations (1- 6) for many of the cells lines that we are reporting on, it was not practicable to use these formulations in a large scale industrial setting. Problems associated with these serum-free formulations included: the incorporation of tissue extracts refined in-house, the need to use Fetal Bovine Serum for cellular attachment and the high cost and availability of growth factors. EX-CELL™501 has overcome these problems by using only the highest grade commercially available chemicals and proteins and the medium was developed specifically to facilitate cellular attachment. EX-CELL™501 includes in its formulation GroPep Pty. Ltd.'s high purity, recombinant growth factors Long R^3 IGF-1 and Long EGF (7, 8). The enhanced biological activity of these growth factors allows for their use at low concentrations, therefore, their use in serum-free media formulations becomes economically feasible.

MATERIALS AND METHODS

Cell Lines:

- All established cell lines where obtained from the ATCC.

Cell Culture:

- All cell culture reagents were supplied by JRH Biosciences. Serum-free adaptation was accomplished by gradual weaning.
- Cell counts were performed using the trypan blue exclusion method. For the Cellular Growth study triplicate 25 cm^2 flasks were seeded at an optimum density in 10 ml of medium with a minimum of three counts.
- Doubling times were calculated by averaging the doubling times from the Cellular Growth study.

Amino Acid Analysis:

- 20 μl of sample was mixed with 2 μl of 1 mM putrescine. This mixture was derivitized with 60 μl o-phthaldehyde and applied to a 4.6 x 100 mm Rainin Microsorb Short-One (#80-200) and eluted with a liner gradient of buffer A: (94.5% 0.078 M sodium acetate: 5% MeOH: 0.5% tetrahydorfuran) to buffer B: (95% MeOH: 5% H_2O).

RESULTS

We have been able to adapt twelve different cell lines, derived from 7 different species into EX-CELL™501 serum-free and have grown many of them for greater then 100 days serum-free (see Table 1). The cell types grown include Embryo Fibroblasts, Kidney cells, Lung Carcinoma cells, Epithelioid Carcinoma cells, and Liver Hepatoblastoma cells.

Table 1. Growth Rates of Serum-Free Adapted Fibroblasts in EX-CELL™501.

EX-CELL™501 Adapted Cell Lines					
Cell Line	Source	Tissue	Pass No.	Serum-Free	+ 1% FBS Growth[1]
NIH 3T3	Mouse	Embryo	30	78%	112%
BALB/c 3T3	Mouse	Embryo	30	73%	133%
Swiss Albino 3T3	Mouse	Embryo	10	92%	127%
PA 317	Mouse	Embryo	8	44%	92%
BHK-21	Hamster	Kidney	30	57%	112%
MDBK	Bovine	Kidney	27	108%	128%
MDCK	Canine	Kidney	20	101%	83%
293	Human	Kidney	11	45%	43%
A 549	Human	Lung	8	fair	57%
SL-29	Chicken	Embryo	4	N.D.[2]	54%
Vero	Monkey	Kidney	25	65%	50%
HeLa	Human	Epithelial	4	N.D.[2]	N.D.[2]

1. Cellular growth rates are expressed as a percentage of growth obtained using ATCC recommended basal media plus 10% FBS.
2. SL-29 and HeLa cell lines are being adapted to serum-free growth and cell numbers have not been determined (N.D.).

Fig.1

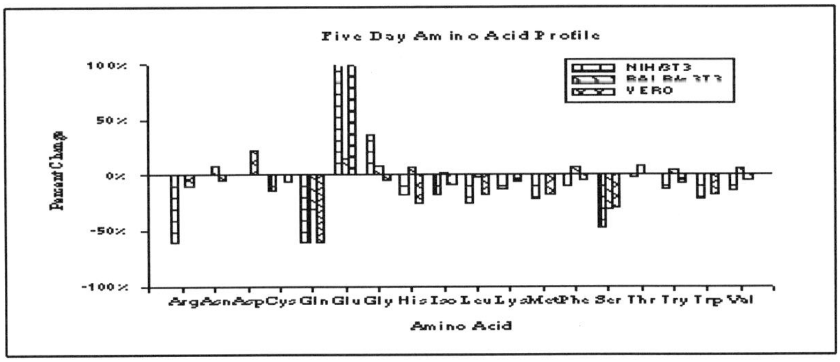

The amino acid utilization and production was examined by HPLC analysis for three cell lines after five days in culture. For all three cell lines tested only Glutamine and Serine were significantly depleted. NIH 3T3 also demonstrated a high degree of utilization of Arginine. Both the NIH 3T3 and VERO cell lines produce excessive quantities of Glutamic Acid (171% and 334% over base line, respectively).

Fig. 2

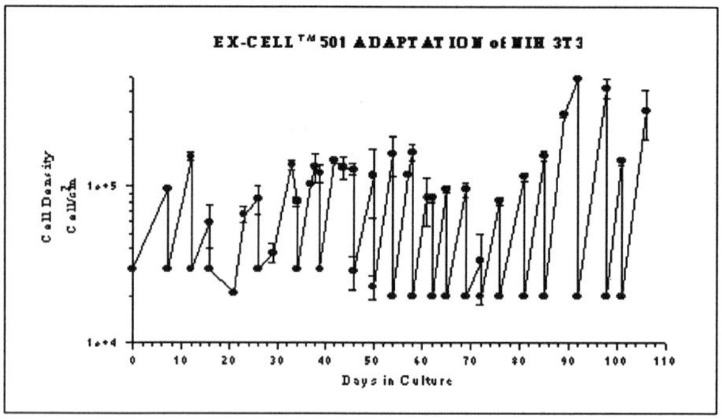

NIH 3T3 cells demonstrate a typical growth adaptation pattern when grown in EX-CELL™501. The early passes (1-5) show variations in final cellular density and growth rates as the cells adapt to a serum-free environment. The later passes demonstrate predictable final cellular densities and growth rates as the cells continue to adapt to EX-CELL™501.

Fig 3.

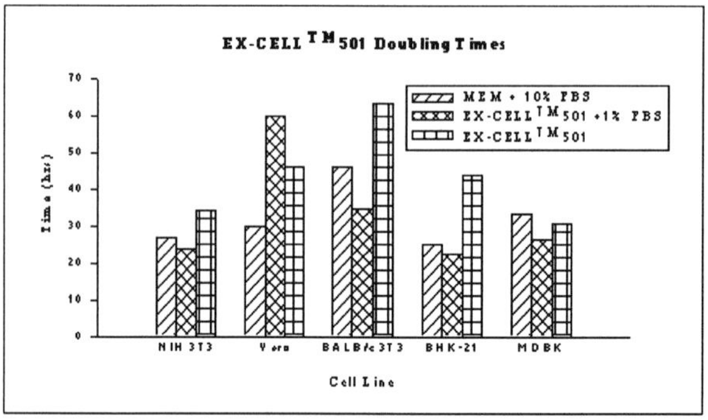

Average doubling times comparisons for five cell lines grown in EX-CELL™501, EX-CELL™501+ 1% FBS, and MEM + 10% FBS. Four of the five cell lines examined demonstrated the best growth in EX-CELL™501+ 1% FBS.

Fig.4

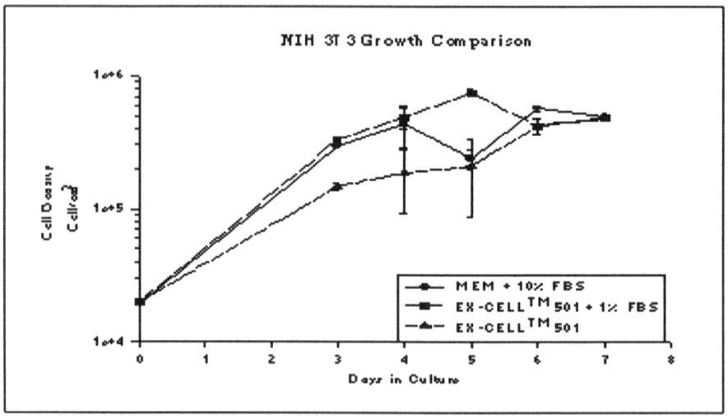

Figures 4.-8. are 5-7 day growth curves with serum-free adapted cell lines grown in EX-CELL™501, EX-CELL™501+ 1% FBS, and MEM + 10% FBS.

Fig. 5

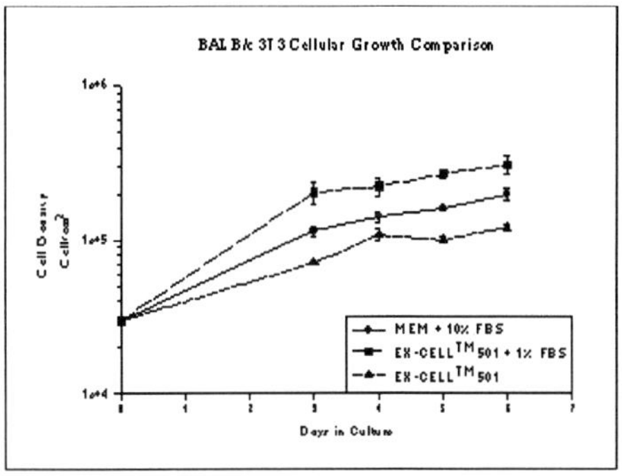

Fig.6

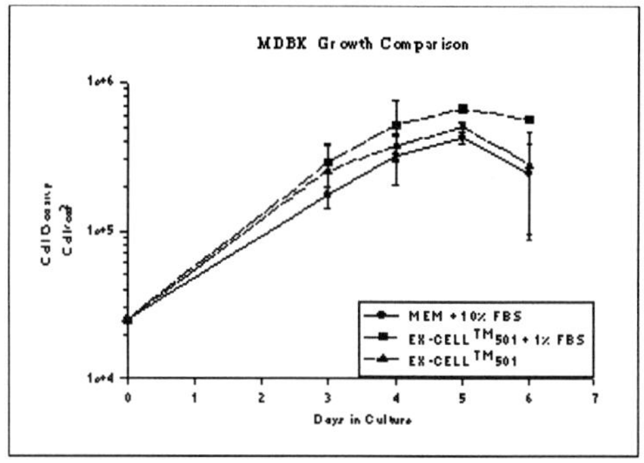

Fig.7

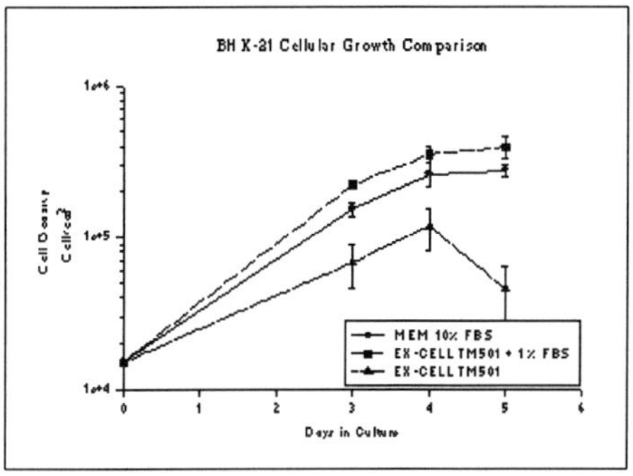

Fig.8

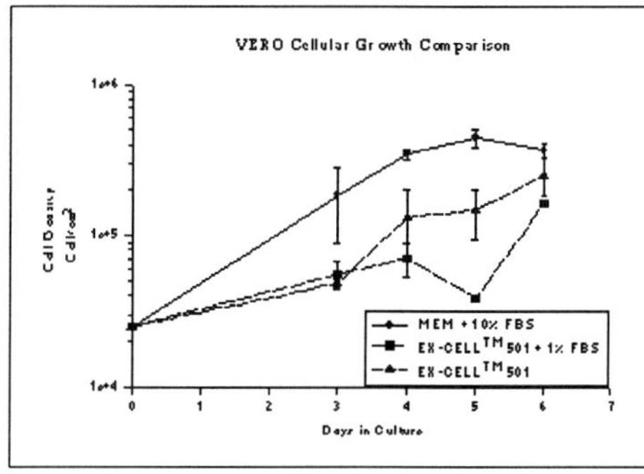

CONCLUSION

- Concentration of amino acids in EX-CELL™501 after 5 days in culture were not growth limiting for the three cell lines tested.
- EX-CELL™501 will support NIH 3T3 cells for long term serum-free culture (>100 days).
- The NIH 3T3 cell line exhibits three distinct phases of adaptation to EX-CELL™501; early (pass 1-4), mid (pass 5-17), and late (pass 18 and higher).
- EX-CELL™501 growth profiles for several cell lines is comparable to MEM + 10% growth profiles.
- EX-CELL™501 + 1% FBS supports higher cellular densities and faster growth rates them MEM + 10% FBS.
- EX-CELL™501 supports the serum-free growth of several fibroblast cell lines derived from a number of species.

REFERENCES

1. Barnes D., Sato G. (1980) Methods of Growth of Cultured Cells in Serum-Free Medium. Analytical Biochemistry 102, 255-270.
2. Bottenstein J., Hayashi I., Hutchings S., Masui H., Mather J., McClure D., Ohasa S., Rizzino A., Sato G., Serrero G., Wolfe R., Wu r. ((1979) The Growth of Cells in Serum-Free Hormone-Supplemented Media. Methods in Enzymology LVIII, 94-109
3. Chiang L., Silnutzer J., Pipas J., Barnes D. (1984) Serum-Free Cell Culture for Growth of NIH 3T3 and 10T1/2 Mouse Embryo Fibroblast Cell Lines, SV40 Virus Propagation, and Selection of SV40-Transformed Cells. In: Methods for Serum-Free Culture of Epithelial and Fibroblastic Cells, pp 265-276, Alan R. Liss, Inc., 150 Fifth Avenue, New York, NY 10011.
4. Taub M. (1984) Growth of Primary and Established Kidney Cell Cultures in Serum-Free Media. In: Methods for Serum-Free Culture of Epithelial and Fibroblastic Cells, pp 3-24, Alan R. Liss, Inc., 150 Fifth Avenue, New York, NY 10011.
5. Traub M., Chuman L., Saier M., Sato G. (1979) Growth of Madin-Darby Canine Kidney Epithelial Cell (MDCK) Line in Hormone-Supplemented, Serum-Free Medium. Proc. Natl. Acad. Sci. 76, No. 7, 3338-3342.
6. Walthall B., Ham R. (1981) Multiplication of Human Diploid Fibroblasts in a Synthetic Medium Supplemented with EGF, Insulin, and Dexamethasone. Exp. Cell Res. 134, 301-311.
7. Francis G., Ross M., Ballard F., Milner S., Senn C., McNeil K., Wallace J., King R., Wells J. (1992) Novel Recombinant Fusion Protein Analogues of Insulin-like Growth Factor (IGF-1) Indicate the Relative Importance of IGF-binding Protein and Receptor Binding for Enhanced Biological Potency. Journal of Molecular Endocrinology 8. 213-223.
8. Xian C., Upton Z., Goddard C., Shoubridge C., McNeil K., Wallace J., Read L., Francis G. (1995) Production of a Human Epidermal Growth Factor Fusion Protein and its Degradation in Rat Gastrointestinal Flushings. Journal of Molecular Endocrinology 15. 1-9.

SPECIAL THANKS

Mara Knapp and Robin Mofle for the Amino Acid Analysis

TRANSFECTION OF HUMAN PLACENTAL ENDOTHELIAL CELLS WITH SV40 T-ANTIGENS

MONIKA SCHÜTZ and PETER FRIEDL
Institut für Biochemie, Technische Hochschule Darmstadt, Petersenstr. 22, D-64287 Darmstadt, Germany

Abstract

Microvascular endothelial cells were isolated from human full-term placenta (Human placental endothelial cells, HPEC). The cells were transfected via lipofection with pRNS1, coding for the SV40 genome. One cell line with largely expanded life span was isolated and kept in culture up to 80 cumulative population doublings. The transfected cell line HPEC-A1 expressed SV40 T-antigens and exhibited endothelial-like characteristics identical to primary cells: uptake of low density lipoprotein, staining with *Ulex europaeus* 1-lectin and indirect immunofluorescence against von Willebrand factor (vWF), α thrombomodulin, and VE-cadherin.

Introduction

The vascular endothelium plays a pivotal role in a number of important physiologic and pathophysiologic events (3). To elucidate specific endothelial functions the studies have mainly relied on the examination of cells *in vitro*. Until now, the most common source of human endothelial cells has been the umbilical cord vein (Human umbilical vein endothelial cells, HUVEC). Despite its biological relevance, little is known about endothelial diversity within the vascular tree, and specific capillary functions remain to be clarified. Studies of human microvascular endothelial cells are hampered by substantial difficulties in isolation and cultivation. Recently we described the preparation of microvascular endothelial cells from human full-term placenta (Human placental endothelial cells, HPEC) (13). Cultures of HPEC and HUVEC provide a valuable system to investigate endothelial characteristics specific for vessel size and not originating from a different tissue environment.

The biochemical and immunological examination of primary cells has a number of limitations, as finite proliferation capability, inability to generate large amounts of cells, and phenotypic instability. In contrast to rodent cells, spontaneous immortalisation of human cells is an extremely rare event. To date, only one cell line derived from human endothelial cells has been reported (16). Another human endothelial cell line has been

M.J.T. Carrondo et al. (eds.), Animal Cell Technology, 749–753.

established by hybridization with tumor cells (2). Several investigators have successfully used the introduction of the large T-antigen to immortalise human endothelial cells (1, 4, 7, 8, 10, 12, 14, 18). So we transformed HPEC with pRNS1, coding for the SV40 genome and the neomycin (G-418) resistence gene (11). Lipofection was used, as an easy and efficient method, previously optimised for human endothelial cells (17).

Materials and Methods

Isolation and cultivation of HPEC

Human placental endothelial cells were prepared and cultivated as previously described (13).

Transfection of HPEC

Transformation of HPEC was performed with plasmid pRNS1 (11) by means of lipofection (17). Briefly, the transfection cocktail consisted of a crude fraction of soyabean lipids (azolectin, Sigma) and 1,2-dimyristyloxypropyl-3-trimethyl-ammonium bromid (DMRIE). The lipid stock solution, containing 1 mg azolectin and 0.73 mg (1.15 mmol) DMRIE in 1 ml sterile deionised water, was sonicated. 96 µl of the lipid mixture and 220 nmol pRNS1 were suspended in 3 ml serum-free culture medium and incubated 30 min at ambient temperature. Logarithmicly growing HPEC were seeded at a density of $1x10^4$ cells/cm^2 in a 75 cm^2-flask. When cultures reached about 50% confluence, growth medium was removed and the cells washed with PBS. 12 ml serum-free culture medium and 3 ml of the liposome-DNA-solution was added. After 5 h the transfection mixture was replaced by culture medium. The day after, HPEC were subcultivated at a *split ratio* of 1:3. Three days later medium was changed to selective medium (supplemented with 200 µg/ml G-418, Gibco). During selection the appearance and growth of cell clones was followed by microscopical observation. When clones reached about 500 to 1000 cells, they were passaged by trypsinisation, gradually propagated and subcultivated. Culture conditions were those of primary HPEC, except for the omission of conditioned medium.

Results

HPEC were transfected with pRNS1 at passage number 4, utilizing lipofection with azolectin and DMRIE. Transfection lead to a dramatic decrease of the cell viability. After two weeks of selection, 8 cell clones emerged and one clone, denoted HPEC-A1, could be subsequently expanded. The basic growth pattern of HPEC-A1 was similar to primary HPEC, but the morphology of transfectants was more heterogenous. Monolayers of HPEC-A1 showed an inhomogeneity in cell size and contact-inhibition of cell growth was less strict (Fig.1). So HPEC-A1 were capable to grow to a higher cell-density in confluence; maximal density of normal HPEC was 6-8x10^4 cells/cm^2, those of transfectants 1-1.1x10^5 cells/cm^2.

HPEC-A1 exhibited endothelial-specific characteristics, identical to primary HPEC (13). Anti-vWF antibodies stained cytoplasmatic granules (Fig.2). α thrombomodulin and VE-cadherin were located at intercellular borders (data not shown). The uptake of acetylated LDL lead to perinuclear staining (Fig.3). There was no binding of *Bandairaea*

Simplicifolia 1-lectin, whereas *Ulex europaeus* 1-lectin stained the perinuclear space (data not shown).

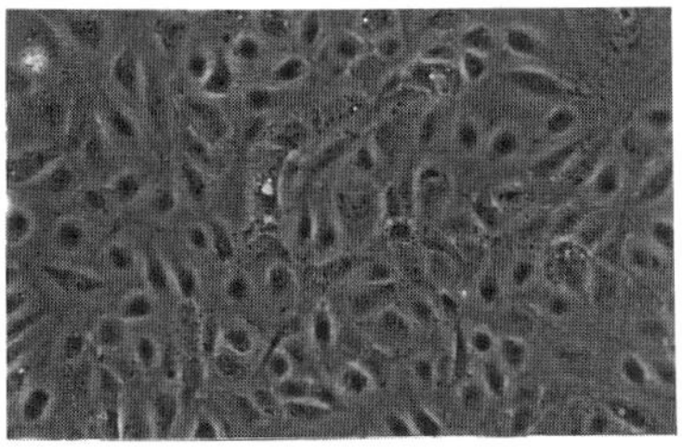

Fig.1. *In vitro* growth pattern of confluent HPEC-A1 monolayers. The cells were elongated and formed expanded spiral structures. x20

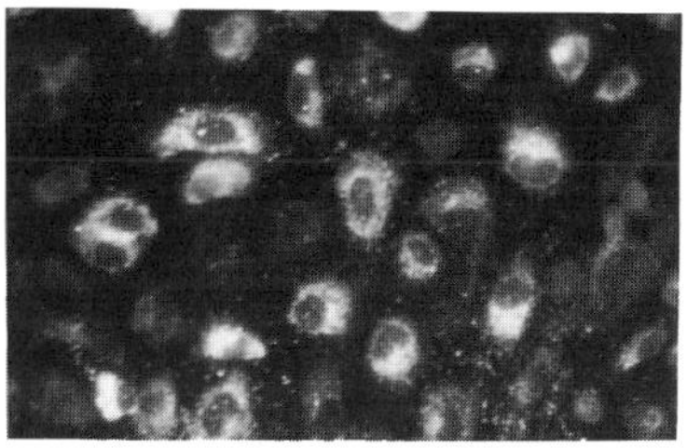

Fig.2. HPEC-A1 displayed positive immunostaining of vWF, accumulated in cytoplasmatic granules. x20

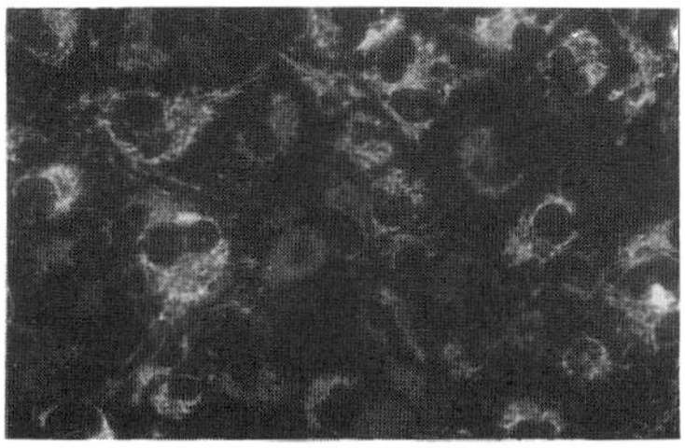

Fig.3. Uptake of Dil-labelled acetylated LDL was demonstrated. x20

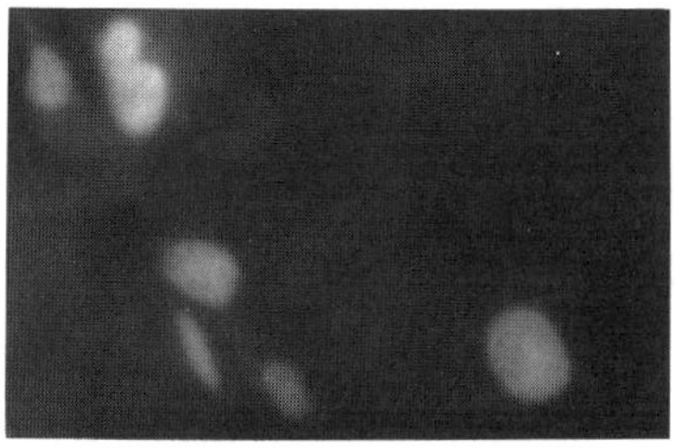

Fig.4. Indirect immunofluorescence revealed the nuclear location of the SV40 T-antigens. x40

The indirect immunostaining of small and large T-antigens lead to a bright nuclear staining (Fig.4). The expression of SV40 early genes enabled HPEC-A1 to overcome senescence of primary HPEC. The introduction of the viral oncogenes resulted in stimulation of cell growth and prolonged cultivation capability up to 80 cumulative population doublings (CPD). But at this age of culture the generation time increased slowly as the cells entered a crisis, finally leading to cell death. So HPEC A1 exhibited a largely expanded, but finite lifespan. Introduction of small and large T-antigens enabled HPEC-A1 to overcome a first mortality stage (M1). But additional genetic events, independent of SV40-early genes seemed to be required to overcome a second mortality stage (M2) for an immortalization (6, 9, 15, 19).

References

1. Ades EW, Candal FJ, Swerlick RA, George VG, Summers S, Bosse DC, Lawley TJ (1992), HMEC-1: Establishment of an immortalized human microvascular endothelial cell line. *J. Inv. Derm.*, **Sept**, 683-690.

2. Edgell CJS, McDonald C, Graham JB (1983), Permanent cell line expressing human factor VIII-related antigen established by hybridization. *Proc. Natl. Acad. Sci. USA*, **80**, 3734-3737.

3. Fajardo LF (1989), The complexity of endothelial cells.
A. J. C. P., **92** (2), 241-250.

4. Fickling SA, Tooze JA, Whitley GStJ (1992), Characterisation of human umbilical vein endothelial cell lines produced by transfection with the early region of SV40.
Exp. Cell Res., **201**, 517-521.

5. Friedl P, Tatje D, Czpla R (1989), An optimized culture medium for human vascular endothelial cells from umbilical cord veins.
Cytotech., **2**, 171-179, 1989.

6. Goletz TJ, Robetorye S, Pereira-Smith OM (1994), Genetic analysis of indefinite division in human cells: Evidence for a common immortalizing mechanism in T and B lymphoid cell lines.
Exp. Cell Res., **215**, 82-89.

7. Hohenwarter O, Zinser E, Schmatz C, Rüker F, Katinger H (1992), Influence of transfected SV40 early region on growth and differentiation of human endothelial cells.
J. Biotech., **25**, 349-356.

8. Hohenwarter O, Jakoubek A, Schmatz C, Katinger H (1994), Expression of SV40 tumour antigens enables human endothelial cells to grow independently from foetal calf serum and exogenous growth factors.
J. Biotech., **34**, 205-211.

9. Hubbard-Smith K, Patsalis P, Pardinas JR, Jha KK, Henderson AS, Ozer HL (1991), Altered chromosome 6 in immortal human fibroblasts.
Mol. Cell. Biol., **12** (5), 2273-2281.

10. Lassalle P, LaGrou C, Delneste Y, Sanceau J, Coll J, Torpier G, Wietzerbin J, Stehelin D, Tonnel AB, Capron A (1992), Human endothelial cells transfected by SV40 T antigens: Characterization and potential use as a source of normal endothelial factors.
Eur. J. Immunol., **22**, 425-431.

11. Litzkas P, Jha KK, Ozer HL (1984) Efficient transfer of cloned DNA into human diploid cells: protoplast fusion in suspension.
Mol. Cell. Biol., **4**, 2549-2552.

12. Sasaguri Y, Yanagi H, Nagase H, Nakano R, Fukuda S, Morimatsu M (1991), Collagenase production by immortalized human aortic endothelial cells infected with simian virus 40.
Virchows Archiv B. Cell. Pathol., **60**, 91-97.

13. Schütz M, Friedl P (1996), Isolation and cultivation of endothelial cells derived from human placenta.
Eur. J. Cell. Biol., submitted.

14. Schwartz B, Vicart P, Delouis C, Paulin D (1991), Mammalian cell lines can be efficiently established in vitro upon expression of the SV40 large T antigen driven by a promoter sequence derived from human vimentin gene.
Biol. Cell., **73**, 7-14.

15. Smith JR, Ning Y, Pereira-Smith OM (1992), Why are transformed cells immortal? Is the process reversible?
Am. J. Clin. Nutr., **55**, 1215S-1221S.

16. Takahashi K, Sawasaki Y, Hata JI, Mukai K, Goto T (1990), Spontaneous transformation and immortalization of human endothelial cells.
In Vitro Cell. Dev. Biol., **25**, 265-274.

17. Teifel M, Heine LT, Milbredt S, Friedl P (1996), Optimization of transfection of human endothelial cells.
Eur. J. Biochem., submitted.

18. Vicart P, Testut P, Schwartz B, Llorens-Cortes C, Perdomo JJ, Paulin D (1993), Cell adhesion markers are expressed by a stable human endothelial cell line transformed by the SV40 large T antigen under vimentin promotor control.
J. Cell. Physiol., **157** (1), 41-51.

19. Wright WE, Pereira-Smith OM; Shay JW (1989), Reversible cellular senescence: Implications for immortalization of normal human diploid fibroblasts.
Mol. Cell. Biol., **9** (7), 3088-3092.

Acknowledgements

We are grateful to the helpful midwifes of the Städtischen Kliniken Darmstadt (Darmstadt/ Germany) for providing human umbilical cord veins and human placenta, H.-G. Kreysch and G.-A. Luckenbach, Merck (Darmstadt/Germany) for their kind cooperation by analysing the integrin expression and doing the flow cytometry. We would like to acknowledge Dr. Bernotat-Danielowski, Merck (Darmstadt/ Germany) for microvascular HMVEC and Prof. Dr. Müller-Berghaus and Dr. Pötsch, Kerckhoff-Klinik (Bad Nauheim/Germany) for the monoclonal antibody TM 6 against α thrombomodulin .

The authors would like to thank the European Commission (DG XII) for granting a bursary to present these results at the 14th ESACT Meeting in Vilamoura, Portugal. M. Schütz was a stipendiate of DFG (grant FR 932/1-1). Diese Arbeit ist Teil der Dr.-Ing. Dissertation von M. Schütz, Technische Hochschule Darmstadt.

Characterization of growth and recombinant protein production of the insect cell line BTI-Tn-5B1-4 (High Five^{TM}) in suspension culture.

Ernesto Chico, Anja Kobold, Bernd Haase and Volker Jäger.
Gesellschaft für Biotechnologische Forschung mbH. Mascheroder Weg 1
D-38124 Braunschweig, Germany.

Abstract

In this work, the insect cell line BTI-Tn-5B1-4 (High Five^{TM}), normally anchorage dependent, was adapted to grow in suspension by a subculturing procedure without supplementation of heparin. Comparative testing of different serum-free media were carried out in batch suspension culture using spinner flasks. Oxygen, asparagine and glutamine were identified as the limiting nutrients in the growth of this cell line in Ex-Cell 401 medium. A recombinant baculovirus, encoding for a human immunoglobulin G, was used for infection of the High Five^{TM} cells during the study of the cell line production capability.
Keywords: Trichoplusia ni, Insect cells, Heparin, Baculovirus expression system, Recombinant immunoglobulin.

Introduction

The high specific recombinant protein productivity of the insect cell line BTI-Tn-5B1-4 infected with Baculovirus has been reported recently [1-4]. The shear sensitivity and the cell-density effect of this cell line have been argued as disadvantages for its use for large scale protein production, in comparison with the *Spodoptera frugiperda* cell lines [2]. Previous investigations [1-3] have been carried out under adherent growth conditions and the attempts to grow this cell line in suspension culture have led to formation of cellular aggregates, which is undesirable because cell viability of the clumped cells may be reduced due to diffusional limitations. To overcome this, heparin has been used to prevent cellular aggregation during adaptation of insect cell lines to suspension culture [5]. The first aim of this work was to establish a procedure to adapt the insect cell line BTI-Tn-5B1-4 (High Five), normally anchorage dependent, to grow in single cell suspension. The influence of heparin on the adaptation process was also evaluated.
Although protein production and secretion capabilities of High Five cells have been studied extensively, few data are available about the growth behaviour, which could be used for process development and optimization. In this work, a characterization of the growth of High Five cells is presented using different commercially available insect cell culture media. Main efforts are made to identify growth limiting compounds in batch suspension culture.

Results and Discussion

Adaptation to growth as single cells.
Two different subculturing strategies were carried out for the adaptation of High Five cells, growing adherent in culture flasks in serum-containing medium, to single cell suspension culture. Cells were either cultured for 6 passages (every 2-3 days) in spinner flasks in serum-containing medium supplemented with different concentrations of

M.J.T. Carrondo et al. (eds.), Animal Cell Technology, 755–760.

heparin (10 and 50 U/ml respectively) or adapted stepwise to serum-free culture conditions in T-flasks and subsequently subcultured in spinner flasks for 6 passages in presence of heparin (200 U/ml). Control cultures, without using heparin, were carried out in parallel for both adaptation procedures. All the experiments were done in 500 ml Techne spinner flasks using 100 ml as working volume. Stirrer speed was adjusted to 90-100 rpm.

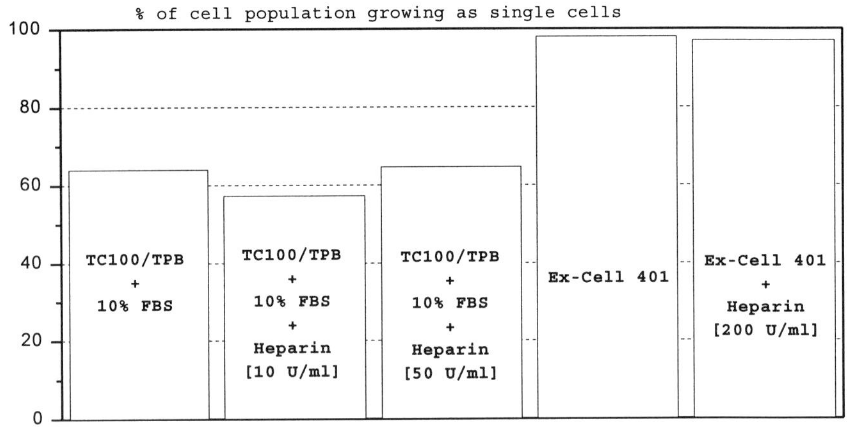

Fig. 1: Results of the adaptation strategies carried out in serum-supplemented medium (TC100/TPB+10% FBS) and serum-free medium (Ex-Cell 401). Total cell number was measured after using a mechanical disaggregation method [6].

Results of the adaptations procedures are presented in Fig. 1. Approximately 40% of the cells grown in medium with serum remained in cellular aggregates until the end of the experiment. On the contrary, subculturing of cells in Ex-Cell 401 serum-free medium (JRH) allowed the adaptation of this cell line to grow in suspension as single cells. Heparin was shown to have no influence on the adaptation process. Adapted High Five cells kept their capability of growing in suspension as single cells even after cryopreservation and subsequent revitalization.

Characterization of High Five Cells in Suspension Culture.

High Five cells, adapted to growth in suspension as single cells and cultured in Ex-Cell 401 medium, were used for comparative studies of different serum-free media and cultivation modes. Batch cultivations were carried out in spinner flasks as described above, using the following media: SIM (Sigma), Express Five (Life Technologies), SF-900 II (Life Technologies) and a 1:1 mixture of SF-900 II and Ex-Cell 401. In addition, a cultivation under controlled disolved oxygen conditions was performed in a 2.4 l membrane-aerated stirred bioreactor [7] using Ex-Cell 401 medium. Figs. 3 and 4 show the main results of these experiments.

Doubling times (*td*) and specific growth rates (μ) of High Five cells grown in different culture media are presented in Fig. 2. Cells behaved very similar in all of the serum free media and cell growth parameters were comparable to those in serum-containing medium. Values of 18-22 h and 0.030-0.035 h^{-1} were obtained for *td* and μ, respectively. Amino acids consumption patterns of High Five cells growing adherently and in suspension culture are shown in Fig. 3. Asparagine and glutamine were almost depleted at the end of the exponential growth phase, which might lead to growth limitation and prevent higher maximum cell densities in batch cultures.

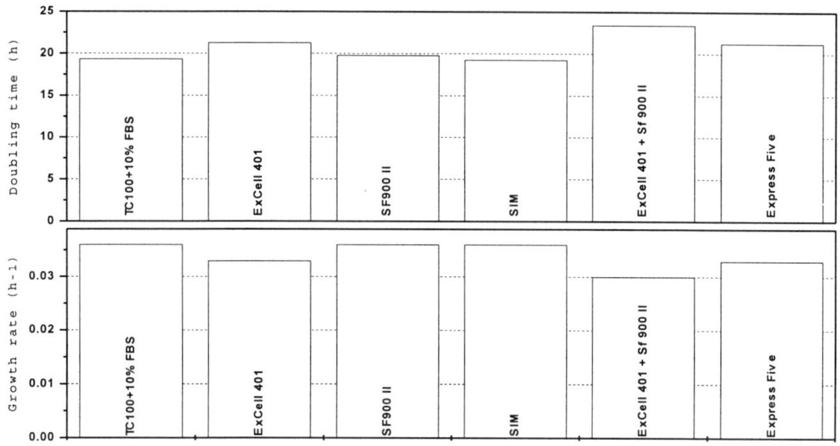

Fig. 2: Growth kinetic parameters of High Five cells in different serum-free media in comparison to serum-supplemented medium. All experiments were done in spinner flasks.

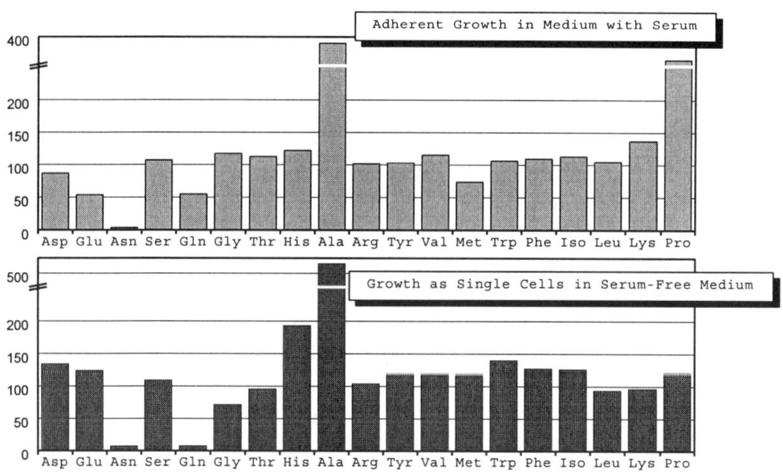

Fig. 3: Amino acids consumption pattern of cells growing adherent in TC100/TPB+10%FBS and cells growing as single cell suspension in Ex-Cell 401 serum-free medium. Results from the end of the exponential growth phase are shown and expressed as % of initial amino acid concentrations.

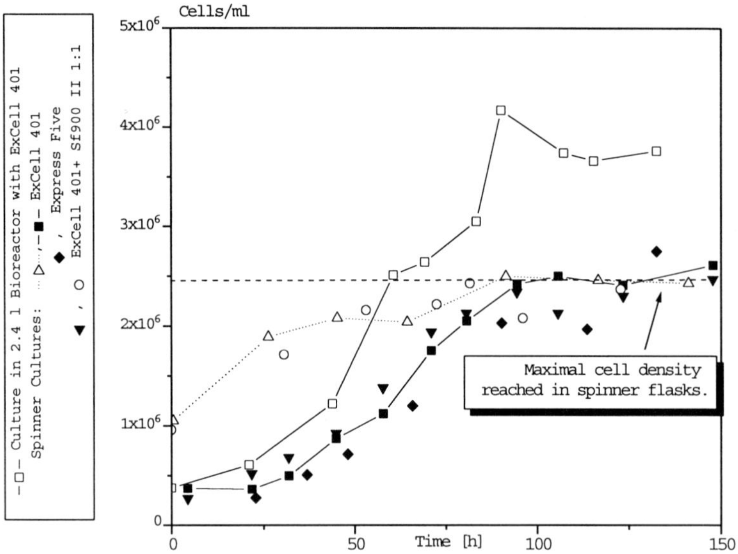

Fig. 4: Cell density profiles under different cultivation conditions.

Cell growth profiles of the cultivations in spinner flasks and bioreactor are presented in Fig. 4. A maximal cell density of 2.5×10^6 cells/ml was reached in spinner flasks for all different serum-free media used. Using a higher cell density as inoculum (1×10^6 cells/ml) did not result in a proportional higher maximal cell yield in spinner flasks. However, none of the measured amino acids was found to be limiting in these cultures at the time of growth cease. To verify wether oxygen could become limiting at this cell density in surface-aerated spinner flasks, a culture experiment was performed using a 2.4 l membrane-aerated bioreactor. Ex-Cell 401 medium was used for this culture and dissolved oxygen concentration was controlled at 55% of air saturation throughout the culture time. Under these conditions, a maximal cell density of $3.5\text{-}4 \times 10^6$ cells/ml was achieved.

These results clearly indicate that surface-aerated spinner cultures with stirrer speeds up to 100 rpm do not supply enough oxygen for the optimal growth of High Five cells. Therefore, this very common production system must be used very carefully for studies of cell growth and recombinant protein production.

Recombinant Protein Production.

A recombinant baculovirus, which was generated using a construct encoding for a human immunoglobulin G, was used for protein production experiments. Cells were infected with a MOI of 0.5 at a cell density of 1×10^6 cells/ml. These experiments were performed in spinner flasks as described above. Infections were carried out using exponentially growing cells after a complete medium exchange. Results in Fig. 5 show that the productivity as well as the final immunoglobulin concentration were very sensibly affected by the experimental conditions. For the infections carried out in a 1:1 mixture of SF-900 II and Ex-Cell 401 media, an increase in the stirrer speed from 90 to 100 rpm resulted in almost three-fold increase of the recombinant antibody

concentration in the supernatant. It can be speculated that an enhanced oxygen transfer rate of the system, due to the sligthly higher stirrer speed, was responsible for the increased recombinant protein productivity, indicating that oxygen could play an important role during virus infection and recombinant protein expression in this cell line. These results are in accordance with findings using Sf9 cells for recombinant protein production, where insufficient oxygen supply substantially reduced the product yield [8].
Under these non-optimized conditions, High Five cells showed comparable or even better production capability in comparison to SF21 cells for the complex recombinant protein expressed.

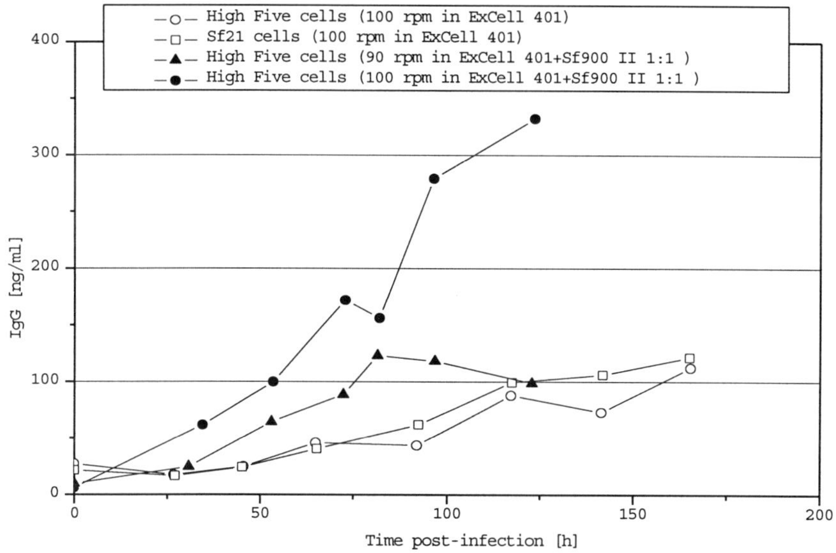

Fig. 5: Comparison of the recombinant protein production after infection of High Five cells and Sf21 cells with the same recombinant baculovirus.

Conclusions

The highly productive cell line BTI-Tn-5B1-4, adapted to growth in single cell suspension, represents a good alternative to *Spodoptera frugiperda* cell lines for the production of complex recombinant proteins using the baculovirus expression system. Further optimization is required to culture this cell line at high cell density conditions. Sufficient supply of asparagine, glutamine and oxygen seem to be the main challenges for such a purpose.

References

1. Wickham TJ, Davis T, Granados RR, Shuler ML and Wood HA (1992) Screening of Insect Cell Lines for the Production of Recombinant Proteins and Infectious Virus in the Baculovirus Expression System. Biotechnol. Prog. 8, 391-396.

2. Wickham TJ and Nemerow GR (1993) Optimization of Growth Methods and Recombinant Protein Production in BTI-Tn-5B1-4 Insect Cells Using the Baculovirus Expression System. Biotechnol. Prog. 9, 25-30.
3. Davis TR, Wickham TJ, McKenna KA, Granados RR, Shuler M and Wood HA (1993) Comparative Recombinant Protein Production of Eight Insect Cell Lines. In Vitro 29A, 388-399.
4. Du RP, Jackson GED, Wyde PR, Yan WY, Wang Q and Ewasyshyn ME (1994) A Prototype Recombinant Vaccine Against Respiratory-syncytial Virus and Parainfluenza Virus-3. Bio/Technology 12, 813-818.
5. Boyce Thompson Inst. Plant. Res. (1994) Preparation of insect suspension cell culture. US Patent 5348877.
6. Moreira JL, Alves PM, Aunins JG, Carrondo MJT (1994) Changes in Animal Cell Natural Aggregates in Suspended Batch Cultures. Appl. Microbiol. Biotechnol. 41, 203-209.
7. Blasey, HD and Jäger V (1991) Strategies to Increase the Efficiency of Membrane Aereated and Perfused Animal Cell Bioreactors by an Improved Medium Perfusion. In: Sasaki R and Ikura K (eds.), Animal Cell Culture and Production of Biologicals, Kluwer Acad. Pub., Dordrecht, pp 61-73.
8. Jäger V and Kobold A (1995) Propagation of *Spodoptera frugiperda* Cells (Sf9) and Production of Recombinant Proteins with the Baculovirus Expression System Using Improved Spinner Flasks with Membrane Aeration. Biotechnol. Techniques 9, 435-440.

Acknowledgements

Authors thank the German Academic Exchange Service (DAAD) for financial support for E. Chico.

CONTROLLING THE GLYCOSYLATION OF RECOMBINANT PROTEINS EXPRESSED IN ANIMAL CELLS BY GENETIC AND PHYSIOLOGICAL ENGINEERING

Damien Lamotte[1], Alex Eon-Duval[2], Giula Acerbis[3], Gianfranco Distefano[3], Lucia Monaco[3], Marco Soria[3], Nigel Jenkins[2], Jean-Marc Engasser[1] & Annie Marc[1]

1 - LSGC / CNRS, BP 451, 54001 Nancy CEDEX, France
2 - De Monfort University - Dpt of Biological Sciences, Leicester, U.K.
3 - San Raffaele Scientific Institute - DIBIT, Milan, Italy

ABSTRACT

The cDNA coding for rat alpha-2,6-sialyltransferase (α2,6-ST) was expressed in a recombinant CHO cell line producing interferon-gamma, using a novel *in vitro* amplification vector. The α2,6-ST was expressed efficiently, 20-30% of the sialic acid pool on interferon-gamma being linked in the α2,6-conformation. Despite the overall level of sialylation was not significantly increased, growth and interferon-gamma production in bioreactor were not compromised by the transfection and expression of the gene encoding for the α2,6-ST.

INTRODUCTION

Animals cells are the most favoured means of producing human therapeutic proteins because of their ability to perform post-translational modifications, the most important being glycosylation. Whilst the fidelity and expression level of the polypeptide blackbone can now be stringently controlled, the oligosaccharide residues on recombinant glycoproteins are highly variable depending on host cell type (1), protein structure and cell cultures conditions (2).

Most recombinant glycoproteins are currently expressed in the CHO cell line (3). Although the CHO-derived glycan structures are similar to those present on natural human proteins, they are not identical. Specifically, the CHO cell line expresses only alpha-2,3-sialyltransferase (α2,3-ST), its alpha-2,6-sialyltransferase (α2,6-ST) being non-functional. As a result, recombinant proteins expressed in CHO cells lack terminal sialic acid linked to core residues via an alpha-2,6 linkage.

We have corrected the α-2,6-ST defect in CHO cells by expressing the α2,6-ST gene in a modified CHO line (CHO 320) producing interferon-gamma as target recombinant glycoprotein. Since the addition of sialic acid is known to be heavily influence by cell culture conditions, the ability to perform alpha-2,6-linkages has been assessed in bioreactor batch culture under maintained culture conditions.

M.J.T. Carrondo et al. (eds.), Animal Cell Technology, 761–765.

MATERIAL AND METHODS

Cell line transfection & selection of clones

The CHO 320 producing interferon-gamma (IFN-γ) was supplied by the Wellcome Foundation Laboratories (now part of the Glaxo-Wellcome Research Group). Then, the α2,6-ST cDNA was transfected into CHO 320 by the *in vitro* amplification expression method (4). Transfection and selection of clones have been described elsewhere in details (5). Briefly, the α2,6-ST cDNA was inserted into the pSfiSV19 vector (donated by Dr J. Paulson) allowing the preparation of long and linear DNA concatenamers. Such molecules contained several copies of the expression unit ligated in tandem and linked to a terminal copy of the neomycin expression unit and were used as such for transfection. This method allowed us to obtain a high proportion of clones isolated displayed good levels of recombinant protein expression. For rapid screening of multiples clones, the fluorescent lectin-FACS analysis of α2,6-ST positive was adopted (6). Two clones, C5 & C6, were chosen for scale-up cultures on the basis of a high level of FACS-positive cells, significant but different α2,6-ST enzyme activity and high level of IFN-γ production.

Bioreactor cultures and samples

Prior to culture in bioreactor, clones C5 & C6 have been progressively adapted to culture in suspension in a serum-based RPMI medium containing FCS (10%), G418 ($450 \mu g.ml^{-1}$) and methotrexate (0.1 mM). CHO 320, C5 & C6 were scale-up in 0.5L plastic shake flask cultures and harvested in exponential phase of growth in order to inoculate 3 litres of medium. A 4L-bioreactor (Incelltech-SGI, France) was run under conditions of constant pH (7.35), pO2 (50% of air saturation), temperature (37°C) and agitation (50 rpm). Viable cells were determined twice daily by trypan blue exclusion combined to a haemocytometer. Supernatants of samples were stored at -20°C and then assayed for IFN-γ concentration. Cell pellets were kept to estimate the intracellular α2,6-ST activity during the cultures. The final supernatant was harvested and filtered on 0,2 μm membrane. IFN-γ was then purified using an immuno-affinity chromatography (2).

Alpha -2,6-sialyltransferase activity and interferon-gamma assays

α-2,6-sialyltransferase activity was estimated on cell lysates (10^7 cells) by a lectin-based microtiter plate assay developed by Mattox *et al.* (7). α2,6-ST activity was expressed per mg cell protein, as determined by the BioRad detergent-compatible protein assay. IFN-γ concentration was assayed by ELISA method calibrated in $IU.L^{-1}$ using an international reference preparation (8).

Determination of sialic acid content

Purified IFN-γ was subjected to total release of all sialic acids by incubation with a non-specific sialidase (*A. ureafaciens*) and partial release with an α-2,3 linkage specific sialidase (*Newcastle disease virus*). Sialic acids were recovered by micro-ultrafiltration. Then, the rpHPLC detection of fluorophore-labelled sialic acids (9) was used to discrimate between α2,6 and α2,3-linked sialic acids. This technique allowed us to resolve the proportion of the antigenic N-glycolylneuraminic acid (NeuGc) in the sialic acid pool.

RESULTS

Growth and IFN-γ production in bioreactor cultures

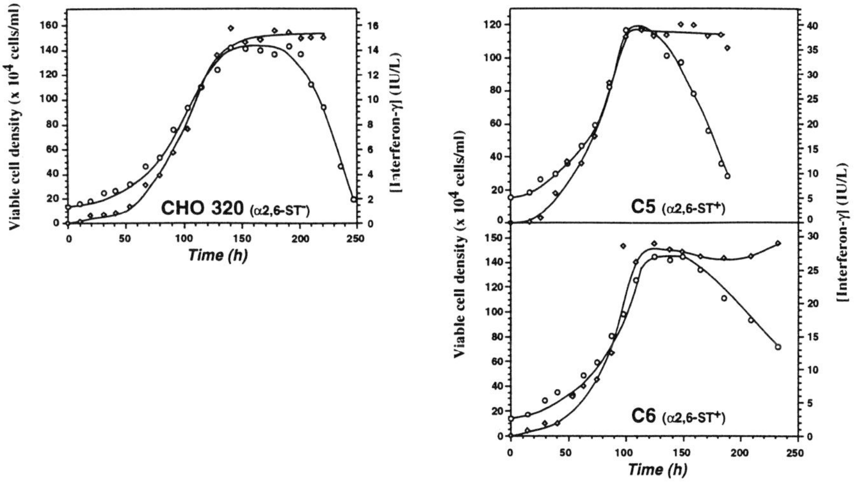

Figure 1: Viable cells (O) and IFN-γ (◇) in stirred batch cultures

Clones C5 & C6 performed well in stirred batch cultures reaching a final cell density slightly similar to the 320 strain. Furthermore, IFN-γ production of clones was 2-3 fold greater (figure 1). Note that high IFN-γ production was a prerequisite at the early step of the clone selection. IFN-γ production and growth were associated in all cases, IFN-γ concentration remaining constant at death phase. The maximum specific rate of cell growth was similar in both modified and non-modified cells (0.020 h^{-1}). Glutamine depletion stopped the growth, glucose depletion occurring ten hours after that of the glutamine (data not shown).
Thus, the transfection of the α2,6-ST cDNA into the parental CHO320 had no harmful effect on cell growth, metabolism and productivity.

Effect of α2,6-ST transfection on amount of α2,6-linkages

Transfection into the parental CHO320 cell line of cDNA coding for α2,6-ST led to significant amount of α2,6-linkages in the oligosaccharide chains of IFN-γ (table1). Thus, the transfected α2,6-ST enzyme restored CHO cell sialylation to a human type. Moreover, intracellular α2,6-ST activity was stable throughout batch culture of the three cell lines. The proportion of sialic acids linked in the α-2,6-conformation seems to be related to the α2,6-ST activity, IFN-γ from C5 being more α2,6-sialylated than that of the C6 cell line.
Nevertheless, the level of α2,6-linked sialic acids in IFN-γ (20-30%) may be the maximum possible for this protein. The competition between α2,3-ST and α2,6-ST toward the termination of the oligosaccharide chains of a protein is depending on this protein. For example, the recombinant β-trace protein has 90% of α2,6-linked sialic acids in a BHK cell modified for the production of α2,6-ST (10). Moreover it is well known that the activity of α2,6-ST and α2,3-ST toward the glycans are modulated by the type of the oligosaccharide chains (bisected, linear, branched, complex) bear by the

protein (11). In the same way, it has been reported a preference for α2,6-sialylation on the Mannose-3-arm rather than on the Mannose-6-arm of antennary chains (10). Finally, in brain, α2,3-ST regulates α2,6-ST, showing an interdependence between the two enzymes in biochemical pathways (12).

	CHO 320 (α2,6-ST$^-$)	C5 (α2,6-ST$^+$)	C6 (α2,6-ST$^+$)
α2,6-ST activity (mU/mg)	0	87.5	19.6
α2,6-ST linkages (%)	0	32	20
NeuGc (%)	6.7	8.7	5.8

Table 1 : α2,6-ST activity, α2,6-ST linkages and NeuGc for the three cell lines

Effect of α2,6-ST transfection on overall sialylation of IFN-γ

The overall molar ratio of sialylation in IFN-γ was similar for the three cell lines (2:1). The expression of α2,6-ST did not improved the overall level of sialylation. Therefore, it appeared that α2,6-ST and α2,3-ST compete for the same intracellular pool of sialic-acid precursor, CMP-sialic acid. Synthesis and transport of CMP-sialic acid are bottlenecks for the full sialylation of recombinant glycoproteins, a high expression of sialyltransferase being necessary but not sufficient.
Concerning the molecular structure of linked sialic acids, the level of α2,6-ST activity did not altered significantly the proportion of NeuGc in IFN-γ (Table 1). Despite some results of other authors (13), α2,6-ST did not transferred preferentially NeuGc (an oncofoetal antigen in humans) onto oligosaccharide chains.

CONCLUSION & PERSPECTIVES

We have created and cultivated two clonal cell lines expressing the transfected α2,6-ST gene which can remain the endogenous gene defect in CHO cells to produce recombinant proteins with a human-type linkage of sialic acids. The overall level of sialylation is not improved by this transfection whereas the proportion of NeuGc is not fortunately increased. Experiments are in progress to express α2,6-ST in a wild type of CHO cell in order to transfect afterwards gene coding for therapeutic recombinant protein. Furthermore, we are testing different culture conditions to increase the overall level of sialylation, glycosylation pathways being greatly influence by culture conditions.

ACKNOWLEDGEMENTS

We thank Glaxo-Wellcome Research Laboratories for the CHO320 cell line and Dr J. Paulson of Cytel Corporation (San Diego-USA) for the cDNA for α2,6-ST. This study is supported by the European Commission through the Biotechnology G-Project (Contract number-BIO-CT-94-3096).

REFERENCES

1. James DC, Freedman RB, Hoare M, Ogonah OW, Rooney BC, Larionov OA, Dobrovolsky VN, Lagutin OV. & Jenkins, N (1995) N-glycosylation of recombinant human interferon-gamma produced in different animal expression systems. Bio/technology 13, 592-596.
2. Castro PML, Ison AP, Hayter PM & Bull AT (1995) The macroheterogeneity of recombinant human interferon-gamma produced by chinese-hamster ovary cells is affected by the protein and lipid content of the culture medium. Biotechnol. Appl. Biochem. 21, 87-100.
3. Hodgson J (1993) Expression systems : a user's guide. Bio/technology 11, 887-893.
4. Monaco L, Tagliabue R, Soria MR & Ulhen M (1994) An *in vitro* amplification approach for the expression of recombinant proteins in mammalian cells. Biotechnol. Appl. Biochem. 20, 157-171.
5. Monaco L, Marc A, Eon-Duval A, Acerbis G, Distefano G, Soria MR, Lamotte D, Engasser J-M & Jenkins N. (1996) Genetic engineering of α2,6-sialyltransferase in recombinant CHO cells and its effect on the sialylation of recombinant interferon-γ. Cytotechnology, *in press.*
6. Lee EU, Roth J and Paulson JC (1989) Alteration of terminal glycosylation sequences on N-linked oligosaccharides of chinese hamster ovary cells by expression of ß-galactoside α2,6-sialyltransferase. J. Biol. Chem. 264, 13848-13855.
7. Mattox S, Walrath K, Ceiler D, Smith DF & Cummings RD (1992) A solid-phase assay for the activity of CMPNeuAc : Gal ß-1-4GlcNac-R α2,6-sialyltransferase. Anal. Biochem. 206, 430-436.
8. Hooker AD, Goldman MH, Markham NH, James DC, Ison AP, Bull AT, Strange PG, Salmon I, Baines AJ & Jenkins N (1995) N-glycans of recombinant human interferon-gamma change during the batch culture of CHO cells. Biotechnol. Bioeng. 48, 639-648.
9. Reuter G & Schauer R (1994) Determination of sialic acids. Methods Enzymol. 230, 168-199.
10. Grabenhorst E, Hoffman A, Nimtz M, Zettlmeissl G & Conradt HS (1995) Construction of stable BHK-21-cells coexpressing human secretory glycoproteins and human Gal(ß-1-4)GlcNAc-r α-2,6-sialyltransferase α-2,6-linked NeuAc is preferentially attached to the Gal(ß-1-4)GlcNAc(ß-1-2)Man(α-1-3)-branch of diantennary oligosaccharides from secreted recombinant ß-trace protein. Eur. J. Biochem. 232, 718-725.
11. Nemansky M, Schiphorst WE & Van den Eijnden DH (1995) Branching and elongation with lactosanoglycan chains of N-linked oligosaccharides result in a shift toward termination with alpha-2,3-linked rather than with alpha-2,6-linked sialic acid residues. *FEBS* Lett. 363, 280-284.
12. Baubichon-Cortay H, Broquet P, George P & Louisot P (1989) Evidence for an O-glycan sialylation system in brain. Characterization of a beta-galactoside alpha 2,3-sialyltransferase from rat brain regulating the expression of an alpha-N-acetylgalactosaminide alpha 2,6-sialyltransferase activity. Eur. J. Biochem. 182, 257-265.
13. Hamamoto T, Kurosawa N, Lee YC & Tsuji S (1995) Donor substrate specificities of gal-ß-1,4GlcNAc α-2,6-sialyltransferase and Gal-ß-1,3-GalNAc α-2,3-sialyltransferase - comparison of N-acetyl and N-glycolylneuraminic acids. Biochimica Et Biophysica Acta-General Subjects. 1244, 223-228.

ESTIMATION OF METABOLIC FLUXES USING MATERIAL BALANCES FOR A CHO CELL LINE

H.Lübben, G. Kretzmer Institut für Technische Chemie der Universität Hannover
Callinstr.3 D-30167 Hannover, Germany
C.Paredes, J.J. Cairo, F.Godia Department d'Enginyeria Quimica Universitat Autonoma de Barcelona 08193 Bellaterra, Barcelona, Spain

ESACT Abstract *:* The metabolic fluxes corresponding to a CHO cell line SSO/A2 producing a human glycoprotein antithrombin IIII have been estimated. The analysis is based on the stoichiometry of the main reactions in the metabolic network, and a number of assumptions based in literature data and previous knowledge of the cell line behavior. Using a weighted least squares technique, the value of the metabolic fluxes is estimated through the minimization of the error between the calculated and experimental specific consumption and production rates for a series of measured extracellular compounds and the cell growth. The results obtaind from the application of this method to the exponential growth phase in batch experiments are reported.

Introduction *:*

Recently metabolic flux analysis had become a powerful tool a large spectrum of applications:
Especially the quantitave analysis of metabolic fluxes enables a better understanding of cell physiology when enviromental conditions are supposed to change cell metabolism. It can be proved, that there is a satisfactory interpretation of experimental data.
The biggest goal to achieve is a way for a better formulated media, because of its overwhelm of nutritions which leads often to an increase of cellular waste products and the uneconomic exploitation of the media. A direct feeding strategy under consideration of the actual demand for each compound would be the solution for this problem. Metabolic flux analysis is a guide for metabolic engeneering, because of the more detailed consideration of the intracellular procedure in contrast to the overall view in biochemical engeneering up to now and therefore attempts to be a further a method for bioprozess optimization.
The analysis is based on the stoichiometry of the main reactions in the metabolic network and a number of assumptions based in literature data and previous knowledge of the cell line behavior.

M.J.T. Carrondo et al. (eds.), Animal Cell Technology, 767–772.

A combination of the stoichiometric matrix with the specific consumption and production rates enables the estimation of the quantitative intracellular fluxes using a least square procedure. It is pointed out that no regulatory information and enzyme kinetic is required. Constructing a metabolic network is a complex task due the highly bifurcated structure of animal cell metabolism and almost negelectable concentrations of most of the intermediates, which make their detection quite difficult. Therefore the mathematical model of the metabolic network was developed in order to establish a relationship between the rates observed extracellulary and the calculated internal fluxes.
Only a comparison of the experimental values with the calculated values can decide if the mathematical solution gives a biological sense.

Materials and Methods

Mathematical background:

Prinziples of the estimation of intracellular fluxes are based upon the work of Stephanopoulos und Wang [1] [2].
There is the folowing fundamental equation :

$$(A^{em} A^{i})^{T} \cdot f = (q^{em} q^{i})^{T}$$

A=stöchiometric matrix of relevant reactions from main metabolic pathways
f= vector of intracellular fluxes
q = vector of specific consumption and production rates
for extracellular measured (em) and internal (i) compounds
stoichiometric matrix A consists of R rows (number of reactions)
and C columns (number of compounds)
compounds are divided into intracellular metabolite (not measurable) and extracellular metabolite (measurable and not measurable)

$$C = c^{ex} + c^{in} = c^{em} + c^{eu} + c^{intr.}$$

$$A = (A^{em} A^{eu} A^{i})$$

vector q is built up by single vectors of a consumed or produced compound, which can be measured or which is assumed to be constant during the investigated period (e.g.intracellular organic acids, coenzymes etc.) and those specific rates of unmeasured compounds (e.g. ammonia and CO_2).

Prinziple steps for the determination of intracellular fluxes

The biggest problems occurred in creating a reaction network with the most appropiate and important reactions.
The constructed network in this case is overdetermined, this means more equations (compounds) than unknowns (fluxes). Least-squares procedure is usually how to deal this kind of matrix. With this approach the solution of the problem consists in finding a set of fluxes which minimize the squares of the error between calculated and experimental q`s.

- Solution of the equation system leads to calculated fluxes
- Establishing the solution und estimation of the calculated q`s
    - Performing a least square procedure :

$$\text{Minimize } X^2 = \sum_{i=1}^{i=c^m+c^i} \left(\sum_{j=1}^{j=r} s_{ij} \cdot f_j - q_i\right)^2$$

A quite important aspect is the normalization of the equations, which relies to the expression weightend least square procedure and prevents that the solution of the least square procedure fits only the major fluxes and neglect the smaller ones.
The general solution may supply the knowledge of so far nonmeasurable metabolite. $(q^i + q^{eu})$

Determination of a relevant growth period

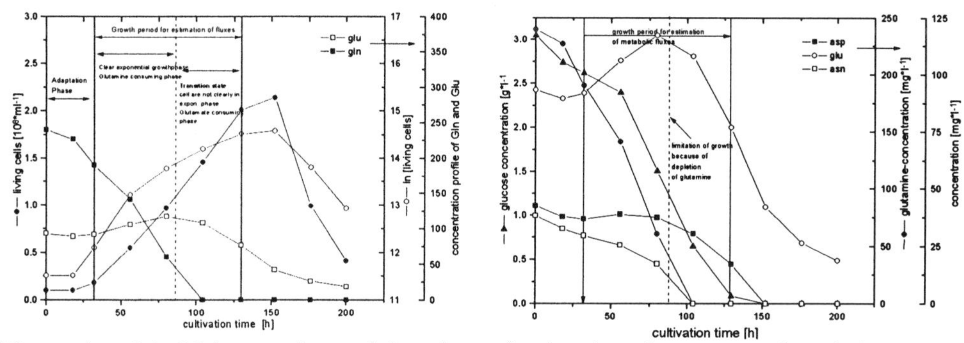

Figure 1 and 2: Main experimental data determinating the relevant growth period

Assumptions

A steady state analysis of cellular metabolism is supposed, assuming that metabolic transients are more rapid than cellular growth rates and dynamic changes in the organism`s enviroment.
Production of new cellullar material is also assumed to be a product of the central metabolism.
The composition of cellular material for the CHO cell line has been taken from an average mammalian cell [3].
Only in exponential growth phase there is no doubt that cells growth without limiting in any nutrition compound. This means for the reaction network all important metabolic pathways (glycolysis or TCA e.g.) are working properly in the reported manner under nonliminary conditions, no additional reactions occur.

Methodological procedure for determination of metabolic fluxes for a CHO cell line

For calculation purposes the data of a CHO batch culture (5Liter Braun fermenter) at 37°C, 50% oxygen saturation, 80RPM in exponential growth phase was chosen was choosen.

Calculation of specific growth rate, specific consumption- and production rates of extracellular measured compounds

Specific growth rate :

$$X_v = X_{vo} * \exp[\mu * t]$$

The specific growth rate was calculated to 0.023 h^{-1}.The calculated cell number was fitted to the experimental one.

Calculating the specific rates :

$$S_i = S_0 + q_S * 10^{-3} * \frac{X_{V0} * \exp((\mu * t) - 1)}{\mu}$$

S_i = measured concentration
q_s = specific consumtion rate
X_{vo} = nitial cell number

The specific rates were estimated by a least square fit of the experimental concentration data to the above equation and are presented in the following table:

Compound	q [nmol/10^6cell*h]	Compound	q [nmol/10^6cell*h]
		Substrates	
glucose	-239,33	glutamate	8,33
glutamine	-35,27	proline	-2,07
arginine	-2,04	histidine	-1,05
asparagine	-4,67	isoleucine	-3,85
serine	-3,68	leucine	-5,02
threonine	-2,84	lysine	-8,56
tyrosine	-1,21	methionine	-1,16
cysteine	-5,97	phenylalanine	-1,66
		valine	-2,07
Products		Cellular material	
ammonia	52,12	DNA	0,0705
alanine	9,4289	RNA	0,2931
glycine	1,0607	lipids	1,8791
aspartate	0	proteines	40,3096
lactate	364.46	sugars	3,1919

Table 1 : Specific consumption and production rates of the external measurable compounds

Results:

In general the metabolic flux analysis data reflect excellent correspondence between experimental data and estimation with the stoichiometric mass bilance. The modell error was just 3.95% compared to the experimental values.

The biggest errors occur in estimation of gaseous products (ammonia and carbondioxide), because of their tendency leaving the fermenter easily.
The biggest problems occur in creating a reaction network with most appropiate reactions:

General solution of the stoichiometric network:

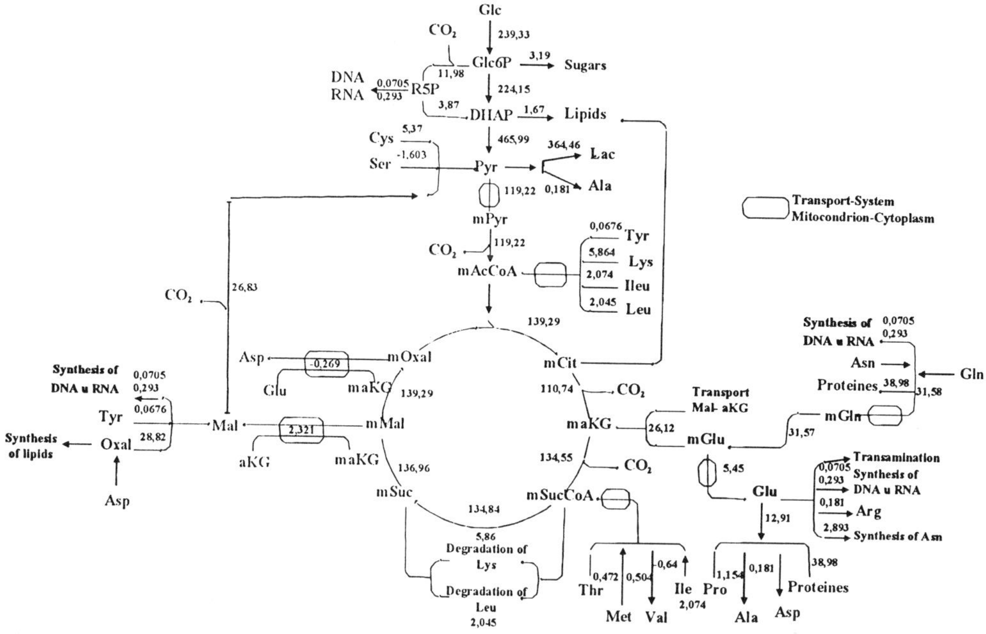

Figure 3: Metabolic map for a CHO cell line with calculated fluxes in $[nmol/10^6\, cells \cdot h]$

Biological significance of the flux data.

In general the data received from metabolic flux analysis show two main aspects: The velocity of the glycolytic pathway is bigger than any other biochemical pathway within CHO SSO/A2 cells. The tricarboxylic acid cycle (TCA) is working more properly compared to other mammalian cell lines.

76% of the initial glucose is converted to lactate (239.3 nmol (1e6 cell h)). 26% of the pyruvate generated in glycolysis (119 nmol/(1e6 cell h)) is entering the TCA compared to 54.77 pyruvate of repectivly 621 nmol(1e6 cell h) pyruvate (9%) for an investigated hybridoma cell line.

Usually in animal cells the conversion from pyruvate to lactate occurs only when there is an limitation in oxygene. For most of cell lines in vitro this conversion take place regardless of the level of oxygene in the culture.Usually glycolysis of in vitro cultured cells takes place at a very high rate. Every conversion of one molecule glucose in the glycolytic pathway generates two molecules of NADH which have to be regenerated due to the catalytic ammounts in the metabolismn. Normally the regeneration of NADH takes place in

the mitochondrion. But this needs a indirect transportsystem, since the mitochondrium membrane is not permeable for the NADH molecule. This so called malate aspartate shuttle has to be working with high velocity otherwise there is only the conversion of pyruvate to lactate which also generates two molecules of NAD^+ and is the main obvious explanation for the large excretion of lactate in many animal cell cultures. For the CHO cell line the decreased lactate production suspects that the malate aspartate shuttle within these cells is slightly functional compared to in any other cell lines.
In the stoichiometric modell especially NAD^+ regeneration reactions in the mitochondrion were treated seperately and an amazing decrease of the absolut error in the least square procedure was achieved compared to a preceding matrix where the mitochondrial reactions were not treated seperatly. It should be acknowleged, however, that all results are based upon a mathematical solution of a stoichiometric modell with a simplification in regard to cellular behavior in vitro.
For validation purposes, labeling experiments with glucose are planned, taking advantage of NMR data which will be applied.

Conclusion

The metabolic fluxes corresponding to a CHO cell line SSO/A2 producing a human glycoprotein, antithrombin IIII, have been estimated using a weighted least squares technique. As a key ascpect the stoichiometic mass and energy balance of this mathematical modell can only be solved by seperation of the reactions that take place in different compartiments within the cell. The biological sence of the obtained mathematical solution has to be proved critically in order to adjust the model to the biochemical behavior of the cells
It has been demonstrated that MFA is an easy but powerful tool which gives a profound knowledge in intracellular behavior of mammalian cells The additional possibility to intervene directly in cell-metabolism enables a further step in process optimization.

References :

[1] Alberts B., D.Bray, J.Lewis, M.Raff, K.Roberts and J.D Watson 1994 Molecular biology of the cell. Garland Publishing, New York
[2] Xie L., and D.I.C: Wang.1994 Stoichiometric analysis of animal cell growth and ist application in medium design. Biotechnology and Bioengineering 43:1164-1164
[3]. Zupke.G, and G. Stephanopoulos Intracellular flux analysis in hybridomas using mass balances and in vitro ^{13}C NM. Biotechnology and Bioengineering 43:1059-1074

Aknowledgements :

The authors would like to thank Direccion General de Investigacion Cientifica y Tecnica and Deutscher Akademischer Austauschdienst (DAAD) for financial support in the Programm Acciones Integradas Hispano-Alemanas

PHYSIOLOGICAL AND KINETIC STUDIES ON TWO HYBRIDOMAS SECRETING MONOCLONAL ANTIBODIES IN BIOREACTORS.

C. HERMIDA [1], A. FERNANDEZ [1], F. GONZALEZ [1], T. BOGGIANO [1], J. VILLAN [1], G. RODRIGUEZ [1], L. BOUZO [1], P. VITON [1], V. SANCHEZ [2] and J.J. ARAGON [2].

1- Centro de Inmunología Molecular. Habana. Cuba.
2- Universidad Autónoma de Madrid. España.

Key words: Hybridoma, kinetic, Glucose, Glutamine, Amino acids.

Abstract

The purpose of this work is to report the fate of glucose and glutamine in two different hybridoma cell lines secreting therapeutic monoclonal antibodies. First, we studied the behaviour of those cell lines in physiological conditions during shorts incubation periods up to 180 min. under aerobic environment. Then, we carried out a kinetic and metabolic study in batch culture using spinners flasks. The specific growth, production and uptake rates were calculated and a comparison was made between those values and the measurements at physiological conditions. Finally, we analyzed the evolution of the amino acid concentration using HPLC.

Introduction

Fermentation of microbial cell cultures is well established nowadays; however, cultivation of mammalian cells often involves sophisticated equipment and culture conditions, besides, complex fermentation processes.
For this reason, the design of production processes for mammalian cell lines requires knowledge for cultivation strategies such as nutrient requirements and waste production (1). The energetic metabolism of a mammalian cell line grown in vitro was analyzed by substrate consumption and products formation rates. Those studies provide an important tool in the scale-up of the production of monoclonal antibodies (MAbs). The aim of this paper was the physiological and kinetical characterization of two hybridoma cell lines, both in serum supplemented medium or serum free medium.

Materials and Methods

Cell lines:

Two hybridomas (ior-t3 and ior-r3, Center of Molecular Immunology, Havana, Cuba) were used. The hybridoma strain ior-t3 secretes monoclonal antibody against a differentiation antigen (CD3) on the

M.J.T. Carrondo et al. (eds.), Animal Cell Technology, 773–778.

surface of mature T lymphocytes, while hybridoma strain ior-r3 secretes monoclonal antibody against epidermal growth factor receptor. The isotype of both Mabs is IgG_2a.

Culture medium:

Hybridomas were maintained in RPMI-1640 medium containing 11.8 mM glucose, 4 mM glutamine, 26 mM sodium bicarbonate, 15 mM HEPES and supplemented with 10 % of fetal calf serum (FCS). For batch experiments different concentrations of glucose, glutamine and FCS were used. DIF medium was used for experiments with serum free media. It was prepared in proportion 1:1 of the basic mediums ISCOVE'S -DMEM and HAM'S-F12 and further supplemented with insulin (10 mg L^{-1}), human transferrin (10 mg L^{-1}), ethanolamine (1.2 µl L^{-1}), glutamine (2 mM) and albumax (a complex mixture of lipids (2), 10 mg L^{-1}). Medium and serum were purchased from Gibco.

Culture conditions:

Physiological studies on ior-t3 hybridoma cell line: The cell suspension was incubated for 90 min under aerobic conditions at 37^{0}C to deplete the intracellular pool of endogenous metabolites. At the end of the incubation, cells were collected by centrifugation, suspended and distributed in aliquots. Additions indicated in each experiment were made and the cells incubated for various periods of time, at 37^{0}C, with gentle shaking.
Kinetic studies were carried on ior-t3 and ior-r3 hybridoma cell lines on batch cultures in 250 ml spinner flasks, at 30 rpm in a 37 ^{0}C incubator. Runnings with FCS were carried on with different concentration (10, 8, and 5 %) while runnings in serum free medium (DIF medium) was carried on with varying glucose concentrations and constant glutamine concentration (DIF-2 and 6 with 17 mM glucose, 2 mM glutamine; DIF-4 and 8 with 34 mM glucose, 2 mM glutamine). Cultures were done in triplicate and samples were taken twice daily. Cell count and viability were determined by the trypan blue dye exclusion method using an hemocytometer.

Analytical methods:

Glucose, lactate, glutamine and ammonia quantification were done on cell free supernatants through enzymatic determinations coupled with redox reactions in 96 plates. Absorbance of pyridine nucleotides was measured at 340 nm using an Organon Teknika reader. ATP, alanine and aspartate were measured by enzymatic techniques in a Shimadzu model (UV3000) dual wavelength spectrophotometer (4). Mabs concentration was evaluated by an indirect ELISA test. Amino acids were determined through HPLC ion exchange using *o*-phthaldialdehyde (OPA) as derivatization method.

Data analysis:

Specific rates were calculated by lineal regression of minimum squares from differential equations which describe the behaviour of kinetics variables (3).

Results and Discussion

The behaviour of ior-t3 hybridoma line in physiological conditions during short incubation periods up to 180 min. under aerobic environment pointed out several results: evaluation of glycolytic flux showed that glucose had not influence in glutamine pathway, however the presence of glutamine decreased the glycolytic flux around 30 %. Same effect of glutamine on glucose metabolism has been studied in ascites tumor cells (4), and it was found that this amino acid strongly inhibit the glycolytic rate and that this inhibition is due to an intense decrease in the flux through the phosphofructokinase reaction. On the other hand, more than 99 % of glucose was converted to lactate. Production of labeled anions (lactate and pyruvate) from [U-^{14}C]-glucose was nearly stoichiometric with glucose consumption; besides, glycolysis was inhibited by 2 mM of glutamine up to 47% during the early 60 min. of incubation and up to 20 % for the later period of incubation (Fig. 1A).

Content of ATP was three times higher when hybridoma cells were incubated in presence of glucose than when incubation was done with glutamine alone (Fig 1B). This fact is not common for other cell lines (5), but it can be explained because of while 99 % of glucose was converted to lactate only around 30 % of glutamine converted to glutamate goes to citric acid cycle.

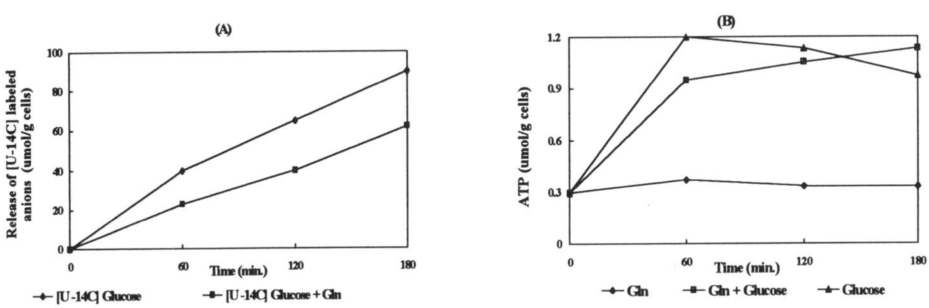

Fig 1. Effect of Glutamine on (A) the release of ^{14}C - labeled anions (lactate and pyruvate) by hybridoma cells, and (B) content of ATP of hybridoma cells (ior-t3).

On the other hand, the presence of glucose did not affect glutamine consumption (Fig. 2A), but slightly reduced glutamate production (Fig. 2B), strongly blocked aspartate production (Fig. 2C) and also increased alanine production (Fig. 2D). This fact could be explained through a piruvate increment which in turn, facilitates alanine transamination and an trapping of oxalacetate by acetyl-CoA which reduced aspartate formation.

In ior-t3 hybridoma cells 31% of glutamate from metabolized glutamine was accumulated in the extracellular medium. From remaining glutamate 36 % was desaminated through GDH pathway while 60 % was transaminated being alanine the main product of transamination.

Only 0.5 % of glucose consumed was oxidated in hybridoma cells. Besides, glucose did not affect [U-^{14}C]-glutamine oxidation (Fig. 3A), but the latter inhibited [6-^{14}C]-glucose oxidation up to 70 % in incubation periods of 120 min (Fig. 3B). CO_2 production from glutamine was almost forty fold higher than obtained from glucose, which pointed out glutamine as the best oxidative substrate.

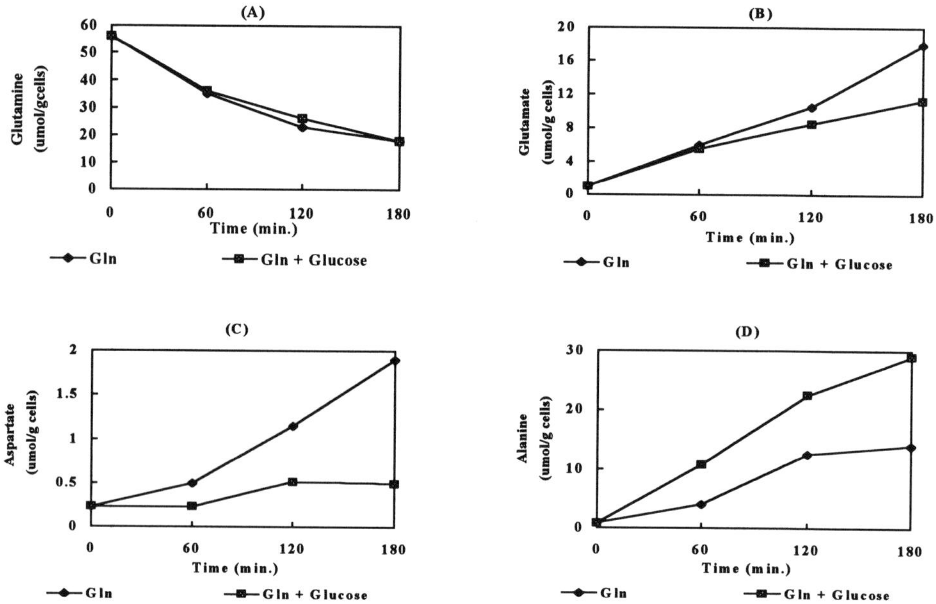

Fig. 2 Metabolic kinetics of (A) Glutamine, (B) Glutamate, (C) Aspartate and (D) Alanine in physiological conditions (ior - t3).

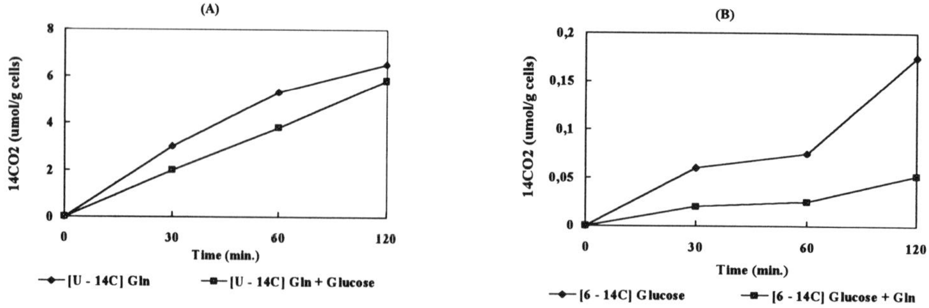

Fig. 3 Production of CO_2 in hybridoma cells (ior - t3) from (A) Glutamine in presence of Glucose and (B) Glucose in presence of Glutamine.

Kinetic studies of ior-t3 and ior-r3 hybridoma lines in batch culture conditions pointed out the following results: glucose was not a limiting nutrient during different culture conditions; its consumption is fast during log phase, reaching a minimum level which never becomes zero. However, glutamine was consumed by both hybridomas in a very similar manner, reaching closely zero levels during the first culture periods and before log phase was over. Ammonia levels were constant through culture time. Similar results were obtained on physiological studies as described above.

Apparent or real specific growth rates were not affected by reduced serum supplements or by serum free media (DIF medium). Both tested hybridoma cell lines had an specific growth rate in a range of 0.025 to 0.035 h^{-1}. Cultures with serum free media showed lag phase extended to more than 70 hours; whereas in control cultures lag phase was around 50 hours.

Ior-t3 clearly showed a closely relation between proliferation and production phases; there was a significative decrease in MAb production rate when plateau phase was reached. For ior-r3 production pattern was different, being higher through the running, meaning a non-growth associated pattern.

Data from Table No. 1 show that reduced serum in culture medium did not affect kinetics coefficients, however, there was a slightly increment in specific glucose consumption rate and specific lactate production rate (q Glucose, q lactate) in serum free media for both hybridomas. Concerning specific glutamine consumption rate and specific ammonia production rate (q Glutamine, q Ammonia) there were no main differences in serum free media as compared to control cultures. However, there was a significant increase in specific immunoglobulin production rate in cultures with serum free media for ior-r3 hybridoma cell line.

Table No. 1 also shows biomass yields from different nutrients and immunoglobulin yield from living cells. Results indicate different behaviors for both hybridomas: a strongly inhibition of biomass yield from glucose or glutamine and a significant increase of immunoglobulins yield for ior-r3, while for ior-t3 biomass yield from glutamine was significantly diminished in serum free medium supplemented with high glucose. These results pointed out better growing conditions for ior-r3 in serum free media.

Cells	Media	q Gluc. x 10^{-7}	q Lact. x 10^{-7}	q Gln. x 10^{-7}	q NH4 x 10^{-8}	q Mab. x 10^{-7}	YX_v/Glu x 10^5	YX_v/Gln x 10^5	$YIgs/X_v$ x 10^5
R3	RPMI 10 %	1.4	1.9	1,7	2,7	5,1	2,1	3,1	1,8
	RPMI 8 %	1,3	2,6	1,0	8,2	1,6	1,5	3,5	0.7
	RPMI 5 %	1,4	1,9	1,0	4,2	3,3	1,5	5,1	1,3
	DIF (2)	4,1	7,1	2,0	6,2	12	0.6	1,5	4,6
	DIF (4)	3,2	5,2	1,0	5,6	15	0.4	6,1	7,3
T3	RPMI 10 %	1,1	1,4	0.9	2,8	8,9	2,1	3,8	3.0
	RPMI 8 %	1,3	2,2	0.9	4,9	9,1	1,7	2,9	4,5
	RPMI 5 %	1,5	2,2	1,1	2,6	5,8	1,7	1,3	5,4
	DIF (6)	3,1	5,2	0.9	2,6	10	1,4	2,6	2,8
	DIF (8)	2,4	3,5	2,0	3,6	12	1,1	5,1	5,3

Table 1 Kinetic coefficients (in pg/cell h) and yields (in pg/cell) from batch cultures in RPMI - 1640 and DIF medium (DIF-2 and 6 with 17 mM glucose, 2 mM glutamine; DIF-4 and 8 with 34 mM glucose, 2 mM glutamine).

Amino acid profiles of both hybridomas cell lines were determined in culture conditions described above. Fig. 4 A shows those amino acids which were produced (mainly alanine) or consumed (mainly glutamine) throughout ior-r3 hybridoma cultures. Besides, amino acids such as histidine, phenylalanine, tyrosine, threonine, and methionine remained constant in serum free medium, while tryptophane, aspartate and asparagine remained constant in RPMI with 10% FCS of ior-r3 hybridoma cultures. Fig. 4 B shows those amino acids which were produced (mainly alanine, glutamate and serine) or consumed (mainly glutamine and arginine) throughout ior-t3 hybridoma cultures. Amino acids that remained constant in serum free medium and control medium were histidine, phenylalanine, tyrosine, threonine, tryptophane, aspartate and methionine.

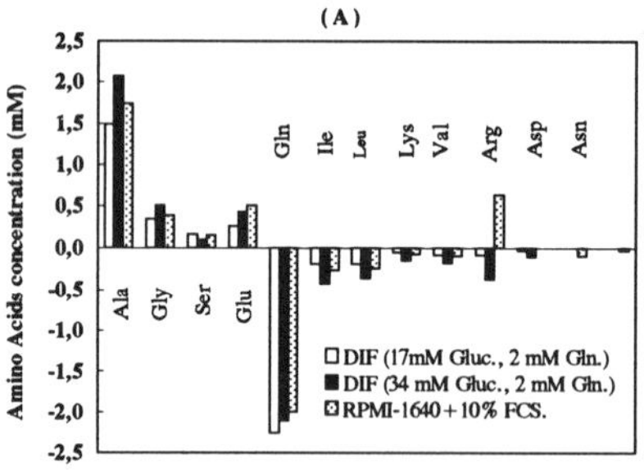

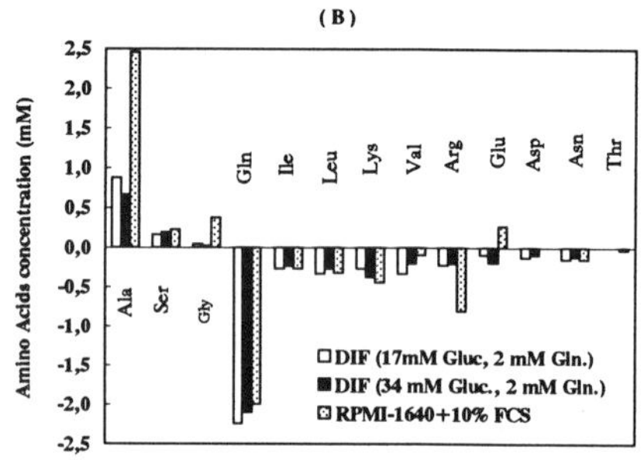

Fig 4. Contents of amino acids in ior-r3 (A) and ior-t3 (B).

These results provide a physiological and kinetical characterization of both hybridomas, which facilitates the scaling-up through hollow fiber bioreactors with therapeutic purposes. Different behaviors were shown for ior-t3 and ior-r3 hybridomas cell lines; data from Table 1 indicates better growing conditions in serum free medium for the latter due to a significant increment of specific immunoglobulins production rate. Concerning amino acids profile, augmented alanine production was found from both physiological studies as for HPLC analysis.

References

1. Buntemeyer H., Stenner A., Seewoster T., Heidemann R., Tebbe H., Wallerius C., Lehmann J. (1995). Comparison of nutrient requirements of mammalian cell lines. In " Animal Cell Technology: developments towards the 21 st Century ". (E.C. Beuvery et al., eds). 117-123.
2. Jager V. et al. (1991). Production of Biologicals from animals cells in culture. "Eds. R. E. Spier, J. B. Griffiths, B. Meignier. Butterworth-Heinemann, Oxford, 1: 155-165.
3. Hu W.S and Wang D.I. (1986). Mammalian cell culture technology: A review from an engineering perspective. In " Mammalian Cell Technology". (William G. Thilly et al, eds). Chapter 8: 167-197.
4 González-Mateos F., Gómez M.E., García-Salguero L., Sánchez V and Aragón J.J. (1993). Inhibition of glycolysis by amino acids in ascites tumor cells. Journal of Biological Chemistry 268: 7809-7817.
5. Kovacevic Z. and J. D. McGivan. (1983). Mitochondrail metabolism of glutamine and glutamate and its physiological significance. Physiological Reviews 6: 591-598.

Hydrodynamic Control of Aggregate Size in Suspended Cultures of Different Animal cells

Carvalhal, A.V.; Moreira, J.L.; Miranda, P.M.; Carrondo, M.J.T.*

ITQB/IBET, Apartado 12, 2780 Oeiras, Portugal,
* also FCT/UNL, 2825 Monte da Caparica, Portugal

Key words: aggregation, size comtrol, hydrodynamic control, BHK, CHO, 293, Vero.

Abstract

Aggregation can only be used as a natural immobilisation system in stirred tanks if aggregate size can be controlled, in order to avoid nutrient and product transport limitations. In this work several cell lines have been studied and their aggregate characteristics compared: BHK, Vero, CHO and 293. 250 ml spinner flasks (Corning) were used varying the agitation rate from 25 to 120 rpm and maintaining the media composition. The aggregates of each cell line showed a different behaviour, particularly in aggregate size, morphology and fraction of single cells in suspension. However, in all cases it was possible to observe a decrease in aggregate size when the agitation rate was increased. A model including the entire range of isotropic eddies was applied to the experimental data, confirming the aggregate size control through the hydrodynamics imposed to the vessel.

Introduction

Some anchorage-dependent animal cells like BHK (Baby Hamster Kidney cells), CHO (Chinese Hamster Ovary cells), 293 (Human embryo kidney cells) and Vero (Monkey African Green kidney cells) have been reported as having the capacity to grow as natural aggregates when cultured in suspension in stirred vessels (Goetghebeur and Hu, 1991; Moreira *et al.*, 1994a; Tolbert *et al.*, 1980).
It has been shown that cells in aggregates as a three-dimensional structure holds three major factors in favour, when compared with microcarriers immobilisation system: cells are less sensitive to damage caused by fluid/mechanical forces (hydrodynamic stress) (Moreira *et al.*, 1994b), lower costs are involved and cell recycling is simpler. Aggregates are more easily retained by sedimentation inside reactors than single cell culture.
To allow this aggregate immobilisation system to be a valid technological possibility it is essential to control their size in order to avoid the formation of necrotic centres and to minimise the zone of nonproliferative cells, and so to maintain a high product quality and productivity.
It has been shown that aggregate characteristics are dependent upon cell line, medium composition and agitation rate (Tolbert *et al.*, 1980), this last factor being the one that plays the major role on the aggregate size (Moreira *et al.*, 1994b, 1995a). This hydrodynamic effect over aggregates has been correlated by a model combining the entire

M.J.T. Carrondo et al. (eds.), Animal Cell Technology, 779–784.

range of isotropic eddies, translating the dependence of aggregate dimensions on the power dissipation rate per unit of mass, to the power of -0.25 (Moreira *et al.*, 1994b, 1995a).
Natural aggregates formed by BHK have been proved to be hydrodynamically controlled (Moreira et al., 1994b). So, the principal aim of this work is to evaluate whether this hydrodynamic control may be considered as valid to other cell lines.
In this work BHK, CHO, Vero and 293 cells grown as aggregates in stirred tanks operated as batch cultures were tested, for different agitation rates (25, 60, 90 and 120 rpm) and for a specific medium culture. The cellular growth, aggregate characteristics, dependence of aggregate size upon power dissipation were also evaluated.

Materials and Methods

Cell line and medium:

BHK 21A7.4 cells were obtained from Dr Hansjorg Hauser (GBF, Braunschwig, Germany) and grown in Dulbecco's Modified Eagle's Medium (DMEM), supplemented with 10% (v/v) foetal bovine serum (FCS), 4.5g/l of glucose, 5 mg/l of puromycin and 1% penicillin/streptomycin. **CHO** cells were obtained from ECACC (no 85050302) and grown in Ham's F12, supplemented with 10% (v/v) foetal bovine serum (FCS), 5g/l of glucose, 4 mM of glutamine, 1 mM of proline and 1% penicilin/streptomycin. **Vero** cells were obtained from ECACC (n 84113001) and grown in DMEM, supplemented with 10% (v/v) foetal bovine serum (FCS), 4.5g/l of glucose and 1% penicillin/streptomycin. **293** cells were obtained from ECACC (n 85120602) and grown in DMEM, supplemented with 10% (v/v) foetal bovine serum (FCS), 4.5g/l of glucose, 1% of non essential amino acids and 1% penicillin/streptomycin.
Cells for spinner inoculum were grown in T-flasks and cultured at an atmosphere of 7% CO_2 and 37 °C.

Bioreaction system:

Bioreaction studies were performed in spinner-flasks (Corning) of 250 cm^3 for the cell lines Vero, 293 and BHK and of 125 cm^3 for CHO cell line (all working volumes). All spinner flasks had a paddle impeller as the stirring system, with a 5.4 cm in diameter and a height of 1.8 cm.
A concentration of 2.10^5 cell/ml was used to inoculate each spinner flask and the medium used was as describe above. An atmosphere of 7% CO_2 and 37 °C was used for all tests. Agitation rates of 25, 60, 90 and 120 rpm were tested for all cell lines.

Sampling:

The sampling method of bioreaction studies in spinner flasks to evaluate aggregates characteristics and cell growth has been previously optimised and described (Moreira *et al.*, 1994a).

Results and Discussion

Cell Growth Behaviour of the Different Cell Lines

In figure 1 maximum cell concentration and average viability of single and adherent cells in aggregates are present for 25 and 120 rpm agitations rates. These results were confirmed by the bioreaction that was performed at 60 and 90 rpm (data not shown).

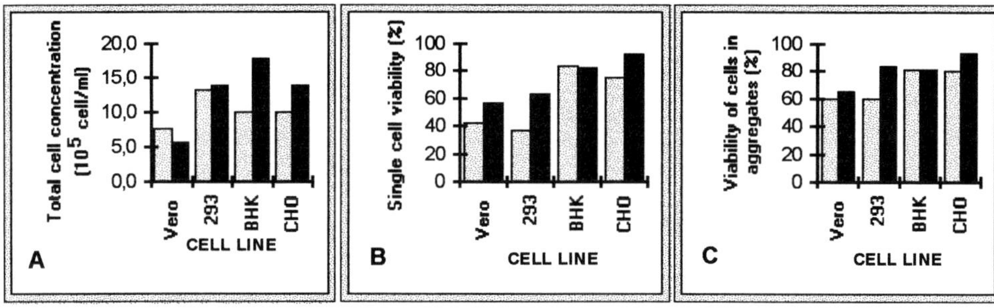

Figure 1: Cellular growth as natural aggregates of cell lines Vero, 293, BHK and CHO at agitation rate of 25 and 120 rpm: (A) Maximum total cell concentration (B) Single cell average viability (C) Average viability of cells in aggregates.

From the results it is possible to observe a different behaviour of the different cells when grown as aggregates: BHK and CHO cells are not strictly anchorage-dependent and thus single cell growth occur, whereas Vero and 293 cells seems to be much more aggregate dependent, with single cell death occurring. This is also confirmed by the different fraction of adherent cells in aggregates (see below) in each case: even at the higher agitation rate of 120 rpm Vero has 76% and 293 87%, of cells in aggregates, whereas BHK and CHO cells have respectively 22 and 13 %.
Since it is known that aggregation is a dynamic process (Moreira *et al.*, 1995b) with cells being exchanged from/to aggregates/single cells, the decrease in cell viability of the aggregates, observed on Vero and 293 cells, is easily explained .

The decrease in cell growth and viability at 25 rpm (figures 1A and 1B) can be explained by oxygen transfer limitations.
The decrease observed for 293 in cell growth and in cell viability either in single cells or aggregates (figures 1B and 1C) is due to the much larger aggregates diameters (see below), which, when associated with the external diffusional limitations at 25 rpm lead to internal diffusional limitations in the aggregates; these in turn, may cause the formation of necrotic centres in the aggregates, situation that does not occur at larger agitation rates or in the aggregates of the other cell lines at 25 rpm.

Cell growth at 120 rpm was smaller than at 60 and 90 rpm only for Vero and 293 cells (data not shown); this confirms the previous discussion that these cell lines are extremely anchorage-dependent since shear stress at 120 rpm is sufficient to increase the fraction of single cells which does not grow and even start to die.

Aggregate Characteristics of the Different Cell Lines

The most important variables to describe aggregate characteristics (fraction of cells in aggregates, aggregate diameter and concentration) are depicted in figure 2.

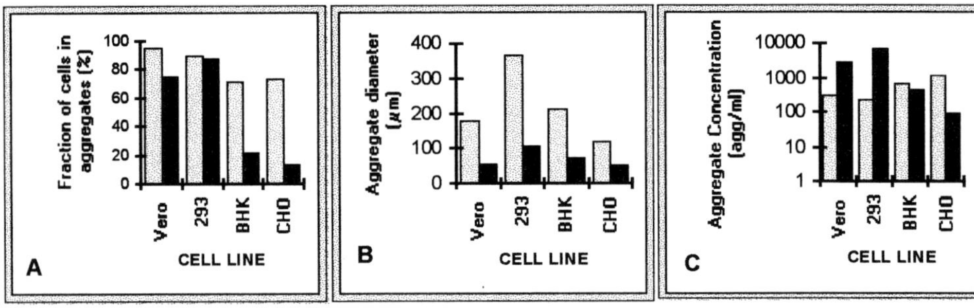

Figure 2: Aggregates characteristics of cell lines Vero, 293, BHK and CHO at agitation rates of 25 and 120 rpm: (A) Average fraction of cells in aggregates (B) Maximum average of aggregate diameter (C) Average aggregate concentration.

As previously indicated, it is possible to observe in figure 2A that Vero and 293 cells grow basically as aggregates whereas BHK and CHO cells prefer to grow in a large fraction as single cells without decrease on cell growth or viability.
In all the cases, however, aggregates size decrease with the agitation rate increase (figure 2B). These results were also clearly confirmed by the data obtained at 60 and 90 rpm.

The increase in aggregate concentration at 120 rpm for Vero and 293 (figure 2C) is not observed for BHK and CHO cell lines since in this last case the fraction of cells in aggregates is very small.

The aggregate concentration at 25 rpm (figures 2C) is due to the large aggregate diameter (figure 2B) and the smaller cell growth (figure 1A). Again, at 60 and 90 rpm, concentration increases with the increase in agitation rate, this being more relevant for Vero and 293 cells since the adhesion fraction is always quite large (data not shown). At 120 rpm a significant decrease in aggregate concentration for BHK and CHO cells is observed (figure 2C) due to the large decrease in the adherent fraction (Figure 2A), since these cells have the ability to grow as single cells in suspension as previously referred to

Hydrodynamic Control of Aggregates

For BHK cells the correlation between aggregate diameter and the hydrodynamic conditions is clearly established- equation 1 (Moreira *et al.*, 1994b).

$$D_{max} = K\varepsilon^{\alpha} \quad (1)$$

where D_{max} is the maximum average aggregate diameter obtained during the batch run, ε is the power input per unit of mass and K and α are empirical parameters.
A model combining the entire range of isotropic eddies has been developed (Moreira *et al.* 1995a), leading to a theoretical prediction law, that shows that in all subranges the dependency of aggregates diameters upon power input is always -0.25
It was possible to observe that this empirical correlation used can be applied to all cell lines regardless of their different behaviours presented in figures 1 and 2.
The results lead to the conclusion that it is possible to control the aggregate size of the cells tested in stirred tanks, with the advantages of this immobilisation system for cell recycle in large particles without the increase in costs inevitable with the use of supports.

Conclusions

The main conclusions that can be drawn from the results in this work are as follows:

• Each cell line shows a different behaviour in cellular growth (growth and viability) and in aggregate characteristics (aggregate size and concentration, fraction of cells in aggregates).

• It was possible to confirm two different kinds of behaviour:
a) Vero and 293 cells are highly anchorage-dependent (very high adherent fractions) and single cells do not grow or even start do die, this being the reason why a decrease in cell growth and viability occurs;
b) BHK and CHO cells are less aggregate dependent and single cells grow in suspension, thus exhibiting high viability even when the agitation rate is low (25 rpm).

• Aggregate characteristics are dependent upon hydrodynamics: with an increase in agitation rate there is a decrease in aggregate diameter and in the adherent fraction.

• It was possible to establish an experimental correlation between maximum aggregate diameter and power dissipation per unit of mass similar to the one obtained by the authors in a recent model where the all range of isotropic eddies was considered.

Notation
Symbols
D_{max} maximum aggregate diameter (μm)
K constant in equation 1
Greek Letters
ε rate of energy dissipation per unit of mass, $cm^2\ s^{-1}$
α empirical parameter of equation 1
η Kolmogoroff's eddy size (cm)
ν kinematic vicosity ($cm^3\ s^{-1}$)
ρ fluid density ($g\ cm^{-3}$)

References

Goetgebeur, S. and Hu, W-S (1991), Appl Microbiol Biotechnol 34: 735-741.
Kawase, Y.; Moo-Young, M. (1990), Chem. Eng. J., 43, 19-41
Moreira, J.L.; Alves, P.M.; Aunins, J.G. and Carrondo, M.J.T. (1994a), Appl. Microbiol Biotechnol 1:203-209.
Moreira, J.L.; Alves, P.M.; Aunins, J.G. and Carrondo, M.J.T. (1994b), Biotechnology and Bioengineering, 46, 351-360.
Moreira, J.L.; Santana, P.C.; Feliciano, A.S.; Cruz, P.E.; Racher, A.J.; Griffiths, J.B. and Carrondo, M.J.T. (1995a), Biotechnol. Prog., 11, 575-583.
Moreira, J.L.; Cruz, P.E. and Santana, P.C. (1995b), Chemical Engineering Science, Vol 50, No. 17, 2747-2764.
Tolbert, W. R.; Hitt, M. M. and Feder, J. (1980), In Vitro 16(6):486-490.

Acknowledgements
The authors acknowledge and appreciate the financial support received from Junta Nacional de Investigação Científica e Tecnológica - Portugal (PBICT/BIO/20333/95)

ANALYSIS OF GLUCOSE AND GLUTAMINE METABOLISM OF HYBRIDOMA CELLS BY CONTINUOUS CULTURE EXPERIMENTS

Anna Sanfeliu, Carles Paredes, Jordi Joan Cairó and Francesc Gòdia

Departament d'Enginyeria Química. Facultat de Ciències. Universitat Autònoma de Barcelona. 08193 Bellaterra. Spain

ABSTRACT.

A series of continuous culture experiments at changing inlet concentrations of glucose and glutamine were carried out for the hybridoma cell line KB 26.5.

The analysis of the steady-states obtained using different glucose and glutamine feed concentrations was done in terms of cell and metabolite concentrations.

The results show the metabolic deregulation existing in hybridoma cells and make it possible to calculate the minimal specific requirements for glucose and glutamine for the cell line used.

INTRODUCTION.

Hybridoma cell cultures tipically produce lactate, ammonia and some amino acids when metabolizing the carbon and energy sources, glucose and glutamine. This byproduct generation represents an inefficient use of carbon source and can inhibit cell growth.

The knowledge of hybridoma phisiology and metabolism can contribute to optimize the culture medium, the strategy of operation and to a better understanding of which are the main substrates or metabolites that need to be monitored and controlled for process optimization.

Continuous culture was chosen to study the objectives of this work because it allows defined and constant environmental conditions.

MATERIALS AND METHODS.

The KB-26.5 murine hybridoma cell line, kindly provided by Laboratorios Knickerbocker SAE (Barcelona), was used in this study. This cell line produces an IgG_3 antibody directed against antigen A_1 of red cells.

The basal medium used in all cultures was DMEM (Sigma D5030) supplemented with sodium bicarbonate 3.7 $g \cdot l^{-1}$ (Probus), phenol red 15 $mg \cdot l^{-1}$, ß-mercaptoethanol 0.391 $mg \cdot l^{-1}$ (Sigma), insulin 0.836 $mg \cdot l^{-1}$ (Novo) and 1% foetal calf serum (FCS, Biological

M.J.T. Carrondo et al. (eds.), Animal Cell Technology, 785–789.

Industries).

Continuous culture was carried out in a spinner flask (Techne) with 250 ml working volume, stirred at 40 rpm and incubated at 37°C in a humidified atmosphere of 5% CO_2 in air (Forma Scientific CO_2 incubator). Continuous culture was initiated by inoculation of cells at $1.5x10^5$ viable cells/ml in the spinner flask using 25 mM glucose and 5 mM glutamine as initial concentrations. The cells were then allowed to grow in a batch fashion and after two days continuous medium flow into and out of the vessel was initiated. The dilution rate used throughout the duration of the experiment (5 months) was 0.0256 h^{-1}. The continuous culture can be divided in two parts: in the first part cells were cultivated with different glutamine concentrations (5, 3, 1.5, 1, 0.75 and 0.4 mM) and a fixed glucose concentration (25 mM) in the feed medium,whereas in the second part cells were cultivated with different glucose concentrations (20, 10, 5, 4, 3, 3.5, 0.75, 1.5, 1 and 0.75 mM) and a fixed glutamine concentration (0.75 mM) in the feed medium.

Viable cell concentration and viability were determined by the the trypan blue exclusion method using an haemocytometer (Neubauer improved, Brand).

Glucose and lactate concentrations were estimated with an YSI 2000 automated glucose and L-lactate analyzer.

Ammonium concentration was determined by a flow injection analysis system.

Glutamine and other amino acids concentrations were measured by HPLC (Hewlet Packard 1090) using a reversed phase column (aminoquant 200 x 2.1 mm) after derivatisation of the samples with OPA and FMOC for the analysis of both primary and secondary amino acids.

RESULTS.

Figure 1 shows the culture response to changes in glutamine feed concentration.

These results indicate that when an excessive amount of glutamine is added to the hybridoma culture it is not efficiently used for cell growth, but rather to produce byproducts such as ammonium, alanine and proline as shown in Table I.

It can be affirmed that glutamine concentrations lower than 0.75 mM are not sufficient to mantain a level of $5x10^5$ cells/ml at the dilution rate used in this study. From these results the minimum specific requirement for glutamine is 30.72 nmols/(10^6 cells·h).

Figure 2 shows the culture response to changes in glucose feed concentration. Similarly to the results obtained with glutamine, when an excessive amount of glucose is added to the hybridoma culture it is not efficiently used for cell growth but converted into a by-product, the lactate.

As shown in Table II, the proline level was negligible and alanine level remained constant for glucose concentrations between 20 and 3 mM. When glucose was reduced from 3 mM to 1.5 mM, the amount of alanine produced was slightly decreased.

Glucose concentrations lower than 1.5 mM are not sufficient to maintain a level of $5x10^5$

cells/ml at the dilution rate employed in this study. From these results the minimum specific requirement for glucose is 76.8 nmols/(10^6 cells·h).

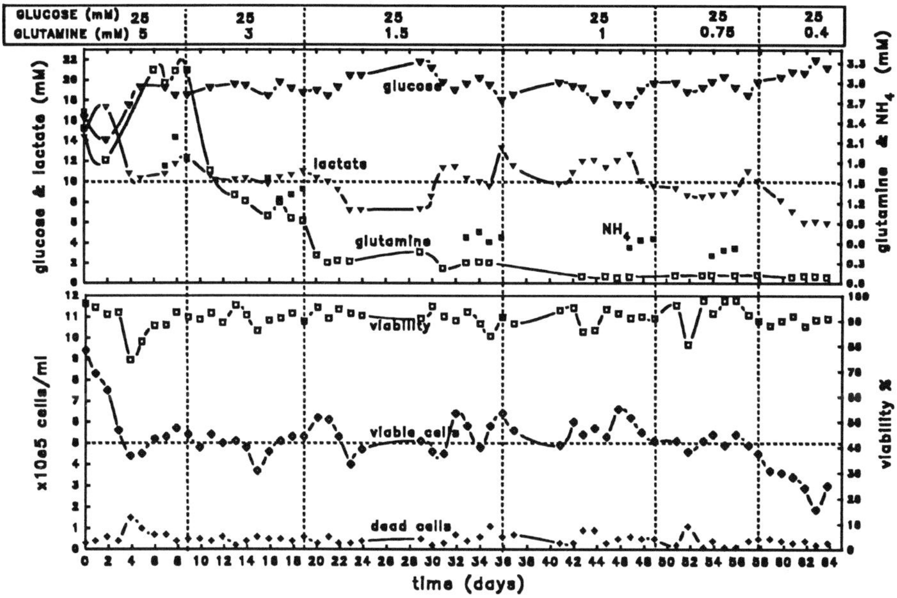

Fig. 1 Culture response to changes in glutamine feed concentration

Table I Steady state metabolite concentrations as a function of different glutamine feed concentrations

GLUCOSE AND GLUTAMINE FEED (mM)	GLUCOSE CONSUMED (mM)	LACTATE PRODUCED (mM)	GLUTAMINE CONSUMED (mM)	AMMONIUM PRODUCED (mM)	ALANINE PRODUCED (mM)	PROLINE PRODUCED (mM)
Gluc 25 mM Gln 5 mM	5.85	11.64	2	1.93	0.99	0.18
Gluc 25 mM Gln 3 mM	6.29	10.71	1.72	1.39	0.79	0.16
Gluc 25 mM Gln 1.5 mM	5.33	9.97	1.14	0.69	0.53	0.05
Gluc 25 mM Gln 1 mM	6.65	10.79	0.80	0.62	0.38	0.04
Gluc 25 mM Gln 0.75 mM	5.12	8.83	0.59	0.48	0.32	0.04

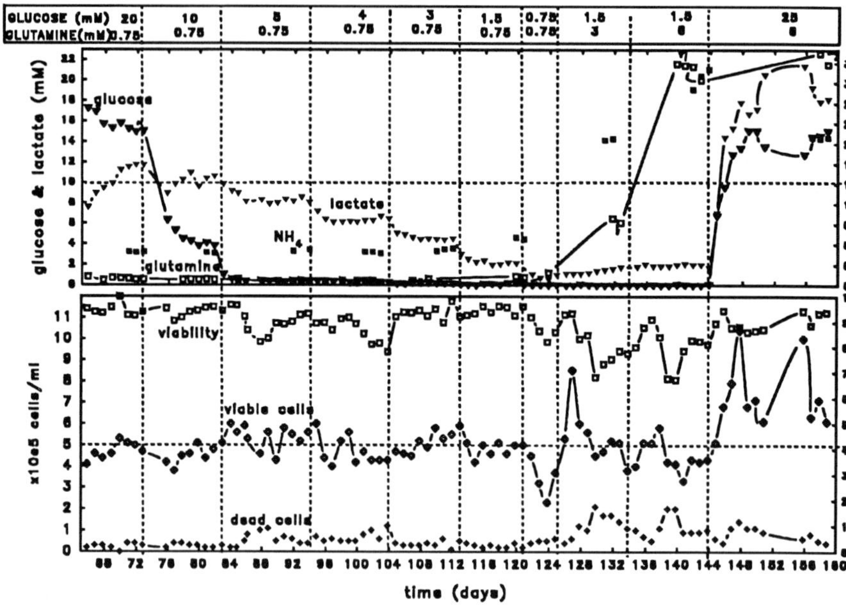

Fig. 2 Culture response to changes in glucose feed concentration

Table II Steady state metabolite concentrations as a function of different glucose feed concentrations

GLUCOSE AND GLUTAMINE FEED (mM)	GLUCOSE CONSUMED (mM)	LACTATE PRODUCED (mM)	GLUTAMINE CONSUMED (mM)	AMMONIUM PRODUCED (mM)	ALANINE PRODUCED (mM)	PROLINE PRODUCED (mM)
Gluc 25 mM Gln 0.75 mM	5.12	8.83	0.59	0.48	0.32	0.04
Gluc 20 mM Gln 0.75 mM	4.987	10.36	0.66	0.49	0.29	0.03
Gluc 10 mM Gln 0.75 mM	5.75	10.12	0.65	0.48	0.29	0.03
Gluc 5 mM Gln 0.75 mM	4.71	8.39	0.70	0.51	0.33	0.02
Gluc 4 mM Gln 0.75 mM	3.92	6.40	0.66	0.49	0.23	0.03
Gluc 3 mM Gln 0.75 mM	2.87	4.46	0.67	0.52	0.25	0.01
Gluc 1.5 mM Gln 0.75 mM	1.55	2.09	0.66	0.69	0.18	0.07

CONCLUSIONS.

The continuous culture has confirmed the metabolic deregulation existing in hybridoma cells. When glucose and glutamine are added in excessive amounts to the culture, cells will rapidly consume them without any apparent benefit, since they are not strictly used for growth, but to produce high levels of lactate, ammonium, and some amino acids (alanine, proline) as waste products. Low concentrations of glucose and glutamine reduce the uptake rates and less byproducts are accumulated.

These results can be explained in terms of the low efficiency in the shuttle system for mitochondrial reoxidation of the large amounts of NADH produced in the glycolysis pathway (Figure 3). As a consequence, NADH should be regenerated via lactate dehydrogenase using piruvate as a substrate. Thus, the pyruvate pool that can be incorporated into TCA cycle is very small and the rapid glutaminolysis unbalances the TCA cycle.

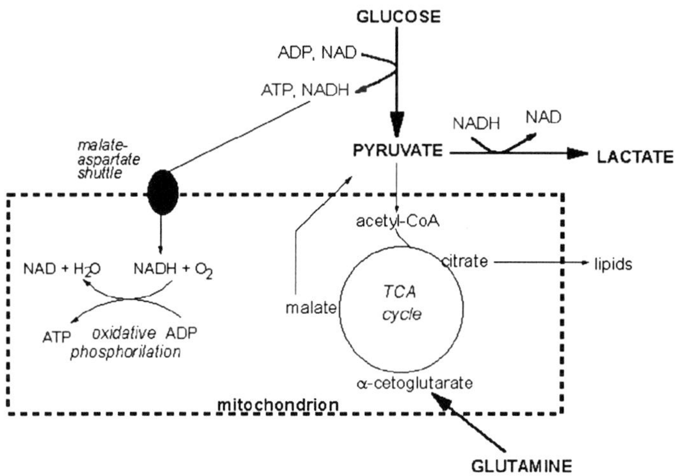

Fig. 3 Scheme of the main pathways of the hybridoma cell metabolism

Additionally, by means of the continuous culture results it has been possible to calculate the minimal specific metabolic requirements for glucose and glutamine. These parameters can be used in the designing of fed-batch hybridoma cultures with an optimal feeding strategy to enable sufficient supply of glucose and glutamine, with the minimal excretion of partly oxidized undesirable end products.

These results suggest a way of improving processes based on hybridoma cell culture, which is to perform metabolic pathway engineering, particularly in glycolysis or glutaminolysis, in order to achieve an efficient use of the substrates and a minimization of the byproduct accumulation.

THE METABOLIC PROFILE OF A MURINE HYBRIDOMA GROWN IN SERUM FREE MEDIUM AT VARIOUS OXYGEN CONCENTRATIONS

D.C.H. JAN, D.A. PETCH, N. HUZEL AND M. BUTLER
Department of Microbiology, University of Manitoba,
Winnipeg, Manitoba, CANADA R3T 2N2

Abstract

A murine hybridoma was grown in continuous cultures at oxygen concentrations in the range 10% to 125% of air saturation. Although some adaptation was required at each step-wise increase in dissolved oxygen (DO) level, the cells could be maintained within this range at high viability at a dilution rate of 1 volume per day. However, an attempt to grow cells at 250% of air saturation was unsuccessful. Adaptation to increased DO involved an induction of the specific activity of 3 anti-oxidant enzymes.

The increase in DO caused changes in energy metabolism. The specific rate of glucose uptake increased at higher DO when a greater proportion of glucose was metabolised anaerobically. Most of the glucose (> 85%) was metabolised via glycolysis. Relatively low flux rates were measured for the pentose phosphate and TCA cycle pathways. Although the specific glutamine utilisation rate increased at higher DO, there was no evidence of a change in the pattern of metabolism. Analysis of the oxygen uptake rate of cell suspensions at steady state under all conditions showed a pronounced Crabtree effect which is manifest by a decrease (up to 40%) in oxygen consumption on addition of glucose.

INTRODUCTION

Although most mammalian cell bioreactor systems are operated at a DO of 30-60%, some mammalian cell lines can be adapted to withstand high oxygen concentrations (1,2) A high level of dissolved oxygen in cultures may enhance the formation of reactive oxygen species that can damage the cell membranes, reduce cell viability and induce cell death (3). However, these effects may be reduced by cellular adaptation which may involve an increase in activity of anti-oxidant enzymes (4). This may extend the operable DO range in bioreactor systems and reduce the requirement for fine set-point control using precision probes and feedback loops.

The purpose of the work reported here was to study the ability of a murine B-lymphocyte hybridoma to adapt and grow at dissolved oxygen levels above those normally used in culture. We report that under continuous culture conditions the cells

M.J.T. Carrondo et al. (eds.), Animal Cell Technology, 791–796.

were able to adapt to DO levels of up to 150% air saturation with step-wise increases in DO from 10% air saturation. This adaptation involved changes in the profile of energy metabolism and the induction of some anti-oxidant enzymes. The hyperoxic conditions were associated with a higher specific antibody production but a lower apparent growth rate.

MATERIALS AND METHODS

Cell line: The murine hybridoma (CC9C10), which secretes a monoclonal antibody (IgG_1) against insulin, was obtained from the ATCC (No HB123).

Culture: The basal medium consisted of DMEM: Ham's F12 (1:1v/v) supplemented with a serum-free formulation (5). Cells were grown in a Celligen (New Brunswick) or LH (210) bioreactor (working volume 1.2-1.5 l). Both bioreactors were operated in a continuous mode with a stirring speed of 100 rpm, a temperature of 37°C and pH of 7.1. A dilution rate of one volume per day was maintained except where indicated to allow for cell adaptation to higher oxygen levels.

Media analysis. Glucose and lactate concentrations were determined by an analyzer (YSI model 27). Glutamine was measured by an enzymatic assay adapted for use on a microplate reader. Ammonia was determined by a specific electrode. Mab concentrations were determined by HPLC using a ProAna Mabs column (Hyclone).

Assays. Metabolic flux rates were determined from $^{14}CO_2$ release from D-[6-^{14}C]-glucose or D-[1-^{14}C]-glucose and ^3H_2O release from D-[3-^{3}H]-glucose as previously described (6). Glutathione peroxidase (GPX), Glutathione S-transferase (GST) and Superoxide dismutase (SOD) activities were determined by spectrophotometric assays. Oxygen uptake rate (OUR) was determined by the rate of decrease of dissolved oxygen concentration using a Biological Oxygen Monitor (YSI 5300).

RESULTS AND DISCUSSION

1. Effect of dissolved oxygen on cell growth

The murine hybridoma (CC9C10) was grown routinely in a simple serum-free formulation developed in our laboratory. However, the lack of serum renders the cultures more susceptible to any potential toxic or stress effects that may arise through changes in culture conditions. In our experiments the DO level was increased step-wise from an initial steady state established in continuous culture at 10% DO. In most cases these increases caused an initial decrease in cell concentration followed by a re-establishment of steady state at the same specific cell growth rate, equivalent to a dilution rate of 1 day^{-1}. During adaptation the dilution rate was temporarily decreased to prevent wash out of the cells.

Steady state was assumed after 5 volume changes of culture and a constant viable cell concentration over 3 days. Steady state growth was attained at levels of DO from 10% up to 125% air saturation (Table 1) at a dilution rate of 1 vol d^{-1} which is equivalent to a specific growth rate (μ) of 0.042 h^{-1}. The % viability was high in all cases and this indicates that the death rate was negligible. However, the significant decrease in viable cell concentration at higher DO values indicates that the cell growth rate and specific yield of cells from the serum-free medium was dependent on DO. However, we failed to maintain cell growth above this level. An increase of DO up to 250% of air saturation caused the viable cell concentration to decline. However, the viable cell concentration was able to recover after the DO was reduced from 250% to 100% of air saturation.

Table 1: The effect of DO on steady state cell and Mab concentration (D=1 d^{-1}).

	Dissolved oxygen level (% air saturation)			
	10%	50%	100%	125%
viable cells (10^6/ ml)	2.15	2.15	1.95	1.55
viability (%)	99.1	99.0	98.0	98.5
Mab (mg/l)	52.3	48.9	41.3	55.4

2. Substrate utilization and product formation

The specific rates of consumption or production of the major substrates and products were determined from the analysis of culture supernatants at each steady state (Table 2). The q_{Glc} value increased at higher levels of DO. The DO level did not affect q_{Gln} significantly in the range 10% to 100%. However, the specific glutamine utilization rate was increased 20% in the culture grown at a DO of 125%.

The ratio of q_{Glc}/q_{Gln} is used as an index of the preference of CC9C10 cells for glucose or glutamine as an energy source. This ratio increased steadily from 2.54 to 4.17 withe a DO increase from 10% to 125%. This indicates an increased preference for glucose rather than glutamine at the higher oxygen concentration.

Table 2. The specific rates of substrate utilization and product formation. Values are means (±SEM) expressed as nmol/ 10^6 cells per min except for Mab which is given as ng/ 10^6 cells per min. n=3.

	Dissolved oxygen level (% air saturation)			
	10%	50%	100%	125%
Glucose	3.71±0.06	4.29±0.03	4.64±0.03	7.34±0.03
Lactate	5.02±0.20	6.07±0.14	6.60±0.08	12.2±0.40
Glutamine	1.46±0.05	1.47±0.03	1.54±0.05	1.76±0.06
Ammonia	0.72±0.01	0.67±0.02	0.66±0.06	0.78±0.01
MAB	16.9±0.76	15.8±2.56	14.7±0.91	24.8±0.62

The specific MAB production rate (q_{Mab}) at 125% DO_2 was significantly higher (57%) than at lower DO. However, the measured Mab concentration did not show a statistically significant variation with change in DO. This is explained by the relatively higher cell concentration at the lower DO and suggests that there is a differential effect of DO on growth rate and q_{Mab}.

Yield coefficients were determined to give an indication of any change in pattern of metabolism. The $Y_{Lac}/_{Glc}$ is a ratio of lactate produced to glucose consumed and gives an indication of the proportion of glucose metabolised anaerobically via glycolysis. This value increased gradually from 1.35 to 1.66 as DO was increased from 10% to 125%. This result suggests that the proportion of glucose metabolised through glycolysis pathway was enhanced at a high oxygen level. The coefficient $Y_{Amm}/_{Gln}$ did not change significantly as the DO increased from 10% to 125%. This suggests that the deamination of glutamine was not affected by dissolved oxygen.

3. Energy metabolism profiles

The flux of glucose through alternative metabolic pathways was determined by short-term radioactive assays. The measured rates were linear for a period of 6 h in all cases and were used to calculate the flux of glucose through the pentose phosphate pathway, glycolysis and TCA cycle as shown previously(6).

The overall rate of glucose utilization increased significantly at higher DO (Table 3). This result was consistent with the increase in q_{Glc} values at higher DO shown in Table 2. The higher glucose uptake rates in Table 3 are explained by the fact that the radioactive assays were conducted in fresh medium containing 17.5 mM glucose. In contrast the q_{Glc} values were determined at the relatively low glucose concentrations (1-6 mM) in cultures at steady state.

Table 3. The metabolic flux of glucose through different pathways (nmole/10^6 cells/min) Each value (±SEM) is based on 6 determinations taken over 6 h. The % of glucose metabolised by each specific pathway.is shown in brackets.

	Dissolved oxygen level (% air saturation)		
Pathway	10%	50%	100%
Glycolysis	2.72±0.19 (85.8%)	6.48±0.29 (89.8%)	7.34±0.97 (90.5%)
PPP	0.451±0.014 (14.2%)	0.740±0.019 (10.3%)	0.771±0.047 (9.5%)
TCA	0.0645±0.0027 (2.03%)	0.0895±0.0027 (1.24%)	0.0169±0.0015 (0.34%)
Total	3.17±0.2	7.22±0.31	8.11±1.02

The increased glucose utilization at higher DO can be explained entirely by an increase in anaerobic metabolism. The flux of glucose through the TCA cycle decreased significantly at high DO, whereas the flux through both glycolysis and the pentose phosphate pathway increased significantly. The proportion of glucose metabolised by each of 3 pathways is indicated by the % values shown in Table 3. A large proportion (> 85%) of glucose was metabolised via glycolysis in all cases and increased at higher DO. On the other hand, the proportion of glucose metabolised via the pentose phosphate and TCA cycle pathways decreased at higher DO.

The oxygen uptake rates (OUR) was determined at 0.22 μmole O_2/ 10^6 cells per h in all cell samples with no apparent significant difference with respect to DO. However, the OUR was sensitive to the addition of substrate to the cell suspension. Supplementation of the sample with glutamine or a mixture of amino acids had no apparent effect on the OUR. Similarly, re-suspension of the cell to the same concentration in PBS had no effect on the OUR. However, supplementation of the original suspension with 17.5 mM glucose or re-suspension of the cells in fresh medium caused an immediate reduction in the OUR by over 40% of the original value. This is indicative of a pronounced Crabtree effect (7) which suggests that the metabolic profile in these cultures is highly sensitive to the availability of glucose which may be a limiting substrate under the conditions used. It also suggests that the oxidative metabolism at steady state is most probably related to glutamine utilisation.

4. Changes in anti-oxidant enzyme activity

The intracellular activity of three anti-oxidant enzymes, glutathione transferase, glutathione peroxidase and superoxide dismutase (GST, GPX and SOD) were determined in cells under 5 levels of DO (Table 4). In all cases the activity increased significantly as values of DO were increased above 10%. For GST and GPX, maximum specific activities were determined at 100% DO with values that were x11 and x18 greater than the corresponding values determined at 10% DO.

Table 4. The effect of DO on the specific activity of 3 anti-oxidant enzymes
Values (unit/ 10^6 cells) are expressed as means (±SEM) n=3. The measurements were performed with cell lysates obtained from cultures at steady state under various levels of DO and by assays described in the Materials and Methods.

	Dissolved oxygen level (% air saturation)				
	10%	50%	100%	125%	150%
GST	27.5±0.9	104±3	313±9	78.2±2.0	75.4±3.2
GPX	2.22±0.01	12.1±1.4	40.8±2.6	1.36±0.06	0.57±0.64
SOD	93.6±3.4	189±49	413±13	420±25	576±26

At higher levels of DO these activities decreased and in the case of GPX the values at 125% and 150% were below those determined at 10% DO. In the case of SOD the activity increased at higher DO levels to a maximum activity at 150% DO. The maximum value was x6 that determined at 10% DO. The increased activities of these ant-oxidant enzymes at DO >10% indicates an adaptive mechanism to counter the inhibitory effects of reactive oxygen species. It may be that at the high DO levels (>150%) in this system, the anti-oxidant activities are overwhelmed by these inhibitory effects. The burden on these enzymes may be greater under the serum-free conditions of the culture because of the lack of antioxidants that may be present in serum. There is also some evidence that the anti-oxidant enzymes themselves may be inactivated by high levels of oxygen-derived free radicals (8).

Acknowledgements
We acknowledge financial support from NSERC, Apotex and Rh Pharmaceuticals.

References

1. Valk van der P, Gille JJP, Oostra, AB, Roubos, EW, Sminia T & Joenje H (1985) Characterization of an oxygen-tolerant cell line derived from Chinese hamster ovary. Cell Tissue Res. 239, 61-68.
2. Oller AR, Buser CW, Tyo MA & Thilly WG (1989) Growth of mammalian cells at high oxygen concentrations. J. Cell Sci., 14, 43-49.
3. Ellis LC (1991) Free Radicals in Tissue Culture: Part IV, Effects on Cells in Culture, Art to Science, 10: 1-5. Hyclone publication.
4. Sullivan SJ, Roberts RJ & Spitz DR (1991) Replacement of media in cell culture alters oxygen toxicity - possible role of lipid aldehydes and glutathione transferases in oxygen toxicity. J Cell Physiol. 147, 427-433.
5. Butler M & Huzel N (1995) The effect of fatty acids on hybridoma cell growth and antibody productivity in serum-free cultures. J. Biotechnol. 39, 165-173.
6. Petch D & Butler M (1994). A profile of energy metabolism in a murine hybridoma: glucose and glutamine utilization. J.Cell Physiol. 161, 71-76.
7. Guppy M, Greiner E & Brand K (1993) The role of the Crabtree effect and an endogenous fuel in the energy metabolism of resting and proliferating thymocytes. Eur. J. Biochem. 212, 95-99.
8. Pigeolet E, Corbisier P, Houbion A, Lambert D, Michiels C, Raes M, Zachary M & Remade J (1990) Glutathione peroxidase, superoxide dismutase and catalase inactivation by peroxides and oxygen-derived free radicals. Mech. Ageing Dev. 51, 283-297.

ALTERNATIVE STRATEGIES FOR THE PRODUCTION OF AN ANTIBODY-CYTOKINE FUSION PROTEIN FROM RECOMBINANT BHK CELLS

Looby D[1], Moreira JL[2], Stacey GN[1], Cruz H[2], Cuffe M[2], Dias E[2], Hayes K[1], House H[1], Beardmore M[1] , Savage A[3], Carrondo MJT[2] and Griffiths JB[1]

1 CAMR, Porton Down, Salisbury, Wiltshire, SP4 0JG, UK
2 IBET, Apatardo 12, 2780 Oeiras, Portugal
3 Dept. of Chemistry, University College Galway, Ireland

Introduction

The choice of scale up strategies for production of recombinant proteins from animal cells is often restricted to suspension cultures in stirred tank bioreactors. The primary aim of the work reported here was to identify alternative strategies for scale up of production, from BHK cells, of a recombinant antibody-cytokine fusion protein from BHK cells. We have used a single recombinant BHK clone to address a wide range of scale up options including batch production using aggregates in stirred systems, and attached cells on microcarriers; and continuous production using a fixed bed porous carrier perfusion system. These modalities have been compared with standard and modified (silicone coated) roller bottles. While cell specific production rate and concentration of product in culture supernatant are key factors in the final choice of production method, further comparative assessments of product quality and feasibility and cost at pilot stage are critical to the final selection of scale up technologies for use in production.

Key words: Bioreactor, BHK, recombinant fusion protein.

Materials and Methods

Cell line: The cell line used was a recombinant BHK clone genetically manipulated to secrete an antibody-cytokine fusion protein.

Culture medium: DMEM (high glucose) was supplemented with 10% FBS and 5ug/ml puromycin as the selective agent.

Cell enumeration: Total and viable cell counts were determined by Trypan blue (0.4% in PBS) dye exclusion and nuclei counts by treating cell suspensions with 0.1M citric acid, 0.1M crystal violet and 1% (V/V) Triton X100.

M.J.T. Carrondo et al. (eds.), Animal Cell Technology, 797–800.

Determination of product level: Product level in culture supernantant was determined by an ELISA employing a capture antibody for IgG heavy chain portion of the fusion protein. A alkaline phosphatase conjugated antibody was then used to give a colourimetric quantifiable reading against an IgG standard. The protocol for this ELISA was standardised between laboratories. NB: Values reported here refer only to ELISA data and do not infer equivalent product quality for each culture system.

The culture systems and their operation:-

Roller bottles: The standard roller bottle used was a 2l ($850cm^2$) polystyrene bottle (Corning Ltd) and the expanded surface area roller bottles were 1l ($450cm^2$) (Textursil, Ashby Scientific Ltd). The inoculum for these vessels was 10^7 cells/bottle. Medium volume for both roller bottle types was 200ml and medium was changed to maintain a glucose level above 5mM, ie. medium was replaced, and product thus harvested, on days 4, 7 and 10 post inoculation.

Spinner Flasks: Wheaton flasks with ball impellers were used for microcarrier cultures. The inoculum was $2.0x10^5$cells/ml for both attached cells and aggregates (1). Cytodex carriers (collagen) were used at a concentration of 3mg/ml and Cultisphere G (gelatin beads) at a concentration of 1mg/ml. Stirrer speeds were 120rpm for aggregates and 60rpm for Cytodex and Cultisphere G.

Stirred tank bioreactors: Stirred tank bioreactors (1l working volume, Applikon) fitted with marine impellers were used for microcarriers and aggregates. The pH was controlled at 7.2 and oxygen (DO) was maintained at 40% saturation with head space air and oxygen on demand. The system was inoculated at $2.0x10^5$cells/ml for microcarriers and aggregates systems. As for spinner cultures Cytodex 3 beads were used at 3mg/ml. Stirrer speeds were 100 and 70rpm for aggregates and microcarriers respectively.

Fixed bed reactor: The fixed bed system was comprised of a reactor vessel (100ml) containing porous borosilicate glass beads (average diameter 5mm, pore volume 60% of bead, pore size 60-300mm) and a reservoir vessel (2l stirred tank bioreactor) for media conditioning (2, 3). Medium was recycled at a flow rate of 0.1l/min through the fixed bed and back to the reservoir. Medium feed rate was adjusted to maintain a glucose concentration above 5mM ie. 1l/day in steady state culture. The pH was controlled at 7.2 and dissolved oxygen at 60% in the reservoir.

Results and Discussion

Comparison of Culture Systems for the Production of an Antibody-Cytokine Fusion Protein from Recombinant BHK Cells:

Culture System	**Cell Type**	**Culture Media Volume**	**React. Active Vol.**	**Max Cell Density $x10^5/ml^b$**	**Max Product Concent. mg/ml^b**	**Mode of Opertn.**
Spinner	Suspension Aggregates	250ml	250ml	9.0	1.0	Batch
Spinner (Cytodex 1)	Attached	250ml	250ml	5.7	1.5	Batch
Stirred Tank	Suspension Aggregates	1000ml	1000ml	6.0	1.0	Batch
Stirred Tank (Cytodex 1)	Attached	1000ml	1000ml	7.0	1.6	Batch
Spinner (Cultispher G)	Attached	250ml	250ml	6.4	1.0	Batch
Fixed Bed	Attached	1000ml	100ml	-	3.8	Contin. perfusion[a]
Roller Bottle (Normal)	Attached	200ml	200ml	4.0	1.5 (Av:1.2)	Repeated Batch 3 harvests
Roller Bottle (Expanded surface area)	Attached	200ml	200ml	20.0	4.0 (Av:2.3)	Repeated Batch 3 harvests

[a] Media feed rate = 1000ml per day in steady state culture

[b] Per ml culture media volume

Average steady state production rate in the fixed bed reactor system was 3.2mg/day and a total of 78mg was produced in 30days.
The cell specific production rate in batch microcarrier spinner culture was 0.67mg/10^6cells/day.

The recombinant cell line (BHK21a) can grow either attached to a substrate or as aggregates. Maximum cell densities and product concentrations were achieved in the fixed bed reactor (3.8ug/ml) and textured silicone coated roller bottles (4.0ug/ml), whereas all the other systems tested achieved between 1.0 and 1.6ug/ml. Similar cell densities were achieved in batch aggregate and microcarrier cultures. The fixed bed perfusion system produced an average of 3.2mg/day in steady state operation, and thus is more productive than batch microcarrier and aggregate cultures.

However, this study investigated only the fixed bed porous glass sphere culture system in continuous perfusion mode and it should be noted that microcarriers (solid and porous) and aggregates can also be retained in perfusion bioreactors and scaled up easily in conventional stirred tank bioreactors. Perfusion systems based on stirred tank bioreactors are more homogenous than fixed bed systems and thus have the potential for improved performance. The fixed bed system is limited in terms of linear scale up (ie. bed height) due to effects of concentration gradients, but does have considerable radial (ie. bed diameter) scale up potential.

References

1) Moreira JL, Alves PM, Feliciano AS, Aunins JG and Carrondo MJT (1995) Serum free and serum containing media for the growth of suspended BHK aggregates in stirred vessels. Enzyme and Microbial Technology 17, 437-444.

2) Looby D and Griffiths JB (1988) Fixed bed porous glass sphere (porosphere) bioreactors for animal cells. Cytotechnology 1, 339-346.

3) Looby D and Griffiths JB (1989) Immobilisation of animal cells in fixed and fluidised porous glass sphere reactors. In: Advances in Animal Cell Biology and Technology for Bioprocesses. Eds: Spier RE, Griffiths JB, Stephenne J and Crooy PJ. pp336-334, Butterworth, Guildford, UK.

OPTIMIZATION OF HUMAN ANTIBODY PRODUCTION BY XENOHYBRIDIZATION OF EBV-TRANSFORMED CELL LINES

E.J.M. Al, R.A. Mijnhart, N.J.J. Dekker, R.J.P. Ouwerkerk, H.M.G. Sillekens and P.C. van Mourik
Central Laboratory of the Netherlands Red Cross Blood Transfusion Service, Department of Biotechnology, P.O.Box 9190, 1006 AD Amsterdam, The Netherlands

Abstract

Two cell lines producing human IgG-antibodies to Hepatitis B Virus Surface Antigen were obtained by EBV-transformation of human blood lymphocytes. Large-scale production for therapeutic purposes was hampered by inconsistent IgG-production and infectious EBV-production. Therefore, xenohybridomas were engineered by electrofusion of the EBV-transformed cell lines with mouse myeloma cell lines. After adaptation to serum-free medium, two clones were obtained and tested in large-scale perfusion fermentation: only one clone yielded consistent antibody production. Properties of the antibodies produced were similar to those from the parental cells; while EBV-DNA was found negative. It is concluded that this procedure yields, although with low efficiency, stable producing and EBV-negative clones.

Introduction

In the past, two cell lines producing human IgG-antibodies to Hepatitis B Virus Surface Antigen (HBsAg) were obtained by EBV-transformation of human blood lymphocytes: CLB-1 and CLB-4 (Ref.1). Pilot production of the antibodies, for human therapeutic applications, was done by large-scale perfusion fermentation (10 l) of the two cell lines. However, inconsistent antibody production was encountered during one of the runs (results not shown). Furthermore, EBV-production by the two cell lines was detected after relevant biosafety testing (not shown). To solve these two problems, xenohybridomas were produced by electrofusion with the murine myeloma cell lines SP2/0 and NS-1 (Ref.2 and 3). The selected cell lines and the antibodies produced were characterized and large-scale fermentation was successfully performed with one clone (Table 2). The results indicate that the applied procedure provides a useable tool for this purpose.

M.J.T. Carrondo et al. (eds.), Animal Cell Technology, 801–806.

Procedure

Electrofusion (according Ref.2) was performed by combining $10x10^6$ viable cells of both the EBV-transformed cell line and the myeloma cell line in 0.2 ml electrofusion medium in the fusion chamber of a ECSI Fusion Porator (Model CFP II). Fusion was induced at 0.3 kV for 5 μsec; electrofusion medium consisted of 0.255 M sucrose - 0.2 mM $CaCL_2$ - 0.2 mM $MgCl_2$. After electrofusion, cells are cultured for one day in culture medium without selection (IMDM-10%FBS). Selection of xenohybridomas was started one day after electrofusion with ouabain, azaserine and thymidine in IMDM with 10% FBS. After cloning by limiting dilution, adaptation of the xenohybridomas to serum-free culture medium (Perfusion medium) was done by diluting FBS-growing clones 1:1 with serum-free medium. Testing for human antibody production was done by a Sandwich-ELISA using mouse monoclonal anti-(human IgG1)-antibodies or mouse monoclonal anti-idiotype antibodies.

DNA fingerprinting was performed using a synthetic, ^{32}P-radiolabelled $(GTG)_5$ oligonucleotide probe after HinfI restriction enzyme digestion, agarose gelelectroforese and Southern blotting. EBV-DNA detection was performed using ^{32}P-radiolabelled LMP and EBNA-2 probes after BamHI restriction enzyme digestion, agarose gelelectroforese and Southern blotting. EBNA (Epstein-Barr Nuclear Antigen) was detected by immunofluorescence microscopy using a human anti-EBNA-positive antiserum.

Perfusion fermentation was performed in a 10 l Bioreactor (Chemap/Braun) equipped with a hollow fibre filtration device for cell retention during harvesting. Serum-free perfusion medium was perfused at a maximum rate of 2 vessel volumes per day. Crude harvest was concentrated and IgG was purified by Protein A Sepharose Chromatography (Pharmacia). Anti-HBsAg-activity was quantified in a parallel line radioimmunoassay (AUSRIA, Abbott). Isoelectric focusing (IEF) of the purified antibodies was performed with the Phastsystem (Pharmacia) using IEF 3-9 gradient gels.

Results and Discussion

Both EBV-transformed cell lines, CLB-1 and CLB-4, were fused to the mouse myeloma cell line SP2/0 as this cell line was tested extensively for biosafety. In addition, NS-1 cell line was used for its improved fusion capacity (Ref.3). Table 1 presents the results of the electrofusion: low numbers of xenohybridomas were obtained and the IgG producing clones were subcloned.

After limiting dilution cloning twice, the stability was at most 92% at which stage adaptation to serum-free culture medium (perfusion medium) was performed. It appeared that after adaptation, stability decreased to less than 40%. Further cloning yielded clones with maximum stability of 90%.

TABLE 1. Results of electrofusions performed.

Fusion	Number of growing clones	Number of IgG-producing clones	Selected clones
CLB-1 x NS-1	13	4	G7
CLB-1 x SP2/0	4	2	H10
CLB-4 x SP2/0	29	7	A6 B8

From two subclones, F-CLB-1.H10 and F-CLB-4.B8, the presence of EBV-DNA was studied by DNA-hybridization. Figure 1 shows that, after BamH1 restriction enzyme digestion, no EBV-DNA was detectable by means of probing with both LMP and EBNA-2. Both genes are clearly present in EBV-transformed cell lines (lanes 1).

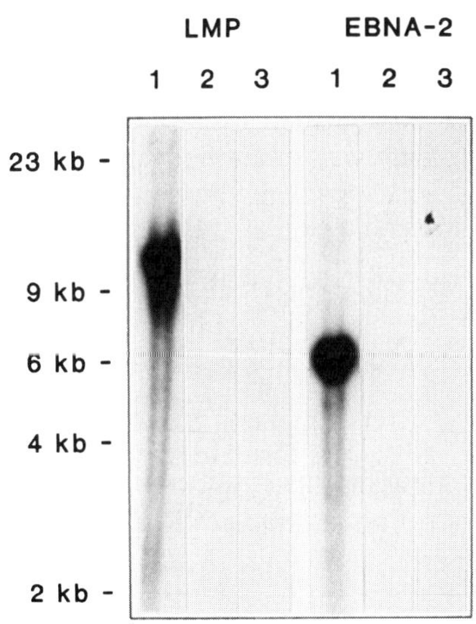

Figure 1. Absence of EBV-DNA in xenohybridomas by DNA-hybridization. Lane 1: CLB-4; lane 2: F-CLB-1.H10; lane 3: F-CLB-4.B8. EBV-probes used are LMP (left panel) and EBNA-2 (right panel) after BamH1 digestion. kb, kilobases.

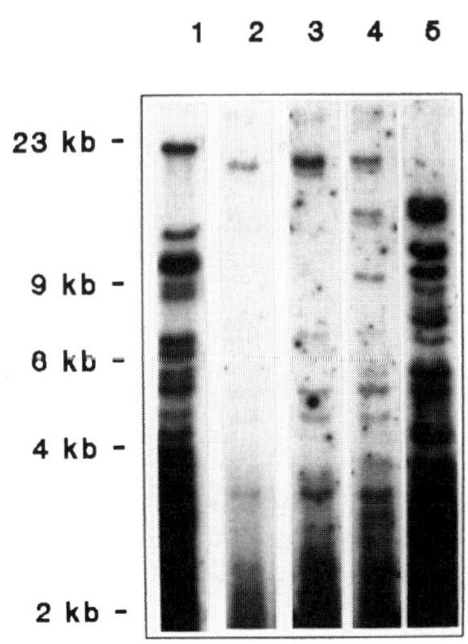

Figure 2. DNA-Fingerprinting analysis. Lane 1: CLB-1; lane 2: F-CLB-1.H10; lane 3: SP2/0; lane 4: F-CLB-4.B8; lane 5: CLB-4. DNA was digested with restriction enzyme Hinf1 and probed with radiolabelled $(GTG)_5$. kb, kilobases.

EBNA expression, indicative for EBV-transformation, was determined by immunofluorescence and found negative in the xenohybridomas.

DNA-fingerprinting was performed to establish the identity of xenohybridomas. Figure 2 shows the $(GTG)_5$-fingerprints of the parent cell lines and the xenohybridomas. It appears that the two xenohybridomas (F-CLB-1.H10 and F-CLB-4.B8) exhibit limited banding similar to the SP2/0 cell line and clearly reduced compared to the EBV-transformed parent lines. This may be caused by a major loss of human DNA from the xenohybridomas.

The F-CLB-4.B8 xenohybridoma cell line was tested for performance during perfusion fermentation growth in a 10 l Bioreactor equipped with hollow fibre for cell retention during harvesting. Perfusion rate was increased from 0 to 2 vessel volumes per day during the first week of fermentation. Figure 3 presents the viable cell number and IgG-production during a run of 1000 hours. It can be seen that maximum viable cell number varies between 10 and 15 x 10^6 per ml. The IgG content in the vessel during the steady state period varied between 40 and 90 mg/L; the total IgG yield after 1000 hours was over 40 grams. A transient decrease in cell number and IgG production was observed after 700 hours of fermentation; this was due to clogging of the hollow fibre device. After replacement of the hollow fibre, cell numbers and IgG-levels returned to maximum values.

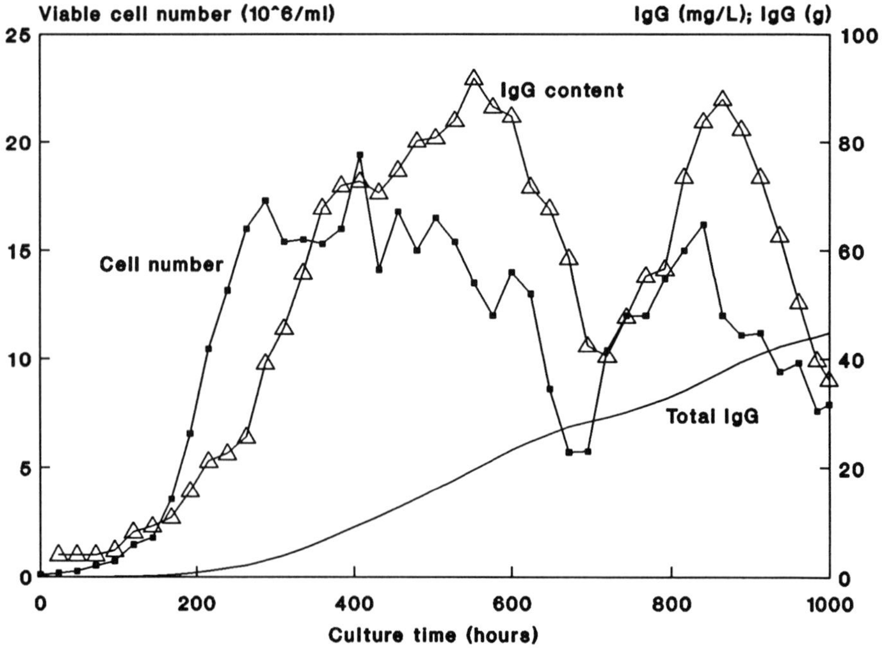

Figure 3. Perfusion fermentation of F-CLB-4.B8.

Table 2 summarizes the most important metabolic parameters of the perfusion fermentation, as calculated as means over a period 20 days of (pseudo-)steady-state fermentation. Result are compared with the data routinely obtained for a mouse-mouse hybridoma in the same perfusion fermentation system: although higher cell numbers are obtained for the xenohybridoma, the IgG-production is low as compared to a mouse-mouse hybridoma. Glucose and glutamine metabolism of the xenohybridoma are similar as for the mouse-mouse hybridoma.

TABLE 2. Metabolic Parameters of Perfusion Fermentation.
Means of steady state values (n=20).

Parameter	Xenohybridoma B8	Mouse-Mouse Hybridoma
Cell number (x10^6/ml)	15 ± 2	8 ± 2
IgG-production (mg/L)	68 ± 15	80 ± 15
IgG-production (mg/L/day)	130 ± 29	160 ± 30
IgG-production (ug/10^6cells/day)	9 ± 2	21 ± 5
Glucose uptake (umole/10^6 cells/day)	2.1 ± 0.3	3.3 ± 0.9
Y (Lactate/Glucose)	1.1 ± 0.2	1.4 ± 0.5
Y (Ammonia/Glutamine)	0.7 ± 0.1	0.8 ± 0.1

After perfusion fermentation, IgG was purified from the concentrated crude harvest by Protein A-Sepharose chromatography. Structural properties of the isolated IgG were determined: anti-idiotype ELISA, IEF and anti-HBsAg specific activity. The isolated IgG exhibited similar properties as the IgG isolated from the parent EBV-line.

Conclusions

1. IgG-producing xenohybridomas have been obtained after electrofusion with murine myeloma cell lines.
2. EBV-DNA is not detectable after xenohybridization of EBV-transformed cell lines.
3. Perfusion fermentation of xenohybridoma was feasible and yielded >40 grams of antibody.
4. Properties of antibody produced by the xenohybridoma are similar to the antibody produced by the parent cell line.

References

1. Stricker EAM, Tiebout RF, Lelie PN, Zeijlemaker WP (1985) A Human Monoclonal IgG1λ Anti-Hepatitis B Surface Antibody. Scand.J.Immunol. 22, 337-343.
2. Kwekkeboom J, De Groot C, Tager JM (1992) Efficient electric field-induced generation of hybridomas without prior activation in vitro. Hum. Antibod.Hybridomas, 3, 48-53.
3. Tiebout RF, Stricker EAM, Oosterhof F, Van Heemstra DJM, Zeijlemaker WP (1985) Xenohybridization of Epstein-Barr Virus-Transformed Cells for the Production of Human Monoclonal Antibodies. Scand.J.Immunol. 22, 691-701.
4. Nurnberg P, Roewer L, neitzel H, Sperling K, Popperl A, Hundrieser J, Poche H, Epplen C, Zichler H, Epplen JT (1989) DNA Fingerprinting with the oligonucleotide probe $(CAC)_5/(GTG)_5$: somatic stability and germline mutations. Hum Genet 84, 75-78.
5. Ring CJA (1994) The B-cell immortalizing functions of Epstein-Barr virus. J.Gen.Virol. 75, 1-13.

Acknowledgements

We acknowledge Biotest Pharma GmbH, Dreieich, Germany, for their fruitful collaboration.

Author Index

Subject Index